AI Foundations and Applications with MATLAB

Ying Bai

AI Foundations and Applications with MATLAB

 Springer

Ying Bai
Department of Computer Science, Engineering & Mathematics
Johnson C. Smith University
Charlotte, NC, USA

This work contains media enhancements, which are displayed with a "play" icon. Material in the print book can be viewed on a mobile device by downloading the Springer Nature "More Media" app available in the major app stores. The media enhancements in the online version of the work can be accessed directly by authorized users.

ISBN 978-3-031-84422-5   ISBN 978-3-031-84423-2   (eBook)
https://doi.org/10.1007/978-3-031-84423-2

This Springer imprint is published by the registered company Springer Nature Switzerland AG
The registered company address is: Gewerbestrasse 11, 6330 Cham, Switzerland

If disposing of this product, please recycle the paper.

*This book is dedicated to my wife, Yan Wang,
and my daughter, Susan Xue Cunningham*

Preface

Today, artificial intelligence (AI) and related technologies have been widely applied in all corners in our real world, and it is one of the hottest topics in most advanced techniques implemented in our society. With the help of AI-related knowledge and technology, our society has become an automatic-control or intelligent-control one to enable all machines to have intelligent learning and decision-making ability to perform all difficult and complicated tasks that are handled by a human being with high-level intelligent ability, even thinking and logic derivation ability, in previous time.

The motivation of developing this book is from a fund, Google TensorFlow College Fund, and I was asked to open a new AI course in my current college based on this fund.

To prepare offering this course, I started my searching trip and tried to find a good textbook, preferably a text with MATLAB as language since it is a powerful language used for AI and I had some experience on it. But unfortunately, I did not find a good book after long time searching. Some books either covered too much pure theoretical knowledge or contained a huge block of codes starting from the first page, which is absolutely a headache to college students. Also, most AI-related books only covered machine learning without touching fuzzy inference system. But the latter is also a key component in the AI study field.

Finally, I decided to build a textbook myself to meet the needs of this course. My desired book is one that should:

1. Provide both fundamental theoretical knowledge and hands-on practices to enable students to study and master the main topics on AI easily and quickly.
2. Meet the needs of both entry-level students and senior-level students to enable them to learn, design, and build professional AI-related projects and apply them in real world effectively and conveniently based on an easy learning language.
3. Cover the most updated and a full spectrum of materials and trends related to AI to enable students to learn and study the latest AI technologies and their applications, including both machine learning and fuzzy inference system (FIS), even the adaptive neuro fuzzy inference system (ANFIS).

4. Provide more practical and actual example projects to enable students to understand and learn AI-related algorithms easily and quickly with their practices. Learning from practicing should be fully reflected in this book.
5. Enable students to learn AI algorithms in an interesting way by providing them a lot of user-friendly GUIs and various tools to make study an enjoyment process.

The outstanding features of this book include, but not limited to, the following:

1. It covers almost all algorithms and implementations for the most popular AI technologies, including fuzzy inference systems, supervised learning, unsupervised learning, neural networks, deep learning, reinforcement learning, and ANFIS.
2. The most latest MATLAB tools related to the current AI, including the Fuzzy Logic Toolbox, Statistics and Machine Learning Toolbox, Curve Fitting Toolbox, Deep Learning Toolbox, Reinforcement Learning Toolbox, and Simulink, are introduced and discussed with real projects.
3. Totally, one hundred and thirty-four (134) real projects, including 77 class projects and 57 lab projects, are involved in this book to enable students and readers to follow up them to understand and strengthen what they learned to build professional AI-related projects efficiently and conveniently.
4. In addition to basic and fundamental discussions and explanations for popular AI-related algorithms, some advanced AI implementations and applications are also provided in Chapters 11 and 12 to set a path and direction for senior students and readers to develop advanced and practical AI applications in easier and quicker ways.
5. To meet the needs of both beginners and senior students and readers on the AI studies, two different learning and developing styles are introduced and analyzed: APPs and library functions provided by MATLAB. The former provide various GUIs and user-friendly tools to enable the entry-level students and readers to quickly and easily build professional AI-related applications with few or no coding lines via those APPs. The latter provide a collection of AI-related library functions to enable senior-level students and readers to develop more complicated and flexible AI applications with large blocks of codes.
6. Both teaching and learning materials are described in the book to help instructors and students to teach and learn AI-related technologies efficiently and easily. The learning materials include 77 class projects, homework questions, exercises, lab projects, and all sample datasets used to build class and lab projects. The teaching materials contain homework solutions, class projects and lab project solutions, teaching slides in PPT, and all sample datasets.
7. It is a good textbook for college students and good reference book for AI programmers, software engineers, and academic researchers.

I sincerely hope that this book can provide useful and practical helps and guides to all students or users who adopted this book, and I will be more than happy to know that all of you would be able to develop and build professional and practical AI-related applications with the help of this book.

Supplementary information: Extra supplementary information can be downloaded from: https://github.com/sn-code-inside/AI-Foundations-and-Applications-with-MATLAB.

Chapter 11 (Case Study Projects on Fuzzy Logic Inference Systems), Chapter 12 (Case Study Projects on Deep Learning) and other Extra supplementary information can be downloaded from: https://github.com/sn-code-inside/AI-Foundations-and-Applications-with-MATLAB.

Charlotte, NC, USA   Ying Bai
October 2024

Acknowledgements

The first and most special thanks go to my wife, Yan Wang, and I could not finish this book without her sincere encouragement and support.

Many thanks should be given to the acquisition editor, Ms. Susan Lagerstrom, and production editor, Ms. Kritheka Elango, who made this book available to public. You could not have found this book in the current book market without their deep perspective and hardworking. The same thanks are extended to the editor team of this book. Without this team's contributions, it would have been impossible for this book to be published.

Thanks should also be extended to the following book reviewers for their precious opinions to the book:

- Dr. Jiang Xie, Professor, Department of Electrical and Computer Engineering at the University of North Carolina at Charlotte
- Dr. Dali Wang, Professor, Department of Physics and Computer Science at Christopher Newport University
- Dr. Nailong Guo, Associate Professor, Department of Mathematics and Computer Science at Benedict College

Finally but not the last, thanks should be forwarded to all the people who supported me to finish this book.

This book is financially supported by Google TensorFlow College Awards granted to Dr. Ying Bai in 2022.

Contents

1 Introduction 1
  1.1 What Is Artificial Intelligence? 1
  1.2 History of Artificial Intelligence 2
  1.3 Basic Structures and Components in Artificial Intelligence 2
    1.3.1 Learning 3
    1.3.2 Reasoning 3
    1.3.3 Problem Solving 6
    1.3.4 Perception 6
    1.3.5 Language Understanding 8
  1.4 Intelligent Agents 8
    1.4.1 Properties and Functions of Intelligent Agents 9
    1.4.2 The Structure of Intelligent Agents 10
    1.4.3 Type of Intelligent Agents 10
    1.4.4 How Intelligent Agents Work 11
  1.5 Popular Implementations of Artificial Intelligence 12
  1.6 Future of Artificial Intelligence 12
  1.7 Risks and Benefits of Artificial Intelligence 13
  1.8 The Book Structure and Description 14
    1.8.1 Outstanding Features About This Book 16
    1.8.2 Class Projects and Lab Projects 17
    1.8.3 Teaching and Learning Materials 18
    1.8.4 How This Book Is Organized and How to Use This Book . 18
    1.8.5 Instructor and Customer Supports 20
  References 21

2 Learning and Decision-Making Process 23
  2.1 Introduction 23
  2.2 Logical Intelligence 24
    2.2.1 Propositional Logic 25
    2.2.2 Inference Systems 28
  2.3 AI Knowledge Cycle 29

2.4   Knowledge-Based Systems   30
     2.4.1   Backward Chaining   31
     2.4.2   Forward Chaining   32
2.5   Knowledge Representations.   33
     2.5.1   What Is Knowledge Representations.   34
     2.5.2   Logical Representation   35
     2.5.3   Semantic Networks   35
     2.5.4   Neural Networks   36
     2.5.5   Cycle of Knowledge Representation in AI   37
     2.5.6   Approaches to Knowledge Representation   38
2.6   Decision Tree.   40
     2.6.1   Entropy.   42
     2.6.2   How to Use Entropy in Decision Tree.   43
     2.6.3   Information Gain   44
     2.6.4   A Real Case Study.   47
2.7   From Decision Tree to Neural Networks.   49
2.8   Chapter Summary   50
References.   55

3   **Fuzzy Logic Inference Systems**   57
3.1   Introduction   57
3.2   Fuzzy Logic Idea   58
3.3   Fuzzy Set   60
     3.3.1   Classical Sets and Operations   60
     3.3.2   Mapping of Classical Sets to Functions.   62
     3.3.3   Fuzzy Sets and Operations.   63
     3.3.4   A Comparison Between the Classical Sets
             and the Fuzzy Sets   65
3.4   Fuzzifications and Membership Functions   66
3.5   Fuzzy Control Rules   68
     3.5.1   Fuzzy Mapping Rules   69
     3.5.2   Fuzzy Implication Rules   70
3.6   Defuzzifications Process and Lookup Table   70
     3.6.1   Mean of Maximum (MOM) Method.   71
     3.6.2   Center of Gravity (COG) Method   72
     3.6.3   The Height Method (HM)   72
     3.6.4   The Online and Offline Output   73
     3.6.5   The Lookup Table   73
3.7   A Typical Architecture of Fuzzy Logic Control System   76
3.8   Implementations of Fuzzy Logic Inference Control System.   78
     3.8.1   MATLAB Development Environment and
             Fuzzy Logic Toolbox™.   79
     3.8.2   Build Fuzzy Logic Control System for Air
             Conditioner with Fuzzy Logic Designer App   82

| | | 3.8.3 | Build Fuzzy Logic Control System for Air Conditioner with Fuzzy Functions. | 95 |

3.8.3   Build Fuzzy Logic Control System for Air
         Conditioner with Fuzzy Functions.   95
3.9   Introduction to Type-2 Fuzzy Inference System   107
      3.9.1   Introduction to Interval Type-2 Fuzzy Inference System . .   109
      3.9.2   Implementations of Interval Type-2
              Fuzzy Inference System with MATLAB   112
3.10 An Example of Interval Type-2 Fuzzy Inference System   116
3.11 A Case Study for Applying an Interval Type-2
      Fuzzy Inference System.   118
      3.11.1 Build IT2 Fuzzy Inference System with
              Fuzzy Logic Designer App   120
      3.11.2 Build the Input and the Output Membership Functions. . . .   121
      3.11.3 Design and Build Control Rules   122
      3.11.4 Build IT2 FIS with MATLAB
              Fuzzy Inference Functions.   125
3.12 Simulation Study for Type-1 and IT2 Fuzzy Logic Control
      Systems   133
      3.12.1 Create a Simulation Block Diagram
              Model for Type-1 FIS and IT2 FIS   134
      3.12.2 Build Each Block by Entering Related Parameters   137
      3.12.3 Perform the Simulation Study for Our Two FIS Models. . .   137
3.13 Chapter Summary   139
References.   147

4   Introduction to Machine Learning   149
4.1   What Is Machine Learning?.   149
4.2   Structure and Component of Machine Learning   150
      4.2.1   Supervised Learning   151
      4.2.2   Unsupervised Learning   153
      4.2.3   Reinforcement Learning   154
      4.2.4   Neural Network   156
      4.2.5   Deep Learning   156
4.3   Machine Learning with MATLAB.   157
      4.3.1   Statistics and Machine Learning Toolbox   158
      4.3.2   Deep Learning Toolbox   159
      4.3.3   Curve Fitting Toolbox   159
4.4   Machine Learning-Related Apps and Functions
      Provided by MATLAB.   160
4.5   Chapter Summary   161
References.   164

5   Introduction to Regression Algorithms   167
5.1   Supervised Learning-Related Apps and Functions   167
5.2   Linear Regression-Related Apps and Functions   167
      5.2.1   Linear Regression   168
      5.2.2   Linear Regression App Example   173

	5.2.3	Regression Learner Functions	181
	5.2.4	Multiple Linear Regression App Modeling	184
	5.2.5	Multiple Linear Regression Function Modeling	192
5.3		Decision Tree-Related Apps and Functions	198
	5.3.1	Regression Tree-Related Apps	200
	5.3.2	Regression Tree-Related Functions	210
5.4		Introduction to Nonlinear Regressions	217
	5.4.1	Nonlinear Regression Algorithms and Models	217
	5.4.2	Nonlinear Regression-Related Apps and Functions	219
5.5		Support Vector Machine-Related App and Functions	219
	5.5.1	Support Vector Machine-Related Apps	221
	5.5.2	Support Vector Machine-Related Functions	225
5.6		Implementations of Nonlinear Regressions	226
	5.6.1	Nonlinear Regression-Related Apps	227
	5.6.2	Nonlinear Regression-Related Functions	240
5.7		K-Nearest Neighbor (KNN) Algorithm	254
	5.7.1	Working Principle of KNN Algorithm	255
	5.7.2	K-Nearest Neighbor (KNN) Regression Algorithm	257
5.8		Introduction to Random Forest Algorithm	260
	5.8.1	Bagging and Bootstrap Aggregation	260
	5.8.2	Navigation from Bagging to Random Forests	261
	5.8.3	Random Forest Algorithms in MATLAB	262
	5.8.4	The RegressionBaggedEnsemble Related Apps	264
	5.8.5	The RegressionBaggedEnsemble Related Functions	268
5.9		A Real Project Example for Stock Prediction Using Regression Algorithm	280
	5.9.1	Introduction to Google Stock Dataset	280
	5.9.2	Build a Sample Project to Train and Test Regression Nonlinear Models	281
5.10		Chapter Summary	285
References			290
6		**Introduction to Classification Algorithms**	**293**
6.1		Introduction to Classifications	293
6.2		Classification-Related Apps in MATLAB	295
6.3		Binary Classification App Example	297
6.4		Classification Classes and Related Functions in MATLAB	302
6.5		Linear Classifications	304
	6.5.1	Binary Linear Classifications and Related Functions	308
	6.5.2	A Case Study for Binary Classification with Heart Disease Prediction	317
6.6		Multiclass Classifications App Example	321
6.7		Multiclass Classification Function Example	325
	6.7.1	Multiclass Classifications with Mixed Type Data in Dataset	330

      6.7.2   Working Principles of Naive Bayes Classifier 333

      6.7.3   A Real Example Project for Naive Bayes
              Classification Model 336

      6.7.4   Working Principles of the Error-Correcting
              Output Code (ECOC) Algorithm 340

      6.7.5   Special Features and Preprocessing for Image
              Classifications 346

  6.8   A Real Project of Classifying Images by Using the ECOC
      Algorithm 350

      6.8.1   Build an Image Classification Project by
              Using the ECOC Algorithm 351

      6.8.2   Build an Evaluation Project to Validate
              the Trained Model categoryClassifier 353

      6.8.3   A Case Study of Using the HOG Features
              to Classify Fruits Images 356

  6.9   Multiclass Classification Function Example for Audio Sounds
      Classifications 359

      6.9.1   What Is Mel-Frequency Cepstral Coefficients (MFCCs) . . 360

      6.9.2   What Is Band Energy Ratio (BER) 361

      6.9.3   MATLAB Classes and Functions Used
              for Audio Signal Classifications 361

      6.9.4   Build Audio Classification Project to
              Classify Digit Sounds 363

      6.9.5   Evaluate the Classification Result for Individual
              Digit Audio Sound 367

  6.10 Multiclass Classification for Animal Sounds Signals 370

      6.10.1 Preprocess the Selected Audio Sounds to Adjust the
              Multichannel Numbers. 370

      6.10.2 Build Audio Classification Project to Classify
              Animal Sounds. 371

  6.11 Chapter Summary 376

  References 382

7   Neural Networks and Deep Learning 385

  7.1   Introduction to Neural Networks 386

  7.2   Structure and Components of a Typical
      Neural Network System. 387

  7.3   The Types of Neural Networks 388

      7.3.1   Feedforward Neural Networks (FNN) 389

      7.3.2   Backpropagation Neural Networks (BNN) 389

      7.3.3   Convolutional Neural Networks (CNN) 390

      7.3.4   Recurrent Neural Networks (RNN). 393

  7.4   How Do Neural Networks Work? 394

  7.5   Introduction to Deep Learning. 396

      7.5.1   Deep Learning Versus Machine Learning 398

  7.6   Deep Learning in MATLAB 399

7.6.1   Deep Learning Related Apps   400
7.6.2   Deep Learning-Related Functions   404
7.6.3   The Workflow and Steps of Deep Learning
         Algorithms on Neural Networks   415
7.6.4   A Practical Decision Between the Transfer
         Learning and New Learning Strategy   420
7.7   Build Practical Example Projects with Deep
       Learning Technology   421
7.7.1   Prediction of Energy Level for Earthquake
         with Deep Learning   421
7.7.2   Build an Earthquake Prediction Project
         with Deep Learning APPS-ANN   422
7.7.3   Build an Earthquake Prediction Project
         with Deep Learning APPS-DND   430
7.7.4   Build an Earthquake Prediction Project
         with Deep Learning Functions   437
7.8   Using Deep Learning Algorithms to Perform Images
       Classifications of Fruits   446
7.8.1   Preprocess the Original Fruits Dataset to Get
         Our Desired Dataset   446
7.8.2   Perform the Classification Process to Get
         Our Deep Learning Model   449
7.8.3   Evaluate the Effectiveness of the Trained
         Fruit Deep Learning Model   453
7.8.4   Using Deep Learning Algorithms to Classify Animals'
         Sounds   455
7.9   Chapter Summary   461
References   467

8   Introduction to Unsupervised Learning   469
8.1   Introduction to Unsupervised Learning Algorithms   470
8.1.1   Exclusive Clustering Algorithm   471
8.1.2   The Overlapping Clustering Algorithm   474
8.1.3   The Hierarchical Clustering   476
8.1.4   The Probabilistic Clustering   477
8.2   The Association Rules   478
8.2.1   A Real Example of Using Support and
         Confidence to Build Association Rules   479
8.3   Unsupervised Learning in MATLAB   481
8.4   Using Neural Net Clustering App to Build
       Unsupervised Learning Models   482
8.4.1   Introduction to Self-Organizing
         Map (SOM) Algorithm   482
8.4.2   Using Neural Net Clustering App to Train a
         Clustering Model Earthquake_Cluster_App   484
8.5   Using MATLAB Functions to Build Clustering Models   487

8.5.1   Exclusive Clustering Algorithm 489

8.5.2   Functions Used for Hierarchical Clustering
Algorithm 497

8.5.3   Functions Used for Overlapping Clustering
(Fuzzy-K) Algorithm 507

8.5.4   Functions Used for Probabilistic
Clustering Algorithm 517

8.5.5   Introduction to Evaluation for Unsupervised
Learning Models 526

8.6   Using MATLAB Functions to Build Association
Rules Models 536

8.6.1   Modify the Diabetes.csv Dataset to Make
It as Our Desired Dataset 537

8.6.2   Build the Diabetes Project to Identify Association
Rules for Some Features 539

8.7   Using MATLAB Functions to Build Apriori
Algorithm Models 541

8.7.1   Introduction to Apriori Algorithm 542

8.7.2   Using MATLAB Function to Build Apriori
Model for Diabetes Dataset 545

8.7.3   Unsupervised Learning and Supervised Learning 548

8.8   Chapter Summary 550

References 554

9   Introduction to Reinforcement Learning 557

9.1   Introduction to Reinforcement Learning Algorithms 558

9.1.1   Components Involved in Reinforcement
Learning Control Systems 558

9.1.2   The Markov Decision Process 559

9.1.3   The Reinforcement Learning Process 561

9.2   Some Basic Algorithms Used in Reinforcement Learning 563

9.2.1   Monte Carlo Methods 564

9.2.2   The Temporal Difference Method 564

9.2.3   The Function Approximation Method 565

9.3   Model-Based Reinforcement Learning 565

9.3.1   Learn the Model 567

9.3.2   Given the Model 567

9.4   Model-Free Reinforcement Learning 568

9.4.1   Policy Optimization and Policy Iteration 569

9.4.2   Q-Learning or Action Value Iteration Algorithm 572

9.4.3   Hybrid RL Algorithms 576

9.5   Reinforcement Learning in MATLAB 577

9.5.1   Procedure to Build Real RL-Related Applications in
MATLAB 578

9.6   Using MATLAB Reinforcement Learning APP to Build Real
Applications 579

9.6.1   Introduction to Reinforcement Learning Designer 579

9.6.2   Build an RL Project with Reinforcement Learning
         Designer.. 581
9.6.3   Three Popular Methods Used to Build Customer
         Environments.. 587
9.6.4   Build an RL Project to Control DC Motor with
         Customer Function Environment......................... 588
9.6.5   Build an RL Project to Control DC Motor with
         Customer Template Environment 596
9.6.6   Build an RL Project to Control DC Motor with
         Customer Simulink Environment........................ 605
9.7   Using MATLAB Reinforcement Learning Functions to
      Build Real Applications... 617
9.7.1   Introduction to MATLAB Reinforcement Learning
         Functions 618
9.7.2   Build a Real RL Project with Reinforcement Learning
         Functions 621
9.7.3   Comparison of Different Agents Based on Their
         Training Results... 624
9.8   Chapter Summary 624
References... 630

10   Introduction to Adaptive Neuro Fuzzy Inference System........... 631
10.1   Introduction to Adaptive Neuro Fuzzy Inference Algorithm..... 631
10.2   The Components and Architecture of an Adaptive Neuro
        Fuzzy Inference System.. 632
10.3   A Real Example of a Two-Input Adaptive Neuro Fuzzy
        Inference System 633
10.4   Adaptive Neuro Fuzzy Inference System in MATLAB......... 637
10.5   Generate Input Datasets and Make It Suitable for
        ANFIS Projects 638
10.5.1   Introduction to Google Stock Price Dataset........... 638
10.5.2   Generate the Training Dataset and Checking Dataset ... 639
10.6   Use Neuro-Fuzzy Designer to Build Our Stock Price
        Prediction Project... 641
10.6.1   Evaluate and Validate the Trained and Checked
          Stock Price Model 646
10.7   Use MATLAB ANFIS Functions to Build Our Stock
        Price Prediction Project 649
10.7.1   Introduction to ANFIS Functions 649
10.7.2   Design and Evaluate Our Stock Price Model
          with MATLAB Functions 650
10.8   Chapter Summary 653
References... 659

Appendix A: Download and Install MATLAB 2023a Software 661

Index.. 665

About the Author

Ying Bai is a Professor in the Department of Computer Science, Engineering and Mathematics at Johnson C. Smith University located at Charlotte, North Carolina. His special interests include artificial intelligence, soft computing, database programming, fuzzy logic controls, automatic and robot controls, as well as robot calibrations. His industry experience includes positions as software and senior software engineers at companies such as Motorola MMS, Schlumberger ATE Technology, Immix TeleCom, and Lam Research. Since 2003, Dr. Bai has published 20 books with publishers such as Prentice Hall, CRC Press LLC, Springer, Cambridge University Press, and Wiley IEEE Press. He has also published more than 70 academic research papers in IEEE Trans. journals and international conferences. Most books are about the microcontroller control and programming, fuzzy logic implementations, classical and modern controls, cross-language interface programming, database programming, and applications.

Chapter 1
Introduction

Today artificial intelligence (AI) and related technologies have been widely applied in all corners of our real world, and it is one of the hottest topics in most advanced techniques implemented in our society. With the help of AI-related knowledge and technology, our society has become an automatic-control or intelligent-control one to enable all machines to have intelligent learning and decision-making ability to perform all difficult and complicated tasks that are handled by human beings with high-level intelligent ability, even thinking and logic derivation ability, in previous time.

To effectively learn and understand this new technology, first we need to have a clear picture about AI. For that purpose, the first question for us is: What is AI?

1.1 What Is Artificial Intelligence?

Artificial Intelligence (AI) is a technology or a method to make computers, computer-controlled devices, or microcontrollers think or work as a human being does. In other words, AI can be considered as an algorithm to be installed in computers to direct the latter to play functions as human beings play.

More directly, AI is just to copy the human being's learning and decision-making abilities or processes and paste them into machines to enable the latter to have similar abilities to perform learning and decision-making functions on various objects in our real world.

In terms of that definition, two basic functions should be adopted by AI: (1) learning ability, which is equivalent to training and checking processes, and (2) decision-making ability, which can be mapped to testing and implementation processes in actual applications.

Y. Bai, *AI Foundations and Applications with MATLAB*, https://doi.org/10.1007/978-3-031-84423-2_1

AI is accomplished by studying the patterns of the human brain and by analyzing the cognitive process, and the outcome of these studies builds intelligent software and systems.

1.2   History of Artificial Intelligence

Here is a list of short history about the AI:

- 1956: John McCarthy coined the term "artificial intelligence" and had the first AI conference.
- 1965: Lotfi A. Zadeh at the University of California at Berkeley developed and reported the first fuzzy logic-related idea, fuzzy sets, which are expanded and evolved to today's fuzzy logic inference systems and widely implemented in the world.
- 1969: Shakey was the first general-purpose mobile robot built. It is now able to do things with a purpose vs. just a list of instructions.
- 1997: Supercomputer **"Deep Blue"** was designed, and it defeated the world champion chess player in a match. It was a massive milestone for IBM to create this large computer.
- 2002: The first commercially successful robotic vacuum cleaner was created.
- 2005–2019: Speech recognition, image and pattern identifications, robotic process automation (RPA), a dancing robot, smart homes, and other innovations make their debut.
- 2020: Baidu releases the Linear-Fold AI algorithm to medical and scientific and medical teams developing a vaccine during the early stages of the SARS-CoV-2 (COVID-19) pandemic. The algorithm can predict the RNA sequence of the virus in only 27 seconds, which is 120 times faster than other methods.

The concept of fuzzy logic and fuzzy semantics is a central component of the programming of artificial intelligence solutions. Due to the faster developments of AI-related technologies, different techniques and methods have been combined or mixed together to get multiple integrations of those techniques. Some fuzzy sets can be embedded into machine learning to get fuzzy-neural works. A typical example of those combinations is the Adaptive Neural Fuzzy Inference System (ANFIS).

1.3   Basic Structures and Components in Artificial Intelligence

Some important techniques or methods involved in AI technology are:

1. Fuzzy Inference System
2. Neural Networks

3. Machine Learning
4. Deep Learning

Generally, five components are considered as basic and fundamental elements, which are:

1. Learning
2. Reasoning
3. Problem Solving
4. Perception
5. Language Understanding

1.3.1   Learning

It can be considered as a try-to-find-correct-way process, which is similar to a close-loop control system. During the learning process, exactly a training process, various inputs are fed into a system to get the related outputs. Then an error-checking function is performed to compare the desired outputs with the actual outputs to obtain differences or errors between them. Then modify the system by tuning some internal parameters or weight factors to repeat that close-loop learning process until the errors between the desired outputs and actual outputs are accepted within a limited range.

An example of this learning process can be described by a driving-car-study process. A student who is the first time to drive a car needs to be trained by a driver tutor to get driving knowledge and technique. The tutor needs to provide feedback for each driving operation based on the student's action, either correct or incorrect, to direct the student to keep correct but avoid incorrect operations to make the student familiar and follow the correct operations. During that process, the student is an operator, and the tutor is a commander providing correct inputs, checking the outputs—student's operations, providing error-checking function, and finally adjusting the student's actions, directing the student to obtain successful driving technique.

1.3.2   Reasoning

Reasoning can be considered as a process of thinking and understanding something in a logical way based on existing knowledge to derive a conclusion or predict something to happen in the future.

Generally, four types of reasoning methods are popularly implemented:

1. Deductive Reasoning
2. Inductive Reasoning

3. Abductive Reasoning
4. Reasoning by Analogy

Reasoning is similar to thinking, but it is exactly a logical and thoughtful way of thinking. This means that a reasoning process needs to use more logical derivation methods or paths to obtain more objective and optimal conclusions.

Three basic components are involved in a reasoning process [1]:

1. Cases
2. Rules
3. Results

These three components play different roles in different reasoning processes.

Generally speaking, a *case* is a special observation or a situation with a condition or a relation connected. A *rule* can be considered as a statement that states that if one condition is true, then the other one would also be true. A *result* is also a special observation or a situation that is similar to a *case*, but the difference is that a *result* depends on a condition that is related to another condition based on the *rule*.

1.3.2.1   Deductive Reasoning

Deductive reasoning uses a case with a rule to derive a result. Deductive reasoning starts with the proposition of a general rule and proceeds from there to a guaranteed specific observation or conclusion. Deductive reasoning moves from the general rule to the specific application: In deductive reasoning, if the original propositions are true, then the observation or conclusion must also be true.

For example, a good student's GPA should be greater than 3.5. Tom is a good student, thus Tom's GPA should be higher than 3.5. Another instance is: New Year is January 1st of each year, today is January 1st, so today is New Year.

1.3.2.2   Inductive Reasoning

Inductive reasoning uses a case with a result or multiple results to derive a rule. In other words, inductive reasoning combines multiple observations with experiential information to get a conclusion. When using a specific set of data or existing knowledge from past experiences to make decisions, you are using inductive reasoning.

For example, a good student's GPA should be greater than 3.5, and Tom's GPA is higher than 3.5, thus Tom is a good student. Similarly, New Year is January 1st of each year, today is New Year, so today is January 1st.

In short summary, both deductive and inductive reasoning bring benefits, but there are some differences that exist between them:

- Deductive reasoning uses theories (cases) and observations (rules) to derive and prove a specific conclusion (result). The goal is to prove a fact.

- Inductive reasoning uses experience (case) and proven observations (results) to predict the outcome (rule). The goal is to guess a possible outcome.

1.3.2.3   Abductive Reasoning

Abductive reasoning uses a result with a rule to derive a case. In fact, abductive reasoning begins with an incomplete set of observations and proceeds to the possible explanation for the set. Abductive reasoning yields the kind of decision-making that does its best with the information at hand, which often is incomplete [2].

Abductive reasoning, unlike deductive and inductive reasoning, produces a plausible conclusion but does not definitively verify it. Abductive conclusions do not eliminate uncertainty or doubt, which is expressed in retreat terms such as "best available" or "most likely" [3].

For example, Tom is a good student and Tom's GPA is higher than 3.5, thus a good student's GPA should be greater than 3.5. Additionally, today is January 1st and it is New Year, so New Year is January 1st.

Symbolically these three types of reasoning can be presented as [1]:

$$A \wedge (A \rightarrow B) \Rightarrow B$$

$$A \wedge B \Rightarrow (A \rightarrow B)$$

$$B \wedge (A \rightarrow B) \Rightarrow A$$

where A represents a case, B means a case or a result, and $\rightarrow$ means a rule.

1.3.2.4   Reasoning by Analogy

Reasoning by analogy involves using a well-known observation to propose something about a less well-understood observation or situation. It can go either from one case to another case or from one rule to another rule.

In other words, analogical reasoning proceeds from the observation that things that are similar in some respects are probably similar in other respects too [4].

For example, Tom is a good student, and you know that another student whose name is Bob and his GPA is greater than 3.5. You may come to the conclusion that Bob is also a good student. Going from "Tom is a good student" to "Bob is also a good student" based on a sense that Tom's GPA is similar to Bob's GPA is an analogy from one rule to another.

Similarly, on each New Year's day, it is a holiday, and on each Christmas Day, people also do not need to work, thus you may consider that Christmas is also a holiday since nobody needs to work on a holiday.

1.3.3 Problem Solving

Problem solving is a process that includes defining and understanding a problem; determining the cause of the problem; selecting alternatives for a solution; and implementing and evaluating a solution.

Define and Understand a Problem

This is the first step for a problem-solving process and the precondition for this step is the symptoms presented by that problem. One may need to identify the root or the real problem via all symptoms.

Determining the Cause of the Problem

Try to find the actual cause or reason why the problem is coming. Multiple possible causes or sources may be involved in a problem, but one needs to identify and determine the main real cause of that problem from those multiple reasons.

Selecting Alternatives for a Solution

There may be multiple possible solutions that exist to solve a problem. To find an optimal and final solution, some researches and solution-shooting methods may be needed via various mediums and resources. Some trade-off may be needed to take care of all aspects of a problem to obtain an optimal solution.

Implementing and Evaluating a Solution

As a final decision or solution is determined, it can be implemented to solve the selected problem. Also, an evaluation process is needed to track, check, and confirm the effectiveness of the selected solution.

Problem-solving methods can also be divided into special purpose and general purpose [5]. A special-purpose method is modified for a particular problem and often exploits very specific features of the situation in which the problem is embedded. In contrast, a general-purpose method is applicable to a wide variety of problems. One general-purpose technique used in AI is means-end analysis, a step-by-step, or incremental, reduction of the difference between the current state and the final goal.

1.3.4 Perception

Perception can be defined as the state of being or process of becoming aware of something through the senses.

Perception is not only the passive receipt for of these signals, but it is also shaped by the recipient's learning, memory, expectation, and attention [6, 7]. Sensory input is a process that transforms this low-level information to higher-level information (e.g., extracts shapes for object recognition) [7]. The process that follows connects a person's concepts and expectations (or knowledge), restorative and selective mechanisms that influence perception.

Perception depends on complex functions of the nervous system, but subjectively seems mostly effortless because this processing happens outside conscious awareness [8].

There are also different types of perception in psychology, including [9]:

- **Person perception** refers to the ability to identify and use social cues about people and relationships.
- **Social perception** is how we perceive certain societies and can be affected by things such as stereotypes and generalizations.

1.3.4.1 How Perception Works

We become more aware of our environment and take our appropriate reactions to it via perception. We also use perception in our communication to identify who are our friends and who are not, and how our loved ones may feel. We use perception in behavior to decide what we think about individuals and groups.

We perceive everything around us continuously, even though we do not typically spend a great deal of time thinking about it. For example, the object that falls on our eye's retinas transforms into a visual image unconsciously and automatically. Small changes in pressure and humidity against our skin, allowing us to feel objects, also occur without any single thought.

1.3.4.2 Perception Process

It would be very helpful if a complete perception process could be provided with detailed operational steps in a sequence. This process or sequence is shown below [8]:

1. **Environmental stimulus**: The world is full of stimuli that can attract attention. Environmental stimulus is everything in the environment that has the potential to be perceived.
2. **Attended stimulus**: The attended stimulus is the specific object in the environment on which our attention is focused.
3. **Image on the retina**: This part of the perception process involves light passing through the cornea and pupil, onto the lens of the eye. The cornea helps focus the light as it enters and the iris controls the size of the pupils to determine how much light to let in. The cornea and lens act together to project an inverted image onto the retina.
4. **Transduction**: The image on the retina is then transformed into electrical signals through a process known as transduction. This allows the visual messages to be transmitted to the brain to be interpreted.
5. **Neural processing**: After transduction, the electrical signals undergo neural processing. The path followed by a particular signal depends on what type of signal it is (i.e., an auditory signal or a visual signal).

6. **Perception**: In this step of the perception process, you perceive the stimulus object in the environment. It is at this point that you become consciously aware of the stimulus.
7. **Recognition**: Perception does not just involve becoming consciously aware of the stimuli. It is also necessary for the brain to categorize and interpret what you are sensing. The ability to interpret and give meaning to the object is the next step, known as recognition.
8. **Action**: The action phase of the perception process involves some type of motor activity that occurs in response to the perceived stimulus. This might involve a major action, like running toward a person in distress. It can also involve doing something as subtle as blinking your eyes in response to a puff of dust blowing through the air.

Considering all the things you perceive on a daily basis. At any given moment, you might see familiar objects, feel a person's touch against your skin, smell the aroma of a home-cooked meal, or hear the sound of music playing in your neighbor's apartment. All of these help make up your conscious experience and allow you to interact with the people and objects around you.

1.3.5   Language Understanding

The final component in development of artificial intelligence is language understanding. To make it simple, language understanding in the context of the development of artificial intelligence can be defined as a set of different system signals that justify their various ways or methods using communication or convention. To this point, as the vast majority of artificial intelligence programs and systems are developed within the English-speaking world, a major component of the creation of many such programs and systems is enabling them to understand the English language. Through this language understanding, software developers are able to ensure that computer programs are able to efficiently execute their respective functions and operations.

So far we have provided a detailed discussion about fundamental components used in the artificial intelligence technology. Next, we need to provide an introduction about how to use those components to perform actual artificial intelligence jobs in our real world. That introduction will cover both components' functions, coordination, and communications among those components. But everything is under the control of a key element, an intelligent agent.

1.4   Intelligent Agents

An intelligent agent can be considered as an algorithm made in a program that can make decisions and perform a special or general task based on its environment, user input, and experiences. These programs can be used to autonomously gather information on a regular, programmed schedule or when prompted by the user in real time [10].

In fact, an intelligent agent (IA) is a core of the artificial intelligence system and it makes all decisions or takes actions automatically based on perceptions or sensors. In other words, an IA is similar to a human being's brain, has a thinking ability, and provides related reactions based on the environment automatically. The ***intelligent*** means that the agent has ability to learn itself during the process of performing general or special tasks.

To make an IA to work automatically, two elements are necessary: Perception unit and Action unit. Perception is done through sensors while actions are initiated through actuators.

Generally, a complete IA is composed of lower level agents and higher level agents. Intelligent agents consist of sub-agents that form a hierarchical structure. Lower-level tasks are performed by these sub-agents.

1.4.1 Properties and Functions of Intelligent Agents

Intelligent agents have the following distinguishing characteristics and properties:

1. They have some level of autonomy to allow them to perform certain tasks on their own.
2. They have a learning ability that enables them to learn even as tasks are carried out.
3. They can interact with other entities such as agents, humans, and systems.
4. New rules can be accommodated by intelligent agents incrementally.
5. They exhibit goal-oriented habits.
6. They are knowledge-based. They use knowledge regarding communications, processes, and entities.

Intelligent agents are often described schematically as an abstract functional system similar to a computer program. Abstract descriptions of intelligent agents are called **abstract intelligent agents** (**AIA**) to distinguish them from their real-world implementations. An **autonomous intelligent agent** is designed to function in the absence of human intervention. Intelligent agents are also closely related to software agents.

In artificial intelligence, an agent can be defined as: A process or a system that can be viewed as perceiving its environment through sensors and acting upon that environment through actuators.

Artificial Intelligence agents perform these functions continuously [11]:

- Perceiving dynamic conditions in the environment
- Acting to affect conditions in the environment
- Using reasoning to interpret perceptions
- Problem-solving
- Drawing inferences
- Determining actions and their outcomes

1.4.2 The Structure of Intelligent Agents

A typical structure of an IA can consist of three main parts: architecture, agent function, and agent program [10].

1. **Architecture**: This refers to machinery or devices that consist of actuators and sensors. The intelligent agent executes on this machinery. Examples include a personal computer, a car, or a camera.
2. **Agent function**: This is a function in which actions are mapped from a certain percept sequence. Percept sequence refers to a history of what the intelligent agent has perceived.
3. **Agent program**: This is an implementation or execution of the agent function. The agent function is produced through the agent program's execution on the physical architecture.

Next, let us take a look at the types of intelligent agents.

1.4.3 Type of Intelligent Agents

There are five main categories of intelligent agents. The grouping of these agents is based on their capabilities and level of perceived intelligence [10].

Simple Reflex Agents

These agents perform actions using the current percept, rather than the percept history. The condition-action rule is used as the basis for the agent function. In this category, a fully observable environment is ideal for the success of the agent function.

Model-Based Reflex Agents

Unlike simple reflex agents, model-based reflex agents consider the percept history in their actions. The agent function can still work well even in an environment that is not fully observable. These agents use an internal model that determines the percept history and effect of actions. They reflect on certain aspects of the present state that have been unobserved.

Goal-Based Agents

These agents have higher capabilities than model-based reflex agents. Goal-based agents use goal information to describe desirable capabilities. This allows them to choose among various possibilities. These agents select the best action that enhances the attainment of the goal.

Utility-Based Agents

These agents make choices based on utility. They are more advanced than goal-based agents because of an extra component of utility measurement. Using a utility function, a state is mapped against a certain measure of utility. A rational agent selects the action that optimizes the expected utility of the outcome.

Learning Agents

These are agents that have the capability of learning from their previous experience.
Learning agents have the following elements:

- **The learning element**: This element enables learning agents to learn from previous experiences.
- **The critic**: It provides feedback on how the agent is doing.
- **The performance element**: This element decides on the external action that needs to be taken.
- **The problem generator**: This acts as a feedback agent that performs certain tasks such as making suggestions (new) and keeping history.

1.4.4   How Intelligent Agents Work

As we discussed in previous sections, an IA can provide its automatic decision and control function with two elements, perception unit and action unit. The former can be achieved via all kinds of sensors, and the latter can be realized with actuators. Figure 1.1 shows an illustration diagram for a simple IA process.

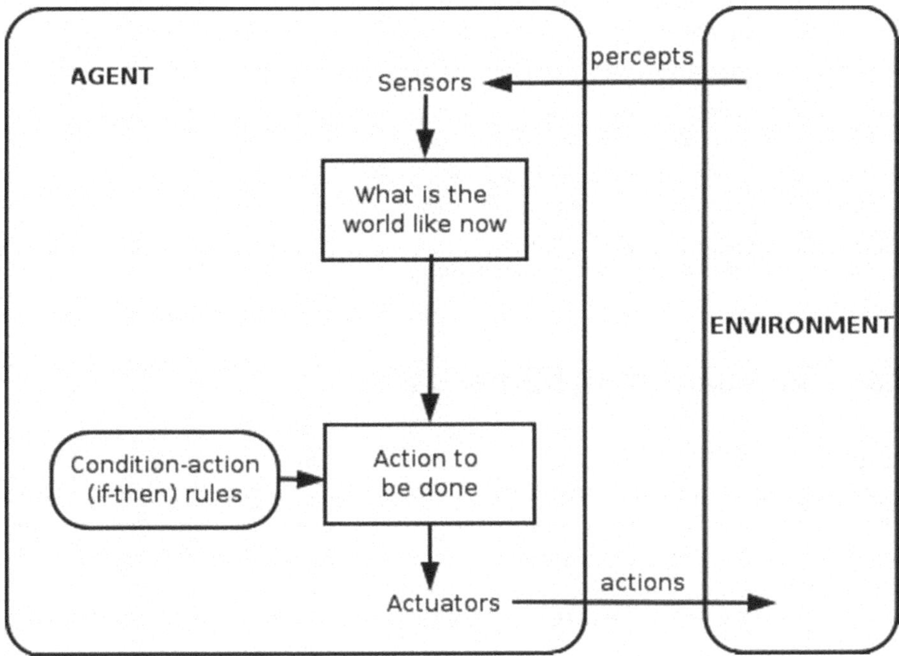

Fig. 1.1 An illustration diagram of IA

The IA can obtain or perceive some knowledge or inputs based on the sensors installed in the related environment. Then a condition-action rule is adopted by the IA to make decisions, derive the related actions, and finally apply actions to the actuators to get the control goal.

Among all those types of agents, the learning agent is a powerful and reliable one due to its auto-learning ability and feedback control feature, to perform jobs to achieve the intelligent control functions.

1.5 Popular Implementations of Artificial Intelligence

As for today, artificial intelligence technology can be implemented in almost all aspects of our world, including but not limited to:

- Manufacturing robots
- Self-driving cars
- Smart assistants
- Healthcare management
- Automated financial investing
- Virtual travel booking agent
- Social media monitoring
- Marketing chatbots

But the most popular implementations can be categorized into the following fields:

- Image processing in medical detection and identifications
- Computer visions
- Image pattern recognitions and inspections
- Automatic control of manufacturing robots
- Self-driving cars and autonomous vehicles
- Big data analysis and data mining
- Intelligent controls on banking and gaming

1.6 Future of Artificial Intelligence

Artificial intelligence (AI) has become an important aspect of the future [12]. This applies equally as well to information technology (IT) as it does many other industries that rely on it. Just a decade ago, AI technology seemed like something straight out of science fiction; today, we use it in everyday life without realizing it—from intelligence research to facial recognition and speech recognition to automation.

AI and machine learning (ML) have taken over traditional computing methods, changing how many industries perform and conduct their day-to-day operations. From research and manufacturing to modernizing finance and healthcare streams, leading AI has changed everything in a relatively short amount of time.

AI and related technologies have had a positive impact on the way the IT sector works. To put it simply, artificial intelligence is a branch of computer science that looks to turning computers into intelligent machines that would, otherwise, not be possible without direct human intervention. By making use of computer-based training and advanced algorithms, AI and machine learning can be used to create systems capable of mimicking human behaviors, provide solutions to difficult and complicated problems, and further develop simulations, aiming to become human-level AI.

According to the statistics, the AI market is expected to reach $190 billion by 2025. By 2021, global spending on cognitive and AI systems will reach $57.6 billion, while 75% of enterprise apps will use AI technologies.

On a more local level, some 83% of businesses say that AI represents a strategic priority, while 31% of creative, marketing, and IT professionals look to invest in AI technologies over the following 12 months. Similarly, some 61% of business professionals point to AI and machine learning as their most significant data initiative over the coming year. In addition, some 95% of business executives who are skilled in using big data also use AI technologies.

1.7   Risks and Benefits of Artificial Intelligence

As a fact, anything has two aspects, good and bad sides, which is true for AI, too. The following facts provide some benefits and risks for artificial intelligence:

Benefits of AI
- Reduction in human error
- Work with high accuracy
- Available 24×7
- Training and operation cost reduction
- Improve processes
- Helping in repetitive jobs
- Digital assistance
- Speed up decision-making
- Daily applications
- New inventions

Risks of AI
- Unsustainability
- Unemployment
- Misuse leading to threats
- Data discrimination
- Making humans lazy
- No emotions
- Lacking out of box thinking
- A future threat to humanity

How Can AI Be Dangerous?
- The AI is programmed to do something devastating.
- The AI is programmed to do something beneficial, but it develops a destructive method for achieving its goal.

1.8 The Book Structure and Description

The book is divided into nine chapters to provide a global and detailed introduction and discussion about artificial intelligence and its implementations in all aspects of our real world. An overview of artificial intelligence and the entire book is provided in Chap. 1. Following that chapter, more detailed discussions about artificial intelligence are given in the following chapters:

- In Chap. 2, a full detailed discussion about a mapping for problem-solving used by human beings and AI is provided to enable readers to understand the similarities and dissimilarities in problem-solving strategies used by both parties.
- Chapter 3 provides detailed discussions about one of the popular AI-related technologies, fuzzy inference system (FIS) and fuzzy logic controls (FLC), with some practical and hands-on example projects. Detailed architectures and structures about FIS and FLC are introduced and explained, including the fuzzification process, membership functions, control rules, fuzzy inference process, and defuzzification process with different methods, such as MOM, COG, HM, and lookup tables. Two major types of FIS, type-I and interval type-II, are discussed with actual projects. Eight (8) real projects are provided in this chapter to help students familiarize them with two MATLAB tools used for building FIS-related applications, including type-I and interval type-II projects, an APP called Fuzzy Logic Designer, and FIS-related functions.
- One of the most fundamental technologies applied in AI, machine learning, is introduced and discussed in Chap. 4. This chapter is crystal important since most popular AI-related techniques are based on machine learning. Details about machine learning algorithms, including supervised learning, unsupervised learning, reinforcement learning, neural networks, and deep learning, are provided in this chapter. Some popular and important machine learning-related toolboxes provided by MATLAB, such as Statistics and Machine Learning Toolbox, Curve Fitting Toolbox, and Deep Learning Toolbox, are also introduced and discussed in detail in this chapter.
- In Chap. 5, detailed introductions and discussions about the regression algorithms are given with eight real projects, which include linear regression, multiple linear regressions, decision trees, nonlinear regressions, support vector machine (SVM), K-Nearest Neighbor (KNN), and random forest. Two different regression tools provided by MATLAB, APP, and library functions are discussed and illustrated with some real project examples. A stock price prediction project is also involved in this chapter to provide an actual application with regression algorithms.

- Chapter 6 provides detailed introductions and discussions about the classification process, including the binary classifications and multi-class classifications. Different classification algorithms, including the Naïve Bayes, Error Correcting Output Code (ECOC), SVM, KNN, decision trees, and random forest, are discussed with 20 real projects. Some special but useful techniques, such as different feature extracting methods, including the histogram of oriented gradients (HOG), the speed-up robust features (SURF), the local binary pattern (LBP), and the bag of features (BOF), are introduced with real projects. Among 20 real projects, two of them are used to classify digit audio signals and animal sounds, one is used to classify fruit images, and others are used to classify diabetes and cars. Both MATLAB APPs and library functions are introduced to help users develop and build professional and actual classification applications.

- Detailed introductions and discussions on neural networks and deep learning are provided in Chap. 7. The typical architecture and components used in neural networks and deep learning algorithms are illustrated with four popular networks, feedforward neural networks, backpropagation neural networks, convolutional neural networks, and recurrent neural networks. The similarities and dissimilarities between machine learning and deep learning are also discussed with some hands-on projects. Two popular MATLAB tools, APPs and library functions, are discussed and illustrated with three real projects, which include a prediction project used to predict possible earthquakes with the APP, Deep Networks Designer, an image classification project used to classify different fruits with functions, and a classification project used to classify animal sounds based on animal audio datasets. Different MATLAB tools, including the APPs and functions, are introduced and applied in those real projects.

- Chapter 8 concentrates discussions on unsupervised learning, including different clustering algorithms, such as exclusive clustering, overlapping clustering, hierarchical clustering, and probabilistic clustering, as well as association rules. Sixteen real projects are involved in this chapter to illustrate how to use MATLAB APPs and library functions to build those unsupervised learning applications. Some suggested new possible evaluation methods are also discussed in this chapter to provide users with certain directions and methods to effectively evaluate unsupervised learning algorithms, which is a challenging topic in AI research and implementation fields today.

- The reinforcement learning technique is discussed in Chap. 9. Some important and useful ideas, such as agent, action, state, reward, environment, policy, state value functions, action value functions, Markov Decision Process (MDP), and Monte Carlo Methods, are introduced and explained in detail with examples. Different reinforcement learning algorithms, such as model-free and model-based, and all related algorithms under those two categories are also introduced. A powerful tool, Reinforcement Learning Toolbox provided by MATLAB, which includes both APPs and a group of functions, is discussed and explained in detail with real project examples. Nine actual projects are involved in this chapter to help users to learn and understand how to design, develop, and build real reinforcement learning application projects and implement them in the real world.

- Chapter 10 provides detailed introductions and explanations about a combination of two different AI techniques, fuzzy inference system (FIS) and neural networks (NN), called adaptive neuro-fuzzy inference system (ANFIS) with solid theoretical discussions and practical implementations. In fact, the FIS provides good decision-making ability for uncertain or vague input data, and the neural networks provide some advantages on learning ability for unknown system. Thus the ANFIS technique combines the advantages of both the FIS and the NN and provides some special merits used for AI applications. Three real projects are involved in this chapter to help users to understand and build their applications with ANFIS to take advantage of both systems.
- Some advanced case study projects related to fuzzy inference systems (FIS) are provided in Chap. 11. Unlike Chap. 3, where only fundamental and basic FIS applications are involved, some advanced real applications with both type-I and interval type-II FIS-related projects are discussed and analyzed in detail, including a type-I and an IT2 FLC used to control DC motors for a laser tracking system (LTS), a decision-making system used to evaluate the optimal military robots, and a control flow control model used to predict the optimal evacuation decision-making plan, an FIS project used to estimate the minimized dose of radiation for pediatrics, and a fuzzy interpolation system used for modeless robots calibrations.
- Chapter 12 contained advanced implementations on deep learning algorithms. Five real implementations are involved in this chapter, which are:

  - Using Deep Network Designer (DND) APP to predict the Euro-US currency exchange rates with a NASDAQ dataset.
  - Using Deep Network Designer (DND) APP and DL-related functions to build a flooding model to predict possible floods in selected areas.
  - Using deep learning-related functions to build a model to identify and classify different types of cars.
  - Using deep learning to classify popular animals based on their images.
  - Building a deep learning model to identify and classify fraud bank checks based on images of selected bank checks.

Due to space limitations, Chaps. 11 and 12 are located on the Springer ftp site in e-book format to save space. Quite a few of real implementation projects are provided in Chaps. 11 and 12 to provide readers with a more clear picture of how to build hands-on projects by using AI technology discussed in the book.

1.8.1   Outstanding Features About This Book

- Covered almost all algorithms and implementations for the most popular AI technologies, including fuzzy inference systems, supervised learning, unsupervised learning, neural networks, deep learning, reinforcement learning, and adaptive neuro-fuzzy inference systems.

- The most latest MATLAB Tools related to the current AI, including the Fuzzy Logic Toolbox, Statistics and Machine Learning Toolbox, Curve Fitting Toolbox, Deep Learning Toolbox, Reinforcement Learning Toolbox, and Simulink, are introduced and discussed with real projects.
- Totally 134 real projects, including 77 class projects and 57 lab projects, are involved in this book to enable students and readers to follow up on them to understand and strengthen what they learned to build professional AI-related projects efficiently and conveniently.
- In addition to basic and fundamental discussions and explanations for popular AI-related algorithms, some advanced AI implementations and applications are also provided in Chaps. 11 and 12 to set a path and direction for senior students and readers to develop advanced and practical AI applications in easier and quicker ways.
- To meet the needs of both beginners and senior students and readers on AI studies, two different learning and developing styles are introduced and analyzed, APPs and library functions provided by MATLAB. The former provided various GUIs and friendly used tools to enable entry-level students and readers to quickly and easily build professional AI-related applications with few or no coding lines via those APPs. The latter provided a collection of AI-related library functions to enable senior-level students and readers to develop more complicated and flexible AI applications with large blocks of codes.
- Both teaching and learning materials are provided by the book to help instructors and students to teach and learn AI-related technologies efficiently and easily. The learning materials include 77 class projects, homework questions, exercises, lab projects, and all sample datasets used to build class and lab projects. The teaching materials contain homework solutions, class projects and lab project solutions, teaching slides PPT, and all sample datasets.

The advantages of using the APPs are that the students and readers do not need to have full knowledge and complete understanding about AI algorithms, and they can study and build projects quickly and easily. This is truly important to entry-level students and readers to avoid some terrible learning curves on AI, and greatly improve and strengthen their learning and studying interests on AI.

The shortcoming for that method is that a lot of details in designing and developing processes may be hidden to students, but this can be compensated by using the Generate Function method in the Export group to convert all their operational steps to the related MATLAB codes to be further studied.

1.8.2 Class Projects and Lab Projects

A huge collection of real projects, exactly 134 projects, is involved in this book to provide readers with actual and hands-on knowledge and practical applications to enable them to understand all topics discussed in the book.

All 134 real projects are divided into two categories:

- **Class Projects**: Totally 77 class projects are involved in the book and distributed in all 12 chapters, an average of 6 projects are covered in each chapter. Each project is illustrated by using a coding program and explained line by line in detail. With the help of these class projects, students and readers can follow them exactly to understand the related AI techniques easily and efficiently.
- **Lab Projects**: Totally 57 lab projects are contained in the book and distributed at the end of each chapter. Approximately, five lab projects are used for each chapter. These lab projects belong to a part of the homework attached at the end of each chapter to enable students and users to develop and build related AI applications by using their hands to improve and strengthen their studied knowledge learning from the classes.

A collection of solutions for all of those lab projects is provided and available for instructors.

1.8.3 Teaching and Learning Materials

A block diagram for distributions of teaching and learning materials is shown in Fig. 1.2.

1.8.4 How This Book Is Organized and How to Use This Book

This book is designed for both college students who are new to AI techniques and professional AI programmers who have some professional experience on this topic.

Chapters 2, 3, and 4 provide the fundamentals of AI structures and components, including popular AI-related algorithms, fuzzy inference system (FIS), and machine learning (ML), which are basics and foundations for AI design and implementations.

Starting from Chap. 5, one of the most important algorithms in machine learning, regressions, is introduced and discussed. For entry-level students and beginners in AI, they only need to study Sects. 5.1–5.7, which is good enough to understand the basics and fundamentals of regression algorithms.

Similar requirements for entry-level students in Chaps. 6 and 7, Sects. 6.1–6.8 and Sects. 7.1–7.7 are good enough for them to understand those foundations on classifications and deep learning as well as neural networks.

Chapters 8 and 9 give a full discussion and analysis of the developments and implementations of unsupervised learning and reinforcement learning algorithms, which are important and basic knowledge on AI applications, and both of them are required for entry-level students and readers.

The Book Related Materials on the Web Sites

FOR INSTRUCTORS:
Teaching materials are available upon request from Springer ftp site:
(https://github.com/sn-code-inside/AI-Foundations-and-Applications-with-MATLAB)

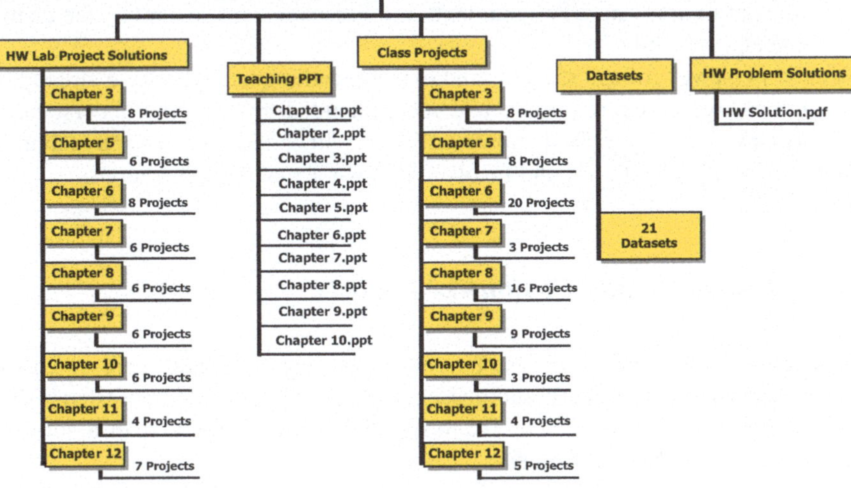

FOR STUDENTS:
Learning materials are free to access via the Springer ftp site:
(https://github.com/sn-code-inside/AI-Foundations-and-Applications-with-MATLAB)

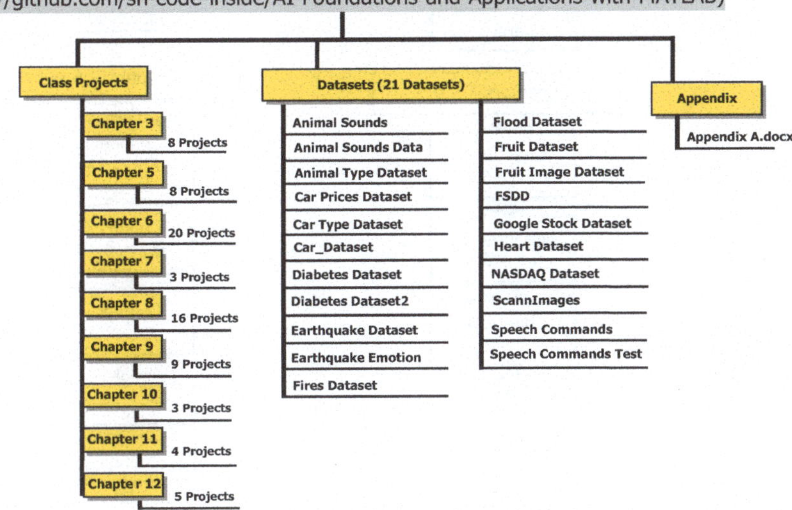

Fig. 1.2 Book-related teaching and learning materials on the Springer site

Chapters 10–12 belong to advanced-level topics on AI, and quite a few advanced and actual AI implementations are involved in those chapters, thus they are good for senior students or experienced readers, but some are challenging for entry-level students or beginners.

Based on the above analysis of this book, this book can be used in two categories such as Level I and Level II, which is shown in Fig. 1.3.

For entry-level undergraduate college students or beginners on AI, it is highly recommended to learn and understand the contents and topics on Level I shown in Fig. 1.3, since those are fundamental knowledge and techniques in AI. Chapters 10–12 are optional to entry-level students and beginners and depend on the instructors' time and schedule.

For senior-level college students and experienced readers who have already had some knowledge and skills in AI, it is recommended to learn and understand the contents and topics in Level II shown in Fig. 1.3 since those topics and techniques belong to the advanced level and involved in most actual AI applications.

1.8.5   Instructor and Customer Supports

The teaching materials for all chapters have been extracted and represented by a sequence of Microsoft PowerPoint files, each file for one chapter. Interested instructors can find them in the folder **Teaching PPT** which is located in a subfolder **Instructors** on the Springer ftp site.

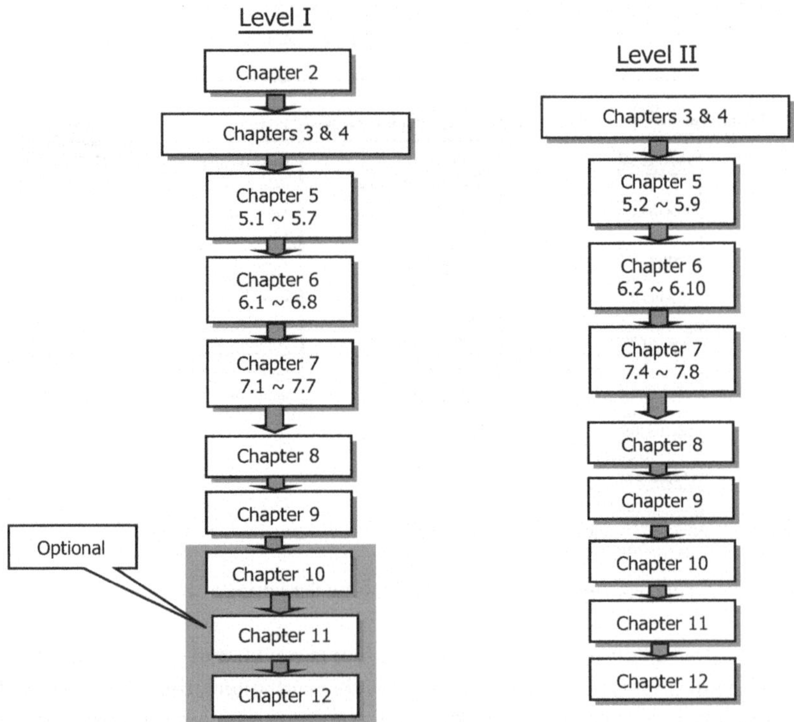

Fig. 1.3 The teaching level organizations of the book

A full homework solution, including homework questions, exercises, and lab projects, is available upon request from the site. E-mail support is also available to readers of this book. When you send an e-mail to us, please provide the following information:

1. A detailed description about your problems, including the error message and debug message as well as the error or debug number if it is provided.
2. Your name and company name.
3. Please send all questions to the e-mail address: ybai@jcsu.edu.

For detailed structure and distribution of all book-related materials, including the teaching and learning materials on the Springer ftp site, refer to Fig. 1.2.

References

1. David Reid & Christine Knipping, *Proof of Mathematics Educations*, E-Book ISBN: 9789460912467, Brill, Jan 2010, https://doi.org/10.1163/9789460912467_009.
2. http://www.butte.edu/departments/cas/tipsheets/thinking/reasoning.html.
3. https://en.wikipedia.org/wiki/Abductive_reasoning.
4. https://www.futurelearn.com/info/courses/logical-and-critical-thinking/0/steps/9166.
5. https://www.britannica.com/technology/artificial-intelligence/Reasoning.
6. Gregory, Richard. "Perception" in Gregory, Zangwill (1987) pp. 598–601.
7. Bernstein, Douglas A. (5 March 2010), *Essentials of Psychology*, Cengage Learning. pp. 123–124, ISBN 978-0-495-90693-3.
8. https://en.wikipedia.org/wiki/Perception#cite_note-mind_perception2-4.
9. https://www.verywellmind.com/perception-and-the-perceptual-process-2795839.
10. https://www.section.io/engineering-education/intelligent-agents-in-ai/.
11. https://www.simplilearn.com/what-is-intelligent-agent-in-ai-types-function-article#the_functions_of_an_artificial_intelligence_agent.
12. https://www.mycomputercareer.edu/news/the-future-of-i-t-and-artificial-intelligence/.

Chapter 2
Learning and Decision-Making Process

2.1 Introduction

Artificial intelligence is exactly a learning and decision-making process to derive or predict the possible results or outputs based on inputs. Four approaches are popularly implemented in this learning and decision-making process:

1. Learning based on experiences
2. Learning based on the input and the output data
3. Learning based on environments
4. Learning based on some statistics or probability observations

In this book, we concentrate on the first three processes.

The first approach can be mapped to the fuzzy inference system (FIS), in which both inputs and control rules are combined together based on the experiences of human beings or some experts to infer the outputs. The second way is to use the machine learning method which is based on the inputs data, either labeled or not, depending on whether the data are labeled by human beings or not labeled. The former is called a supervised learning system or model and the latter is referenced as an unsupervised learning model. The third one is similar to a reinforcement learning process, in which an agent is learned or trained by its environment.

Basically, three big categories of intelligent techniques are covered by AI systems:

1. Fuzzy Inference System (FIS)
2. Machine Learning (ML) Algorithms
3. Deep Learning (DL) Algorithm

The FIS is developed based on the logic intelligence method, and ML contains a group of intelligent algorithms to train different ML models by using input-output data pairs to derive the ideal models, which can be used to work as an independent

© The Author(s), under exclusive license to Springer Nature
Switzerland AG 2025
Y. Bai, *AI Foundations and Applications with MATLAB*,
https://doi.org/10.1007/978-3-031-84423-2_2

and intelligent machine to do decision making based on the further new input data. The DL is just a subset of the ML and it only uses a neural network with multiple layers (deep) and nodes as a model to be trained (learning) by using input-output data pairs. The trained neural network can then work as a robot or intelligent machine to make decisions based on the current input data.

The ML can be further divided into four sub-categories:

1. Supervised Learning Process
2. Unsupervised Learning Process
3. Semi-supervised Learning Process
4. Reinforcement Learning Process

Due to the space limitations, we will limit our discussions on three of them: Supervised Learning, Unsupervised Learning, and Reinforcement Learning.

Both (1) and (2) belong to the so-called off-line or open-loop learning process, and (4) can be categorized as a so-called online or closed-loop learning process. The reason for that is, both supervised and unsupervised learning models need to be trained first to get ideal or intelligent models. After that those trained models can work as intelligent machines to do decision-making themselves based on the inputs only without further learning process. But reinforcement learning is different, and it still needs to be trained in real time based on the current input data or environments via a learning process. A typical example of this kind of model is a robot or an auto-driving automobile.

However, some relationships exist between machine learning and neural networks. Generally, a neural network is only a structure or a component that can be used as a foundation of a model to be trained.

As we know, ML is composed of various algorithms that can be used to train different ML models based on those algorithms. Those algorithms can be categorized into two groups, as we mentioned, supervised learning and unsupervised learning process. Figure 2.1 shows a functional block diagram for a general AI system.

It can be found from Fig. 2.1 that the regression algorithm, especially the logistic regression, is one of the ML algorithms. This algorithm can be further derived to become a neural network, and the detailed derivation process will be given in Chap. 4.

Let us have our discussions about some intelligence concepts, starting with logical intelligence.

2.2   Logical Intelligence

Logic intelligence is also called logic-mathematical intelligence, and it means a kind of ability to reason, recognize, and logically analyze complicated problems in a logical way. Any person with this logic intelligence is also good at scientific investigations, identifying relationships between different things, and understanding complex and abstract ideas in effective and efficient ways.

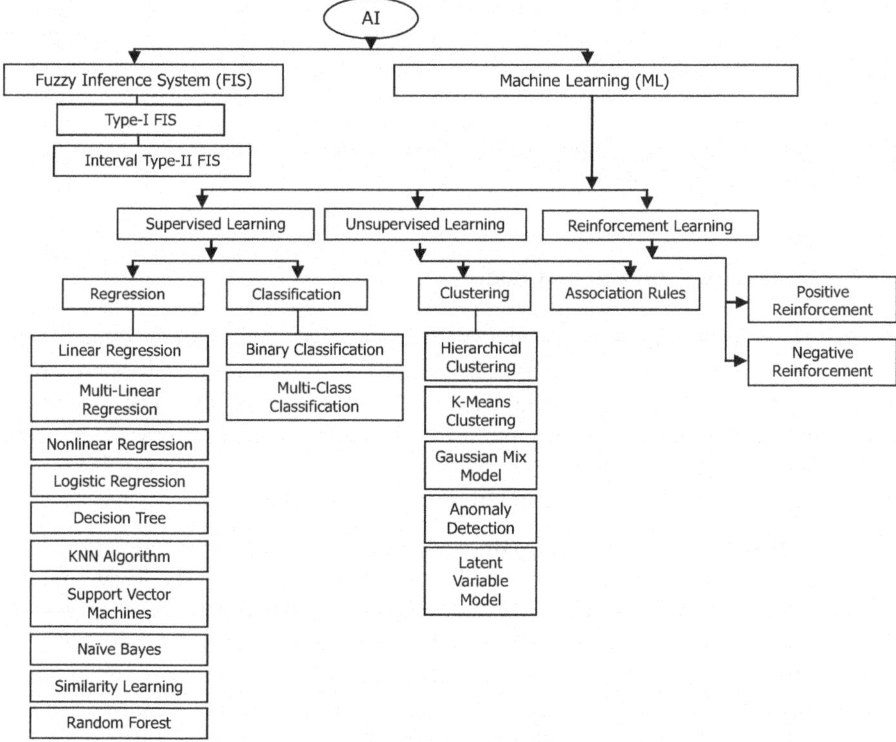

Fig. 2.1 Functional block diagram of a typical AI system

Some typical characteristics of logical intelligence include, but are not limited to [1]:

- Easily calculate mathematical sums in their heads
- Be terrific problem-solvers
- Identify and recognize patterns
- Draw conclusions from facts
- Pose logical arguments
- Be good at playing chess
- Connect different pieces of information

In logical intelligence, propositional logic is a typical component of logic intelligence, and let us take a closer look at it.

2.2.1 Propositional Logic

Generally, propositional logic handles propositions and statements without any reasoning process [2]. In other words, propositional logic (PL) is the simplest form of logic where all statements are made by propositions. In fact, a proposition is a declarative statement that is resulted in either true or false. It is a kind of technique of knowledge representation in logical and mathematical form [3]. Some popular examples of propositions are:

- Tom is a young boy.
- Washington D.C. is the capital of the United States.
- Peach is a kind of fruit.

2.2.1.1   Basics of Propositional Logic

Each of the propositions is assigned a Boolean value of either **true** or **false** [4]. In digital application areas, for example, computer logic gates, these values are given by the binary representations 1 (true) and 0 (false). Where true (1) can be mapped to an analog value of 3.5 V or above and false (0) represents an analog voltage value of 0.7 V or below for TTL logic.

Some basic properties that existed in propositions include:

- Propositional logic is also called Boolean logic as it works on 0 and 1.
- In propositional logic, we use symbolic variables to represent the logic, and we can use any symbol for representing a proposition, such as A, B, C, P, Q, and R.
- Propositions can be either true or false, but they cannot be both.
- Propositional logic consists of an object, relations or functions, and logical connectives.
- These connectives are also called logical operators.
- Propositions and connectives are the basic elements of the propositional logic.
- Connectives can be said as a logical operator, which connects two sentences.
- A proposition formula that is always true is called tautology, and it is also called a valid sentence.
- A proposition formula that is always false is called contradiction.

In propositional logic, there are some relationships among propositions and those relationships are called logic operators and can be represented by connectives. Totally five connectives are existed and they are:

- Negation
- Conjunction
- Disjunction
- Conditional
- Bi-conditional

Table 2.1 shows these connectives and their operational symbols.

The order or the precede of these connectives is arranged from the top (most higher priority) to the bottom (least priority) in Table 2.1.

All of those logic operators or connectives can be represented by a table called **Truth Table**, in which the input and the output mapping are described. The function or definition of each connective is summarized below:

Table 2.1 Five connectives

Connectives	Symbol	Operations
Negation	—	Not
Conjugation	$\wedge$	And
Disjunction	$\vee$	Or
Conditional	$\rightarrow$	If……Then
Bi-Conditional	$\leftrightarrow$	If and only If

1. **Negation**: It is equivalent to inversion or flip-over of a logic variable. For example, a negation of logic variable A is equal to $\overline{A}$. The truth table representation of a negation operator is

A	$\overline{A}$
F	T
T	F

2. **Conjugation**: It is equivalent to an *AND* operation, generally it works for more than one logic variable. For example, if we have two variables, A and B, the conjugation operation for those two logic variables is $X = A \wedge B$. It means that the output X will be true only when both variables are true. The truth table for a conjugation operator $X = A \wedge B$ is

A	B	X
F	F	F
F	T	F
T	F	F
T	T	T

3. **Disjunction**: It is equivalent to an *OR* operation, and regularly it works for more than two variables. For example, if we have two variables, A and B, the disjunction operation for those two variables is $X = A \vee B$, which means that the output X will be true if any of them is true. The truth table for a disjunction operator $X = A \vee B$ is

A	B	X
F	F	F
F	T	T
T	F	T
T	T	T

4. **Conditional**: It means that logic decision operation generally works for more than two variables. For example, with two logic variables, A and B, if A is true, then B is also true, and it is represented as $A \rightarrow B$, but if A is false, then we do not need to take care of B, which can be true or false. In that case the result will be true. The truth table for $A \rightarrow B$ is

A	B	$A{\rightarrow}B$
F	F	F
F	T	T
T	F	F
T	T	T

5. **Bi-Conditional**: This is a bi-directional operator, and it means that if A is true if, and only if, B is true. It is indicated as $A \leftrightarrow B$, and means that the result is true; either both A and B are true or both A and B are false. The truth table for $A \leftrightarrow B$ is

A	B	$A{\rightarrow}B$
F	F	T
F	T	F
T	F	F
T	T	T

2.2.2   Inference Systems

By using different truth tables shown in the above sections, we can perform some simple logical analysis to derive the desired conclusion based on logic inputs. However, this method may encounter some difficulties when the logic system becomes too complicated and contains too many rows. For example, for n input variables, the related truth table needs 2^n rows, which makes the situation too tough to be analyzed. In that case, we may need to use some inference systems to derive the conclusions.

The so-called inference system in AI refers to the process of reasoning and making decisions based on available information or input data. It involves deriving new knowledge or conclusions from existing knowledge or data. In other words, it is the process of going beyond the information provided to make predictions or draw conclusions based on that information.

In AI, an inference system is similar to a reasoning system and it can be categorized into four types (refer to Sect. 1.3.2 in Chap. 1):

- **Deductive Inference System**
- **Inductive Inference System**
- **Abductive Inference System**
- **Inference System by Analogy**

Deductive inference system involves reasoning from general principles to specific conclusions, while inductive inference system involves inferring general principles or rules based on specific observations or data [5].

One of the most popular and vital inference systems is the fuzzy inference system (FIS). A typical FIS combines the input data represented as a sequence of membership functions (MFs) and various rules that are based on experiences and previous knowledge to infer the final outputs.

For example, to figure out how much tip to pay after a lunch or dinner in a restaurant, an FIS inference reasoning process can be adopted. Two factors (inputs) can be used to determine the amount of tip (output) that should be paid; food quality and service, and all of them can be fuzzificated as three membership functions:

Food	Service	Tip
Delicious	Excellent	Great
Average	Good	Average
Rancid	Poor	Cheap

With three variables, totally we should be able to cover nine rules. To make things simple, here we only use three rules:

1. If the food is rancid or the service is poor, then the tip is cheap.
2. If the service is good, then the tip is average.
3. If the food is delicious and the service is excellent, then the tip is great.

When all of these input variables represented as membership functions are combined with these three rules, the fuzzy inference system can be triggered to infer the related output which is also a fuzzy variable. A Center of Gravity (COG) method could be used to convert the output, tip, from a fuzzy variable to a crisp variable.

Anyway, using knowledge-based systems and other AI techniques can help enhance the intelligence of machines and enable them to perform a wide range of tasks. Before we can dig deeper into AI-related knowledge, let us first have a global picture of the relationship and operational sequence of knowledge used in AI, or AI knowledge cycle.

2.3 AI Knowledge Cycle

The AI knowledge cycle is a process that involves the acquisition, representation, and utilization of knowledge by AI systems. It consists of several stages, which include the following:

1. Data Collection: This stage involves gathering relevant data from various sources such as sensors, databases, or datasets including the original or raw collected data.
2. Data Preprocessing: The collected data then needs to be cleaned, filtered, normalized, or transformed into a suitable format for analysis.
3. Knowledge Base: Knowledge-based systems are computerized systems that emulate human reasoning.
4. Knowledge Representation: This stage involves encoding the data into a format that an AI system can use. This can include symbolic representations, such as

knowledge graphs or ontologies, membership functions, or numerical representations, such as feature vectors.

5. Knowledge Inference: Once the data has been represented, an AI system can use this knowledge to make predictions or decisions. This involves applying machine learning algorithms or other inference techniques to the data.
6. Knowledge Evaluation: This stage involves evaluating the accuracy and effectiveness of the knowledge that has been inferred. This can involve testing the AI system on known examples or other evaluation metrics.
7. Knowledge Utilization: Finally, the knowledge acquired and inferred can be used to perform various tasks, such as natural language processing, image recognition, or decision-making.

The AI knowledge cycle is a continuous process, as new data is constantly being generated, and the AI system can learn and adapt based on this new information. By following this cycle, AI systems can improve their performance and perform a wide range of tasks more effectively.

2.4 Knowledge-Based Systems

Knowledge-based systems are computerized systems that emulate human reasoning. Such systems are built with specific knowledge in certain domains of application and operate in a way similar to that of a human expert. They were created in an attempt to capture and emulate human intelligence in symbolic form, usually in a set of *if-then* rules. Given an input, the system triggers a corresponding rule to produce a response [6].

Knowledge-based systems (KBS) typically have three components, which include [7]:

1. **Knowledge base**: A knowledge base is an established collection of information and resources, generally a dataset. The system uses this as its repository for the knowledge it uses to make decisions.
2. **Inference engine**: An inference engine processes the knowledge throughout the system. It acts similarly to a human brain to derive or infer a response by combining relevant information based on the requests.
3. **User interface**: The user interface is how the knowledge-based system appears to users on the computer. This allows users to interact with the system and submit requests.

Generally, the KBS can be divided into the following five types [7]:

1. **Case-based systems**: Case-based systems use case-based reasoning. This involves reviewing past knowledge of similar situations. Based on what it finds, the knowledge-based system provides solutions that are effective in those given situations.

2. **Expert systems**: Expert systems are one of the most common types of knowledge-based systems. These systems mimic human experts' decision-making processes, making them helpful for complex analyses, calculations, and predictions. In addition to presenting solutions, they provide specific explanations for the problems they are solving.

3. **Hypertext manipulation systems**: Hypertext manipulation systems store knowledge by linking text to other texts and by using hypertext. Hypertext refers to a network of discrete blocks of information interconnected as a way to store data. This type of system allows you to access many types of data easily.

4. **Intelligent tutoring systems**: Intelligent tutoring systems are knowledge-based systems specifically designed to support learning. These systems provide users with personalized feedback and instructions based on their performance and inquiries. As such, they are often used in education, allowing students to learn more and have a personalized learning experience without direct intervention from a teacher.

5. **Rule-based systems**: Rule-based systems rely on human-made, hard-coded rules. It uses these rules to analyze and manipulate data to achieve specific outcomes. This may involve using *IF-THEN* rules, which establish that if a user makes a certain request, then the system delivers a certain outcome.

Among these five types, the Rule-based system is one of the most popular types and it has been widely applied in AI today. Here we will concentrate our study on two popular inference engines for rule-based systems.

2.4.1 Backward Chaining

Backward chaining is the logical process of inferring unknown truths from known conclusions by moving backward from a solution to determine the initial conditions and rules [8]. Backward chaining is often applied in artificial intelligence (AI) and may be used along with its counterpart, forward chaining.

In AI, backward chaining is used to find the conditions and rules by which a logical result or conclusion was reached. An AI might utilize backward chaining to find information related to conclusions or solutions in reverse engineering applications.

As a goal-driven and top-down form of reasoning, backward chaining usually employs a depth-first search strategy by starting from a conclusion, result, or goal and going backward to infer the conditions from which it resulted. Backward chaining traces back and looks through a rules table. In the rules table, it seeks out any actions that are specified in *if-then* statements, applying logic to determine which of the possible actions would have caused the end result.

In summary, the backward chaining engine works as an expert system and tries to find or derive the reason or the answer to the question, *"why this happened?"* based on the conclusion.

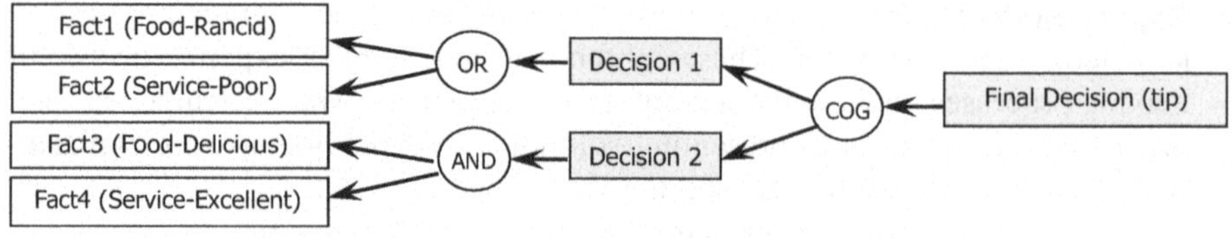

Fig. 2.2 Functional block diagram of a backward chaining process

Still, we try to use the tip payment in a restaurant as an example to illustrate the operational steps and process for the backward chaining method.

Three working rules for the tip payment process are:

1. If the food is rancid or the service is poor, then the tip is cheap.
2. If the service is good, then the tip is average.
3. If the food is delicious and the service is excellent, then the tip is great.

As shown in Fig. 2.2, the backward chaining engine tries to find out in which or what conditions the final result or decision could be obtained. This derivation process starts from the final result and performs a backward investigation or backward chaining to try to find the cause or reason for the current result.

Starting from the final decision or the amount of the paid tip, backward chaining could derive two possible decisions, Decision 1 and Decision 2, respectively. Both decisions were made by combining inputs, food and service, and the *if-then* rules via the inferring reasoning process. Furthermore, the values of two input variables, food and service, can also be derived based on rules 3 and 1 shown above.

Based on the output of the COG, the backward chaining could determine the part or the ratio of each decision. If the tip is great, Decision 2, rule 3 could be identified, and furthermore both input values, food is delicious and the service is excellent, can also be determined. Similarly, if the tip is cheap, Decision 1, rule 1 can be derived and the values of two inputs, food is rancid or the service is poor, could also be determined.

A very similar algorithm or process to backward chaining is the backpropagation algorithm, which is widely implemented in the supervised learning process in machine learning. In fact, as a machine-learning algorithm, backpropagation performs a backward pass for a neural network to adjust the ML model's parameters, aiming to minimize the mean squared error (MSE), and further to get an optimally trained ML model. More detailed discussions about the backpropagation algorithm will be given in Chap. 5.

2.4.2   Forward Chaining

Opposite to backward chaining, forward chaining is used to break down the logic sequence and work through it from beginning to end by attaching each step after the previous one is solved [8]. As a matter of fact, forward chaining is the logical process of inferring unknown truths from known data and moving forward using determined conditions and rules to find a result or a solution [9].

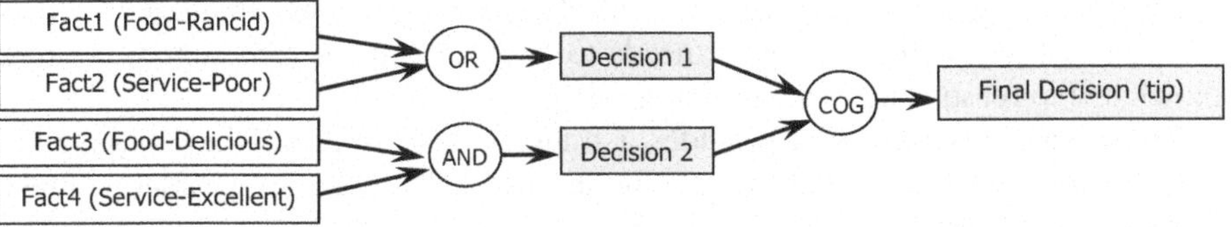

Fig. 2.3 Functional block diagram of a forward chaining process

Both backward chaining and forward chaining are two main reasoning methods and have been widely applied in the supervised learning process in machine learning since both processes use an inference engine and can be described logically as repeated applications of propositional logic processes.

A functional block diagram of using forward chaining to process our tip payment is shown in Fig. 2.3. Three working rules for the tip payment process are:

1. If the food is rancid or the service is poor, then the tip is cheap.
2. If the service is good, then the tip is average.
3. If the food is delicious and the service is excellent, then the tip is great.

Starting from the input variables, food and service, depending on their real values represented as membership functions, related rules can be triggered. If the food is rancid or the service is poor, rule 1 is selected and the first decision, Decision 1, could be derived. Similarly, if the food is delicious and the service is excellent, rule 3 is triggered and the second decision, Decision 2, should be made. After the first inputs layer and the second decision layer, the forward chaining can continue to the next step to derive the final tip or output layer via the COG method.

The name "forward chaining" comes from the fact that the inference engine starts with the data and reasons its way to the answer, as opposed to backward chaining, which works the other way around. In the derivation, the rules are used in the opposite order as compared to backward chaining. In this example, rule 2 is not used in determining the final tip.

Because the data determines which rules are selected and used, this method is called data-driven, in contrast to goal-driven backward chaining inference. The forward chaining approach is often employed by expert systems, such as a fuzzy inference system (FIS).

One of the advantages of forward-chaining over backward-chaining is that the reception of new data can trigger new inferences, which makes the engine better suited to dynamic situations in which conditions are likely to change.

2.5 Knowledge Representations

We have provided a detailed discussion about knowledge-based systems (KBS) with three key components in the last section. In this section, we will concentrate on how to present and use those components to perform reasoning jobs to derive the desired results.

Knowledge representation is a crucial element of Artificial Intelligence. It is believed that an intelligent system needs to have an explicit representation of its knowledge to reason and make decisions [10].

Knowledge representation provides a framework for representing, organizing, and manipulating knowledge that can be used to solve complex problems, make decisions, and learn from data.

2.5.1 What Is Knowledge Representations

Knowledge representation is a fundamental concept in artificial intelligence (AI) that involves creating models and structures to represent information and knowledge in a way that intelligent systems can use. The goal of knowledge representation is to enable machines to reason about the world like humans, by capturing and encoding knowledge in a format that can be easily processed and utilized by AI systems.

There are various approaches to knowledge representation in AI, as shown in Fig. 2.4, which include [10]:

1. **Logical representation**: This involves representing knowledge in a symbolic logic or rule-based system, which uses formal languages to express and infer new knowledge.
2. **Semantic networks**: This involves representing knowledge through nodes and links, where nodes represent concepts or objects, and links represent their relationships.
3. **Frames**: This approach involves representing knowledge in the form of structures called frames, which capture the properties and attributes of objects or concepts and the relationships between them.
4. **Ontologies**: This involves representing knowledge in the form of a formal, explicit specification of the concepts, properties, and relationships between them within a particular domain.

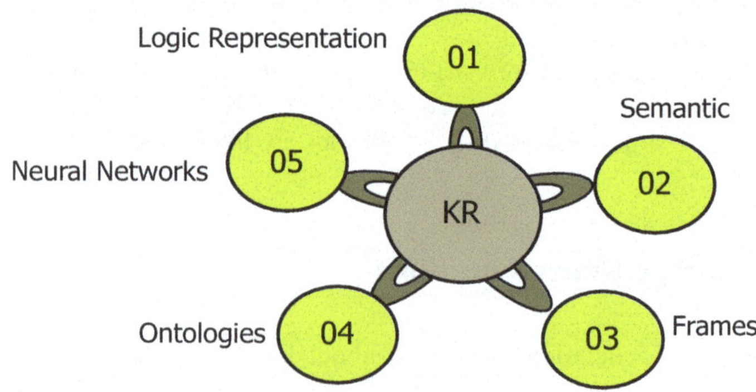

Fig. 2.4 Functional connection among five knowledge representations

5. **Neural networks**: This involves representing knowledge in the form of patterns or connections between nodes in a network, which can be used to learn and infer new knowledge from data.

In this section, we only concentrate on three of them—logical representation, semantic networks, and neural networks, since all of them are popular and widely implemented in today's AI studies and researches.

2.5.2   Logical Representation

Logical representation is a language with some **definite rules** that deal with propositions and has no ambiguity in representation. It represents a conclusion based on various conditions and lays down some important **communication rules**. Also, it consists of precisely defined syntax and semantics which supports sound inference. Each sentence can be translated into logics using syntax and semantics [11].

Syntax decides how we can construct legal sentences in logic and determines which symbol we can use in knowledge representation, and how to write those symbols. Semantics are the rules by which we can interpret the sentence in the logic, and they assign a meaning to each sentence.

Advantages of using logical representation include:

- Logical representation helps to perform logical reasoning.
- This representation is the basis for the programming languages.

  Disadvantages of using logical representation include:

- Logical representations have some restrictions and are challenging to work with.
- This technique may not be very natural, and inference may not be very efficient.

Refer to Sect. 2.2 to get more details about the logical representation.

2.5.3   Semantic Networks

Semantic networks work as an **alternative** to **predicate logic** for knowledge representation [11]. In Semantic networks, you can represent your knowledge in the form of graphical or block networks. This network consists of nodes representing objects and arcs which describe the relationship between those objects. Also, it categorizes the object in different forms and links those objects. This representation consists of two types of relations, as shown in Fig. 2.5:

1. IS-A relation (Inheritance)
2. Kind-of-relation

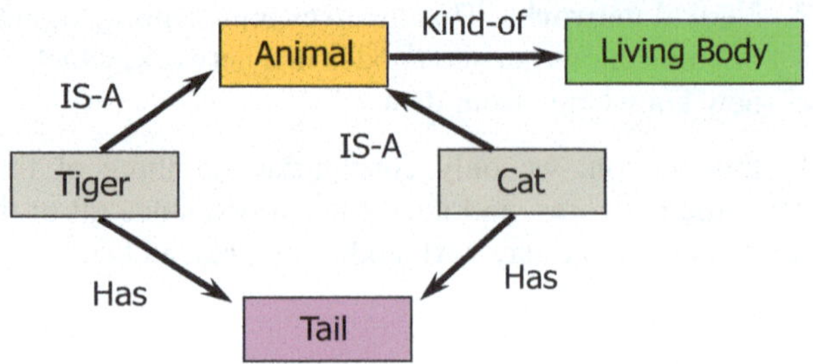

Fig. 2.5 Two types of relations in semantic networks

The animal is a kind of or belongs to a living body class (kind-of-relation), at the same time, both tiger and cat are animals (IS-A relation). However, both animals have tails.

Advantages of using semantic networks include:

- Semantic networks are a natural representation of knowledge.
- It conveys meaning in a transparent manner.
- These networks are simple and easy to understand.

Disadvantages of using semantic networks include:

- Semantic networks take more computational time at runtime.
- Also, these are inadequate as they do not have any equivalent quantifiers.
- These networks are not intelligent and depend on the creator of the system.

2.5.4　Neural Networks

This involves representing knowledge in the form of patterns or connections between nodes in a network, which can be used to learn and infer new knowledge from input data.

A neural network is a method in artificial intelligence that teaches computers to process data in a way that is inspired by the human brain [12]. It is a type of machine learning process, called deep learning, which uses interconnected nodes or neurons in a layered structure that resembles the human brain. It creates an adaptive system that computers use to learn from their mistakes and improve continuously. Thus, artificial neural networks attempt to solve complicated problems, like summarizing documents or recognizing faces, with greater accuracy.

A structural and components diagram of a typical neural network is shown in Fig. 2.6.

A neural network can be considered as a foundation or a structure of a model that is composed of a sequence of layers, and each layer contains some nodes or neurons that are similar to cells in the human brain, which are connected by a group of nervous systems. All kinds of different signals can be detected, transferred, analyzed, and identified by those layers, exactly by those nodes, from input such as sensors, to the output to make decisions or take actions to respond to those signals.

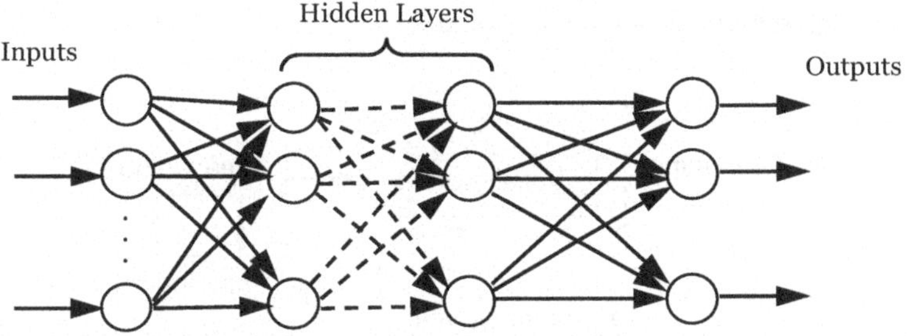

Fig. 2.6 A typical structure and components of a neural network

Neural networks containing a limited number of layers, such as 2 or 3, are called shallow neural networks or shallow learning systems. But those neural networks that include multiple layers, says above 5, are called deep neural networks or deep learning models. The layers between the input and the output layers are called hidden layers since they cannot be seen or observable from outside.

Regularly, neural networks need to be trained by input-output data to learn and understand the required functions or actions assigned by human beings. The trained neural network models are called deep learning models. A more detailed discussion about neural networks and related training-checking methods will be discussed in Chap. 7.

2.5.5 Cycle of Knowledge Representation in AI

The knowledge representation plays an important role in artificial intelligence studies. In fact, Artificial Intelligent Systems (AIS) usually consist of various components to display their intelligent behavior. Some of these components include [11]:

1. Perception
2. Learning
3. Knowledge Representation and Reasoning
4. Planning
5. Execution

In Sect. 1.3 in Chap. 1, we have provided some introductions about basic components used in AI study. The difference between those five components and the above components is that here both components, Planning and Execution, are combined together to map to the Problem-Solving shown in Sect. 1.3 in Chap. 1.

Figure 2.7 shows a functional block diagram to show the different components of the system and how it works.

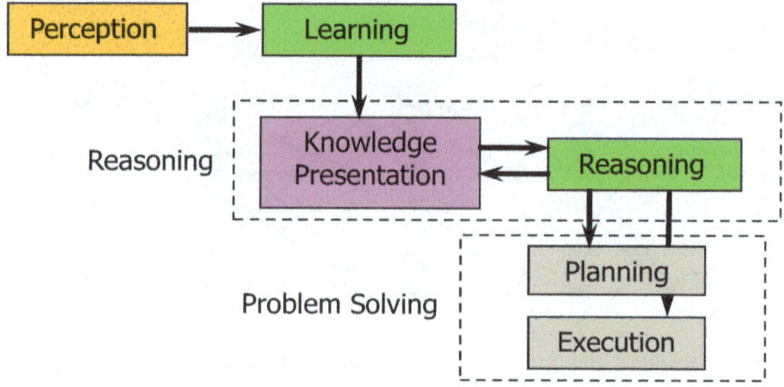

Fig. 2.7 Functional block diagram of the cycle of knowledge representation

2.5.6   Approaches to Knowledge Representation

Various different types of knowledge representation methods have been widely implemented in AI studies. Basically, there are four types of them:

1. Simple relational knowledge
2. Inheritable knowledge
3. Inferential knowledge
4. Procedural knowledge

Simple relational knowledge is one of the simplest types of knowledge representations. For example, a relational database or a dataset can be considered as a kind of relational knowledge representation. This type of knowledge representation is used in database systems where the relationship between different entities is represented. Due to its simplicity, there is a low opportunity for this kind of representation to work as an inference engine.

Table 2.2 shows a typical data table, **Student** table, in a relational database **CSE_DEPT**. All students in this table are connected or related by the **student_id**, which is a primary key in this table. The unique relationship between this table and other tables is identified by that primary key, **student_id** that may be a foreign key in other tables. This table provides a simplest knowledge representation for all students.

Inheritable knowledge refers to knowledge acquired by an AI system through learning and can be transferred or inherited by other AI systems. This knowledge can include models, rules, or other forms of knowledge that an AI system learns through training or experience [10].

In this approach, all data must be stored in a hierarchy of classes.

With this kind of knowledge representation, an AI system can inherit knowledge from other systems to allow it to learn faster and avoid repeating mistakes that have already been made. Inheritable knowledge also allows for knowledge transfer across domains and allows an AI system to apply knowledge learned in one domain to another.

Table 2.2 An example of database: student table

student_id	student_name	gpa	credits	major	schoolYear	email	simage
A78835	Andrew Woods	3.26	108	Computer Science	Senior	awoods@college.edu	NULL
A97850	Ashely Jade	3.57	116	Info System Engineering	Junior	ajade@college.edu	NULL
B92996	Blue Valley	3.52	102	Computer Science	Senior	bvalley@college.edu	NULL
H10210	Holes Smith	3.87	78	Computer Engineering	Sophomore	hsmith@college.edu	NULL
T77896	Tom Erica	3.95	127	Computer Science	Senior	terica@college.edu	NULL

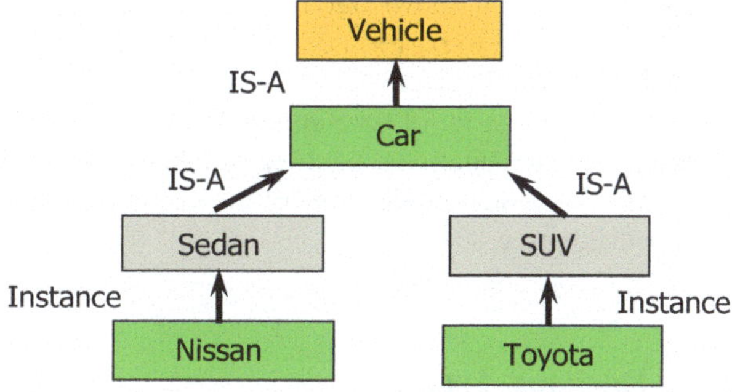

Fig. 2.8 An example of inheritable knowledge representation

In this kind of representation, we use boxed nodes to represent objects and their values and use arrows that point from objects to their values. Figure 2.8 shows an example of inheritable knowledge representation.

Where the top root of this architecture is Vehicle class, which is an abstract description to all automobiles. Under the top class, the Car is a child class with two types of cars, Sedan and SUV, respectively. To instantiate two types of classes, Sedan and SUV, two instances or objects, Nissan and Toyota, are generated. The relations between the top three layers of classes are **IS-A**, but the relationships between Sedan and SUV classes and their derived objects are **Instances**.

Inferential Knowledge refers to the ability to draw logical conclusions or make predictions based on available data or information. In artificial intelligence, inferential knowledge is often used in machine learning algorithms, where models are trained on large amounts of data and then used to make predictions or decisions about new data [10].

As we mentioned in our above discussions, a knowledge base system should contain a knowledge base and an inference engine that can perform inferring function to predict or estimate the conclusion based on the input data in the knowledge base.

For example, as we mentioned in Sect. 1.3.2.1 in Chap. 1, a good student's GPA should be greater than 3.5. Tom is a good student, thus Tom's GPA should be higher than 3.5. Another instance is: New Year is January 1st of each year, today is January 1st, so today is New Year.

Procedural Knowledge refers to the knowledge or instructions required to perform a specific task or solve a problem. This knowledge is often represented in algorithms or rules dictating how a machine processes data or performs tasks.

Procedural knowledge is also known as imperative knowledge, performative knowledge, or practical knowledge. An example of procedural knowledge could be learning to ride a bicycle or learning the driving process [10]. Other examples of procedural knowledge include recipes, science experiments, assembly manuals, or instructions for playing games.

Thus, procedural knowledge is an important aspect of artificial intelligence, and it allows machines to perform complicated tasks and make decisions based on specific instructions.

We have spent a lot of time and space discussing about the knowledge base and knowledge representations, next let us take care of how to make decisions for the given knowledge and model. One of the most powerful tools used in decision analysis is decision tree.

2.6 Decision Tree

A decision tree is a decision support hierarchical model that uses a tree-like model of decisions and their possible consequences, including chance event outcomes, resource costs, and utility. It is one way to display an algorithm that only contains conditional control statements.

Decision trees are commonly used in operations research, specifically in decision analysis [13], to help identify a strategy most likely to reach a goal, but are also a popular tool in machine learning.

A decision tree is a flowchart-like structure in which each internal node represents a *test* on an attribute (e.g., whether a coin flip comes up heads or tails), each branch represents the outcome of the test, and each leaf node represents a class label or a decision (decision taken after computing all attributes). The paths from root to leaf represent classification rules. The tree structure comprises a root node, branches, internal nodes, and leaf nodes, forming a hierarchical, tree-like structure.

In decision analysis, a decision tree and the closely related influence diagram are used as a visual and analytical decision support tool, where the expected values or expected utility of competing alternatives are calculated. A decision tree consists of three types of nodes [14]:

1. Decision nodes—typically represented by squares
2. Test or chance nodes—typically represented by circles
3. End nodes—typically represented by triangles

An example of making decisions for a possible trip by using a decision tree is shown in Fig. 2.9.

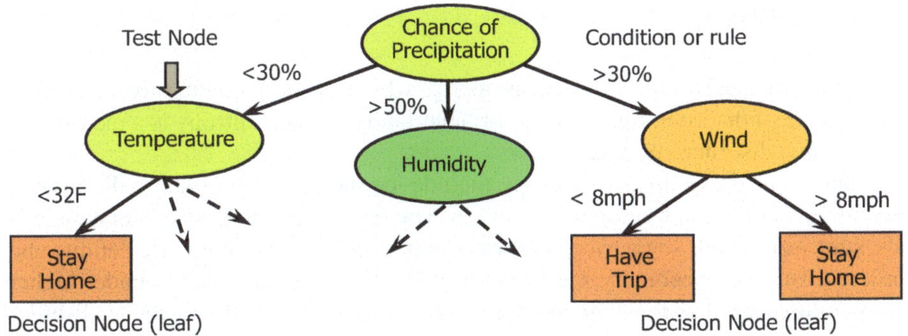

Fig. 2.9 An example of a decision tree to make decision for a trip

To make a decision on whether to have a trip based on the weather is described by this tree. All circle or ellipse nodes in the tree represent the related test or chance nodes, such as temperature, humidity, and wind. The numbers or logic representations between test nodes are conditions or rules, and the square nodes represent the decision nodes, such as *stay home* or *have trip* at the bottom.

The decision tree can be considered as a non-parametric supervised learning algorithm, which is utilized for both classification and regression tasks.

The most popular terminologies used in a decision tree include:

- **Root Node**—It is the node present at the beginning or the top of a decision tree from this node the population starts dividing according to various features. In Fig. 2.9, the top node *Chance of Precipitation* is a root node.
- **Test Nodes**—The nodes we get after splitting the root nodes are called test or chance nodes. Three nodes in Fig. 2.9, *Temperature*, *Humidity*, and *Wind*, belong to test nodes.
- **Decision or Leaf Nodes**—The nodes where further splitting is not possible are called leaf nodes or decision nodes. Three nodes in Fig. 2.9, *Stay Home, Have Trip*, and *Stay Home*, are three decision nodes.
- **Sub-tree**—Just like a small portion of a graph is called a sub-graph, a sub-section of this decision tree is called a sub-tree.
- **Pruning**—It is nothing but cutting down some nodes to stop over-fitting.

Decision trees are upside down which means the root is at the top and then this root is split into various several nodes. Decision trees are nothing but a bunch of *if-else* statements in layman's terms. It checks if the condition is true and if it is then it goes to the next node attached to that decision [15].

The goal of machine learning is to decrease uncertainty or disorders from the dataset, and for this, we use decision trees.

Now you must be thinking how do I know what should be the root node? What should be the decision node? When should I stop splitting? To decide this, there is a metric called *Entropy* which is the amount of uncertainty in the dataset.

2.6.1 *Entropy*

Entropy is a measurable physical property, which is most commonly associated with a state of disorder, randomness, or uncertainty. In fact, entropy is a measure of uncertainty, disorder, or randomness.

Some typical examples of entropy include ice melting, salt or sugar dissolving, making popcorn, and boiling water for tea. The molecules on those original materials, such as ice, salt, sugar, rice, and water under room temperature, are definite and ordered. But they become some disorder after those physical actions under higher temperatures or dissolved in the water. To measure or estimate those disorder degrees, entropy is a good candidate and tool.

Most implementations and applications of using entropy are related to risk assessment or estimations, especially in financial studies, including stocks and investments.

To better understand entropy, we had better starting a real simple example.

During the summer break, a group of students want to decide to have their vacation in some interesting spots to have fun with that period of time. At the time, two options are available: Paris and London. Totally we assume that there are 11 students and they can select one of them by voting. After all students provided their voting, the result is: 6 selected Paris and 5 selected London, and the voting results are almost equal.

Which city should be selected as the target spot now? It is hard to answer that question and make the final decision since the voting for both of them is equal.

This is exactly what we call a disorder since we have an equal number of votes for both cities, and we cannot really decide which spot students should visit. It would have been much easier if the votes for Paris were 8 and for London it was 3. Here we could easily say that the majority of votes are for Paris hence everyone will be going to that city. This situation is similar to a decision tree with the output being either a Yes or a No.

The general mathematical formula for entropy can be described as follows [16]:

$$E(S) = \sum_{i=1}^{n} - p_i \log_2 p_i \tag{2.1}$$

where p_i is simply the probability of an element or a class i in our dataset. For simplicity's sake let us say we only have two classes, a positive class and a negative class. Therefore i here could be either (+) or (−). So if we had a total of 100 data points in our dataset with 30 belonging to the positive class and 70 belonging to the negative class then $p_{(+)}$ would be 3/10 and $p_{(-)}$ would be 7/10, yes, it is pretty straightforward.

If we want to calculate the entropy of the above classes in this example, using the formula above we can get the following result:

$$E(S) = -\frac{3}{10} \times \log_2\left(\frac{3}{10}\right) - \frac{7}{10} \times \log_2\left(\frac{7}{10}\right) \cong 0.88$$

The calculated entropy is approximately 0.88, which is considered a relatively higher entropy. A high level of disorder generally means a low level of purity or a high level of impurity. Regularly, entropy is measured between 0 and 1. Of course, this depends on the number of classes in the real dataset. Entropy can also be greater than 1 but it means the same thing, a very high level of disorder. For the sake of simplicity, the examples in this section will have entropy between 0 and 1.

Now let us consider our example above, the entropy in our case can be calculated as below:

$$E(S) = -p_{(+)} \log_2 p_{(+)} - p_{(-)} \log_2 p_{(-)} \tag{2.2}$$

where

$p_{(+)}$ is the probability of positive class.
$p_{(-)}$ is the probability of negative class.
S is the subset of the training example.

Next, let us discuss how to use entropy to estimate the degree of uncertainty or randomness for a node in decision tree or decision analysis.

2.6.2 How to Use Entropy in Decision Tree

As we discussed above, entropy is basically used to measure the impurity of a node. Exactly, impurity is the degree of randomness or uncertainty; it indicates how random or how uncertain our data are. A so-called *pure-sub-split* means that either you should be getting a *Yes*, or you should be getting a *No* based on the rules or conditions.

Now we have 6 students selected Paris (*Yes*) and 5 selected London (*No*) initially, after the first split the left node gets 4 *Yes* and 3 *No* whereas the right node gets 3 *Yes* and 1 *No*, as shown in Fig. 2.10.

We see here the split is not pure since we can still see some negative classes in both nodes. In order to make a decision tree, we need to calculate the impurity of each split, and when the purity is 100%, we make it as a leaf node.

To check the impurity of feature 2 and feature 3, we need the help of the entropy formula to calculate the impurity degree for each feature.

For feature 2, we have

$$E(S_2) = -\frac{4}{7} \times \log_2\left(\frac{4}{7}\right) - \frac{3}{7} \times \log_2\left(\frac{3}{7}\right) = 0.9852$$

For feature 3, we have

$$E(S_3) = -\frac{3}{4} \times \log_2\left(\frac{3}{4}\right) - \frac{1}{4} \times \log_2\left(\frac{1}{4}\right) = 0.8113$$

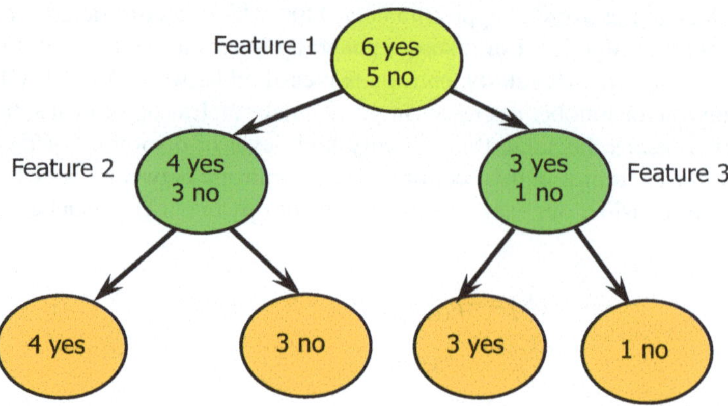

Fig. 2.10 An example of using entropy in decision tree

From this example, it is clear that the right node (feature 3) has low entropy or more purity than the left node (feature 2) since the right node has a greater number of *Yes* and it is easy to decide from here.

We can get a conclusion from this example, which is: the higher the entropy, the lower the purity and the higher the impurity.

Now we have gotten some basics about how to measure disorder or randomness degree for given nodes. Next, we need to find a method to measure the reduction of this disorder in our target variable or class for giving additional information about it. This method is called Information Gain.

2.6.3   Information Gain

Information gain measures the reduction of uncertainty given some feature and it is also a deciding factor for which attribute should be selected as a decision node or root node. This method can be mathematically written as (Information Gain = IG(Y, X)):

$$IG(Y,X) = E(Y) - E(Y|X)$$
(2.3)

where $E(Y|X)$ is the entropy of Y given X and $E(Y)$ is just the entropy of Y.

By simply subtracting the entropy of Y given X from the entropy of just Y, we can calculate the net reduction of uncertainty about Y given an additional piece of information X about Y. This is called Information Gain. The greater the reduction in this uncertainty, the more information is gained about Y from X.

In fact, the so-called information gain is the difference between the original entropy $E(Y)$ and the modified entropy $E(Y|X)$ after some other additional factor, such as X, is added. Exactly we want to get reduced entropy compared with the original entropy for the input data Y. If that difference is a positive number, which means that the entropy is really reduced and the uncertainty is also reduced, it results in an increment on certainty. From this point of view, we say that we get a gain or an improvement on our information system.

Let us build an example to illustrate the Information Gain (IG) more clearly.

Assume we have a traveling club with 30 members. The dataset is to predict whether the person will go on an overseas tour or not. Let us say 16 people go on the tour and 14 people do not.

Two features are involved to predict whether a member will go on the tour or not.

1. Feature 1 is the amount of the **Money**, which takes two values *high* and *low*.
2. Feature 2 is **Time**, which takes three values *little*, *some*, and *enough*.

Let us take a closer look at how our decision tree will be made using these two features. We will use the information gained to decide which feature should be the root node and which feature should be placed after the split.

Based on the decision tree shown in Fig. 2.11, let us calculate the entropy for the parent node and the special entropies under some additional conditions. First, the entropy for the root node is:

$$E(Root) = -\frac{16}{30} \times \log_2\left(\frac{16}{30}\right) - \frac{14}{30} \times \log_2\left(\frac{14}{30}\right) = 0.99$$

The entropy of our root variable is close to 1, at maximum disorder due to the even split about equally between class labels *Go to tour* and *Not go to tour*. Our next step is to calculate the entropy of our target variable by giving additional information about the money amount to support the tour. For this, we will calculate the entropy for the root via each value of money amount and add them using a weighted average of the proportion of observations that end up in each value:

$$E(Root|Money = high) = -\frac{12}{13} \times \log_2\left(\frac{12}{13}\right) - \frac{1}{13} \times \log_2\left(\frac{1}{13}\right) \cong 0.39$$

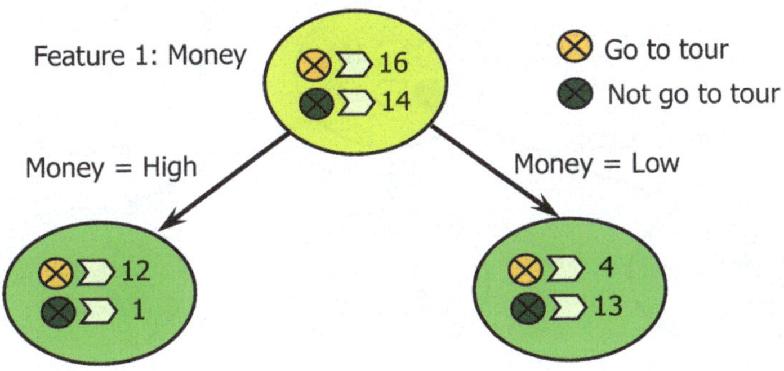

Fig. 2.11 Decision tree for feature 1: Money

$$E\left(Root|Money = low\right) = -\frac{4}{17} \times \log_2\left(\frac{4}{17}\right) - \frac{13}{17} \times \log_2\left(\frac{13}{17}\right) \cong 0.79$$

To get the weighted average of the entropy of each node, we need to do as below:

$$E\left(Root|Money\right) = \frac{13}{30} \times 0.39 + \frac{17}{30} \times 0.79 \cong 0.62$$

Now we have the value of $E(Root)$ and $E(Root|Money)$; next we can get the difference between the original entropy and the modified entropy, or the information gain:

$$IG\left(Root, Money\right) = E\left(Root\right) - E\left(Root|Money\right) = 0.99 - 0.62 = 0.37$$

Our root entropy was near 0.99 and after checking this value of information gain, we can get a conclusion that the entropy of the dataset will decrease by 0.37 if we make *Money* as our root node in this decision tree.

In a similar way, we can do this with the other feature *Time* and calculate its information gain.

The decision tree for feature 2, Time, is shown in Fig. 2.12. First, let us calculate the entropy for each node:

$$E\left(Root\right) = -\frac{16}{30} \times \log_2\left(\frac{16}{30}\right) - \frac{14}{30} \times \log_2\left(\frac{14}{30}\right) = 0.99$$

$$E\left(Root|Time = Little\right) = -\frac{6}{8} \times \log_2\left(\frac{6}{8}\right) - \frac{2}{8} \times \log_2\left(\frac{2}{8}\right) \cong 0.81$$

$$E\left(Root|Time = Some\right) = -\frac{4}{10} \times \log_2\left(\frac{4}{10}\right) - \frac{6}{10} \times \log_2\left(\frac{6}{10}\right) \cong 0.97$$

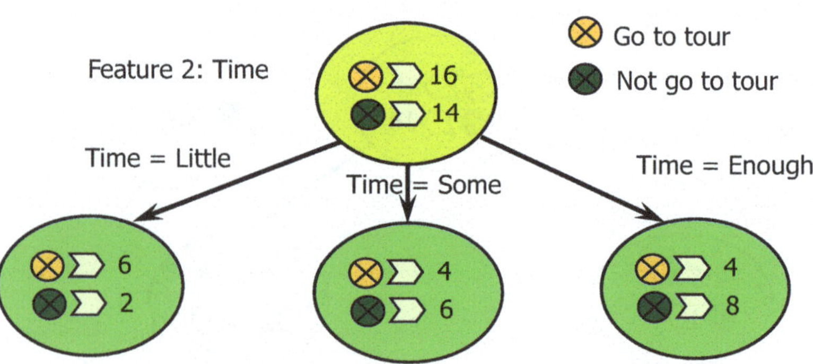

Fig. 2.12 Decision tree for feature 2: Time

$$E\left(Root | Time = Enough\right) = -\frac{4}{12} \times \log_2\left(\frac{4}{12}\right) - \frac{8}{12} \times \log_2\left(\frac{8}{12}\right) \cong 0.92$$

The weighted average of entropy of each node can be calculated as below:

$$E\left(Root | Time\right) = \frac{8}{30} \times 0.81 + \frac{10}{30} \times 0.97 + \frac{12}{30} \times 0.92 \cong 0.91$$

The value of $E(Root)$ and $E(Root | Time)$, as well as the information gain, will be:

$$IG\left(Root, Time\right) = E\left(Root\right) - E\left(Root | Time\right) = 0.99 - 0.91 = 0.08$$

By comparing the values of two information gains, we can see that the *Money* feature gives more reduction, which is 0.37 than that of the *Time* feature, which is 0.08. Hence, we should select the feature that has the highest information gain and then split the node based on that feature.

Using the result of this example, *Money* will be our root node and we should do the same for sub-nodes. Here we can see that when the money is *high*, the entropy is low and hence we can assume or derive that a person will definitely go to the tour if he/she has high funds, but what if the money is low? We will again split the node based on the new feature, which is *Time*.

Next, let us have a real case study to illustrate the entropy and information gain more clearly.

2.6.4   A Real Case Study

Suppose an auto insurance company will determine the possible insurance premium or quota for a given number of customers' insurance applications. A background check and analysis may be performed based on the background-question surveys for customers. Based on those answers, the insurance company can determine the related auto quota and provide appropriate premium costs. The first step to do those analyses is to develop or build a table based on the facts provided by customers. A typical table is shown in Table 2.3.

The table contains the relationship between the number of traffic accidents that happened in the past 5 years and the risk estimations. Three levels of traffic accidents, named *Less*, *Normal*, and *More*, are defined. Two levels are built for the risk, named *Low* and *High*.

Our target variable is *Risk* which can take on two values *Low* and *High* and we only have one feature called *Accident Number* which can take on values *Low*, *Mid*, and *High*. There are a total of 14 observations: 7 of them belong to the *Low* class and 7 belong to *High* Class. So it is an even split by itself.

Table 2.3 Customer survey feedback table

Accident Number	Risk Level		Total
	Low	*High*	
Less	4	0	4
Normal	2	3	5
More	1	4	5
Total	**7**	**7**	**14**

Checking across the top row we can find that there are four observations that have values of *Low* for the feature risk level. Furthermore, we can also see how our target variable is split for *Low* in risk. For observations that have a value *Low* for their risk, there are four that belong to the *Less* class, two belong to the *Normal* class, and only one belongs to the *More* class. I can similarly figure out such values for other values of *Risk Level* from the table.

For this case study, we will use this survey table as a starting point to calculate the entropy of our target variable, *Risk Level*, by itself and then calculate the entropy of our target variable given additional information about the feature, *Accident Number*. This will enable us to calculate how much additional information the *Accident Number* can provide for our target variable *Risk Level*.

First, let us calculate the entropy for our target variable *Risk Level* as below:

$$E\left(Risk\ Level\right) = -\frac{7}{14}\log_2\left(\frac{7}{14}\right) - = -\frac{7}{14}\log_2\left(\frac{7}{14}\right) = 1$$

It can be seen that the entropy of our target variable is 1, it is a maximum disorder or impurity due to the even split between class labels, *Low* and *High*. Based on this result, it is hard to make any meaningful decision. Therefore, we need to perform our next step to calculate the entropy of our target variable given additional information about *Risk Level*.

For this purpose, we need to calculate the entropy for *Risk Level* for each value of *Accident Number*, and add them using a weighted average of the proportion of observations that end up in each value. Let us do those calculations one by one shown below:

$$E\left(Risk\ Level|Accident\ Number = Less\right) = -\frac{4}{4}\log_2\left(\frac{4}{4}\right) - \frac{0}{4}\log_2\left(\frac{0}{4}\right) = 0$$

$$E\left(Risk\ Level|Accident\ Number = Normal\right) = -\frac{2}{5}\log_2\left(\frac{2}{5}\right) - \frac{3}{5}\log_2\left(\frac{3}{5}\right) = 0.97$$

$$E\left(Risk\ Level|Accident\ Number = More\right) = -\frac{1}{5}\log_2\left(\frac{1}{5}\right) - \frac{4}{5}\log_2\left(\frac{4}{5}\right) = 0.72$$

The weight average is:

$$E\left(Accident\ Number|Risk\ Levl\right)=\frac{4}{14}\times0+\frac{5}{14}\times0.97+\frac{5}{14}\times0.72=0.60$$

Now we can calculate two different information gains, one is for the original *Risk Level* and the other is for the changed *Risk Level* after adding additional *Accident Number* classes, to see how much the impurity or disorder value can be reduced:

$$\text{Information Gain}:\text{IG}\left(Risk\ Level,Accident\ Number\right)=1-0.60=0.40$$

It can be found that the information gain has been reduced from the original value of 1 to the current value of 0.40. Similarly, the impurity or disorder level is also reduced by the same amount. Relatively, the purity level has been increased.

2.7   From Decision Tree to Neural Networks

In Sect. 2.6.3, we provided a real example to illustrate the entropy and information gain and their implementations in detail. In that example, two groups of club members provided two options for one overseas trip; among them 16 people voted to go on the tour but 14 people did not. Two features are involved to predict whether a member will go on the tour or not, money and time. In terms of those two features, two decision trees were built, as shown in Figs. 2.11 and 2.12.

Now let us have a closer look at those figures. Assume that now we try to combine both voting results with two decision trees together to perform our decision-making process. Also, we like to add two features, money and time, as two inputs to this new decision tree.

To make our decision tree more straightforward without any confusion, we can rotate both decision trees shown in Figs. 2.11 and 2.12 counter-clockwise by 90°. Our rotated two decision trees or our new decision tree is shown in Fig. 2.13.

It can be found from Fig. 2.13 that this new decision tree is very similar to a neural network in structure. Each test node can be mapped to a node or a neuron in the neural network, and each rule or condition can be considered as a weight factor w_{ij} in neural network. More leaf nodes could be developed if we want to split the tree into more layers.

In fact, neural networks did do a similar decision-making process as decision trees did, but the difference is that the neural networks must be trained, which is called the learning process by using input-output data pairs to identify all weighted factors' values. Then the trained neural networks can perform the desired decision-making process to do some intelligent jobs. Another difference is that the nodes in the neural networks may have multiple rules for different nodes at the next layers simultaneously.

More detailed discussions about neural networks will be given in Chap. 7.

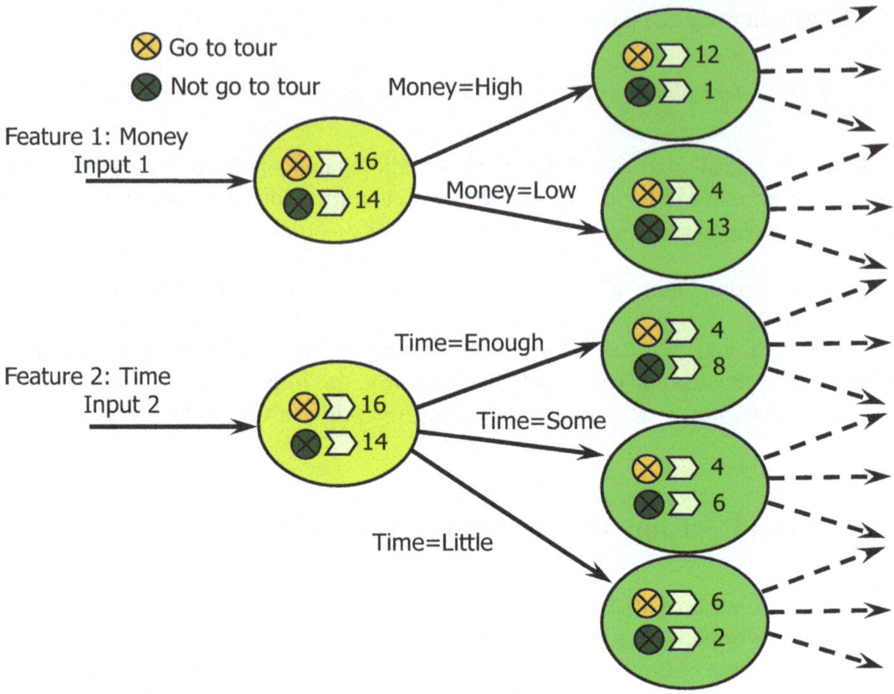

Fig. 2.13 An example of decision tree to neural networks

2.8 Chapter Summary

In this chapter, three basic and popular intelligent techniques in AI systems are introduced, which include

1. Fuzzy Inference System (FIS)
2. Machine Learning (ML) Algorithms
3. Deep Learning (DL) Algorithm

Among them, the first strategy belongs to the category of learning based on experiences, but the second and the third techniques can be considered as a category of learning based on the input and the output data.

In fact, an artificial intelligence system can be considered as an intelligent machine that can perform a learning and decision-making process. A completed artificial intelligence (AI) system is composed of two basic components: fuzzy inference system (FIS) and machine learning (ML) process.

The FIS can be further divided into three sub-components: Type-I FIS, Interval Type-II FIS, and higher-level type FIS.

ML can be categorized into three sub-groups: Supervised Learning, Unsupervised Learning, and Reinforcement Learning processes.

In order to have a clear picture and solid understanding about those AI components and their working procedures, some fundamental and basic AI-related concepts and operational principles are discussed starting from Sect. 2.2.

These concepts and operational principles include:

1. Logical Intelligence
2. Propositional Logic
3. Knowledge Cycle
4. Knowledge-Based System
5. Knowledge Presentation
6. Decision Tree

Most popular elements and features related to learning and decision-making process are also introduced, such as forward chaining, backward chaining, simple relational knowledge, inheritable knowledge, inferential knowledge, procedural knowledge, entropy, and information gain.

A mapping discussion between the decision trees and the neural networks is given in the last section to provide an illustration to show a similarity between a decision tree and a neural network in structure.

When finishing this chapter, students will be able to:

- Understand the foundation of AI and related components as well as basic operational procedures for learning and decision-making process.
- Understand the basic knowledge-based system, knowledge presentations, and knowledge cycle procedure.
- Understand the decision tree architecture and decision-making process.
- Analyze simple knowledge systems by building decision tree, calculating entropy and information gains to reduce the impurity or disorder degree.
- Get a mapping relationship between the decision trees and neural networks.

Home Works

I. True/False Selections

_____1. Artificial intelligence is exactly a learning and making decision process to derive or predict the possible results or outputs based on inputs.

_____2. Two approaches are popularly implemented in learning and decision-making process, learning based on knowledge and learning based on experience.

_____3. Machine learning includes supervised learning, unsupervised learning, semi-supervised learning, and reinforcement learning.

_____4. The deep learning is a subset of the ML and it only uses a neural network with multiple layers and nodes as a model to be trained by using input-output data pairs.

_____5. Logic intelligence is logic-mathematical intelligence, and it means a kind of ability to reason, recognize, and logically analyze complicated problems in a logical way.

_____6. Propositional logic handles propositions and statements with some reasoning process.

_____7. Conjugation is equivalent to an *AND* operation; generally, it works for more than one logic variable.

_____8. Knowledge-based systems typically have two components, knowledge base and inference engine.

_____9. A neural network is a method in artificial intelligence that teaches computers to process data in a way that is inspired by the human brain.

_____10. Entropy is a measure of uncertainty, disorder, or randomness for some variables.

II. Multiple Choices

1. Knowledge-based systems typically have three components, which include _____.

   (a) Knowledge base, rule base, and reasoning
   (b) Rule base, inference engine, and reasoning
   (c) Reasoning, knowledge cycle, and user interface
   (d) Knowledge, inference engine, and user interface

2. The following approaches are involved in knowledge representation in AI _________.

   (a) Logical representation, semantic networks, ontologies
   (b) Ontologies, semantic networks, frames
   (c) Frames, neural networks, knowledge base
   (d) None of those

3. Four types of knowledge representation methods have been widely implemented in AI studies, they are ________.

   (a) Simple relational, logical, procedural, and inferential knowledge
   (b) Inheritable, inferential, seasoning, and procedural knowledge
   (c) Procedural, inheritable, logical, and inferential knowledge
   (d) Inferential, inheritable, procedural, and simple relational knowledge

4. A decision tree consists of three types of nodes, they are ________.

   (a) Decision nodes, chance nodes, and end nodes
   (b) Terminal nodes, starting nodes, and middle nodes
   (c) Root nodes, end nodes, and test nodes
   (d) None of them

5. Entropy is a measurable physical property of a state for ________.

   (a) Disorder
   (b) Randomness
   (c) Uncertainty
   (d) All of them

6. Which of the following is a correct equation for information gain ______?

   (a) $IG(X,Y) = E(Y) - E(X)$
   (b) $IG(Y,X) = E(X) - E(Y)$
   (c) $IG(Y,X) = E(Y) - E(Y|X)$
   (d) $IG(Y,X) = E(X) - E(X|Y)$

7. Some typical examples of entropy include ____________.

   (a) Ice melting
   (b) Salt or sugar dissolving
   (c) Boiling water for tea
   (d) All of them

8. Backward chaining is the logical process of inferring unknown truths from ________ conclusions by moving backward from a solution to determine the ___________.

   (a) Unknown, known input
   (b) Unknown, truths
   (c) Known, unknown input
   (d) Unknown, unknown output

9. Forward chaining is the logical process of inferring unknown truths from _________ and moving forward using determined conditions and rules to find a ___________.

   (a) Unknown inputs, known output
   (b) Known inputs, unknown output
   (c) Known inputs, known output
   (d) Unknown inputs, unknown output

10. Entropy is a measurable physical property, which is most commonly associated with a state of disorder, randomness, or uncertainty. The ________ the entropy value, the ______ the uncertainty of a variable.

    (a) Lower, less
    (b) Lower, lower
    (c) Higher, higher
    (d) Higher, lower

III. Exercises

1. Provide a basic description about artificial intelligence, including the definition and components involved.
2. List three basic categories of intelligent techniques that are covered by AI systems.
3. List four sub-categories under machine learning.
4. Represent the propositional logic A ∧ B ∨ C to a truth table (using either T/F or 1/0 in the truth table).

5. Given the following truth table. Convert it back to a propositional logic representation.

A	B	C	X
F	F	F	F
F	F	T	F
F	T	F	F
F	T	T	T
T	F	F	F
T	F	T	T
T	T	F	F
T	T	T	T

6. Represent the propositional logic A $\wedge$ (B $\vee$ C) to a truth table (using either T/F or 1/0 in the truth table).
7. A weather forecast indicates that there is 70% rain but 30% with no rain. Calculate the entropy value for the uncertainty of raining $E(Rain)$.
8. Table 2.4 shows the relationship between a target variable *Humidity* and a feature *Rain*. Determine the entropy values for the target variable *Humidity*. Also calculate the entropy for *Humidity* for each value of *Rain*, and add them using a weighted average of the proportion of observations that end up in each value. Finally get the information gains to check whether the disorder or uncertainty level can be reduced.
9. Table 2.5 shows the relations between the target variable *Fires* and a feature of forest fires, *Wind*. The target variable *Fires* has been categorized into three levels: *No*, *Small*, and *Large*. The feature *Wind* is also divided into three classes; *Weak*, *Normal*, and *Strong*. Calculate the original entropy value for the target variable *Fires*. Also derive the entropy for *Fires* on each class on the *Wind*. Get the information gain to confirm your result.

Table 2.4 Weather forecast for raining

Rain	Humidity		Total
	Low	High	
No	4	0	4
Little	2	3	5
Much	1	4	5
Total	7	7	14

Table 2.5 Risk for forest fires

Wind	Fires			Total
	No	Small	Large	
Weak	5	1	0	6
Normal	1	0	4	5
Strong	0	5	6	11
Total	6	6	10	22

References

1. https://www.multiplenatures.com/insight-posts/logical-intelligence.
2. R. Neapolitan & X. Jiang, Artificial Intelligence with an Introduction to Machine Learning, CRC Press, 2nd Edition, ISBN: 13:978-1-138-50238-3, 2018.
3. https://www.javatpoint.com/propositional-logic-in-artificial-intelligence.
4. https://brilliant.org/wiki/propositional-logic/.
5. https://www.scaler.com/topics/inference-rules-in-ai/.
6. Peter C.Y. Chen, Aun-Neow Poo, *Engineering, Artificial Intelligence in*, Editor(s): Hossein Bidgoli, Encyclopedia of Information Systems, Elsevier, 2003, Pages 141–155, ISBN 9780122272400.
7. https://www.indeed.com/career-advice/career-development/what-is-knowledge-based-system.
8. https://www.techtarget.com/whatis/definition/backward-chaining?Offer=abt_pubpro_AI-Insider.
9. https://www.techtarget.com/whatis/definition/forward-chaining?Offer=abt_pubpro_AI-Insider.
10. https://www.scaler.com/topics/knowledge-representation-in-ai/.
11. https://www.edureka.co/blog/knowledge-representation-in-ai/#:~:text=Logical%20representation%20is%20a%20language,down%20some%20important%20communication%20rules.
12. https://aws.amazon.com/what-is/neural-network/#:~:text=A%20neural%20network%20is%20a,that%20resembles%20the%20human%20brain.
13. Von Winterfeldt, Detlof; Edwards, Ward (1986). "Decision trees". *Decision Analysis and Behavioral Research*. Cambridge University Press. pp. 63–89. ISBN 0-521-27304-8.
14. Kamiński, B.; Jakubczyk, M.; Szufel, P. (2017). "A framework for sensitivity analysis of decision trees". *Central European Journal of Operations Research*. **26** (1): 135–159. doi:https://doi.org/10.1007/s10100-017-0479-6. PMC 5767274.PMID 29375266.
15. https://www.analyticsvidhya.com/blog/2021/08/decision-tree-algorithm/#:~:text=A%20decision%20tree%20is%20a%20tree%2Dlike%20structure%20that%20represents,based%20on%20the%20weather%20conditions.
16. https://towardsdatascience.com/entropy-how-decision-trees-make-decisions-2946b9c18c8.

Chapter 3
Fuzzy Logic Inference Systems

3.1 Introduction

As we discussed in Chap. 2, fuzzy inference system (FIS) is one of the major components in artificial intelligence (AI), and it plays an important role in logical intelligence and inferring process in the AI study field. Depending on its function, it can be divided into three categories:

1. Type-I FIS
2. Type-II FIS including the Interval Type-II FIS
3. Type-III FIS

The number of types is equivalent to the dimensions of FIS implemented in applications. The higher the type number, the more complicated in FIS. Due to the complex degree, in this book, we only take care of the first two types, exactly Type-I and Interval Type-II FIS.

In this chapter, we will provide detailed and completed discussions about this technology.

Fuzzy logic or fuzzy inference systems are not new terminologies and they have been developed and applied in all aspects of our world for a long time.

The idea of fuzzy logic was invented by professor L. A. Zadeh of the University of California at Berkeley in 1965 [1]. This invention was not well recognized until Dr. E. H. Mamdani, who is a professor at London University, applied the fuzzy logic in a practical application to control an automatic steam engine in 1974 [2], which is almost 10 years after the fuzzy theory was invented. Then in 1976, Blue Circle Cement and SIRA in Denmark developed an industrial application to control cement

---

Supplementary Information The online version contains supplementary material available at https://doi.org/10.1007/978-3-031-84423-2_3.

Y. Bai, *AI Foundations and Applications with MATLAB*,
https://doi.org/10.1007/978-3-031-84423-2_3

kiln [3]. That system began to operate in 1982. More and more fuzzy implementations have been reported since the 1980s, including those applications in industrial manufacturing, automatic controls, automobile productions, banks, hospitals, libraries, and academic education. Fuzzy logic techniques have been widely applied in our society today.

Starting with Type-I FIS, before we can continue our discussion about this topic, we first need to have a definite answer for those fuzzy-related terminologies, such as what is fuzzy logic? What is a fuzzy inference system? What is the meaning of the terminology fuzzy?

3.2 Fuzzy Logic Idea

Generally speaking, a fuzzy logic idea is similar to the human being's feeling and inference or derivation process. For example, at one working day's morning after you get up, your mom might tell you: Being slowly when driving on your way to work in highway! How slow is it? Your mom did not tell you the exact driving speed such as 50 MPH or 60 MPH in the highway at all, but you have a 100% understanding and an idea of what is the meaning of "slowly" your mom meant. Yes, this is a fuzzy idea and this is the meaning of the terminology fuzzy.

Fuzzy idea and fuzzy logic are so often utilized in our routine life that nobody even pays any attention to them. For instance, to answer some questions in certain surveys, most time one could answer with "Not very satisfied" or "Somehow satisfied," which are also fuzzy or ambiguous answers. Exactly in what degree is one satisfied or dissatisfied with some service or product for those surveys? These vague answers can only be created and implemented by human beings, but not by machines. Is it possible for a computer to directly answer those survey questions as human beings did? Absolutely it is impossible without the help of AI technology. Computers can only understand either "0" or "1," and "HIGH" or "LOW." Those data are called crisp or classic data and can be processed by all digital machines.

Is it possible to allow computers to handle those ambiguous data with the help of human beings? If it is, how can computers and machines handle those vague data? The answer to the first question is yes. But to answer the second question, we need some fuzzy logic technique (FLT) and knowledge of fuzzy inference system (FIS).

To implement FLT in a real application, one needs the following three steps:

1. Fuzzifications—Convert classical data or crisp data into the fuzzy data or Membership Functions (MFs).
2. Fuzzy Inference Process—Combine Membership Functions with the control rules together to derive the fuzzy output.
3. Defuzzifications—Use different methods to calculate each associated fuzzy output, either online or offline, and keep those outputs or put them into a table: Lookup Table. Convert fuzzy outputs into crisp outputs or pick up the outputs from the lookup table based on the current input during an application.

As mentioned before, all machines can process crisp or classical data such as either "0" or "1." In order to allow machines to handle the vague language input such as "Somehow satisfied," the crisp input and output must be converted to linguistic variables with fuzzy components. For instance, to control an air conditioner system, the input temperature and the output rotation speed of control motors must be converted to the associated linguistic variables such as "HIGH," "MEDIUM," "LOW" and "FAST," "MEDIUM," or "SLOW." The former is corresponding to the input temperature and the latter is associated with the rotation speed of the operating motor. Besides those conversions, both the input and the output must also be converted from the crisp data to the fuzzy data. All of these jobs are performed by the first step—fuzzifications.

In the second step, to begin the fuzzy inference process, one needs to combine the Membership Functions with the control rules together to derive the fuzzy control output, and keep those outputs or arrange those outputs into a table called the lookup table. The control rule is the core of the fuzzy inference process, and those rules are directly related to human beings' intuition and knowledge. For example, in an air conditioner control system, if the temperature is too high, the heater should be turned off, or the heat-driving motor should be slowed down, which is the human beings' intuition or common sense.

At the third step, different methods, such as Center of Gravity (COG) or Mean of Maximum (MOM), are utilized to either convert the associated fuzzy control outputs to the crisp outputs or pick up each control output from the lookup table.

During a real application, a control output can be either directly calculated or selected from the lookup table developed from the last step based on the current input. Furthermore, that control output should be converted from the linguistic variable back to the crisp variable and output to the control operator. This process is involved in step 3 called the defuzzification process.

In most cases, the input variables are more than one dimension for real applications. Thus, one needs to perform fuzzifications or develop Membership Function for each dimensional variable separately. Perform the same operation if the system has multiple output variables.

Summarily, a fuzzy inference system is a process of performing crisp-fuzzy-crisp for inputs on a real system. The original input and the terminal output must be crisp variables, but they should be fuzzy variables during the fuzzy inference process. The reason why one needs to change the crisp to a fuzzy variable is: from the point of view of fuzzy control or human beings' intuition, no absolutely crisp variable exists in our real world. Any physical variable may contain some uncertain components. For instance, if someone says: the temperature here is high. This high temperature contains some middle and even low temperature components. From this point of view, fuzzy control uses universal or global components, not just a limited range of components, as the classical variables did.

With the rapid development of the fuzzy technologies, different fuzzy control strategies have been developed based on different classical control methods, such as the PID-fuzzy control [4], sliding-mode fuzzy control [5], neural fuzzy control, adaptor fuzzy control [6], and phase-plan mapping fuzzy control [7]. More and more new fuzzy control strategies or combined crisp and fuzzy control techniques have been developed and applied to all aspects of our society today.

This chapter starts with an introduction to the fuzzy logic control and the fuzzy terminology. The following sections provide a detailed description of the fuzzy inference process and the architecture of the fuzzy logic control. Section 3.2 discusses the fuzzy sets and crisp sets. The fuzzifications and membership function are represented in Sect. 3.3. Fuzzy control rule and defuzzifications are discussed in Sects. 3.4 and 3.5, respectively. Some related combined fuzzy control methods are provided in the rest of the sections of this chapter.

3.3 Fuzzy Set

The concept of the fuzzy set is only an extension of the concept of a classical or crisp set. Fuzzy set is actually a broader set compared with the classical or crisp set. The classical set only considers a limited number of degrees of membership such as "0" or "1," or a range of data with limited degrees of membership. For instance, if a speed is defined as a crisp high, its range must be between 70 MPH and higher and it has nothing to do with 60 or even 50 MPH. But fuzzy set will take care of a much broader range for this high speed. In other words, fuzzy set will consider a much larger speed range such as from 20 MPH to higher speeds as a high speed. The exact speed to which the 20 MPH can contribute to that high speed depends on the membership function. This means that the fuzzy set uses a universe of discourse as its base and it considers an infinite number of speeds of membership in a set. In this way, the classical or crisp set can be considered as a subset of the fuzzy set.

To get better understanding about this fuzzy set, let us start our discussion from a crisp set.

3.3.1 Classical Sets and Operations

A classical or crisp set is a collection of objects in a given range with a sharp boundary. An object can either belong to the set or not belong to the set. For example, we assume to create a faculty set or a faculty collection F with nine faculty members x_1, $x_2 \ldots \ldots x_9$ in a college.

$$F = \left\{ x_1, x_2, x_3, x_4, x_5, x_6, x_7, x_8, x_9 \right\} \tag{3.1}$$

In general, the entire object of discussion F is called a universe of discourse, and each member x_i is called an element. Assuming that elements x_1–x_4 belong to the department of computer science, which can be considered as another set A. The elements x_1–x_3 are under age 50, which can be considered as a set B. Therefore, the following relations exist:

$$X \in F$$
$$A = \left\{ x_1, x_2, x_3, x_4 \right\} \in F \tag{3.2}$$
$$B = \left\{ x_1, x_2, x_3 \right\} \subset A$$

It can be found that all elements in set B belong to set A, or the set A contains set B. In this case, set B can be considered as a subset of set A and can be expressed as $B \subset A$. The relationship among different sets we discussed can be described in Fig. 3.1. The basic classical set operations include complement, intersection, and union, which are represented as:

Complement of A (A^C)

$$A^C(x) = 1 - A \tag{3.3}$$

Intersection of A and B ($A \cap B$)

$$A \cap B = A(x) \cap B(x) \tag{3.4}$$

Union of A and B ($A \cup B$)

$$A \cup B = A(x) \cup B(x) \tag{3.5}$$

The representations of those classical set operations are shown in Fig. 3.2.

It is clear that either an element belongs to a set or does not belong to that set in the classical set and its operation. There is a sharp boundary between different elements for different sets and cannot be mixed with each other. But for the fuzzy set, it has different laws.

Fig. 3.1 The classical set of faculty members in a college

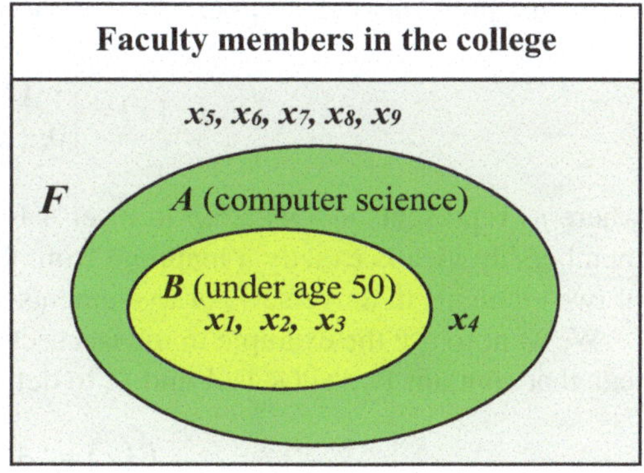

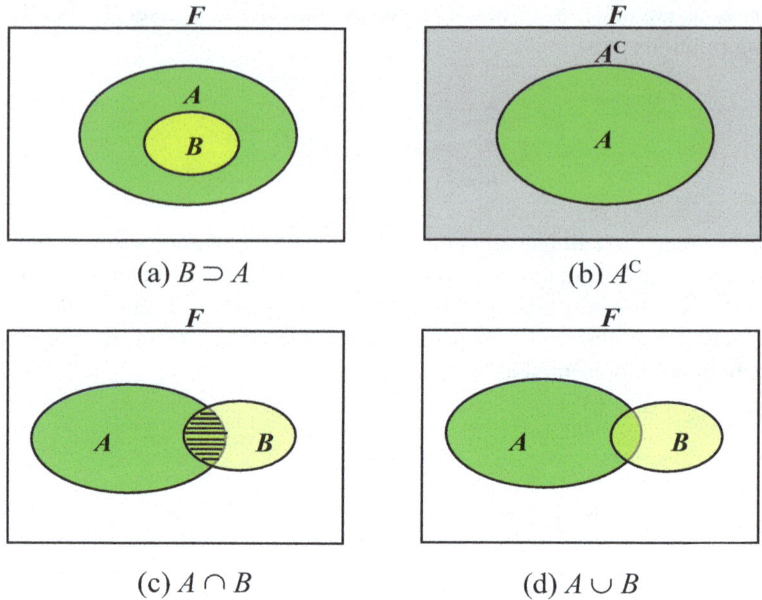

Fig. 3.2 Classical set and its operations. (**a**) $B \supset A$, (**b**) A^C, (**c**) $A \cap B$, (**d**) $A \cup B$

3.3.2 Mapping of Classical Sets to Functions

The classical set we discussed in the last part can be represented or mapped into certain functions. This means that we can relate set-theoretic forms to function-theoretic representations, and map elements in one universe of discourse to elements or sets in another universe. For the classical or crisp sets, this mapping is very easy and straightforward. Assume that X and Y are two different universes of discourse. If an element x belongs to X and it corresponds to an element y belonging to Y, the mapping between them can be expressed as

$$\mu_A(x) = \begin{cases} 1, & (x \in A) \\ 0, & (x \notin A) \end{cases} \tag{3.6}$$

where μ_A represents *membership* in a set A for the element x in the universe. This membership idea is exactly a mapping from the element x in the universe X to one of two elements in universe Y, or to elements 0 or 1.

We want to use the example in the last section, a set of faculty members in a college that contains two subsets A and B, to define a new set $P(x)$ as

$$P(x) = \{A, B\} \tag{3.7}$$

and mapping these two sets from the universe of discourse F to another universe Y with two elements, 0 or 1.

$$M\big(P(x)\big) = \big\{ (1,1,1,1,0,0,0,0,0),(1,1,1,0,0,0,0,0,0) \big\} \tag{3.8}$$

Now we can define operations of two classical sets, A and B, on the universe F as follows:

$$\text{Union}: A \cup B = \mu_A(x) \cup \mu_B(x) = \max\big(\mu_A(x),\mu_B(x)\big)$$

$$\text{Intersection}: A \cap B = \mu_A(x) \cap \mu_B(x) = \min\big(\mu_A(x),\mu_B(x)\big) \tag{3.9}$$

$$\text{Complement}: A^C = F \setminus A$$

Based on the discussions about the classical sets and the fuzzy sets, we should have some basic knowledge and understanding about both of them. Next let us take a look at the fuzzy set operations.

3.3.3 Fuzzy Sets and Operations

As we mentioned in the Sect. 3.3.1, the classical set has a sharp boundary, which means that a member either belongs to that set or not. Also this classical set can be mapped to a function with two elements, 0 or 1, as we discussed in the last section. For example, in Sect. 3.3.1, we defined the faculty member in the department of computer science as set A. A faculty either fully belongs to that set ($\mu_A(x) = 1$) if one is a faculty of the computer science department or has nothing to do with set A ($\mu_A(x) = 0$) if he is not a faculty in that department. This mapping is straightforward with a sharp boundary without any ambiguous. In other words, this fully can be mapped as a member of set A with a degree of 1, and not belonging to can be mapped as a member of set A with a degree of 0. This mapping is similar to a black-and-white binary categorization.

Compared with classical sets, a fuzzy set allows members to have a much smoother boundary. In other words, fuzzy set allows a member to belong to or not belong to a set with some partial degree. For instance, still we use the driving speed as an example. The speed can be divided into three categories: LOW (0–30 MPH), MID (30–70 MPH), and HIGH (70–100 MPH) from the point of view of the classical set, which is shown in Fig. 3.3a.

In the classical set, any speed can only be categorized into one subset, either LOW, MID, or HIGH, and the boundary is crystal clear. But in the fuzzy set such as shown in Fig. 3.3b, these boundaries become vague or smooth. One speed can be categorized into two or maybe even three subsets simultaneously. For example, the speed 40 MPH can be considered to belong to LOW to a certain degree, say $0.45°$, but at the same time, it can belong to MID with about $0.55°$. Another interesting thing is the speed 50 MPH, which can be considered to belong to LOW and HIGH at around $0.2°$ and belong to MID at almost $1°$. The dashed line in Fig. 3.3b represents the classical set boundary.

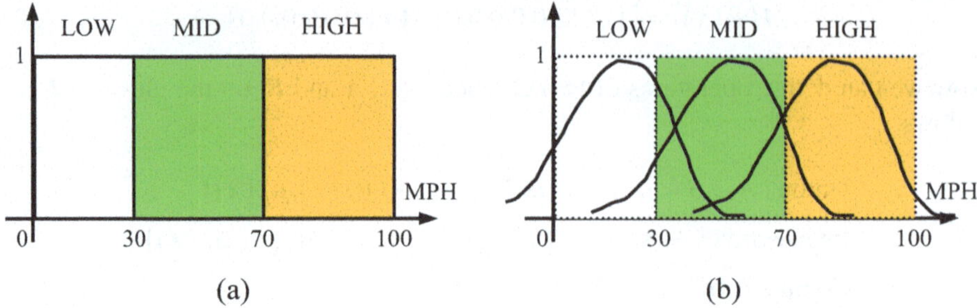

Fig. 3.3 The representations of the classical and fuzzy sets. (**a**) Classical speed set, (**b**) fuzzy speed set

It can be seen that a fuzzy set contains elements which have varying degrees of membership in the set, and this is contrasted with the classical or crisp sets because members in a classical set cannot be other members unless their membership is full or complete in that set. A fuzzy set allows a member to have a partial degree of membership, and this partial degree membership can be mapped into a function or a universe of membership values. Assume that we have a fuzzy set A, and if an element x is a member of that fuzzy set A, that mapping can be denoted as.

$$\mu_A(x) \in [0,1] \quad (A = (x, |\mu_A(x), |x \in X) \tag{3.10}$$

Suppose we have a fuzzy subset B with an element x that has a membership function of $\mu_B(x)$. When the universe of discourse X is discrete and finite, this mapping can be expressed as

$$B = \frac{\mu_B(x_1)}{x_1} + \frac{\mu_B(x_2)}{x_2} + \ldots = \sum_i \frac{\mu_B(x_i)}{x_i} \tag{3.11}$$

When the universe X is continuous and infinite, the fuzzy set B can be represented as

$$B = \int \frac{\mu_B(x)}{x} \tag{3.12}$$

As we discussed in the previous section for the classical set operations, the basic fuzzy set operations also include intersection, union, and complement, and those operations are defined as

$$\text{Union}: A \cup B = \mu_A(x) \cup \mu_B(x) = \max(\mu_A(x), \mu_B(x))$$

$$\text{Intersection}: A \cap B = \mu_A(x) \cap \mu_B(x) = \min(\mu_A(x), \mu_B(x)) \tag{3.13}$$

$$\text{Complement}: A^C = F \setminus A$$

where A and B are two fuzzy sets and x is an element in the universe of discourse, X. The graphical illustrations of those operations are shown in Fig. 3.4.

According to the definition of the fuzzy set operations, a fuzzy set union operation is exactly equivalent to selecting the maximum member from those members in the sets, and the intersection operation is to select the minimum member from the sets.

3.3.4   A Comparison Between the Classical Sets and the Fuzzy Sets

In Sect. 3.3.1, we defined a college faculty set F, a faculty set for the computer science department, set A, and a set of faculty members who are under age 50, or set B. Now we can make a clear comparison between the classical and fuzzy sets using that example.

According to the classical or crisp set theory, set B includes only three members who are under age 50 (x_1, x_2, x_3), so the boundary between the faculty who are under age 50 and the faculty who are above age 50 is crystal clear, as shown the solid line in Fig. 3.5. But for the fuzzy set theory, set B contains not only those three members, x_1, x_2, x_3, but also contains some other members in some varying degrees. The actual age distribution for all 9 faculty members is displayed on the x-axis in Fig. 3.5.

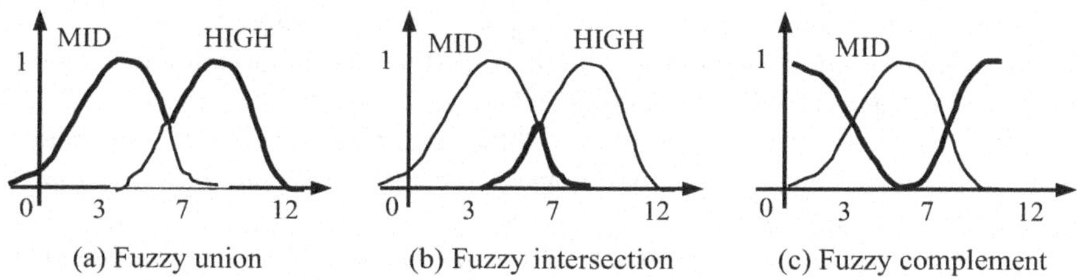

(a) Fuzzy union   (b) Fuzzy intersection   (c) Fuzzy complement

Fig. 3.4 Fuzzy set operations. (**a**) Fuzzy union, (**b**) fuzzy intersection, (**c**) fuzzy complement

Fig. 3.5 The comparison between the classical sets and the fuzzy sets

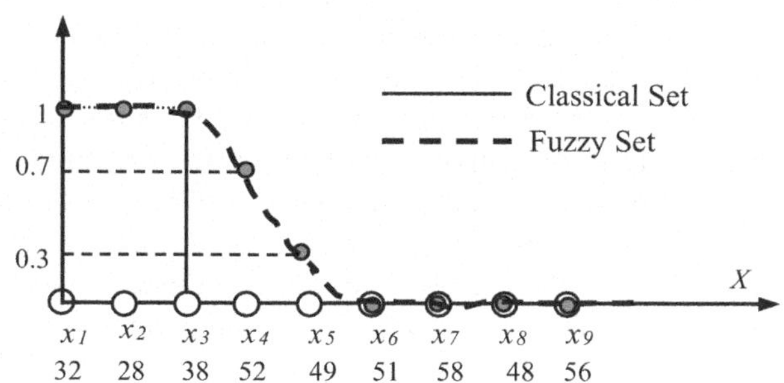

It can be found that members x_4 and x_5 do not belong to the set B in the view of crisp set, but they can be considered as partial members in a partial membership of $0.7°$ and $0.3°$, respectively. To map this to a membership function, it can be expressed as

$$\mu_B(x) = \frac{1}{x_1} + \frac{1}{x_2} + \frac{1}{x_3} + \frac{0.7}{x_4} + \frac{0.3}{x_5}$$

where the operator "−" is called a separator, and "+" is the *OR* operator. All other members whose membership is 0 are omitted from this function.

We have provided a detailed and complete discussion about the crisp or classical set and the fuzzy set; next let us concentrate on the fuzzification process or the membership functions.

3.4 Fuzzifications and Membership Functions

Fuzzy set is a useful and powerful tool and allows us to represent objects or members in a vague or ambiguous way. It also provides a way that is similar to the human beings' concepts and thinking process. However, the fuzzy set itself cannot lead to any useful and practical products until the fuzzy inference process is applied. To implement fuzzy inference to a real product to solve an actual problem, as we discussed before, three consecutive steps are needed, which are: fuzzification, fuzzy inference, and defuzzification.

Fuzzification is the first step prior to applying a fuzzy inference system. Most variables that existed in our real world are crisp or classical variables. We need to convert those crisp variables (both input and output) to the fuzzy variables and then apply fuzzy inference system to process those data to obtain the desired output. Finally, those fuzzy outputs need to be converted back to crisp variables to complete the desired control objectives.

Generally, two processes are involved in fuzzification: derive the membership functions for both input and output variables, and convert and represent them with linguistic variables. This process is equivalent to converting or mapping classical sets to fuzzy sets with related degrees.

In practice, membership functions can have various types, such as the triangular waveform, trapezoidal waveform, Gaussian waveform, bell-shaped waveform, sigmoidal waveform, and S-curve waveform. The exact type depends on the actual applications. For those systems that need significant dynamic variation in a short period of time, a triangular or a trapezoidal waveform should be utilized. For those systems that need very high control accuracy, a Gaussian or S-curve waveform should be selected.

To illustrate the process of the fuzzifications, we still use the air conditioner as an example we discussed in Sect. 3.2. Assume that we have an air conditioner control system that is under the control of a heater motor. If the temperature is high, the heater control motor should be stopped, and if the temperature is low, that heater motor should be turned on and sped up, which are common senses.

Regularly, the normal temperature range is from 30 F° to 90 F°. This range can be further divided into three subranges or subsets, which are:

Low temperature: 30–50 F°, 40 F° is a center
Mid temperature: 40–80 F°, 60 F° is a center
High temperature: 60–90 F°, 75 F° is a center

Convert those three temperature ranges to the related linguistic variables: LOW, MID, and HIGH, which are corresponding to three temperature ranges listed above.

The membership function of these input temperatures is shown in Fig. 3.6a. To make our study simple, a trapezoidal waveform is utilized for the type of the membership function. A crisp low temperature can be considered as a mid temperature with some degree in this fuzzy membership function representation. For instance, a 45 F° belongs to both a LOW and a MID in 0.5°. Some terminologies used for the membership function are also shown in Fig. 3.6.

The so-called *support* for a fuzzy set, says LOW, is the set of elements whose degree of membership in LOW is greater than 0. This support can be expressed in a function form as

$$Support\left(\text{LOW}\right) = \left\{\, x \in T \,|\, \mu_{\text{LOW}}\left(x\right) > 0 \,\right\} \tag{3.14}$$

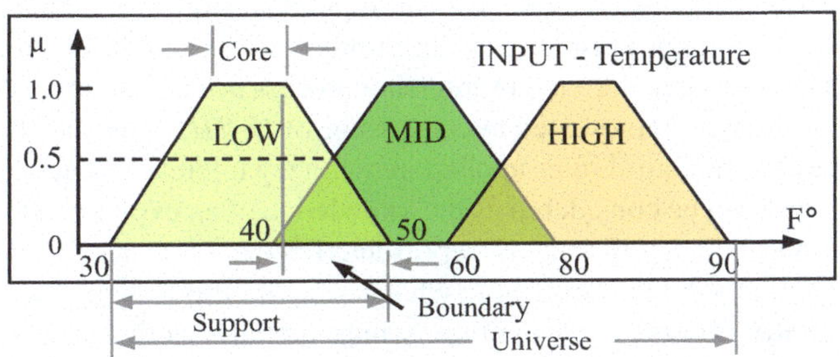

(a) Membership Function for the input - temperature.

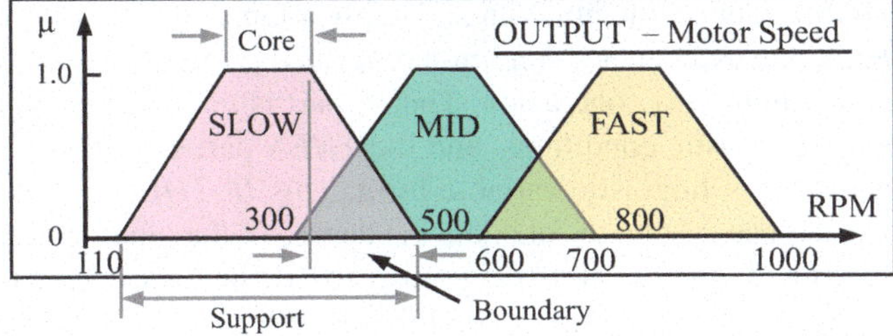

(b) Membership Function for the output – motor rotating speed.

Fig. 3.6 Membership function for input and output. (**a**) Membership function for the input–temperature. (**b**) Membership function for the output–motor rotating speed

It can be concluded that the support of a fuzzy set is a classical set.

The *core* of a fuzzy set is the set of elements whose degree of membership is equal to 1 in that set, which is also equivalent to a crisp set. The *boundary* of a fuzzy set indicates the range in which all elements whose degree of membership is between 0 and 1 (0 and 1 are excluded) in that set.

The membership function for the output, the rotating speed of the heater motor, is also shown in Fig. 3.6b. Similarly, the normal motor rotating speed is ranged from 10 revolutions per minute (RPM) to 90 RPM. This range can be further divided into three subranges or subsets, which are:

SLOW rotating speed: 100–500 RPM, 300 RPM is a center
MID rotating speed: 300–800 RPM, 500 RPM is a center
FAST rotating speed: 600–1000 RPM, 800 RPM is a center

After the membership functions for both input and output are defined, the next step is to define the fuzzy control rules prior to applying the fuzzy inference system.

3.5 Fuzzy Control Rules

As we discussed in Sect. 2.1, fuzzy inference system (FIS) is a branch of AI study, and it belongs to a category of learning based on experiences. From the knowledge base, we can obtain all knowledge sources, such as all input data stored in a database or a dataset. By using the inference system or algorithm, we can derive or predict the desired outputs. However, that deriving or inferring process needs both input data and related rules, or experiences represented by a sequence of *IF-THEN* format. Those are typical representations of control rules applied in the fuzzy inference system.

Fuzzy control rule can be considered as the knowledge of an expert in any related field of application. The fuzzy rule is represented by a sequence of the form *IF-THEN*, leading to algorithms describing what action or output should be taken in terms of the currently observed information, which includes both input and feedback if a closed-loop control system is applied. The law to design or to build a set of fuzzy rules is based on the human beings' knowledge or experience, which is dependent on each different actual situation and application.

A fuzzy *IF-THEN* rule associates a condition described using linguistic variables and fuzzy sets to an output or a conclusion. The *IF* part is mainly used to capture knowledge by using the elastic conditions, and the *THEN* part can be used to give the conclusion or output in linguistic variable form. This *IF-THEN* rule is widely used by the fuzzy inference system to calculate the degree of the input data matched to the condition of a rule. Figure 3.7 illustrates a way to calculate the degree between a fuzzy input T (temperature) and a fuzzy condition LOW. Here we still use the air conditioner system as an example.

This calculation can also be represented by the function

$$M\left(T,\text{LOW}\right) = Support\min\left(\mu_T\left(x\right),\mu_{\text{LOW}}\left(x\right)\right) \qquad (3.15)$$

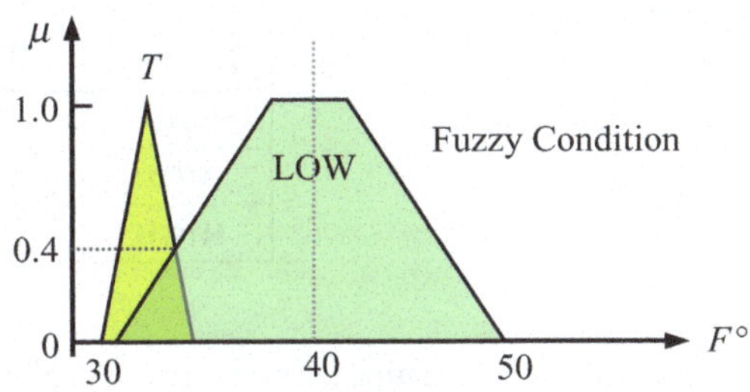

Fig. 3.7 Match a fuzzy input with a fuzzy condition

Two types of fuzzy control rules are widely utilized for most real applications. One is fuzzy mapping rules and the other lis called fuzzy implication rules.

3.5.1   *Fuzzy Mapping Rules*

Fuzzy mapping rule provides a functional mapping between the input and the output using linguistic variables. The foundation of a fuzzy mapping rule is a fuzzy graph, which describes the relationship between the fuzzy input and the fuzzy output. In some real applications, it is very hard to derive a definite relationship between the input and the output; also the relationships between those inputs and outputs are very complicated even when they were developed. Fuzzy mapping rule is a good solution under those situations.

Exactly, fuzzy mapping rule works a similar way as human intuitions or insights, and each fuzzy mapping rule only approximates a limited number of elements of the function, so the entire function should be approximated by a set of fuzzy mapping rules. When using our air conditioner system as an example, a fuzzy mapping rule can be derived as.

IF the temperature is LOW, *THEN* the rotating speed of the heater motor should be FAST.

For other input temperatures, different rules could be built and developed.

In most actual applications, the input variables are commonly more than one dimension. For example, in our air conditioner system, two input variables are involved: the current temperature and the change rate of the temperature. The fuzzy control rules should also be extended to allow multiple inputs to be considered to derive the output. Table 3.1 is an example of fuzzy control rules applied in a real air conditioner system. Where T is the current temperature and $\dot{T}$ is the change rate of the temperature.

The row and column represent two inputs, the temperature input and the change rate of the temperature input, and those inputs are related to *IF* parts in *IF-THEN* rules. The control output can be considered as a third-dimensional variable that is located at the cross point of each row (temperature) and each column (change rate of the temperature), and that conclusion is associated with the *THEN* part in *IF-THEN* rules. For example, when the current temperature is LOW, and the current

Table 3.1 An example of fuzzy rules

$\dot{T}$ \ T	LOW	MID	HIGH
LOW	FAST	MID	MID
MID	FAST	SLOW	SLOW
HIGH	MID	SLOW	SLOW

change rate of the temperature is also LOW, the rotating speed of the heater motor should be FAST to increase the temperature as soon as possible. This can be represented by the *IF-THEN* rule as

IF the temperature is LOW, and the change rate of the temperature is also LOW, *THEN* the output–rotating speed of the heater motor should be FAST.

All other related control rules can also be developed by following the similar strategy, which is very similar to human beings' intuition and common senses.

For this air conditioner example, totally nine (9) rules are built, as shown in Table 3.1. Both input variables, T and $\dot{T}$, can be determined from each row and column, and the output of each rule is derived from the intersection of each related row and column. For those applications that need higher control accuracy, the input and output should be divided into smaller segments, and more fuzzy rules should be adopted and applied.

3.5.2 *Fuzzy Implication Rules*

Fuzzy implication rule describes a generalized logic implication relationship between inputs and outputs. The foundation of a fuzzy implication rule is the narrow sense of fuzzy logic [8]. The fuzzy implication rules are related to classical two-valued logic and multiple-valued logic.

Still using the air conditioner system as an example, the implication is:

IF the temperature is LOW, *THEN* the heater motor should be FAST.

Based on this implication and a fact, if the temperature is HIGH; the result that the heater motor should be slow down or SLOW, could be inferred.

3.6 **Defuzzifications Process and Lookup Table**

The control output derived from the FIS by a combination of input, output membership functions (MFs), and fuzzy rules is still a vague or a fuzzy variable, and this process is called fuzzy inference. To make that fuzzy output available to a crisp variable and apply it in all real applications, a defuzzification process is needed.

The defuzzification process is exactly to convert the fuzzy output to the crisp or classical output to the control objective. A point to be noted is, the fuzzy derivation or output is still a linguistic variable, and this linguistic variable needs to be converted to the crisp variable via the defuzzification process. Generally, three defuzzification techniques are commonly used, which are: Mean of Maximum method, Center of Gravity method, and the Height method.

Let us have a closer look at those methods.

3.6.1   Mean of Maximum (MOM) Method

The Mean of Maximum (MOM) defuzzification method computes the average of those fuzzy outputs that have the highest degrees. For example, the fuzzy output is: the rotating speed of the heater motor $x^{'}$ is FAST. By using the MOM method, this defuzzification can be expressed as

$$\mathrm{MOM}\left(\mathrm{FAST}\right) = \frac{\displaystyle\sum_{x^{'} \in T} x^{'}}{T} \quad T = \left\{ x^{'} \mid \mu_{\mathrm{FAST}}\left(x^{'}\right) = \mathrm{Support}\ \mu_{\mathrm{FAST}}\left(x\right) \right\} \qquad (3.16)$$

where T is the set of output $x^{'}$ which has the highest degrees in the set FAST.

A graphic representation of the MOM method is shown in Fig. 3.8a.

A disadvantage of using the MOM method is that it does not consider the entire shape of the output membership function, but it only takes care of the points that have the highest degrees in that function. For some membership functions that have different shapes but same highest degrees, this method will get the same result, which may not be accurate.

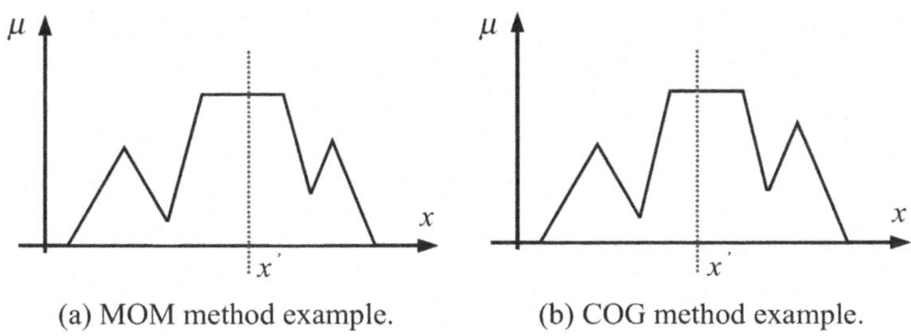

(a) MOM method example.   (b) COG method example.

Fig. 3.8 Graphic representation of defuzzification techniques. (**a**) MOM method example. (**b**) COG method example

3.6.2   *Center of Gravity (COG) Method*

The Center of Gravity method (COG) is one of the most popular defuzzification techniques and is widely implemented in most actual applications. This method is very similar to the formula used for calculating the center of gravity in physics. The weighted average of the membership function or the center of gravity of the area bounded by the membership function curve is computed to get the most crisp value of the fuzzy quantity. For example, for one fuzzy output: the rotating speed of the heater motor x is FAST. The COG output can be represented as

$$\mathrm{COG}\left(\mathrm{FAST}\right) = \frac{\sum_{x} \mu_{\mathrm{FAST}}\left(x\right) \times x}{\sum_{x} \mu_{\mathrm{FAST}}\left(x\right)} \tag{3.17}$$

If x is a continuous variable, the defuzzification result should be

$$\mathrm{COG}\left(\mathrm{FAST}\right) = \frac{\int \mu_{\mathrm{FAST}}\left(x\right) x dx}{\int \mu_{\mathrm{FAST}}\left(x\right) dx} \tag{3.18}$$

A graphic representation of the COG method is shown in Fig. 3.8b.

3.6.3   *The Height Method (HM)*

The HM defuzzification method works fine only for the case where the output membership function is an aggregated union result of symmetrical functions [9]. The process of this method can be divided into two steps. First, the consequent membership function F_i can be converted into crisp consequent $x = f_i$, where f_i is the center of gravity of F_i. Then the COG method is applied to the rules with crisp consequents, which can be expressed as

$$x = \frac{\sum_{i=1}^{M} w_i f_i}{\sum_{i=1}^{M} w_i} \tag{3.19}$$

where w_i is the degree to which the ith rule matches the input data. The advantage of using this method is its simplicity. Therefore, many neuro-fuzzy models use this kind of defuzzificationmethod to reduce the complex of calculations.

3.6.4   *The Online and Offline Output*

The final output result of the FIS is a crisp output variable used to directly control the system target to obtain a perfect or a desired objective. This kind of result can be obtained in two different formats: online or offline result, in all real applications.

The so-called online result can be obtained by using any defuzzification method we discussed above based on the current inputs. The advantage of using this online method is that the control output has real-time response to the current inputs and the output can be computed or calculated immediately following the inputs. In other words, the output has higher controllability in control accuracy.

The shortcoming of the online method is that the responding speed is relatively slower since it needs longer time to perform that defuzzification process in real time due to the complex of the defuzzification algorithm, which is a time-consuming process.

On the other hand, the offline output does not need to be calculated in real time, and it does not need the real-time inputs. In fact, all offline results can be calculated via a defuzzification method in advance or prior to real applications based on the inputs intervals, that is to divide inputs into smaller intervals or smaller segments equally in terms of the range of the inputs. Then all outputs are stored into a table (called lookup table) and each output is set to a cell in the table with the input row and input column as indices for that output.

The advantage of using the offline method is that the responding speed is faster since it does not need to perform any real-time calculation for the defuzzification process, while the disadvantage by using that method is the control accuracy since each output is picked up from the lookup table and it is not calculated based on the current input.

Next let us have a closer look at the lookup table.

3.6.5   *The Lookup Table*

When a lookup table is used, the terminal product of the defuzzification process is a lookup table. In fact, the defuzzification process needs to be performed for each subset of the membership function, both inputs and outputs. For instance, in our air conditioner system, one needs to perform the defuzzification for each subset of temperature input such as LOW, MID, and HIGH based on the associated fuzzy rules. The defuzzification result for each subset can be stored into an associated cell in the lookup table according to the current temperature and temperature change rate. Here we use the air conditioner system as an example to illustrate the defuzzification process and the creation of the lookup table.

To make this illustration simple, we make two assumptions: (1) Assume that the membership function of the change rate of the temperature can be described as in Fig. 3.9. (2) Only four rules are applied to this air conditioner system, which are:

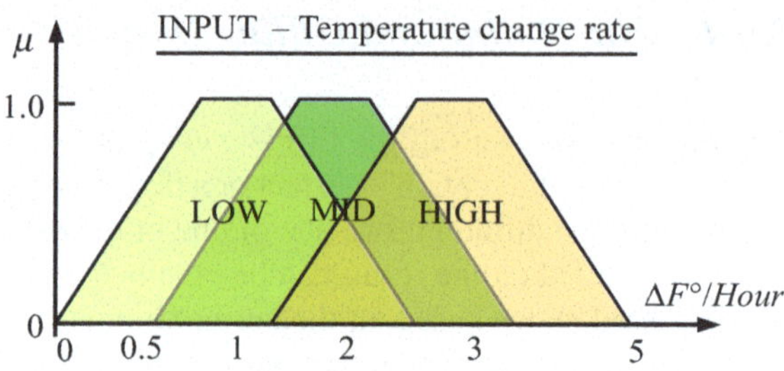

Fig. 3.9 The membership function of the change rate of the temperature

1. *IF* temperature is LOW, and the change rate of the temperature is LOW, *THEN* the rotating speed of the heater motor should be FAST.
2. *IF* temperature is MID, and the change rate of the temperature is MID, *THEN* the rotating speed of the heater motor should be SLOW.
3. *IF* temperature is LOW, and the change rate of the temperature is MID, *THEN* the rotating speed of the heater motor should be FAST.
4. *IF* temperature is MID, and the change rate of the temperature is LOW, *THEN* the rotating speed of the heater motor should be MID.

Based on the assumption made above for the membership function and fuzzy rules, we can illustrate this defuzzification process using a graph. Four (4) fuzzy rules can be interpreted as functional diagrams, as shown in Fig. 3.10.

Based on this example, consider the current input temperature is 35 F° and the change rate of the temperature is 1 F° per hour.

From Fig. 3.10, it can be found that the points of intersection between the temperature values of 35 F° and the graph in the first column (temperature input T) have the membership functions of 0.6, 0.8, 0.5, and 0.8. Similarly, the second column (change rate of temperature ΔT) shows that a temperature change rate of 1 F° per hour has the membership functions of 1.0, 0.4, 0.4, and 1.0. The fuzzy output for the four rules is the intersection of the paired values obtained from the graph, or the AND result between the temperature input and the temperature change rate input. According to Eq. (3.13), this operation result should be: min (0.6, 1.0), min (0.8, 0.4), min (0.5, 0.4), and min (0.8, 1.0), which produce 0.6, 0.4, 0.4, and 0.8, respectively.

The membership functions representing the control adjustment are weighted according to the input change and different control contributions as shown in Fig. 3.11.

Now, for a pair of temperature and temperature change rate, four sets of fuzzy outputs exist. To determine the crisp value of action to be taken from these contributions, one can either choose the maximum value using the MOM method or use the Center of Gravity method (COG). In this example, the COG method is used and the action is given by the center of the summed area, which is contributed by the different fuzzy outputs. Furthermore, the COG method gives a more reliable lookup table compared with the MOM operation.

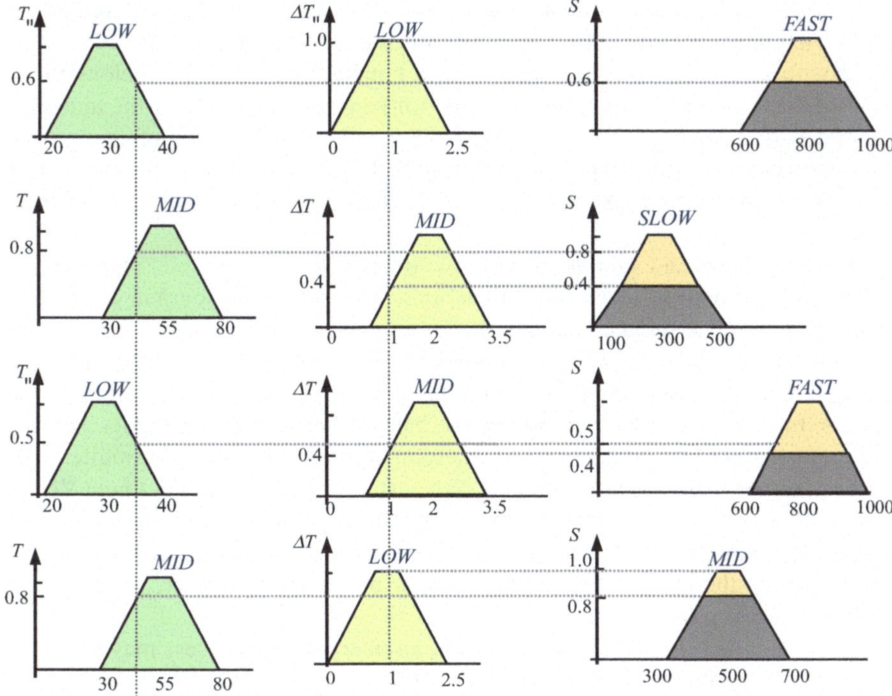

Fig. 3.10   An explanation of the fuzzy output calculation

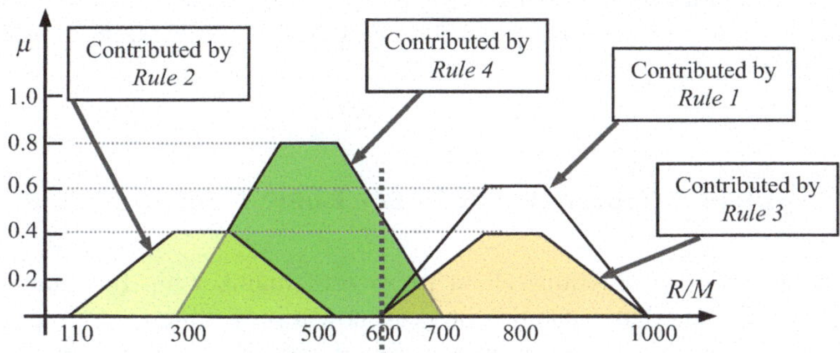

Fig. 3.11   Determination of the fuzzy output by using Center of Gravity (COG) method

Thus, for a temperature of 35 F° and a change rate of temperature 1 F° per hour, the fuzzy output element y for this input pair is

$$y = \frac{0.6 \times 800 + 0.4 \times 300 + 0.4 \times 800 + 0.8 \times 500}{0.6 + 0.4 + 0.4 + 0.8} = 600 R / M \qquad (3.20)$$

This defuzzified output is a crisp or classical value, and should be entered into a cell located in a table called lookup table. Since this fuzzy output element is associated with a temperature input pair with a temperature of 35 F° (belonging to LOW in the temperature membership function) and a change rate of the temperature of 1 F°/H (also belonging to LOW in the membership function of change rate of the temperature), this output value should be located in the cross point between the LOW temperature T (row) and the LOW change rate $\dot{T}$(column) as shown in Table 3.2.

Table 3.2 shows an example of our lookup table. To fill this table, one needs to use the defuzzification technique to calculate all other fuzzy output values and locate them to the associated cells in the lookup table as we did above. Generally, the dimensions of the fuzzy rules should be identical to the dimensions of the lookup table, just as shown in this example. To obtain more accurate control accuracy for fuzzy output element values, we can further divide the inputs, say the temperature input and change rate of the temperature, into multiple smaller subsets to get finer membership functions. For instance, here we can define a VERY LOW subset that covers 20–30 F° for the temperature input, a LOW subset that contains a range of 30–40 F°, and so on. Performing the same process to the change rate of the temperature and the heater motor speed output, we can get a much finer lookup table.

For higher control accuracy applications, an interpolation process may be added after the lookup table to obtain finer output [10].

When implementing a Type-I fuzzy logic technique in a real system, the lookup table can be stored in a computer's memory, and the fuzzy output can be obtained based on the current inputs, exactly based on the current row and column of two inputs. As we mentioned, in practice, there are two ways to calculate the fuzzy output value using fuzzy inference process in real control applications: offline and online methods.

3.7 A Typical Architecture of Fuzzy Logic Control System

Based on our above discussions about fuzzy sets, membership functions, fuzzy inference system, and defuzzification process, now it is the time for us to provide a complete fuzzy control system to illustrate the role of each component and a global

Table 3.2 An example of a lookup table

$\dot{T}$ \ T	LOW	MID	HIGH
LOW	600	?	?
MID	?	?	?
HIGH	?	?	?

picture for its control function. Combined with the discussions we made in the previous sections, a structure or architecture of fuzzy logic control system is given in this part.

A complete closed-loop fuzzy logic control system used for general motor control is shown in Fig. 3.12. This FLC is equivalent to a Proportional-Derivative (PD) controller. The inputs are error and error rate, which are combined together by block M to input to the fuzzy inference system. The FIS can also be a lookup table derived based on the membership function of inputs, the output, and the fuzzy control rules. A control gain factor G is used to tune the output of the FIS or the lookup table to obtain different output values. A feedback signal is obtained from the output of the system.

The input r is set to a desired speed S, and that input will be compared with the feedback speed coming from the output–motor rotating speed, which is called a closed-loop control strategy, and the difference of that comparison will be fed into the input of the fuzzy controller that is composed of an FIS with some additional elements, such as control gain or filters.

Both input error e and the change rate of the input error Δe will work together as a combination of inputs, which are under the control of a multiple switch M, to be fed to the controller.

The FIS will infer and derive a desired output, a related voltage, based on the control rules and fuzzy sets, and feed it to the motor to get a desired rotating speed of that motor, further adjusting the rotating speed. A tachometer attached on the motor works as a sensor to convert the rotating speed back to an associated voltage to be fed back to the input.

In most real applications, the input is not a desired speed, but it is a voltage that is mapped to the desired speed. The feedback variable coming from the sensor or the tachometer is also a voltage proportional to the rotating speed of the motor. Therefore, the inputs and the output to the comparator are voltages since the motor is generally driven by voltages.

So far we should have some basic and solid knowledge about fuzzy inference and control systems. To get a better understanding of those topics, we need to develop some actual systems by implementing fuzzy logic controllers on those systems.

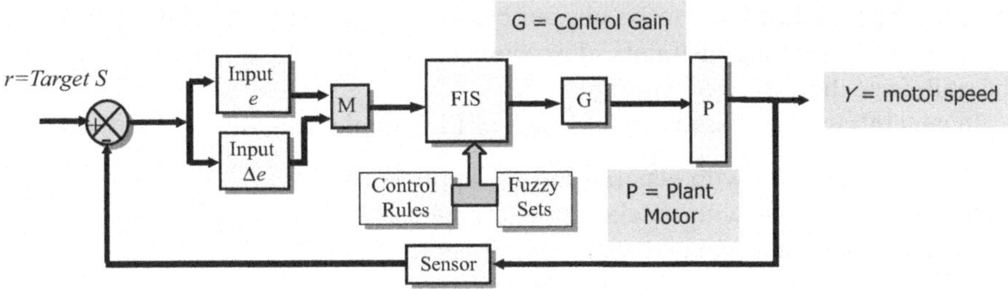

Fig. 3.12 The block diagram of a fuzzy control system

3.8 Implementations of Fuzzy Logic Inference Control System

Generally speaking, all real control systems can be categorized into two groups based on the function and controllability

- Linear Controller Systems
- Nonlinear Control Systems

However, historically speaking, all control systems can also be defined as another two categories based on the control history:

- Classical or Conventional Control Systems
- Modern Control Systems

A common feature of the conventional control system, such as a Proportional-Integral-Derivative (PID) control system, is that the control algorithm is analytically described by mathematical equations. In general, the synthesis of such control algorithms requires a formalized analytical description of the controlled system by a mathematical model [11]. The main advantage of fuzzy logic is its capability to express the knowledge in a linguistic way, and allow a system to be described by simple, human-friendly control rules [12, 13]. This alternative linguistic approach leads to a meaningful reduction of the computational burden. Generally, the fuzzy logic solutions are not aimed at achieving the computational precision as traditional techniques did, but aim at finding acceptable solutions in a shorter time. The natural applications of fuzzy logic control (FLC) are those systems whose mathematical model is unknown or too difficult to be derived, or very complex and time-varying, and where human experience can play an important role on those systems.

As in other control applications, fuzzy logic implementation could be done either with general-purpose systems or dedicated systems, and many solutions have been proposed for both alternatives. The structure and relative simplicity of fuzzy processing algorithms naturally leads to straightforward implementation in dedicated hardware structures. However, the majority of fuzzy logic implementations reported in literatures use general-purpose hardware due to the fact that such an implementation involves a low start-up cost with a well-defined control algorithm such as FLC. An optimal solution for the full application range of fuzzy logic does not exist. Different approaches have to be chosen based on the features of the application: complexity, real-time constraints, development time, quantities, and other factors that influence the choice of a design.

In general, we can classify four classes of FLC implementation alternatives:

1. Software solutions with general-purpose applications
2. Software solutions with special-purpose applications
3. Hardware solutions using dedicated fuzzy processors
4. Hardware solutions using fuzzy ASICs

Due to the space limitation, in this chapter, we only concentrate our study on the first two parts. We start with some general analyses of some real FLC applications with a few of powerful tools provided by MATLAB, such as MATLAB Fuzzy Logic Toolbox™ and MATLAB Simulink©. This lays a foundation for the following section, such as software special-purpose applications.

Prior to starting our real FLC implementations, first let us have some knowledge about MATLAB developing environments and related toolboxes.

3.8.1 *MATLAB Development Environment and Fuzzy Logic Toolbox*™

MATLAB provided a good and friendly developing environment for all FLC applications, which includes Fuzzy Logic Toolbox™ and a simulation tool called Simulink. These tools enable users to build and develop some general and specific FLC applications easily and efficiently in a short time. More important, MATLAB Fuzzy Logic Toolbox™ software provides support for two types of fuzzy inference systems:

1. Mamdani systems
2. Sugeno systems

These two types of FISs have different advantages and shortcomings, and one needs to select one of them based on the actual applications. For both Mamdani and Sugeno FISs, one can create either Type-1 or Type-2 fuzzy inference system applications.

Moreover, MATLAB Fuzzy Logic Toolbox™ provided two major developing modes to allow both beginning and sophisticated students to build their professional applications quickly and easily. These two developing modes include:

1. Fuzzy Logic Designer App Mode
2. Fuzzy Logic Functions Mode

Let us have a closer look at these two types and two modes.

3.8.1.1 Mamdani Fuzzy Inference Systems

Mamdani fuzzy inference was first introduced as a method to create a control system by synthesizing a set of linguistic control rules obtained from experienced human operators [14]. In a Mamdani system, the output of each rule is a fuzzy set. The FIS we discussed in Sect. 3.6.5 belongs to the Mamdani FIS.

Since Mamdani systems have more intuitive and easier-to-understand rule bases, they are well-suited to expert system applications where the rules are created from human expert knowledge, such as medical diagnostics, automatic closed-loop control systems, and robotic control systems.

3.8.1.2 Sugeno Fuzzy Inference Systems

Sugeno fuzzy inference system, which is also referred to as Takagi-Sugeno-Kang fuzzy inference, uses *singleton* output membership functions that are either constant or a linear function of the input values. The defuzzification process for a Sugeno system is more computationally efficient compared to that of a Mamdani system since it uses a weighted average or weighted sum of a few data points rather than compute a centroid of a two-dimensional area [15].

Unlike control rules defined in the Mamdani FIS, each control rule defined in a Sugeno system operates as a branch shown in Fig. 3.13, which shows a two-input system with input values x and y.

Each rule generates two results:

1. z_i—Rule output value, which is either a constant or a linear function of the input values

$$z_i = a_i x + b_i y + c_i \tag{3.21}$$

   Here, x and y are the values of input 1 and input 2, and a_i, b_i, and c_i are constant coefficients. For a zero-order Sugeno system, z_i is a constant ($a = b = 0$).
2. w_i—Rule firing strength derived from the rule antecedent

$$w_i = And\left[F_1(x), F_2(y) \right] \tag{3.22}$$

   Here, $F_1(...)$ and $F_2(...)$ are the membership functions for inputs 1 and 2, respectively.

   The output of each rule is the weighted output level, which is the product of w_i and z_i.

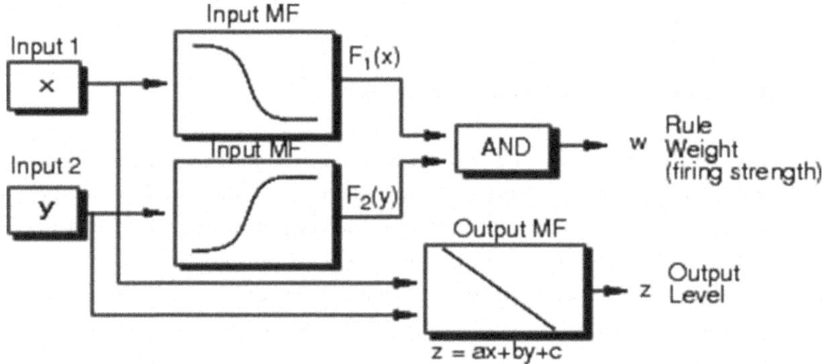

Fig. 3.13 A functional block diagram of Sugeno FIS control system

The easiest way to visualize first-order Sugeno systems (a and b are nonzero) is to think of each rule as defining the location of a moving singleton. That is, the singleton output spikes can move around in a linear fashion within the output space, depending on the input values. The rule firing strength then defines the size of the singleton spike.

The final output of the system is the weighted average over all rule outputs:

$$\text{Final output} = \frac{\sum_{i=1}^{N} w_i z_i}{\sum_{i=1}^{N} w_i} \tag{3.23}$$

where N is the number of rules.

3.8.1.3   MATLAB Fuzzy Logic Designer App Mode

The Fuzzy Logic Designer App mode allows users or students to design, test, and tune a fuzzy inference system (FIS) for modeling complex system behavior. Using this App, you can design Mamdani and Sugeno FIS applications with Type-I and Type-II FIS applications.

The most important and interesting point of using this App is that the App provided a sequence or a group of Graphic User Interfaces (GUIs) to enable students, especially beginning students who have not much knowledge about FLC, to quickly and easily build a fuzzy control system with a few of minutes without any coding!

Yes, that is true. By opening the related GUIs in that App, selecting desired membership functions and control rules, the App can produce the ideal outputs as you want. Everything you need to do is to select your control parameters by clicking and selecting some key data items. Also you can get and adjust the outputs by selecting different inputs as you like, and check and inspect the 3D output envelopes in real time.

As the fast development of Fuzzy Logic Toolbox, a new Fuzzy Logic Designer App has been released by MathWorks® recently. Compared with the original version, this new App allows users to build both Type-I and Type-II FIS applications in an easier and convenient way. Also some better and friendly GUIs are presented in the new version to help users to build FIS implementations quickly and effectively.

To activate the original and the new Fuzzy Logic Designer App, different commands should be typed in the Command window, and they are:

- **fuzzy** (for original App),
- **fuzzyLogicDesigner** (for new App).

To catch up with new technologies, we prefer to use the new version App to build our project.

We will use our air conditioner as an example to illustrate how to use Fuzzy Logic Designer App to build an FLC or FIS application to perform an open-loop control to the heater motor in the next section.

The good thing about using this App mode is that the students' learning curve for FLC can be significantly reduced, and the students' learning interests can also be greatly increased. This mode is especially useful to the beginning students.

A good thing must come with some bad thing, which is also true for this App mode. A shortcoming of using this mode is that a lot of function codes are built by the App itself and those codes are hidden and covered by the system, and cannot seen by the users, which blocks students and unable them to get more details about the coding functions.

To solve that disadvantage, the second mode, the Fuzzy Logic Function Mode, can effectively help experienced students to build more specific FLC applications with a lot of blocks of codes.

3.8.1.4 MATLAB Fuzzy Logic Functions Mode

This mode includes all FLC-related functions that are packaged in some system libraries located at different namespaces.

To use those functions, users or students need to have a solid understanding and knowledge about FLC and the detailed features of those functions. Unlike the App mode, in which all fuzzy features and components can be presented and accessed with different GUIs, in the function mode, all features and components must be built directly by using MATLAB codes, which are very similar to C codes. Moreover, users can even build the FIS with customer functions, which means that when users build a fuzzy inference system (FIS), they can replace the built-in membership functions or inference functions with their custom functions. They can create an FIS that uses these custom functions in the Fuzzy Logic Designer App and at the MATLAB command line.

Due to the similarity between the MATLAB codes and C codes, the learning curve is also lower since most students should have learned or studied the C language in other courses.

Next let us build our real fuzzy logic control project by using our air conditioner system as an example to illustrate the detailed operational steps. First let us start to build that project with the Fuzzy Logic Designer App mode.

3.8.2 Build Fuzzy Logic Control System for Air Conditioner with Fuzzy Logic Designer App

To make things simple and easy, we like to build our project with a simple closed-loop control system by using Fuzzy Logic Designer App (FLDA). The functional block diagram of this control system is shown in Fig. 3.14.

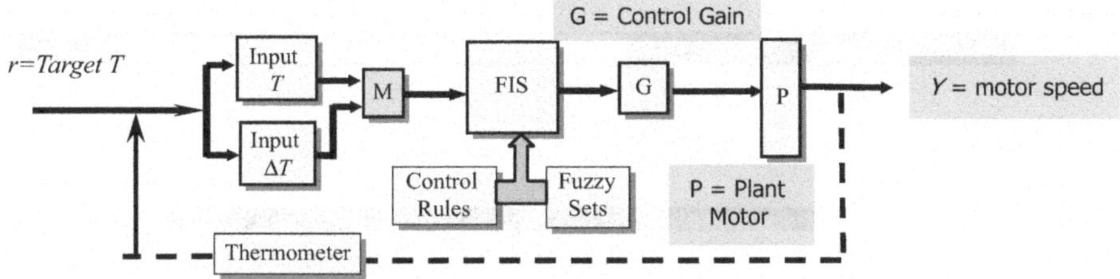

Fig. 3.14 Functional block diagram of FLC for air conditioner

The desired or the target input temperature T is divided into two parts: input temperature T and the change rate of input temperature ΔT, and both of them are fed into the FIS with a membership functions (MFs) format. The FIS calculates the desired output, the rotating speed of the motor Y, via the input MFs and the fuzzy control rules, and drives the motor with certain speed to adjust the room temperature. The thermometer works as a sensor to provide feedback.

The software we need to use includes MATLAB, Fuzzy Logic Toolbox, and Simulink. Be sure to download and install these into your computer. Refer to Appendix A to complete the downloading and installing process for those components and tools.

Let us first build membership functions for two inputs, T and ΔT, and the output that is the rotating speed of the motor S.

3.8.2.1   Build Membership Functions for Both Inputs and the Output

First let us create a folder to hold our project. Open the File Explorer and create a new folder under the root as, **C:\AI Projects\Chapter 3**, and this is the location where all our AI projects will be stored in this chapter.

Open the MATLAB by double-clicking on its shortcut from the desktop and perform the following operational steps to create our new project:

1. On the opened MATLAB, as shown in Fig. 3.15, in the Command window, type **fuzzyLogicDesigner**, and press the **Enter** key to open the Fuzzy Logic Designer App.
2. The opened Fuzzy Logic Designer App is shown in Fig. 3.16. With this App, different types of FIS can be built, including:

   (a) Custom FIS
   (b) FIS from Data
   (c) Mamdani Type-I FIS
   (d) Mamdani Type-II FIS
   (e) Sugeno Type-I FIS
   (f) Sugeno Type-II FIS

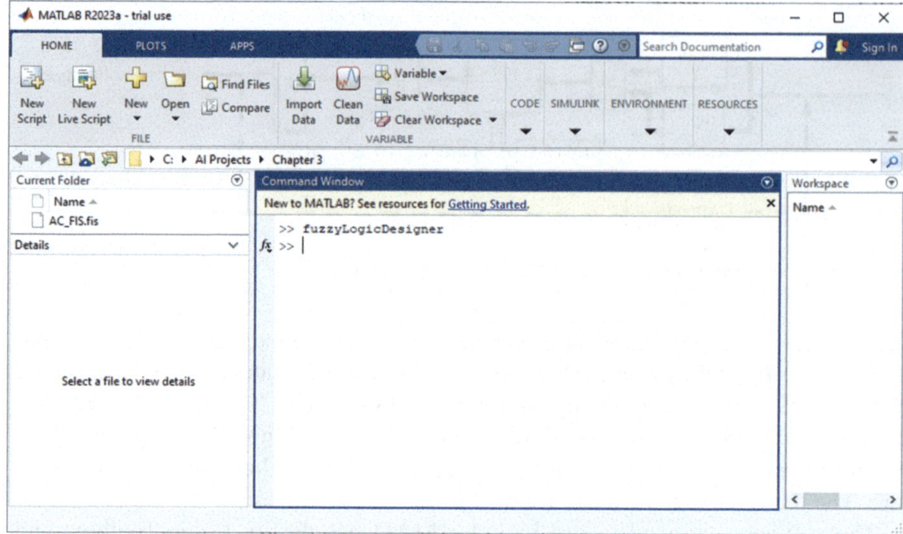

Fig. 3.15 The opened MATLAB window

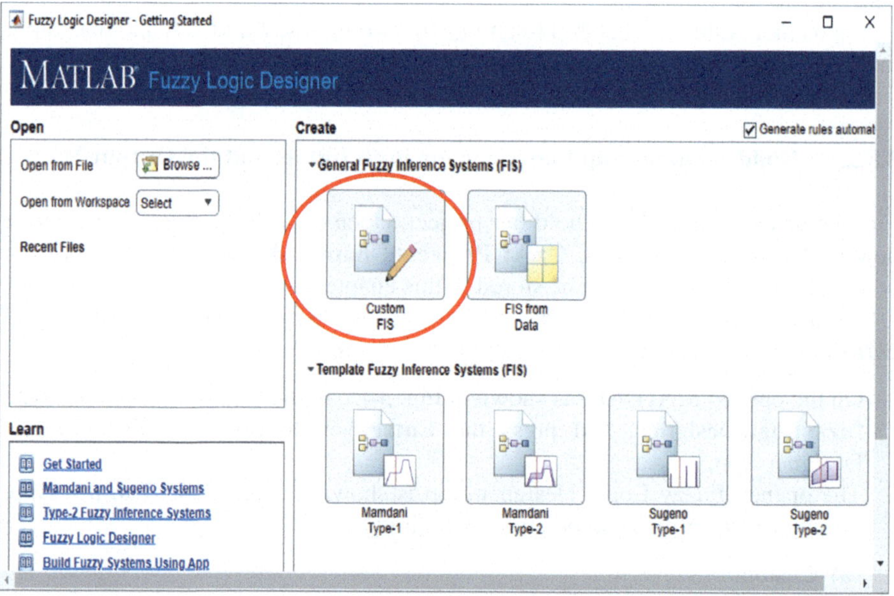

Fig. 3.16 The opened Fuzzy Logic Designer App

3. This App also allows users to open an existing FIS, either from a file saved in the hard disk or from the Workspace. Just click on the **Browse** button on the upper-left panel, you can select and open an existing FIS, or click on the drop-down arrow on the **Select** combo box on the upper-left panel to open an existing FIS from the Workspace.

4. Some predefined templates for different types of FIS can make the FIS building process faster and easier, such as from the **Mamdani Type-I FIS** to **Sugeno Type-II FIS** displayed under the **Template Fuzzy Inference System (FIS)** shown in Fig. 3.16. To build a more special FIS, such as our air conditioner FIS, we prefer to use the **Custom FIS** mode.
5. Click on the **Custom FIS** icon to open the Custom System wizard, which is shown in Fig. 3.17, to enter our FIS information. The name for our FIS is **AC_FIS**, and we need two inputs, **Temp** and **Temp_Rate** and one output, **Motor_Speed**. Enter those pieces of information to the related box, and your finished Custom System wizard is shown in Fig. 3.17. Click on the **OK** button to continue.
6. An initial configuration for our FIS is shown in Fig. 3.18.

Fig. 3.17 The opened Custom System wizard

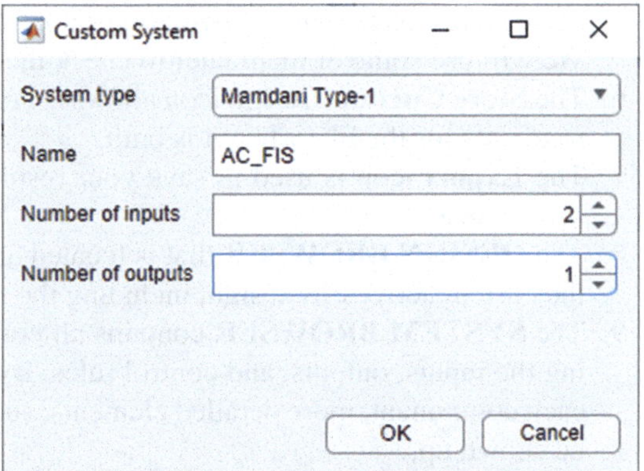

Fig. 3.18 An initial configuration for our FIS

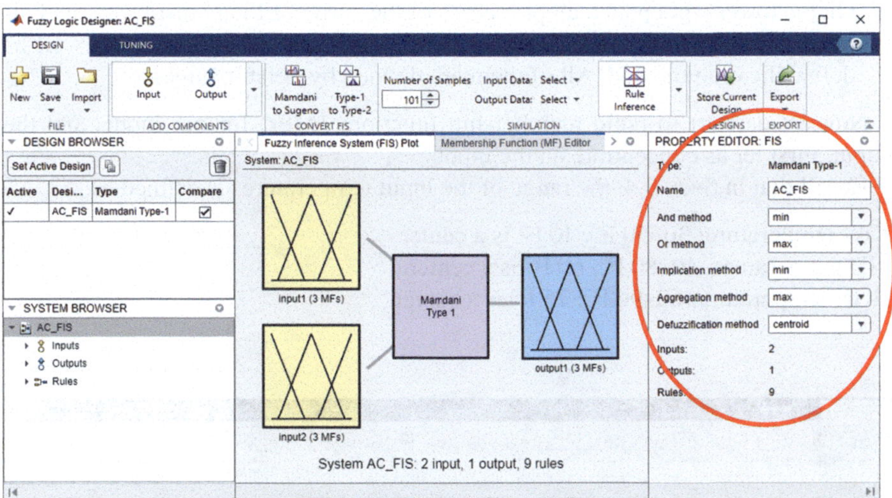

Let us have a closer look at this wizard to get us familiar with all useful functions provided by this wizard starting from the top menu items, which is shown in Fig. 3.19.

1. Starting from the left, three buttons, **New**, **Save**, and **Import**, are used to create a new FIS, save an FIS to a file or load an FIS from a file or from the Workspace.
2. With the help of the second left group, **ADD COMPONENTS**, you can add more input or output variables, control rules, into the current FIS.
3. The function on the third group allows you to perform conversion operations between different types of FIS.
4. The next group function allows users to perform some simulations by selecting testing input or output data, and the **Number of Samples** is used to set up the number of samples for inputs or outputs.
5. The function of the next group is to enable users to display some performance results, such as the rules distributions, control envelops, and error distributions. We will use some of them later to check the performance of our FIS.
6. The **Store Current Design** icon allows users to save the current designing environments for the FIS when it is built.
7. The **Export** icon is used to save your resulted FIS or simulation result to the Workspace.
8. The **DESIGN BROWSER** that is located at the mid-left panel is used to show the current active FIS design, including the name and type of the activated FIS.
9. The **SYSTEM BROWSER** contains all components built in your FIS, including the inputs, outputs, and control rules. By clicking on each arrow in front of each component, more detailed elements, such as all MFs and control rules, can be shown up.
10. The center panel contains the configuration of the current FIS, including all inputs, output, number of MFs for all variables, and number of rules.
11. On the lower-right panel, all properties of the current FIS are displayed, including the fuzzy set operators, implication and aggregation method, as well as defuzzification method. All of them are defined by default values.

Now let us start to build membership functions (MFs) for our inputs and the output. First let us concentrate on the inputs.

Recall that in Sect. 3.4, the range of the input temperatures is defined as:

LOW temperature: 30–50 F°, 40 F° is a center
MID temperature: 40–80 F°, 60 F° is a center
HIGH temperature: 60–90 F°, 75 F° is a center

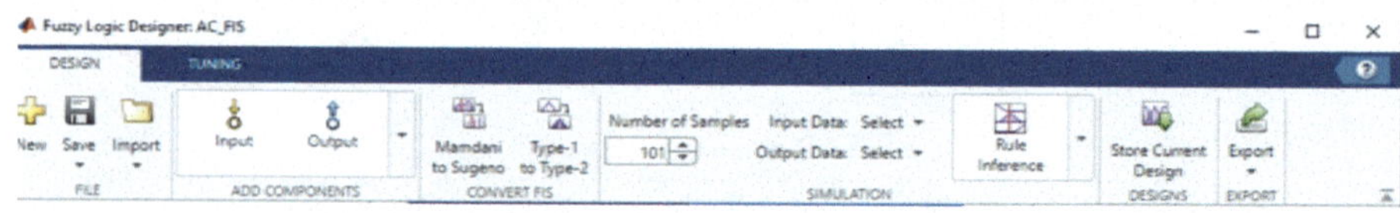

Fig. 3.19 The top menu items on the App

The range of the change rate of the input temperatures is from 0 to 5 (Fig. 3.9):

LOW change rate: 0.0–2.5 F°/H
MID change rate: 0.5–3.5 F°/H
HIGH change rate: 1.5–5.0 F°/H

Perform the following operations to set up MFs for our input variables first.

1. On the opened Fuzzy Logic Designer App, click on the **Input1(3 MFs)** icon to open its editor in the lower-right corner, which is shown in Fig. 3.20.
2. Enter **Temp** into the **Name** box under the **PROPERTY EDITOR: INPUT** panel and press the **Enter** key on the keyboard.
3. Change the input universe as [30 90] by entering them into the **Range** box, and click on the **Evenly Distribute MFs** button, as shown in Fig. 3.20.

Next let us set up the parameters for our three MFs, **LOW**, **MID**, and **HIGH** in the **Temp** input variable. Perform the following operations to complete this setup:

1. Keep the **Temp** variable icon selected, the properties of our three MFs are displayed in a three-line table, as shown in Fig. 3.20. Set the name for three MFs as **LOW**, **MID**, and **HIGH** by entering them one by one into the **Name** column in that table.
2. Double-click on the **Triangular** under the **Type** column for the **LOW** in MF, and select the **Generalized bell** by clicking on it. Do the same thing for the **HIGH** in MF, but select **Gaussian** for the **MID** in MF.

Your finished setup for three MFs of **Temp** variable is shown in Fig. 3.20.

Perform similar operations to set up three membership functions for the second variable **Temp_Rate**. The only differences are the Name (**Temp_Rate**) and the **Range** (0–5).

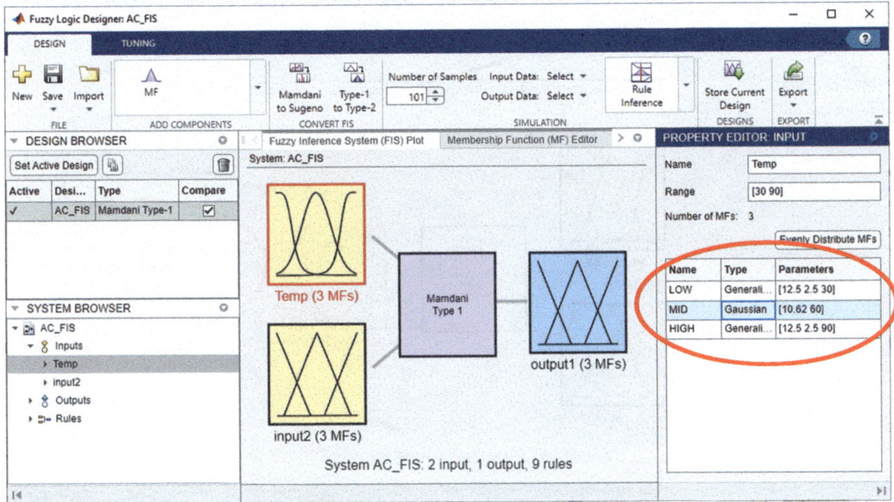

Fig. 3.20 The finished setup for three MFs on Temp variable

Finally let us build MFs for the output, **Motor_Speed**, to complete our membership functions definitions. The names and the range for three variables of the output, **Motor_Speed**, are:

SLOW rotating speed: 100–500 RPM, 300 RPM is a center
MID rotating speed: 300–800 RPM, 500 RPM is a center
FAST rotating speed: 600–1000 RPM, 800 RPM is a center

Perform the following operations to set up MFs for our output variable:

1. On the opened our FIS project, **AC_FIS**, click on the **output1** icon to open its **PROPERTY EDITOR: OUTPUT** wizard. You can open that project if you just start a new MATLAB window by clicking on the link **AC_FIS** that is located under the **Recent Files** in the **Open** window at the upper-left panel.
2. Then enter **Motor_Speed** into the **Name** box, [**1001000**] into the **Range** box, and click on the **Evenly Distribute MFs** button.
3. Set the name for three MFs as **SLOW**, **MID**, and **FAST** by entering them one by one into the **Name** column on that table.
4. Double-click on the **Triangular** under the **Type** column for the **SLOW** in MF, and select the **Generalized bell** by clicking on it. Do the same thing for the **FAST** in MF, but select **Gaussian** for the **MID** in MF.
5. Your finished Fuzzy Logic Designer should match the one that is shown in Fig. 3.21.

One point to be noted is the **Range** for each variable. That *range* is the entire universe or the length in which three MFs whose values are not 0 are covered (refer to Fig. 3.6).

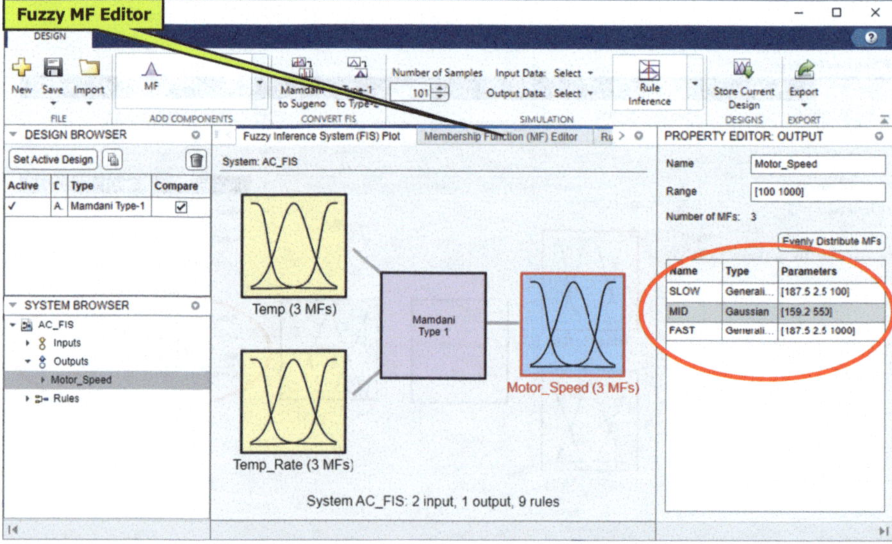

Fig. 3.21 The finished setup for three MFs on output–Motor_Speed

Before we can continue to build our FIS for this air conditioner control system, we had better save this FIS to make it safe. Click on the **Save** icon under the **DESIGN** menu, and select our project **AC_FIS** to save it to a desired folder. Click **Yes** to the popup message box to update it if an original project is already in there.

You can also save this FIS to the Workspace by using the **Export** icon on the top menu bar. MATLAB Workspace is a template used to enable users to store any variables used in the current project and reuse them in the future. To do that saving, just click on the drop-down arrow on the **Export** icon and select the item **Export Fuzzy Inference System to Workspace**. When the Export wizard appears, check the checkbox in front of our FIS, **AC_FIS**, and click on the **Export** button.

> **Tips**
>
> **When saving a FIS to the Workspace or to the hard disk, it is different for the file extension. For Workspace, you do not need to add any extension, such as .fis after the FIS's name. But when you save the FIS to the hard disk, you must add the extension, .fis, after the FIS's name.**

Before we can continue to build the control rules for our FIS, we like to have a closer look at some other interesting properties and functions provided by this Designer to enable us to build more professional and specific features for our MFs.

3.8.2.2 Use Membership Function Editor to Build Professional and Specific MFs

The Fuzzy Logic Designer App provided an MF specified editor to enable users to build more professional and specific MFs based on the practical considerations on various applications. In this section, we like to use our air conditioner FIS to illustrate how to use that editor to build more professional and flexible MFs.

To open that editor, just select one of our input or output variables, such as **Temp**, by clicking on that variable. Then click on the **Membership Function (MF) Editor** tab, as shown in Fig. 3.21. The opened MF Editor is shown in Fig. 3.22.

Now you can tune or modify any MF as you like, such as moving or shifting any MF in either left or right direction to change its range, or clicking and holding one square icon on an MF, the **MID** MF in our case, and shifting that square icon to adjust the core, support, and boundary of that MF. You can even get nonsymmetry MF by using this editor if you need. Figure 3.23 shows an example of the tuning result for our **Temp** variable with the help of this editor.

To recover our **Temp** variable to the original one with the previous MFs, just click on the **Evenly Distribute MFs** button. Next let us build the control rules for this system.

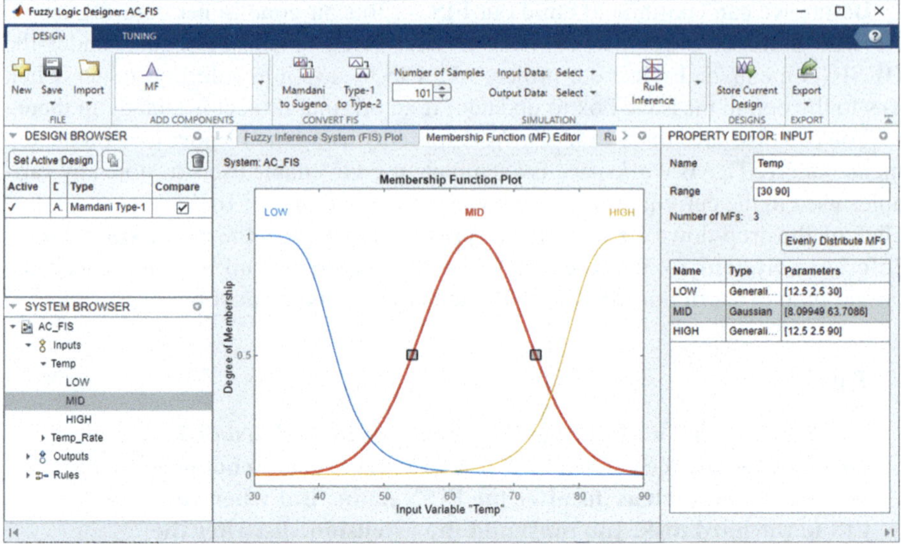

Fig. 3.22 The opened MF Editor

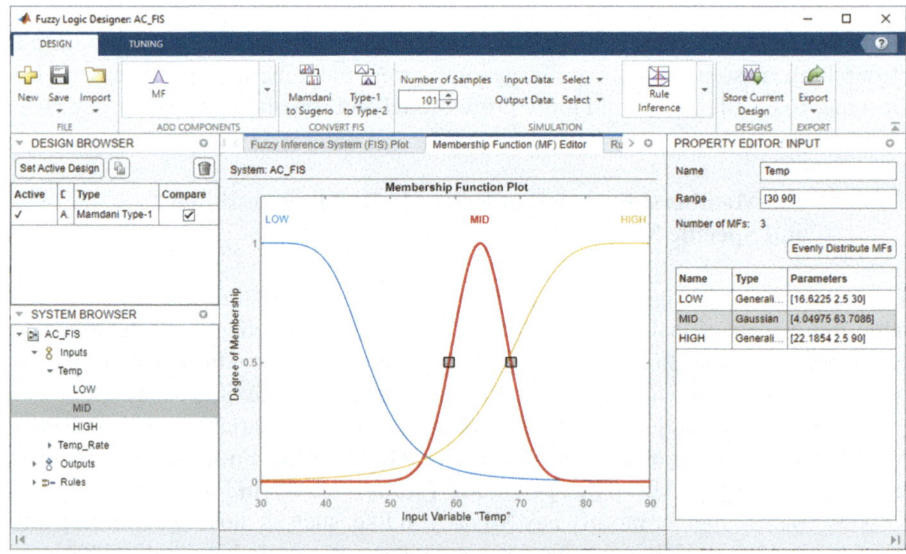

Fig. 3.23 A tuning example of our Temp variable with the MF Editor

3.8.2.3   Build the Control Rules for the Air Conditioner Control FIS

As we did in Sect. 3.5, we have nine fuzzy control rules for this air conditioner control system, which are shown in Table 3.1. For your convenience, we show that table here again in Table 3.3.

Table 3.3 An example of fuzzy rules

ΔT \ T	LOW	MID	HIGH
LOW	FAST	MID	MID
MID	FAST	SLOW	SLOW
HIGH	MID	SLOW	SLOW

Exactly, a total of nine control rules can be presented as:

$$
\begin{aligned}
&1.\ \text{If the } T \text{ is LOW and the } \Delta T \text{ is LOW, the motor speed is FAST.}\\
&2.\ \text{If the } T \text{ is LOW and the } \Delta T \text{ is MID, the motor speed is FAST.}\\
&3.\ \text{If the } T \text{ is LOW and the } \Delta T \text{ is HIGH, the motor speed is MID.}\\
&4.\ \text{If the } T \text{ is MID and the } \Delta T \text{ is LOW, the motor speed is MID.}\\
&5.\ \text{If the } T \text{ is MID and the } \Delta T \text{ is MID, the motor speed is SLOW.}\\
&6.\ \text{If the } T \text{ is MID and the } \Delta T \text{ is HIGH, the motor speed is SLOW.}\\
&7.\ \text{If the } T \text{ is HIGH and the } \Delta T \text{ is LOW, the motor speed is MID.}\\
&8.\ \text{If the } T \text{ is HIGH and the } \Delta T \text{ is MID, the motor speed is SLOW.}\\
&9.\ \text{If the } T \text{ is HIGH and the } \Delta T \text{ is HIGH, the motor speed is SLOW.}
\end{aligned}
\tag{3.24}
$$

To build those rules in MATLAB Fuzzy Logic Designer App, open the MATLAB and the Designer. You can load our fis file, **AC_FIS.fis**, from the location or the folder we saved before by using the **Import** icon menu item in the top bar. Just click on the drop-down arrow of that **Import** icon and select the item, Import Fuzzy Inference System from File. Then browse to the folder where our FIS, **AC_FIS.fis**, is stored, in our case it is **C:\AI Projects\Chapter 3**, and click on the **Open** button to load it.

Now click on the **Rules Editor** menu item, as shown in Fig. 3.24, to open the Rule Editor, which is also shown in Fig. 3.24.

Based on all rules shown in Eq. (3.24), we can build each rule in this Rule Editor by selecting different values, such as **LOW, MID**, and the output **HIGH**, from the related rule in the center panel.

To set up the first rule, just click on it from the center panel. Immediately you can find that the first rule has been displayed in the **PROPERTY EDITOR: RULE** window at the right-hand side, as shown in Fig. 3.25.

To set up this rule, keep it selected and go to the **PROPERTY EDITOR: RULE** window. Select each value of MF for each variable by clicking on the drop-down arrow and clicking the desired value based on Eq. (3.24). For example, the first rule is:

1. If the T is LOW and the ΔT is LOW, the motor speed is FAST.

Perform the following operations to set it up:

1. Along the **Temp** row, click on the drop-down arrow on the second combo box and select the **LOW** for the **Temp** variable.

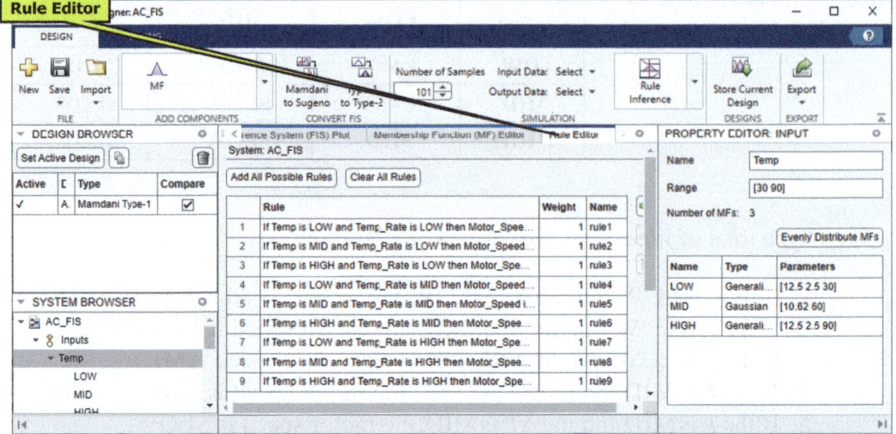

Fig. 3.24 The opened Rule Editor

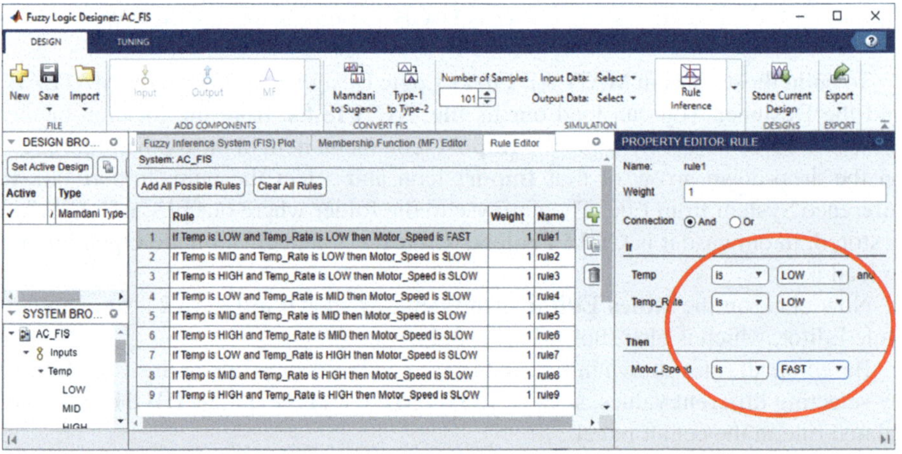

Fig. 3.25 The finished setting up for the first rule

2. Along the **Temp_Rate** row, click on the drop-down arrow on the second combo box and select the **LOW** for the **Temp_Rate** variable.
3. Along the **Motor_Speed** row, click on the drop-down arrow on the second combo box and select the **FAST** for the **Motor_Speed** variable.

Immediately you can find that the first rule shown in the center panel has been changed and now it is identical to your setting-up values in the **PROPERTY EDITOR: RULE** window.

In a similar way, by following Eq. (3.24) we can build all other control rules. The rule numbers are identical to those numbers in our Eq. (3.24). The finished Rule Editor is shown in Fig. 3.26.

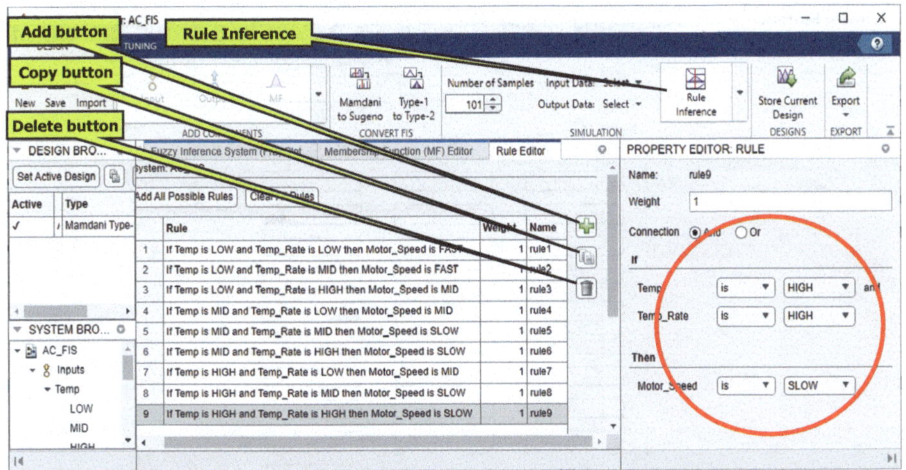

Fig. 3.26 The finished Rule Editor

As shown in Fig. 3.26, you can modify any rule you have built, including to delete or copy any of them by clicking the related button. To do any change, just select that rule first by clicking on it, and then click on any desired button to do what you want. Three buttons are available for three related modifications; Plus (+), Copy, and Delete, as shown in Fig. 3.26.

To perform either a delete or copy function for any rule, just select that rule first by clicking on it and click on the delete or the copy button. To add a new rule, just click on the (+) button and enter your new rule from the **PROPERTY EDITOR: RULE** window.

You can also add all possible rules or clear all rules with two buttons, **Add All Possible Rules** and **Clear All Rules**, as shown in Fig. 3.26.

At this point, we have completed designing and building our fuzzy logic control system for our air conditioner system. Next we can check and test our design by using related tools. It is recommended to save this FIS prior to testing our project.

3.8.2.4   Check and Test the Air Conditioner Control System

In the opened Designer, click on the **Rule Inference** icon, as shown in Fig. 3.26, on the menu bar to open the Rule Inference Viewer. Here we can check and test the functions of our FIS for the air conditioner system. The opened Rule Inference Viewer is shown in Fig. 3.27.

You can also test this FIS at any other inputs values in real time to obtain the output immediately. It is so convenient and quick, yes, it is!

Furthermore, you can have a look at the 3D input-output relationship to have a global view of our control performance. To do that, just click on the **Control Surface** icon on the top bar, as shown in Fig. 3.27, to open the Surface Viewer wizard, as shown in Fig. 3.28.

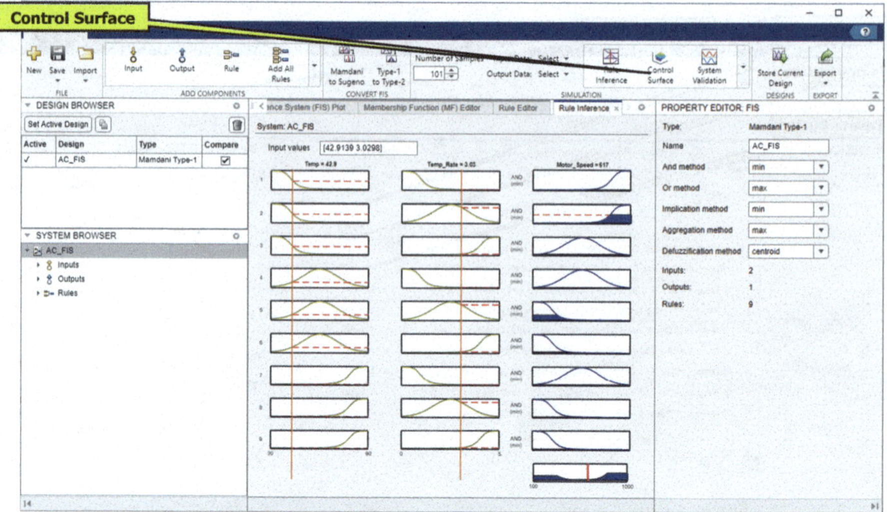

Fig. 3.27 The opened Rule Inference Viewer

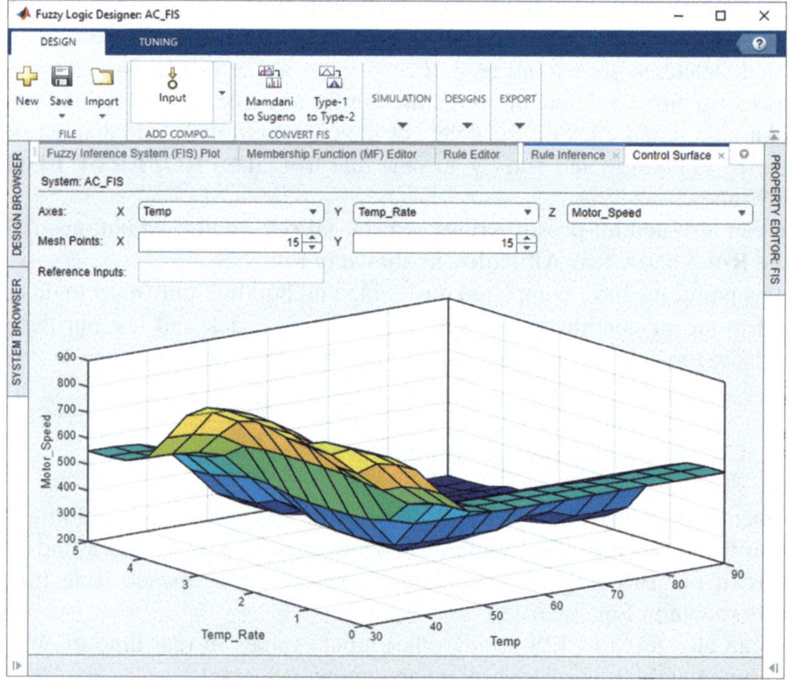

Fig. 3.28 The opened Control Surface wizard

We have successfully completed building our air conditioner system with Fuzzy Logic Designer App, next let us build the same project with the Fuzzy Logic Functions.

3.8.3   Build Fuzzy Logic Control System for Air Conditioner with Fuzzy Functions

The fuzzy control system built for our air conditioner in the last section is very easy and quick with no significant learning curves, even with no coding process. Yes, that is true. With the help of the Fuzzy Logic Designer App, a lot of or maybe all coding processes are provided by the MATLAB engine itself. The good point is that the students' learning curves can be greatly reduced and the students' learning interests on FIS can also be greatly increased. But the bad point is that all coding processes are hidden to students, and students may not be able to understand what really happened behind that App as well as the working principle and detailed building process for the FIS.

To solve that problem, we will provide detailed introductions and discussions about those knowledge and coding process behind that App in this section by using the other mode provided by MATLAB, Fuzzy Logic Functions Mode.

Unlike Fuzzy Logic Designer App, in which a lot of GUIs and toolboxes are provided to help users easily and quickly build a fuzzy control system, building a control system with Fuzzy Logic Functions is a totally different story. In the Fuzzy Functions mode, no GUI is provided and users must develop the control system with codes, exactly in MATLAB codes which are very similar to the programming language C.

3.8.3.1   MATLAB Script and Live Script Files

In order to build MATLAB® codes, two editors are provided by MathWorks®, Script file and Live Script file. The so-called Script file is similar to a coding editor file to enable users to create and enter their source codes, which will be compiled later to become the binary codes to be executed. The Script file has an extension **.m**. The Live Script file and live functions are interactive documents that combine MATLAB code with formatted text, equations, and images in a single environment called the Live Editor. In addition, live scripts store and display output alongside the code that creates it. The Live Script file has an extension **.mlx**.

In this section, we only concentrate on how to use Script Editor to generate the MATLAB source file or codes.

First let us create a folder to hold our project. Open the File Explorer and create a new folder under the root as, **C:\AI Projects\Chapter 3**, and this is the location where all our AI projects will be stored in this chapter.

Open the MATLAB by double-clicking on its shortcut from the desktop and perform the following operational steps to create our new project:

1. Click on the **New Script** icon that is located in the upper-left corner under the **Home** menu item to open a new script file. A new untitled script file is generated.
2. First let us save our file by clicking the drop-down arrow on the **Save** icon and clicking on the **Save** item to save our project. Browse to our default folder **C:\AI Projects\Chapter 3**, and enter **AC_Func.m** into the **File name** box, and click on the **Save** button.

Regularly, to build a fuzzy control system with functions, we need to perform the following operations:

1. Create an FIS variable by using function **mamfis()** or **sugfis()** to create a new Type-1 Mamdani system or Type-1 Sugeno system.
2. Use **addInput()** and **addOutput()** functions to add input variables or output variables into the FIS system created above.
3. You can configure the input variable properties using dot notation. For example, specify the name and range of the variable. **Fis.Inputs(1).Name = "service"**; **fis.Inputs(1).Range = [0 10]**; You can also add the input variable property using the **addInput()** method.
4. You can use **plotmf()** function to display your generated or added MFs.
5. Use **addMF()** function to add MFs into an added variable in step 2.
6. Use **addRule()** function to add control rules into your FIS system.
7. Use **showrule()** function to display all rules in text format.
8. Use **genfis()** function to generate a Type-1 FIS object based on input and output data.
9. Use **writeFIS()** function to save the generated FIS to the hard disk as a file.
10. Use **gensurf()** function to check the envelope of the generated FIS.
11. Use **plotfis()** function to plot the structure of the generated Type-1 FIS.
12. After an FIS is built and you want to use it, use **readfis()** function to get it.
13. Use **evalfis()** function to calculate the output for a given Type-1 FIS based on the current inputs.
14. Some other functions, including **removeInput()**, **removeOutput()**, and **removeMF()**.
15. Some other functions include **convertfis()**, which is used to convert an old FIS to a current one, **convertToSugeno()**, which is used to convert a Mamdani FIS to a Sugeno FIS, and **convertToType1()** and **convertToType2()**, as well as **defuzz()** function.

All of these functions are assembled and categorized in Table 3.4 based on their purposes.

It can be found from Table 3.4 that most functions, about 87% of them, can be applied for both types, either Type-1 or Type-2 FIS. Three of them can only work for Type-1 FIS.

Next let us build the membership function for two inputs and output variables.

Table 3.4 Most popular MATLAB functions used in Fuzzy Logic Toolbox

Purpose	Function	Arguments
Create a new FIS Outline (Type-1)	mamfis()	fis = mamfis("Name", "acfis");
	sugfis()	fis = sugfis("Name", "acfis");
Add Input Variables	addInput()	fis = addInput(fis,[0 10],"Name","food");
Add Input MFs	addMF()	fis = addMF(fis,"food","gaussmf",[0 1],"Name","rancid");
Add Output Variables	addOutput()	fis = addOutput(fis,[0 30],"Name","tip");
Add Output MFs	addMF()	fis = addMF(fis,"tip","trimf",[0 5 10],"Name","cheap");

Add Control Rules	addRule()	ruleList = [1 1 1 1 2;
		    2 0 2 1 1;
		    3 2 3 1 2];
		
		fis = addRule(fis,ruleList);

Plot MFs	plotmf()	plotmf(fis,'input',1);
Display Rules	showrule()	showrule(fis); show rule in symbolic, verbose or indexed.
Create a new FIS Object (Type-1)	genfis()	fis = genfis(inputData, outputData, options);
Save a FIS to a File	writeFIS()	writeFIS(fis, fileName);
Generate FIS Output Surface	gensurf()	gensurf(fis);
Display FIS in structure mode	plotfis()	plotfis(fis);
Load FIS from a file	readfis()	fis = readfis('tipper');
Evaluate a FIS Based on inputs	evalfis()	output = evalfis(fis, input, options);
Options set for evalfis() function	evalfisOptions()	opt = evalfisOptions(Name,Value);
Remove input variable from FIS	removeInput()	fisOut = removeInput(fisIn,inputName);
Remove output variable from FIS	removeOutput()	fisOut = removeOutput(fisIn,outputName);
Remove MFs from a FIS	removeMF()	fis = removeMF(fis,"service","poor");
Convert an old FIS to a new FIS (Type-1)	convertfis()	fisNew = convertfis(fisOld);
Convert Mamdani FIS to Sugeno FIS	convertToSugeno()	sugenoFIS = convertToSugeno(mamdaniFIS);
Convert Type-2 FIS to Type-1 FIS	convertToType1()	fisT1 = convertToType1(fisT2);
Convert Type-1 FIS to Type-2 FIS	convertToType2()	fisT2 = convertToType2(fisT1);
Defuzzify MFs	defuzz()	output = defuzz(x, mf, method);

3.8.3.2 Create Fuzzy Inference System for our Project

As we discussed in the last section, the first step to build our fuzzy logic control system is to generate an FIS by using MATLAB functions. Thus first let us create our FIS using MATLAB function **mamfis()** since we want to use the Mamdani system to create a Type-1 type FIS.

Open MATLAB and our new Script file **AC_Func.m** we built in the last section. Perform the following operational steps to create this new FIS:

1. Type a comment line at the top to indicate the project name and function, as shown in Fig. 3.29.
2. Generate our FIS by calling the function **mamfis()** with the name of the FIS, **acfis**, as shown in step **5** in Fig. 3.29.

Next let us continue to add input and output variables to our FIS by calling related functions shown in Table 3.4.

```
1  % Name: AC_Func.m
2  % Func: Build a FIS for an AC system with MATLAB Functions
3  % Date: August 24, 2023
4
5  acfis = mamfis("Name", "acfis");
```

Fig. 3.29 The initial codes to generate our FIS

```
1  % Name: AC_Func.m
2  % Func: Build a FIS for an AC system with MATLAB Functions
3  % Date: August 24, 2023
4
5  acfis = mamfis("Name", "acfis");
6  acfis = addInput(acfis,[30 90],"Name","Temp");
7  acfis = addInput(acfis,[0 5],"Name","Temp_Rate");
8  acfis = addOutput(acfis,[100 1000],"Name","Motor_Speed");
```

Fig. 3.30 Add two input and one output variables into our FIS

3.8.3.3 Add Input and Output Variables to Our Fuzzy Inference System

Prior to adding input and output variables into our project, first let us familiarize some popular MATLAB functions in the Fuzzy Logic Toolbox™ and their functionalities as well as their working groups. Table 3.4 shows the most popular and important functions included in the Fuzzy Logic Toolbox™. More detailed introductions and implementations of those functions will be given by real example coding when we build our air conditioner project later.

First let us add two input variables, **Temp** and **Temp_Rate**, into our FIS. To do that, we need to call the function **addInput**() shown in Table 3.4.

Just move the cursor under the first coding line we just typed, and enter the codes shown in steps **6–8** in Fig. 3.30. The previous coding line has been highlighted to differentiate from the new coding lines.

The arguments of the function **addInput**() include four parts: the first one is out generated FIS, **acfis**, the second is the range of the input variable represented by the starting and the ending values embedded in a bracket [], the third part is a name indicator to the input variable and it must be written as **"Name"**, and the last or the fourth part is the actual name of our input variable, **"Temp"**. Both the name indicator and the real name must be enclosed by the quotation marks "" since that is required by MATLAB coding syntax.

To determine the actual interval values for the ranges of our input and output variables, such as **Temp**, **Temp_Rate** and **Motor_Speed**, refer to Sect. 3.8.2.1 or Figs. 3.20 and 3.21.

To add our output variable into the FIS, perform a similar coding job, and your finished code should match one that is shown in step **8** in Fig. 3.30.

Table 3.5 Types of membership function used in Fuzzy Logic Toolbox

Membership Function Type	Descriptions
"gbellmf"	Generalized bell-shaped membership function
"gaussmf"	Gaussian membership function
"gauss2mf"	Gaussian combination membership function
"trimf"	Triangular membership function
"trapmf"	Trapezoidal membership function
"linsmf"	Linear s-shaped saturation membership function
"linzmf"	Linear z-shaped saturation membership function
"sigmf"	Sigmoidal membership function
"dsigmf"	Difference between two sigmoidal membership functions
"psigmf"	Product of two sigmoidal membership functions
"zmf"	Z-shaped membership function
"pimf"	Pi-shaped membership function
"smf"	S-shaped membership function
"constant"	Constant membership function for Sugeno output membership functions
"linear"	Linear membership function for Sugeno output membership functions
String or character vector	Name of a custom membership function in the current working folder or on the MATLAB path. Custom output membership functions are not supported for Sugeno systems.
Function handle	Handle to a custom membership function in the current working folder or on the MATLAB path. Custom output membership functions are not supported for Sugeno systems.

3.8.3.4   Add Membership Functions to Input and Output Variables in FIS

Now let us add related membership functions to our input and output variables. Starting from our first input variable, **Temp**.

Based on our discussions in Sect. 3.8.2.1, three MFs are assigned to our first input variable, they are, LOW, MID, and HIGH. Both LOW and the HIGH MFs have the **Generalized bell** type, but the HIGH has a **Gaussian** type. According to Table 3.4, we need to use **addMF()** function to add those MFs into our FIS.

One issue with this adding MFs function is the type of MFs. We can select this type from a combo box when we build those MFs using the Fuzzy Logic Designer App. However, no App or GUI is available when we add those MFs with the function mode. MATLAB defined those MF types with a group of type names, which are shown in Table 3.5, to solve this problem.

Now we can add our MFs into our input and output variables by using **addMF()** function with the help of type names defined in Table 3.5.

Enter the codes shown in steps **9–20** in Fig. 3.31 into our script file. The previous coding parts have been highlighted.

Let us have a closer look at this piece of newly added codes to see how it works.

```
1  % Name: AC_Func.m
2  % Func: Build a FIS for an AC system with MATLAB Functions
3  % Date: August 24, 2023
4
5  acfis = mamfis("Name", "acfis");

6  acfis = addInput(acfis,[30 90],"Name","Temp");
7  acfis = addInput(acfis,[0 5],"Name","Temp_Rate");
8  acfis = addOutput(acfis,[100 1000],"Name","Motor_Speed");

9  % add 3 MFs to input variable Temp
10 acfis = addMF(acfis,"Temp","gbellmf",[12.5 2.5 30],"Name","LOW");
11 acfis = addMF(acfis,"Temp","gaussmf",[10.62 60],"Name","MID");
12 acfis = addMF(acfis,"Temp","gbellmf",[12.5 2.5 90],"Name","HIGH");

13 subplot(3,1,1);
14 plotmf(acfis,"input",1);

15 % add 3 MFs to input variable Temp_Rate
16 acfis = addMF(acfis,"Temp_Rate","gbellmf",[1.042 2.5 2.776e-17],"Name","LOW");
17 acfis = addMF(acfis,"Temp_Rate","gaussmf",[0.8847 2.5],"Name","MID");
18 acfis = addMF(acfis,"Temp_Rate","gbellmf",[1.042 2.5 5],"Name","HIGH");
19 subplot(3,1,2);
20 plotmf(acfis,"input",2);
```

Fig. 3.31 Add three MFs to our input variables

1. This is a comment line used to indicate the purpose of the following coding lines.
2. When using **addMF()** to add MFs, the format is: **fis_name, variable_nam, mf_type, range, mf_name_title, mf_name**. All of those arguments must be a string covered by quotation marks "" except the **fis_name** and the range.
3. The **mf_type** we used for **LOW** and **HIGH** is **"gbellmf"**, which is equivalent to **Generalized bell** type.
4. But it is **Gaussian** type or **"gaussmf"** for the MF **MID**.
5. The coding lines 13 and 14 are used to plot our first three MFs to confirm it. Similar function is provided in coding lines 19 and 20 to confirm our MFs for the second variable, **Temp_Rate**.
6. The coding lines 16–18 are used to set three MFs to our second input variable, **Temp_Rate**. To determine the actual interval values for the ranges of our input variables, such as **Temp** and **Temp_Rate**, refer to Sect. 3.8.2.1 or Fig. 3.20.

Finally let us enter the codes shown in steps 21–26 in Fig. 3.32 to define three MFs for our output variable, **Motor_Speed**. As for the range of three MFs, refer to Fig. 3.21.

One point to be noted when using **addMF()** function to add MFs into our variables is the argument name and its order. Also to be noted is the case of letters used in all arguments since MATLAB is a case-sensitive programming language. Make sure to use **"Name"**, instead of **"name"**, to indicate each name of MF to be added into variables.

Now run our script by clicking on green-color **Run** button on the top menu bar, the running result is shown in Fig. 3.33. Next let us generate and add control rules into our FIS.

```matlab
1  % Name: AC_Func.m
2  % Func: Build a FIS for an AC system with MATLAB Functions
3  % Date: August 24, 2023
4
5  acfis = mamfis("Name", "acfis");
6  acfis = addInput(acfis,[30 90],"Name","Temp");
7  acfis = addInput(acfis,[0 5],"Name","Temp_Rate");
8  acfis = addOutput(acfis,[100 1000],"Name","Motor_Speed");
9  % add 3 MFs to input variable Temp
10 acfis = addMF(acfis,"Temp","gbellmf",[12.5 2.5 30],"Name","LOW");
11 acfis = addMF(acfis,"Temp","gaussmf",[10.62 60],"Name","MID");
12 acfis = addMF(acfis,"Temp","gbellmf",[12.5 2.5 90],"Name","HIGH");
13 subplot(3,1,1);
14 plotmf(acfis,"input",1);
15 % add 3 MFs to input variable Temp_Rate
16 acfis = addMF(acfis,"Temp_Rate","gbellmf",[1.042 2.5 2.776e-17],"Name","LOW");
17 acfis = addMF(acfis,"Temp_Rate","gaussmf",[0.8847 2.5],"Name","MID");
18 acfis = addMF(acfis,"Temp_Rate","gbellmf",[1.042 2.5 5],"Name","HIGH");
19 subplot(3,1,2);
20 plotmf(acfis,"input",2);
21 % add 3 MFs to output variable Motor_Speed
22 acfis = addMF(acfis,"Motor_Speed","gbellmf",[187.5 2.5 100],"Name","SLOW");
23 acfis = addMF(acfis,"Motor_Speed","gaussmf",[159.2 550],"Name","MID");
24 acfis = addMF(acfis,"Motor_Speed","gbellmf",[187.5 2.5 1000],"Name","FAST");
25 subplot(3,1,3);
26 plotmf(acfis,"output",1);
```

Fig. 3.32 Add three MFs to our output variable

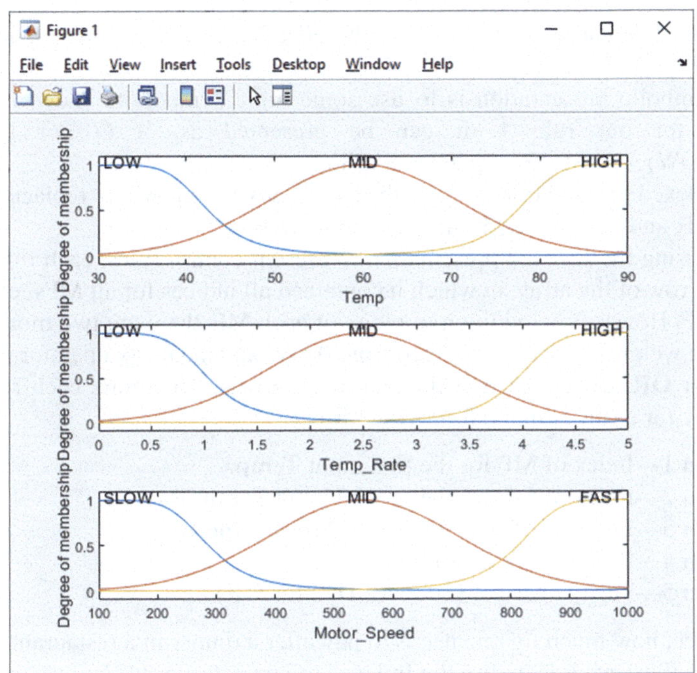

Fig. 3.33 Our added MFs for input and output variables

3.8.3.5 Generate and Add Control Rules into Our Fuzzy Inference System

Recall that when we discussed our control rules for this air conditioner system in Sect. 3.8.2.3, totally nine control rules can be used in our system:

1. If the T is LOW and the ΔT is LOW, the motor speed is FAST.
2. If the T is LOW and the ΔT is MID, the motor speed is FAST.
3. If the T is LOW and the ΔT is HIGH, the motor speed is MID.
4. If the T is MID and the ΔT is LOW, the motor speed is MID.
5. If the T is MID and the ΔT is MID, the motor speed is SLOW.   (3.25)
6. If the T is MID and the ΔT is HIGH, the motor speed is SLOW.
7. If the T is HIGH and the ΔT is LOW, the motor speed is MID.
8. If the T is HIGH and the ΔT is MID, the motor speed is SLOW.
9. If the T is HIGH and the ΔT is HIGH, the motor speed is SLOW.

As we know, all control rules have three different presentation ways:

1. Verbose
2. Symbolic
3. Indexed

The so-called **Verbose** way is to describe all rules as a standard text format, as shown in Eq. (3.25) above.

The symbolic presentation is to use some logic symbols to show a rule. For example, for our rule 1, it can be presented as, if $(T == \text{LOW})$ & $(\Delta T == \text{LOW}) = > (\textbf{Motor\_Speed} = \text{FAST})$.

The Indexed presentation is to use the each index for one MF to replace the name of each MF, and describe each rule in an index way.

When using the Indexed presentation to present control rules, each rule is composed of a row of the array, in which it contained all indices for all MFs, each index for one MF. However, in addition to index of each MF, there are two more indices, such as the weight of each rule ranged from 0 to 1, and the fuzzy operator type, such as **AND** or **OR**, are attached at the end of each rule. Therefore, each row of the index array for a rule is in the following format:

1. Column 1—Index of MF for the first input **Temp**.
2. Column 2—Index of MF for the second input **Temp_Rate**.
3. Column 3—Index of MF for the output **Motor_Speed**.
4. Column 4—Rule weight (from **0** to **1**).
5. Column 5—Fuzzy operator (1 for **AND**, 2 for **OR**).

For instance, how much tip one needs to pay after a dinner in a restaurant is a good example to illustrate how to use the Indexed to present rules [16].

Two categories or variables could be used as inputs, **Service** and **Food**, and one variable, **Tip**, can work as an output. They are defined as below:

Service with three MFs:	**Poor, Good, Excellent**	MF Index: 1, 2, 3
Food with two MFs:	**Rancid, Delicious**	MF Index: 1, 0, 3
Tip with three MFs:	**Cheap, Average, Generous**	MF Index: 1, 2, 3

Three control rules are used for these variables and MFs:

1. If (Service is poor) or (Food is rancid), then (Tip is cheap).
2. If (Service is good), then (Tip is average).
3. If (Service is excellent) or (Food is delicious), then (Tip is generous).

These three rules can be presented by using Indexed format as below:

   ruleList = [1 1 1 1 2 ⟸ The last number 2 indicates that the operator is OR.

   2 0 2 1 1; ⟸ The last number 1 indicates that the default operator is AND.

   3 2 3 1 2]; ⟸ The last number 2 indicates that the operator is OR.

The weight used for all three rules is full weight, which is 1 on the fourth column.

For our air conditioner application, three MFs for the first variable **Temp** are, **LOW**, **MID**, and **HIGH**. The related indices for them are 1 for **LOW**, 2 for **MID**, and 3 for **HIGH**, respectively. Similar index values for the second variable **Temp_Rate**. For the output variable, **Motor_Speed**, three index values are: 1 for **SLOW**, 2 for **MID**, and 3 for **FAST**.

Regularly, we prefer to use either Verbose or Indexed way to do our coding process. Based on the Indexed representation format, our nine control rules can be written as a matrix since we used the whole weight for each rule (1) and the **AND** (1) operator:

$$Rulelist = \begin{bmatrix} 1 & 1 & 3 & 1 & 1; \\ 1 & 2 & 3 & 1 & 1; \\ 1 & 3 & 2 & 1 & 1; \\ 2 & 1 & 2 & 1 & 1; \\ 2 & 2 & 1 & 1 & 1; \\ 2 & 3 & 1 & 1 & 1; \\ 3 & 1 & 2 & 1 & 1; \\ 3 & 2 & 1 & 1 & 1; \\ 3 & 3 & 1 & 1 & 1 \end{bmatrix}; \tag{3.26}$$

In real MATLAB coding process, we can use the Verbose format to setup our rule as:

```
rule1 = 'If Temp is LOW and Temp_Rate is LOW then Motor_Speed
is FAST';
rule2 = 'If Temp is LOW and Temp_Rate is MID then Motor_Speed
is FAST';
rule3 = 'If Temp is LOW and Temp_Rate is HIGH then Motor_Speed
is MID';
ruleList = char(rule1, rule2, rule3);
```

If we use the Indexed format to set up our control rules as shown in Eq. (3.26) above.

To save space and make it simple, we will use the Indexed format to build our control rules in this project. Enter the codes shown in Fig. 3.34 into our project code window.

```
1   % Name: AC_Func.m
2   % Func: Build a FIS for an AC system with MATLAB Functions
3   % Date: August 24, 2023
4
5   acfis = mamfis("Name", "acfis");
6   acfis = addInput(acfis,[30 90],"Name","Temp");
7   acfis = addInput(acfis,[0 5],"Name","Temp_Rate");
8   acfis = addOutput(acfis,[100 1000],"Name","Motor_Speed");
9   % add 3 MFs to input variable Temp
10  acfis = addMF(acfis,"Temp","gbellmf",[12.5 2.5 30],"Name","LOW");
11  acfis = addMF(acfis,"Temp","gaussmf",[10.62 60],"Name","MID");
12  acfis = addMF(acfis,"Temp","gbellmf",[12.5 2.5 90],"Name","HIGH");
13  subplot(3,1,1);
14  plotmf(acfis,"input",1);
15  % add 3 MFs to input variable Temp_Rate
16  acfis = addMF(acfis,"Temp_Rate","gbellmf",[1.042 2.5 2.776e-17],"Name","LOW");
17  acfis = addMF(acfis,"Temp_Rate","gaussmf",[0.8847 2.5],"Name","MID");
18  acfis = addMF(acfis,"Temp_Rate","gbellmf",[1.042 2.5 5],"Name","HIGH");
19  subplot(3,1,2);
20  plotmf(acfis,"input",2);
21  % add 3 MFs to output variable Motor_Speed
22  acfis = addMF(acfis,"Motor_Speed","gbellmf",[187.5 2.5 100],"Name","SLOW");
23  acfis = addMF(acfis,"Motor_Speed","gaussmf",[159.2 550],"Name","MID");
24  acfis = addMF(acfis,"Motor_Speed","gbellmf",[187.5 2.5 1000],"Name","FAST");
25  subplot(3,1,3);
26  plotmf(acfis,"output",1);
27  % add control rules to our FIS
28  rulelist = [ 1    1    3    1    1;
29               1    2    3    1    1;
30               1    3    2    1    1;
31               2    1    2    1    1;
32               2    2    1    1    1;
33               2    3    1    1    1;
34               3    1    2    1    1;
35               3    2    1    1    1;
36               3    3    1    1    1 ];
37  acfis = addRule(acfis, rulelist);
38  acfis = writeFIS(acfis);
39  % display the surface of our FIS
40  ac_fis = readfis('acfis');
41  subplot(1,1,1);
42  gensurf(ac_fis);
```

Fig. 3.34 The codes for adding control rules and displaying surface of the FIS

Let us have a closer look at this piece of codes step by step to see how it works and add nine control rules into our FIS.

1. First we need to generate our control rule matrix with nine rules or nine rows, and each row vector is mapped to a rule with five columns. The first row is: the index of the **LOW** for the first variable **Temp** (1), the index of the **LOW** for the second variable **Temp_Rate** (1), and the index of the **FAST** for the output variable **Motor_Speed** (3). This rule will be used as a whole weight (1), and an **AND** operator is used for the first and the second input variables (1). The resulted row vector is: **1 1 3 1 1**;
2. From lines 29 to 36, the rest eight rules or rows are defined in a similar way.
3. The function **addRule**() is used to add our defined nine control rules into our FIS.

   To save our FIS, the function **writeFIS**() is called to do that saving. One point to be noted is that no file name is accompanied with this FIS when calling that function. In this case, a dialog box will be displayed to allow or ask users to select the location and enter a valid file name to do this saving. This function provided a flexibility to save this FIS.
4. The codes between lines 40 and 42 are used to display a surface of our FIS. A trick is that you cannot directly use our FIS, **acfis**, to do this displaying with the function **gensurf**(); instead, you must use **readfis**() function to load our saved FIS and then call **gensurf**() to do that displaying. The function **subplot(1, 1, 1)** is used to try to avoid plotting that surface on the third plot as we did in line 25

Now run our project and close the three MFs, and enter a valid name for our FIS to save it. The generated surface for our FIS is shown in Fig. 3.35.

3.8.3.6   Evaluate and Test Our Developed FIS with Selected Inputs

One can test and confirm our developed FIS, **acfis**, with any appropriate inputs, **Temp** and **Temp_Rate**, with different values, and check the outputs, **Motor_Speed**. The function **evalfis**() should be used for that purpose.

The syntax of using that function is, **output = evalfis(fis, input);**

The input can be a single input pair or a matrix that contained multiple input pairs. For example, the coding line **output = evalfis(acfis, [70, 2.5]);** can generate an output of **243.3114**.

Generate another new MATLAB Script file named **Eval_ACFIS.m** and enter the codes shown in Fig. 3.36 into that file. Let us have a closer look at this piece of codes to see how it works.

1. First we need to load our developed FIS file, **ac_fis**, by calling the function **readfis**() from our current folder. We did save that file in the folder, **C:\AI Projects\ Chapter 3**. You may need to use MATLAB **cd** command to find and get the current folder if you save that file into different folder in your machine. For example, if you saved your FIS file into a location **C:\My_Project\My_Folder\ac_fis. fis**, you need to enter one coding line as, **cd 'C:\My_Project\My_Folder'**, to allow machine to access that folder and file.

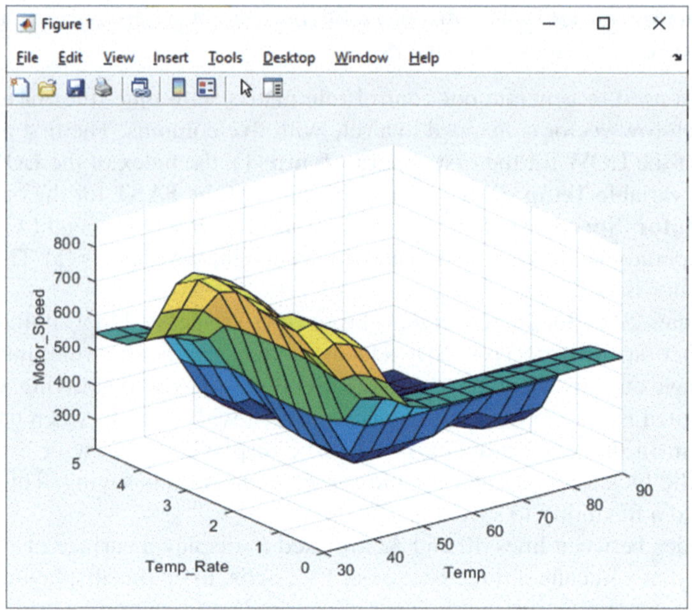

Fig. 3.35 The displayed surface of our FIS

```
% Evaluate our air conditioner FIS
% Eval_ACFIS.m
% August 29, 2023
1  fis = readfis('ac_fis');
2  output = evalfis(fis, [70 2.5])
3  input = [38 0.6; 42 1.2; 48 1.5; 52 1.8; 67 3.2; 70 3.3; 78 4.1; 45 5.0; 89 4.8;
           82 3.6; 72 4.2; 69 1.6; 58 2.3; 48 2.2; 59 3.3; 67 4.1; 79 3.5; 81 3.8];

4  output = evalfis(fis, input)

5  plot(output, 'b-o', 'LineWidth',2);
   grid;
   legend('output', 'Location', 'northeast');
```

Fig. 3.36 The codes for the FIS evaluation process

2. First we test our FIS by using an input pair, **Temp = 70** and **Temp_Rate = 2.5**, by calling the function **evalfis()**. The output should be printed in the Command window since we did not add any semicolon,**;**, after that coding line.
3. Now we use a matrix to contain multiple input pairs to get more outputs. To do that, first we need to generate an input matrix.
4. The **evalfis()** is called to run that function to perform evaluation for our input matrix. The output is also a matrix.
5. The output is plotted to display our results, as shown in Fig. 3.37.

At this point, we have provided a completed introduction and discussion about how to use fuzzy functions to build our air conditioner control system. Compared

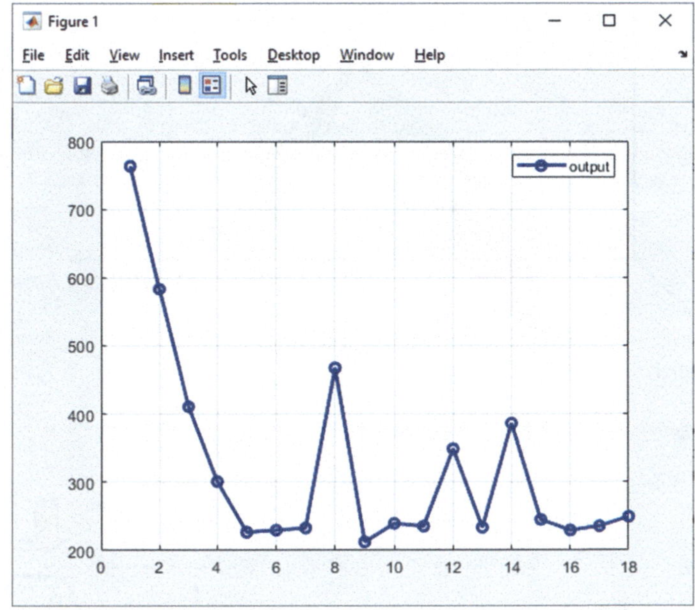

Fig. 3.37 The evaluation result for our FIS

with the same project we built in Sect. 3.8.2, where the Fuzzy Logic Designer App is used to build our air conditioner system, it provides more flexibility and control-lability in building our project by using the fuzzy functions. However, more coding processes are involved into that development, which may be a challenging issue to the beginning students who have never any coding experience.

In the following sections, we will do more discussions on a more complicated topic, Type-2 fuzzy inference system.

3.9   Introduction to Type-2 Fuzzy Inference System

As we mentioned at the beginning at this chapter, the so-called type number is exactly equivalent to the dimension of fuzzy inference system applied in our real world. Type-1 is for 1D and Type-2 is for 2D FIS. However, you will see that those definitions are untrue, especially for the Type-2 FIS.

Generally, the so-called Type-1 or Type-2 FIS is exactly determined by the dimensions of MFs. For a Type-1 FIS, its MFs have a single- or one-dimensional membership value; therefore, they can only model the degree of membership in a given linguistic set, but they cannot model uncertainty in the degree of membership.

The membership function of a general Type-2 fuzzy set, $\tilde{A}$, is a three-dimensional function (Fig. 3.38), where the third dimension is the value of the membership

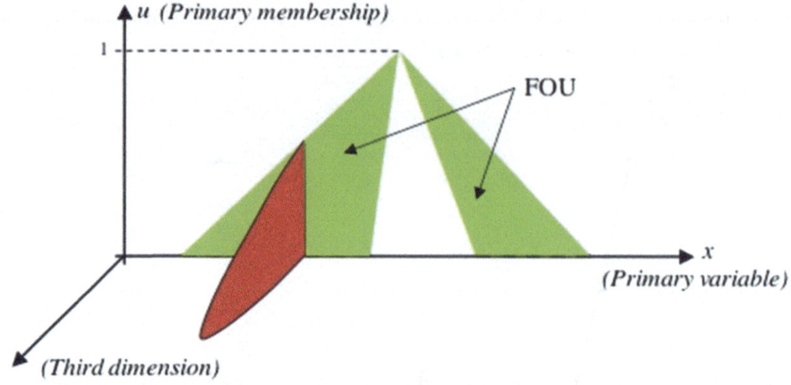

Fig. 3.38 A general type-2 FIS is a 3D system

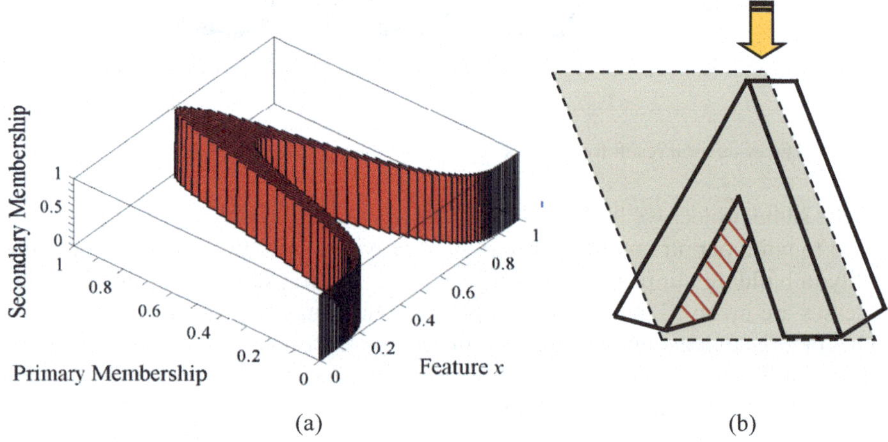

Fig. 3.39 Some actual type-2 MFs

function at each point on its two-dimensional domain that is called its **Footprint of Uncertainty** (FOU).

It can be found from Fig. 3.38 that a Type-2 FIS is a 3D system with more complicated MFs, and it is very difficult to be mapped to a mathematical equation. To simplify that case, a so-called Interval Type-2 FIS (IT2FIS) is introduced, which is exactly a 2D system and can be described in a 2D coordinate system.

Figures 3.39a, b shows some MFs used for actual type-2 FIS and they are 3D MFs. In fact, the MFs in an IT2FIS are cut along the x-y plan to get an intersection or an interval for the uncertainty along that plane. In this way, the complicated 3D system can be reduced to a 2D system, as shown in Fig. 3.40a, which is a cutting-off plane of an MF in Fig. 3.39b.

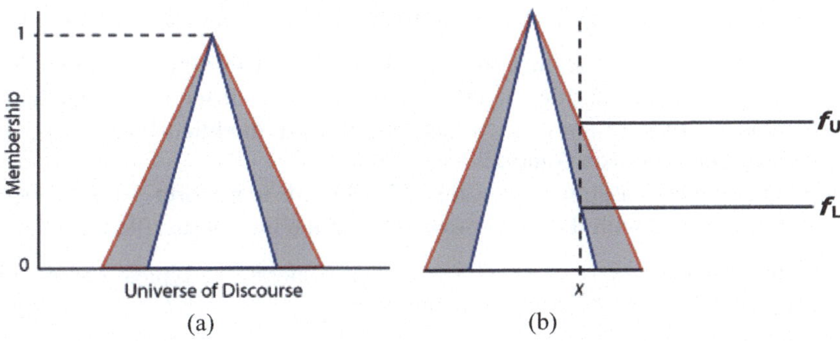

Fig. 3.40 The interval type-2 MFs

The interval type-2 MF now has two boundaries, the upper and the lower boundary, as shown in Figure 3.40a, which are called upper MF and lower MF, respectively. Now if you try to get the actual values of MFs, a range of values, not a single point value, are provided by type-2 MFs. As shown in Fig. 3.40b, there are two values for that MF, the value in the upper membership function (f_U) and the value in the lower membership function (f_L).

The IT2FIS still uses the input and output membership functions, combined with the control rules, to derive the outputs. However, the fuzzy sets used in the interval type 2 fuzzy logic or the membership grades in each membership function are not crisp values, but another fuzzy sets.

Due to its relatively simple structure and computational process, we will limit our discussions on the interval type-2 fuzzy inference system in this chapter. Let us have a closer look at the structure and working principles of the interval type-2 FIS.

3.9.1   Introduction to Interval Type-2 Fuzzy Inference System

Unlike developing process for type-1 FIS, the value of MFs used in an interval type-2 FIS is not a single value; instead, it is another modified fuzzy set represented by a range of or an interval of values between 0 and 1. After combining those MFs with control rules, the inferred results are also fuzzy sets. In order to get the crisp output values, the outputs of an IT2FIS need one more process to reduce that type-2 fuzzy set to a type-1 set, and then any converting method, such as COG, can be applied to convert that set to crisp output values.

Another difference between the general type-2 fuzzy system and the interval type-2 fuzzy system is that in the former system, the membership degrees are pure fuzzy sets, but the membership degrees are a set of crisp values with a range of 0–1 or an interval for the latter.

Thus, based on the above facts, a conclusion can be derived, and it is as follows:

1. Similar to type-1 FIS, the interval type-2 FIS still utilizes input and output MFs, combining with control rules, to infer the outputs. The difference is that the outputs in an IT2FIS are not crisp values, but they are modified fuzzy sets with a range of values between 0 and 1.
2. In order to get crisp output values, the IT2FIS needs to perform a type-reduction process combined with the COG method to get the final defuzzified results.

Assume the input vector is $x' = (x'_1, x'_2, \ldots, x'_i)$. Generally, a typical computation process in an IT2FIS involves the following steps:

1. Compute the membership of x_i' on each X_i^n to get the Lower Membership Function (LMF) and Upper Membership Function (UMF), $\{\underline{\mu}(x'), \overline{\mu}(x')\}$.
2. Compute the firing interval of the nth rule, $F^n(x')$ which is

$$F^n(x') = \left[\underline{\mu}(x_1') \times \underline{\mu}(x_2') \times \ldots \underline{\mu}(x_I'), \overline{\mu}(x_1') \times \overline{\mu}(x_2') \times \ldots \overline{\mu}(x_I') \right] = \left[\underline{f}^n, \overline{f}^n \right]$$

3. Perform type-reduction to combine $F^n(x')$ and the corresponding rule consequents with the center-of-sets type-reducer:

$$Y_{\cos}(x') = \bigcup_{\substack{f^n \in P^n(x') \\ y^n \in Y^n}} \frac{\sum_{n=1}^{N} f^n y^n}{\sum_{n=1}^{N} f^n} = [y_l, y_r] \tag{3.27}$$

4. Compute the defuzzified output as

$$y = \frac{y_l + y_r}{2} \tag{3.28}$$

where y_l and y_r are called switch points from where the UMF will be changed to the LMF or vice versa. In step 3, in order to calculate y_l and y_r, the following equations are used:

$$\begin{aligned}
y_l &= \min_{k \in [1,N-1]} \frac{\sum_{n=1}^{k} \overline{f}^n y^n + \sum_{n=k+1}^{N} \underline{f}^n y^n}{\sum_{n=1}^{k} \overline{f}^n + \sum_{n=k+1}^{N} \underline{f}^n} \equiv \frac{\sum_{n=1}^{L} \overline{f}^n y^n + \sum_{n=L+1}^{N} \underline{f}^n y^n}{\sum_{n=1}^{L} \overline{f}^n + \sum_{n=L+1}^{N} \underline{f}^n} \\[2ex]
y_r &= \min_{k \in [1,N-1]} \frac{\sum_{n=1}^{k} \underline{f}^n \overline{y}^n + \sum_{n=k+1}^{N} \overline{f}^n \overline{y}^n}{\sum_{n=1}^{k} \underline{f}^n + \sum_{n=k+1}^{N} \overline{f}^n} \equiv \frac{\sum_{n=1}^{R} \underline{f}^n \overline{y}^n + \sum_{n=R+1}^{N} \overline{f}^n \overline{y}^n}{\sum_{n=1}^{R} \underline{f}^n + \sum_{n=R+1}^{N} \overline{f}^n}
\end{aligned} \tag{3.29}$$

The switch points, y_l and y_r, can be determined using the Karnik-Mendel (**KM**) algorithms. Refer to Appendix B to get more details about the KM algorithm.

$$\underline{y}^L \leq y_l \leq \underline{y}^{L+1}$$
$$\overline{y}^R \leq y_r \leq \overline{y}^{R+1} \tag{3.30}$$

Figure 3.41 shows an example of using Karnik-Mendel algorithm to calculate switch points with seven interval type-2 MFs. The main idea of the KM algorithm is to find the switch points for y_l and y_r. For $n \leq L$, the upper membership grades are used to calculate y_l; for $n > L$, the lower membership grades are used. This will ensure y_l be the minimum (Fig. 3.41a).

Similar process to y_r, the y_r should be the maximum of $Y_{\cos}(x')$.

Since y^n increases from the left to the right along the horizontal axis, one should choose a small weight (LMF grade) for y^n on the left and a large weight (UMF grade) for y^n on the right.

The KM algorithm finds the switch point y_r. For $n \leq R$, the lower membership grades are used to calculate y_r; for $n > R$, the upper membership grades are used (Fig. 3.41b). This will ensure y_r be the maximum.

Figure 3.42 shows the functional block diagram of an Interval Type-2 FIS [17]. It is similar to Type-1 FIS, but the major difference is that at least one of the fuzzy sets in the rule base is an IT2 fuzzy set. The outputs of the inference engine are IT2 fuzzy sets, and a type-reducer is needed to convert them into a Typr-1 fuzzy set before defuzzification can be started.

It can be found from Fig. 3.42 that the only difference between type-1 and interval type-2 FIS is: the input and output MFs in an IT2FIS are IT2 fuzzy sets, and a type-reducer is needed to convert IT2 fuzzy sets to type-1 fuzzy set, and furthermore to get crisp outputs with the help of COG method. Some fundamental operations in the type-2 fuzzy system are union (3.31), intersection (3.32), and complement (3.33) [18].

The union for interval type-2 fuzzy sets $\tilde{A}$ and $\tilde{B}$ is:

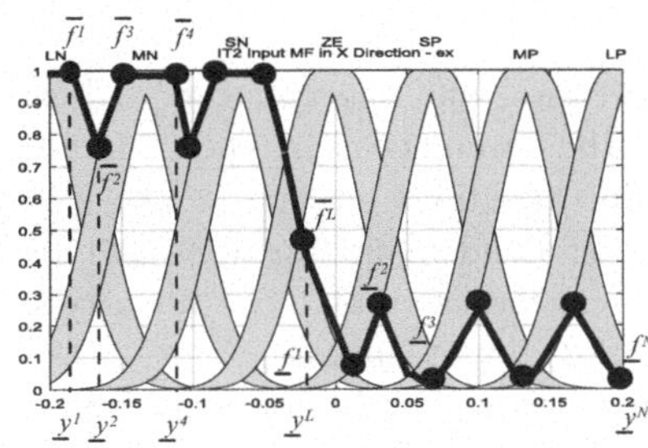

(a) Compute y_l – switch from UMF to LMF.

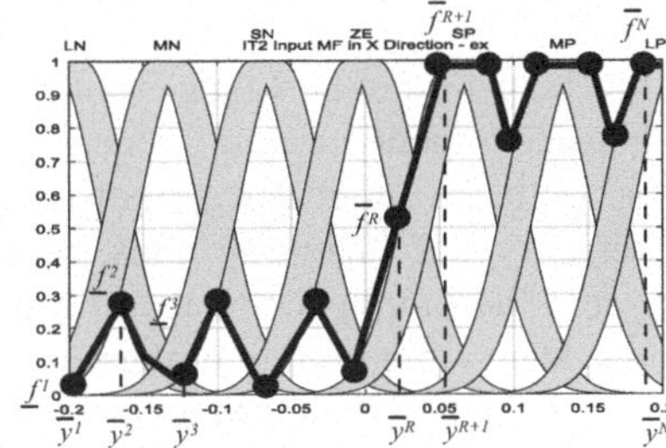

(b) Compute y_r – switch from LMF to UMF.

Fig. 3.41 An example of using Karnik-Mendel algorithm to get switch points. (**a**) Compute y_l – switch from UMF to LMF. (**b**) Compute y_r – switch from LMF to UMF

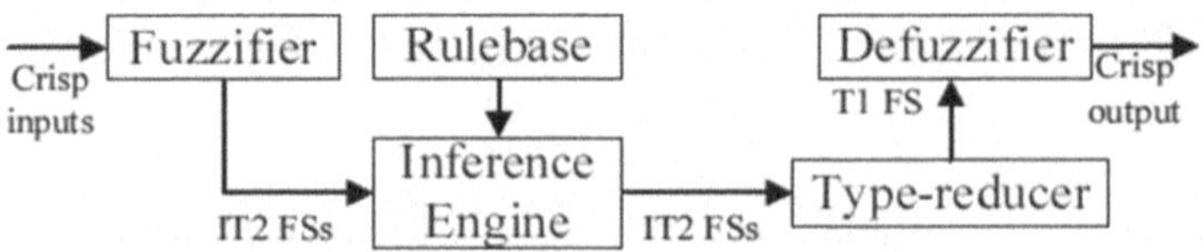

Fig. 3.42 A functional block diagram of a typical IT2FIS

$$\tilde{\tilde{A}} \cup \tilde{\tilde{B}} = \left\{ \int_{x \in X} \mu_{\tilde{\tilde{A}}}(x) \cup \mu_{\tilde{\tilde{B}}}(x) / x \right\} = \left\{ \int_{x \in X} \left[\int_{a \in \left[\underline{\mu}_{\tilde{A}(x)} \vee \underline{\mu}_{\tilde{A}(x)}, \overline{\mu}_{\tilde{A}(x)} \vee \overline{\mu}_{\tilde{B}(x)} \right]} 1 / \alpha \right] / x \right\} \qquad (3.31)$$

The intersection for interval type-2 fuzzy sets $\tilde{A}$ and B is:

$$\tilde{\tilde{A}} \cap \tilde{\tilde{B}} = \left\{ \int_{x \in X} \mu_{\tilde{\tilde{A}}}(x) \cap \mu_{\tilde{\tilde{B}}}(x) / x \right\} = \left\{ \int_{x \in X} \left[\int_{\alpha \in \left[\underline{\mu}_{\tilde{A}(x)} \wedge \underline{\mu}_{\tilde{B}(x)}, \overline{\mu}_{\tilde{A}(x)} \wedge \overline{\mu}_{\tilde{B}(x)} \right]} 1 / \alpha \right] / x \right\} \qquad (3.32)$$

The complement for interval type-2 fuzzy sets $\tilde{A}$ and B is:

$$\neg \tilde{\tilde{A}} = \left\{ \int_{x \in X} \mu_{\neg \tilde{\tilde{A}}}(x) / x \right\} = \left\{ \int_{x \in X} \left[\int_{\alpha \in \left[1 - \overline{\mu}_{\tilde{A}}(x), 1 - \underline{\mu}_{\tilde{A}}(x) \right]} 1 / \alpha \right] / x \right\} \qquad (3.33)$$

3.9.2 Implementations of Interval Type-2 Fuzzy Inference System with MATLAB

With the help of Fuzzy Logic Toolbox™ software, one can create both type-2 Mamdani and Sugeno fuzzy inference systems. The difference is: in type-2 Mamdani systems, both the input and output membership functions are type-2 fuzzy sets. However, in type-2 Sugeno systems, only the input membership functions are type-2 fuzzy sets. The output membership functions are the same as those for a type-1 Sugeno system—constant or a linear function of the input values.

To create interval type-2 Mamdani and Sugeno systems at the command line, use command **mamfistype2** and **sugfistype2** objects, respectively. These objects have the same parameters as the type-1 **mamfis** and **sugfis** objects along with an additional **TypeReductionMethod** parameter.

Another way to create an interval type-2 fuzzy inference system is to convert an existing type-1 system, such as one created using the **genfis()** function that can only be used to generate type-1 FIS (refer to Table 3.4). To do so, use the **convertTo-Type2()** function (refer to Table 3.4).

Once an interval type-2 fuzzy inference system is generated, one can:

1. Evaluate the fuzzy system using the **evalfis()** functions.
2. Simulate the fuzzy system using the **Fuzzy Logic Controller** block.
3. Tune the parameters of the fuzzy system using the **tunefis()** function.
4. Deploy the fuzzy system as described in **Deploy Fuzzy Inference Systems**.

One can also create an interval type-2 fuzzy inference system using the **Fuzzy Logic Designer** App, as we discussed in Sect. 3.8.1.3.

Now let us have a closer look at how an IT2FIS performs its inferring process to derive the crisp outputs based on IT2 fuzzy sets.

3.9.2.1   Fuzzy Inference Process for Interval Type-2 Fuzzy Systems

For interval type-2 fuzzy inference systems, input values are fuzzified by finding the corresponding degree of membership in both the UMFs and LMFs from the rule antecedent. Doing so generates two fuzzy values for each type-2 membership function as we discussed in Sect. 3.9. For example, the fuzzification in Fig. 3.38b shows the membership value in the upper membership function (f_U) and the lower membership function (f_L).

Next, a range of rule firing strengths is found by applying the fuzzy operator to the fuzzified values of the interval type-2 membership functions, as shown in Fig. 3.43a. The maximum value of this range (w_U) is the result of applying the fuzzy operator to the fuzzy values from the UMFs. The minimum value (w_L) is the result of applying the fuzzy operator to the fuzzy values from the LMFs.

For a Mamdani system, the implication method clips (min implication) or scales (prod implication) the UMF and LMF of the output type-2 membership function using the rule firing range limits. This process produces an output fuzzy set for each rule. Figure 3.43b shows the output fuzzy set (dark gray region) produced by applying min implication to the UMF (red) and LMF (blue).

For an interval type-2 Sugeno system, the output level z_i for the ith rule is computed in the same manner as for a type-1 Sugeno system.

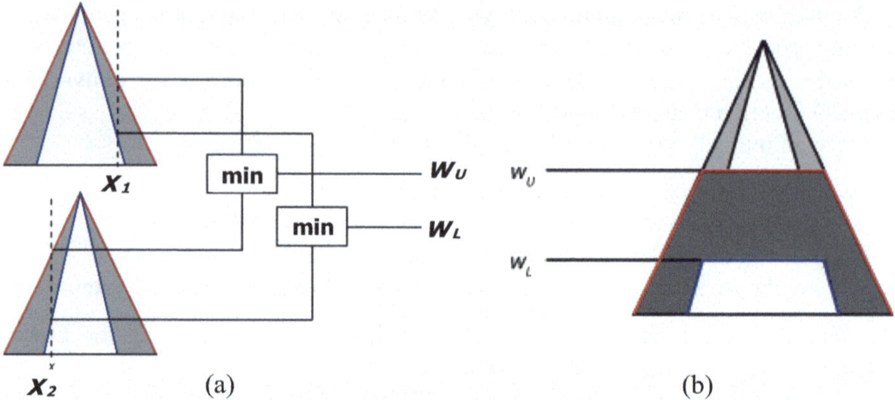

Fig. 3.43 Fuzzy inference process for IT2FIS

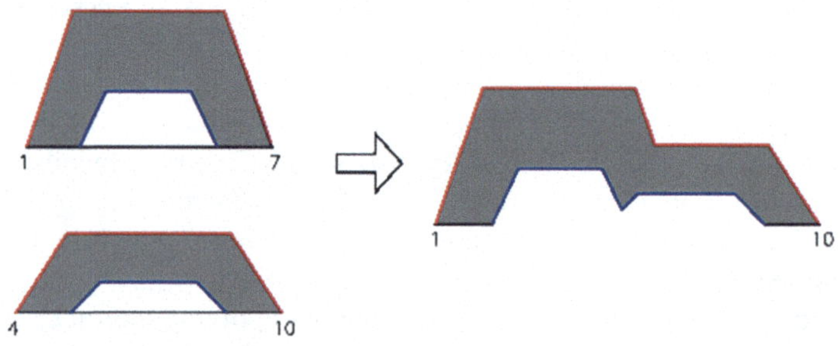

Fig. 3.44 An example of aggregation of two type-2 fuzzy sets

$$z_i = c_0^i + \sum_{j=1}^{M} c_j^i x_j \tag{3.34}$$

Here, j is the input index, x_j is the value of the jth input variable, and the c terms are the upper membership function parameters.

Unlike a type-1 Sugeno system, the rule firing strengths are not used to process the consequences of each rule. Instead, the output level and rule firing strengths are used during the aggregation process.

The goal of the aggregation stage is to derive a single type-2 fuzzy set from the rule output fuzzy sets. For a type-2 Mamdani system, the software finds an aggregate type-2 fuzzy set by applying the aggregation method to the UMFs and LMFs of the output fuzzy sets of all the rules.

Figure 3.44 shows the aggregation of two type-2 fuzzy sets (the outputs for a two-rule system) using max aggregation.

3.9.2.2 Type Reduction and Defuzzification for Interval Type-2 FIS

To obtain the final crisp output value for the inference process, the aggregate type-2 fuzzy set is first reduced to an interval type-1 fuzzy set, which is a range with lower limit c_L and upper limit c_R. This interval type-1 fuzzy set is commonly referred to as the centroid of the type-2 fuzzy set. In theory, this centroid is the average of the centroids of all the type-1 fuzzy sets embedded in the type-2 fuzzy set. In practice, it is not possible to compute the exact values of c_L and c_R. Instead, iterative type-reduction methods are used to estimate these values.

For a given aggregate type-2 fuzzy set, as shown in Fig. 3.45, the approximate values of c_L and c_R are the centroids of the following type-1 fuzzy sets (green). A point to be noted is that the lower limit c_L and upper limit c_R are exactly equivalent to y_l and y_r as we discussed in Sect. 3.9.1 above. They are called switch points used to switch from the upper MFs to the lower MFs or vice versa.

Mathematically, these centroids are found using the following equations [19]:

$$c_L = \frac{\sum_{i=1}^{L} x_i \bar{\mu}(x_i) + \sum_{i=L+1}^{N} x_i \underline{\mu}(x_i)}{\sum_{i=1}^{L} \bar{\mu}(x_i) + \sum_{i=L+1}^{N} \underline{\mu}(x_i)}$$

$$c_R = \frac{\sum_{i=1}^{R} x_i \underline{\mu}(x_i) + \sum_{i=R+1}^{N} x_i \bar{\mu}(x_i)}{\sum_{i=1}^{R} \underline{\mu}(x_i) + \sum_{i=R+1}^{N} \bar{\mu}(x_i)} \tag{3.35}$$

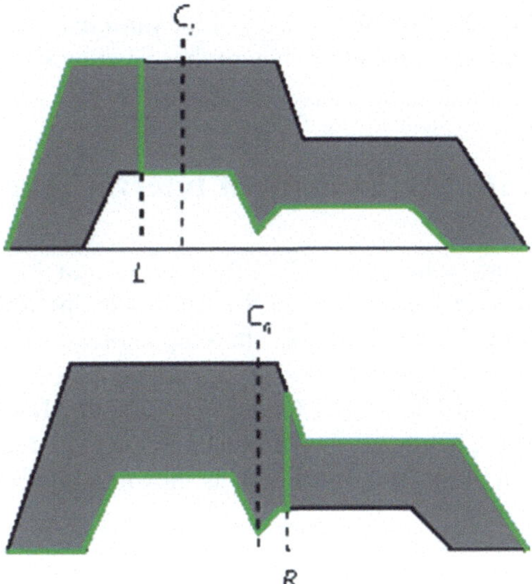

Fig. 3.45 An illustration of estimating the switch points

Table 3.6 Type-reduction methods

Type Reduction Method	TypeReduction Property Value	Description
Karnik-Mendel (KM) [20]	"karnikmendel"	Original KM Method
Enhanced Karnik-Mendel (EKM) [21]	"ekm"	Modified KM Method
Iterative algorithm with stop condition (IASC) [22]	"iasc"	Iterative improvement to brute force method
Enhanced IASC (EIASC) [23]	"eiasc"	Improved version of IASC algorithm

where:

- N is the number of samples taken across the output variable range, specified using **evalfisOptions**.
- x_i is the ith output value sample.
- $\bar{\mu}(x_i)$ is the upper membership function.
- $\underline{\mu}(x_i)$ is the lower membership function.
- L and R are *switch points* that are estimated by the various type-reduction methods. For a list of supported methods, refer to Table 3.6.

For both Mamdani and Sugeno systems, the final defuzzified output value (y) is the average of the two centroid values from the type reduction process.

$$y = \frac{c_L + c_R}{2} \tag{3.36}$$

It can be found that Eq. (3.35) is identical to Eqs. (3.29), and (3.36) is equivalent to (3.28) in Sect. 3.9.1.

Fuzzy Logic Toolbox software supports four (4) built-in type-reduction methods, as shown in Table 3.6. These algorithms differ in their initialization methods, assumptions, computational efficiency, and terminating conditions.

To set the type-reduction method for an interval type-2 fuzzy system, set the **TypeReduction** property of the **mamfistype2** or **sugfistype2** object. In general, the computational efficiency of these methods improved as you move down the table.

3.10   An Example of Interval Type-2 Fuzzy Inference System

To make the above IT2 FIS more meaningful, we try to use an example to illustrate how an IT2 FLC or FLS performs its inference function based on input MFs and control rules to derive the output fuzzy sets, and furthermore to obtain the crisp fuzzy output by using a type-reducer process.

Consider an IT2 FIS with two inputs, **Input1** and **Input2**, and one output, **output**. Each input domain contains two IT2 fuzzy sets, **West (W)** and **East (E)**, and **North (N)** and **South (S)**, as shown in Fig. 3.46.

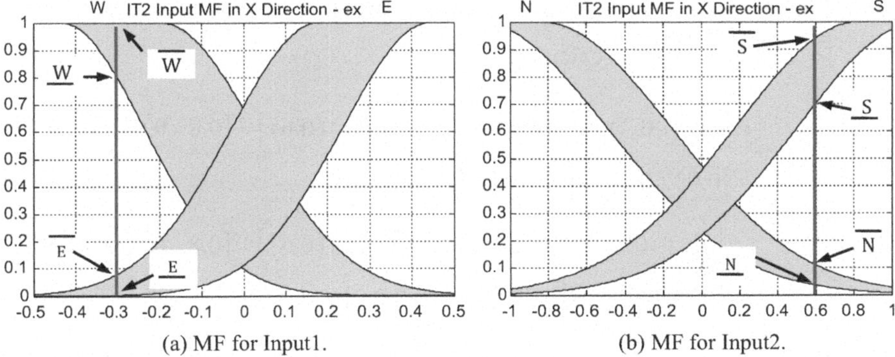

(a) MF for Input1.   (b) MF for Input2.

Fig. 3.46 Input membership functions for IT2FIS example system. (**a**) MF for Input1. (**b**) MF for Input2

Table 3.7 Control rules and consequents of the IT2 FIS

Input2 / Input1	N	S
W	$Y^1 = [\underline{y}^1, \overline{y}^1] = [-0.6, -0.3]$	$Y^2 = [\underline{y}^2, \overline{y}^2] = [-0.2, 0.2]$
E	$Y^3 = [\underline{y}^3, \overline{y}^3] = [0.2, 0.6]$	$Y^4 = [\underline{y}^4, \overline{y}^4] = [0.4, 0.8]$

Both fuzzy sets on two inputs are described by using the Gaussian type-2 curve, as shown in Fig. 3.46. Four control rules are applied to this IT2 FIS, and they are:

1. IF Input1 is W and Input2 is N, the output is Y^1.
2. IF Input1 is W and Input2 is S, the output is Y^2.
3. IF Input1 is E and Input2 is N, the output is Y^3.
4. IF Input1 is E and Input2 is S, the output is Y^4.

The complete relationship between control rules and the corresponding outputs is shown in Table 3.7. The output values on each intersection cell are calculated based on the following four (4) equations, from (1) to (4).

Suppose we have an input vector Input = {**Input1, Input2**} = {−0.3, 0.6}, based on the above four control rules, the firing intervals of the four IT2 Fuzzy Sets are:

$$\left[\mu_{\underline{W}}\left(-0.3\right), \mu_{\overline{W}}\left(-0.3\right) \right] = \left[0.8, 1.0\right] \tag{3.37}$$

$$\left[\mu_{\underline{E}}\left(-0.3\right), \mu_{\overline{E}}\left(-0.3\right) \right] = \left[0.0, 0.08\right] \tag{3.38}$$

$$\left[\mu_{\underline{N}}\left(0.6\right), \mu_{\overline{N}}\left(0.6\right) \right] = \left[0.04, 0.1\right] \tag{3.39}$$

$$\left[\mu_{\underline{S}}\left(0.6\right), \mu_{\overline{S}}\left(0.6\right) \right] = \left[0.7, 0.95\right] \tag{3.40}$$

According to step (2) in Sect. 3.9.1, the firing intervals of the four rules $F^n(\mathbf{x}') = \left[\underline{f^n}\ \overline{f^n}\right]$ can be calculated as:

1. $\left[\underline{f^1}, \overline{f^1}\right] = \left[\mu_{\underline{W}}(-0.3) \times \mu_{\underline{N}}(0.6), \mu_{\overline{W}}(-0.3) \times \mu_{\overline{N}}(0.6)\right] = [0.8 \times 0.04, 1.0 \times 0.1]$
$= [0.032, 0.1] \rightarrow \left[\underline{y^1}, \overline{y^1}\right] = [-0.6, -0.3]$

2. $\left[\underline{f^2}, \overline{f^2}\right] = \left[\mu_{\underline{W}}(-0.3) \times \mu_{\underline{S}}(0.6), \mu_{\overline{W}}(-0.3) \times \mu_{\overline{S}}(0.6)\right] = [0.8 \times 0.7, 1.0 \times 0.95]$
$= [0.56, 0.95] \rightarrow \left[\underline{y^2}, \overline{y^2}\right] = [-0.2, 0.2]$

3. $\left[\underline{f^3}, \overline{f^3}\right] = \left[\mu_{\underline{E}}(-0.3) \times \mu_{\underline{N}}(0.6), \mu_{\overline{E}}(-0.3) \times \mu_{\overline{N}}(0.6)\right] = [0.0 \times 0.04, 0.08 \times 0.1]$
$= [0.0, 0.008] \rightarrow \left[\underline{y^3}, \overline{y^3}\right] = [0.2, 0.6]$

4. $\left[\underline{f^4}, \overline{f^4}\right] = \left[\mu_{\underline{E}}(-0.3) \times \mu_{\underline{S}}(0.6), \mu_{\overline{E}}(-0.3) \times \mu_{\overline{S}}(0.6)\right] = [0.0 \times 0.7, 0.08 \times 0.95]$
$= [0.0, 0.076] \rightarrow \left[\underline{y^4}, \overline{y^4}\right] = [0.4, 0.8]$

Based on the KM algorithms, we find that $L = 1$ and $R = 2$ for this example. Thus,

$$y_l = \frac{\overline{f^1} y^1 + \underline{f^2} y^2 + \underline{f^3} y^3 + \underline{f^4} y^4}{\overline{f^1} + \underline{f^2} + \underline{f^3} + \underline{f^4}}$$
$$= \frac{0.1 \times (-0.6) + 0.56 \times (-0.2) + 0 \times 0.2 + 0 \times 0.4}{0.1 + 0.56} = -0.2606$$

$$y_r = \frac{\underline{f^1} \overline{y^1} + \underline{f^2} \overline{y^2} + \overline{f^3} \overline{y^3} + \overline{f^4} \overline{y^4}}{\underline{f^1} + \underline{f^2} + \overline{f^3} + \overline{f^4}}$$
$$= \frac{0.032 \times (-0.3) + 0.56 \times 0.2 + 0.008 \times 0.6 + 0.076 \times 0.8}{0.032 + 0.56 + 0.008 + 0.076} = 0.2485$$

Therefore, the final crisp output of this IT2 FLS y is

$$y = \frac{y_l + y_r}{2} = \frac{-0.2606 + 0.2485}{2} = -0.0061$$

Next let us use a real case study to illustrate how to design and implement a real IT2 Fuzzy Logic Controller (FLC) to control a DC motor system.

3.11   A Case Study for Applying an Interval Type-2 Fuzzy Inference System

To make things simple, suppose we have a DC motor system with a transfer function

$$G(s) = \frac{6520}{s(s+430.6)} e^{-0.005s} \tag{3.41}$$

Figure 3.47 shows a functional block diagram for this DC motor closed-loop control system. The inputs include the input error **ex** and error rate **dex**, and both of them are fed into an A/D converter to become digital variables. A multiplexer is used to combine them into the IT2 FIS to obtain the output. After a type-reducer, a crisp output variable can be obtained and fed into a D/A converter to become analog voltage to control the DC motor that is represented by its transfer function.

The error input voltage is coming from a tachometer that is working as a sensor to detect the actual motor rotation speed and convert that speed back to the related voltage. Exactly, that sensor can transfer the motor rotation velocity (*machine steps/ second*) to the related error voltages.

Two inputs are applied to this control system, the error **ex** and the error rate **dex**. The control gain K is used to adjust and tune the fuzzy logic control output value.

The required parameters we need to design an IT2 fuzzy logic controller (FLC) are the input and output variable ranges for this DC motor system. Table 3.8 shows the input and the output ranges for this motor control system.

To represent the error input and error rate input using fuzzy sets, a set of linguistic variables is chosen to represent a 5-*degree* of error, 5-*degree* of error rate, and 5-*degree* of control output.

Membership functions are constructed to represent the input and output in which degree belongs to a different membership set or linguistic variable set.

The membership functions of the input error, the error rate, and the controller output can be defined as a set of linguistic variables. The units for both the input error and error rate are *Voltage*, and the unit for output u is *Machine steps/second*.

The linguistic variables for 5-degree MFs are defined as follows:

- LN—Large Negative
- SN—Small Negative

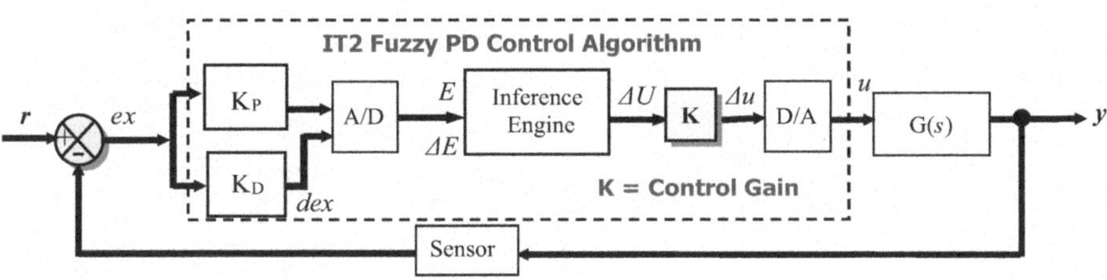

Fig. 3.47 IT2 fuzzy logic control system block diagram

Table 3.8 Ranges of the input-output variables

Variables	Variables Ranges	Units
Input Error	-0.3 ~ 0.3	V
Input Error Rate	-0.15 ~ 0.15	V
Motor Output Speed	-600 ~ 600	Machine Steps/s

- ZE—Zero
- SP—Small Positive
- LP—Large Positive

In this section, we will discuss how to design and implement an interval type-2 FLC for this DC motor system and compare the control performances for both types of FLCs with a MATLAB Simulation process. The value in the exponent part, **0.005 s**, exactly is equivalent to a pure time delay, which is 5 *ms* after a motor identification process is performed.

First let us build an IT2 FLC for this DC motor system with Fuzzy Logic Designer App, exactly with an IT2 Fuzzy Logic Toolbox App.

3.11.1 Build IT2 Fuzzy Inference System with Fuzzy Logic Designer App

Open the MATLAB and type **fuzzyLogicDesigner** in the Command window to open the Fuzzy Logic Designer App. Click on the **Mamdani Type-2** icon to open the fuzzy inference system designer, as shown in Fig. 3.48.

Then click on the **Mamdani Type-2** box located at the center of the Designer, as shown in Fig. 3.48, and click on the **Save** button in the upper-left corner on the top bar to save our FIS to a desired folder on your machine, such as **C:\AI Projects\ Chapter 3**, and name the project as **IT2_Fuzzy.fis**. Then click on the **Save** button to save the project.

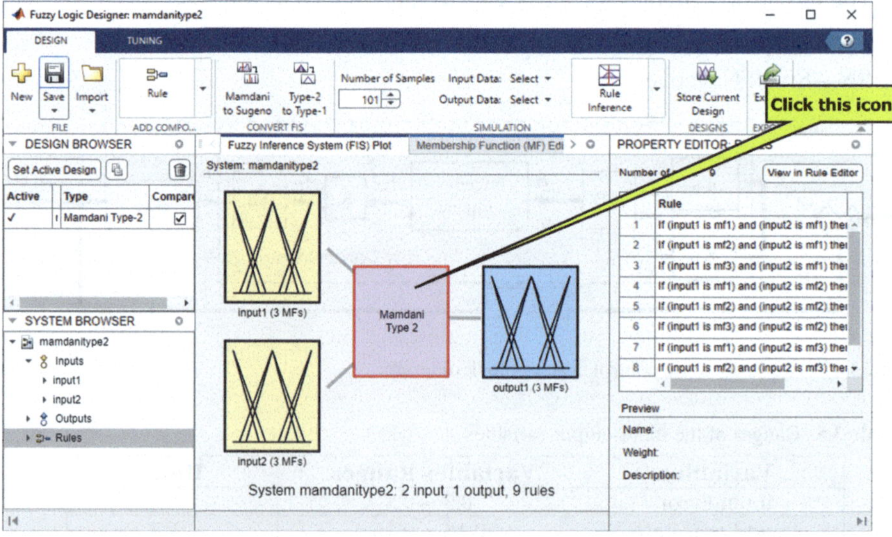

Fig. **3.48** The opened fuzzy logic App designer

Next let us build the membership functions for both inputs and the output variables.

3.11.2   Build the Input and the Output Membership Functions

For this motor control system, we have two input variables, input position error **ex** and input position error change rate **dex**, and one output **yit2**.

Perform the following operational steps to build both inputs and the output membership functions:

1. Click on the **input1 (3MFs)** box and then click on the **Membership Function (MF) Editor** tab to open the MFs Editor wizard, as shown in Fig. 3.49.
2. Go to the **SYSTEM BROWSER** located at the lower-left corner, and right-click on the first MF **mf1** under the **input1** variable, and select the **Delete** item to remove it. Perform a similar operation to remove the other two MFs, **mf2** and **mf3**, respectively.
3. Now click on the **MF** icon on the top bar to add our first MF with the default Triangular type to the **input1** variable.
4. Go to the **PROPERTY EDITOR:INPUT** window on the right, change the variable name to **ex** by entering it into the **Name** box, as shown in Fig. 3.49. Also change the **Range** for this MF to [−0.3 0.3].
5. Change the name of this MF to **ZE** by entering it into the **Name** box, and double-click on the line under the **Type** column and select the **Gaussian** as the type for this MF.
6. In a similar way, add the other four MFs, named **LN** (Large Negative), **SN**, (Small Negative), **SP** (Small Positive), and **LP** (Large Positive). The type for both **SN** and **SP** are **Gaussian** type, but the type for **LN** and **LP** are **Sigmoid**, as shown in Fig. 3.49.

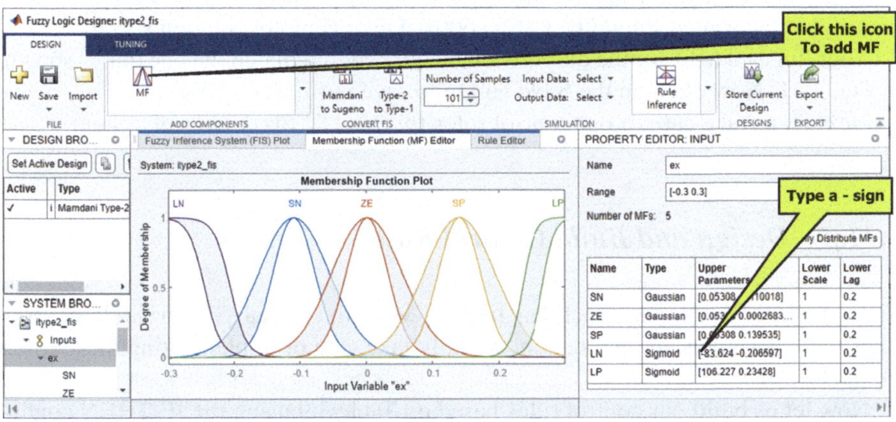

Name	Type	Upper Parameters	Lower Scale	Lower Lag
SN	Gaussian	[0.05308 ...10018]	1	0.2
ZE	Gaussian	[0.053... 0.0002683...]	1	0.2
SP	Gaussian	[0.05308 0.139535]	1	0.2
LN	Sigmoid	[-83.624 -0.206597]	1	0.2
LP	Sigmoid	[106.227 0.23428]	1	0.2

Fig. 3.49 The finished MFs for input variable ex

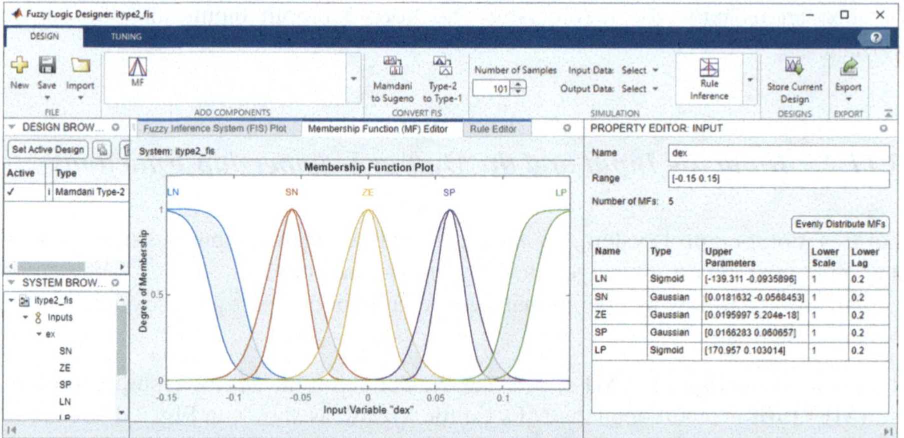

Fig. 3.50 The completed MFs for the second input variable dex

Now click on the **Evenly Distribute MFs** button to make all five MFs to be displayed in a symmetry way. Also one needs to click on the small square cell to reduce the width of each MF and move each of them to left or right. Your finished MF for the input **ex** is shown in Fig. 3.49.

A point is that you may need to type a negative sign in the left bond of the **Range** for **LN** to get a required and correct Sigmoid MF.

Now click on the **Fuzzy Inference System (FIS) Plot** tab to return to the Designer wizard. Then click on the **input2 (3MFs)** icon to build MFs for our second input variable **dex**.

Perform similar operational steps from 1 to 6 above to setup MFs for **dex** variable with the same five MFs and the same names. The only difference is the Range, which is [−0.15 0.15]. The finished MFs for the second variable **dex** should match the one that is shown in Fig. 3.50.

In a similar way, set up five MFs for our output variable **yit2** with the same names, but the Range should be [−600,600], which is motor rotation speed.

Your finished MFs for the output variable **yit2** should match the one that is shown in Fig. 3.51. Now click on the **Save** button to save our FIS.

Next let us take care of our control rules for this IT2 fuzzy control system.

3.11.3   Design and Build Control Rules

Since we have two inputs, each has five (5) MFs; thus, totally we have 25 control rules to be designed. In this section, we only take care of implementing these rules in this IT2 FIS.

Now let us build our control rules based on Table 3.9 using the IF-THEN conditions. In Table 3.9, each column represents the **ex** value and each row represents the

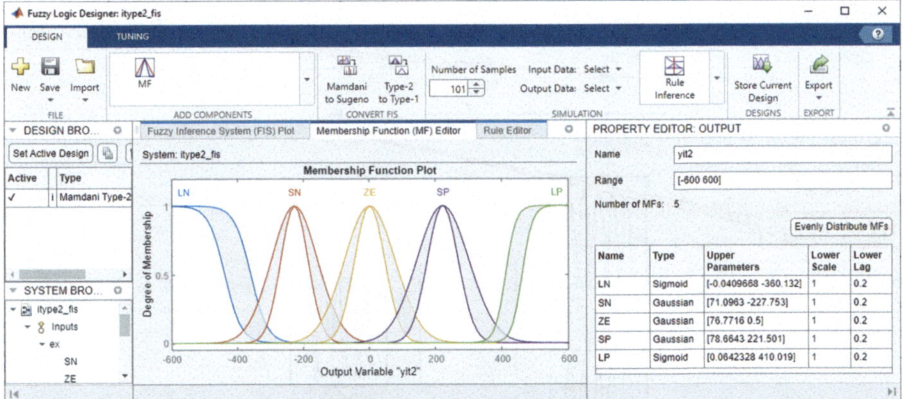

Fig. 3.51 The completed MFs for the output variable yit2

Table 3.9 IT2 FIS control rules

dex \ ex	LN	SN	ZE	SP	LP
LP	ZE	SP	LP	LP	LP
SP	SN	ZE	SP	LP	LP
ZE	LN	SN	ZE	SP	LP
SN	LN	SN	SN	ZE	SP
LN	LN	LN	SN	SN	ZE

dex value. The intersection cell for each column and each row is the related output value.

For example, in the first column and the first row as well as the related intersection cell in Table 3.9, the control rule is:

IF the **ex** is LN and the **dex** is LP, THEN the output **U** is ZE.

This is making sense since the motor rotating speed should be zero if the input error **ex** is LN and the error rate **dex** is LP.

Now open the MATLAB and enter **fuzzyLogicDesigner** into the Command window to open the Fuzzy Logic Designer App, and open our IT2 FIS, **IT2_Fuzzy**, by clicking on it from the **Open** panel located at the upper-left corner. Then (1) click on the **Rule** icon, as shown in Fig. 3.52, and (2) the **Rule Editor** tab, to open the Rule Editor, as shown in Fig. 3.53.

Now go to the **PROPERTY EDITOR:RULE** wizard that is located at the right and set up our control rules based on Table 3.8. Starting from the first rule, perform the following operational steps to set it up:

1. Select the LN from the **ex** row and LP from the **dex** row, as shown in Fig. 3.52, and select ZE from the **yit2** row under the **Then** tab.
2. Click on the green-color + button, as shown in Fig. 3.53, to add this first rule into our IT2 FIS.

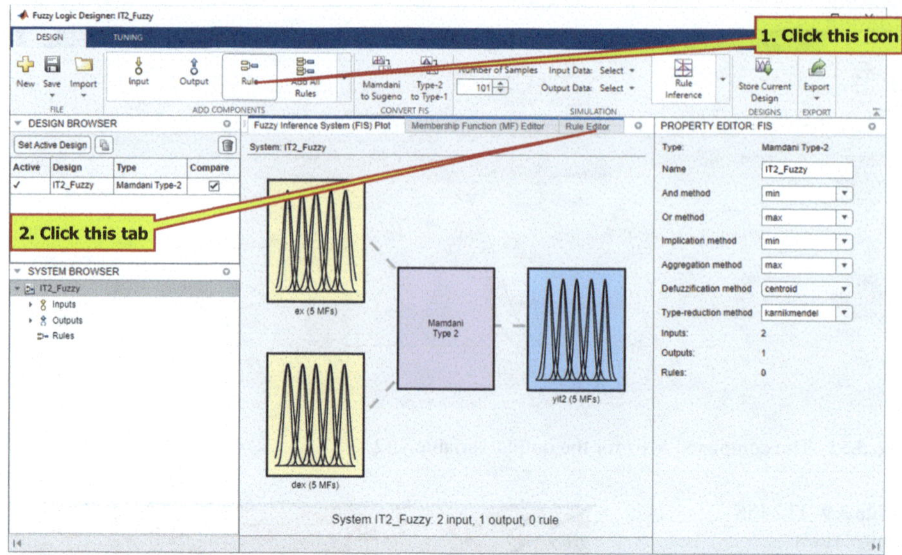

Fig. 3.52 Open the Rule Editor by clicking on the Rule item

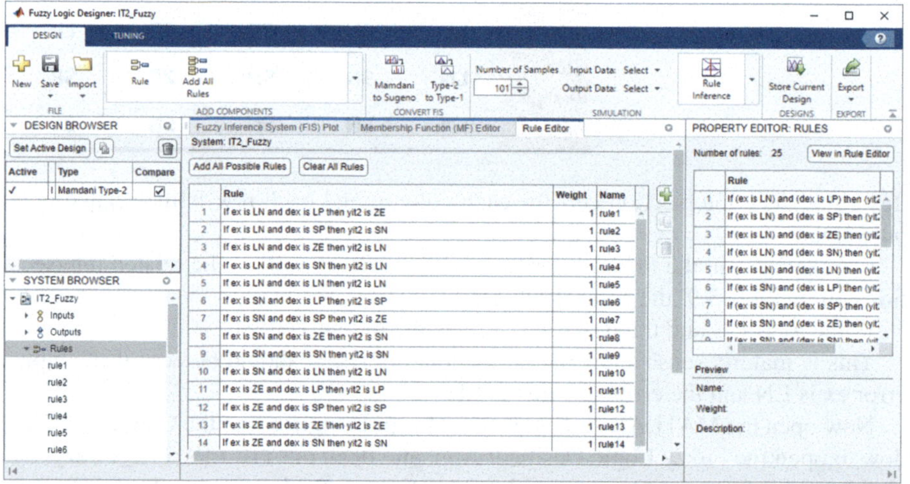

Fig. 3.53 The finished 25 control rules

3. In a similar way, continue to set up for our second rule based on Table 3.8.
4. With the same operation, add all 25 rules into our IT2 FIS.

The finished 25 control rules should match the one that is shown in Fig. 3.53. Now click on the **Save** button to save our IT2 FIS. Click on the **Yes** button if a message box is displayed to remind you to over-write the original FIS file.

If you click on the **Rule Inference** button on the top, as shown in Fig. 3.54, the final control rules combined with all inputs and the output are displayed, as shown

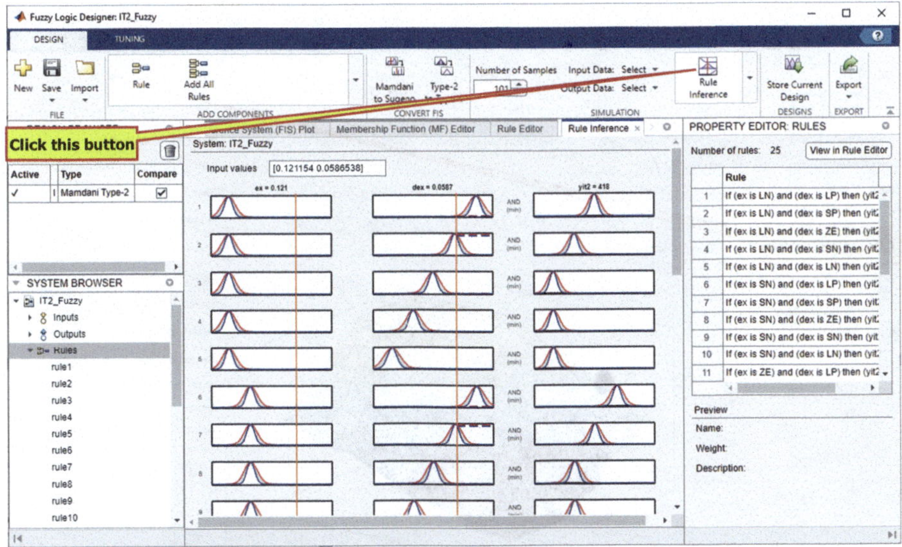

Fig. 3.54 The opened Rule Inference wizard

in Fig. 3.54. You can now change or modify each or both input values by moving the vertical bar on inputs columns to obtain the output in real time if you like. An example is shown in Fig. 3.54 with the following inputs and output values:

- Input **ex** = 0.121
- Input **dex** = 0.0587
- Output **yit2** = 418

Click on the drop-down arrow at the **Rule Inference** box, as shown in Fig. 3.55, and select the **Control Surface** item, and one can get the control envelope or surface for our IT2 FIS, which is shown in Fig. 3.55.

So far we built our IT2 FIS control system by using the Fuzzy Logic Designer App method. Next let us build our IT2 FIS with MATLAB Functions method, which is more professional with more flexibility to build an FLC system, but it will involve a lot of coding jobs with more difficulties.

3.11.4   Build IT2 FIS with MATLAB Fuzzy Inference Functions

Recall that we provided a very detailed discussion about the most popular Fuzzy Inference Functions in Table 3.4 in Sect. 3.8.3.2. However, a different function should be used to create a new Mamdani interval type-2 fuzzy inference system, which is **mamfistype2**(). We need to call this function to generate our new IT2 FIS.

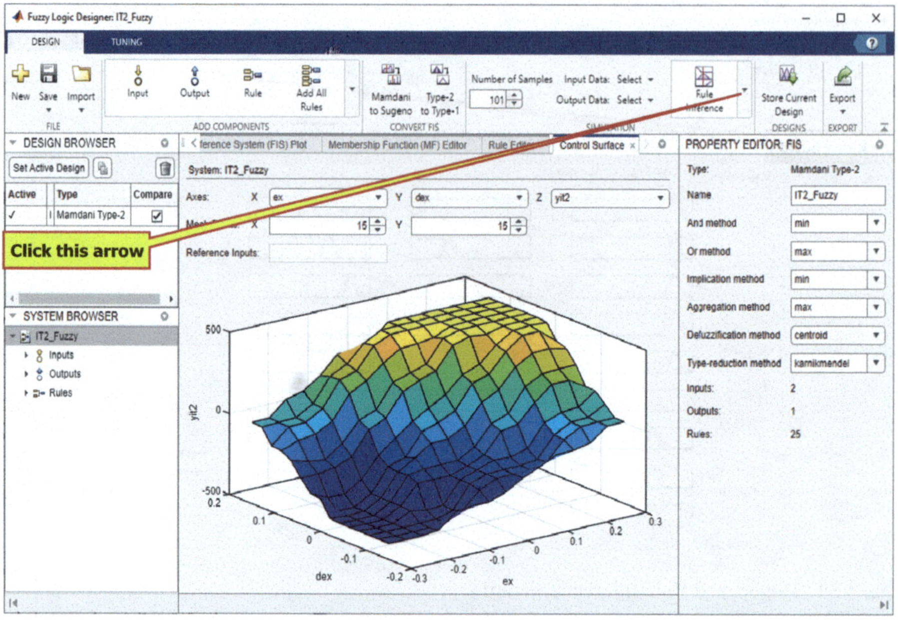

Fig. 3.55 The displayed control surface for our IT2 FIS

First let us create that fuzzy inference system by calling the function **mamfistype2()**.

3.11.4.1 Generate a New Mamdani Interval Type-2 Fuzzy Inference System

Open the MATLAB by double-clicking on its shortcut from the desktop and perform the following operational steps to create our new project:

1. Click on the **New Script** icon that is located in the upper-left corner under the **Home** menu item to open a new script file. A new untitled script file is generated.
2. Let us save our file by clicking the drop-down arrow on the **Save** icon and click on the **Save** item to save our project. Browse to our default folder **C:\AI Projects\ Chapter 3**, and enter **IT2_Func.m** into the **File name** box, and click on the **Save** button.
3. Enter the codes shown in Fig. 3.56 to generate our IT2 FIS with three variables.

Let us have a closer look at this piece of codes to see how it works.

1. First a Mamdani Interval Type-2 FIS is generated with the name of **IT2_Func**. The exact name should be **IT2_Func.fis**, but it is unnecessary to provide the **.fis** extension following the FIS's name when creating this type of FIS with this function, and MATLAB will automatically add that extension for you.

```
1  % Name: IT2_Func.m
2  % Build an IT2 FIS with Fuzzy Inference Functions
3  % Y. Bai
4  % Sept 7, 2023

5  % generate a new IT2 FIS
6  fis = mamfistype2(("Name", "IT2_Func");
7  fis = addInput(fis,[-0.3 0.3],"Name","ex");
8  fis = addInput(fis,[-0.15 0.15],"Name","dex");
9  fis = addOutput(fis,[-600 600],"Name","yit2");
```

Fig. 3.56 The codes used to generate our new IT2 FIS with three variables

2. From coding lines 7 to 9, two input variables and one output variable are added into our generated FIS in step 6 above with each related variable range and name

3.11.4.2   Add Membership Functions to Our IT2 FIS

Add five MFs to the input variable **ex** by entering the codes shown in Fig. 3.57. The previous codes are highlighted. Let us have a look at this piece of codes to see how it works.

1. The function **addMF()** is called to add five MFs into our first input variable **ex**. Refer to our previous project **IT2_Fuzzy.fis** we discussed in Sect. 3.11.2, exactly in Fig. 3.49 to get ranges for those five MFs. The types of three MFs are **Gaussian** forms, and a type-code **gaussmf** is used, but for the types of MFs, **LN** and **LP** are **Sigmoid** with the type-code of **sigmf**. Refer to Table 3.5 to get more details about the type-code for each type.
2. In addition to types, range, and name, one can also add the **LowScale** and **LowerLag** parameters. However, those parameters are not necessary and they are optional. Refer to Fig. 3.49 to get values for those parameters.
3. These three coding lines are used to test our MFs with **plot()** function.

In a similar way, enter the codes shown in Fig. 3.58 into this Script to add five MFs for our second input and the output variables. Let us have a closer look at this piece of codes to see how it works.

1. The coding lines between 21 and 25 are used to add five MFs into our second input variable **dex**. They are similar to those codes from lines 11 to 15, and the only differences are the range of the MFs and the name of variable.
2. The coding lines between 27 and 31 are used to add five MFs into our output variable **yint2**. They are similar to those codes from lines 11 to 15, and the only differences are the range of the MFs and the name of variable.
3. The coding lines between 33 and 38 are used to plot five MFs for the second input and the output variables.

Now run our project by clicking the green-color **Run** button on the top bar; the running result is shown in Fig. 3.59. Our five MFs are set up correctly.

```
1   % Name: IT2_Func.m
2   % Build an IT2 FIS with Fuzzy Inference Functions
3   % Y. Bai
4   % Sept 7, 2023

5   % generate a new IT2 FIS
6   fis = mamfistype2(("Name", "IT2_Func");
7   fis = addInput(fis,[-0.3 0.3],"Name","ex");
8   fis = addInput(fis,[-0.15 0.15],"Name","dex");
9   fis = addOutput(fis,[-600 600],"Name","yit2");

10  % add 5 MFs to input variable ex
11  fis = addMF(fis,"ex","sigmf",[-83.624 -0.206597],"Name","LN");
12  fis = addMF(fis,"ex","gaussmf",[0.0332566 -0.107229],"Name","SN");
13  fis = addMF(fis,"ex","gaussmf",[0.0424661 -0.00361446],"Name","ZE");
14  fis = addMF(fis,"ex","gaussmf",[0.0342799 0.1084340],"Name","SP");
15  fis = addMF(fis,"ex","sigmf",[106.227 0.23428],"Name","LP",'LowerScale',1,'LowerLag',0.2);

16  % plot 5 MFs for the first input variable ex
17  subplot(3,1,1);
18  plotmf(fis,"input",1);
19  grid;
```

Fig. 3.57 The codes for adding five MFs for the first input variable

```
1   % Name: IT2_Func.m
2   % Build an IT2 FIS with Fuzzy Inference Functions
3   % Y. Bai
4   % Sept 7, 2023

5   % generate a new IT2 FIS
6   fis = mamfistype2(("Name", "IT2_Func");
7   fis = addInput(fis,[-0.3 0.3],"Name","ex");
8   fis = addInput(fis,[-0.15 0.15],"Name","dex");
9   fis = addOutput(fis,[-600 600],"Name","yit2");

10  % add 5 MFs to input variable ex
11  fis = addMF(fis,"ex","sigmf",[-83.624 -0.206597],"Name","LN");
12  fis = addMF(fis,"ex","gaussmf",[0.0332566 -0.107229],"Name","SN");
13  fis = addMF(fis,"ex","gaussmf",[0.0424661 -0.00361446],"Name","ZE");
14  fis = addMF(fis,"ex","gaussmf",[0.0342799 0.1084340],"Name","SP");
15  fis = addMF(fis,"ex","sigmf",[106.227 0.23428],"Name","LP",'LowerScale',1,'LowerLag',0.2);

16  % plot 5 MFs for the first input variable ex
17  subplot(3,1,1);
18  plotmf(fis,"input",1);
19  grid;

20  % add 5 MFs to input variable dex
21  fis = addMF(fis,"dex","sigmf",[-139.311 -0.0935896],"Name","LN");
22  fis = addMF(fis,"dex","gaussmf",[0.0181632 -0.0487952],"Name","SN");
23  fis = addMF(fis,"dex","gaussmf",[0.0171399 5.204e-18],"Name","ZE");
24  fis = addMF(fis,"dex","gaussmf",[0.0166283 0.0542169],"Name","SP");
25  fis = addMF(fis,"dex","sigmf",[170.957 0.103014],"Name","LP",'LowerScale',1,'LowerLag',0.2);

26  % add 5 MFs to output variable yit2
27  fis = addMF(fis,"yit2","sigmf",[-0.0409668 -360.132],"Name","LN");
28  fis = addMF(fis,"yit2","gaussmf",[64.0868 -199.594],"Name","SN");
29  fis = addMF(fis,"yit2","gaussmf",[61.5948 0.5],"Name","ZE");
30  fis = addMF(fis,"yit2","gaussmf",[66.1541 221.501],"Name","SP");
31  fis = addMF(fis,"yit2","sigmf",[0.0642328 410.019],"Name","LP",'LowerScale',1,'LowerLag',0.2);

33  subplot(3,1,2);
34  plotmf(fis,"input",2);
35  grid;

36  subplot(3,1,3);
37  plotmf(fis,"output",1);
38  grid;
```

Fig. 3.58 The finished adding MFs to the input and the output variables

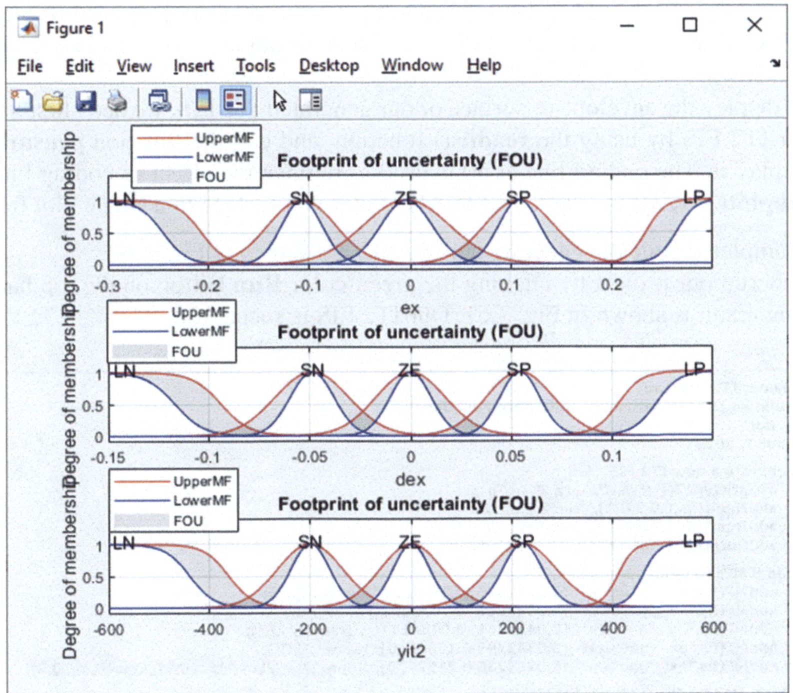

Fig. 3.59 The added five MFs to our IT2_Func.fis

3.11.4.3   Build the Control Rules for Our IT2 Fuzzy Inference System

Refer to Sect. 3.8.3.5 to get more details about how to define each rule based on its index, weight, and operations.

For our current IT2 FIS, we have five MFs for two input variables and one output variable. Therefore, the index for each MF for all three variables can be defined as below:

- **LN**: 1, **SN**: 2, **ZE**: 3, **SP**: 4, **LP**: 5.

Based on Table 3.9, we can represent our 25 control rules as index format with weight as 1 and using the **AND** (1) operation. The resulting control rules are:

Enter this **Rulelist** with the associated codes into our Script file, as shown in Fig. 3.58. Let us have a closer look at this piece of newly added codes to see how it works.

1. The generated **Rulelist** that is presented as the index format is declared based on the MF index and the control rules defined in Table 3.9. Here in order to save space, we divided this rule into different segments and separated them with semi-colons.
2. The generated Rulelist is added into our IT2 FIS by calling the **addRule**() function.

3. Then the generated IT2 FIS is saved to the hard disk as a file named **itype2_fis.fis**. A point to be noted is that the extension **.fis** is not needed when saving an FIS in this way.

4. To display the envelope or surface of our generated IT2 FIS, we need first to load our IT2 FIS by using the **readfis**() function, and call the function **gensurf**() to display it. The coding line at 50 is used to remove the effect of coding line 36, **subplot(3, 1, 3);** to enable MATLAB to plot this surface in a single plot frame.

The completed codes to generate our IT2 FIS are shown in Fig. 3.60.

Now run our project by clicking the green-color **Run** button on the top bar; the running result is shown in Fig. 3.61. Our IT2 FIS is successful.

```
1   % Name: IT2_Func.m
2   % Build an IT2 FIS with Fuzzy Inference Functions
3   % Y. Bai
4   % Sept 7, 2023
5   % generate a new IT2 FIS
6   fis = mamfistype2(("Name", "IT2_Func");
7   fis = addInput(fis,[-0.3 0.3],"Name","ex");
8   fis = addInput(fis,[-0.15 0.15],"Name","dex");
9   fis = addOutput(fis,[-600 600],"Name","yit2");
10  % add 5 MFs to input variable ex
11  fis = addMF(fis,"ex","gaussmf",[0.0332566 -0.213253],"Name","LN");
12  fis = addMF(fis,"ex","gaussmf",[0.0332566 -0.107229],"Name","SN");
13  fis = addMF(fis,"ex","gaussmf",[0.0424661 -0.00361446],"Name","ZE");
14  fis = addMF(fis,"ex","gaussmf",[0.0342799 0.1084340],"Name","SP");
15  fis = addMF(fis,"ex","gaussmf",[0.0383730 0.2192770],"Name","LP",'LowerScale',1,'LowerLag',0.2);
16  % plot 5 MFs for the first input variable ex
17  subplot(3,1,1);
18  plotmf(fis,"input",1);
19  grid;
20  % add 5 MFs to input variable dex
21  fis = addMF(fis,"dex","sigmf",[-139.311 -0.0935896],"Name","LN");
22  fis = addMF(fis,"dex","gaussmf",[0.0181632 -0.0487952],"Name","SN");
23  fis = addMF(fis,"dex","gaussmf",[0.0171399 5.204e-18],"Name","ZE");
24  fis = addMF(fis,"dex","gaussmf",[0.0166283 0.0542169],"Name","SP");
25  fis = addMF(fis,"dex","sigmf",[170.957 0.103014],"Name","LP",'LowerScale',1,'LowerLag',0.2);
26  % add 5 MFs to output variable yit2
27  fis = addMF(fis,"yit2","sigmf",[-0.0409668 -360.132],"Name","LN");
28  fis = addMF(fis,"yit2","gaussmf",[64.0868 -199.594],"Name","SN");
29  fis = addMF(fis,"yit2","gaussmf",[61.5948 0.5],"Name","ZE");
30  fis = addMF(fis,"yit2","gaussmf",[66.1541 221.501],"Name","SP");
31  fis = addMF(fis,"yit2","sigmf",[0.0642328 410.019],"Name","LP",'LowerScale',1,'LowerLag',0.2);
33  subplot(3,1,2);
34  plotmf(fis,"input",2);
35  grid;
36  subplot(3,1,3);
37  plotmf(fis,"output",1);
38  grid;
39  Rulelist = [ 1    5    3    1    1; 1    4    2    1    1; 1    3    1    1    1; 1    2    1    1    1;
40               1    1    1    1    1; 2    5    4    1    1; 2    4    3    1    1; 2    3    2    1    1;
41               2    2    2    1    1; 2    1    1    1    1; 3    5    5    1    1; 3    4    4    1    1;
42               3    3    3    1    1 ; 3    2    2    1    1; 3    1    2    1    1; 4    5    5    1    1;
43               4    4    5    1    1; 4    3    4    1    1; 4    2    3    1    1; 4    1    2    1    1;
44               5    5    5    1    1; 5    4    5    1    1; 5    3    5    1    1; 5    2    4    1    1;
45               5    1    3    1    1];
46  fis = addRule(fis, Rulelist);
47  it2_fis = writeFIS(fis, 'itype2_fis');
48  % display the surface of ou FIS
49  it_fis = readfis('itype2_fis');
50  subplot(1,1,1);
51  gensurf(it_fis);
```

Fig. 3.60 The completed codes to generate our IT2 FIS

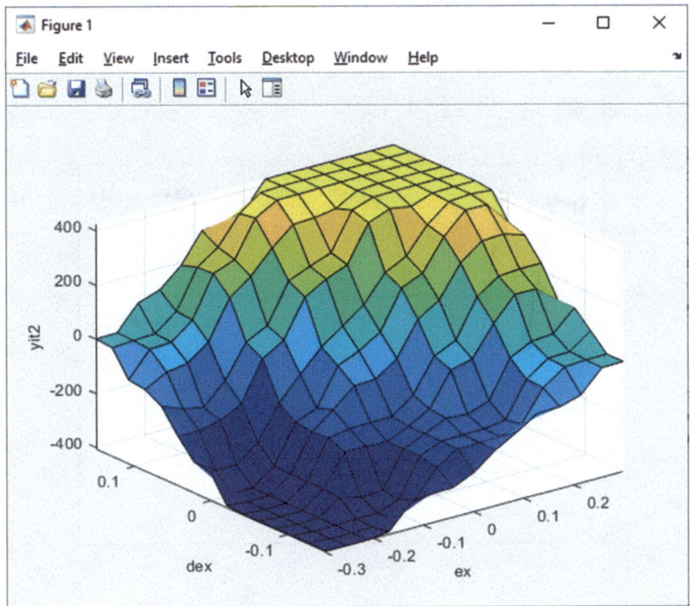

Fig. 3.61 The displayed surface of our generated IT2 FIS

```
1  % Evaluate our interval type-2 FIS
2  % Eval_IT2FIS.m
3  % September 10, 2023
4  fis = readfis('itype2_fis');
5  output = evalfis(fis, [0.2 0.05])
6  input = [0.2 0.05; 0.21 0.06; 0.02 0.02; 0.28 0.10; -0.23 0.08; -0.12 -0.02; 0.07 -0.03; 0.16 0.041;
7          0.19 0.14; 0.25 -0.08; 0.05 -0.09; 0.13 -0.13; -0.27 -0.11; 0.23 -0.15; 0.22 -0.08; 0.09 -0.09;
8          0.17 -0.041; 0.07 0.02; -0.21 0.13];
9  output = evalfis(fis, input)
10 plot(output, 'b-o', 'LineWidth',2);
11 grid;
12 legend('output', 'Location', 'northeast');
```

Fig. 3.62 The codes for the IT2 FIS evaluation process

3.11.4.4   Evaluate and Test Our Developed IT2 FIS with Selected Inputs

One can test and confirm our developed IT2 FIS, **itype2_fis**, with any appropriate inputs, **ex** and **dex**, with different values, and check the outputs, **yit2**. The function **evalfis()** should be used for that purpose. The syntax of using that function is, **output = evalfis(fis, input);**

The input can be a single pair or a matrix that contained multiple input pairs. For example, the coding line **output = evalfis(itype2_fis, [0.2, 0.05]);** can generate an output of **342.9356**.

Generate another new MATLAB Script file named **Eval_IT2FIS.m** and enter the codes shown in Fig. 3.62 into that file. Let us have a closer look at this piece of codes to see how it works.

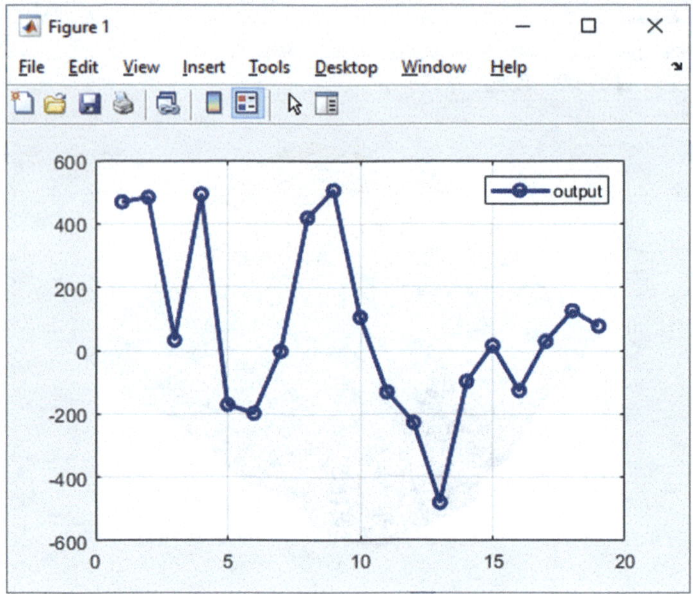

Fig. 3.63 The testing result for our IT2 FIS

1. First we need to load our developed FIS file, **it2_fis**, by calling the function **readfis**() from our current folder. We did save that file in the folder, **C:\AI Projects\Chapter 3**. You may need to use MATLAB **cd** command to find and get the current folder if you save that file into different folder in your machine. For example, if you have saved your FIS file into a location **C:\My_Project\ My_Folder\it2_fis.fis**, you need to enter one coding line as, **cd "C:\My_Project\ My_Folder"**, to allow machine to access that folder and file.
2. First we test our IT2 FIS by using an input pair, **ex = 0.20** and **dex = 0.05**, by calling the function **evalfis**(). The output should be printed in the Command window since we did not add any semicolon (**;**), after that coding line.
3. Now we use a matrix to contain multiple input pairs to get more outputs. To do that, first we need to generate an input matrix.
4. The **evalfis**() is called to run that function to perform evaluation for our input matrix. The output is also a matrix.
5. The output is plotted to display our results, as shown in Fig. 3.63.

At this point, we have provided a completed introduction and discussion about how to use fuzzy functions to build our air conditioner control system. Compared with the same project we built in Sect. 3.8.2, where the Fuzzy Logic Designer App is used to build our air conditioner system, it provides more flexibility and controllability in building our project by using the fuzzy inference functions.

Next let us concentrate our discussions on a simulation study for this IT2 FIS. We also like to perform some comparison studies to compare a type-1 FIS with our IT2 FIS to show readers some differences between these two kinds of controllers.

3.12   Simulation Study for Type-1 and IT2 Fuzzy Logic Control Systems

In order for us to perform a comparison study between a type-1 and a type-2 fuzzy inference system or logic controller, we need to generate a type-1 fuzzy inference system or a logic controller. Of course, one can modify the codes in Fig. 3.60 to get a type-1 FIS. Exactly one only needs to replace the coding line 6 with a type-1 Mamdani FIS by using the coding line, **fis = mamfis**(). Another way is to use the function **convertToType1**() to convert our current IT2 FIS to a type-1 FIS.

Perform the following operations to create a new Script file **ConvTIFIS.m** and enter the codes shown in Fig. 3.64 into that file. Let us have a closer look at this piece of codes to see how it works.

1. First we need to load our generated IT2 FIS by calling the function **readfis**() and assign it to a new local variable fisT2.
2. The codes between lines 5 and 7 are used to confirm the original IT2 FIS by plotting the first input MF.
3. The function **convertToType1**() is called to convert the current IT2 FIS to a related type-1 FIS with the IT2 FIS as an argument.
4. The converted type-1 FIS is saved to the current folder on the hard disk with the name of **t1_fis.fis**.
5. The codes between lines 10 and 12 are used to check the converting result by plotting the first input MFs of the converted type-1 FIS.

Run the project now by clicking on the green-color **Run** button, and the running result is shown in Fig. 3.65.

Now let us use MATLAB Simulink to perform a simulation study for our type-1 and type-2 FIS and compare their control performances.

```
1   % convert an ITS FIS to a type-1 FIS
2   % Name: ConvT1FIS.m
3   % September 10, 2023

4   fisT2 = readfis('itype2_fis');

5   subplot(2, 1, 1);
6   plotmf(fisT2,"input",1);
7   grid;

8   fisT1 = convertToType1(fisT2);
9   t1_fis = writeFIS(fisT1, 'type1_fis');

10  subplot(2, 1, 2);
11  plotmf(fisT1,"input",1);
12  grid;
```

Fig. 3.64 The completed codes to convert IT2 FIS to a type-1 FIS

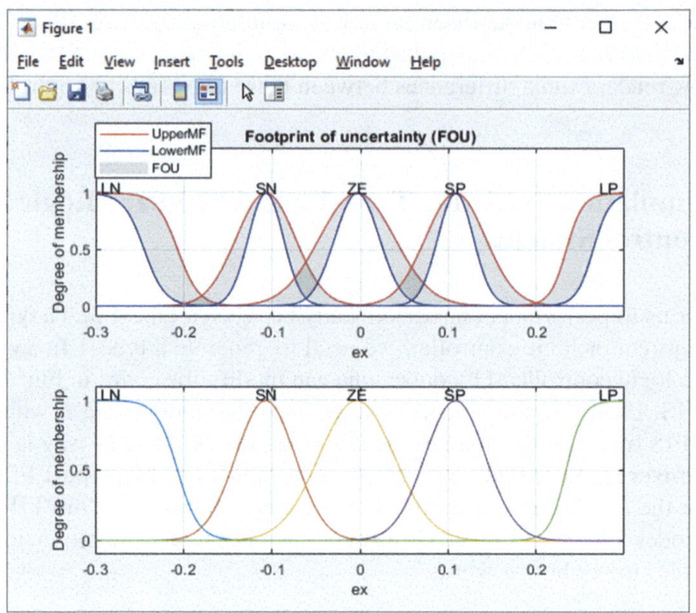

Fig. 3.65 The testing result for converted type-1 FIS

3.12.1 Create a Simulation Block Diagram Model for Type-1 FIS and IT2 FIS

First let us create a Simulink® model named **t1_it2_fis_comp.slx**.

Open MATLAB® Simulink® by typing **simulink** in the MATLAB® Command window after the prompt symbol **>>** to open the **Simulink Library Browser**, as shown in Fig. 3.66. Then we need to build this Simulink model by adding different blocks into this model container.

To create a new model, click on the **Blank Model** icon, as shown in Fig. 3.66, to start it. A new **untitled** model is generated, as shown in Fig. 3.67, with no block since it is a blank model now.

All Simulink blocks are located in the Simulink Library, and we need to open those blocks by clicking on the **Library Browser**, as shown in Fig. 3.67. Then expand the **Simulink** item that is located under the **Library** tab in the **Library Browser** to open a completed list for all simulation blocks available to us, as shown in Fig. 3.67.

To build our model, we need the following blocks that are located at the different related library classes as listed below:

1. **Gain** and **MUX** (Multiplexer) blocks under the **Commonly Used Blocks** class.
2. Transfer Function (**Transfer Fcn**), **Transport Delay** and **Derivative** blocks under the **Continuous** class.

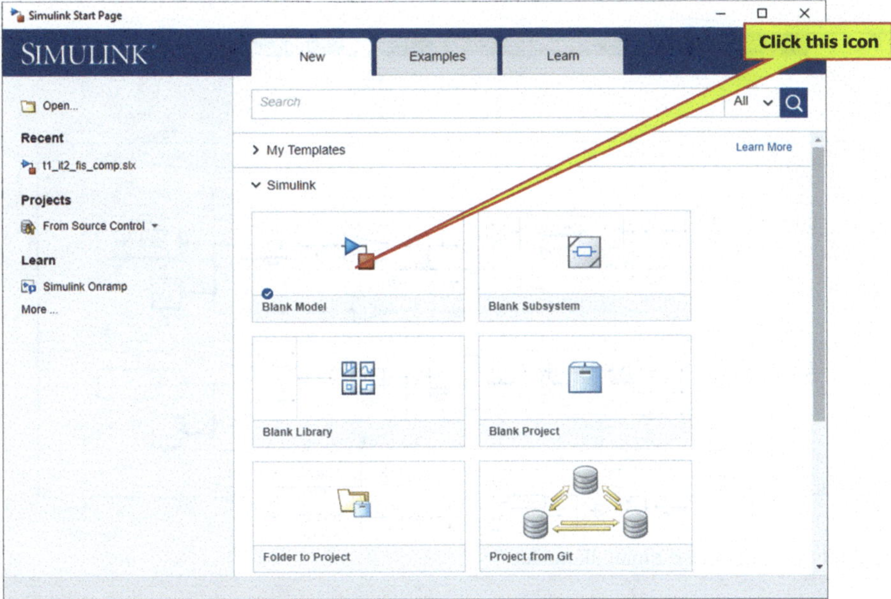

Fig. 3.66 The opened Simulink Start Page

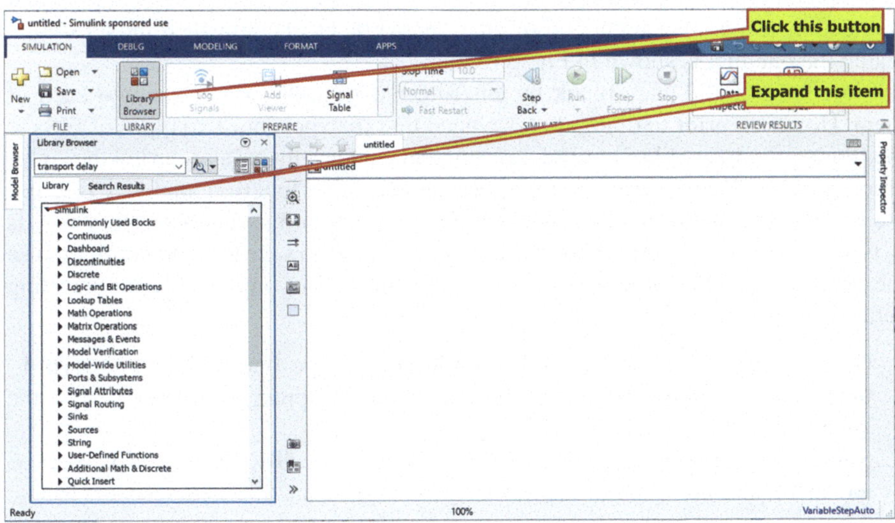

Fig. 3.67 The opened new untitled model

3. **Subtract** block under the **Math Operations** class.
4. **Scope** and **To Workspace** blocks under the **Sinks** class.
5. **Pulse Generator** block under the **Sources** class.
6. **Fuzzy Logic Controller** block under the **Fuzzy Logic Toolbox** class.

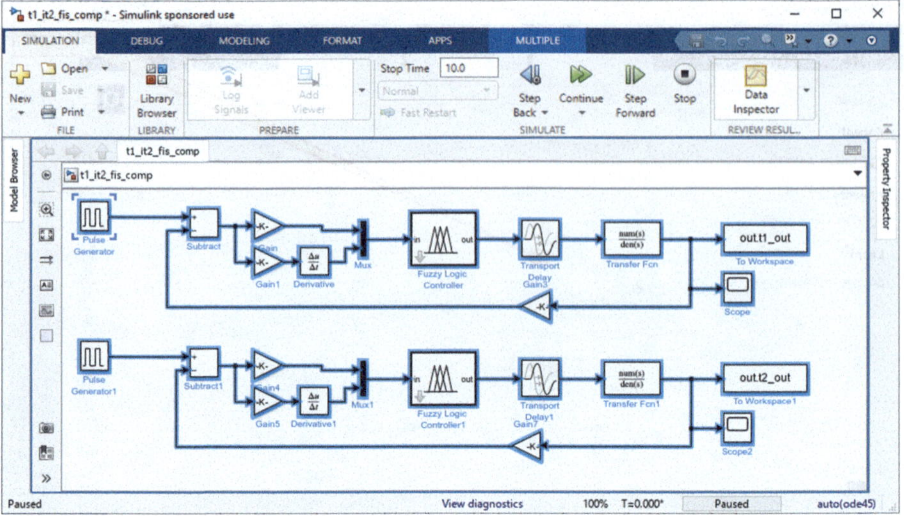

Fig. 3.68 The completed Simulink model

Expand each of these classes in the **Library** list on the left and click on each of blocks, and drag it to our model space on the right pane. Make connections between each related block by clicking on one side of the block, dragging and stopping that connection at the other input end of the selected block. The finished model should match the one that is shown in Fig. 3.68.

Now let us build each block by entering our desired parameters into each of them. However, prior to doing that, we need first to export our two types of FISs to the Workspace to enable Simulink to recognize and use them later during the simulation process. There are two ways to do that exporting function: one way is to use the Fuzzy Logic Designer App with the **Export** function and another easy way is to type commands directly in the Command window. To save time, we prefer to use the second way. Perform the following operations to complete these exporting actions:

1. On the opened MATLAB Command window, type one coding line, **type1_ fis = readfis('type1_fis')**, after the cursor **>>** and press the **Enter** key on the keyboard.
2. Similarly, type another coding line, **itype2_fis = readfis('itype2_fis')**, after the cursor **>>** and press the **Enter** key on your keyboard.

Immediately one can find that two types of FIS, **type1_fis** and **itype2_fis**, have been exported to the Workspace window on the right.

3.12.2   Build Each Block by Entering Related Parameters

To set up parameters for each block, one needs to double-click on each of them from our model window, as shown in Fig. 3.68, to open its Parameter wizard, and enter the related parameters shown in Table 3.10 into the associated box.

For both types of FIS, all parameters are identical with one exception, which is the Gain for input ex. That value is **8.80** for type-1 FIS, but it is changed to **10.80** for type-2 FIS for this simulation process. The **Transfer Fcn** is a Laplace dynamic model for our DC motor, which is identified by using MATLAB Identification Toolbox. The transport delay is 5 *ms* that is also identified by above Toolbox.

For all other parameters in all blocks, keep them with no changes, or in other words, keep the default values for them.

For the simulation parameters, such as **Stop Time** that is shown on the top box, keep the default value of 10 s, too.

3.12.3   Perform the Simulation Study for Our Two FIS Models

Now let us start to perform our simulation study for our two FISs. First we need to open our two output monitors, **Scope** and **Scope1**, which are connected to our outputs for two fuzzy control systems, to watch the control performance in real time.

Table 3.10 Setup parameters for each block

Block	Parameters
Pulse Generator	Amplitude: 0.3, Period: 4, Pulse Width: 50%, Phase delay: 0
Pulse Generator1	Amplitude: 0.3, Period: 4, Pulse Width: 50%, Phase delay: 0
Transport Delay	Time delay: 0.005, Initial output: 0, Initial buff size: 1024, Pade order: 0
Transport Delay1	Time delay: 0.005, Initial output: 0, Initial buff size: 1024, Pade order: 0
Transfer Fcn	Numerator coefficients: [6520], Denominator coefficients: [1 430 0]
Transfer Fcn1	Numerator coefficients: [6520], Denominator coefficients: [1 430 0]
Gain	8.80
Gain1	0.01
Gain2	0.0005 (feedback sensor used to transfer motor speed 600 to input voltage of 0.3 V: 0.3/600 = 1/2000 = 0.0005)
Gain3	10.80
Gain4	0.01
Gain5	0.0005 (feedback sensor used to transfer motor speed 600 to input voltage of 0.3 V: 0.3/600 = 1/2000 = 0.0005)
Scope	Main|Sample time: 0.001, Axes scaling: Auto
Scope2	Main|Sample time: 0.001, Axes scaling: Auto
Fuzzy Logic Controller	FIS name: 'type1_fis.fis'
Fuzzy Logic Controller1	FIS name: 'itype2_fis.fis'
To Workspace	Variable name: t1_out, Sample time: 0.005
To Workspace1	Variable name: t2_out, Sample time: 0.005

To do that, double-click on both **Scopes** to open them, and drag them to the appropriate location on your screen.

Now click on the green-color **Run** button to start our simulation study. The simulation model is first compiled and then the simulation is executed.

The simulation results are shown in Fig. 3.69a (response of type-1 FIS) and 3.69b (response of IT2 FIS). The reason we selected the **Pulse Generator** as the input for our control system is that it can provide both rising and falling edges for a step-like signal to test the dynamic function of our system.

To get a better comparison result, we can collect those response data from the Workspace and use **plot**() function to plot both of them in a single graph. An easy way to do that is to create a new Script file named **Plot_t1_it2_output.m** and enter the codes shown in Fig. 3.70 into that file as a new project. Let us have a closer look at this piece of codes to see how it works.

1. The **plot**() function is called to plot the outputs for both FIS. The output data for both FIS are stored in a folder **out** under the Workspace window since we used two **To Workspace** blocks in our Simulink model to send both outputs to the Workspace.

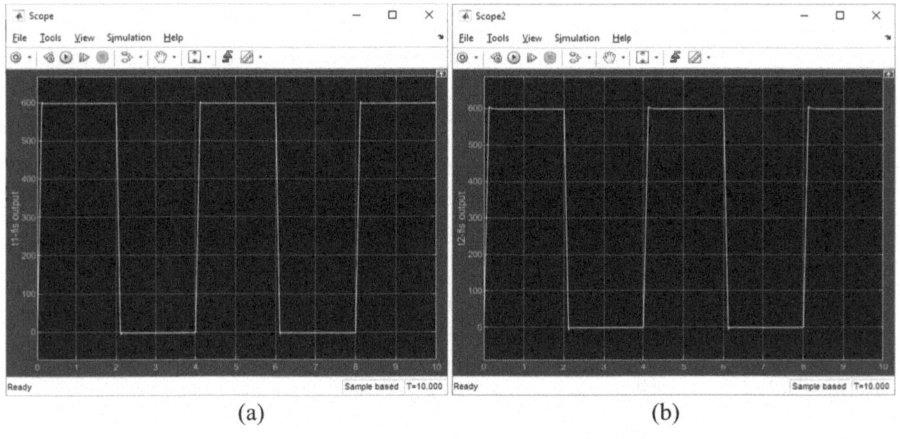

(a) (b)

Fig. 3.69 The simulated results

```
1  % Func:   Plot both FIS
2  % Name: Plot_t1_it2_output.m
3  % Date:   Sept. 19, 2023

4  plot(out.t1_out.Time, out.t1_out.Data, out.t2_out.Time, out.t2_out.Data, 'LineWidth',2);
5  grid;
6  legend('Type-1 FIS', 'IT2 FIS', 'Location', 'southeast');

   % END
```

Fig. 3.70 The codes to plot the outputs for both FIS

2. A point to be noted is that under the **out** folder, two groups of output data can be found, **t1_out** and **t2_out**, which represent the output data for type-1 and interval type-2 FIS. Each group data contained two data columns: the first one is the **Time** and the second one is the responding **Data**; both of them are 2001 by 1 array.
3. The **legend**() function is used to display an illustration for both trajectories.

Run that Script file to plot the responses for both FIS, and the comparison result is shown in Fig. 3.71. It can be found that there is almost no significant difference between the outputs of two FIS, but this result is only suitable for our study case.

It can also be found that both responses for pulse inputs are almost perfect with very short rising time, very small overshoot, very short falling time, and very small valley value. This kind of result absolutely cannot be achieved by using any traditional controller, such as PID controller. This case study also confirmed that the FIS can provide much better performances over conventional controllers.

3.13 Chapter Summary

The main topics covered by this chapter include fuzzy inference systems, fuzzy logic control systems, and some typical implementations of those systems via a few real projects.

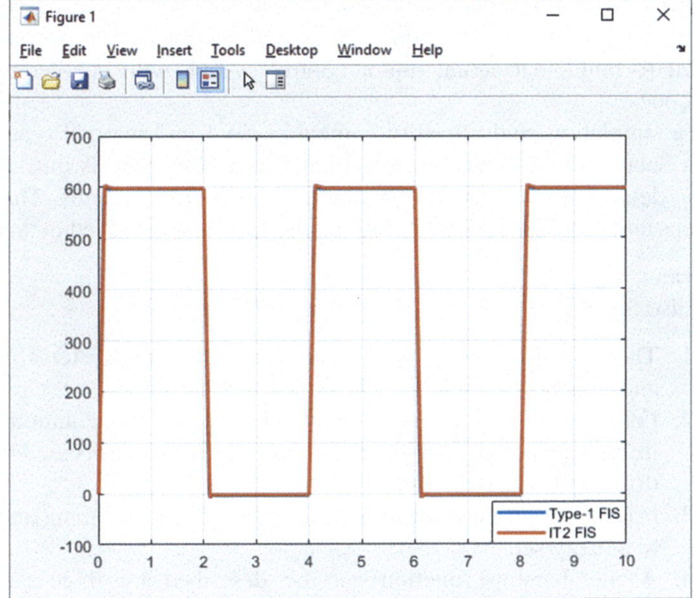

Fig. 3.71 The comparison between the outputs of type-1 and IT2 FIS

Starting from Sect. 3.2, the fuzzy idea is first introduced and discussed by using some routine life examples. The design and implementation processes of using FIS are also involved in that section.

To get more details about the FIS, some basic and fundamental knowledge and terminologies are introduced in Sect. 3.3, such as fuzzy sets and classical sets as well as their operations, membership functions, control rules, and defuzzifications. Some popular and powerful defuzzification technologies, such as the mean of maximum (MOM) method, center of gravity (COG) method, and height method (HM), are also discussed in detail with actual examples. Both offline and online FIS are introduced with lookup tables and real-time defuzzification process. A typical architecture of fuzzy logic control system is given in Sect. 3.7.

Starting from Sect. 3.8, a fully introduction and discussion about a real developing tool or environment, MATLAB and related Toolboxes, including Fuzzy Logic Toolbox, and two typical fuzzy inference systems, Mamdani and Sugeno, are discussed. Two developing modes provided by the MATLAB Fuzzy Logic Toolbox, Fuzzy Logic Design App mode and Fuzzy Logic Function mode, are also introduced and illustrated with a real project, air conditioner control system, with detailed building steps. A completed introduction for the most popular fuzzy logic functions built in the Fuzzy Logic Toolbox is provided in Sect. 3.8.3.2.

A detailed discussion of another important type of FIS, interval type-2 FIS, is given in Sect. 3.9. Some important ideas, such as upper membership functions and lower membership functions, Karnik-Mendel (KM) algorithms, and type-reducer, are also covered in that part.

A case study for applying an interval type-2 fuzzy inference system to a real DC motor control system is introduced and discussed in Sect. 3.11. Both design modes, Fuzzy Logic Designer App mode and Fuzzy Logic Functions mode, are applied and implemented to build that actual motor control system with detailed steps and MATLAB codes.

Finally, a simulation study used to compare type-1 and interval type-2 FIS is provided in Sect. 3.12. A powerful and popular tool, Simulink, is introduced and discussed in detail with two DC motor closed-loop control systems. The detailed building steps and procedure for the Simulink model is also involved in that section.

Home Works

I. True/False Selections

      ______1. There are two kinds of popular fuzzy inference systems, Mamdani and Sugeno.

      ______2. Four operational steps are involved in the implementation of fuzzy inference system, fuzzifications, fuzzy inference process, defuzzifications, and type reduction.

      ______3. Friday belongs to weekends, and this is a fuzzy statement and belongs to a fuzzy set.

      ______4. A membership function can be described by three parameters, Universe, Support, and Boundary.

_____5. Lookup tables are terminal products of the defuzzification process and they are output of offline fuzzy inference systems.

_____6. Two implementation modes, Fuzzy Logic Designer App mode and Fuzzy Functions mode, are widely used in Fuzzy Logic Toolbox.

_____7. MATLAB Fuzzy Logic Toolbox provided two major types of fuzzy inference systems, type-1 and interval type-2 FIS.

_____8. By using MATLAB functions, **mamfis()** and **sugfis()**, one can generate any new fuzzy inference systems for both type-1 and type-2 FIS.

_____9. A fuzzy control rule can be presented by three different ways, Verbose, Symbolic, and Indexed.

_____10. Unlike a type-1 FIS, an interval type-2 FIS needs to provide a type-reducer to reduce the outputs from a fuzzy set to the crisp values.

II. Multiple Choices

1. Which of the following statements is not a fuzzy statement _____

   (a) Today is a nice day
   (b) Come here quickly
   (c) Go home right now
   (d) The service is terrible

2. Which of the following operations is a correct one for fuzzy set $A \cap B$ _____

   (a) $\max(\mu_A(x), \mu_B(x))$
   (b) $\min(\mu_A(x), \mu_B(x))$
   (c) $A(x) \cap B(x)$
   (d) None of these

3. Which of the following methods is not a method used in defuzzification process _____

   (a) Mean of Maximum (MOM) Method
   (b) The Height Method (HM)
   (c) Center of Gravity (COG) Method
   (d) Lookup Table

4. To save a generated FIS to a file in hard disk, which of the following methods should be called _____

   (a) genfis()
   (b) savefis()
   (c) storefis()
   (d) writeFIS()

5. Which of the following methods is used to load an FIS from the workspace ________

   (a) readfis()
   (b) loadfis()
   (c) genfis()
   (d) All of them

6. Which of the following is a correct format a fuzzy control rule can be represented ______

   (a) Verbose
   (b) Indexed
   (c) Symbolic
   (d) All of them

7. After using the KM algorithm, the switch points, y_l and y_r, can be obtained from an IT2 FIS. The defuzzified or crisp output can be calculated as____________

   (a) $y_l + y_r$
   (b) $y_l - y_r$
   (c) $(y_l + y_r)/2$
   (d) $(y_l - y_r)/2$

8. To save or export a fuzzy inference system (**.fis**) to the MATLAB workspace, which of the following methods is correct ___________

   (a) writeFIS("filename");
   (b) load("filename");
   (c) save("filename");
   (d) Using the **Export** button in the Fuzzy Logic Designer App

9. To send the simulation results to the Workspace, ________ block should be used and this block is located at the ___________ class in Simulink class library.

   (a) Workspace, Commonly Used Blocks
   (b) Workspace, Sources
   (c) To Workspace, Sinks
   (d) To Workspace, Commonly Used Blocks

10. To perform a simulation study for a fuzzy inference system, one needs to build a simulation ________ by adding various blocks that are located at Simulink ________.

   (a) Class, Model library
   (b) Model, Model library
   (c) Model, Function library
   (d) Model, Class library

III. Exercises

1. List three operations for a fuzzy inference system to control a closed-loop control system.
2. List three methods used to perform the defuzzification process.
3. Explain the differences between a type-1 FIS and an interval type-2 FIS.

IV. Lab Projects

1. Using MATLAB® Fuzzy Logic Toolbox to design a type-1 FLC for a heater-fan control system shown in Fig. 3.72. The room temperature is controlled by a heater-fan. If the room temperature is low or too cold, the heater should be turned on to heat the room. Otherwise, if the temperature is high or too hot, the heater-fan should be turned off to cool down the room. A thermometer works as a feedback sensor. The generated FIS is named **LAB3_1.fis**.

   The definition of the membership functions for the input temperature **Temp** and the output heater-fan rotating speed **Fan_Speed** is given in Fig. 3.73.

   Four control rules are applied to this FIS:

   (a) IF temp is Cold, THEN the fan_speed is High.
   (b) IF temp is Cool, THEN the fan_speed is Med.
   (c) IF temp is Warm, THEN the fan_speed is Low.
   (d) IF temp is Hot, THEN the fan_speed is Zero.

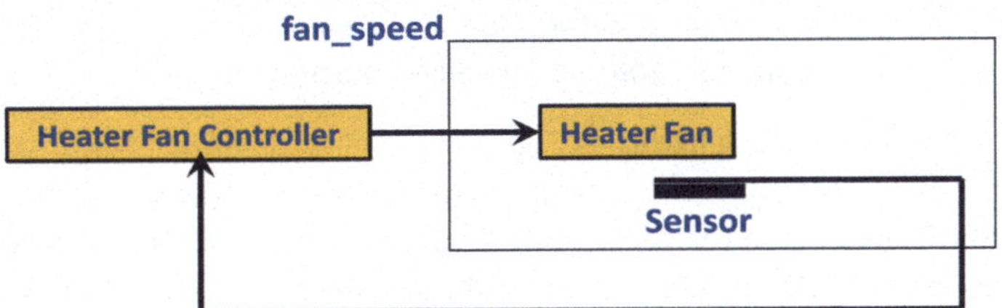

Fig. 3.72 A heater-fan control system

Fig. 3.73 Definition of the MF for the input temperature

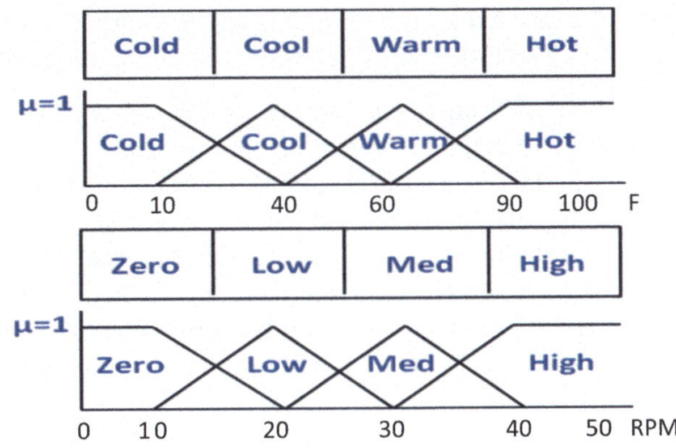

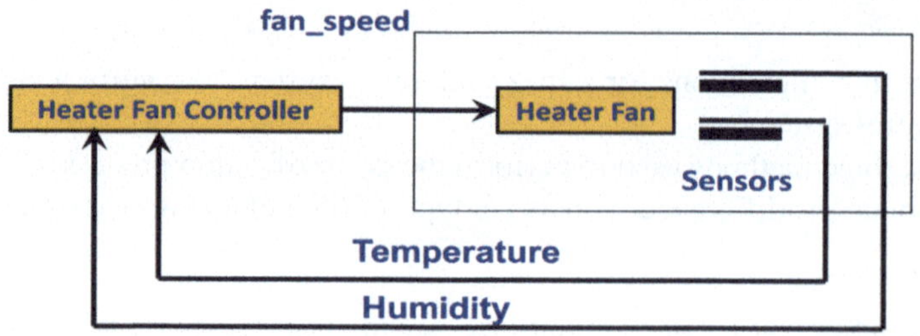

Fig. 3.74 The modified heater-fan control system

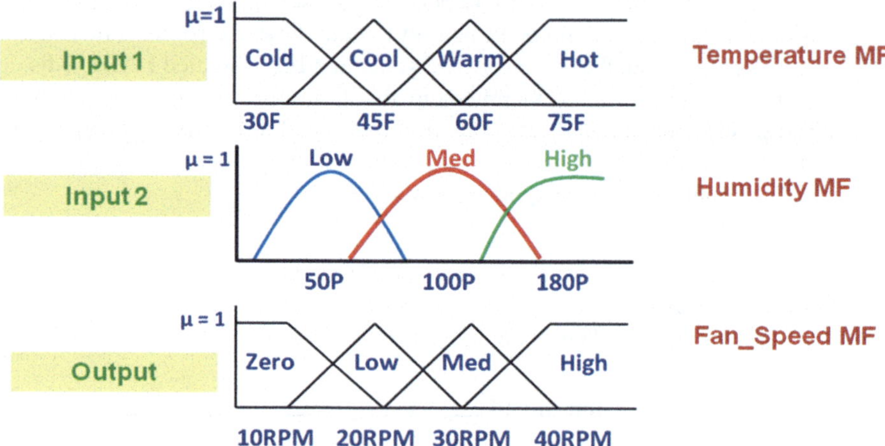

Fig. 3.75 MFs for the inputs and the output

Show the design result by plotting the rule inference and control surface.

2. Using MATLAB® Fuzzy Logic Toolbox to design a type-1 FLC for a modified heater-fan control system shown in Fig. 3.74. A thermometer and a humidity detector work as the feedback sensors. The generated FIS is named **LAB3_2.fis**

The MFs for both inputs and the output are shown in Fig. 3.75.
The following 12 control rules are implemented in this FIS:

(a) IF Temp is *cold* and Humidity is *high*, THEN fan_speed is *high*.
(b) IF Temp is *cool* and Humidity is *high*, THEN fan_speed is *med*.
(c) IF Temp is *warm* and Humidity is *high*, THEN fan_speed is *low*.
(d) IF Temp is *hot* and Humidity is *high*, THEN fan_speed is *zero*.
(e) IF Temp is *cold* and Humidity is *med*, THEN fan_speed is *med*.
(f) IF Temp is *cool* and Humidity is *med*, THEN fan_speed is *low*.
(g) IF Temp is *warm* and Humidity is *med*, THEN fan_speed is *zero*.
(h) IF Temp is *hot* and Humidity is *med*, THEN fan_speed is *zero*.
(i) IF Temp is *cold* and Humidity is *low*, THEN fan_speed is *med*.

(j) IF Temp is *cool* and Humidity is *low*, THEN fan_speed is *low*.

(k) IF Temp is *warm* and Humidity is *low*, THEN fan_speed is *zero*.

(l) IF Temp is *hot* and Humidity is *low*, THEN fan_speed is *zero*.

3. Redo Lab Project 2 by using MATLAB Fuzzy Logic Toolbox to build an interval type-2 FIS. The created IT2 FIS is named **LAB3_3.fis**.

4. Redo Lab Project 1 with MATLAB® Fuzzy Logic functions to create a new FIS named **LAB3_4.fis**.

5. Redo Lab Project 3 with MATLAB® Fuzzy Logic functions to create a new IT2 FIS named **LAB3_5.fis**.

6. For a given DC motor system plant with a transfer function as

$$G(s) = \frac{2000}{s(s+220)}$$

Design an interval type-2 FIS, **LAB3_6.fis**, for this control system using either Fuzzy Logic Designer App or Fuzzy functions method with the following given conditions:

(a) Two variables, input error **err** and input error rate **err_rate**, work as input voltages to this system.

(b) One output is the motor rotating speed **m_sp** (revolution per minute—RPM).

(c) The ranges for three variables, **err**, **err_rate**, and **m_sp**, are shown in Table 3.11.

(d) For both **err** and **err_rate**, five MFs can be designed as: Large Negative (**LN**), Small Negative (**SN**), Zero (**ZE**), Small Positive (**SP**), and Large Positive (**LP**). Gaussian and Sigmoid curves should be used to get more accurate controls.

(e) For the output **m_sp**, five MFs can be considered as: Large Negative (**LN**), Small Negative (**SN**), Zero (**ZE**), Small Positive (**SP**), and Large Positive (**LP**). Gaussian and Sigmoid curves can be used.

(f) Design 25 rules based on Table 3.9 or the common sense for this FLC system.

7. Using MATLAB® Simulink to perform a simulation study for Lab Project 6. Show the simulation model **LAB3_7.slx** and simulation results by sending the results to the Workspace, and plotting the pulse responses for the pulse generator as input.

   Hint1: To get desired simulation results, some Gain factors must be adjusted appropriately. Gain for the input **error** should be around 30, and

Table 3.11 Ranges of the input-output variables

Variables	Variables ranges	Units
Input Error (err)	−0.2–0.2	V
Input Error Rate (err_rate)	−0.05–0.05	V
Motor Output Speed (m_sp)	−800–800	RPM

Table 3.12 Ranges of the input-output variables

Variables	Variables ranges	Units
Input Error (err)	−0.4–0.4	V
Input Error Rate (err_rate)	−0.08–0.08	V
Motor Output Speed (m_sp)	−500–500	RPM

the Gain for the input **error_rate** should be around 1, and the Gain for the feedback sensor should be 0.2/800 = 1/4000 = 0.00025.

Hint2: The Pulse Generator should be set: Amplitude: 0.2, Period: 4, Pulse Width: 50%.

8. For a given fan motor control system, the transfer function of the plant is

$$G(s) = \frac{5000}{s(s+650)}$$

Design an interval type-2 FIS, **LAB3_8.fis**, for this control system using either Fuzzy Logic Designer App or Fuzzy functions method with the following given conditions:

(a) Two variables, input error **err** and input error rate **err_rate**, work as input voltages to this system.
(b) One output is the motor rotating speed **m_sp** (revolution per minute—RPM).
(c) The ranges for three variables, **err**, **err_rate**, and **m_sp**, are shown in Table 3.12.
(d) For both **err** and **err_rate**, five MFs can be designed as: Large Negative (**LN**), Small Negative (**SN**), Zero (**ZE**), Small Positive (**SP**), and Large Positive (**LP**). Gaussian and Sigmoid curves should be used to get more accurate controls.
(e) For the output **m_sp**, five MFs can be considered as: Large Negative (**LN**), Small Negative (**SN**), Zero (**ZE**), Small Positive (**SP**), and Large Positive (**LP**). Gaussian and Sigmoid curves can be used.
(f) Design 25 rules based on Table 3.9 or the common sense for this FLC system.

9. Using MATLAB® Simulink to perform a simulation study for Lab Project 8. Build the simulation model with the name **LAB3_8.slx**. Show the simulation results by sending the results to the Workspace and plotting the pulse responses for the pulse generator as input.

Hint1: To get the desired simulation results, some Gain factors must be adjusted appropriately. Gain for the input **err** should be around 150, and the Gain for the input **err_rate** should be around 0.001, and the Gain for the feedback sensor should be 0.4/500 = 0.0008.

Hint2: The Pulse Generator should be set: Amplitude: 0.4, Period: 4, Pulse Width: 50%.

References

1. Zadeh L. A. (1965) Fuzzy Sets. Intl J. Information Control 8:338–353.
2. Mamdani E. H., Assilion S (1974) An Experiment in Linguistic Synthesis With a Fuzzy Logic Controller, Intl J. Man-Machine Stud 7:1–13.
3. Holmblad L. P., Ostergaard J. J. (1982) Control of Cement Kiln by Fuzzy Logic, Gupta M. M, Sarchez E, Fuzzy Information and Decision Processes, North Holland, pp. 389–399.
4. Meng Joo Er and Ya Lei Sun, "Hybrid Fuzzy Proportional Integral Plus Conventional Derivative Control of Linear and Nonlinear Systems", IEEE Trans. On Industrial Electronics, Vol. 48, No. 6 December 2001, pp. 1109-1117.
5. J. C. Wu and T. S. Liu, "A Sliding-Mode Approach to Fuzzy Control Design", IEEE Trans. On Control Systems Technology, Vol. 4 2, March 1996, pp. 141–151.
6. Feng-Yi Hsu and Li-Chen Fu, "Intelligent Robot Deburring Using Adaptive Fuzzy Hybrid Position/Force Control", IEEE Trans. On Robots and Automation, Vol. 16, No. 4, August 2000, pp. 325–334.
7. H. X. Li and H. B. Gatland, "A New Methodology for Designing a Fuzzy Logic Controller", IEEE Trans. On Sys., Man, and Cybernetics, Vol. 25, No. 3, March 1995, pp. 505–512.
8. John Yen and Reza Langari (1999) Fuzzy Logic – Intelligence, Control, and Information, Prentice Hall.
9. Mohammad Jamshidi, Nader Vadiee and Timothy J. Ross (1993), Fuzzy Logic and Control, Prentice Hall.
10. Ying Bai, Hanqi Zhuang and Zvi. S Roth, "Fuzzy Logic Control to Suppress Noises and Coupling Effects in a Laser Tracking System", *IEEE Trans on Control Systems Technology*, Vol. 13, No. 1, January 2005, pp. 113–121.
11. R. Yager, D. Filev, *Essentials of Fuzzy Modeling and Control*, John Wiley & Sons, 1994.
12. A. Costa, A. De Gloria, P. Faraboschi, A. Pagni, G. Rizzotto, "Hardware Solutions for Fuzzy Control", Proceedings of the IEEE, Volume 83, Issue 3, March 1995, page(s): 422–434.
13. J. M. Mendel, "Fuzzy Logic Systems for Engineering: A Tutorial", Proceedings of the IEEE, Volume 83, Issue 3, March 1995, page(s): 345–377.
14. Mamdani, E.H., and S. Assilian. "An Experiment in Linguistic Synthesis with a Fuzzy Logic Controller". *International Journal of Man-Machine Studies* 7, no. 1 (January 1975): 1–13. https://doi.org/10.1016/S0020-7373(75)80002-2.
15. Sugeno, Michio, ed. *Industrial Applications of Fuzzy Control*. Amsterdam; New York: New York, N.Y., U.S.A.: North-Holland; Sole distributors for the U.S.A. and Canada, Elsevier Science Pub. Co., 1985.
16. Fuzzy Logic Toolbox™ User Guide, MathWorks®, pp. 1–22.
17. Dongrui Wu, "A Tutorial on Interval Type-2 Fuzzy Sets and Systems", University of Southern California, USA 2012.
18. ITT/UABC, Interval Type-2 Fuzzy Logic Toolbox For Use with MATLAB®, Tijuana Institute of Technology and Baja California Autonomous, University, Tijuana Campus, Mexico, 2008.
19. Mendel, Jerry M., Hani Hagras, Woei-Wan Tan, William W. Melek, and Hao Ying. *Introduction to Type-2 Fuzzy Logic Control: Theory and Applications*. Hoboken, New Jersey: IEEE Press, John Wiley & Sons, 2014.
20. Karnik, Nilesh N., and Jerry M. Mendel. "Centroid of a Type-2 Fuzzy Set." *Information Sciences* 132, no. 1–4 (February 2001): 195–220. https://doi.org/10.1016/S0020-0255(01)00069-X.
21. Wu, D. and J.M. Mendel. "Enhanced Karnik-Mendel algorithms." *IEEE Transactions on Fuzzy Systems* 17 (2009): 923–934.
22. Duran, K., H. Bernal, and M. Melgarejo. "Improved iterative algorithm for computing the generalized centroid of an interval type-2 fuzzy set," *Annual Meeting of the North American Fuzzy Information Processing Society* (2008): 190–194
23. Wu, D. and M. Nie. "Comparison and practical implementations of type-reduction algorithms for type-2 fuzzy sets and systems." *Proceedings of FUZZ-IEEE* (2011): 2131–2138

Chapter 4
Introduction to Machine Learning

As we discussed in Chap. 2, machine learning is one of the key components in the artificial intelligence (AI) study field. Basically, AI is composed of two major components, fuzzy inference system (FIS) and machine learning (ML). We have provided detailed discussions about the former, and in this chapter, we will concentrate on the latter. But first let us try to answer a question, what is machine learning?

4.1 What Is Machine Learning?

Machine learning is closely related to artificial intelligence, and exactly it is a part of AI techniques. Why we call it as artificial intelligence, not animal intelligence? The answer is that those intelligence abilities are related to or owned by human beings only. Only human beings can have those kinds of high-level abilities to sense and process received data to make final decision-making via a logic and derivation procedures. Therefore, the so-called AI is to copy those sensing, logic and derivation processes, or abilities from human beings' brains and paste them into some machines to enable them to have similar intelligent abilities to perform desired tasks as human beings do. In other words, the purpose of AI is to transfer those human beings' intelligent abilities to machines to enable them to handle all jobs intelligently.

Now another question comes, how to do those copying and pasting jobs? In other words, how to transfer those human beings' intelligent abilities to machines? While, that is the jobs of machine learning are supposed to do. Machine learning is to use different algorithms to train various ML models to fulfill those intelligence-transferring jobs. Therefore, from this point of view, ML is a foundation tool for AI, or ML provides various ways to train different models to make AI possible in practice.

Y. Bai, *AI Foundations and Applications with MATLAB*,
https://doi.org/10.1007/978-3-031-84423-2_4

Next let us have a more detailed discussion about machine learning and its components.

4.2   Structure and Component of Machine Learning

As we discussed in Chap. 2, machine learning is composed of various algorithms that can be used to train different ML models based on those algorithms. Those algorithms mainly can be categorized into two fundamental groups, supervised learning and unsupervised learning process. Figure 4.1 shows a functional block diagram for a general machine learning system.

In fact, machine learning can be further divided into the following groups of techniques:

1. Supervised Learning
2. Unsupervised Learning
3. Reinforcement Learning

Under those three groups, different methodologies and algorithms are involved with each of them. Let us first have a closer look at all of them to provide a clear picture of this topic.

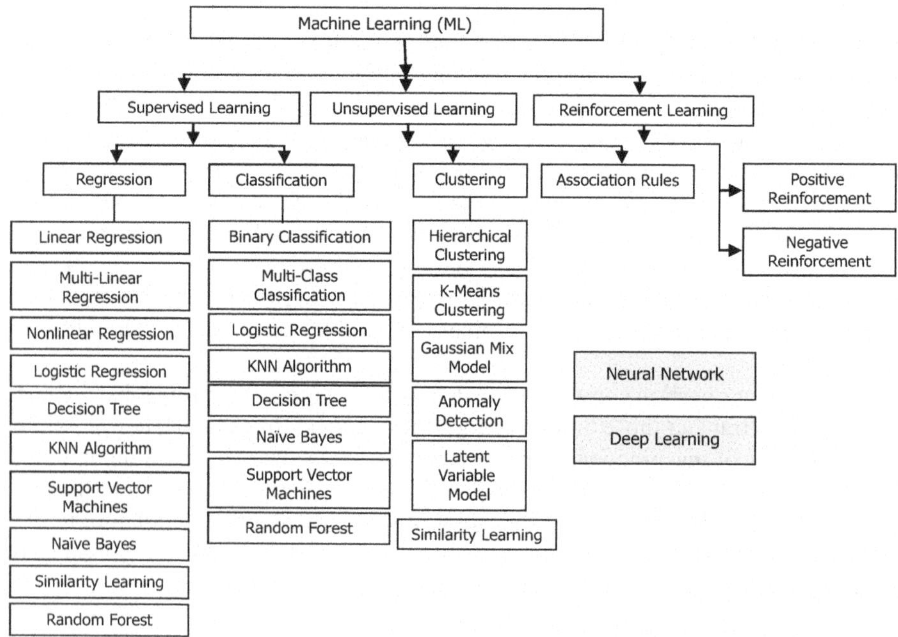

Fig. 4.1 Functional block diagram of a typical machine learning system

4.2.1   Supervised Learning

The so-called supervised learning refers to a training process by using a group of labeled input-output pairing data to predict or classify outcomes accurately for an ML model. As labeled input data is fed into an ML model, it adjusts its weights until the model has been fitted to produce output data closely or approximately equal to the labeled output accordingly, which occurs as part of the cross-validation process. Supervised learning helps organizations solve a variety of real-world problems at scale, such as classifying spam in a separate folder from your inbox.

In fact, supervised learning uses a training dataset to teach models to yield the desired output. This training dataset includes known or labeled inputs and correct outputs, which allow the model to learn over time. The algorithm measures its accuracy through the loss function, adjusting until the error has been sufficiently minimized.

As shown in Fig. 4.1, supervised learning can also be divided into two categories in algorithms: **Regression** and **Classification**.

In fact, it is hard to distinguish or provide a clear boundary between a regression and a classification algorithm since both of them provide similar mapping functions. For example, some regression algorithms, such as Logistic Regression, KNN algorithm, Decision Tree, and Naïve Bayes, are used for the classification functions. However, some other algorithms under the regression category, such as Support Vector Machines and Random Forest, could perform either regression or classification functions.

In addition, some other algorithms, such as Similarity Learning, belong to the unsupervised learning category and it can provide clustering functions.

Thus, based on the above facts, in order to provide a crisp boundary to distinguish the regression and classification algorithm, we can use the relationship of inputs and outputs applied to two algorithms. The difference between the two algorithms can be described as below:

- The regression algorithm can be considered as a mapping function and it maps the inputs to the continuous outputs, either linear or nonlinear.
- The classification algorithm can also be considered as a mapping function and it can map the inputs to discrete outputs, either binary or multiclass.

4.2.1.1   Regression Algorithm

The term *regression* comes from the Latin term *regressus*, which means to go back or to return to something. The term *regression* was first introduced by a scientist named Francis Galton to explain the relationship between the height of the father and the height of the son [1]. Regression analysis is mainly used to look back on the historical data to find patterns and use them for prediction.

In fact, regression is used to understand the relationship between dependent and independent variables for a model. It is commonly used to make projections or predictions based on the previous and current input-output dataset for a model, such as the dangerous degrees of forest fire to be occurred for some given weather conditions, flooding possibility for current and historical raining-drop amount, and stock market values for historical stock values.

Linear and logistic regression algorithms are relatively simple, and they are used to build and easy to interpret even for technical and nontechnical problems. They are mostly intuitive and take relatively little time to execute. Regression and logistic regression are the two most basic and essential algorithms in machine learning.

The relation between linear and logistic regression is the fact that they use labeled input and output datasets to make predictions. However, the main difference between them is how they are being used. Linear Regression is used to solve Regression problems whereas Logistic Regression is used to solve Classification problems [2].

4.2.1.2   Classification Algorithm

Classification is defined as the process of recognition, understanding, and grouping of objects and ideas into preset categories. With the help of these precategorized training datasets, classification in machine learning programs leverages a wide range of algorithms to classify future datasets into respective and relevant categories.

Classification algorithms used in machine learning utilize input training data for the purpose of predicting the likelihood or probability that the data that follows will fall into one of the predetermined categories [3].

The classification algorithm can be divided into two categories, binary and multiclass, which is determined by the number of outputs. For the binary classification purpose, only one output value, such as **true** or **false**, would be classified or derived. However, multiple output results could be classified based on input dataset for the multiclass classification algorithm.

Some typical multiclass classifications include to identify or classify a certain type of animals from given groups of animals, either images or voices, and identify a certain type of cars from given groups of vehicles in images. A typical binary classification is to identify or detect a personal or a business bank checks, or to classify whether a product is in good condition or not.

In short, classification can be considered as a kind of **Pattern Recognition**. Here, classification algorithms applied to the training data find the same pattern, such as similar number sequences, words or sentiments, and the like, in future data sets.

In summary, classification is a kind of categorization process using labeled input and output in a dataset to identify a single or a group of discrete output values.

4.2.2   *Unsupervised Learning*

Unlike regression and classification algorithms, in which some labeled input and output data must be used to train the ML model, the unsupervised learning algorithms do not need any labeled data to perform those kinds of training jobs. Instead, they only use input data values to group or cluster a similar class of input data and categorize them into different target groups based on the given clustering criteria.

Unsupervised learning uses self-learning algorithms and they learn without any labels or prior training. Instead, the model is given raw, unlabeled data and has to infer its own rules and structure the information based on similarities, differences, and patterns without explicit instructions on how to work with each piece of data [4].

Unsupervised learning algorithms are better suited for more complex processing tasks, such as organizing large datasets into clusters. They are useful for identifying previously undetected patterns in data and can help identify features useful for categorizing data.

Under the unsupervised learning umbrella, there are two significant algorithms called clustering and association rules.

4.2.2.1   Clustering

Cluster analysis or clustering is the task of grouping a set of objects in such a way that objects in the same group or called a *cluster* are more similar (in some sense) to each other than to those in other groups or called clusters. It is a main task of exploratory data analysis, and a common technique for statistical data analysis, used in many fields, including pattern recognition, image analysis, information retrieval, bioinformatics, data compression, computer graphics, and machine learning [5].

Dividing the data into clusters or subsets in such a way that within a specific cluster all the elements are similar to each other and they are dissimilar from elements in other clusters. In market research, cluster analysis is used for grouping people with similar properties to run targeted market campaigns. In the insurance industry, cluster analysis is used for identifying groups of people with similar properties so that the premium rate can be fixed [1].

Cluster analysis itself is not one specific algorithm, but the general task to be solved. It can be achieved by various algorithms that differ significantly in their understanding of what constitutes a cluster and how to efficiently find them. Popular notions of clusters include groups with small distances between cluster members, dense areas of the data space, intervals, or particular statistical distributions.

Clustering can therefore be formulated as a multiobjective optimization problem. The appropriate clustering algorithm and parameter settings that include parameters such as the distance function to use and a density threshold or the number of expected clusters depend on the individual data set and intended use of the results. Cluster analysis as such is not an automatic task, but an iterative process of knowledge discovery or interactive multiobjective optimization that involves trial and failure. It is often necessary to modify data preprocessing and model parameters until the result achieves the desired properties.

4.2.2.2 Association Rules

Association rule learning is a rule-based machine learning method for discovering interesting relations between variables in large databases. It is intended to identify strong rules discovered in databases using some measures of interestingness [6]. In any given transaction with a variety of items, association rules are meant to discover the rules that determine how or why certain items are connected.

Based on the concept of strong rules, Rakesh Agrawal, Tomasz Imieliński, and Arun Swami [7] introduced association rules for discovering regularities between products in large-scale transaction data recorded by point-of-sale (POS) systems in supermarkets. For example, the rule **{onions, potatoes} = {burger}** found in the sales data of a supermarket would indicate that if a customer buys onions and potatoes together, they are likely to also buy hamburger meat. Such information can be used as the basis for decisions about marketing activities such as, e.g., promotional pricing or product placements [8].

4.2.3 Reinforcement Learning

Reinforcement learning (RL) is an interdisciplinary area of machine learning and optimal control concerned with how an intelligent agent ought to take actions in a dynamic environment in order to maximize the cumulative reward. Reinforcement learning is one of three basic machine learning paradigms, alongside supervised learning and unsupervised learning.

Reinforcement learning differs from supervised learning in not needing labeled input/output pairs to be presented, and in not needing sub-optimal actions to be explicitly corrected. Instead, the focus is on finding a balance between exploration of uncharted territory and exploitation of current knowledge [9].

The environment is typically stated in the form of a Markov decision process (MDP) because many reinforcement learning algorithms for this context use dynamic programming techniques [10]. The main difference between the classical dynamic programming methods and reinforcement learning algorithms is that the latter do not assume knowledge of an exact mathematical model of the Markov decision process and they target large Markov decision processes where exact methods become infeasible [11].

In fact, Reinforcement Learning (RL) can be considered as a decision-making process. It is about learning the optimal behavior in an environment to obtain maximum reward. This optimal behavior is learned through interactions with the environment and observations of how it responds, similar to children exploring the world around them and learning the actions that help them achieve a goal [12].

In the absence of a supervisor, the learner must independently discover the sequence of actions that maximize the reward. This discovery process is akin to a trial-and-error search process. The quality of actions is measured by not just the immediate reward they return, but also the delayed reward they might fetch. As it can learn the actions that result in eventual success in an unseen environment without the help of a supervisor, reinforcement learning is a very powerful algorithm in machine learning study.

There are two types of Reinforcement Learning algorithms and they are widely implemented in our world.

4.2.3.1   Positive Reinforcement

Positive reinforcement learning algorithm is defined as when an event, which occurs due to a particular behavior, increases the strength and frequency of behaviors. In other words, it has a positive effect on behaviors.

In short words, some advantages of reinforcement learning include:

1. Maximizes Performance.
2. Sustain Change for a long period of time.
3. Too much Reinforcement can lead to an overload of states which can diminish the results.

4.2.3.2   Negative Reinforcement

Negative Reinforcement Learning algorithm is defined as strengthening of behavior because a negative condition is stopped or avoided. Some advantages of negative reinforcement learning include [13]:

1. Increases behaviors
2. Provides defiance to a minimum standard of performance
3. Provides enough to meet up the minimum behaviors

Regularly, reinforcement learning elements contain the following components:

1. Policy
2. Reward function
3. Value function
4. Model of the environment

4.2.4   Neural Network

Neural networks, also called artificial neural networks (ANNs) or simulated neural networks (SNNs), are a subset of machine learning and are the backbone of deep learning algorithms. They are called **neural** because they mimic how neurons in the brain signal one another [14].

Neural networks are composed of nodes and layers—an input layer, one or more hidden layers, and an output layer. Each layer contains multiple nodes and each node is an artificial neuron that connects to the next, and each has a weight and threshold value. When one node's output is above the threshold value, that node is activated and sends its data to the network's next layer. If it is below the threshold, no data passes along.

In fact, neural network can be considered as a special machine learning model, but it is different from all other machine learning models since it uses a neural network as a body to mimic the human being's brain to improve its analyzing and derivation abilities to provide better performance compared with other models.

The learning algorithm applied to a neural network can be either supervised, unsupervised, or reinforcement. Thus, the neural network block shown in Fig. 4.1 is a floating body.

4.2.5   Deep Learning

Deep learning is a subset of machine learning, which is essentially a neural network with three or more layers. These neural networks attempt to simulate the behavior of the human brain, albeit far from matching its ability, allowing it to *learn* from large amounts of data. While a neural network with less layers (≤3), which is called shallow learning, can still make approximate predictions, but additional hidden layers can help to optimize and refine for accuracy [15].

As we mentioned, neural networks are the backbone of deep learning algorithms, which means that if machine learning uses neural networks as models, a deep learning algorithm is coming.

In fact, deep learning involves the use of complex models, neural networks, which exceed the capabilities of all other machine tools or algorithms such as logistic regression and support vector machines, to provide much better performances or some function approximations with smaller errors.

Deep learning can use supervised learning algorithm to perform predictions as regression did and classify images as classification algorithm did by using labeled input and output data.

On the other hand, deep learning can be used to cluster objects or images based on their similarities and dissimilarities. One can extract special features with a trained neural network (deep learning), then deploy an unsupervised algorithm such as K-Means clustering.

Additionally, autoencoders are neural nets that can be used for image compression and reconstruction via a latent space representation of compressed data, and it outputs whatever is inputted. Therefore, those autoencoders can be considered as self-supervised learning neural nets.

The reinforcement learning with neural networks can be used and is the methodology behind **DeeMind** and its victory in the game **Go**.

In summary, deep learning can be used for supervised, unsupervised, self-supervised, or reinforcement, and it depends on the neural network that is used. For this reason, the Deep Learning block shown in Fig. 4.1 is a floating block, which means that it can be used for any learning study category.

Due to some similarities among different algorithms used for supervised, unsupervised, and reinforcement learning studies, we will concentrate on our discussions for those algorithms under different categories to avoid any possible duplications.

We will go through most of those algorithms in this chapter; first let us have a clear and complete picture about related tools provided by MATLAB in machine learning to design and build various applications.

4.3   Machine Learning with MATLAB

MATLAB is a powerful platform and provides a huge selection of machine learning, including designing, developing, simulating, testing, and deploying all different algorithms for machine learning-related models. Compared with other platforms and languages, such as Python, R, and Java, MATLAB provided some super-developing tools and environments to meet the needs of both beginners and experienced developers in AI studies.

Fundamentally, MATLAB provided two major groups of tools for machine learning developers, and they are:

1. User-friendly app used to build and develop machine learning-related projects in easy and fast ways.
2. Function library used to design, build, and develop machine learning-related projects in more professional and flexible ways.

The first group contains many selected tools with user-friendly graphic user interfaces (GUIs) to enable readers or users to design and develop their desired machine learning applications more easily and quickly with some basic knowledge about machine learning. It is so easy and convenient to build a machine learning application for users so that even they do not need to type any piece of coding line.

For some experienced readers or users, the second group is a good selection for them. MATLAB categorizes different machine learning algorithms into different subsets, and each of those subsets can be considered as a library with a collection of functions. Readers need to call those functions to build their applications without any GUI available.

For most beginners or readers who are new to machine learning, the first group is a better choice since they do not need much and deep knowledge in machine learning, and they can easily and conveniently design and build some practical machine learning-related projects by using App with GUIs. However, for experienced developers, the second group would be a better choice since they can design and build more professional applications by calling related functions to make their development process more flexible and practical.

Comparably speaking, the first group or the App method enables readers or students to be able to design and build their applications more easily and quickly. With the help of that App, the study and learning curve on machine learning can be greatly reduced and the students' learning interests can be greatly increased and improved, but a shortcoming is that some detailed knowledge behind the App and GUIs are hidden from students.

On the other hand, by using the functions in different libraries, readers and students can design and build their applications in more professional and flexible ways, but they need to learn and study more and deep knowledge and techniques in machine learning.

All of those machine learning Apps and functions are located at different or related Toolboxes, which include:

- Statistics and Machine Learning Toolbox
- Deep Learning Toolbox
- Curve Fitting Toolbox

4.3.1   Statistics and Machine Learning Toolbox

The Statistics and Machine Learning Toolbox™ provides functions and Apps to describe, analyze, and model data. One can use descriptive statistics, visualizations, and clustering for exploratory data analysis, fit probability distributions to data, generate random numbers for Monte Carlo simulations, and perform hypothesis tests. Regression and classification algorithms enable users to draw inferences from data and build predictive models either interactively, using the Classification and Regression Learner apps, or programmatically, using AutoML.

For multidimensional data analysis and feature extraction, the toolbox provides principal component analysis (PCA), regularization, dimensionality reduction, and feature selection methods that let you identify variables with the best predictive power.

The toolbox provides supervised, semi-supervised, and unsupervised machine learning algorithms, including Support Vector Machines (SVMs), boosted decision trees, k-means, and other clustering methods. One can apply interpretability techniques such as partial dependence plots and LIME, and automatically generate C/C++ code for embedded deployment. Many toolbox algorithms can be used on data sets that are too big to be stored in memory.

4.3.2   Deep Learning Toolbox

Deep Learning Toolbox™ provides a framework for designing and implementing deep neural networks with algorithms, pretrained models, and Apps. One can use convolution neural networks (ConvNets, CNNs) and long short-term memory (LSTM) networks to perform classification and regression on image, time-series, and text data. One can also build network architectures such as generative adversarial networks (GANs) and Siamese networks using automatic differentiation, custom training loops, and shared weights.

With the Deep Network Designer App, you can design, analyze, and train networks graphically. The Experiment Manager App helps you manage multiple deep learning experiments, keep track of training parameters, analyze results, and compare code from different experiments. One can visualize layer activations and graphically monitor training progress.

Users can import networks and layer graphs from TensorFlow™ 2, TensorFlow-Keras, and PyTorch®, the ONNX™ (Open Neural Network Exchange) model format, and Caffe. One can also export Deep Learning Toolbox networks and layer graphs to TensorFlow 2 and the ONNX model format. The toolbox supports transfer learning with DarkNet-53, ResNet-50, NASNet, SqueezeNet, and many other pretrained models.

4.3.3   Curve Fitting Toolbox

Curve Fitting Toolbox™ provides an App and functions for fitting curves and surfaces to data. The toolbox lets you perform exploratory data analysis, preprocess and postprocess data, compare candidate models, and remove outliers. One can conduct regression analysis using the library of linear and nonlinear models provided or specify their own custom equations. The library provides optimized solver parameters and starting conditions to improve the quality of your fits. The toolbox also supports nonparametric modeling techniques, such as splines, interpolation, and smoothing.

After creating a fit, one can apply a variety of postprocessing methods for plotting, interpolation, and extrapolation; estimating confidence intervals; and calculating integrals and derivatives.

In summary, all Apps and functions related to machine learning and deep learning are included in those three Toolboxes. There are some duplicated Apps and functions among those three Toolboxes; thus, we will concentrate our discussions on the top two of them in the book.

In fact, some Apps and functions related to machine learning and deep learning are combined or mixed together in some Toolboxes, and we will show them in the following discussions.

First let us concentrate our study on machine learning-related Apps provided by MATLAB.

4.4   Machine Learning-Related Apps and Functions Provided by MATLAB

The Statistics and Machine Learning Toolbox contains all the tools necessary to extract knowledge from large datasets. It provides functions and Apps to analyze, describe, and model data. Starting exploratory data analysis becomes a breeze with the descriptive statistics and graphs contained in the toolbox. Furthermore, fitting probability distributions to data, generating random numbers, and performing hypothesis tests will be extremely easy. Finally, through the regression and classification algorithms, we can draw inferences from data and build predictive models.

For data mining, the Statistics and Machine Learning Toolbox offers feature selection, stepwise regression, Principal Component Analysis (PCA), regularization, and other dimensionality reduction methods that allow the identification of variables or functions that impact your model.

Most popular supervised and unsupervised machine learning algorithms, including Support Vector Machines (SVMs), decision trees, K-Nearest Neighbor (KNN), K-Means, K-Medoids, hierarchical clustering, Gaussian Mixture Models (GMM), and hidden Markov models (HMM), are also involved in this toolbox. Through the using of such algorithms and calculations on datasets that are too large to be stored in memory can be correctly executed [16].

Exactly, MATLAB Statistics and Machine Learning Toolbox combined nine (9) Apps together, and they contain machine learning and deep learning Apps in one package. These nine Apps include (Fig. 4.2):

1. Classification Learner
2. Deep Network Designer
3. Deep Network Quantizer
4. Experiment Manager
5. Neural Net Clustering
6. Neural Net Fitting
7. Neural Net Pattern Recognition
8. Neural Net Time Series
9. Regression Learner

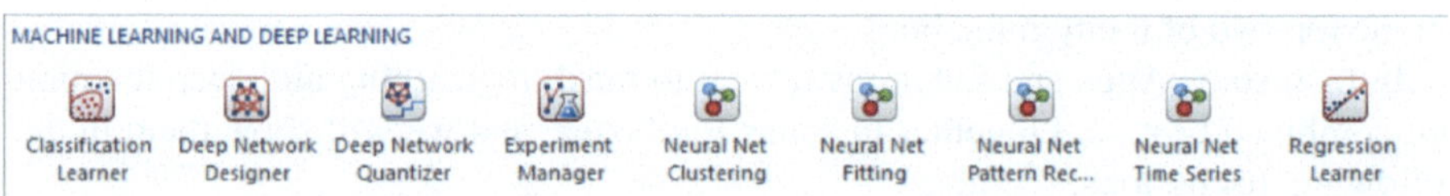

Fig. 4.2 The Apps categorized under the Statistics Machine Learning Toolbox

Due to space limitation and availability, we cannot provide a very detailed discussion for all of them. Instead, we will select some important Apps that are closely related to machine learning in the following sections.

4.5   Chapter Summary

This chapter provides an overview of machine learning and related algorithms closely related to supervised, unsupervised, and reinforcement learning technologies.

The definition about machine learning is given in Sect. 4.1. Following that, a detailed description about major components and structure used in machine learning is provided. A global picture about machine learning is discussed with most popular algorithms widely implemented in supervised and unsupervised learning as well as reinforcement learning.

Three fundamental components are involved under the machine learning umbrella, which include:

1. Supervised Learning
2. Unsupervised Learning
3. Reinforcement Learning

Under the supervised learning category, two components are also involved, and they are:

1. Regression
2. Classification

Multiple different algorithms and models are covered by these two components, such as linear regression, multilinear regression, nonlinear regression, logistic regression, decision tree, KNN, SVM, naive Bayes, and random forest. In fact, all of those algorithms can be applied and implemented by both regression and classification processes. The difference between these two components is the output or response values; for the former, the outputs are continuous or numeric values, but for the latter, the outputs are classes or categorical values.

Comparably speaking, both supervised and unsupervised learning techniques belong to open-loop control systems, but the reinforcement learning belongs to closed-loop control system. The reason for that is both supervised and unsupervised learning models need to be trained by input data stored in a dataset. During the training process, it used a closed-loop control mode since it needs feedback to check and adjust the error between the desired and accrual responses. But after the training process, both models will no longer need any feedback and they work as some open-loop systems.

However, the reinforcement learning model is different. This means that it needs the training process, but it still needs the feedback from the response to detect and check the error between the input and the output in real time to make sure that the output is correct. Thus, it is a closed-loop control system even after its training process.

In addition to the above three major models, some submodels are involved in the machine learning, which include the neural networks and deep learning model. A neural network is just a structure to be trained by using different algorithms, and it becomes a deep learning model when it is trained by using any machine learning algorithm.

To help users to design and build their desired machine learning projects, three important and popular MATLAB Toolboxes, Statistics and Machine Learning Toolbox, Deep Learning Toolbox, and Curve Fitting Toolbox, are introduced and discussed with more details to enable users to complete their designs and developments easily and conveniently.

Home Works

I. True/False Selections

_____1. AI is a technology used to transfer human beings' intelligent abilities to machines to enable them to handle all jobs intelligently.

_____2. Generally, machine learning is to use different algorithms to train various ML models to perform some intelligence jobs as human beings do.

_____3. Machine learning contains two categories: supervised learning and unsupervised learning.

_____4. Supervised learning is composed of two categories: regressions and classifications.

_____5. It is easy to distinguish or provide a clearly boundary between a regression and a classification algorithm since both of them provide different mapping functions.

_____6. Some popular algorithms, such as SVM, Random Forest, KNN algorithm, Decision Tree, and Naïve Bayes, can be used for both regression and classification functions.

_____7. Neural networks and deep learning belong to unsupervised learning algorithms.

_____8. Regression analysis is mainly used to look back on the historical data to find patterns and use them for prediction.

_____9. Classification is defined as the process of recognition, understanding, and grouping of objects and ideas into preset categories.

_____10. Reinforcement learning differs from supervised learning in not needing labeled input/output pairs to be presented, and in not needing suboptimal actions to be explicitly corrected.

II. Multiple Choices

1. Machine learning contains the following algorithms ____________.

   (a) Regressions and classifications
   (b) Supervised learning and unsupervised learning
   (c) Linear and multiclass classifications
   (d) Supervised learning, unsupervised learning, and reinforcement learning

2. The supervised learning contained the following algorithms, __________.

   (a) Clustering and reinforcement learning
   (b) K-Means clustering and positive reinforcement learning
   (c) Neural networks and deep learning
   (d) Binary classifications and k-means clustering

3. The unsupervised learning includes the following algorithms __________.

   (a) Clustering and association rules
   (b) Negative reinforcement and K-Nearest Neighbors
   (c) Naïve Bayes and kernel approximation
   (d) All of them

4. Clustering is a task of __________________.

   (a) Grouping a set of objects in such a way that objects in the same group or called a *cluster* are more similar (in some sense) to each other than to those in other groups or called clusters
   (b) Dividing the data into clusters or subsets in such a way that within a specific cluster all the elements are similar to each other and they are dissimilar from elements in other clusters
   (c) Being formulated as a multiobjective optimization problem
   (d) All of them

5. Reinforcement learning contains the following components __________________.

   (a) Policy
   (b) Reward function
   (c) Value function
   (d) All of them

6. MATLAB provides all machine learning-related tools and supports in __________________.

   (a) Statistics and Machine Learning Toolbox and Deep learning toolbox
   (b) Statistics and Machine Learning Toolbox, Deep learning and curve fitting toolbox
   (c) Curve Fitting Toolbox and Signal processing toolbox
   (d) Fuzzy logic toolbox and Deep learning toolbox

7. MATLAB provided two major groups of tools for machine learning, __________________.

   (a) Apps and function library
   (b) Toolboxes and functions
   (c) GUIs and Apps
   (d) None of them

8. Deep learning belongs to _______________________.

   (a) An unsupervised learning algorithm
   (b) A supervised learning algorithm
   (c) Both a and b
   (d) Reinforcement algorithm

9. The Statistics and Machine Learning Toolbox contains _______________.

   (a) Apps for supervised learning tools and functions
   (b) Apps for unsupervised learning tools and functions
   (c) Apps and functions for supervised and unsupervised learning
   (d) All of them

10. Generally, Deep Learning can be considered as_____________________________.

   (a) A subset of machine learning, and it is a neural network with three or more layers
   (b) Involving the use of complex models, neural networks, which exceed the capabilities of all other machine tools or algorithms to provide much better performances
   (c) A supervised learning algorithm to perform predictions as regression did and classify images as classification algorithm did by using labeled input and output data
   (d) All of them

III.　　Exercises

1. Explain the similarity and dissimilarity between the regression and the classification algorithms.
2. Provide an explanation of why MATLAB provided a better platform for AI applications compared with other platforms or languages.
3. Describe the relationship between the neural networks and deep learning.

References

1. Venkata Reddy Konasani & Shailendra Kadre, *Machine Learning and Deep Learning Using Python and TensorFlow*, McGraw Hill, 1st Edition, ISBN-13: 978-1260462296, Feb, 2021.
2. https://www.kdnuggets.com/2022/03/linear-logistic-regression-succinct-explanation.html#:~:text=The%20relation%20between%20Linear%20and,used%20to%20solve%20Classification%20problems.
3. https://www.simplilearn.com/tutorials/machine-learning-tutorial/classification-in-machine-learning#what_is_classification.
4. https://cloud.google.com/discover/what-is-unsupervised-learning#:~:text=Unsupervised%20learning%20in%20artificial%20intelligence,any%20explicit%20guidance%20or%20instruction.

5. https://en.wikipedia.org/wiki/Cluster_analysis.
6. Piatetsky-Shapiro, Gregory (1991), *Discovery, analysis, and presentation of strong rules*, in Piatetsky-Shapiro, Gregory; and Frawley, William J.; eds., *Knowledge Discovery in Databases*, AAAI/MIT Press, Cambridge, MA.
7. Agrawal, R.; Imieliński, T.; Swami, A. (1993). "Mining association rules between sets of items in large databases". *Proceedings of the 1993 ACM SIGMOD international conference on Management of data—SIGMOD '93*. p. 207.
8. https://en.wikipedia.org/wiki/Association_rule_learning.
9. Kaelbling, Leslie P.; Littman, Michael L.; Moore, Andrew W. (1996). "Reinforcement Learning: A Survey". *Journal of Artificial Intelligence Research*. **4**: 237–285.
10. van Otterlo, M.; Wiering, M. (2012). "Reinforcement Learning and Markov Decision Processes". *Reinforcement Learning*. Adaptation, Learning, and Optimization. Vol. 12. pp. 3–42. doi:https://doi.org/10.1007/978-3-642-27645-3_1. ISBN 978-3-642-27644-6.
11. Li, Shengbo (2023). *Reinforcement Learning for Sequential Decision and Optimal Control* (First ed.), Springer Verlag, Singapore. pp. 1–460. doi:https://doi.org/10.1007/978-981-19-7784-8. ISBN 978-9-811-97783-1.
12. https://www.synopsys.com/ai/what-is-reinforcement-learning.html#:~:text=Definition,enviro nment%20to%20obtain%20maximum%20reward.
13. https://www.geeksforgeeks.org/what-is-reinforcement-learning/.
14. https://www.ibm.com/blog/ai-vs-machine-learning-vs-deep-learning-vs-neural-networks/.
15. https://www.ibm.com/topics/deep-learning.
16. https://subscription.packtpub.com/book/data/9781788398435/1/011v11sec13/ statistics-and-machine-learning-toolbox.

Chapter 5
Introduction to Regression Algorithms

The so-called supervised learning technique is to use input-output pair as a data source to train different models with different algorithms to obtain optimal models to perform related predications for a set of given or new data.

5.1 Supervised Learning-Related Apps and Functions

As we discussed in Sect. 4.2 in Chap. 4, Supervised Learning contains two groups, regression and classification algorithms. However, under those groups, various algorithms can be utilized by either group. In other words, some algorithms can perform regression function, but at the same time, those algorithms can also perform classification function. In order to distinguish these groups, we will provide our discussions for both groups. First let us start our discussions about Apps used for supervised learning, exactly for linear regression and logistic regression.

5.2 Linear Regression-Related Apps and Functions

MATLAB provides a powerful App, which is called Regression Learner. The function of this App is to interactively train, validate, and tune regression models with different algorithms.

By using this App, one can train different regression models, including linear regression models, regression trees, Gaussian process regression models, support vector machines, kernel approximation, ensembles of regression trees, and neural

Supplementary Information The online version contains supplementary material available at https://doi.org/10.1007/978-3-031-84423-2_5.

network regression models. In addition to training models, you can also explore your data, select features, specify validation schemes, and evaluate results. You can export a model to the workspace to use the model with new data or generate MATLAB® code to learn about programmatic regression.

Training a model in Regression Learner consists of the following two parts:

1. Validated Model: Train a model with a validation scheme. By default, the app protects against over-fitting by applying cross-validation. Alternatively, you can choose holdout validation. The validated model is visible in the App.
2. Full Model: Train a model on full data, excluding test data. The App trains this model simultaneously with the validated model. However, the model trained on full data is not visible in the App. When you choose a regression model to export to the workspace, Regression Learner exports the full model.

As we mentioned, a regression is a statistical algorithm that relates a dependent variable, which is an input, to one or more independent or explanatory variables, such as outputs.

A regression model is able to show whether changes observed in the dependent variable are associated with changes in one or more of the output variables.

In fact, a regression process or algorithm is to use a least-squares approach to fit a best-fit line and check how the data is dispersed around that line. The least-squares technique is used to make the sum of squares created by a mathematical fitting function as small as possible.

The flowchart shown in Fig. 5.1 illustrates a popular and common workflow for training regression models in the Regression Learner App.

Among all regression algorithms, the linear and multilinear regression algorithms are most popular and simpler techniques.

5.2.1　Linear Regression

The linear regression establishes a linear relationship between two variables, input and output, based on a line of best fitting, and it is thus graphically depicted using a straight line with the slope defining how the change in one variable impacts a change in the other. The y-intercept of a linear regression relationship represents the value of one variable, such as output y, when the value of the other, such as input x, is zero [1].

An example implementation of linear regression algorithm is to predict or estimate the number of the occurred traffic accidents and the population of a selected

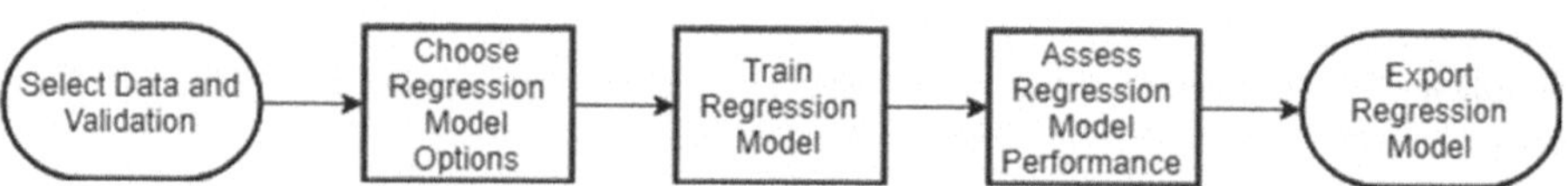

Fig. 5.1 Flowchart of workflow for training regression models

state in the United States. Roughly this should be a linear relationship between them, as shown in Fig. 5.2.

This example shows how to perform simple linear regression using the accidents dataset. The example also shows you how to calculate the coefficient of determination R^2 to evaluate the regressions. The accidents dataset, **accidents**, contains data for fatal traffic accidents in the United States.

Linear regression models the relation between a dependent, or response, variable y and one or more independent, or predictor, variables $x_1,...,x_n$. Simple linear regression considers only one independent variable using the relation.

$$y = \beta_0 + \beta_1 x + \epsilon \tag{5.1}$$

where β_0 is the y-intercept, β_1 is the slope or regression coefficient, and ϵ is the error term.

Start with a set of n observed values of x and y given by (x_1, y_1), (x_2, y_2), ..., (x_n, y_n). Using the simple linear regression relation, these values form a system of linear equations. Represent these equations in matrix form as

$$\begin{bmatrix} y_1 \\ y_2 \\ \cdot \\ \cdot \\ \cdot \\ y_n \end{bmatrix} = \begin{bmatrix} 1 & x_1 \\ 1 & x_2 \\ \cdot & \cdot \\ \cdot & \cdot \\ \cdot & \cdot \\ 1 & x_n \end{bmatrix} \begin{bmatrix} \beta_0 \\ \beta_1 \end{bmatrix} + \begin{bmatrix} \epsilon_1 \\ \epsilon_2 \\ \cdot \\ \cdot \\ \cdot \\ \epsilon_n \end{bmatrix} \tag{5.2}$$

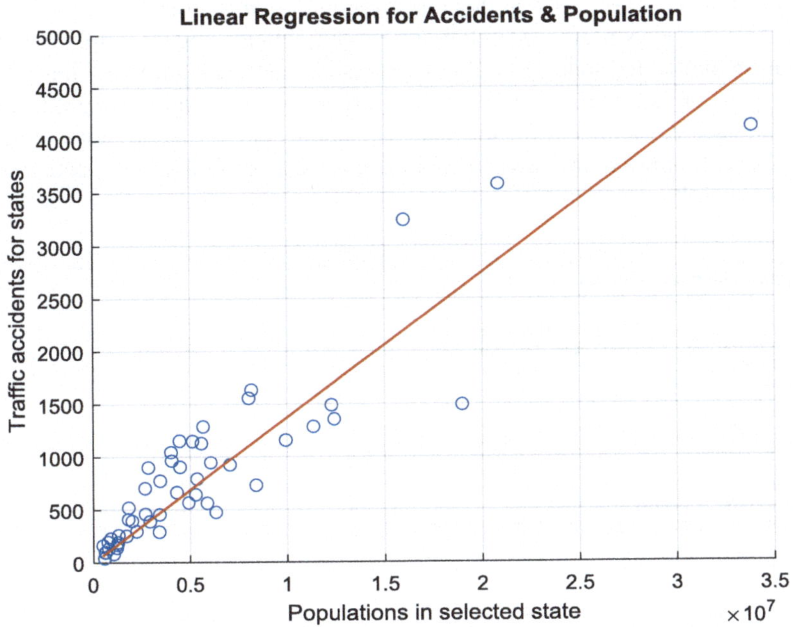

Fig. 5.2 A linear relationship between the traffic accidents and population

Without considering the error and y-intercept, the relation is now $Y = XB$. In MATLAB, one can find B using the **mldivide** operator as $B = X\backslash Y$.

The dataset **accidents** can be found in the US Department of Transportation and it covered US traffic accidents and fatalities in 2004. The data covers all 50 states and the District of Columbia. The dataset is composed of 51 rows and 17 columns with a size of 8 KB.

To get the distribution relationship between the traffic accident numbers and related state, the codes shown in Fig. 5.3 can be used. Let us have a closer look at that piece of codes to see how it works.

1. First the dataset **accidents** is loaded into the workspace. The exact path for that dataset should have been set up in your local folder as the Toolbox is installed.
2. The **hwydata** is a variable holder used to get two columns data stored in the **accidents** dataset, columns 14 and 4, which are state's population and the traffic accident numbers occurred in that state, and convert them to double in data type.
3. The retrieved two columns data are converted to long in format.
4. The vector or matrix division is performed to get the regressed slope β_1, and the result is displayed in the Command window.
5. The regressed result is calculated based on the linear relation.
6. The function **scatter(X, Y)** is used to create a scatter plot with markers at the locations specified by X and Y. This plot shows the actual input and output values, or real state and actual traffic accidents occurred on that state.
7. The regressed result is also plotted, and it is used to compare with the real values.
8. Related labels or title are added to make displaying result more illustrative.

Now let us have a closer look at this linear regression algorithm.

Figure 5.4 shows a detailed relation between actual accident numbers and the population in selected state, as well as regressed linear result without y-intercept β_0. The purpose of using this linear regression algorithm is to find a straight line to fit all actual accident numbers for the selected state as closely as possible. However, due to the different variations on actual data, we cannot find an ideal line to fit all data points without any error.

```
% testing plot linear regression
1  load accidents
2  x = hwydata(:,14);      %Population of states
   y = hwydata(:,4);       %Accidents per state
3  format long
4  b1 = x\y

5  yCalc1 = b1*x;
6  scatter(x,y)
   hold on
7  plot(x,yCalc1)
8  xlabel('Populations in selected state')
   ylabel('Traffic accidents for states')
   title('Linear Regression for Accidents & Population')
   grid on
```

Fig. 5.3 Detailed codes used to calculate and display the regression result

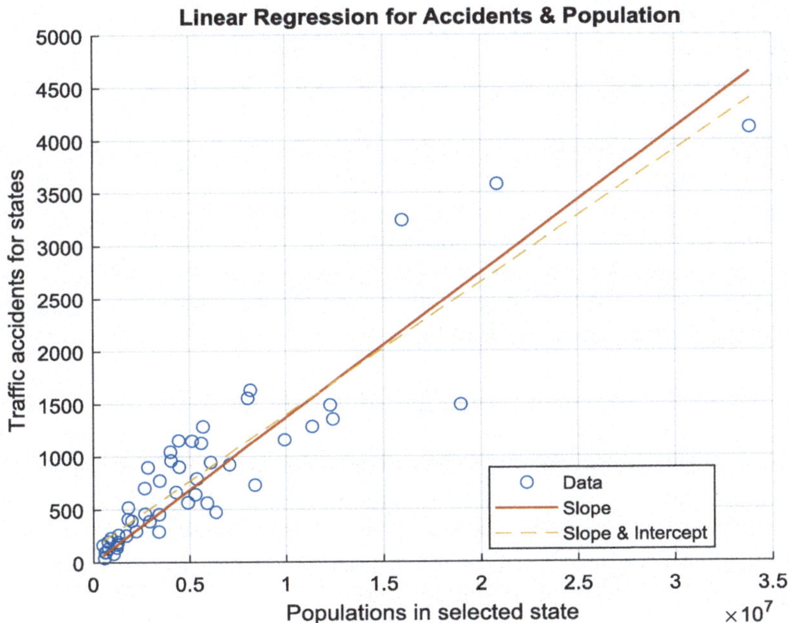

Fig. 5.4 The comparison result for two fitting methods

Now we can improve the fit shown in Fig. 5.2 by including a y-intercept β_0 in the model as $y = \beta_0 + \beta_1 x$. Calculate β_0 by padding x with a column of ones and using the \ operator. Adding the following codes into the Script shown in Fig. 5.3:

```
xx = [ones(length(x), 1) x];          % Function length() is to get
the number of input x, which is row.
b = xx\y
```

```
Run the Script, the result is:
```

```
b = 1.0e+02 *

   1.427120171726538
   0.000001256394274
```

This result represents the relation $y = \beta_0 + \beta_1 x = 142.7120 + 0.0001256x.$

To compare this modified fitting with the first one, we can use the following codes to plot both fitting results, as shown in Fig. 5.4.

```
yCalc2 = xx*b;
plot(x,yCalc2,'----')
legend('Data','Slope','Slope & Intercept','Location','best');
```

Figure 5.5 is a duplication of Fig. 5.4 by adding errors. In some points, the errors between the actual data and the regression result are very small, such as those points with small populations, but the errors become significantly larger on some other points, such as points with bigger populations, ε_n, ε_{n-1}, ε_{n-2}, and ε_{n-3}. To solve this kind of problem, we need to use the least-squares algorithm to make all errors as small as possible. This so-called as-small-as-possible operation can be realized with the following math equation (refer to Eq. 5.1),

$$\min \in = \min \sum \left(y - \left(\beta_0 + \beta_1 x \right) \right)^2 \tag{5.3}$$

where ε, y, and x are vectors with n as the number of input and output data.

We already have the data on x and y. With that, we need to find the values of β_0 and β_1, which will minimize this sum of squares of errors. Now we are working with an optimization problem. We want to find optimal values for β_0 and β_1 so that Min $\sum (y - (\beta_0 + \beta_1 x))^2$ is the minimum value.

Now we have obtained two fitting results, one has y-intercept but another does not. How do we know which one is better than another one? In other words, how to evaluate those fitting results with what criterion?

A good answer to that question is to use R^2 method or called Residuals (coefficient of determination) to evaluate the error between the actual and regressed values.

In fact, Residuals are the difference between the *observed* values of the response (dependent) variable and the values that a model *predicts*. When you fit a model that is appropriate for your data, the residuals approximate independent random errors. That is, the distribution of residuals ought not to exhibit a discernible pattern.

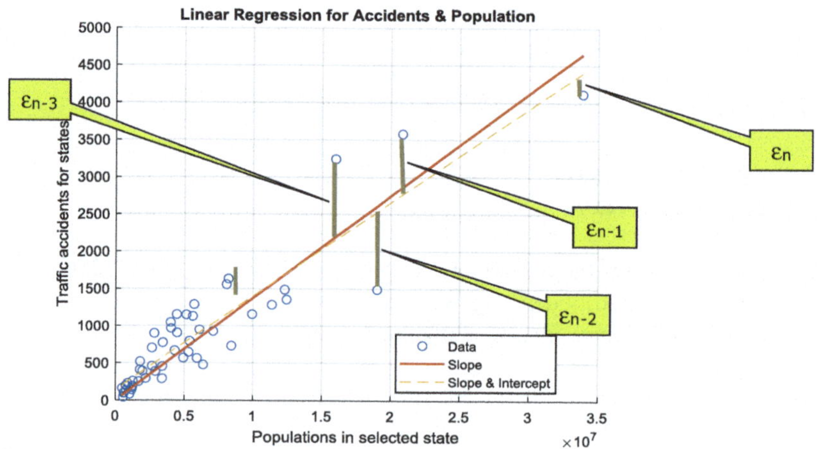

Fig. 5.5 Regression line used to fit the data in accidents dataset

Producing a fit using a linear model requires minimizing the sum of the squares of the residuals. This minimization yields a least-squares fit. You can gain insight into the *goodness* of a fit by visually examining a plot of the residuals. If the residuals plot has a certain pattern, it means that the residual data points do not appear to have a random scatter, while the randomness indicates that the model does not properly fit the data.

Evaluate each fit you make in the context of your data. For example, if your goal of fitting the data is to extract coefficients that have physical meaning, then it is important that your model reflect the physics of the data. Understanding what your data represents, how it was measured, and how it is modeled is important when evaluating the goodness of fit.

One measure of goodness of fit is the *coefficient of determination*, or R^2 (called R-Square). This statistic indicates how closely values you obtained from fitting a model match the dependent variable the model is intended to predict. Statisticians often define R^2 using the residual variance from a fitted model.

R^2 is one measure of how well a model can predict the data, and falls between 0 and 1. The higher the value of R^2, the better the model is at predicting the data.

Where $\hat{y}$ represents the calculated values of y and $\bar{y}$ is the mean of y, R^2 is defined as

$$R^2 = 1 - \frac{\sum_{i=1}^{n}\left(y_i - \widehat{y}_i\right)^2}{\sum_{i=1}^{n}\left(y_i - \bar{y}\right)^2} \tag{5.4}$$

By using Eq. (5.4), we can calculate two R^2 for two fitting lines, and they can be expressed by using MATLAB codes as:

```
Rsq1 = 1 - sum((y - yCalc1).^2)/sum((y - mean(y)).^2)      % Rsq1 =
0.822235650485566


Rsq2 = 1 - sum((y - yCalc2).^2)/sum((y - mean(y)).^2)      % Rsq2 =
0.838210531103428
```

One can see that R_{sq2} is better than R_{sq1}.

Next let us build a real example to illustrate how to use MATLAB Apps to perform regression algorithm for an actual project.

5.2.2   Linear Regression App Example

Now let us use a real project example to illustrate how to use Linear Regression App, Regression Learner, to estimate or predict the DC motor rotation speed (response) based on motor input voltage (predictor).

As we know, a DC motor is driven by input DC voltage and its rotating speed is proportional to the input voltage. Ideally, this input and output relation can be a linear function represented as:

$$y = Kx + b \tag{5.5}$$

where K is the slope and b is the intercept on the y-axis. In other words, y is the output or the motor rotating speed and x is the input DC voltage.

In order to estimate this linear function, exactly to find matched fitting parameters, K and b, we can use linear regression algorithm to build a regression model for this motor system.

In this study, we try to use a motor dataset, **Motor.xlsx**, which is an Excel file with two columns data, input motor voltage (V) and output motor rotating speed (RPM). Totally there are 100 rows or input data, and some of them will be used as training data, checking data, and validation data to build a linear regression model. Now we will use the Regression Learner App to build this linear regression model.

Perform the following operations to build this model:

1. Open MATLAB R2023a or later version and click the **App** tab, click on the drop-down arrow to find the **Regression Learner** App icon under the **MACHINE LEARNING AND DEEP LEARNING** category, as shown in Fig. 5.6. Click it to open this App.
2. By using the Regression Learner App, one can generate a new model or open an existing model. For our case, we need to create a new model. There are two ways to generate a new model:

   (a) **From Workspace**—locate the dataset from workspace.
   (b) **From File**—locate the dataset from a file.

   We prefer to use the second way since it is simple and easy, and you do not need to save your dataset into the workspace.

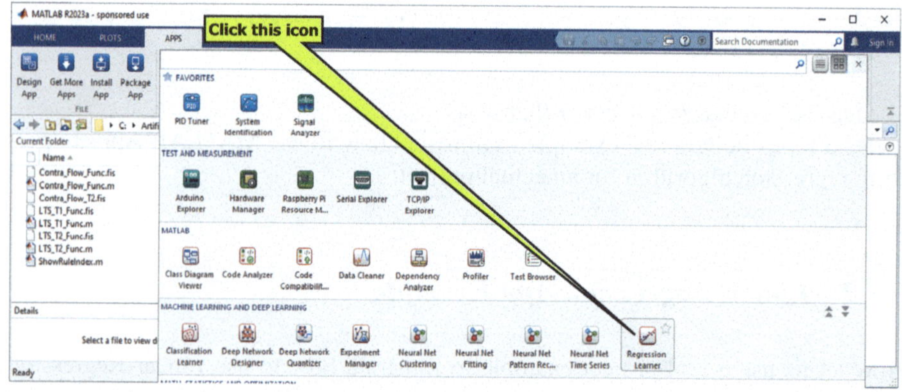

Fig. 5.6 The opened Statistics and Machine Learning App wizard

3. Click on the drop-down arrow in the **New Session** icon and select the **From File** item.

4. On the opened Select File explorer, browse to the folder where our motor dataset, **Motor.xlsx**, is stored. One can find this dataset from the Springer ftp site under the **Students\Datasets** folder. In our case, it is **C:\Artificial Intelligence Book\Students\Datasets**. Select that dataset and click on the **Open** button to open it.

5. Click the green-color check, as shown in Fig. 5.7, to import this dataset into our model.

6. In the next wizard, as shown in Fig. 5.8, we need to check and configure our imported dataset to make it as our desired source with the correct settings to perform this linear regression fitting process.

7. In Fig. 5.8, all details about our dataset are displayed, including the dataset name, the output variable, **Response**, which is motor speed, and a single input or called **Predicator**, under the **Data set** group. All configurations related to this regression process are shown under the **Validation** group on the right.

8. The so-called **Validation** is a process used to examine the predictive accuracy of the fitted models. It estimates model performance on new data compared to the training data, and helps you choose the best model. In fact, three kinds of validation methods can be used:

   (a) Cross-Validation
   (b) Holdout Validation
   (c) Resubstitution Validation

   The cross-validation can automatically divide all data in dataset into various folders to perform training, checking, and testing function as the regression is

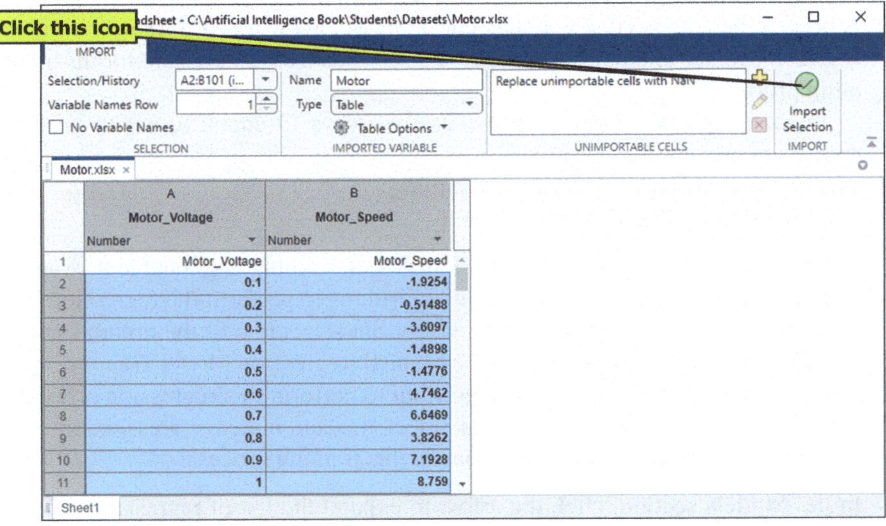

Fig. 5.7 The opened dataset Motor.xlsx

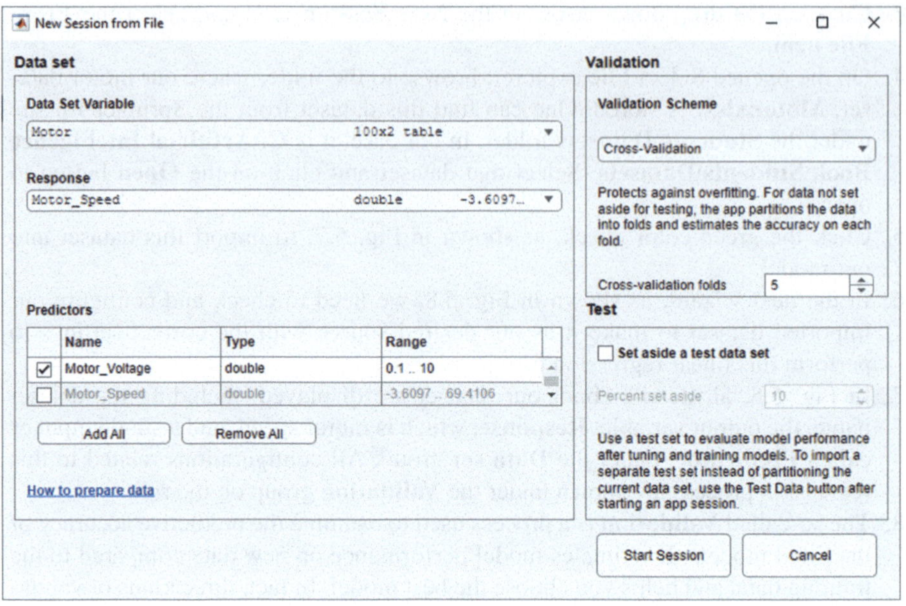

Fig. 5.8 The completed setup for imported dataset

performed. The holdout validation can also be used to train, check, and test your model, but the dataset must be separated into three parts manually by users. The third method does not provide any testing function.

For our validation purpose, we prefer to use cross-validation method.

9. Keep all default settings under the **Validation** group with no changes, as shown in Fig. 5.8, and click on the **Start Session** button to begin the regression process. A point to be noted is that you may check the **Set aside a test data set** checkbox if you like to use another or an additional set of data as testing data for this modeling process.

   In fact, regression learner provides two ways to enable users to perform regression process:

   (a) Automated Regression Model Training
   (b) Manual Regression Model Training

   The so-called Automated Regression Model Training is to enable the regression learner to select and try different algorithms to perform the regression process and display all possible results. Users can select one of the optimal results as their target model. However, for the Manual Regression Model Training, users can select one specific or desired algorithm to perform this regression process.

For the testing purpose, first let us select the automated regression training method. Perform the following operations in this training process:

1. In the **Models** section, click the arrow to expand the list of regression models. Select **All Quick-To-Train**, as shown in Fig. 5.9. This option trains all the model presets that are fast to fit.

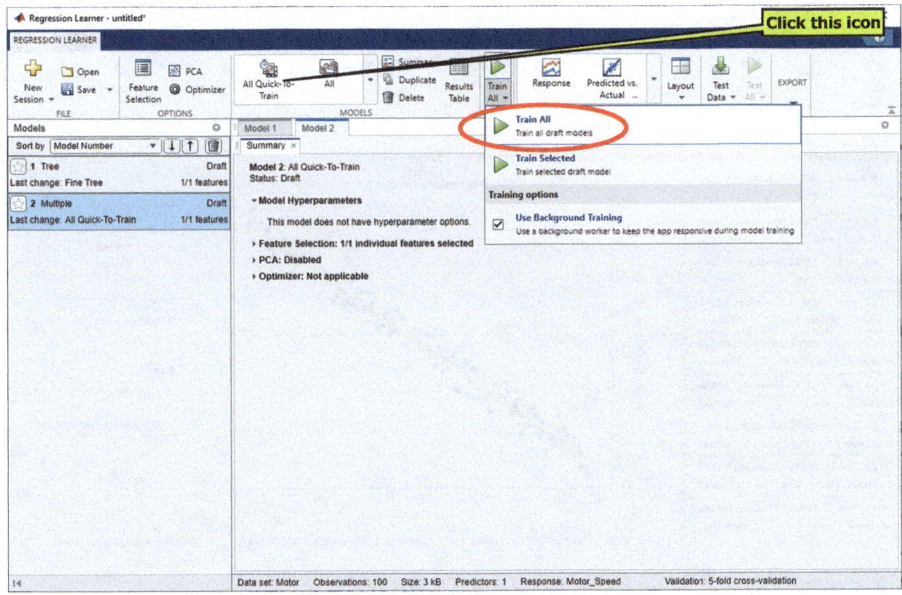

Fig. 5.9 The finished Start Session result

2. In the **Train** section, click on the drop-down arrow on the **Train All** and select **Train All**, as shown in Fig. 5.9.
3. The automated regression model training process starts, and all possible results or models are displayed in Fig. 5.10. The default resulted model is the optimal one with the validation method as Root Mean Square Error (RMSE), which is the second possible algorithm listed as 2.1 in Fig. 5.10.
4. As shown in Fig. 5.10, some other possible algorithms are shown for this regression process. Each algorithm is evaluated with a related validation error. Relatively speaking, the algorithm 2.1 has the smallest RMSE value.
5. Now click on the **Predicted vs. Actual** icon, as shown in Fig. 5.10, to show a comparison between the regression result and the actual data values, which is shown in Fig. 5.11. A good fitting should have all actual data points around the predicted line as closely as possible, which is true for our situation.
6. You can also check this regression model training with a summary and table format by clicking on the **Summary** and **Results Table** icons, respectively, as shown in Fig. 5.11.

It can be found from Fig. 5.11 that all different decision tree algorithms, including the coarse tree, medium tree, and fine tree, are trigged and used to train each related model. The linear regression algorithm is also involved into this all training processes. The standard to assess and evaluate these algorithms is the Root Mean Square Error (RMSE) resulted in the validation process.

Among them, the best or the smallest RMSE value is 3.0088, which is obtained by using the linear regression algorithm, compared with all other algorithms. For

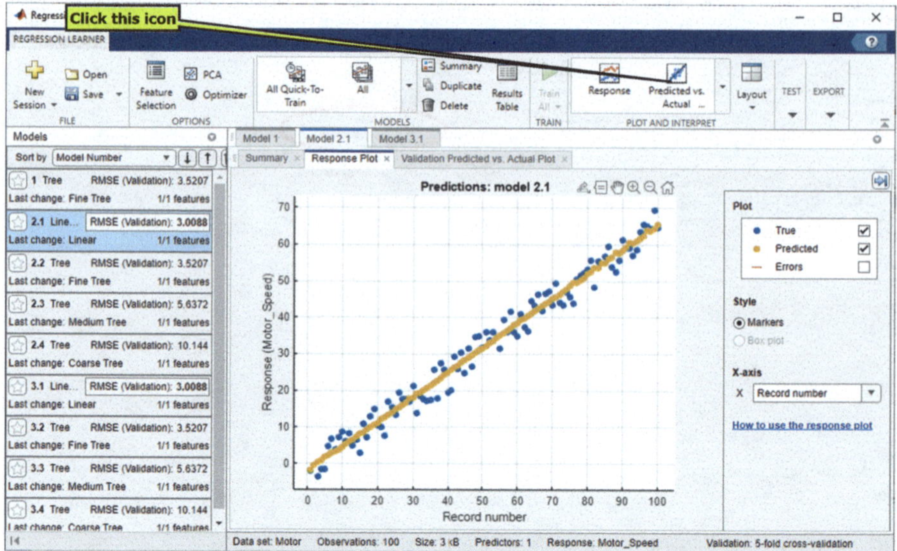

Fig. 5.10 The training result for the linear regression model

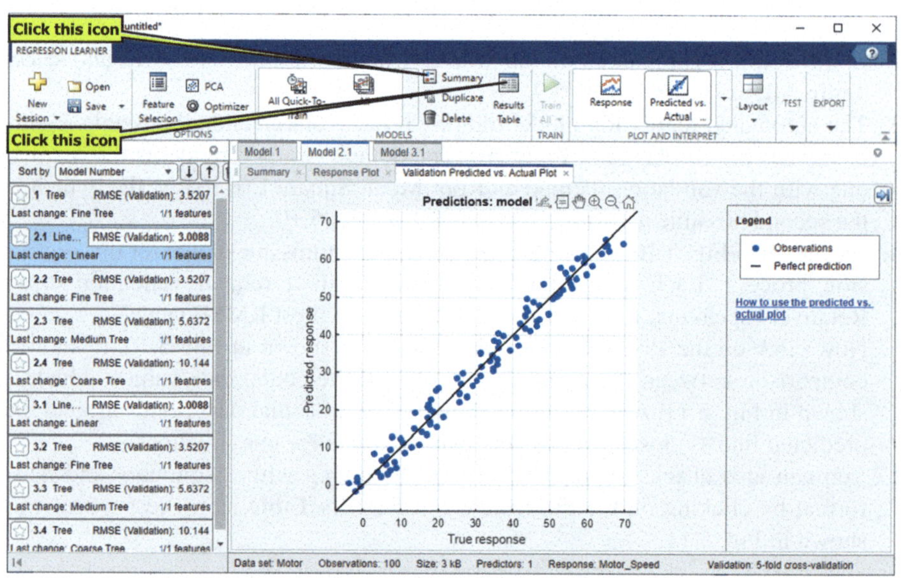

Fig. 5.11 A comparison between the regression result and the actual data

all other decision tree algorithms, the fine tree method provided the best result with an RMSE value 3.5207. The medium tree gave a good result with an RMSE value 5.6372, and the coarse tree released the worst result with an RMSE value 10.144.

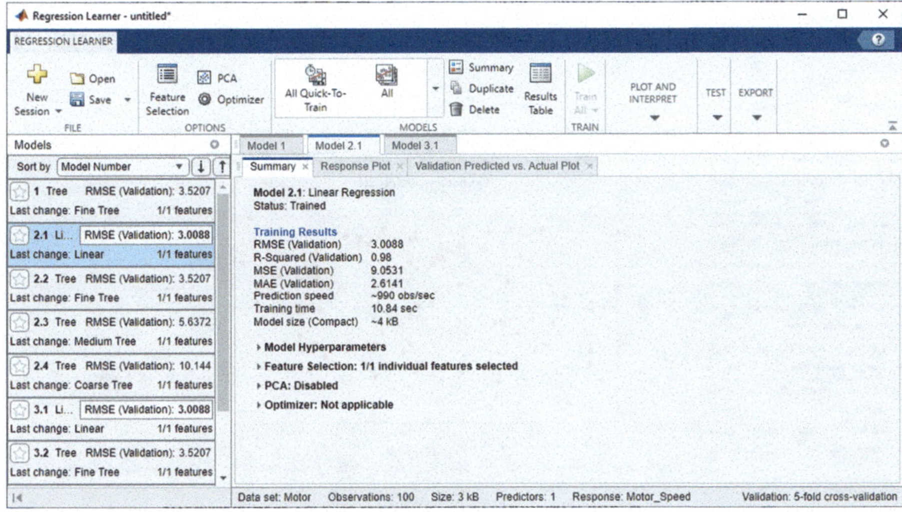

Fig. 5.12 The regression model training summary

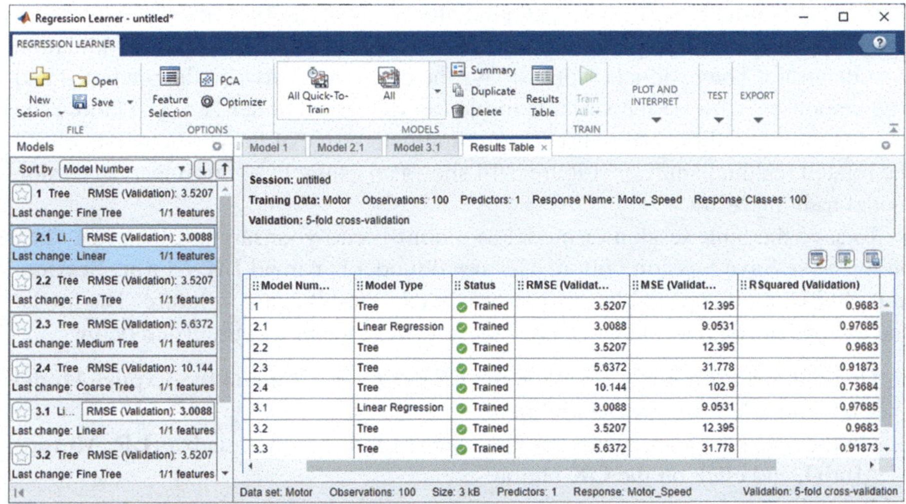

Fig. 5.13 The regression model training result in table format

The advantage of using Train-All method is that the method provides a good and easy way to enable users to try and test all training algorithms by one click, and compare them to select the optimal one.

The regression model training summary and result table are shown in Figs. 5.12 and 5.13. To view more details of training results for any other algorithms as shown in Fig. 5.12, users can click on the desired algorithm on the left pane one by one to show them.

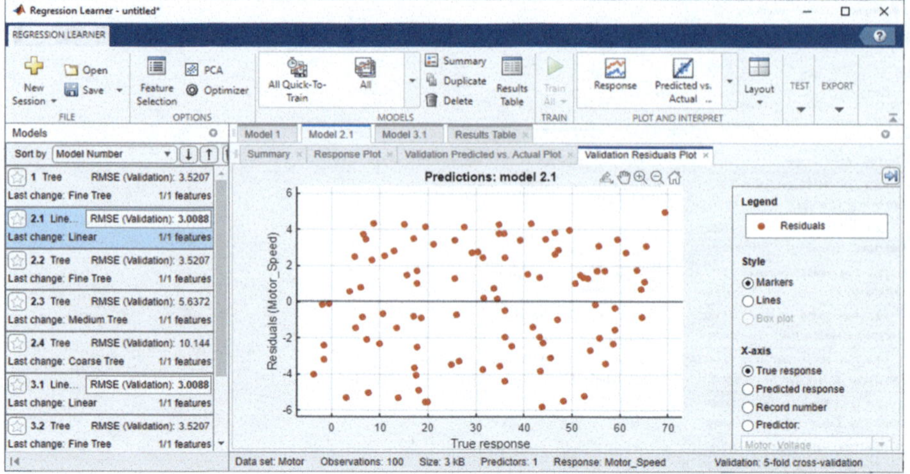

Fig. 5.14 The residual analysis

One can also use the residuals plot to check model performance. To view the residuals plot after training a model, click the arrow in the **Plot and Interpret** section to open the gallery, and then click **Residuals (Validation)** in the Validation Results group. The residuals plot displays the difference between the predicted and true responses. Choose the variable to plot on the x-axis under X-axis. Choose the true response, predicted response, record number, or one of the predictors, to check the related residual analysis. Figure 5.14 shows an example of our linear regression model training result.

To save the built session or model as a **.mat** file to your target folder, just click on the **Save\Save Session** button. The saved model is named **LR_Motor_Session. mat**, and it can be opened later when you need to use it again in future. The model can also be stored into the workspace by following the procedure listed below:

1. Click on the drop-down arrow on the **EXPORT** icon and on the **Export Model** icon, and select **Export Model** item.
2. On the opened wizard, enter a desired name for this model, such as **LR_Motor_ Model**, and click on the **OK** button.
3. Immediately the exported model can be found in the workspace.

It can be found that it is so easy and convenient to enable users to use this App to perform desired linear regression by using different algorithms, and compare their performances in one GUI.

After finishing this model building process via regression learner with linear regression algorithm, one can find that this process is so easy and simple, and enable us to train, check, and build this model quickly and easily. However, any good thing may bring some bad points, which is true for our project. The main issue is when users use this App to build this model, a lot of detailed coding and developing processes are missed. How can we allow users to know and use detailed codes to make this process more professionally and flexibly? The answer is to use functions provided by MATLAB library.

5.2.3   Regression Learner Functions

In Statistics and Machine Learning Toolbox, MATLAB provided a huge collection for various functions used to help readers to design, build, and develop regression models. By using those functions, more professional and flexible models can be developed. Some popular and important functions provided by MATLAB are shown in Table 5.1.

Now let us use the function **fitlm()** to illustrate how to fit our motor control data via a linear regression model. Perform the following operations to build this regression model:

1. Open MATLAB and create a new Script file named **LR_Motor_Func.m**.
2. Enter the codes shown in Fig. 5.15 into this file.

Let us have a closer look at this piece of codes to see how it works.

1. First the name of the dataset file to be used is given. The point to be noted is that a full name of the used dataset must be provided including the path. You need to use your specified path to replace our path if you saved the dataset in different folder.
2. A MATLAB function, **readtable()**, is used to retrieve all data from the dataset, and assign them to a local variable **ds**. This function is an updated one, and it is

Table 5.1 Some popular and important regression fitting functions

Function Name	Descriptions
fitlm()	Fit linear regression model
fitnet()	Fit a neural network
fitnlm()	Fit nonlinear regression model
fitrnet()	Train neural network regression model
fitcnet()	Train neural network classification model
fitglm()	Create generalized linear regression model
fitrlinear()	Fit linear regression model to high-dimensional data

```
% Using functions to perform linear regression
% Input: fire dataset
% Output: Regressed model

1  fileName = 'C:\\Artificial Intelligence Book\\Students\\Datasets\\Motor.xlsx';
2  ds = readtable(fileName);
3  m_mdl = fitlm(ds, 'linear')
4  plot(m_mdl);
   grid;
```

Fig. 5.15 The codes used for the fitting linear regression model

```
m_mdl =

Linear regression model:
   Motor_Speed ~ 1 + Motor_Voltage

Estimated Coefficients:
                    Estimate        SE        tStat              pValue
                    ________      _______    ________           _________

   (Intercept)      -2.0105       0.59775    -3.3635            0.0010992
   Motor_Voltage     6.7485       0.10276    65.671             7.9284e-83

Number of observations: 100, Error degrees of freedom: 98
Root Mean Squared Error: 2.97
R-squared: 0.978,  Adjusted R-Squared: 0.978
F-statistic vs. constant model: 4.31e+03, p-value = 7.93e-83
```

Fig. 5.16 The running result of linear regression algorithm

used to replace some older functions, such as **xlsread()** that is not recommended, to retrieve a text file, an excel sheet, a HTML, or a Word file.

3. The linear fitting function **fitlm()** is called with the dataset file and fitting type as arguments to build the linear regression model. This function has multiple constructors with one, two, three, or four input arguments. The default type is linear if not clearly indicated for the regression type. When this function is executed without any semi-colon at the end, all regression model parameters will be displayed in the Command window.
4. The **plot()** function is called to plot the regression model with a comparison with the original or actual data distribution.

Now save this Script file as **LR_Motor_Func.m** and run it. The running result is shown in Command window, as shown in Fig. 5.16. The Coefficient property includes these columns:

- **Estimate**: Coefficient estimates for each corresponding term in the model. For example, the estimate for the constant term (intercept) is -2.0105.
- **SE**: Standard error of the coefficients.
- **tStat**: t-statistic for each coefficient to test the null hypothesis that the corresponding coefficient is zero against the alternative that it is different from zero, given the other predictors in the model. Note that **tStat = Estimate/SE**. For example, the t-statistic for the intercept is $-2.0105/0.59775 = -3.36345$.
- **pValue**: p-value for the t-statistic of the two-sided hypothesis test.

The summary statistics of the model shown at the bottom include:

- **Number of observations**: Number of rows without any NaN values. For example, Number of observations is 100 means that no NaN value existed for any observation, where the number of rows in X is 100.
- **Error degrees of freedom**: $n-p$, where n is the number of observations, and p is the number of coefficients in the model, including the intercept. For example, the model has 2 predictors, including the intercept. So the Error degree of freedom is $100-2 = 98$.

- **Root mean squared error:** Square root of the mean squared error, which estimates the standard deviation of the error distribution.
- **R-squared and Adjusted R-squared:** Coefficient of determination and adjusted coefficient of determination, respectively. For example, the R-squared value suggests that the model explains approximately 98% of the variability in the response variable Y.
- **F-statistic vs. constant model:** Test statistic for the F-test on the regression model, which tests whether the model fits significantly better than a degenerate model consisting of only a constant term.
- **p-value:** p-value for the F-test on the model. For example, the model is significant with a p-value of 0.

The plotting result that is used to create an added variable plot (partial regression leverage plot) for the whole model is shown in Fig. 5.17. The fitting equation is $y = -2.0105 + 6.7485x$.

The fitting equation can be derived from the running result shown in Fig. 5.16. In the **Estimate** column, the intercept value $b = -2.0105$ and the slope value $K = 6.7485$ represent the values for the estimated linear equation $y = Kx + b$.

In fact, when linear regression algorithm performs fitting process, by default it always uses the data in the last column in the input dataset as the response or output data. Therefore, one may need to arrange the output data in the dataset in the last column, and this may be a trick for the beginners.

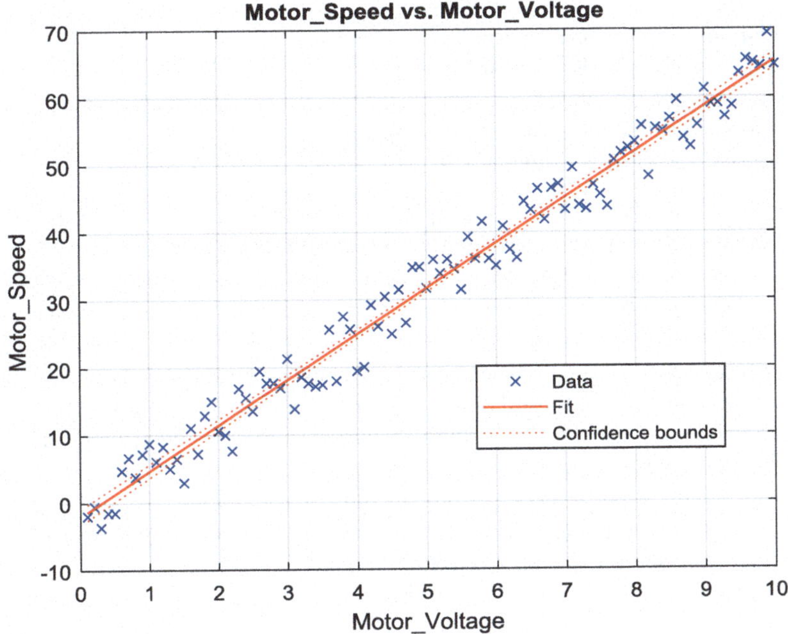

Fig. 5.17 The plotted regression model result

5.2.4 Multiple Linear Regression App Modeling

Multiple linear regression (MLR), also known simply as multiple regression or multivariate regression, is a statistical technique that uses several explanatory variables to predict the outcome of a response variable. The goal of multiple linear regression is to model the linear relationship between the explanatory (independent) variables and response (dependent) variables. In essence, multiple regression is the extension of ordinary least-squares (OLS) regression because it involves more than one explanatory variable [2].

The differences between linear and multiple linear regression algorithms include [3]:

- Multiple regression is a broader class of regressions that encompasses linear and nonlinear regressions with multiple explanatory variables.
- Whereas linear regression only has one independent variable impacting the slope of the relationship, multiple regression incorporates multiple independent variables.

A typical multiple linear regression algorithm can be expressed by using the following equation:

$$y = \beta_0 + \beta_1 x_1 + \beta_2 x_2 + \ldots + \beta_n x_n + \varepsilon \tag{5.6}$$

where for all observations,

- y = response or dependent variable
- x_i = input or independent variables
- β_0 = y-intercept
- β_i = slope coefficients for each input variable
- ε = model's error or residuals

The purpose or the function of a multiple linear regression algorithm is to estimate or predict all of those parameters for Eq. (5.6), including y-intercept β_0, slope coefficients β_i, and residuals ε based on a set of given inputs and output variables to make the difference between the regressed response and the actual output as smaller as possible.

In other words, the so-called multiple linear regression models are to derive the response variable based on more than one predictor variable. You can perform multiple linear regressions with or without the Linear Model object, or by using the Regression Learner App or functions.

Now let us use a real project example to illustrate how to use Linear Regression App, Regression Learner, to estimate or predict the dangerous degree of forest fires to be occurred based on various environment conditions.

The Canadian Forest Fire Danger Rating System (CFFDRS) is the national system for rating fire danger in Canada. The Canadian Forest Fire Weather Index (FWI) System is a subsystem of the CFFDRS and has been in usage since 1970. The FWI

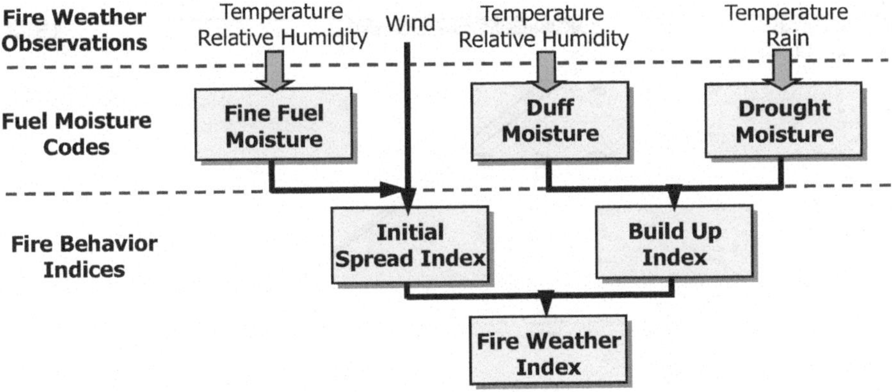

Fig. 5.18 The structure of Fire Weather Index (FWI) system

System is composed of six components with two functional layers, as shown in Fig. 5.18, three fuel moisture codes, and three fire behavior indexes.

On the top layer, three fuel moisture codes, Fine Fuel Moisture Code (FFMC), Duff Moisture Code (DMC), and Drought Code (DC), are used to provide some quantitative descriptions for fuel moisture and drought degree on the studied forest areas. On the bottom layer, three index components, Initial Spread Index (ISI), Build Up Index (BUI), and Fire Weather Index (FWI), provide the qualitative descriptions with Fire Behavior Indices to indicate the possibility or the dangerous degree of fire to be occurred.

The FWI is a combination of ISI and BUI, and is a numerical rating of the potential frontal fire intensity. In effect, it indicates fire intensity by combining the rate of fire spread with the amount of fuel being consumed. In fact, the FWI value is a key parameter or an indicator and it could be used to indicate how dangerous and in what degree a forest wildfire would be occurred.

In this study, we try to use a popular fire dataset, **forestfires.csv**, which is provided by **UCI Machine Learning Repository** [4]. However, some modifications are needed for that dataset since there is no FWI value in that original dataset. A modified dataset, **MFire_Database.xls**, is generated by adding an additional column, FWI, into the dataset. It contains 14 independent variables working as inputs, such as temperature, wind, rainfall, FFMC, ISI, BUI, and DC, and one dependent variable FWI as the output.

Totally there are 518 rows or input data, and some of them will be used as training data, checking data, and validation data to build a multiple linear regression model.

Perform the following operations to build this model:

1. Open MATLAB R2023a or later version and click the **App** tab, click on the drop-down arrow to find the **Regression Learner** App icon under the **MACHINE LEARNING AND DEEP LEARNING** category, as shown in Fig. 5.19. Click it to open this App.

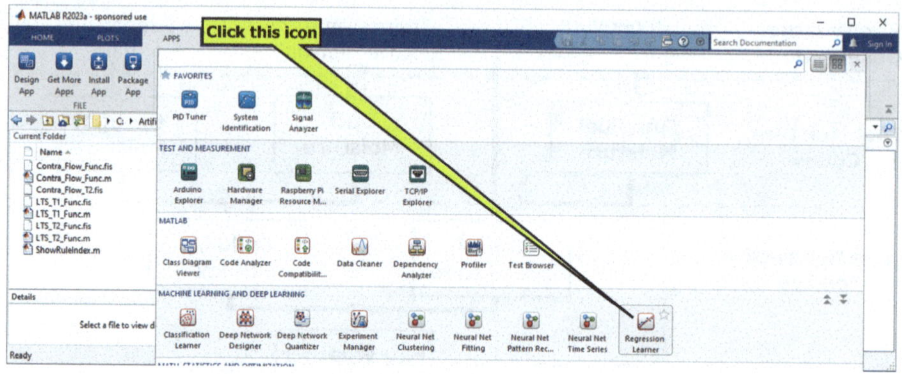

Fig. 5.19 The opened Statistics and Machine Learning App

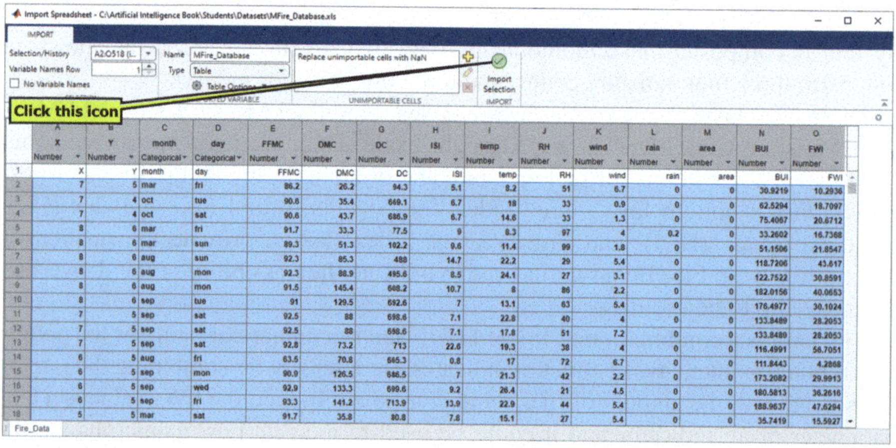

Fig. 5.20 The opened dataset MFire_Database.xls

2. By using the Regression Learner App, one can generate a new model or open an existing model. For our case, we need to create a new model. There are two ways to generate a new model:

   (a) **From Workspace**—locate the dataset from workspace.
   (b) **From File**—locate the dataset from a file.

   We prefer to use the second way since it is simple and easy, and you do not need to save your dataset into the workspace.

3. Click on the drop-down arrow in the **New Session** icon and select the **From File** item.

4. On the opened Select File explorer, browse to the folder where your modified dataset, **MFire_Database.xls**, is stored. One can find this dataset from the Springer ftp site under the **Students\Datasets** folder. In our case, it is **C:\ Artificial Intelligence Book\Students\ Datasets**. Select that dataset and click on the **Open** button to open it. The opened dataset is shown in Fig. 5.20.

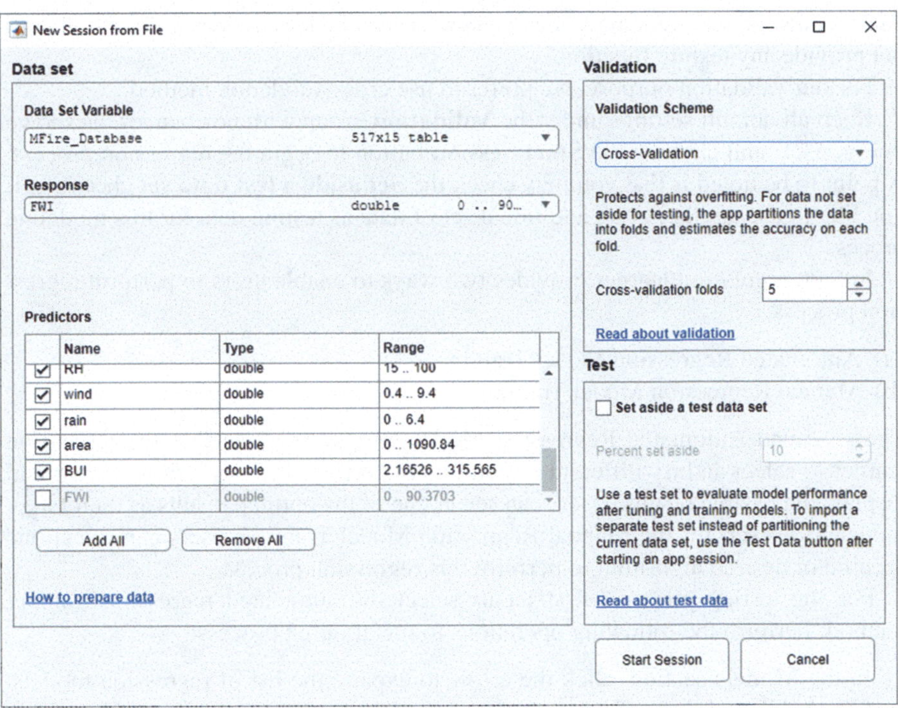

Fig. 5.21 The completed setup for imported dataset

5. Click on the green-color check mark, as shown in Fig. 5.20, to import this dataset to our model.
6. In next wizard, as shown in Fig. 5.21, we need to check and configure our imported dataset to make it as our desired source with the correct settings to perform this linear regression fitting process.
7. In Fig. 5.21, all details about our dataset are displayed, including the dataset name, the output variable, **Response**, which is FWI, and all 14 inputs or called **Predicators**, under the **Data set** group. All configurations related to this regression process are shown under the **Validation** group on the right.

The so-called **Validation** is a process used to examine the predictive accuracy of the fitted models. It estimates model performance on new data compared to the training data, and helps you choose the best model. In fact, three kinds of validation methods can be used:

(a) Cross-Validation
(b) Holdout Validation
(c) Resubstitution Validation

The cross-validation can automatically divide all data in dataset into various folders to perform training, checking, and testing function as the regression is performed. The holdout validation can also be used to train, check, and test your model, but the dataset must be separated into three parts manually by users. The third method does not provide any testing function.

For our validation purpose, we prefer to use cross-validation method.

Keep all default settings under the **Validation** group with no changes, as shown in Fig. 5.21, and click on the **Start Session** button to begin the regression process. A point to be noted is that you may check the **Set aside a test data set** checkbox if you like to use another or an additional set of data as testing data for this modeling process.

In fact, regression learner provides two ways to enable users to perform regression process,

(a) Automated Regression Model Training.
(b) Manual Regression Model Training.

The so-called Automated Regression Model Training is to enable the regression learner to select and try different algorithms to perform the regression process and display all possible results. Users can select one of the optimal results as their target model. However, for the Manual Regression Model Training, users can select one specific or desired algorithm to perform this regression process.

For the testing purpose, first let us select the automated regression training method. Perform the following operations to this training process:

1. In the **Models** section, click the arrow to expand the list of regression models. Select **All Quick-To-Train**, as shown in Fig. 5.22. This option trains all the model presets that are fast to fit.
2. In the **Train** section, click on the drop-down arrow on the **Train All** and select **Train All**, as shown in Fig. 5.22.

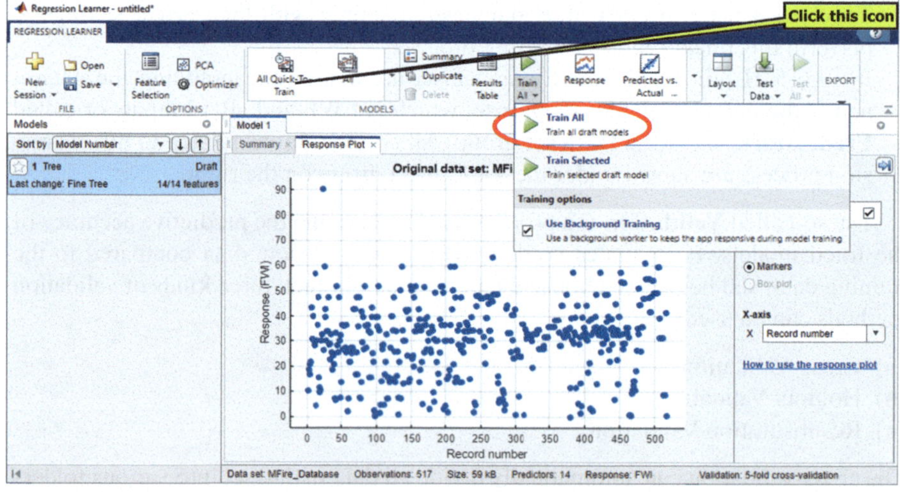

Fig. 5.22 The finished Start Session result

3. The automated regression model training process starts, and all possible results or models are displayed in Fig. 5.23. The default resulted model is the optimal one with the validation method as Root Mean Square Error (RMSE), which is the second possible algorithm listed as 2.1 in Fig. 5.23.

4. As shown in Fig. 5.23, five possible algorithms are shown for this regression process. Each algorithm is evaluated with a related validation error. Relatively speaking, the algorithm 2.2, that is a Tree mapping algorithm, has the smallest RMSE value.

5. Still keep selecting our linear regression algorithm since we are discussing it, and click on the **Predicted vs. Actual** icon, as shown in Fig. 5.23, to show a comparison between the regression result and the actual data values, which is shown in Fig. 5.24. A good fitting should have all actual data points around the predicted line as closely as possible, which is true for our situation.

6. You can also check this regression model training with a summary and table format by clicking on the **Summary** and **Results Table** icons, respectively, as shown in Fig. 5.24.

The regression model training summary and result table are shown in Figs. 5.25 and 5.26.

Under the X-axis, choose the true response, predicted response, record number, or one of the predictors, to check the related residual analysis. Figure 5.27 shows an example for our linear regression model training result.

To save the built session or model as a **.mat** file to your target folder, just click on the **Save\Save Session** button. The saved model is named **MLR_Fire_Session.**

Fig. 5.23 The training result for the linear regression model

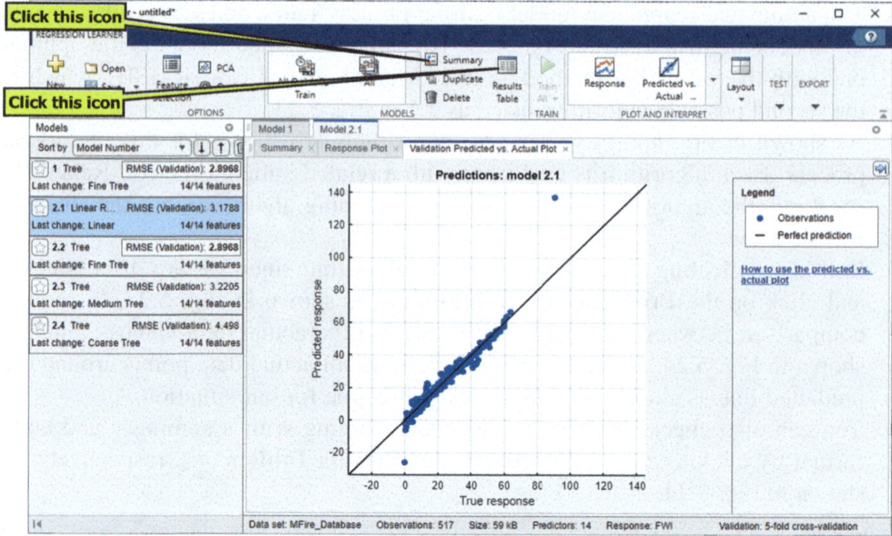

Fig. 5.24 A comparison between the regression result and the actual data

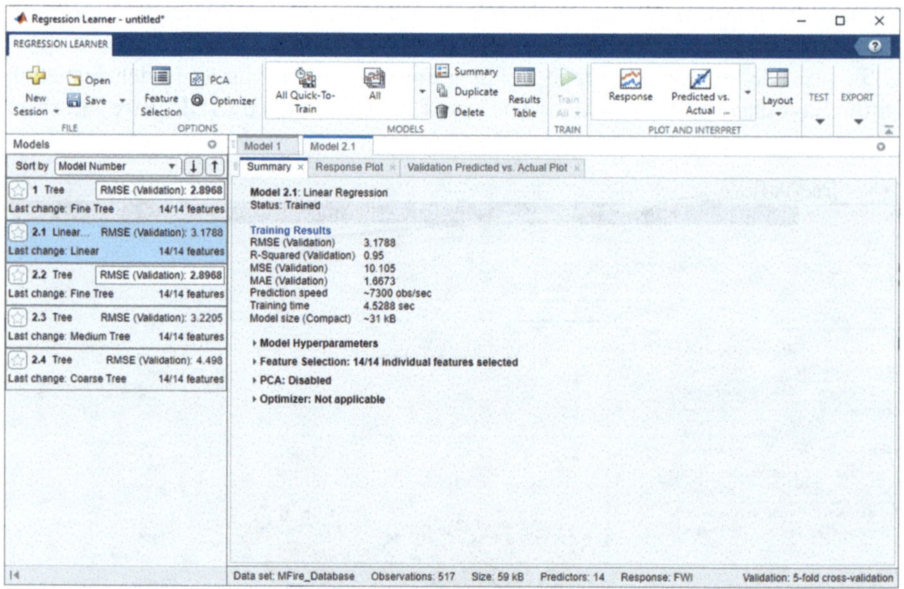

Fig. 5.25 The regression model training summary

mat and it can be opened later when you need to use it again in future. The model can also be stored into the workspace by following the procedure listed below:

1. Click on the drop-down arrow on the **EXPORT** icon and on the **Export Model** icon, and select **Export Model** item.

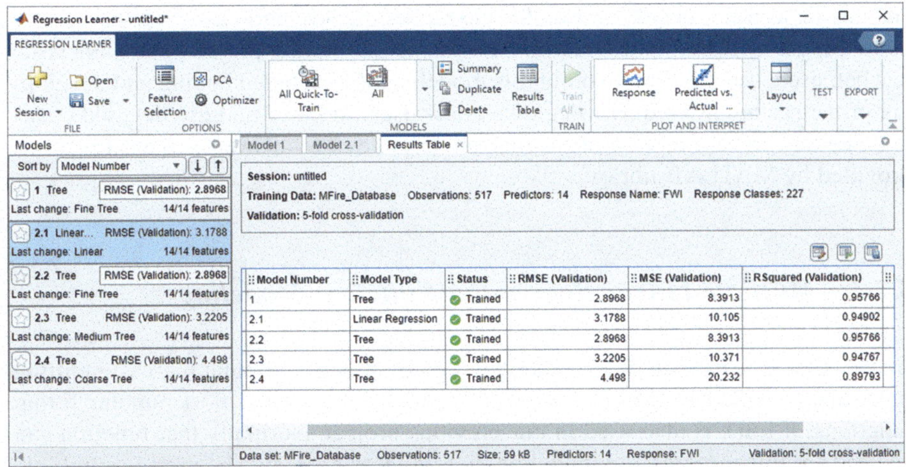

Fig. 5.26 The regression model training result in table format

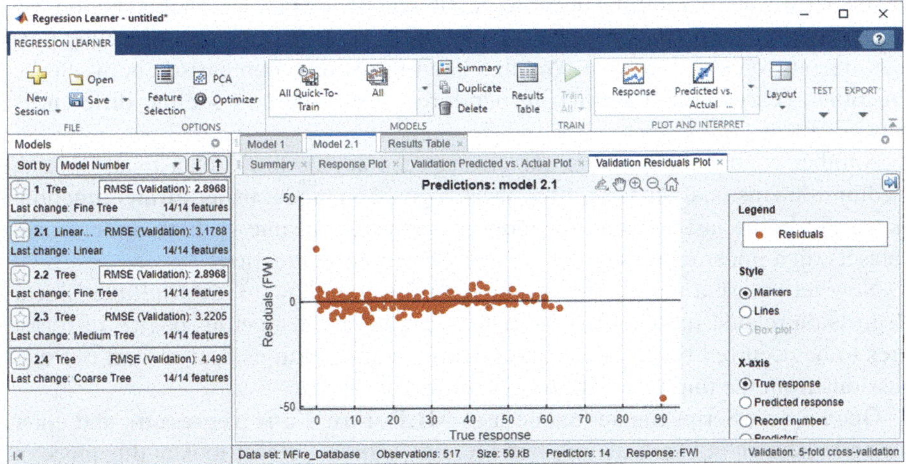

Fig. 5.27 The residual analysis

One can use the residuals plot to check model performance. To view the residuals plot after training a model, click the arrow in the **Plot and Interpret** section to open the gallery, and then click **Residuals (Validation)** in the Validation Results group. The residuals plot displays the difference between the predicted and true responses. Choose the variable to plot on the *x*-axis.

2. On the opened wizard, enter a desired name for this model, such as **MLR_Fire_ Model**, and click on the **OK** button. Immediately the exported model can be found in the workspace.

After finishing this model-building process via regression learner with multiple linear regression algorithm, one can find that this process is so easy and simple, and enables us to train, check, and build this model quickly and easily. However, any good thing may bring some bad points, which is true for our project. The main issue is when users use this App to build this model, a lot of detailed coding and developing processes are missed. How can we allow users to know and use detailed codes to make this process more professional and flexible? The answer is to use functions provided by MATLAB library.

5.2.5 Multiple Linear Regression Function Modeling

As we discussed in Sect. 5.2.3, exactly in Table 5.1, some popular regression fitting functions provided by MATLAB are discussed. One of the most popular fitting functions, **fitlm()**, is also used in our previous project. Normally that function can be used for linear regression modeling, but, in fact, it can also be used for multiple linear regression models.

In addition to that function, MATLAB provided another function, **regress()**, to be specially used for multiple linear regression modeling process. In this section, we try to use it as our main function to handle multiple linear regression modeling process for our forest fire project. We also like to provide a comparison between the functions, **regress()** and **fitlm()**, in our forest fire project to see the differences between them.

Another point to be noted is, for greater accuracy on low-dimensional through medium-dimensional datasets, fit a linear regression model using **fitlm()** function, as we did in the last section. For reduced computation time on high-dimensional datasets, fit a linear regression model using **fitrlinear()** function.

Now let us use a real project example to illustrate how to use Multiple Linear Regression related functions to estimate or predict the dangerous degree of forest fires to be occurred based on various environment conditions. Let us start our discussions from the implementation of function **regress()**.

Open a new Script file and name it as **MLR_Fire_Func_regress.m**, and enter the codes shown in Fig. 5.28 into that file. Let us have a closer look at this piece of codes to see how it works.

```
  % using regress() function to do multiple linear regression
1 load MFire_Database.mat ds

2 x1 = ds.FFMC;
  x2 = ds.DMC;
  x3 = ds.temp;
  x4 = ds.wind;
3 y = ds.FWI;

4 X = [ones(size(x1)) x1 x2 x3 x4];
5 [b, ~, ~, ~, stats] = regress(y, X)
```

Fig. 5.28 The codes used for regression modeling with regress() function

1. The fire dataset **MFire_Database** has two versions, **.xls** and **.mat**. The former is an Excel file and the latter is a MATLAB data file. The reason for those versions is to make our development processes more convenient and flexible.

   In this project, we used the second version, **MFire_Database.mat**. This dataset can be found from the Springer site under the folder **Students\Datasets**. Make sure to copy and paste that dataset into one of your local folders, and build this project in the same folder. Otherwise, you may need to use MATLAB Command **cd** to indicate your folder if you store this dataset in another location in your machine. First we need to load our dataset.

2. Because we do not want to use all input variables to do this multiple linear regression process; thus, we only select four of them, **FFMC**, **DMC**, **temp**, and **wind**, as our inputs. Since that dataset was saved in a dataset format and its original name is **ds**, thus we need to use that dataset name with dot operator to access each variable.

3. The output or dependent variable is **FWI** and is assigned to a response variable y.

4. Four input or independent variables are built as a table with M intercepts (value = 1) as the first column, so totally this table contains 5 columns. The number of rows M can be determined by the function **ones(size(x1))**, which is a $M \times 1$ table or a vector with M rows whose number is the size of **x1**. In fact, all columns in the dataset have the same size or same number of rows.

5. The multiple linear regress function **regress()** is executed with input and output tables. The returned components by calling this function include the estimates for five coefficients stored in the vector **b** with some statistics result, including the residuals, F-Statics and p-value, and an estimate of the error variance.

Save this file as **MLR_Fire_Func_regress.m**, and run it. The running result is shown below:

```
b =
       -76.2683
         0.9145
         0.0829
         0.6476
         0.6534
 stats = 0.6124   202.2406     0.0000     77.2664
```

This running result is equivalent to a multiple linear regression estimated equation,

$$y = -76.2683 + 0.9145x1 + 0.0829x2 + 0.6476x3 + 0.6534x4 \tag{5.7}$$

The returned residual value R^2 is 0.6124 and the p-value of 0.0000 is less than the default significance level of 0.05; a significant linear regression relationship exists between the response y and the predictor variables in X.

```
% Using fitlm() function to perform multiple linear regression
% Input: fire dataset
% Output: Regressed model
1  load MFire_Database.mat ds

2  x1 = ds.FFMC;
   x2 = ds.DMC;
   x3 = ds.temp;
   x4 = ds.wind;
3  y = ds.FWI;

4  X = [ones(size(x1)) x1 x2 x3 x4];
5  f_mdl = fitlm(X, y)
6  plot(f_mdl);
   grid;
```

Fig. 5.29 The codes used for the regression modeling with fitlm() function

Now let us use **fitlm**() function to do a similar multiple linear regression modeling and make a comparison between this function with the function **regress**() we used in the previous project.

Open another new Script file and name it as **MLR_Fire_fitlm.m**, and enter the codes shown in Fig. 5.29 into that file. Let us have a closer look at this piece of codes to see how it works.

1. First, as we did for the last project, we need to load our fire dataset as a **.mat** file with the original dataset name **ds**.
2. Then we need to reassign four independent or input variables (columns), such as **FFMC, DMC, temp**, and **wind**, as inputs to train this regression model.
3. Then choose **FWI** as the response or output variable y.
4. Organize the input table X as five columns and intercept with four selected input variables.
5. The linear regress function, **fitlm**(), is called to perform this multiple linear regress modeling process. Unlike this function used in the previous project, this time we used another constructor for this function, in which both the input and the output variables work as two arguments.
6. A functional plotting is performed to see a possible regression result. In fact, for this kind of multiple linear regression process, the plotting should be a multidimensional graphic result, which is too complicated to be printed out. Here we just use it to do some basic illustration.

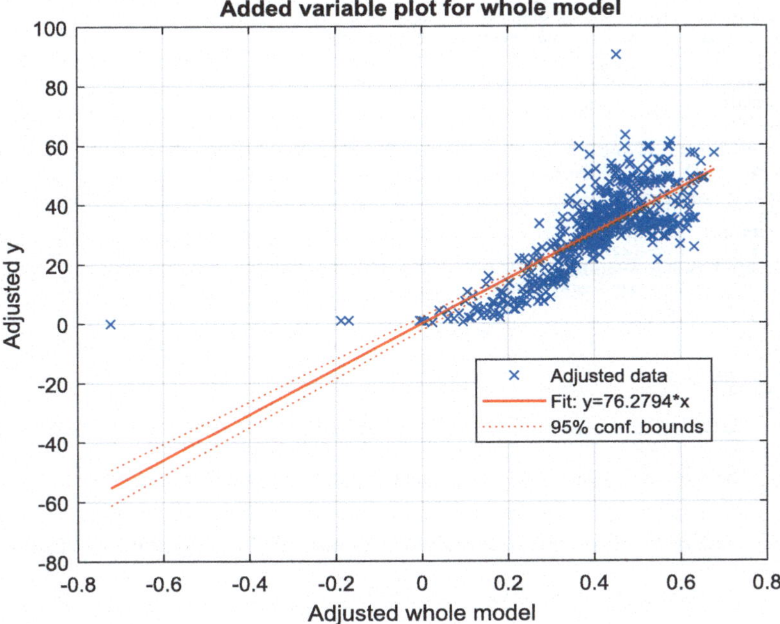

Fig. 5.30 The multiple linear regression result with fitlm() function

Now run this file and the running result is shown below. The plotting result is shown
in Fig. 5.30.

```
f_mdl = Linear regression model: y ~ 1 + x1 + x2 + x3 + x4 + x5
Estimated Coefficients:
                Estimate          SE              tStat             pValue
                ________          ________        ________          ________

(Intercept)  0                 0                 NaN               aN
x1             -76.268          6.7546           -11.291           1.5059e-26
x2             0.9145           0.080034          11.426           4.3866e-27
x3             0.082857         0.0070261         11.793           1.4781e-28
x4             0.64759          0.081295          7.9659           1.0737e-14
x5             0.65341        w 0.22251           2.9366           0.0034679


Number of observations: 517, Error degrees of freedom: 512
Root Mean Squared Error: 8.79
R-squared: 0.612,  Adjusted R-Squared: 0.609
F-statistic vs. constant model: 202, p-value = 6.66e-104
```

```
% using regress() function to do multiple linear regression
% The dataset MFire_Database.mat should be in desired folder

     load MFire_Database.mat ds
 1   x1 = ds.DC;
     x2 = ds.BUI;
 2   y = ds.FWI;

 3   X = [ones(size(x1)) x1 x2];
     [b, ~, ~, ~, stats] = regress(y,X)

 4   scatter3(x1,x2,y,'filled')
 5   hold on
 6   x1fit = min(x1):5:max(x1);
     x2fit = min(x2):5:max(x2);
 7   [X1FIT,X2FIT] = meshgrid(x1fit, x2fit);
 8   YFIT = b(1) + b(2)*X1FIT + b(3)*X2FIT;

 9   mesh(X1FIT,X2FIT,YFIT)
10   xlabel('DC')
     ylabel('BUI')
     zlabel('FWI')
11   view(40,30)
12   hold off
```

Fig. 5.31 The modified codes used to plot a 3D plotting for multiple linear regression

The regressed model can be equivalent to a multiple linear equation expressed as:

$$y = -76.268 + 0.9145x1 + 0.082857x2 + 0.64759x3 + 0.65341x4 \qquad (5.8)$$

Compared with Eq. (5.7), it can be found that two regression methods or two regression functions derive a similar result. Generally, we prefer to use the second function **fitlm()** since it provides more flexibility and more detailed results.

In Fig. 5.30, it only shows a simple relation between the model and the original input data. To get a much clearer and straightforward mapping between some inputs and the output, we can use a three-dimensional plotting to illustrate a multiple linear regression modeling result, but the number of inputs must be limited to two variables with the third dimension as the output.

Now let us modify the codes shown in Fig. 5.28 by changing some codes shown in Fig. 5.31 to try to provide a complete description among inputs-output relationship in a 3D format. Two input variables, **DC** and **BUI**, which are located at columns 7 and 14 in the original dataset, are used for this fitting. The output variable is still **FWI**.

Let us have a closer look at this piece of attached codes to see how it works.

1. After loading the dataset, two input variables, **DC** and **BUI**, are selected and assigned to two new variables, x_1 and x_2, respectively. Of course, you can select any other two input variable for this fitting purpose.

2. The response or output variable, **FWI**, is selected and assigned to the output variable y.

3. For our plotting purpose, only two input or independent variables are built as a table with 2 intercepts (value = 1) as the first column, so totally this table contains 3 columns. The number of rows M can be determined by the function **ones(size(x1))**, which is a $M \times 1$ table or a vector with M rows whose number is the size of x_1. In fact, all columns in the dataset have the same size or same number of rows.

4. The MATLAB function **scatter3()** is used to plot a 3D plotting with circles at the locations specified by 3D coordinates, say x, y, and z. The input argument **'filled'** indicates that those circles should be filled with solid colors.

5. The Command **hold on** in MATLAB is used to retain plots in the current axes so that new plots added to the axes do not delete existing plots. Because we need to perform multiple plot actions to complete this kind of complicated operation, we need to use that command to hold our current and previous plotting with no changes.

6. Also we need to determine the lower and upper limits for both input variables in two axes on this plotting; both **min()** and **max()** functions are used for those purposes. Depending on those limits, the interval or increment amount on both variables' axes can be determined. In our case, a value of 5 is selected.

7. The function **[X, Y] = meshgrid(x, y)** is used to return 2D grid coordinates based on the coordinates contained in vectors x and y. X is a matrix where each row is a copy of x, and Y is a matrix where each column is a copy of y. The grid represented by the coordinates X and Y has **length(y)** rows and **length(x)** columns.

8. The regression estimation is executed by running a multiple linear regression equation represented by a linear combination of two inputs with an intercept constant.

9. The function **mesh()** is used to create a mesh plot, which is a three-dimensional surface that has solid edge colors and no face colors. The function plots the values in matrix Z as heights above a grid in the x-y plane defined by X and Y. The edge colors vary according to the heights specified by Z.

10. Three labels used to display on three axes, **DC**, **BUI**, and **FWI**, are assigned to three label functions.

11. For your convenience, the function **view(az, el)** is used to set the azimuth and elevation angles of the camera's line of sight for the current axes. In our case, we used 40-degree as the angle of rotation around the z-axis, and 30-degree as the minimum angle between the line of sight and the x-y plane.

12. The command **hold off** sets the hold state to off so that new plots added to the axes clear existing plots and reset all axes properties.

Save this Script file as a name of **Plot_MLR.m** to your local folder in your machine.
   Now run this project and the regression model parameters are shown below, and the 3D plotting is shown in Fig. 5.32.

```
Plot_MLR

b =

   10.3874
    0.0078
    0.1124

stats =

   0.4845   241.5541      0.0000   102.3621
```

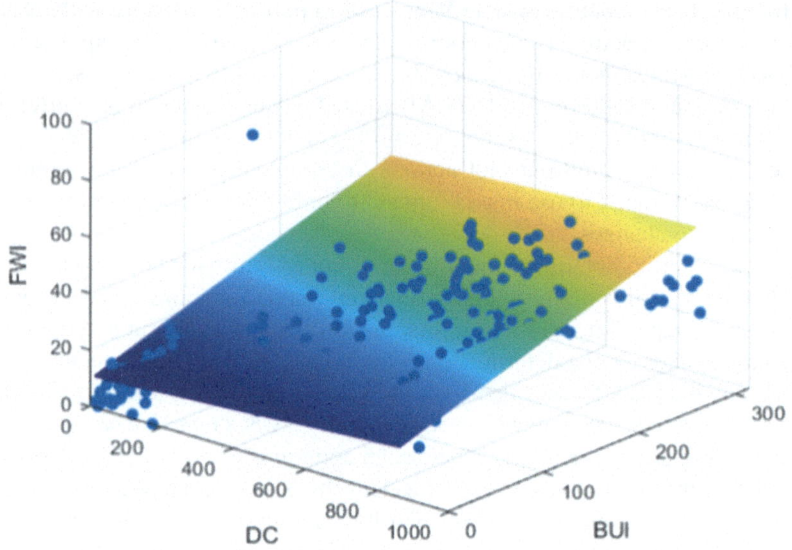

Fig. 5.32   The finished 3D plotting for two input variables

The regression estimation is equivalent to a linear equation,

$$y = 10.3874 + 0.0078\,x1 + 0.1124\,x2 \qquad (5.9)$$

The returned residual value R^2 is 0.4845 and the p-value of 0.0000 is less than the default significance level of 0.05; a significant linear regression relationship exists between the response y and the predictor variables in X.

5.3   Decision Tree-Related Apps and Functions

As we discussed in Sect. 2.6 in Chap. 2, a decision tree is a decision support hierarchical model that uses a tree-like model of decisions and their possible consequences, including chance event outcomes, resource costs, and utility. It is one way to display an algorithm that only contains conditional control statements.

In decision analysis, a decision tree and the closely related influence diagram are used as a visual and analytical decision support tool, where the expected values or expected utility of competing alternatives are calculated. A decision tree consists of three types of nodes:

1. Decision nodes—typically represented by squares
2. Test or chance nodes—typically represented by circles
3. End nodes—typically represented by triangles

Regularly, decision trees can be divided into two major types, named

- Regression Trees
- Classification Trees

The term classification and regression tree (CART) analysis is an umbrella term used to refer to either of the above procedures, first introduced by Breiman et al. in 1984 [5]. Trees used for regression and trees used for classification have some similarities, but also some differences, such as the procedure used to determine where to split [5].

Basically speaking, Regression Tree analysis is when the predicted outcome can be considered as a real number, such as the price of a house, or a patient's length of stay in a hospital. While Classification Tree analysis is when the predicted outcome is the class or a discrete value to which the data belongs [6].

In this section, we will concentrate our discussions on regression trees only. As for the classification trees, we will discuss it on the following sections.

A typical regression tree is to evaluate or assess the quality for some selected products, such as vehicles, or simply some popular cars. Figure 5.33 shows a regression tree used to evaluate the quality of some selected cars based on certain specifications.

In Fig. 5.33, the evaluation criteria include two specifications:

- Number of cylinders—x_1
- Acceleration—x_2

This tree is used to predict the response based on two predictors, x_1 (number of cylinders) and x_2 (acceleration). To predict, start at the top node. At each node, check the values of the predictors to decide which branch to follow. When the branches reach a leaf node, the response (evaluation score) is set to the value corresponding to that node.

The Statistics and Machine Learning Toolbox provided App and functions for both types of decision trees, regression and classification tree. First let us concentrate our discussions on the regression tree-related Apps and functions.

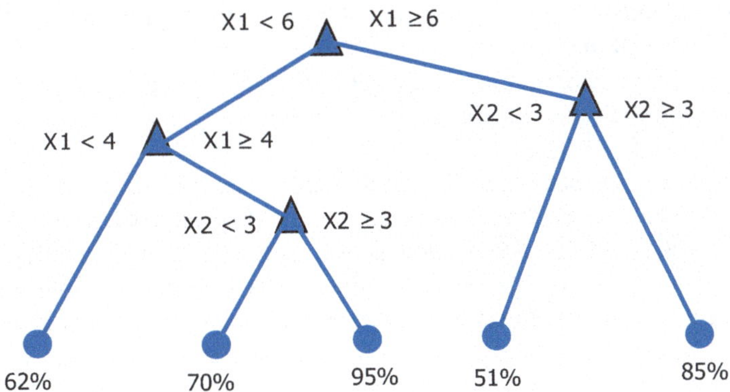

Fig. 5.33 A typical regression tree used to evaluate the quality of cars

5.3.1 Regression Tree-Related Apps

All decision tree-related Apps are involved in the Regression Learner App, which includes regression trees and classification trees. In this section, we limited our discussions on the regression trees Apps; the regression tree-related functions will be discussed in the next section.

To make our discussions easier and simpler, we will use a dataset **carbig** that is provided by MATLAB. This dataset contains characteristics of different big car models produced from 1970 through 1982, and it includes the following eight variables with eight columns:

1. Acceleration—**Acceleration**
2. Number of cylinders—**Cylinders**
3. Engine displacement—**Displacement**
4. Engine power—**Horsepower**
5. Model year—**Model_Year**
6. Weight—**Weight**
7. Country of origin—**Origin**
8. Miles per gallon—**MPG**

The highlighted items above are the actual columns' names on the dataset. In fact, the dataset contains 406 records in 406 rows, which is a 406×8 matrix.

For convenience purpose, we changed the name for that dataset and made it as an Excel format as **Cartable.xls**. One can find this dataset in the Springer ftp site under the folder **Students\Datasets**. You can save this dataset to one of your local folders and use it later. Now let us build our project **RT_Car_App** with Regression Learner and use regression tree to estimate our optimal model.

Open MATLAB and Regression Learner by clicking on the **APPs** tab and **Regression Learner** icon. Perform the following operations to build this model:

1. Click on the drop-down arrow on **New Session** icon, then select **From File** item to create a new regress tree session from our dataset, **Cartable.xls**.
2. On the opened **Select File to Open** wizard, browse to the folder where the dataset **Cartable.xls** is located, select it and click on the **Open** button. The opened dataset is shown in Fig. 5.34.
3. All eight variables (columns) in the dataset are displayed. Click on the **Import Selection** button (green-color circle) shown in Fig. 5.34 to import this dataset into our model.
4. The imported dataset with all default regression configurations is shown in Fig. 5.35. The dataset name is shown under the **Data Set Variables** box, which is a 406×8 table. The **Response** has been selected as the last column, **MPG**, whose data type is double. Since we do not need to change those default settings, just click on the **Start Session** button to begin the regress tree modeling process. The default validation option is cross-validation, to protect against over-fitting. Regression Learner creates a plot of the response with the record number on the x-axis, as shown in Fig. 5.36.

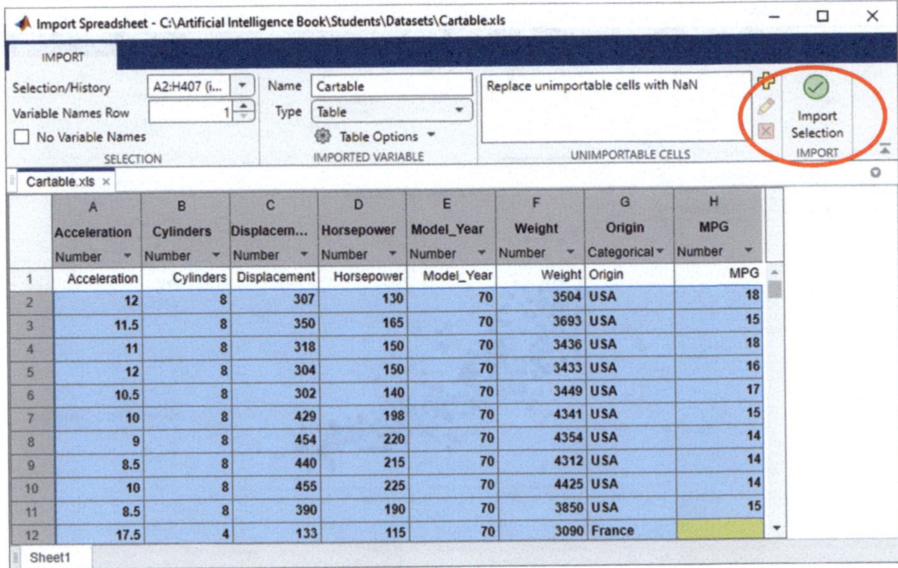

Fig. 5.34 The opened dataset Cartable.xls

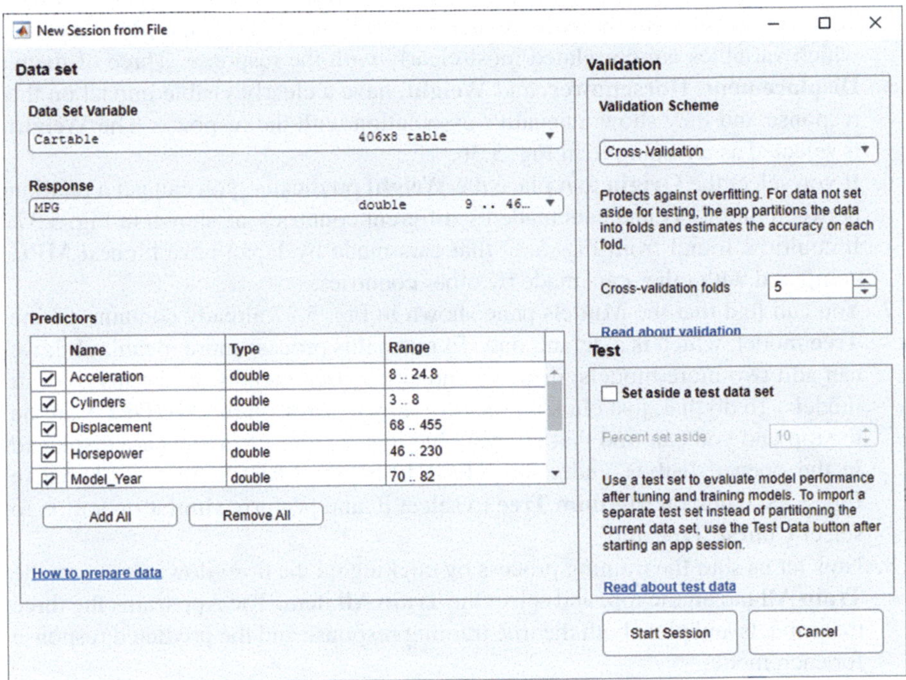

Fig. 5.35 The imported dataset and default settings for regression tree model

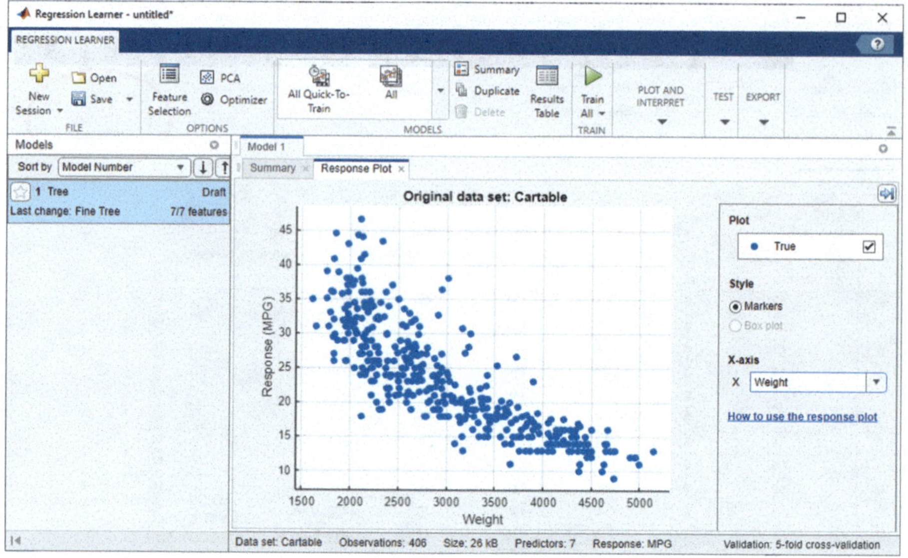

Fig. 5.36 The plotted relation between MPG and weight of the car

5. To visualize the relation between different predictors and the response, select different variables in the **X** list under **X-axis** to the right of the plot. Observe which variables are correlated most clearly with the response. Three of them, **Displacement**, **Horsepower**, and **Weight**, have a clearly visible impact on the response and they show a negative association with the response. The **Weight** is selected as an example in Fig. 5.36.

6. If you select the **Origin** to replace the **Weight** predicator, you can get a relation between the **MPG** and cars made by different countries, as shown in Fig. 5.37. It could be found from Fig. 5.37 that cars made by Japan have highest **MPG** compared with other cars made by other countries.

7. You can find that the **Models** pane shown in Fig. 5.37 already contains a **Fine Tree** model, which is a default one. To make this process more meaningful, we can add two more models, medium and coarse tree models, to the list of draft models. To do that, just click on the drop-down arrow on the **MODELS** tab on the top and you can find that all available regression tree models are displayed in the opened listbox under the **REGRESSION TREES** title, as shown in Fig. 5.38. Click on **Medium Tree** to select it, and perform similar operation to select **Coarse Tree**, too.

8. Now let us start the training process by clicking on the drop-down arrow on the **Train All** tab on the top, and select the **Train All** item. The App trains the three tree models and plots both the true training response and the predicted response for each model.

9. In the **Models** pane, click on the drop-down arrow on the right of **Sort by** box and check the **RMSE (Validation)** (validation root mean squared error) of the

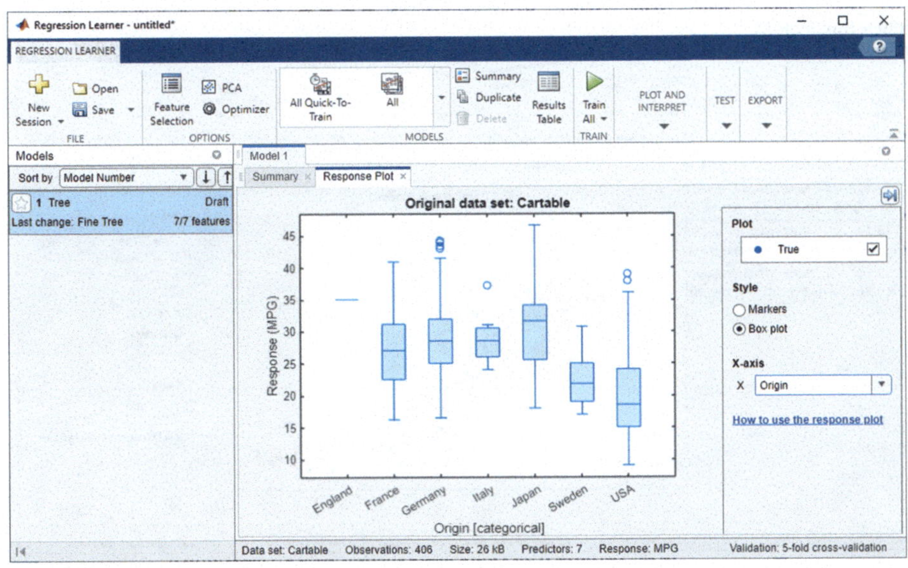

Fig. 5.37 Relation between MPG and cars made by different countries

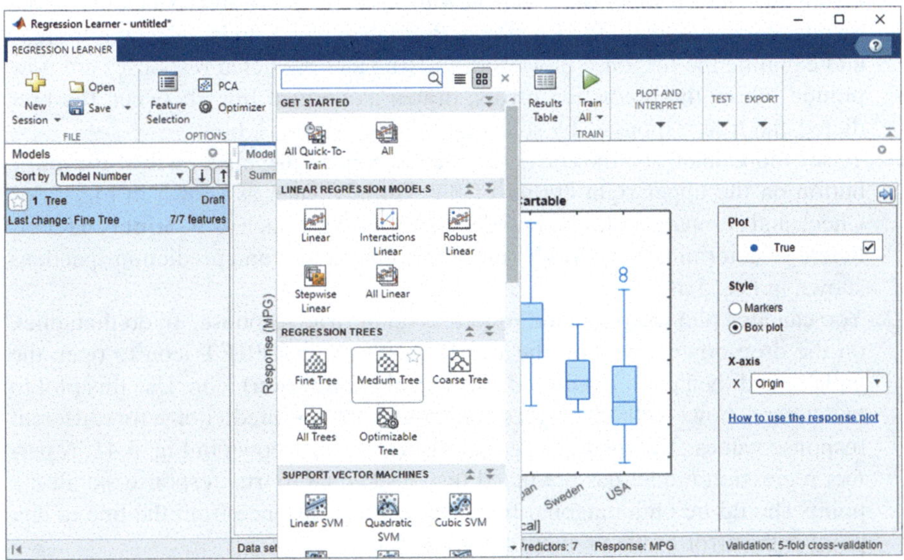

Fig. 5.38 Open the Medium and Coarse tree models

models. The best score is highlighted in a box. The **Fine Tree** and the **Medium Tree** have similar RMSEs (3.3133 and 3.3217), while the **Coarse Tree** is less accurate (3.8621), as shown in Fig. 5.39.

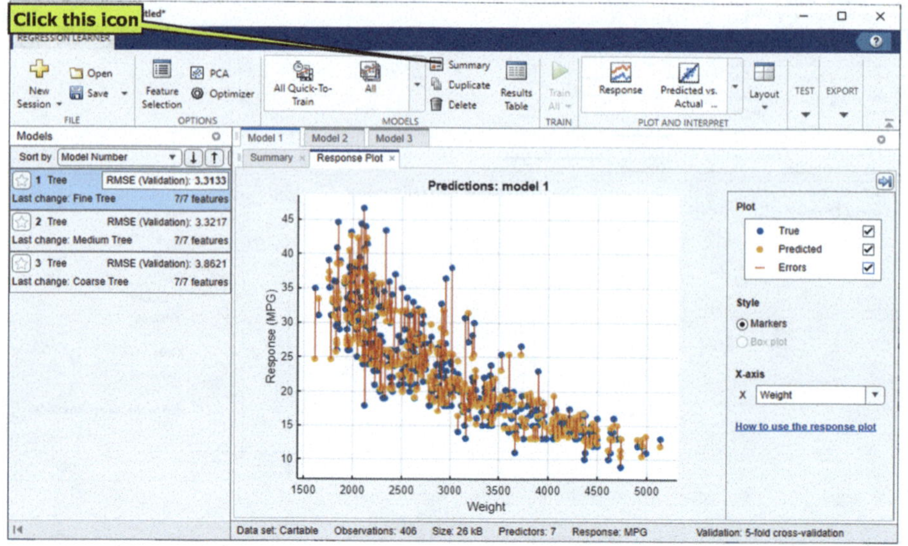

Fig. 5.39 The selected best training result with the smallest RMSE value

10. Choose a desired model, such as **Medium Tree**, in the **Models** pane to view the results of that model. In the **Response Plot** tab, under **X-axis**, select the **Weight** and examine the response plot. Both the true and predicted responses are now plotted. Show the prediction errors, drawn as vertical lines between the predicted and true responses, by selecting the **Errors** check box.

11. To see more details on the currently selected model, just click on the **Summary** button on the upper right corner of the **Models** pane, as shown in Fig. 5.39. Check and compare additional model characteristics, such as **R-Squared** (coefficient of determination), **MAE** (mean absolute error), and prediction speed, as shown in Fig. 5.40.

12. You can also plot the predicted response versus true response. To do that, click on the drop-down arrow on the **PLOT AND INTERPRET** icon to open the gallery, and then click **Predicted vs. Actual (Validation)** icon. Use this plot to understand how well the regression model makes predictions for different response values. The displayed comparison result is shown in Fig. 5.41. A perfect regression model has predicted response equal to true response, so all the points should lie on a diagonal line. The vertical distance from the line to any point is the error of the prediction for that point.

13. You can also compare the responses among multiple models used for the training process. To do that, just click on the drop-down arrow on the **Layout** button and select **Compare models** item to open two models panel, as shown in Fig. 5.42. Both responses for model1 and model3 are displayed in that opened panel. Click on the hide option button, as shown in Fig. 5.42, to enlarge both responses.

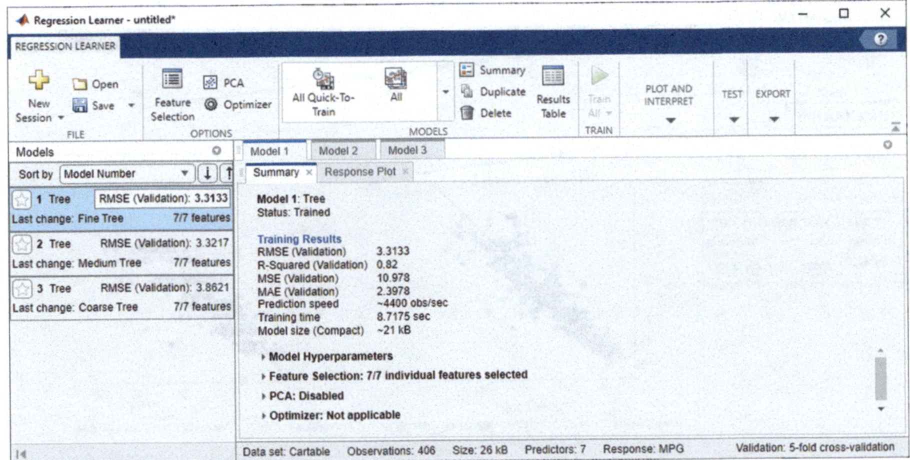

Fig. 5.40 Summary information of the selected regression tree modeling

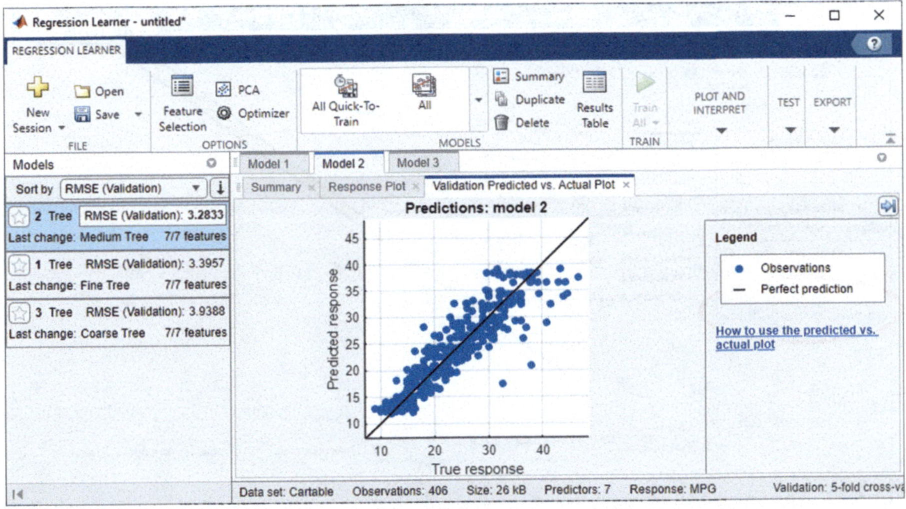

Fig. 5.41 The comparison between predicted response and true response

14. To try to improve the tree models, click on the drop-down arrow on the **MODELS** pane to open all trees listbox, and select **All Trees** icon, as shown in Fig. 5.43.

15. A **Multiple** tree model is added into our **Models** pane at the bottom in the left pane, which is indicated by a circle mark in Fig. 5.43.

16. To improve the tree models, click on the **Feature Selection** icon in the **OPTIONS** tab group to open the **Default Feature Selection** wizard, as shown in Fig. 5.44. In our case, there are two options available, **MRMR** and **F Test**,

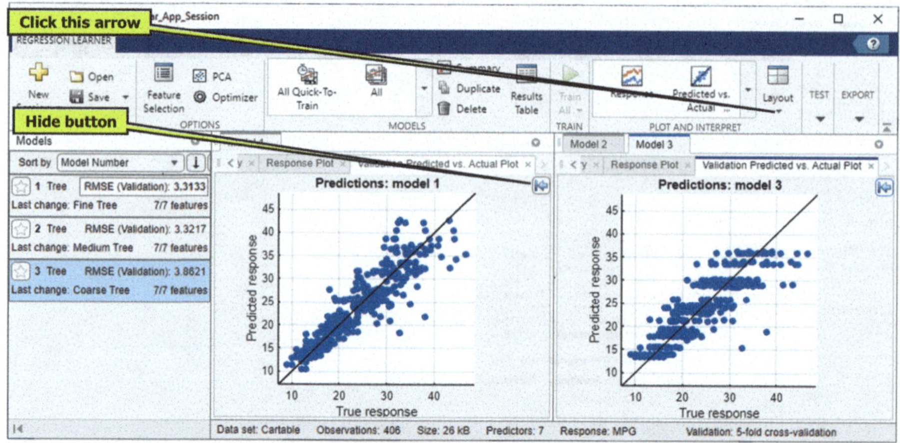

Fig. 5.42 The opened multiple models panel used to compare two models

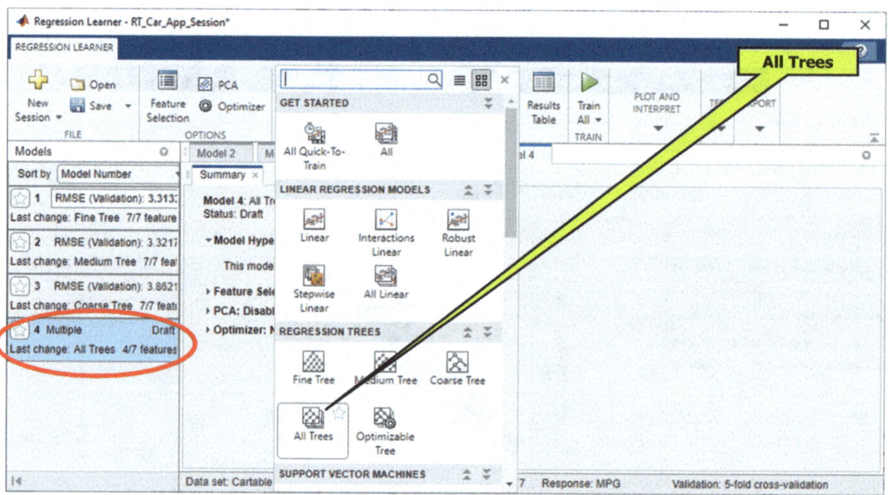

Fig. 5.43 The opened all trees listbox

which are feature ranking algorithms to rank the acceleration and country of origin predictors the lowest. The **RReliefF** option is disabled because the predictors include a mix of numeric and categorical variables.

17. Under **Feature Ranking Algorithm**, click **F Test**. Under **Feature Selection**, keep the default selection, **Select the highest ranked features**, to avoid bias in the validation metrics. Specify to keep 4 of the 7 features for model training, as shown in Fig. 5.44.

18. Click **Save and Apply** button and the App applies the feature selection changes to the current draft model and any new models created using the **Models** gallery.

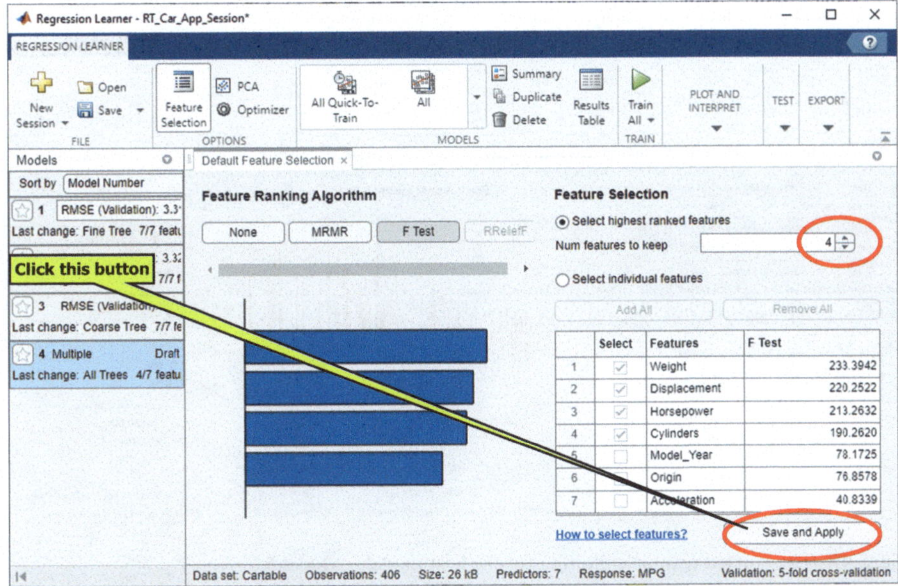

	Select	Features	F Test
1	✓	Weight	233.3942
2	✓	Displacement	220.2522
3	✓	Horsepower	213.2632
4	✓	Cylinders	190.2620
5	☐	Model_Year	78.1725
6	☐	Origin	76.8578
7	☐	Acceleration	40.8339

Fig. 5.44 The opened Feature Selection wizard

19. Now we can train the tree models using the reduced set of features. Click on the drop-down arrow in the **Train All** tab and select the **Train All** icon to start the training process.

20. Observe three new models in the **Models** pane. These models are the same regression trees as before but trained using only 4 of 7 predictors. The App displays how many predictors are used. To check which predictors are used, click a model in the **Models** pane, and note the check boxes in the expanded **Feature Selection** section of the model **Summary** tab.

21. To check the training errors, click on the drop-down arrow in the **PLOT AND INTERPRET** pane and select the **Residuals** icon. The residuals plot is shown in Fig. 5.45. Check the **Lines** radio button under the **Style** to display residuals in the line format. You can try to use different variables under the **X-axis** to check different responses.

22. Relatively speaking, the predicted response with the best model, Model 5.1, is not as good as that of Model 1 since only 4 predicators are used for the selected feature models, Models 5.1–5.3.

23. You can export a full or compact version of the selected model to the Workspace. To do that, click on the drop-down arrows on the **Export** icon and the **Export Model** icon and select the **Export Model** item. Enter **RT_Car_Model** into the model name box, and click on the **OK** button. To exclude the training data and export a compact model, clear on the check in the **Include training data in the exported model** checkbox. You can still use the compact model for making predictions on new data. The Command Window displays information about the results.

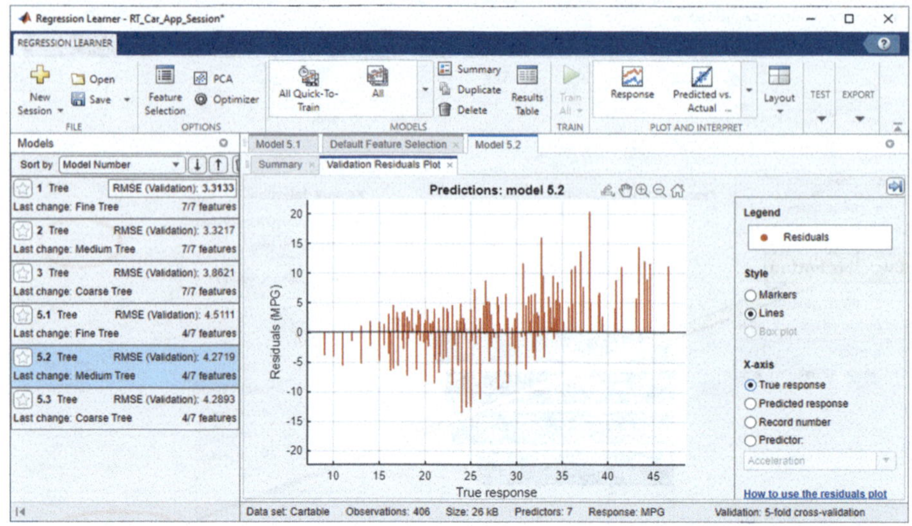

Fig. 5.45 The opened residuals plotting for the best model—Model 5.2

Now that we have finished the training for a regression tree model and exported that trained model to Workspace, next we need to test or evaluate this model with some new data tables. After you export a model to the workspace from Regression Learner, or run the code generated from the App, you get a trained model structure that you can use to make predictions using new data. The structure contains a model object and a function for prediction. The structure enables you to make predictions for models that include principal component analysis (PCA).

To do that model testing or model predicting correctly, the following two points must be kept in mind:

1. A new testing data table must be generated first with the same variable names and format as those in the training data table. Also the order of those variables must be the same as that in the training table.
2. A special function, **predictFcn()**, must be used and it is exported with the trained model together to the Workspace as you exported the model. The calling format for that function is

```
yfit = trainedModel.predictFcn(T);
```

where **yfit** is the testing or predicting result, and the **trainedModel** should be replaced by your actual trained model's name. In our case, it is **RT_Car_Model**.

Based on the above two points, let us build our testing codes to do our predicting in terms of a new data table **T**. Create a new Script file and enter the codes shown in Fig. 5.46 into that file.

```
% Create a new data table to test trained regression tree model with pediction function
1  Acceleration = 18;
   Cylinders = 8;
   Displacement = 320;
   Horsepower = 150;
   Model_Year = 70;
   Weight = 3804;
   Origin = "USA";
2  T = table(Acceleration, Cylinders, Displacement, Horsepower, Model_Year, Weight, Origin)
3  yfit = RT_Car_Model.predictFcn(T)
```

Fig. 5.46 The codes used to test or predict the result based on the trained model

Let us have a closer look at that piece of codes to see how it works.

1. First let us build a new data table **T** with seven input variables. The point to be noted is that all variable names must be identical to those used in the training table. A set of new testing data values are assigned to those testing variables.
2. A new testing data table **T** is generated and the order of all data columns must be same as that in the training data table.
3. The **predictFcn()** is executed with our new table as argument to estimate or predict the result based on our trained model.

Save that file as **Test_Trained_Model.m** and run it. The running result is displayed in the Command window as below:

```
T = 1×7 table

      Acceleration      Cylinders      Displacement      Horsepower
Model_Year     Weight      Origin

      ____________      _________      ____________      __________

________      ______      ______

           18                 8                                320
150                  70             3804        "USA"


yfit = 13.7500
```

You can also use a partial data dataset to test this trained model. For example, you can use a partial dataset **cartable** that only contained the first 6 rows of data from the original dataset **carsmall** as the testing dataset. An example coding file **Test_Trained_Model_M.m** can be found in the Springer ftp site at the folder **Students\Class Projects\Chapter 5\Regression Trees Project**.

Next let us take care of building regress trees model with functions.

5.3.2 Regression Tree-Related Functions

The regression tree-related function is categorized into different classes in MATLAB. The base class or called super-class is **CompactRegressionTree**, which is a compact version of a regression tree of class **RegressionTree** that is a derived class based on the base class. The compact version does not include the data for training the regression tree. Therefore, you cannot perform some tasks with a compact regression tree, such as cross-validation. But one can use a compact regression tree to make predictions for new data after it is trained.

The **RegressionTree** is a derived or a child class of **CompactRegressionTree** class. In fact, a RegressionTree class is a decision tree used for binary splits for regression. An object of class RegressionTree can be used to perform all regression tree functions, including to predict responses for new data with the **predict()** method. The object contains the data used for training, so can compute re-substitution predictions.

To generate or create an object of class RegressionTree, one needs to use the function **fitrtree()**, which is a powerful and complicated function with lot of properties. By using this function to create objects to perform regression tree modeling, more advantages and flexibilities are provided compared with those in Regression Learner, but more complicated setup and configuration processes are also accompanied, which may be a headache or challenge to beginners in machine learning studies.

To make discussion simple and easy, we try to introduce only some popular and useful properties to make this study easier to reduce the learning curve. Table 5.2 shows the most popular properties used for the object of the RegressionTree class.

In addition to these properties, the class also contains some functions. Some popular and important functions are shown in Table 5.3.

The function **fitrtree()** contains five different constructors, which include:

1. **tree = fitrtree(Tbl, ResponseVarName)**
2. **tree = fitrtree(Tbl, formula)**
3. **tree = fitrtree(Tbl, Y)**
4. **tree = fitrtree(X,Y)**
5. **tree = fitrtree(___, Name, Value)**

The **Tbl** is a data table used to train the model, and **X** is the predictor (input) data matrix, **Y** is the response (output) data vector. Depending on the syntax of top four constructors, all arguments need to have different values, as described below.

- If **Tbl** contains the response variable, and you want to use all remaining variables in **Tbl** as predictors, then specify the response variable by using **ResponseVarName**.

Table 5.2 Some popular and important properties

Properties	Descriptions
CategoricalPredictors	Categorical predictor indices, specified as a vector of positive integers. It contains index values indicating that the corresponding predictors are categorical. The index values are between 1 and p, where p is the number of predictors used to train the model. If none of the predictors are categorical, then this property is empty ([]).
Children	An n-by-2 array containing the numbers of the child nodes for each node in tree, where n is the number of nodes. Leaf nodes have child node 0.
ModelParameters	Object holding parameters of tree.
NumObservations	Number of observations in the training data, a numeric scalar. It can be less than the number of rows of input data X when there are missing values in X or response Y.
NodeSize	An n-element vector size of the sizes of the nodes in tree, n is the number of nodes.
NumNodes	The number of nodes n in tree.
PredictorNames	A cell array of names for the predictor variables, in the order in which they appear in X.
ResponseName	A character vector that specifies the name of the response variable (Y).
RowsUsed	An n-element logical vector indicating which rows of the original predictor data (X) were used in fitting. If the software uses all rows of X, then RowsUsed is an empty array ([]).
W	The scaled weights, a vector with length n, the number of rows in X.
X	A matrix or table of predictor values. Each column of X represents one variable, and each row represents one observation.
Y	A numeric column vector with the same number of rows as X. Each entry in Y is the response to the data in the corresponding row of X.

Table 5.3 Some popular and important class functions

Functions	Descriptions
crossval()	Create a cross-validated decision tree.
cvloss()	Regression error by cross validation.
Loss()	Regression error.
Predict()	Predict responses using regression tree.
predictorImportance()	Estimates of predictor importance for regression tree.
view()	View regression tree.

- If **Tbl** contains the response variable, and you want to use only a subset of the remaining variables in **Tbl** as predictors, then specify a formula by using **formula**.
- If **Tbl** does not contain the response variable, then specify a response variable by using **Y**. The length of the response variable and the number of rows in **Tbl** must be equal.

One more point to be emphasized for the argument **formula** in the second constructor is that it can be either a character vector or a string scalar in the form of "**Y ~ x1 + x2 + x3**" format. In this form, **Y** represents the response variable, and **x1**, **x2**, and **x3** represent the predictor variables.

To specify a subset of variables in **Tbl** as predictors for training the model, use a formula. If you specify a **formula**, then the software does not use any variables in **Tbl** that do not appear in formula.

The variable names in the formula must be valid variable names used in **Tbl** and valid names defined by MATLAB. A piece of example codes of using the formula to create a new regression tree model is:

```
load carsmall
        cartable = table(Acceleration, Cylinders, Horsepower,
Mfg, MPG)
  tree = fitrtree(cartable, 'MPG~Acceleration+Mfg')
```

where both **Acceleration** and **Mfg** are valid variable names used in the table **cartable**, and the response **MPG** is also a valid name used in that table.

As for the argument in the fifth constructor, one needs to specify options using one or more name-value pair in addition to any of the input argument combinations in previous syntaxes. For example, you can specify observation weights or train a cross-validated model.

In order to get a more detailed and clearer picture about those name-value pair, we need to discuss them with more spaces. Some popular and useful name-value pairs are shown in Table 5.4. An example of using this kind of name-value pair in a regression tree function can be expressed as:

Table 5.4 Some popular and important name-value pairs

Name-Value Pair	Descriptions
CategoricalPredictors	Each entry in the vector is an index value indicating that the corresponding predictor is categorical. The index values are between 1 and p, where p is the number of predictors used to train the model. Example: 'CategoricalPredictors','all'.
MaxDepth	Maximum tree depth, specified as the comma-separated pair consisting of 'MaxDepth' and a positive integer.
MinParentSize	Minimum number of branch node observations, specified as the comma-separated pair consisting of 'MinParentSize' and a positive integer value. Each branch node in the tree has at least MinParentSize observations.
PredictorNames	Predictor variable names, specified as a string array of unique names or cell array of unique character vectors. By default, PredictorNames contains the names of all predictor variables.
ResponseName	Response variable name, specified as a character vector or string scalar.
SplitCriterion	Split criterion, specified as the comma-separated pair consisting of 'SplitCriterion' and 'MSE', meaning mean squared error. Default value is MSE.
Weights	Observation weights, specified as the comma-separated pair consisting of 'Weights' and a vector of scalar values or the name of a variable in Tbl.
CrossVal	Cross-validation flag, specified as the comma-separated pair consisting of 'CrossVal' and either 'on' or 'off'.
MaxNumSplits	Maximal number of decision splits (or branch nodes), specified as the comma-separated pair consisting of 'MaxNumSplits' and a positive integer.
MinLeafSize	Minimum number of leaf node observations, specified as the comma-separated pair consisting of 'MinLeafSize' and a positive integer value
NumVariablesToSample	Number of predictors to select at random for each split, specified as the comma-separated pair consisting of 'NumVariablesToSample' and a positive integer value.

```
load carsmall
tree = fitrtree([Weight, Cylinders],MPG,...
                                    'CategoricalPredictors',2,'MinPar
entSize',20,...
                                    'PredictorNames',{'W','C'},
'ResponseName', 'MPG', 'CrossVal', 'on', 'KFold', 5)
```

where two predictors or input variables, **Weight** and **Cylinders**, are used as training source and the response or output variable is **MPG**. The following name-value pairs are used:

- The first name-value pair, **CategoricalPredicators**, is used with a value of 2, which indicates that two categorical predicators are used for the tree.
- The second name-value, **MinParentSize**, is to indicate the size of parent as 20.
- The third name-value pair, **PredictorNames**, shows two predictors' names as **'W'** and **'C'**.
- The fourth name-value pair, **ResponseName**, indicates the response name is **'MPG'**.
- The fifth name-value pair, **CrossVal**, is **'on'** to allow the fitrtree to grow a cross-validated decision tree with 10 folds.
- The sixth name-value pair, **KFold**, is to override the above cross-validation setting by using one **'KFold'** as 5.

Now let us use an example project to build a regression tree with the function **fitrtree()** to illustrate the building process.

Create a new Script file and name it as **RT_Car_Func.m**, and enter the codes shown in Fig. 5.47 into that file. In this piece of codes, three example regression trees are generated, **tree**, **tree1**, and **tree2**.

```
% Using fitrtree() function to create regression tree model
% The dataset is carsmall
% Three different models are generated, tree, tree1 and tree2
1  load carsmall
   X = [Acceleration, Weight, Cylinders, Horsepower];
2  tree = fitrtree(X, MPG,...
                   'CategoricalPredictors',4, 'MinParentSize',20,...
                   'PredictorNames',{'A', 'W','C', 'H'}, 'ResponseName', 'MPG', 'CrossVal','on')
3  tree1 = fitrtree([Weight, Acceleration], MPG,...
                   'CategoricalPredictors',2, 'MinParentSize',20,...
                   'PredictorNames',{'W','A'},'ResponseName', 'MPG')
4  XX = [4000 8; 4000 15; 4000 20];
5  MPGpred = predict(tree1, XX)
6  XXX = [Cylinders, Horsepower, Displacement];
7  tree2 = fitrtree(XXX, MPG, 'CrossVal', 'on')
8  numBranches = @(x)sum(x.IsBranch);
   tree2NumSplits = cellfun(numBranches, tree2.Trained);
9  figure;
   histogram(tree2NumSplits)
10 view(tree2.Trained{1},'Mode','graph')
```

Fig. 5.47 Example codes for generate regression tree with function fitrtree()

The first generated **tree** is used to illustrate how to use selected predictors and response variable with name-value pairs to create a regression tree. The second tree is used to call the function **predict()** to evaluate the generated tree, **tree1**, with a new predictor vector. The third tree, **tree2**, is used to plot a histogram of the number of imposed splits on the trees. The number of imposed splits is one less than the number of leaves. Finally, a complete tree is displayed with the **view()** function.

Let us have a closer look at this piece of codes to see how it works.

1. First we need to load our dataset **carsmall** that is a sample cars dataset with 100×10 cells. Totally 10 variables are included in the dataset in 10 columns. For the first tree, we only need to use four of them, **Acceleration**, **Weight**, **Cylinders**, and **Horsepower**, and get them into an input vector **X**.
2. Build our first tree by calling the function **fitrtree()** with **X** as predictors or input variables, **MPG** as the response variable with some name-value pairs.
3. To evaluate a trained regression tree, **tree1**, first we need to create that tree. For this purpose, we only need to select two input variables, **Weight** and **Acceleration**, as our predictors, and **MPG** as our response variable with some selected name-value pairs.
4. To test our trained model, **tree1**, we used three sets of inputs as our testing predictors to predict the mileage-per-gallon (**MPG**) of 4000-pound cars with 8, 15, and 20 **Acceleration**s. Put them into a 3×2 matrix **XX**.
5. Now run the function **predict()** with our trained tree, **tree1**, and testing matrix as input to calculate the predicted results. The results should be displayed in the Command window as this Script is executed.
6. To plot a histogram of the number of imposed splits on the trees, consider three input variables, **Cylinders**, **Horsepower**, and **Displacement** as predictors and **MPG** as response variable, and put them into a 100×3 matrix named **XXX**.
7. Execute the function **fitrtree()** to build and train our target model with cross-validation operation being active.
8. First we need to get how many branches are in the trained tree by using an anonymous function **@(x)sum()**, where the x is the trained tree model **tree2**. Then the function **cellfun()** is called to get all numbers of splits.
9. A new figure window is generated by using the **figure** command with default property values. The resulting figure is the current figure. The **histogram()** function is called to plot a histogram of the number of imposed splits on the trees.
10. Finally the function **view()** is executed to plot the entire tree.

Now run this Script file and the running result is shown below on the Command window:

```
tree1 =

  RegressionTree
    PredictorNames: {'W'  'A'}
      ResponseName: 'MPG'
 CategoricalPredictors: 2
 ResponseTransform: 'none'
    NumObservations: 94


  Properties, Methods

MPGpred =

    14.6875
    11.0000
    15.5417

tree2 =


  RegressionPartitionedModel
    CrossValidatedModel: 'Tree'
    PredictorNames: {'x1'  'x2'  'x3'}
      ResponseName: 'Y'
    NumObservations: 94
      KFold: 10
    Partition: [1×1 cvpartition]
 ResponseTransform: 'none'
```

The histogram of the number of imposed splits on the trees and the final view of the entire tree are shown in Figs. 5.48 and 5.49. In Fig. 5.49, three predictors, **Cylinders**, **Horsepower**, and **Displacement**, are utilized to build a regression tree with the 10 folders (**KFold** = 10) for cross-validation setting.

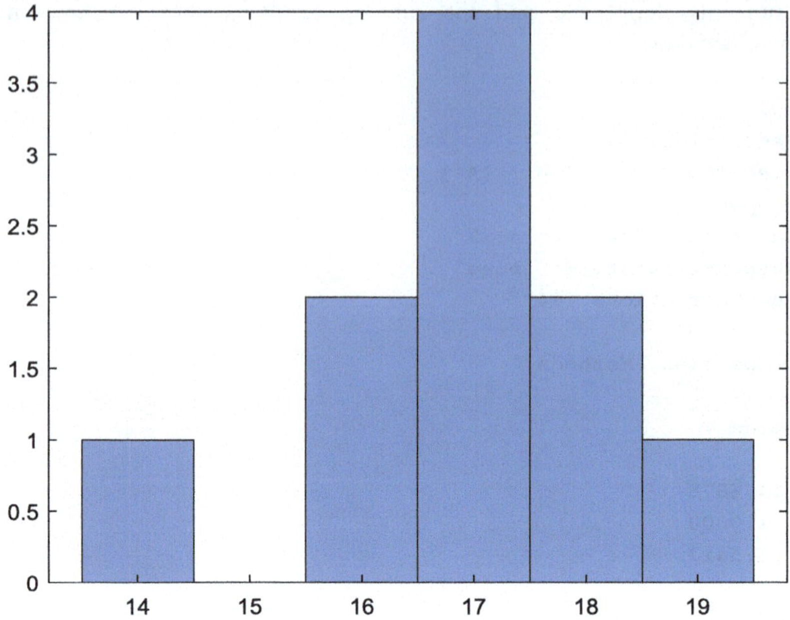

Fig. 5.48 The histogram of the number of imposed splits on the trees

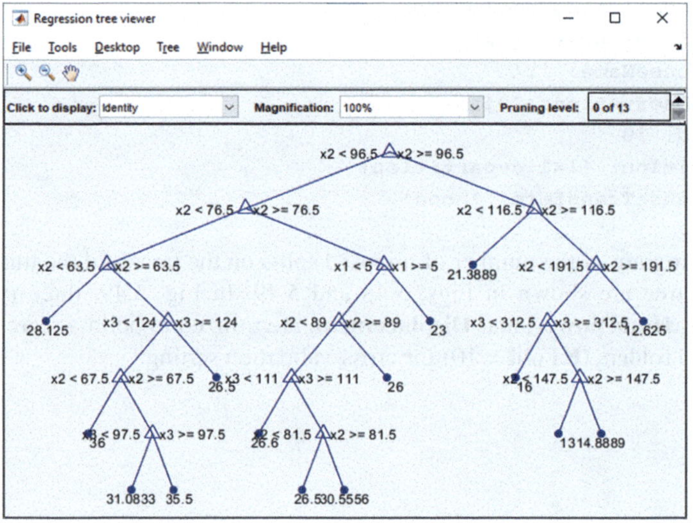

Fig. 5.49 The final view of entire tree

5.4   Introduction to Nonlinear Regressions

Nonlinear regression is a statistical technique that helps describe nonlinear relationships in experimental data. Nonlinear regression models are generally assumed to be parametric, where.

the model is described as a nonlinear equation. Typically machine learning methods are used for nonparametric nonlinear regression.

Parametric nonlinear regression models the dependent variable called the response as a function of a combination of nonlinear parameters and one or more independent variables called predictors. The model can be univariate (single response variable) or multivariate (multiple response variables).

The parameters can take the form of an exponential, trigonometric, power, or any other nonlinear function. To determine the nonlinear parameter estimates, an iterative algorithm is typically used [7].

$$y = f(X,\beta) + \varepsilon \tag{5.10}$$

where β is the nonlinear parameter estimation to be computed and ε represents the error terms.

There are some popular nonlinear regression algorithms and models widely implemented in various applications. Let us have a clear picture about those algorithms and models first.

5.4.1   Nonlinear Regression Algorithms and Models

Some popular algorithms used for fitting a nonlinear regression include:

- Decision Trees
- Support Vector Machine
- Polynomial Regression
- Artificial Neural Network
- Gauss-Newton Algorithm
- Gradient Descent Algorithm
- Levenberg-Marquardt Algorithm

Decision Trees (DT)
A detailed discussion about using regression decision tree or regression tree to build related models to estimate or predict the desired response based on inputs has been given in the last section. Refer to that section to get more details for that kind of algorithm.

Support Vector Machine (SVM)
A Support Vector Machine (SVM) is a kind of supervised learning algorithm used for many classification and regression problems, including linear or nonlinear models. We will have a detailed discussion about SVM in the next section.

Artificial Neural Network (ANN)
ANNs are powerful models that consist of a group of interconnected nodes or neurons organized in different layers. By adjusting the weights and biases of the network during the training process, ANNs can learn complex nonlinear relationships between inputs and outputs.

Polynomial Regression (PLR)
It is a regression analysis in which the relationship between the independent variable or input and the dependent variable or target is modeled as an nth-degree polynomial. In polynomial regression, the input data is transformed by adding polynomial terms of different degrees.

Gauss-Newton Algorithm (G-N)
Gauss-Newton is a calculating method for solving nonlinear least squares problems. It is a first-order approximation method that does not require the second-order term, the Hessian matrix, in the Taylor series expansion [8].

Gradient Descent Algorithm (GD)
Gradient Descent (GD) is an iterative first-order optimization algorithm used to find a local minimum/maximum of a given function. This method is commonly used in *machine learning* (ML) and *deep learning* (DL) to minimize a cost/loss function (e.g., in a linear regression). Due to its importance and ease of implementation, this algorithm is usually taught at the beginning of almost all machine learning courses [9].

Levenberg-Marquardt Algorithm
Levenberg-Marquardt is a popular alternative to the Gauss-Newton method of finding the minimum of a function that is a sum of squares of nonlinear functions [10].

Due to the similarity among the last three algorithms, we will concentrate our study only on one of them to avoid duplications.

In addition to those nonlinear regression algorithms, there are some popular regression models used for nonlinear regression processes, and these models include:

1. Transformable Nonlinear Models
2. Polynomial Models
3. Exponential Models
4. Logarithmic Models
5. Fourier Models
6. Gaussian Models
7. Rational Models

Next let us have a closer look at these models and related functions.

5.4.2 Nonlinear Regression-Related Apps and Functions

In the above seven popular models, the first model, **Transformable Nonlinear Models**, means that some nonlinear regression problems can be moved to a linear domain by a suitable transformation of the model formulation. We will not cover that model in this part due to its simplicity.

For models 2–5, MATLAB provided both an App named **Curve Fitter** and some functions, such as **fit()**, **fitnlm()**, and **nlinfit()**, to help readers to build powerful nonlinear regression models to predict response based on input data easily and quickly. We will discuss those models in detail in the following sections.

As for models 6–7, the **Curve Fitter** also provided some related Apps named **Gaussian** and **Rational** to help users to build nonlinear regression models to find desired matching response values. The function **fit()** can be used to build either a **Gaussian** or a **Rational** model to replace those Apps to perform more professional functions for nonlinear regression applications.

We will have our discussions about these algorithms and models in the following sequences:

1. We start our discussion about Support Vector Machine algorithm in the next section.
2. Followed by SVM, some popular Apps and functions related to nonlinear regression will be discussed with all seven models defined in Curve Fitter App and related functions.
3. The Artificial Neural Network algorithm will be discussed in Chap. 7.
4. Due to the similarity among the last three algorithms, we will concentrate our study only on one of them to avoid duplications.

Now let us start our discussion on Support Vector Machine algorithm.

5.5 Support Vector Machine-Related App and Functions

A Support Vector Machine (SVM) is a kind of supervised learning algorithm used for many classification and regression problems, including signal processing, medical applications, natural language processing, and speech and image recognition [11].

The objective of the SVM algorithm is to find a hyperplane that, to the best degree possible, separates data points of one class from those of another class. **Best** is defined as the hyperplane with the largest margin between the two classes, represented by plus versus minus in the Fig. 5.50. Margin means the maximal width of the slab parallel to the hyperplane that has no interior data points. Only for linearly separable problems can the algorithm find such a hyperplane; for most practical problems, the algorithm maximizes the soft margin allowing a small number of misclassifications.

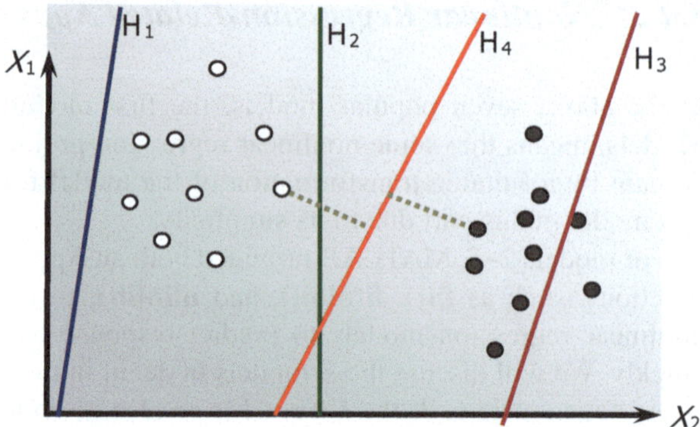

Fig. 5.50 Different hyperplanes used to classify two classes

In geometry, a hyperplane is a generalization of a two-dimensional plane in three-dimensional space to mathematical spaces of arbitrary dimension. Like a plane in space, a hyperplane is a flat hypersurface, a subspace whose dimension is one less than that of the ambient space [12].

In fact, a so-called hyperplane is to use a single line or an intersection line to present a single plan or one dimension in a multiple dimensional space or Euler space in mathematics. Regularly the SVM is used for classification problems, but it can also be used in regression issues. The SVM is mainly used in unsupervised learning, such as clustering process. Let us use a simple example to illustrate how to use SVM to perform clustering process.

In Fig. 5.50, two classes are represented by two groups of white and black data points; each of them can be considered as a p dimensional vector (a list of p-numbers), and we want to know whether we can separate such points with a $(p{-}1)$-dimensional hyperplane. This is called a linear classifier. Each hyperplane can be mapped to a single line, H_i, $i = 1$–4. The purpose of SVM is to find or build a hyperplane in between two classes of datasets to indicate which class it belongs to. The challenge is to train the machine to understand structure from data and mapping with the right class label; for the best result, the hyperplane has the largest distance or largest margin to the nearest training data points of any class.

As shown in Fig. 5.50, four hyperplanes are created for these two classes of data points. Hyperplanes H_1 and H_3 did not separate two groups of data into two classes at all. H_2 did separate two groups of data into two classes without the largest margin between them. Only H_4 separates two groups of data into two classes with the largest distance; therefore, H_4 is the desired hyperplane created and used by support vector machine.

Now let us have a little deeper discussion about the support vector machine with hyperplane. Suppose we have a training dataset of **n** points, (x_1, y_1), ... (x_n, y_n), and these data points belong to two classes. The condition is, $y_i = 1$ if related x_i belong to one class, otherwise $y_i = -1$. We want to find the **maximum-margin hyperplane** that divides the group of points x_i for which $y_i = 1$ from the group of points for which $y_i = -1$, which is defined so that the distance between the hyperplane and the nearest point x_i from either group is maximized.

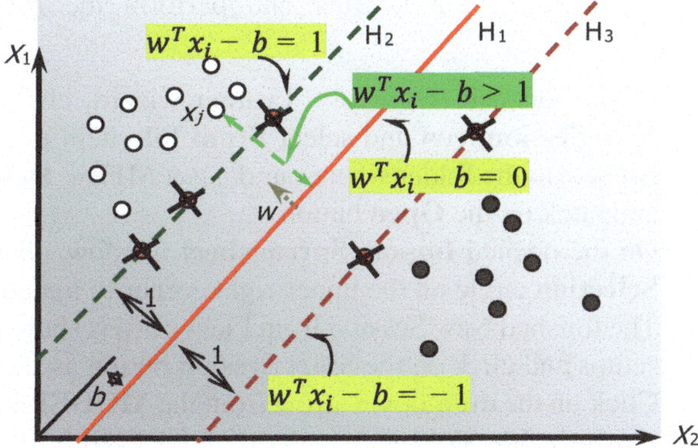

Fig. 5.51 An illustration for some hyperplanes with support vectors

A hyperplane can be written as the set of points x_i satisfying

$$w^T x_i - b = 0 \qquad (5.11)$$

where w is the normalized vector to the hyperplane.

As shown in Fig. 5.51, two parallel hyperplanes (H_2 and H_3) that separate the two classes of data are built, so that the distance between them is as large as possible. The region bounded by these two hyperplanes is called the **margin**, and the maximum-margin hyperplane (H_1) is the hyperplane that lies halfway between them. With a normalized or standardized dataset, these hyperplanes can be described by the equations

$w^T x_i - b = 1$ (H_2—points on or above this boundary is of one class, with label 1)

and

$w^T x_i - b = -1$ (H_3—points on or below this boundary is of the other class, with label -1)

The points on two boundary hyperplanes are called support vectors, which are indicated by **x** marks. An example point x_j is shown in Fig. 5.51. The vertical distance between the point x_j and the hyperplane $w^T x_i - b = 0$ is $w^T x_j - b$, which is greater than 1. Thus, that point belongs to the top class with $y_i = 1$. The w is a normalized vector with magnitude of 1 and it only provides a direction that is vertical to the hyperplane $w^T x_i - b = 0$.

Now let us concentrate our discussions on the support vector machine-related Apps, exactly, use the Regression Leaner to develop regression tree models.

5.5.1   *Support Vector Machine-Related Apps*

We like to use the forest fire dataset to build this kind of regression tree model with support vector machine algorithm.

Open Regression Learner and perform the following operations to create this model.

1. On the opened Regression Learner wizard, click on the drop-down arrow on **New Session** icon and select **From File** item to open the **Select File** window. Browse to the folder where our dataset **MFire_Database.xls** is located, select it, and click on the **Open** button.
2. On the opened **Import Spreadsheet** window, click on the green-color **Import Selection** circle on the upper-right corner to load that dataset into the App.
3. The finished New Session from Fie wizard is shown in Fig. 5.52. Keep all default setups and click on the **Start Session** button as shown in Fig. 5.52 to continue.
4. Click on the drop-down arrow from the **MODELS** pane on the top and select the **Optimizable SVM** item under the **SUPPORT VECTOR MACHINS** group, as shown in Fig. 5.53.
5. On the opened **Summary** group, select some **desired** check boxes for the hyperparameters that you want to optimize. All the check boxes for the available hyperparameters are selected by default. For this example, just clear the **Optimize** check boxes for **Kernel function** and **Standardize data**. By default, the App disables the **Optimize** check box for **Kernel scale** whenever the kernel function has a fixed value other than **Gaussian**. Select a **Gaussian** kernel func-

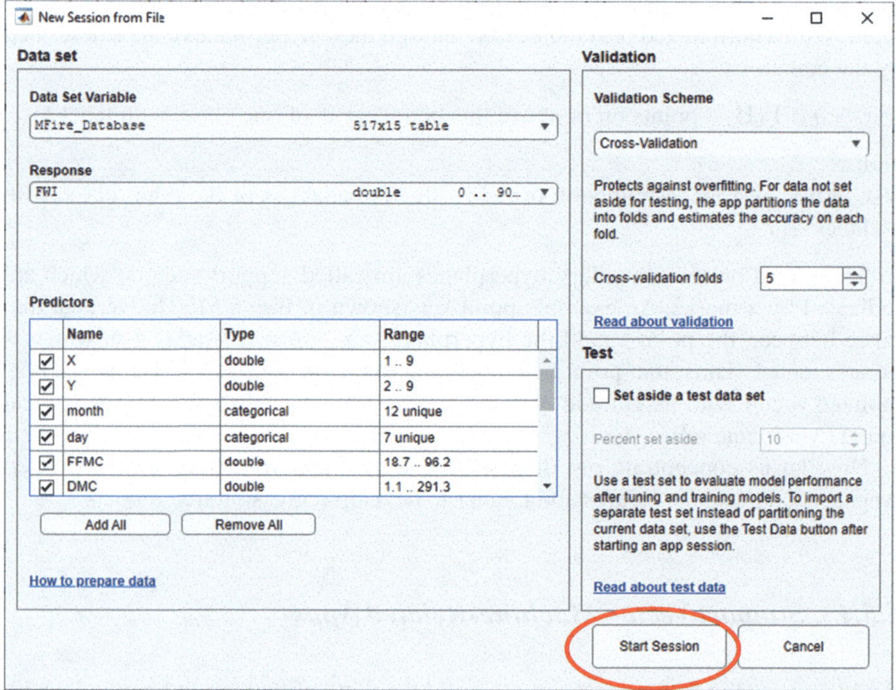

Fig. 5.52 The finished New Session from File wizard

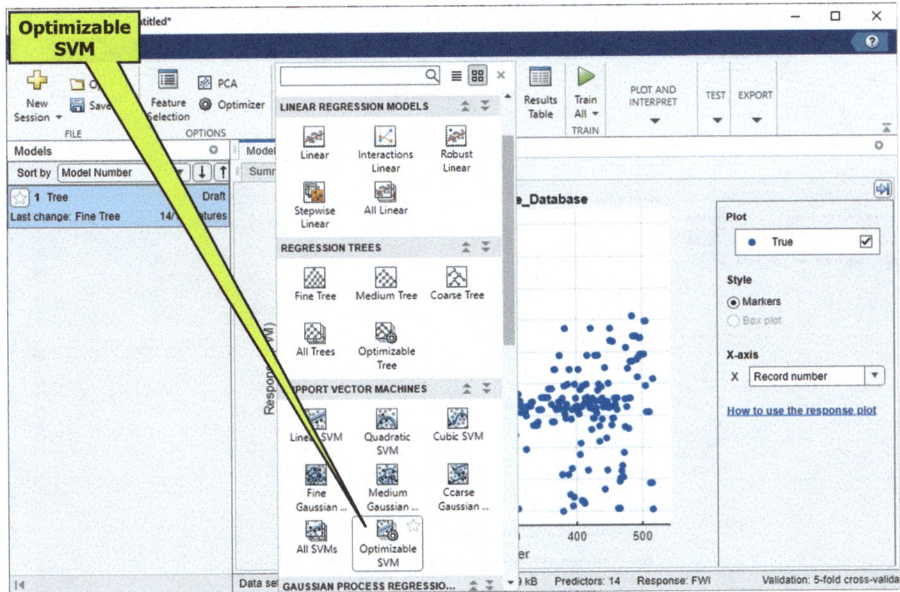

Fig. 5.53 The selected Optimizable SVM algorithm

tion, and select the **Optimize** check box for **Kernel scale**. Your finished **Summary** parameters list is shown in Fig. 5.54.

6. Click on the drop-down arrow on the **Train All** and select **Train Selected** item. The App displays a **Minimum MSE Plot** as it runs the optimization process. For each iteration, the App tries a different combination of hyperparameter values and updates the plot with the minimum validation mean squared error (**MSE**) observed up to that iteration, indicated in dark blue. When the App completes the optimization process, it selects the set of optimized hyperparameters, indicated by a red square.

7. On the displayed training resulted plot, select the **BUI** item from the X box under the X-axis group to display a relation between the predicated support vectors and data points for two classes, which is shown in Fig. 5.55.

The App lists the optimized hyperparameters in both the **Optimization Results** section and the **Model Hyperparameters** section of the model by clicking on the **Summary** tab, as shown in Fig. 5.56. In general, the optimization results are not reproducible.

To export the trained model to the MATLAB workspace, click on the drop-down arrow on the **Export** section and the **Export Model**, and select **Export Model**, enter our model name, **SVM_Fire_Model**, into the Variable name box, then click on the **OK** button. Immediately you can find that model is added into the Workspace.

Alternatively, you can generate a piece of MATLAB codes that train a regression model with the same settings used to train the SVM model in the App. On the **Regression Learner** tab, expand the **Export** section, click on the **Generate**

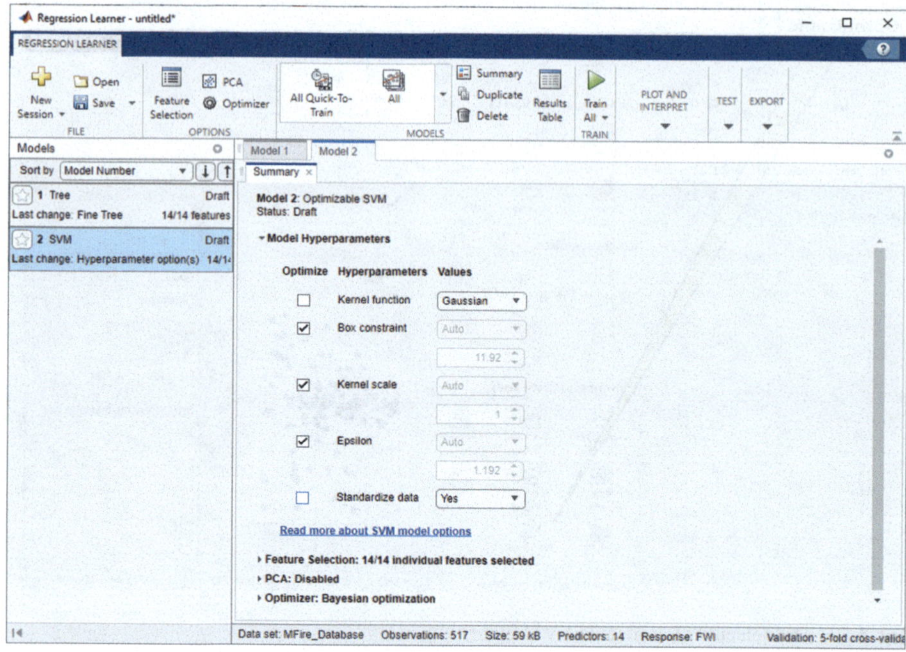

Fig. 5.54 The selected Optimizable parameters

Fig. 5.55 The training response result using the SVM algorithm

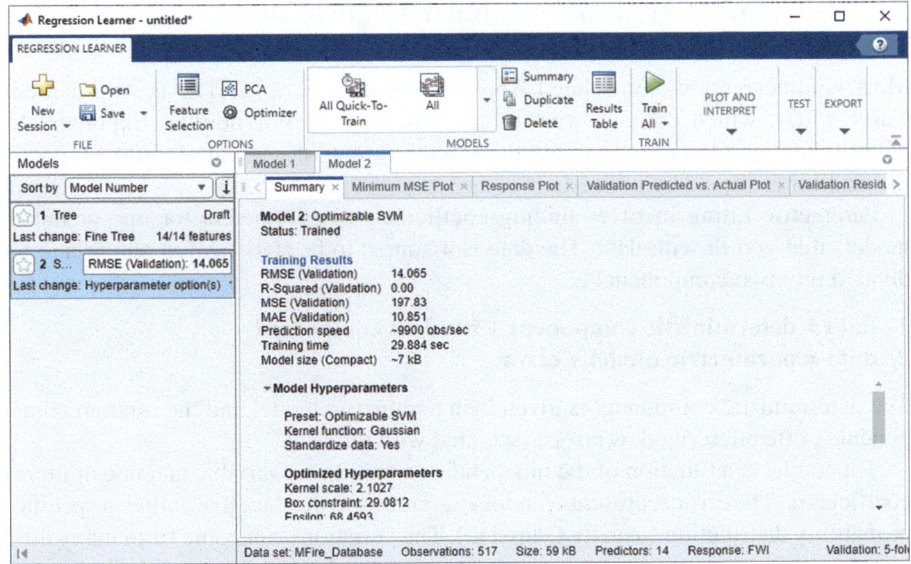

Fig. 5.56 The training results

Function item. The App generates codes from your session and displays the file in the MATLAB Editor. The file defines a function that accepts predictor and response variables, trains a regression model, and performs cross-validation. Change the function name to **Trained_SVM_Func** and save the file as **Trained_SVM_Func.m**.

You can train an SVM model by using that function later.

5.5.2 *Support Vector Machine-Related Functions*

Since most popular applications of using support vector machine algorithms are for classifications, either binary or multiple class classifications, we will provide very detailed discussions about using support vector machine functions later when we introduce the classification applications.

Alternatively, you can use support vector machine-related function exported from the Regression Learner shown in the last section to handle SVM-related applications.

5.6 Implementations of Nonlinear Regressions

Most nonlinear regression-related Apps are involved in a MATLAB App named Curve Fitter, which contains various models, such as Polynomial, Exponential, Algorithmic, and Fourier. To apply those models, one of popular algorithms, parametric fitting, should be utilized.

Parametric fitting involves finding coefficients or parameters for one or more models that you fit with data. The data is assumed to be statistical in nature and is divided into two components:

1. **data = deterministic component + random component**
2. **data = parametric model + error**

The deterministic component is given by a parametric model and the random component is often described as error associated with the data.

The model is a function of the independent or predictor variable and one or more coefficients. The error represents random variations in the data that follow a specific probability distribution (usually Gaussian). The variations can come from many different sources, but are always present at some level when you are dealing with measured data. Systematic variations can also exist, but they can lead to a fitted model that does not represent the data well.

The model coefficients often have physical significance. For example, suppose we collected data that corresponds to a voltage discharging process to a RC circuit, and we want to get estimations about voltage discharging level on the capacitor. The law of an RC circuit states that the voltage discharging level across the capacitor is exponentially proportional to the time with a time constant $\tau = RC$. Therefore, the model to use in the fit is given by

$$y = y_0 e^{-\frac{t}{\tau}} \tag{5.12}$$

where y_0 is the magnitude of the input voltage at time $t = 0$, and τ is the time constant. The data can be described by

$$\text{data} = y_0 e^{-\frac{t}{\tau}} + \text{error} \tag{5.13}$$

Both y_0 and τ are coefficients that are estimated by the fit. Because the data contains some error, the deterministic component of the equation cannot be determined exactly from the data. Therefore, the coefficients and time constant will have some uncertainty associated with them. If the uncertainty is acceptable, then you are done fitting the data. If the uncertainty is not acceptable, then you might have to take steps to reduce it either by collecting more data or by reducing measurement error and collecting new data and repeating the model fit.

With other problems where there is no theory to dictate a model, you might also modify the model by adding or removing terms, or substitute an entirely different model.

As we mentioned, four popular models can be used with the parametric fitting algorithm. Thus, let us have a clear picture about Curve Fitting App and related parametric library models.

5.6.1   Nonlinear Regression-Related Apps

As we discussed in Sect. 5.4.1, seven popular nonlinear regression models are involved in the Curve Fitter App, which are.

1. Transformable Nonlinear Models
2. Polynomial Models
3. Exponential Models
4. Logarithmic Models
5. Fourier Models
6. Gaussian Models
7. Rational Models

In this section, we will concentrate our discussions on six of them without the first one, transformable nonlinear model, due to its simplicity.

The Curve Fitter App provides a flexible graphic user interface where you can interactively fit curves and surfaces to data and view plots. With the Curve Fitter app, you can:

1. Create, plot, and compare multiple fits.
2. Use linear or nonlinear regression, interpolation, smoothing, and custom equations.
3. View goodness-of-fit statistics, display confidence intervals and residuals, remove outliers, and assess fits with validation data.
4. Automatically generate code to fit and plot curves and surfaces, or export fits to the workspace for further analysis. Export a curve or surface fit to a Simulink lookup table.

First let us have a quick introduction about these models.

Polynomial Models
Polynomial models used for curves fitting is to use a polynomial equation to approximate a curve, and all coefficients p_i can be estimated by

$$y = \sum_{i=1}^{n+1} p_i x^{n+1-i} \tag{5.14}$$

where $n + 1$ is the *order* of the polynomial, n is the *degree* of the polynomial, and $1 \leq n \leq 9$. The order gives the number of coefficients to be fit, and the degree gives the highest power of the predictor variable.

In this section, polynomials are described in terms of their degree. For example, a third-degree (cubic) polynomial is given by

$$y = p_1 x^3 + p_2 x^2 + p_3 x + p_4 \tag{5.15}$$

Polynomials are often used when a simple empirical model is required. You can use the polynomial model for interpolation or extrapolation, or to characterize data using a global fit. For example, the temperature-to-voltage conversion for a Type J thermocouple in the 0 to $760°$ temperature range is described by a seventh-degree polynomial.

The main advantages of using polynomial fits include reasonable flexibility for data that is not too complicated, and they are linear, which means the fitting process is simple. The main disadvantage is that high-degree fits can become unstable. Additionally, polynomials of any degree can provide a good fit within the data range, but can diverge wildly outside that range. Therefore, exercise caution when extrapolating with polynomials.

When you fit with high-degree polynomials, the fitting procedure uses the predictor values as the basis for a matrix with very large values, which can result in scaling problems. To handle this, you should normalize the data by centering it at zero mean and scaling it to unit standard deviation. Normalize data by selecting the **Center and scale** check box in the Curve Fitter App.

Exponential Models

The Curve Fitter App provides a one-term and a two-term exponential model as given by

$$y = ae^{bx} \ \ or \ \ y = ae^{bx} + ce^{dx} \tag{5.16}$$

Exponentials are often used when the rate of change of a quantity is proportional to the initial amount of the quantity. If the coefficient associated with b and/or d is negative, y represents exponential decay. If the coefficient is positive, y represents exponential growth.

For example, a single radioactive decay mode of a nuclide is described by a one-term exponential. Where a is interpreted as the initial number of nuclei, b is the decay constant, x is time, and y is the number of remaining nuclei after a specific amount of time passes. If two decay modes exist, then you must use the two-term exponential model. For the second decay mode, you add another exponential term to the model.

Examples of exponential growth include contagious diseases for which a cure is unavailable, and biological populations whose growth is uninhibited by predation, environmental factors, and so on.

Logarithmic Models

A logarithmic model has a steep initial period of growth before continuing to grow at a slower rate. Logarithmic models are used in a variety of applications, such as studies of population growth and signal processing. Curve Fitting App supports the logarithmic models as described below:

$$\text{Natural log}: y = a \times \log(x) + b$$
$$\text{Common log}: y = a \times \log 10(x) + b \qquad (5.17)$$
$$\text{Binary log}: y = a \times \log 2(x) + b$$

In the above equations, a is a scaling parameter and b is the value for y when $x = 1$. You can convert between the different logarithmic models using the change of base formula $log_j(x) = log_i(x)/log_i(j)$, where j is the base of the model to convert to, and i is the base of the model being converted. The logarithmic model can be used to fit to your data depending on the type of problem you are solving.

Fourier Models

A Fourier series describes a periodic function as a sum of sine and cosine functions. You can separate an arbitrary periodic function into simple components by using a Fourier series. These components are easy to integrate, differentiate, and analyze. For this reason, Fourier series are often used to approximate periodic functions or signals.

Fourier series are represented in several forms. Curve Fitting App uses the trigonometric Fourier series form, such as

$$y = a_0 + \sum_{i=1}^{n} a_i \cos(i\omega x) + b_i \sin(i\omega x) \qquad (5.18)$$

where a_0 models a constant or an intercept term in the data and it is associated with the $i = 0$ cosine term, ω is the fundamental frequency of the signal, and n is the number of terms or harmonics. Curve Fitting App or Toolbox supports Fourier series regression for $1 \leq n \leq 8$.

Gaussian Models

The Gaussian model is used to try to fit a Gaussian-like curve to experimental data by Marquardt-Levenberg nonlinear least squares minimization algorithm. The fitting function has a Gaussian form as

$$y = \sum_{i=1}^{n} a_i e^{-\left(\frac{x - b_i}{ci}\right)^2} \qquad (5.19)$$

where a is the amplitude, b is the position of the center of the peak, c is related to the peak width or standard deviation, n is the number of peaks to fit, and $1 \leq n \leq 8$.

Gaussian peaks are encountered in many areas of science and engineering. For example, Gaussian peaks can be used to describe line emission spectra and chemical concentration assays.

Rational Models

Rational models are called *rational functions*, which means that they can be represented by ratios of two polynomials given as

$$y = \frac{\sum_{i=1}^{n+1} a_i x^{n+1-i}}{x_m + \sum_{i=1}^{m} b_i x^{m-1}} \tag{5.20}$$

where n is the degree of the numerator polynomial, and m is the degree of the denominator polynomial. Curve Fitting App supports rational models with $0 \le n \le 5$ and $1 \le m \le 5$. The coefficient associated with x^m is always 1, which makes the numerator and denominator unique when the polynomial degrees are the same.

Here rationals are described in terms of the *degree of the numerator/the degree of the denominator*. For example, a quadratic/cubic rational equation is given by

$$y = \frac{a_1 x^2 + a_2 x + a_3}{x^3 + b_1 x^2 + b_2 x + b_3} \tag{5.21}$$

Like polynomials, rationals are often used when a simple empirical model is required. The main advantage of rationals is their flexibility with data that has a complicated structure. The main disadvantage is that they become unstable when the denominator is around 0.

Next let us use some real examples to illustrate how to use Curve Fitter App to perform nonlinear regression fitting with these models.

We start our discussion with a sample dataset, **nl_motor.xls**, which is a simple dataset used to provide a relation between the input DC voltages and motor rotation speed, and this dataset can be found from the Springer ftp site under a folder **Students\Datasets**. Copy and paste it to one of your local folders to be used later.

Now open MATLAB and type command **curveFitter** in the Command window to open the Curve Fitter App. The default fitting model is **Polynomial** when you open that App. Perform the following operations to complete this fitting:

1. Click the **Select Data** icon, as shown in Fig. 5.57, to open the **Select Fitting Data** wizard.
2. Change the **Fit name** to **motor_fit**, select **inputVol** as the **X data** and **motorSP** as **Y data**. Your finished wizard should match the one that is shown in Fig. 5.58. Click on the **Close** button.
3. Select **2** as the **Degree** and **Off** as the **Robust** boxes on the right pane, as shown in Fig. 5.57. The fitting result is shown in Fig. 5.57.

It can be found from Fig. 5.57 that the fitting parametric equation is,

$$f(x) = p_1 x^2 + p_2 x + p_3 = -0.3452 x^2 + 12.8301 x + 124.5188 \tag{5.22}$$

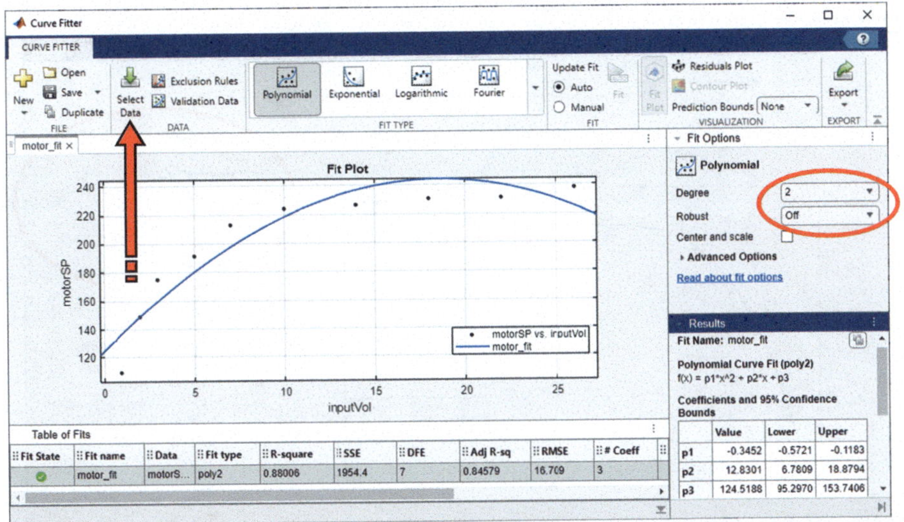

Fig. 5.57 The finished setup and fitting result

Fig. 5.58 The completed select data wizard

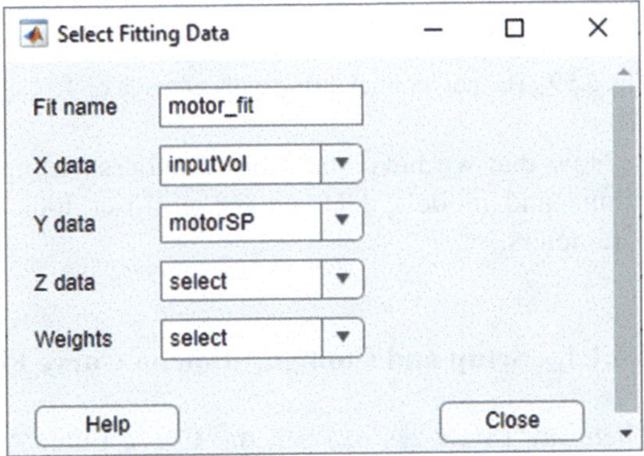

Each estimated coefficient or parameter has both lower and upper bounds shown in the lower-right corner in Fig. 5.57.

To improve the fitting accuracy, we can select higher-degree polynomials. To do that, replace 2 with 3 or 4 from the **Degree** box shown in Fig. 5.57. You can see that the fitting results are greatly improved, as shown in Fig. 5.59, but the polynomials become more complicated with higher order and degree.

It can be found from Fig. 5.59 that the fitting parametric equation is

$$f(x) = p_1 x^4 + p_2 x^3 + p_3 x^2 + p_4 x + p_5 \tag{5.23}$$

which is

$$f(x) = -0.0022 x^4 + 0.1503 x^3 - 3.5613 x^2 + 38.0895 x + 81.1754$$

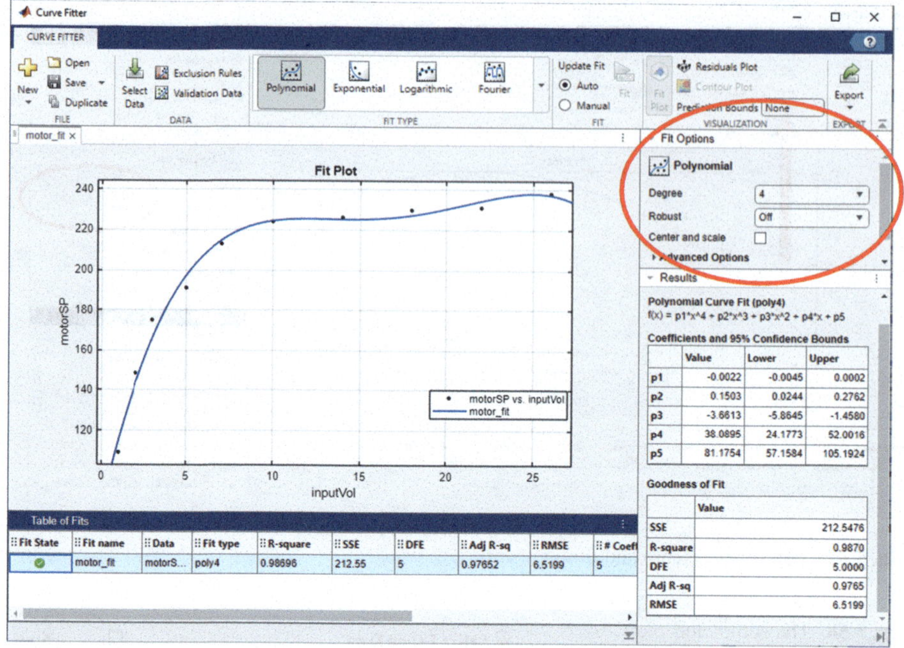

Fig. 5.59 The polynomial fitting with a degree of 4

Now that we have some basic understanding about nonlinear regression algorithms and models, let us have a closer look at some setup and configuration parameters.

5.6.1.1   Setup and Configuration on Curve Fitter App

There are two ways to open the Curve Fitter App: one way is to type command **curveFitter** in the Command window and the other way is to open that App by going to the **APPS** tab, and select the **Curve Fitter** from the opened **APPS** wizard under the **MATH, STATISTICS AND OPTIMIZATION** group. The precondition of using this App is that the Curve Fitter App or Toolbox must have been installed on your machine since that App is involved in the Curve Fitter Toolbox.

The following setups and configurations are included with this App,

1. Center and Scale Data
2. Fit Options and Optimized Starting Points

    (a) Fitting Method and Algorithm—**Method**, **Robust** and **Algorithm** (Table 5.5)
    (b) Finite Differencing Parameters—**DiffMinChange** and **DiffMaxChange** (Table 5.5)

Table 5.5 Some popular and important fit options

Fit Options	Descriptions
Method	Method used to perform fitting function. Here is the Nonlinear Least Square.
Robust	Robust linear least-squares fitting method, specified as the comma-separated pair consisting of 'Robust' and one of these values: 'LAR' is the least absolute residual method.'Bisquare' is the bisquare weights method. Available when the Method is LinearLeast-Squares or NonlinearLeastSquares.
Algorithm	Algorithm to be used for the fitting procedure, specified as the comma-separated pair consisting of 'Algorithm' and either 'Levenberg-Marquardt' or 'Trust-Region'.
DiffMinChange	Minimum change in coefficients for finite difference gradients, specified as the comma-separated pair consisting of 'DiffMinChange' and a scalar.
DiffMaxChange	Maximum change in coefficients for finite difference gradients, specified as the comma-separated pair consisting of 'DiffMaxChange' and a scalar.
MaxFuncEvals	Maximum number of evaluations of the model allowed, specified as the comma-separated pair consisting of 'MaxFunEvals' and a scalar.
MaxIter	Maximum number of iterations allowed for the fit, specified as the comma-separated pair consisting of 'MaxIter' and a scalar. Available when the Method is NonlinearLeastSquares.
TolFun	Termination tolerance on the model value, specified as the comma-separated pair consisting of 'TolFun' and a scalar. Available when the Method is NonlinearLeastSquares.
TolX	Termination tolerance on the coefficient values, specified as the comma-separated pair consisting of 'TolX' and a scalar. Available when the Method is NonlinearLeastSquares.
Coefficients	All coefficients used for fitting method and model with symbols and values.
StartPoint	Initial values for the estimated coefficients, specified as the comma-separated pair consisting of 'StartPoint' and a vector. Find the order of the entries for coefficients in the vector value by using the coeffnames() function. See fit() for examples. If no start points (the default value of an empty vector) are passed to the fit function, starting points for some library models are determined heuristically. For rational and Weibull models, and all custom nonlinear models, the toolbox selects default initial values for coefficients uniformly at random from the interval (0,1). As a result, multiple fits using the same data and model might lead to different fitted coefficients. To avoid this, specify initial values for coefficients with a vector value for the StartPoint property.
Lower	Lower bounds on the coefficients to be fitted, specified as the comma-separated pair consisting of 'Lower' and a vector. The default value is an empty vector, indicating that the fit is unconstrained by lower bounds. If bounds are specified, the vector length must equal the number of coefficients. Find the order of the entries for coefficients in the vector value by using the coeffnames() function. See fit() for examples. Individual unconstrained lower bounds can be specified by -Inf.
Upper	Upper bounds on the coefficients to be fitted, specified as the comma-separated pair consisting of 'Upper' and a vector. The default value is an empty vector, indicating that the fit is unconstrained by upper bounds. If bounds are specified, the vector length must equal the number of coefficients. Find the order of the entries for coefficients in the vector value by using the coeffnames() function. See fit() for examples. Individual unconstrained upper bounds can be specified by +Inf.

(c) Fit Convergence Criteria—**MaxFuncEvals**, **MaxIter**, **TolFun**, and **TolX** (Table 5.5)

(d) Coefficients Parameters—**Coefficients**, **StartPoint**, **Lower**, and **Upper** (Table 5.5)

Let us discuss them one by one in the following sections.

Center and Scale Data

Most fits in this App provide the **Center and scale** option in the **Fit Options** pane, as shown in Fig. 5.59. When you select this option, the App refits the model with the data centered and scaled, which is a good assistance to some data with significant variations.

To alleviate numerical problems with variables of different scales, normalizing the input data or predictor data is a good starting point. For example, suppose your surface fit inputs are engine speed with a range of 500–4500 RPM and engine load percentage with a range of 0–100%. Then, **Center and scale** option can generally improve the fit because of the great difference in scale between those two inputs. However, if your inputs are in the same units or similar scale, then **Center and scale** is less useful. When you normalize inputs with this option, the values of the fitted coefficients change when compared to the original data.

But if you are fitting a curve or surface to estimate coefficients, or the coefficients have physical significance, you may need to disable or clear the **Center and scale** check box. The plots in the Curve Fitter App always use the original scale, regardless of the **Center and scale** status.

To use this option with function format, you need to center and scale the data before fitting, just create the options structure by using the **fitoptions**() function with **options.Normal** specified as 'on'. Then, use the fit function with the specified options. Alternatively, you can use the name-value pair, such as **'Normilize'**, **'on'** as arguments on the **fitoptions**() function.

Fit Options and Optimized Starting Points

To make this Fit Option more sense, we need to use the second model, Exponential, to help us to understand it better.

On the opened App, as shown in Fig. 5.59, click on the **Exponential** icon on the **FIT TYPE** pane to open that model for our motor system. Select **2** as the number of items from the **Number of terms** box, and the fitting result is shown in Fig. 5.60.

Regularly, you can specify fit options interactively in the **Fit Options** pane. All fits except **Interpolant**, **Smoothing Spline**, and **Lowess** have configurable fit options. The available options depend on the fit you select, such as linear, nonlinear, or nonparametric fit. The options described here are available for nonlinear models.

Lower and **Upper** coefficient constraints are the only fit options available in the **Fit Options** pane for **Polynomial** fits.

Nonparametric fits, such as **Interpolant**, **Smoothing Spline**, and **Lowess** fits, do not have **Advanced Options**. The **Coefficient Constraints** values are for the census data.

All popular options and related values are listed in Table 5.5.

To evaluate the fitting result, under the fitting curve displayed in Fig. 5.59, some fitting errors, such as R-Square, SSE, DFE, and RMSE, are shown in the **Table of Fits**. These errors are also grouped and displayed in a table named **Goodness of Fit** located at the lower-right corner. Compared between the Polynomial and Exponential models, the latter provides a better fitting result based on those error values.

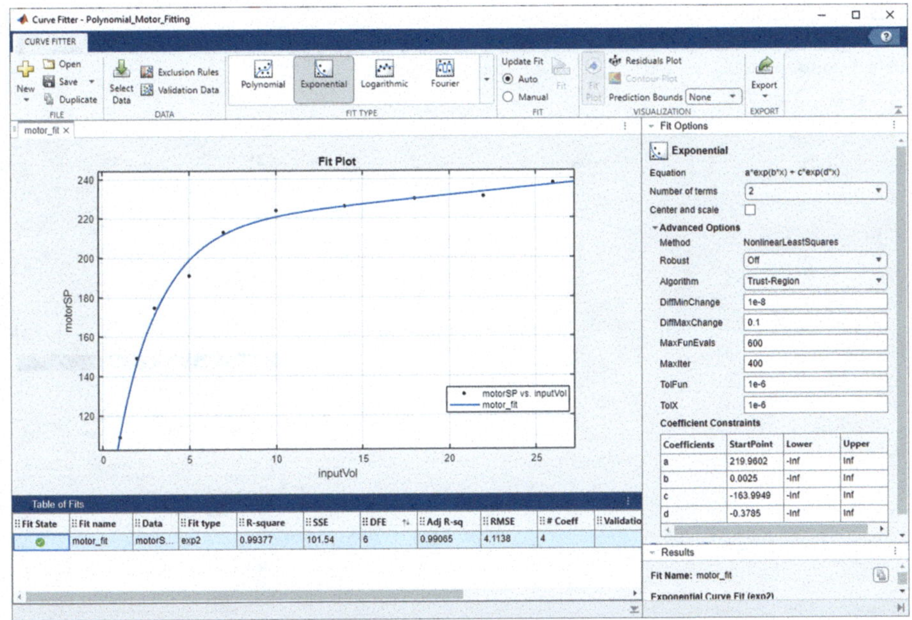

Fig. 5.60 The fitting result for Exponential model

Now we can try to use other models, such as Logarithmic and Fourier, to perform fitting for our motor system. Just click on each related model on the top to do those fitting functions. The fitting results for both models are shown in Figs. 5.61 and 5.62.

Change the **Robust** option to **LAR** for the Logarithmic model and select **2** as number items for the Fourier model in order to get the desired fitting results.

Comparably speaking, among all four models, the fitting performances for our motor system can be ordered as the following sequence starting from the best to the worst,

1. Exponential
2. Polynomial
3. Fourier
4. Logarithmic

Next let us continue to test the Gaussian and Rational models for our motor system.

In the opened Curve Fitter App, click on the drop-down arrow from the **FIT TYPE** pane on the top and select the **Gaussian** model icon. Then set 3 as the **Number of Terms** on the **Fit Options** pane on the right, and the fitting result is shown in Fig. 5.63.

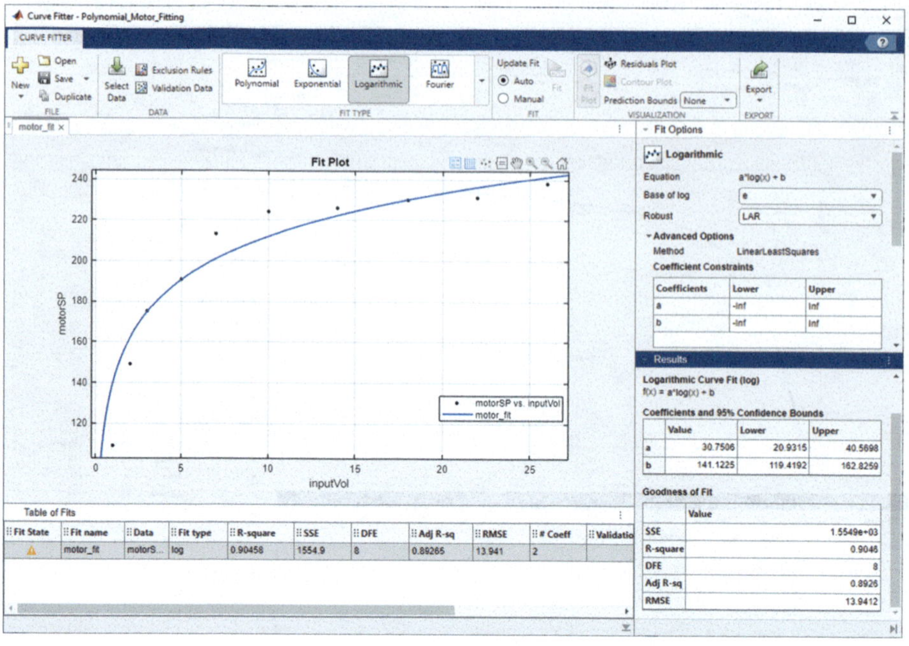

Fig. 5.61 The fitting result with Logarithmic model

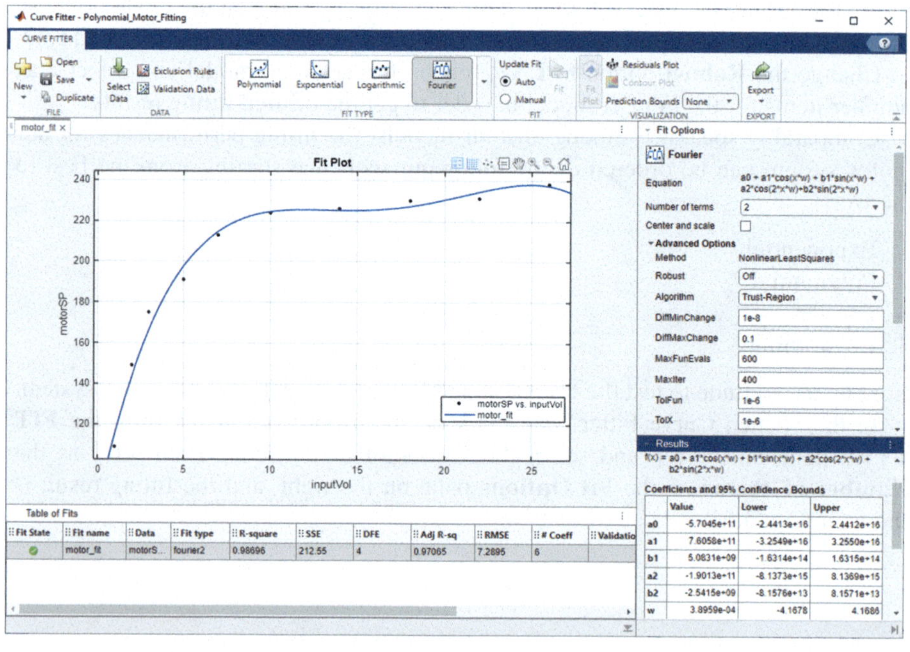

Fig. 5.62 The fitting result with Fourier model

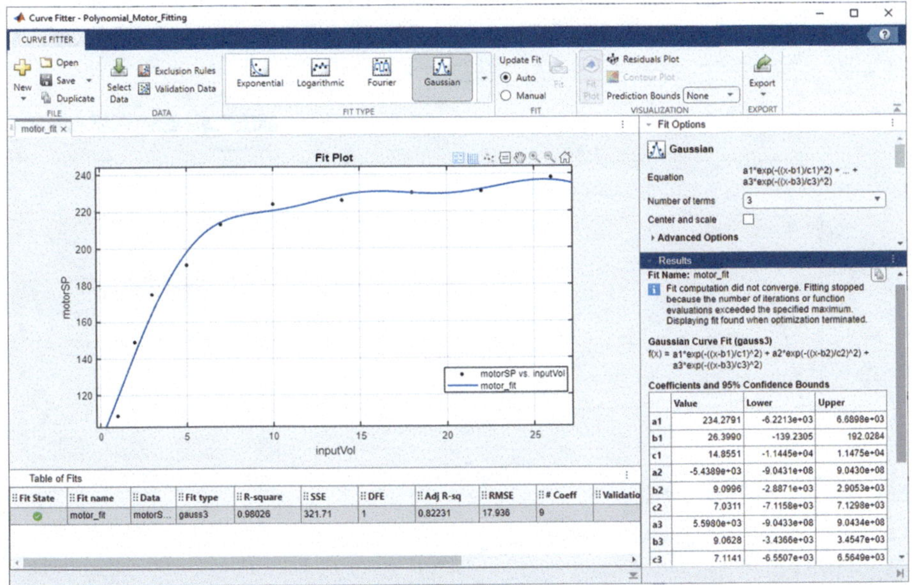

Fig. 5.63 The fitting result with Gaussian model

Similarly, select the **Rational** model icon and change the **Numerator degree** to 2 and **Denominator degree** to 4 on the Fit Options pane. Also check the **Center and scale** checkbox to start this fitting. The fitting result is shown in Fig. 5.64.

Relatively speaking, the Rational model has better performance than that of the Gaussian model. However, one important issue is that the Rational model becomes unstable as the numerator or denominator degree is increased, and the fitting result is different for each time.

5.6.1.2   Save and Export Fitting Results to File and Workspace

You can save your fitted model with results to a file or to the MATLAB Workspace.

To save the current curve fitting session to a session file, on the **Curve Fitter** tab in the **File** section, click on the drop-down arrow on the **Save** item and select **Save Session**. The session file is extended with **.sfit** and it contains all the fits and variables in your session and remembers your layout. In our case, the file name is **curve_motor_fitting.sfit**. You can use **Open** item in the Curve Fitter App to open your saved fitting session as you need it later.

To reserve our fitting models with results to Workspace, you can export the selected fitting model with results to the Workspace. On the **Curve Fitter** tab, in the **Export** section, click the drop-down arrow on the **Export** icon and select **Export to Workspace**. The fit is saved as a MATLAB object and the associated fit results are saved as structures, as shown in Fig. 5.65.

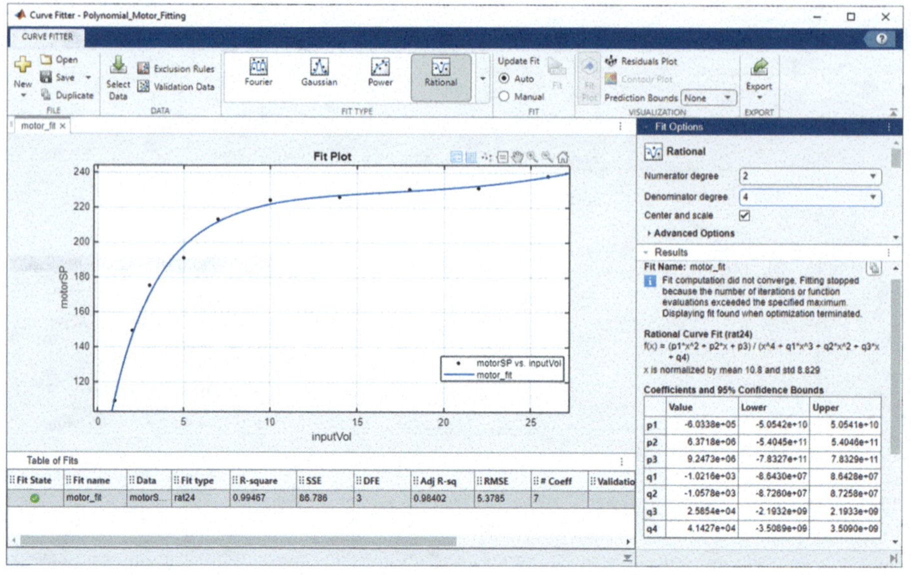

Fig. 5.64 The fitting result with Rational model

Fig. 5.65 The Save Fit to Workspace wizard

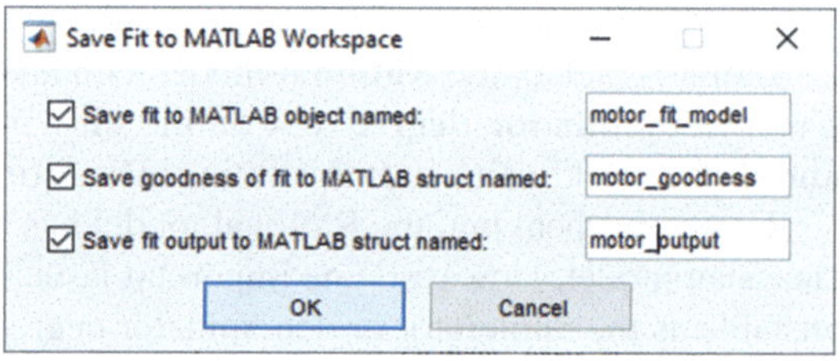

For our case, we changed the names for model, goodness, and output as shown in Fig. 5.65. Click on the **OK** button to complete this exporting action.

In addition to exporting the models with results to Workspace, you can also export the current selected fit to a figure to be utilized later. Moreover, you can even convert the current fitting to MATLAB codes and save them in a new MATLAB function called **createFit()**, which can be used later without touching Curve Fitter. To do that, just click on the drop-down arrow on the **Export** icon and select the item **Generate Codes**.

An example of those generated codes with that function is shown in Fig. 5.66, in which an **Exponential** model is adopted.

5.6.1.3 Evaluate the Fitted Model with New Date

If you want to evaluate the fitted model with a set of new input data, you can build a piece of simple codes to do that job. A piece of sample codes used to evaluate our fitted model, **curve_ motor_fitting.sfit**, is shown in Fig. 5.67. The name of this

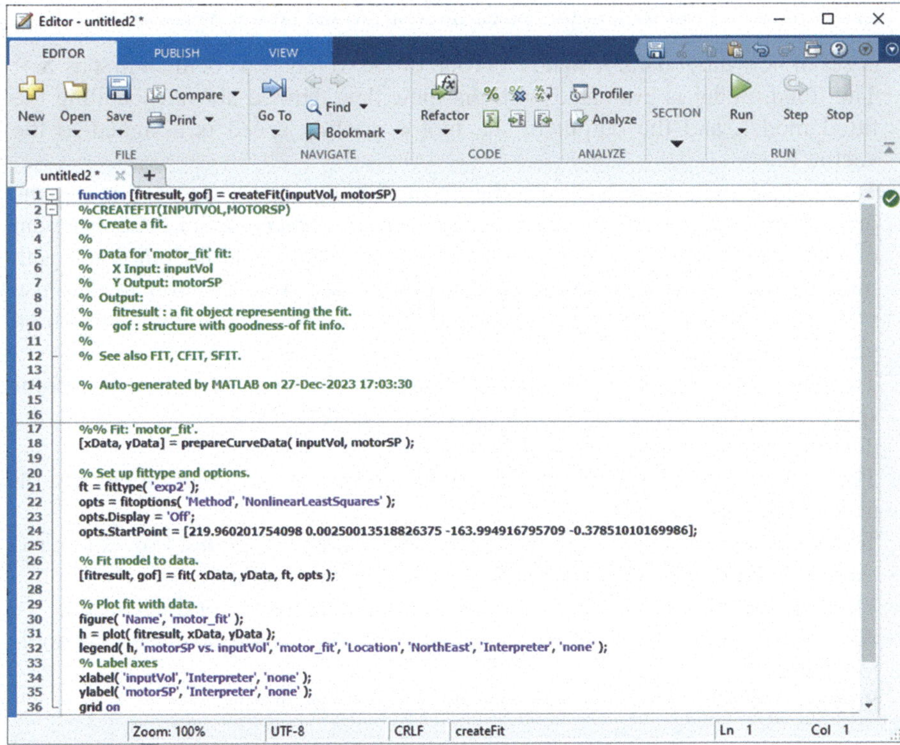

Fig. 5.66 The converted MATLAB codes for the Exponential fitting model

```matlab
% Test to evaluate fitting model that is exported from Curve Fitter App
% Name: eval_model_code.m
% Modle name = motor_fit_model
% Input vector = [1 3 8 9 12 15 18]
% Prior to run this file, the fitted model, motor_fit_model, must have been  exported to Workspace.
X = [1 3 8 9 12 15 18]
Y = motor_fit_model(X)
T = readtable("C:\\Artificial Intelligence Book\\Students\\Datasets\\nl_motor.xls")

cVol = T.inputVol
cMSP = T.motorSP
plot(motor_fit_model, cVol, cMSP)
hold on
plot(motor_fit_model, X, Y,"k+")
hold off
grid;
legend(["data","","extrapolated data","fitted curve"], ...
   "Location","southeast")
xlabel('Input Voltage (V)');
ylabel('Motor Speed (RPM)');
```

Fig. 5.67 The codes used to evaluate the fitted model with new data

script file is **eval_model_code.m** and it can be found from the Springer ftp site at the folder **Students\Class Projects\Chapter 5\Nonlinear Regression Apps**.

A key point is that the fitted model must have been exported to the Workspace prior to running this script file. Otherwise, an error may be encountered as you run it.

Let us have a closer look at this piece of codes to see how it works.

1. A set of new input data, which is a group of DC voltages, is defined first as **X**.
2. The fitted model is evaluated with that new data defined above by calling the fitted model, and the output or the motor rotating speed is assigned to the vector **Y**.
3. To compare the tested data with the original fitting curve, we need to get the original motor input-output data by using the **readtable**() function. The argument of that function is the original data file name. A full name including the path must be used since our data file is stored in that location. You may need to use your actual path + file name if you saved that data file at a different location on your machine.
4. Two data columns on the original data table, input voltages and output motor speed, are assigned to two variables, **cVol** and **cMSP**, respectively, by using the dot operator.
5. The original fitting curve is plotted based on the original input-output data in terms of the fitted model.
6. To enable us to do the second plotting over the first one, the **hold on** command must be used to keep the original plot in there without being refreshed.
7. Then the second plot with our new data is plotted on the top of the first plot.
8. After two plots are done, the command **hold off** can be used to release any holding action.
9. Some illustrations are displayed with the **legend**() function.

Now run this piece of codes to do evaluation for the fitted model and comparison between original and new data. The running result is shown in Fig. 5.68. Where the circle dots are the original data and the + dots are new testing data.

Next let' concentrate our discussion on nonlinear regression-related functions.

5.6.2 Nonlinear Regression-Related Functions

The Curve Fitter App is exactly involved in a MATLAB Toolbox, called Curve Fitting Toolbox, in which all fitting-related Apps and functions are included. In this book, we only concentrate on the curve fitting but provide a little about surface fitting.

To perform curve or surface fitting with programming codes or called programmatic curve or surface fitting, we need to be familiar with curve and surface fitting objects and related object functions, which are included in the MATLAB Curve Fitting Toolbox.

In this section, we will focus our discussion on,

- How to create curve and surface fitting objects programmatically.
- How to use their object functions to manipulate the fits programmatically.

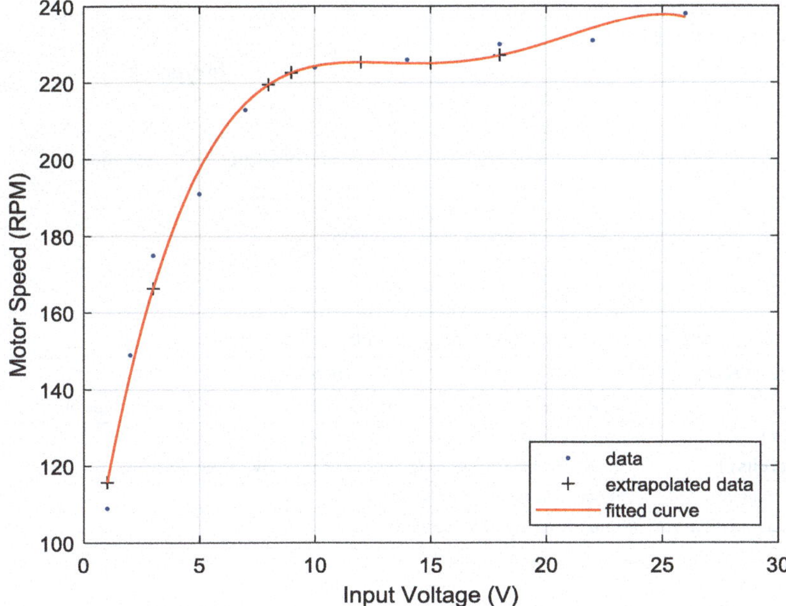

Fig. 5.68 The evaluation and comparison result for our fitted model

After finishing this section, you will be able to use the Curve Fitting Toolbox objects and object functions at the MATLAB command line or to write MATLAB programs for curve and surface fit applications.

5.6.2.1   The Curve Fitting Toolbox Objects and Object Functions

In MATLAB programming, all workspace variables are **Objects** of a particular **Class**. Some typical MATLAB classes include **double**, **char**, **function_handle**, and **Data types**. You can also create custom MATLAB classes, as you did in object-oriented programming.

Object functions are functions that operate exclusively on objects of a particular class. The **Data types** class packs together objects and object functions so that the object functions operate exclusively on objects of their own type, and not on objects of other types. A clearly defined encapsulation of objects and object functions is the goal of object-oriented programming.

Curve Fitting Toolbox software provides you with new MATLAB data type objects for performing curve fitting:

1. **fittype**—Objects allow you to encapsulate information describing a parametric model for your data. Object functions allow you to access and modify that information.

Fig. 5.69 Relations among fittype objects and cfit and sfit objects

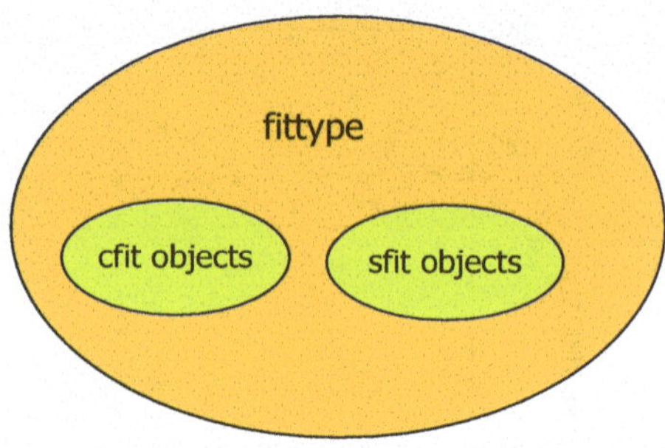

Table 5.6 Some popular and important Create Fits object functions

Create Fit Functions	Descriptions
`fit()`	Create a new **cfit** object by using the model type specified by the **fittype** object.
`fittype()`	Create or construct fit types by specifying library model or function.
`fitOptions()`	Create or modify fit options for the specified library model with additional options.
`prepareCurveData()`	Prepare data inputs for curve fitting.
`prepareSurfaceData()`	Prepare data inputs for surface fitting

2. **Cfit** and **sfit**—Two subtypes of **fittype**, for curves and surfaces. Objects capture information from a particular fit by assigning values to coefficients, confidence intervals, fit statistics, etc. Object functions allow you to postprocess the fit through plotting, extrapolation, integration, etc.

The relations among these objects are shown in Fig. 5.69.

As shown in Fig. 5.69, the **cfit** is a derived or a child class of **fittype**; thus, **cfit** inherits all **fittype** class functions. In other words, you can apply **fittype** object functions for both **fittype** and **cfit** objects, but the **cfit** object functions can only be used by the **cfit** objects. The same is true for the **sfit** objects.

All object functions can be divided either based on **Create Fits** and **Work with Fits**, or **Fit Type** object and **Curve Fit** object. To make it simple, here we just categorize them based on the former. Tables 5.6 and 5.7 show most popular and important Create Fits object functions and Work with Fits object functions. Where the functions with * marks are the Curve Fit object functions, and all other functions belong to Fit Type object functions.

5.6.2.2   Three Popular Nonlinear Fitting Functions in Curve Fitting Toolbox

Generally, as we mentioned in the previous section, MATLAB provided three major fitting functions that can be used to perform nonlinear regression process:

Table 5.7 Some popular and important work with Fits object functions

Work with Fits Functions	Descriptions
argnames()	Return input argument names of cfit, sfit, or fittype object.
category()	Return the fit category of the cfit, sfit, or fittype object.
coeffnames()	Return the coefficient (parameter) names of the cfit, sfit, or fittype object .
coeffvalues()[*]	Return the values of the coefficients, or parameters, of the cfit or sfit object.
feval()	Evaluate cfit, sfit, or fittype object.
formula()	Return the formula of the cfit, sfit, or fittype object fun as a character array.
get()	Return or display all property names and values of the fit options structure options.
islinear()	Determine if cfit, sfit, or fittype object is linear, returns 1 = linear, 0 = not.
numargs()	Return the number of input arguments of the cfit, sfit, or fittype object function.
numcoeffs()	Returns the number of coefficients of the cfit, sfit, or fittype object function.
plot()[*]	Plot the fit.
probvalues()[*]	Return the values of the problem-dependent (fixed) parameters of cfit object function as a row vector.
set()	Assign values to the fit options structure.
type()	Returns the custom or library name of the cfit, sfit, or fittype object function as a character array.

1. **fit**() function—Used for linear and nonlinear models.
2. **fitnlim**() function—Used for nonlinear models.
3. **nlinfit**() function—Used for nonlinear models.

The relations between above object functions and these three fitting functions are: all object functions discussed above are used to assist these three major fitting functions to perform the desired fitting process. Let us have a closer look at these three functions with related functions.

fit() Function

This function is used to create a new **cfit** object by using the model type specified by the **fittype** object. Regularly you do not need to create a **cfit** object directly unless if you want to assign values to coefficients and problem parameters of a **fittype** object *without* performing a fit.

There are six different constructors when calling this function to create a new **cfit** object:

1. **fitobject = fit(x, y, fitType):** Create the fit to the data in x and y with the model by **fitType**.
2. **fitobject = fit([x,y], z, fitType):** Create a surface fit to the data in vectors x, y, and z.
3. **fitobject = fit(x, y, fitType, fitOptions):** Create a fit to the data using the algorithm options specified by the **fitOptions** object.
4. **fitobject = fit(x,y,fitType,Name=Value):** Create a fit to the data using library model **fitType** with additional options by one or more **Name = Value** pair arguments. Use **fitOptions** to display available property names and default values for the specific library model.

5. **[fitobject,gof] = fit(x,y,fitType)**: Return goodness-of-fit statistics in the structure **gof**.
6. **[fitobject,gof,output] = fit(x,y,fitType)**: Return fitting algorithm information in the structure `output`.

Two important arguments, **fitType** and **fitOptions**, are key functions in this function, and need more discussions about them.

The function **fittype()** is used to construct fit types for a fitting function by.

- Specifying library model names.
- Specifying a MATLAB expression.
- Specifying an anonymousFunction.

There are five major constructors for this class:

1. **aFittype = fittype(libraryModelName)**: Create the **fittype** object for the model specified by a `libraryModelName`.
2. **aFittype = fittype(expression)**: Create a fit type for the model specified by a MATLAB expression.
3. **aFittype = fittype(expression,Name,Value)**: Construct the fit type with additional options specified by one or more **Name-Value** pair arguments.
4. **aFittype = fittype(anonymousFunction)**: Create a fit type for the model specified by an `anonymous function`.
5. **aFittype = fittype(anonymousFunction,Name,Value)**: Construct the fit type with additional options specified by one or more **Name-Value** pair arguments.

Table 5.8 lists some popular model names used for **fittype()** function.

An example code line by using the function **fittype()** to construct a fit type for model **poly3** is,

```
f = fittype('poly3')
f = Linear model Poly3: f(p1,p2,p3,p4,x)= p1*x^3 + p2*x^2 +
p3*x + p4
```

The function **fitOptios()** is similar to **fittype()**, but it provides some additional options to enable users to construct the fitting model with more flexibilities. There are seven major constructors for this class:

1. **fitOptions = fitoptions**: Create the default fit options object **fitOptions**.
2. **fitOptions = fitoptions(libraryModelName)**: Create the default fit options object for the library model.

Table 5.8 Some popular model name used for fittype() function

Library Model Name	Descriptions
`'poly1'`	Linear polynomial curve, `Y = p1*x+p2`
`'poly2'`	Quadratic polynomial curve, `Y = p1*x^2+p2*x+p3`
`'poly3'`	Cubic polynomial curve, `Y = p1*x^3+p2*x^2+...+p4`
`...upto'poly9'`	Higher order polynomial curve, `Y = p1*x^9+p2*x^8+...+p10`
`'weibull'`	Distribution model, `Y = a*b*x^(b-1)*exp(-a*x^b)`
`'exp1'`	Exponential model, `Y = a*exp(b*x)`
`'exp2'`	Exponential model, `Y = a*exp(b*x)+c*exp(d*x)`
`'fourier1'`	Fourier model, `Y = a0+a1*cos(x*p)+b1*sin(x*p)`
`'fourier2'`	Fourier model, `Y = a0+a1*cos(x*p)+b1*sin(x*p)+a2*cos(2*x*p)+b2*sin(2*x*p)`
`'fourier3'`	Fourier model, `Y = a0+a1*cos(x*p)+b1*sin(x*p)+...+a3*cos(3*x*p)+b3*sin(3*x*p)`
`...upto'fourier8'`	Fourier model, `Y = a0+a1*cos(x*p)+b1*sin(x*p)+...+a8*cos(8*x*p)+b8*sin(8*x*p)`
	Where `p = 2*pi/(max(xdata)-min(xdata))`.
`'gauss1'`	Gaussian model, `Y = a1*exp(-((x-b1)/c1)^2)`
`'gauss2'`	Gaussian model, `Y = a1*exp(-((x-b1)/c1)^2)+a2*exp(-((x-b2)/c2)^2)`
`'gauss3'`	Gaussian model, `Y = a1*exp(-((x-b1)/c1)^2)+...+a3*exp(-((x-b3)/c3)^2)`
`...upto'gauss8'`	Gaussian model, `Y = a1*exp(-((x-b1)/c1)^2)+...+a8*exp(-((x-b8)/c8)^2)`
`'log'`	Logarithmic model, `Y = a*log(x)+b`
`'log10'`	Base-10 Logarithmic model, `Y = a*log10(x)+b`
`'log2'`	Base-2 Logarithmic model, `Y = a*log2(x)+b`
`'power1'`	Power model, `Y = a*x^b`
`'power2'`	Power model, `Y = a*x^b+c`
`'rat02'`	Rational model, `Y = (p1)/(x^2+q1*x+q2)`
`'rat21'`	Rational model, `Y = (p1*x^2+p2*x+p3)/(x+q1)`
`'rat55'`	Power model, `Y = (p1*x^5+...+p6)/(x^5+...+q5)`
`'lowess'`	Local linear regression (surface)
`'logistic4'`	Four-parameter logistic curve

3. **fitOptions = fitoptions(libraryModelName,Name,Value)**: Create fit options for the specified library model with additional options specified by one or more **Name-Value** pair.

4. **fitOptions = fitoptions(fitType)**: Get the fit options object for the specified fitType. Use this syntax to work with fit options for custom models.

5. **fitOptions = fitoptions(Name,Value):** Create fit options with additional options specified by one or more **Name-Value** pair arguments.

6. **newOptions = fitoptions(fitOptions,Name,Value)**: Modify the existing fit options object **fitOptions** and return updated fit options in **newOptions** with new options specified by one or more **Name-Value** pair arguments.

7. **newOptions = fitoptions(options1,options2)**: Combine the existing fit options object option1 and options2 in **newOptions**.

The available options can be divided into different categories, such as

1. Options for all fitting methods
2. Options for interpolations
3. Options for smoothing
4. Options for linear and nonlinear Least-Squares

All of these options are specified as **Name-Value** pair to be included in the arguments of related calling functions. For our purpose, options 2 and 3 are excluded and we only pay more attention to options 1 and 4. Table 5.5 lists all options for option 4 and some options for option 1. Here some additional fitting methods are shown in Table 5.9.

An example of using the function **fitoptions**() to construct a fitting model is,

```
    options = fitoptions('poly3','Normalize','on','Robust','LAR','Lo
wer',[0 -Inf 0 0 -Inf 0])
    options = llsqoptions with properties:


          Lower: [0 -Inf 0 0 -Inf 0]
          Upper: []
          Robust: 'LAR'
          Normalize: 'on'
          Exclude: []
          Weights: []
        Method: 'LinearLeastSquares'
```

fitnlim() Function

Similar with the function **fit**(), the function **fitnlm**() can also create, construct and fit a nonlinear regression model, but the model is specified by a model function, not a library model name. In most applications, the model function is specified by an anonymous function represented by a function handle. Another difference is that the

Table 5.9 Some additional options for fitting methods

Fitting Method	Description
'NearestInterpolant'	Nearest neighbor interpolation
'LinearInterpolant'	Linear interpolation
'PchipInterpolant'	Piecewise cubic Hermite interpolation (curves only)
'CubicSplineInterpolant'	Cubic spline interpolation
'BiharmonicInterpolant'	Biharmonic surface interpolation
'SmoothingSpline'	Smoothing spline
'LowessFit'	Lowess smoothing (surfaces only)
'LinearLeastSquares'	Linear least squares
'NonlinearLeastSquares'	Nonlinear least squares

function **fit()** can be used to fit both linear and nonlinear models, but the function **fitnlm()** is used to fit nonlinear models.

Three constructors are available for this fit model creation:

1. **mdl = fitnlm(tbl,modelfun,beta0)**: Fit the model specified by **modelfun** to variables in the table or dataset array **tbl**, and return the nonlinear model **mdl**. The **fitnlm()** estimates model coefficients using an iterative procedure starting from the initial values in **beta0**.
2. **mdl = fitnlm(X,y,modelfun,beta0)**: Fit a nonlinear regression model using the column vector y as a response variable and the columns of the matrix **X** as predictor variables.
3. **mdl = fitnlm(__, modelfun,beta0,Name,Value)**: Fit a nonlinear regression model with additional options specified by one or more **Name-Value** pair arguments.

Two arguments, **modelfun** and **beta0**, are needed for our more attentions.

The **modelfun** is used to define a model and most time it is specified by an anonymous function represented as a function handle. The **beta0** is the initial value for all coefficients on that anonymous equation.

In fact, the **modelfun** is a functional form of the model, specified as either of the following:

1. In the function handle, **@modelfun** or **@(b,x)modelfun**, b is a coefficient vector with the same number of elements as **beta0**, and x is a matrix with the same number of columns as **X** or the number of predictor variable columns on the table **tbl**. **modelfun(b,x)** returns a column vector that contains the same number of rows as x. Each row of the vector is the result of evaluating **modelfun** on the corresponding row of x. In other words, **modelfun** is a *vectorized* function, one that operates on all data rows and returns all evaluations in one function call. **Modelfun** should return real numbers to obtain meaningful coefficients.
2. Character vector or string scalar formula in the form $y \sim f(b_1, b_2,..., b_j, x_1, x_2,..., x_k)$, where f represents a scalar function of the scalar coefficient variables $b_1,..., b_j$ and the scalar data variables $x_1,...,x_k$. The variable names in the formula must be valid MATLAB identifiers.

Figure 5.70 shows a piece of sample codes to illustrate how to use **modelfun** and **beta0**. Let us have a closer look at this piece of codes to see how it works.

1. A sample dataset **carbig** is used and loaded into the Workspace.
2. Three columns, **Horsepower**, **Weight**, and **MPG**, are selected and the first two columns work as inputs and the last one works as the response or output variable. All of those columns are fed into a table **tbl**.
3. A **modelfun** is created with five coefficients, b_1–b_5, and the first and the second columns in **tbl** work as input variables. The definition **@(b, x)** indicates that the

```
% An example of using modelfun() to perform a nonlinear fitting
1  load carbig
2  tbl = table(Horsepower, Weight, MPG);
3  modelfun = @(b,x)b(1) + b(2)*x(:,1).^b(3) + ...
                    b(4)*x(:,2).^b(5);
4  beta0 = [-50 500 -1 500 -1];
5  mdl = fitnlm(tbl, modelfun, beta0)
```

Fig. 5.70 A piece of sample codes used to illustrate the modelfun and beta0

model function is composed of the coefficients b and input variables x, both are matrix. Following that definition is the real equation.

4. The initial input vector is defined by **beta0 = [−50,500−1500−1],** which is equivalent to assign values to b_1–b_5, respectively.
5. The **fitnlm()** function is executed with the **modelfun** and **beta0** to fit that model.

The actual model is defined as: **$MPG \sim b_1 + b_2*Horsepower^{\wedge}b_3 + b_4*Weight^{\wedge}b_5$.**

nlinfit() Function

This function is very similar to the function **fitnlm()** constructed with the second constructor format. But this function only returns a vector of estimated coefficients for the fitted model. Totally four different constructors can be used to create this fitting model:

1. **beta = nlinfit(X,Y,modelfun,beta0)**: Return a vector of estimated coefficients for the nonlinear regression of the responses in **Y** on the predictors in **X** using the model specified by **modelfun**. The coefficients are estimated using iterative least squares estimation, with initial values specified by **beta0**.
2. **beta = nlinfit(X,Y,modelfun,beta0,options)**: Fit the nonlinear regression using the algorithm control parameters in the structure **options**. You can return any of the output arguments in the previous syntaxes.
3. **beta = nlinfit(__, Name,Value)**: Use additional options specified by one or more Name-Value pair arguments. For example, you can specify observation weights or a nonconstant error model. You can use any of the input arguments in the previous syntaxes.
4. **[beta,R,J,CovB,MSE,ErrorModelInfo] = nlinfit(__)**: Additionally return the residuals **R**, the Jacobian of modelfun **J**, the estimated variance-covariance matrix for the estimated coefficients **CovB**, an estimate of the variance of the error term **MSE**, and a structure containing details about the error model **ErrorModelInfo**.

where the input argument, **options**, needs to be paid more attention since all other input arguments have been discussed in above functions.

The **options** is a structure and it can be created and set by using a function **statset('funName')**. The argument **funName** is a valid function name defined in the MATLAB Statistics and Machine Learning Toolbox. Different functions have different options structures. Some popular fitting functions include, but are not limited to, **evfit, fit, fitnlm, nlinfit,** and **fitnet**.

Table 5.10 shows some popular options values and weight function names. Table 5.11 lists popular **Value-Name** pairs with related parameter values.

A piece of example codes that is used to test the **nlinfit()** function with all four constructors is shown in Fig. 5.71. Let us have a closer look at that piece of codes to see how it works.

1. This example uses a dataset **carbig** as the data source, so first just load it to Workspace.
2. Only two columns, **Horsepower** and **Weight**, are selected as the inputs, and the **MPG** is chosen as the response variable.
3. The selected inputs are fed to a new array variable **X**, and output is fed to **Y**.
4. The **modelfun** is defined based on a function handle with five coefficients and two input columns, the first and the second columns.
5. The coefficients are initialized by assigning five initial values to **beta0**. If all initial values are zeros, a warning may be given but it still works as the codes are running.
6. The first **nlinfit()** function is called and the returned coefficients are assigned to the vector **beta1**, which can be displayed in the Command window.
7. A new options structure is made by calling the function **statset()**. The argument of this function is a name of a fitting function. In our case, it is **nlinfit**.

Table 5.10 Some popular values for the options structure and weight functions

Options Value	Description
`'DerivStep'`	Relative difference for finite difference gradient, vector
`'FunValCheck'`	Indicator for whether to check for invalid values, 'on' or 'off'
`'MaxIter'`	Maximum number of iterations, positive integer, 100 default
`'RobustWgtFun'`	Weight function (see below for weight functions table for available funs)
`'TolFun'`	Termination tolerance on residual sum of squares, 1e-8 default
`'TolX'`	Termination tolerance on estimated coefficients, 1e-8 default
Weight Function	**Description**
`'andrews'`	$w=I[|r|<\pi]\times\sin(r)/r$
`'bisquare'`	$w=I[|r|<1]\times(1-r^2)^2$
`'fair'`	$w=1/(1+|r|)$
`'huber'`	$w=1/\max(1,|r|)$
`'logistic'`	$w=\tanh(r)/r$
`'welsch'`	$w=\exp\{-r^2\}$

Table 5.11 Some popular Value-Name pairs and related parameter values

Value-Name Pair	Description
`ErrorModel`	Form of error terms, 'constant', 'proportional', 'combined'
`ErrorParameters`	Initial estimates for error model parameters
	'constant' = a, 1, 'proportional'=b, 1, 'combined'=a,b, [1,1]
`Weights`	Observation weights, a vector or a function handle

```
% testing nlinfit() function
1  load carbig
2  tbl = table(Horsepower,Weight,MPG);
3  X = [tbl.Horsepower tbl.Weight];
   Y = tbl.MPG;
4  modelfun = @(b,x)b(1) + b(2)*x(:,1).^b(3) + ...
                  b(4)*x(:,2).^b(5);
5  beta0 = [-50 500 -1 500 -1];
   %beta0 = [0 0 0 0 0];
6  beta1 = nlinfit(X, Y, modelfun, beta0)
7  opts = statset('nlinfit');
8  opts.MaxIter = 150;
   opts.RobustWgtFun = 'bisquare';
   opts.TolFun = 1e-9;
9  beta2 = nlinfit(X, Y, modelfun, beta0, opts)
10 beta3 = nlinfit(X, Y, modelfun, beta0, opts, "ErrorModel","constant")
11 [beta4, R, J, CovB, MSE, ErrorModelInfo] = nlinfit(X, Y, modelfun, beta0)
```

Fig. 5.71 A piece of example codes to test nlinfit() function

8. An option value, **MaxIter**, is modified to 150, which means that the maximum number of iterations is set to 150 (default is 100). The **Weight Function** is selected as **bisquare** and assigned to the **RobustWgFun** option. Also the termination tolerance on residual sum of squares is set to 1e-9.
9. The second **nlinfit()** function is executed with the additional options.
10. The third **nlinfit()** function is called and it is used to test the **Name-Value** pairs, such as **ErrorModel** with a value of **constant**.
11. The fourth **nlinfit()** is tested with the returned parameters. All of those returned parameters will be displayed in the Command window as this piece of codes runs.

Next let us use some of these fitting functions to build fitting models for our simple motor system as examples to illustrate how to utilize these functions to fit our motor system.

5.6.2.3   Build Nonlinear Fitting Models with Fitting Functions

We start our project with a sample dataset, **nl_motor.xls**, which is a simple dataset used to provide a relation between the input DC voltages and motor rotation speed, and this dataset can be found from the Springer ftp site under a folder **Students\ Datasets**. Copy and paste it to one of your local folders to be used later.

Create a new MATLAB Script file and name it **motor_nlfit_func.m** and enter the codes shown in Fig. 5.72 into that file. Let us have a closer look at this piece of codes to see how it works.

1. First we need to load our simple motor dataset and assign it to a local table variable **T**. A point to be noted is the location of that dataset. You must use a **full path\\dataset_name** to get it. Also you need to use your actual path to replace our path if you saved that dataset at different folder in your machine.

```
% Use nonlinear fit functions to set nonlinear model for motor system
% Name: motor_nlfit_func.m
% Dataset is nl_motor.xls
% Output: Fitted model

% Use fit() function...
1    T = readtable("C:\\Artificial Intelligence Book\\Students\\Datasets\\nl_motor.xls")

2    cVol = T.inputVol;
     cMSP = T.motorSP;
3    ft = fittype('exp1')
4    fo = fitoptions('Method','NonlinearLeastSquares',...
                     'Lower',[0,0],...
                     'Upper',[Inf, max(cVol)],...
                     'StartPoint',[1 1]);
5    mdl_1 = fit(cVol, cMSP, ft, fo)

     % Use fitlnm() function.... Define the model as Y = a*log(x) + b
6    ModelFunc = @(b, x) b(1)*log(x(:, 1)) + b(2)
7    beta0 = [30 100];                        % Refer to Figure 5.63
8    mdl_2 = fitnlm(cVol, cMSP, ModelFunc, beta0)

     % Use nlinfit() function.... Define model as poly4: Y = p1*x^4 + p2*x^3 + p3*x^2 + p4*x + p5
9    ModelFunc = @(b, x) b(1)*x(:, 1).^4 + b(2)*x(:, 1).^3 + b(3)*x(:, 1).^2 + b(4)*x(:, 1) + b(5);
10   beta0 = [0 0 -4 20 50];                   % Refer to Figure 5.61
     %beta0 = [0 0 0 0 0];
11   mdl_3 = nlinfit(cVol, cMSP, ModelFunc, beta0)

     % Use fitlnm() function..., Define the model as Y = a* exp(-((x - b)/c)^2)
12   ModelFunc = @(b, x) b(1) * exp(-((x(:, 1) - b(2))/b(3)).^2);
13   beta0 = [200, 10, 20];                     % Refer to Figure 5.65
14   mdl_4 = fitnlm(cVol, cMSP, ModelFunc, beta0)
```

Fig. 5.72 The codes used to perform nonlinear regression with functions

2. Extract two columns, **inputVol** and **motorSP**, from the dataset as input and response variable, and assign them to two local variables **cVol** and **cMSP**, respectively.

3. Construct the fitting type as **exp1** model ($Y = a*exp(b*x)$) with the **fittype()** function.

4. Using **fitoptions()** function to add additional options for this fitting, such as **Method**, **Lower**, and **Upper** bounds, as well as **StartPoint[]**, by using **Name-Value** pairs.

5. The fitting function **fit()**, who can handle both linear and nonlinear regression fitting, is called to perform this fitting action to derive a desired model **mdl_1**. The estimated model with related parameters will be displayed in the Command window as the project runs since we did not attach any semicolon (;) after that coding line.

6. Next we can test the function **fitnlm()** with a logarithmic model by building a **ModelFunc** with the function handle. Refer to Fig. 5.61 to get that actual model format.

7. Set up certain initial values for two parameters in vector variable **beta0**. Refer to Fig. 5.61 to get two possible initial values based on their estimated values (**37.8334 124.718**).

8. The function **fitnlm()** is executed with the input **X**, output **Y**, **ModelFunc**, and **beta0**. The model, **mdl_2**, with the estimated parameters will be displayed in the Command window.

9. Next we like to test the fitting function **nlinfit()** with a polynomial model that has a degree of 4, **poly4**. Refer to Fig. 5.59 and Eq. (5.23) to get more details about the format of that polynomial equation. The **ModelFunc** is constructed with a function handle by following that equation format. One point to be noted is how to calculate the power of vectors. A good way to do that is to use **dot** (.) and $\wedge$ operators. For example, in our model function, $b(1)*x(:, 1)^4$, it must be presented as $b(1)*x(:, 1).\wedge 4$, where $\wedge$ is a power symbol and $\wedge 4$ means $()^4$. Sine $x(:, 1)$ is a vector, thus a **dot** (.) operator must be used to enable a vector to be powered by another vector or a scale factor.

10. The initial values for those estimated parameters must be provided and assigned to the vector variable **beta0**. Refer to Fig. 5.59 to get more details about those values.

11. The fitting function **nlinfit()** is called with the input, output, model function and **beta0** to get the estimated model. The estimated model, **mdl_3**, with related parameters will be displayed in the Command window as the project runs.

12. Finally another model function constructed as a Gaussian model is tested with the **fitnlm()** function. Refer to Fig. 5.63 to get more details about the Gaussian model equation.

13. The initial values for three estimated parameters are given and assigned to the **beta0**. One point to be noted is the Gaussian equation and three parameters shown in Fig. 5.63. In this test, we only need to use the top equation with three parameters, a_1, b_1, and c_1, shown in Fig. 5.63.

14. The fitting function **fitnlm()** is executed to get the estimated model, **mdl_4**. The estimated result, including the estimated parameters, will be displayed in the Command window as the project runs.

Now run this piece of codes or this project, the running results are shown in Fig. 5.73.

Next let us do some evaluations to check the effectiveness of using these fitting methods.

5.6.2.4 Evaluate the Fitted Model with New Date Via Functions

As we did in Sect. 5.6.1.3, we can use some new data to check the performance of those fitting algorithms for our motor system.

Create a new Script file name it **eval_fit_func.m** and enter the codes shown in Fig. 5.74 into that file. Let us have a closer look at that piece of codes to see how it works.

1. First we need to load our simple motor dataset and assign it to a local table variable **T**. A point to be noted is the location of that dataset. You must use a **full path\\dataset_name** to get it. Also you need to use your actual path to replace our path if you saved that dataset in a different folder in your machine.

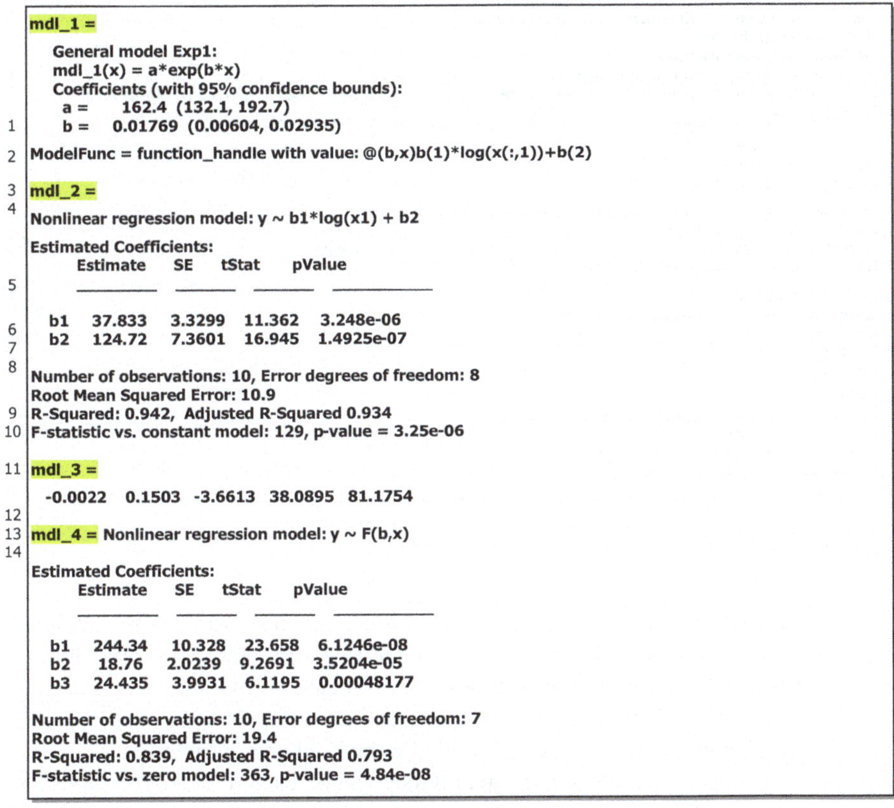

Fig. 5.73 The running results of testing nonlinear regression with functions

2. Extract two columns, **inputVol** and **motorSP**, from the dataset as input and response variable, and assign them to two local variables **cVol** and **cMSP**, respectively.

3. Create a new input data vector, input voltages, and assign it to the local variable **X**.

4. Construct the model function as a logarithmic ($Y = a*log(x) + b$) with the function handle, and perform a nonlinear regression fitting. The fitted model is returned to **mdl_2**.

5. The function **feval**() is executed to perform an evaluation for the newly generated data vector **X** based on the fitted model, and the evaluation result is assigned to the variable **Y**.

6. Now plot the original fitted model with the original input-output data. That original fitted model, **motor_fit_model**, is obtained from a regression nonlinear App method in Sect. 5.6.1.1. The purpose of this plotting is to compare the new fitted model with the original one.

```
% Evaluate nonlinear fit functions for motor system
% Name: eval_fit_func.m
% Dataset is nl_motor.xls
% Output: Fitted model performance
% Prior to run this file, the fitted model, motor_fit_model, must have been  exported to Workspace.
1  T = readtable("C:\\Artificial Intelligence Book\\Students\\Datasets\\nl_motor.xls")

2  cVol = T.inputVol;
   cMSP = T.motorSP;
3  X = [1 3 8 9 12 15 18];

   % Use fitlnm() function.... Define the model as Y = a*log(x) + b
4  ModelFunc = @(b, x) b(1)*log(x(:, 1)) + b(2)
   beta0 = [30 100];                         % Refer to Figure 5.63
   mdl_2 = fitnlm(cVol, cMSP, ModelFunc, beta0)

5  Y = feval(mdl_2, X)
6  plot(motor_fit_model, cVol, cMSP)
7  hold on
8  plot(X, Y,"k+")
9  hold off
   grid;

10 legend(["data","","extrapolated data"], ...
    "Location","southeast")
   xlabel('Input Voltage (V)');
   ylabel('Motor Speed (RPM)');
```

Fig. 5.74 The codes used to evaluate the fitting models

7. A **hold on** command is executed to keep the original plotting to be displayed and this command enables us to plot another trajectory on the original plotting.
8. The new fitting model with new input and evaluated output data is plotted with the + symbols to distinguish from the original ones.
9. A **hold off** command is executed to release the hold on command.
10. Some legends and labels are set up and they can be used to make the plot more meaningful.

Now run this script file and the running result is shown in Fig. 5.75.

It can be found from that figure that the estimated model is close to the original model, and this can be confirmed by checking the fitted curves and response data points marked with "+."

Next let us have a quick discussion about the K-Nearest Neighbor (KNN) algorithm used in regression nonlinear applications.

5.7 K-Nearest Neighbor (KNN) Algorithm

The K-Nearest Neighbors (KNN) algorithm is a robust and intuitive machine learning method employed to tackle classification and regression problems. By capitalizing on the concept of similarity, KNN predicts the label or value of a new data point by considering its K closest neighbors in the training dataset [13].

K-Nearest Neighbors (KNN) is a nonparametric machine learning algorithm that can be used for both classification and regression tasks. In the context of regression,

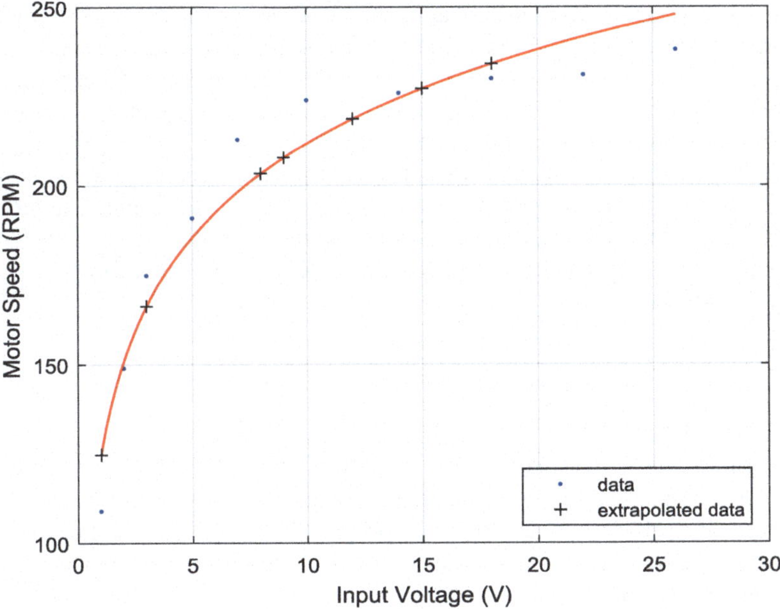

Fig. 5.75 The running result of the evaluation process

KNN is often referred to as "K-Nearest Neighbors Regression" or "KNN Regression." It is a simple and intuitive algorithm that makes predictions by finding the K nearest data points to a given input and averaging their target values for numerical regression or selecting the majority class for classification [14].

5.7.1 Working Principle of KNN Algorithm

The K-Nearest Neighbors (KNN) algorithm operates on the principle of similarity, where it predicts the label or value of a new data point by considering the labels or values of its K-nearest neighbors in the training dataset.

To make predictions, the algorithm calculates the distance between each new data point in the test dataset and all the data points in the training dataset. The Euclidean distance is a commonly used distance metric in K-NN, but other distance metrics, such as Manhattan distance or Minkowski distance, can also be used depending on the problem and data. Once the distances between the new data point and all the data points in the training dataset are calculated, the algorithm proceeds to find the K-nearest neighbors based on these distances. The specific method for selecting the nearest neighbors can vary, but a common approach is to sort the distances in ascending order and choose the K data points with the shortest distances.

After identifying the K-nearest neighbors, the algorithm makes predictions based on the labels or values associated with these neighbors. For classification tasks, the majority class among the K neighbors is assigned as the predicted label for the new data point. For regression tasks, the average or weighted average of the values of the K neighbors is assigned as the predicted value.

Let **X** be the training dataset with **n** data points, where each data point is represented by a d-dimensional feature vector X_i and Y be the corresponding labels or values for each data point in **X**. Given a new data point x, the algorithm calculates the distance between x and each data point X_i in **X** using a distance metric, such as Euclidean distance:

$$distance\left(x,X_i\right) = \sqrt{\sum_{j=1}^{d}\left(x_j - X_{ij}\right)^2} \qquad (5.24)$$

The algorithm selects the K data points from **X** that have the shortest distances to x. For classification tasks, the algorithm assigns the label y that is most frequent among the K-nearest neighbors to x. For regression tasks, the algorithm calculates the average or weighted average of the values y of the K-nearest neighbors and assigns it as the predicted value for x [13].

A graphical illustration of applying KNN algorithm on nonlinear regression process is shown in Fig. 5.76.

In Fig. 5.76, the white circles are original data and the black circles are estimated or predicted data. The model is trained based on the original input-output data pairs

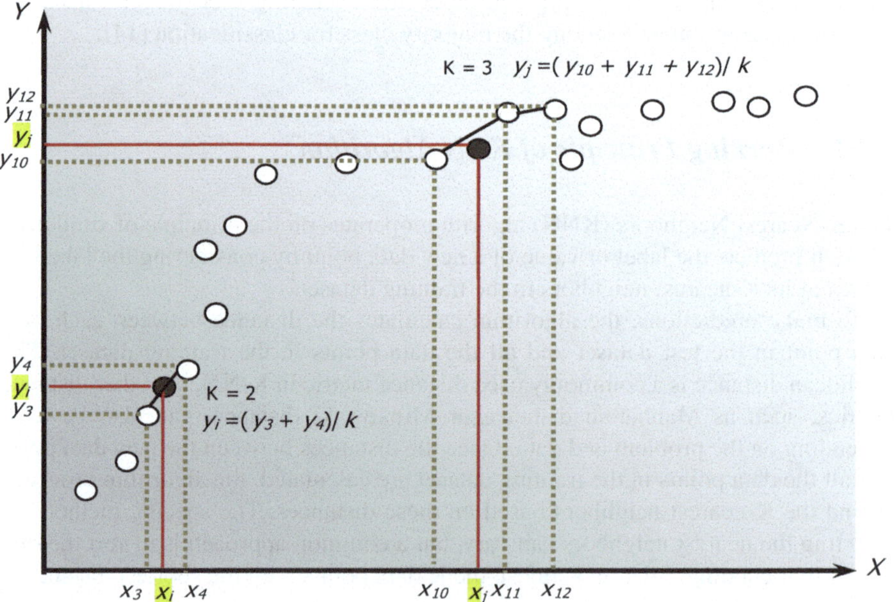

Fig. 5.76 A graphic illustration of using KNN regression for nonlinear estimation

and then it can be used to predict the outputs based on new input data. Here two examples with two different numbers of neighbors, **K**, are selected for this illustration, **K** = 2 and **K** = 3, respectively.

For **K** = 2, the new estimated output y_i for a new input x_i is an average of two or **K**-nearest neighbor points, (x_3, y_3) and (x_4, y_4). Similarly, when **K** = 3, the predicted new output y_j based on another new input x_j is also an average of three or **K**-nearest neighbor points, (x_{10}, y_{10}), (x_{11}, y_{11}), and (x_{12}, y_{12}).

Regularly, KNN algorithm is widely implemented in classifications even it can be used for both classifications and regressions. Due to that possible reason, MATLAB did not provide any related App and function for the KNN algorithm. In this section, we need to use a customer-made function, **K-Nearest Neighbor (KNN) Regressor** [15], to assist us to build KNN models.

5.7.2   K-Nearest Neighbor (KNN) Regression Algorithm

In this section, we only concentrate our study and discussion on KNN Regression nonlinear fitting problems.

First let us have a closer look at the operational procedure by using the KNN Regression algorithm. Totally five operational steps are involved as below [16]:

1. **Data Collection**: Starting with a dataset that includes both input features and target values. In regression tasks, the target values are continuous and represent the output data you want to predict.
2. **Choosing the Number of Neighbors (K)**: You need to choose the number of nearest neighbors, **K**, which will be used to make predictions. This is a hyperparameter that you can tune based on the characteristics of your data. A small **K** (e.g., 1 or 3) may lead to noisy predictions, while a large **K** may lead to overly smoothed predictions.
3. **Distance Metric**: KNN relies on a distance metric (e.g., Euclidean distance) to measure the similarity between data points. Different distance metrics can be used depending on the nature of your data.
4. **Prediction**: When you want to make a prediction for a new input data point, KNN calculates the distance between this point and all other data points in the dataset. It then selects the **K** data points with the smallest distances.
5. **Regression Prediction**: For regression, the predicted value for the new data point is the average of the target values of the **K**-nearest neighbors. This could be a simple arithmetic mean.

In fact, the **K-Nearest Neighbor (KNN) Regressor** is a function named **kNNeighborsRegressor**() and it is defined inside a class with the same name. The syntax of this function is,

kNNeighborsRegressor(k, metric, weights)
This function has three inputs, *k*, *metric*, and *weights*. The meaning of these inputs is:

- **k: Number of Neighbors**, is a positive integer ranged between 1 and inf.
- *metric*: **Distance Metric**, indicates the desired distance measured method to be used from three of them, Euclidean Distance, Manhattan Distance, and Minkowski Distance.
- *weights*: **Weight Function**, indicate the desired weight method to be used from two of them, **uniform** and **distance**. The model fitting trajectory is uniformly distributed among all data points, which is called **uniform**. The model fitting trajectory is distributed based on the distance between each data point and the trajectory, which is called **distance**.

Now let us use our motor system as a data source to illustrate how to use KNN Regression algorithm to fit the regression nonlinear model. To make things simple and complete, we develop our fitting model codes and the evaluation codes together and plot both results together to compare both results.

Now create a new Script file and name it **knn_fit_motor.m**. Then enter the codes shown in Fig. 5.77 into that file. Let us have a closer look at this piece of codes to see how it works.

1. First we need to load our simple motor dataset and assign it to a local table variable **T**. A point to be noted is the location of that dataset. You must use a **full path\\dataset_name** to get it. Also you need to use your actual path to replace our path if you saved that dataset in a different folder in your machine.
2. Extract two columns, **inputVol** and **motorSP**, from the dataset as input and response variables, and assign them to two local variables **cVol** and **cMSP**, respectively.
3. Create a new testing data vector, input voltages, and assign it to the local variable **Xnew**. A point to be noted is that this input vector must be a column vec-

```
% Test KNN Regression for our DC motor system
% Name: knn_fit_motor.m
% Dataset is nl_motor.xls
%
1   T = readtable("C:\\Artificial Intelligence Book\\Students\\Datasets\\nl_motor.xls");

2   cVol = T.inputVol
    cMSP = T.motorSP
3   Xnew = [1 1.2 1.1 1.3 1.5 1.8 2.1 2.0 2.2 2.8 3.0 3.1 7.2 7.5 8.1 8.3 8.7 8.8 9.5 9.1 9.3 12.2 15.5 15.7 16.1...
            16.8 18.2 19 20.3 20.8 21.1 21.8 22]';

4   K = 5;
5   metric = 'euclidean';
    weights = {'uniform', 'distance'};

6   for i = 1:2
7       mdl = kNNeighborsRegressor(K, metric, weights(i));
8       mdl = mdl.fit(cVol, cMSP);
9       Ypred = mdl.predict(Xnew);
10      subplot(2,1,i)
11      plot(cVol,cMSP,'o',Xnew,Ypred);
        grid;
12      legend('data','prediction')
        title(strcat('kNNeighborsRegressor (K = 5, metric = "euclidean", weights = "', weights(i), '")'))
    end

13  cd 'C:\\Artificial Intelligence Book\\Students\\Class Projects\\Chapter 5\\KNN Project';
14  saveas(gcf, 'Regression_results', 'jpeg');
```

Fig. 5.77 The detailed codes used to fit and evaluate the KNN model

tor; thus, do not forget to add a transposed symbol, a single quota ', in the superscript of the new vector. Another point to be noted is the length of this new testing vector, which is relatively longer with more data points involved. The reason for that is that we try to get a good fitting.

4. The default number of neighbor is 5, and we like to keep it for this project.

5. The default metric is **Euclidean**, but the weights contain two methods, **uniform** and **distance**. The reason for that is that we try to compare both methods later by plotting both fitting models to see the differences between them.

6. A **for**() loop is utilized to build two models based on two weights methods, and plot both fitting results to enable us to see the differences between both models.

7. The customer-built KNN function, **kNNeighborsRegressor**(), is called with three arguments, and the estimated model structure is returned and assigned to the local variable **mdl**.

8. Then the original input-output data are used with the **fit**() function to get the estimated model **mdl**.

9. To use the estimated model to predict the output for a new input data vector **Xnew**, the **predict**() function that belongs to that customer-built KNN function is executed, and the predicted outputs are returned and assigned to the local variable **Ypred**.

10. To plot two or more graphs in one sheet, we need to use the **subplot**() function, which has a format as: the number of rows, the number of columns, and the number of the current graph.

11. Then the **plot**() function is executed to plot both original input-output data points marked by **'o'** and the predicted input-output data points on two graphs.

12. Some legends and labels are set to assist and make the displaying results clearer.

13. In addition to plotting both fitting results, we can also save those results as images with different format, such as png, tif, tiff, jpg, and jpeg. In order to save those images to our desired folder, we can use MATLAB Command **cd** to set up our destination folder.

14. Using **saveas**() function to save our graphs as an image with desired file name, **Regression_results**, and format **jpeg**. The first argument of this function is **gcf**, which is a default variable name and it indicates to save the current figure as a graph.

A point to be noticed is that you may need to save this customer-built function file, **kNNeighborsRegressor.m**, at the same folder as this Script file is located to enable the Script file to be run smoothly.

Now run this piece of codes, and the running result is shown in Fig. 5.78.

The detailed codes for the customer-build function file, **kNNeighborsRegressor.m**, can be found from the Springer ftp site, exactly under a folder **Students\Class Projects\Chapter 5\KNN Project**.

Next let us have our discussions concentrated on another supervised machine learning algorithm, random forest algorithm.

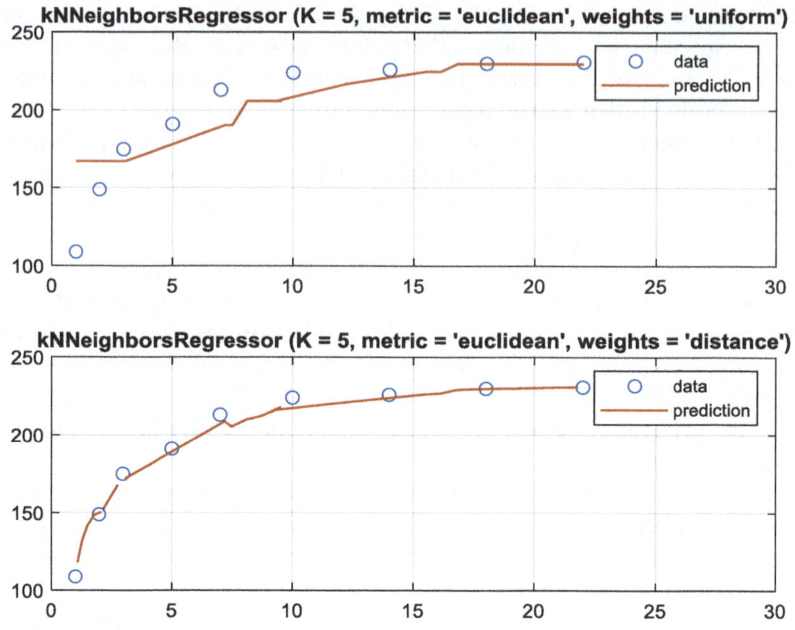

Fig. 5.78 The running result of the KNN Regression algorithm

5.8   Introduction to Random Forest Algorithm

Random forests or random decision forests is an **ensemble learning** method for classification, regression, and other tasks that operate by constructing a multitude of decision trees at training time. For classification tasks, the output of the random forest is the class selected by most trees [17]. For regression tasks, the output of the random forest is a continuous data value.

Basically the random forest algorithm is based on decision tree algorithm, especially on the multiple decision trees. In fact, ensemble learning methods are made up of a set of classifiers, e.g., decision trees, and their predictions are aggregated to identify the most popular result. The most well-known ensemble methods are **bagging**, also known as **bootstrap aggregation**, and **boosting**.

5.8.1   Bagging and Bootstrap Aggregation

The training algorithm for random forests applies the general technique of bootstrap aggregating, or bagging, to tree learners. Given a training set $X = x_1, ..., x_n$ with responses $Y = y_1, ..., y_n$, bagging repeatedly with B times selects a random sample with replacement of the training set and fits trees to these samples [17]:

For $b = 1, ..., B$:

1. Sample, with replacement of n training examples from X, Y; call these X_b, Y_b.
2. Train a classification or regression tree f_b on X_b, Y_b.

After training, predictions for unseen samples x' can be made by averaging the predictions from all the individual regression trees on x':

$$\hat{f} = \frac{1}{B}\sum_{b=1}^{B} f_b\left(x'\right) \tag{5.25}$$

or by taking the plurality vote in the case of classification trees.

This bootstrapping procedure leads to better model performance because it decreases the variance of the model, without increasing the bias. This means that while the predictions of a single tree are highly sensitive to noise in its training set, the average of many trees is not, as long as the trees are not correlated. Simply training many trees on a single training set would give strongly correlated trees (or even the same tree many times, if the training algorithm is deterministic); bootstrap sampling is a way of de-correlating the trees by showing them different training sets.

Additionally, an estimate of the uncertainty of the prediction can be made as the standard deviation of the predictions from all the individual regression trees on x':

$$\sigma = \sqrt{\frac{\sum_{b=1}^{B}\left(f_b\left(x'\right)-\hat{f}\right)^2}{B-1}} \tag{5.26}$$

The number of samples-trees, B, is a free parameter. Typically, a few hundred to several thousand trees are used, depending on the size and nature of the training set. An optimal number of trees B can be found using cross-validation, or by observing the *out-of-bag error*: the mean prediction error on each training sample x_i, using only the trees that did not have x_i in their bootstrap sample [18]. The training and test error tend to level off after some number of trees have been fit.

An illustration of training a random forest model is shown in Fig. 5.79 [17].

As shown in Fig. 5.79, a training dataset that is a 250×100 matrix is randomly sampled with replacement n times. Then, a decision tree is trained on each sample. Finally, for prediction, the results of all n trees are aggregated to produce a final decision.

5.8.2   *Navigation from Bagging to Random Forests*

The above procedure describes the original bagging algorithm for trees. Random forests also include another type of bagging scheme: they use a modified tree learning algorithm that selects, at each candidate split in the learning process, a random subset of the features. This process is sometimes called **feature bagging**. The reason for doing this is the correlation of the trees in an ordinary bootstrap sample: if

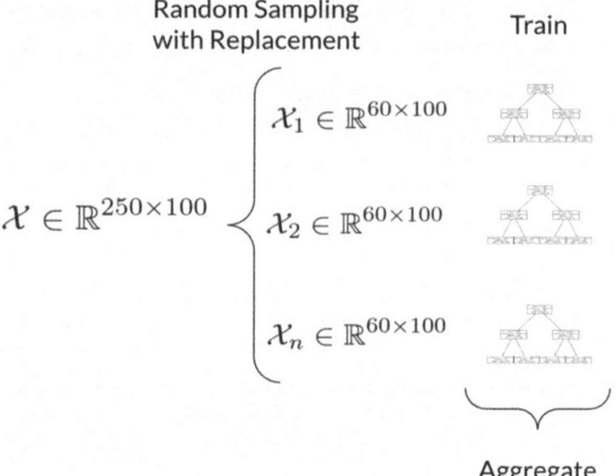

Fig. 5.79 An illustration of training a random forest model

one or a few features are very strong predictors for the response variable (target output), these features will be selected in many of the B trees, causing them to become correlated.

In summary, the random forest algorithm is an extension of the bagging method as it utilizes both bagging and feature randomness to create an uncorrelated forest of decision trees. Feature randomness, also known as **feature bagging** or the **random subspace** method, generates a random subset of features, which ensures low correlation among decision trees. This is a key difference between decision trees and random forests [19].

Random forest algorithms have three main hyperparameters, which need to be set before training. These include *node size*, the *number of trees*, and the *number of features* sampled. From there, the random forest classifier can be used to solve for regression or classification problems.

5.8.3 *Random Forest Algorithms in MATLAB*

In MATLAB, all random forest-related components are included in the Statistic and Machine Learning Toolbox. Essentially, three (3) objects are involved in that Toolbox,

1. **ClassificationBaggedEnsemble** object created by the **fitcensemble()** function for classification.
2. **RegressionBaggedEnsemble** object created by the **fitrensemble()** function for regression.
3. **TreeBagger** object created by the **TreeBagger()** function for classification and regression.

In this section, we only concentrate our study on the second and the third objects, **TreeBagger** and **RegressionBaggedEnsemble**, and leave the first one in the classification part discussed later.

All of these objects can be created based on some related classes listed below:

- **CompactRegressionEnsemble class** (Super class)
- **RegressionEnsemble class** (Child class of CompactRegressionEnsemble class)
- **RegressionBaggedEnsemble class** (Child class of RegressionEnsemble class)

The CompactRegressionEnsemble class is a compact version of a regression ensemble class RegressionEnsemble. The compact version does not include the data for training the regression ensemble. Therefore, you cannot perform some tasks with a compact regression ensemble, such as cross-validation. Use a compact regression ensemble for making regressions predictions of new data.

The RegressionEnsemble class is a child or derived class of CompactRegressionEnsemble class. The RegressionEnsemble combines a set of trained weak learner models and data on which these learners were trained. It can predict ensemble response for new data by aggregating predictions from its weak learners. You need to use the function **fitrensemble**() to create a new RegressionEnsemble object.

The RegressionBaggedEnsemble class is a child or derived class of RegressionEnsemble class. It combines a set of trained weak learner models and data on which these learners were trained. It can predict ensemble response for new data by aggregating predictions from its weak learners. You need to use the function **fitrensemble**() to create a new RegressionBaggedEnsemble object.

A TreeBagger object is an ensemble of bagged decision trees for either classification or regression. Individual decision trees tend to over-fit. *Bagging*, which stands for bootstrap aggregation, is an ensemble method that reduces the effects of over-fitting and improves generalization. You can create a new TreeBagger object by calling the function **TreeBagger**(). The **TreeBagger**() function grows every tree in the TreeBagger ensemble model using bootstrap samples of the input data. Observations not included in a sample are considered **out-of-bag** for that tree. The function selects a random subset of predictors for each decision split by using the random forest algorithm. By default, the **TreeBagger**() function grows classification decision trees. To grow regression trees, specify the **Name-Value** pair for the argument **Method** as **regression**.

The **Name-Value** pairs used for the **Method** argument are different when calling functions **fitrensemble**() and **TreeBagger**(). Table 5.12 shows these values.

Table 5.12 Value-Name pairs used for function fitrensemble() and TreeBagger()

Name-Value Pair - fitrensemble()	Description
Bag	Binary and multiclass classification and regression
LSBoost	Least Squares Boosting - Regression
Name-Value Pair - TreeBagger()	Description
classification (default)	Classification decision trees
regression	Regression decision trees

To make things simple, we only discuss two types of random forest-related algorithms, RegressionBaggedEnsemble and TreeBagger.

5.8.4 The RegressionBaggedEnsemble Related Apps

The Statistics and Machine Learning Toolbox provided both Apps and Functions to perform random forest-related estimations with regression ensemble and tree bagger algorithms. First let us have a closer look at these Apps.

All regression ensemble and random forest-related Apps are involved in the **Regression Learner** App, which includes the **Boosted Trees**, **Bagged Trees**, and **All Ensembles**.

Now we like to use another customer-built dataset, **nl_motor2.xls**, to illustrate how to use these Apps to perform regression ensemble and random forest estimations to build our desired tree models. This dataset is similar to the dataset, **nl_motor.xls**, but has more data points. You can find that dataset from the Springer ftp site under a folder **Students\Datasets**. Copy and paste it to one of your local folders to make it used later.

Perform the following operations to build this model,

1. Open the **Regression Learner** and click on the drop-down arrow on the **New Session** icon, and select **From File** item to open the **Select File to Open** wizard. Browse to your local folder where the dataset **nl_motor2.xls** is located, select it, and click on the **Open** button.
2. Click on the green-color **Import Selection** button on the upper-right corner on the opened App to get all data from the dataset.
3. Keep all default settings on next wizard and click on the **Start Session** button. The session is started and the input-output data pairs are displayed based on the **Record number**.
4. Select our input **inputVol** from the x combo box under the **X-axis**, our input-output data pair trajectory is displayed, as shown in Fig. 5.80.
5. Now select the training model by clicking on the drop-down arrow on the **MODELS** pane, as shown in Fig. 5.80, to open a list of all available training model trees.
6. Scroll down until you find the model named **All Ensembles** under the **ENSEMBLES OF TREES** group. Select it to open this training model on the left pane.
7. Then click on the drop-down arrow on the **Train All** tab on the top and select the **Train Selected** item to begin the training process. The training result is displayed as the training process is completed.
8. Select our input, **inputVol**, from the x combo box under the **X-axis**, the trained result is shown in Fig. 5.81. Check the **Errors** box to show the errors between the true values and estimated values.

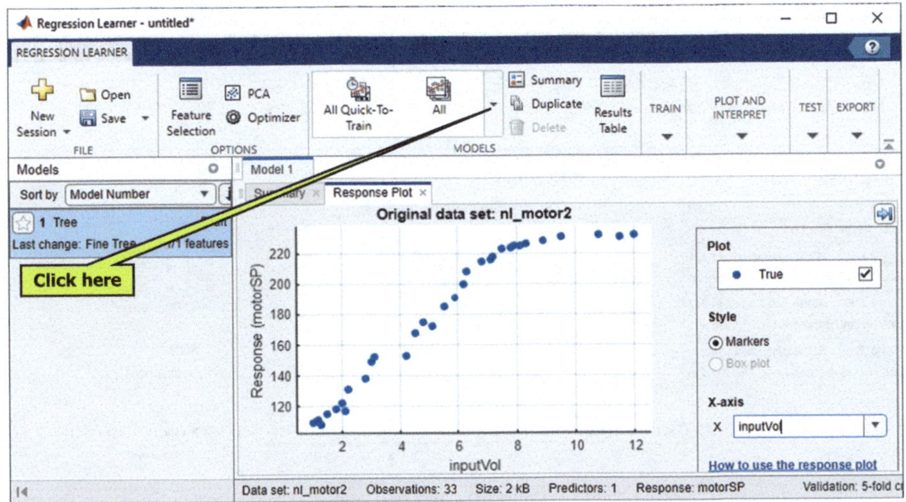

Fig. 5.80 The collected input-output data from the dataset

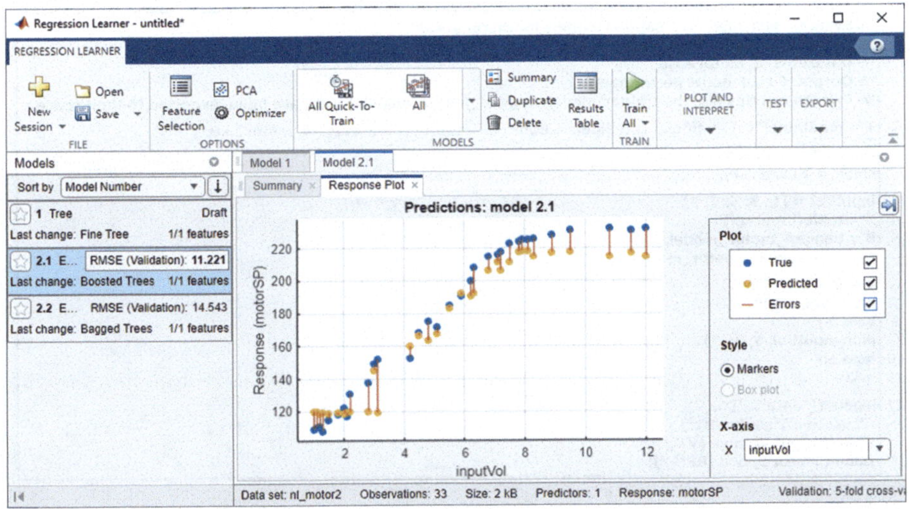

Fig. 5.81 The trained result with All Ensembles model

9. For testing purpose, you can select the **Bagged Trees** model under the **ENSEMBLES OF TREES** group to redo this training to compare its training result with that of the **All Ensembles** model. The training result in **Bagged Trees** model is shown in Fig. 5.82.

10. Click on the drop-down arrow on the **Export** icon and the **Export Model** item, and click on the **Export Model** item to export our trained model tree to the Workspace. Change the name to **begged_motor_model** and click on the **OK** button.

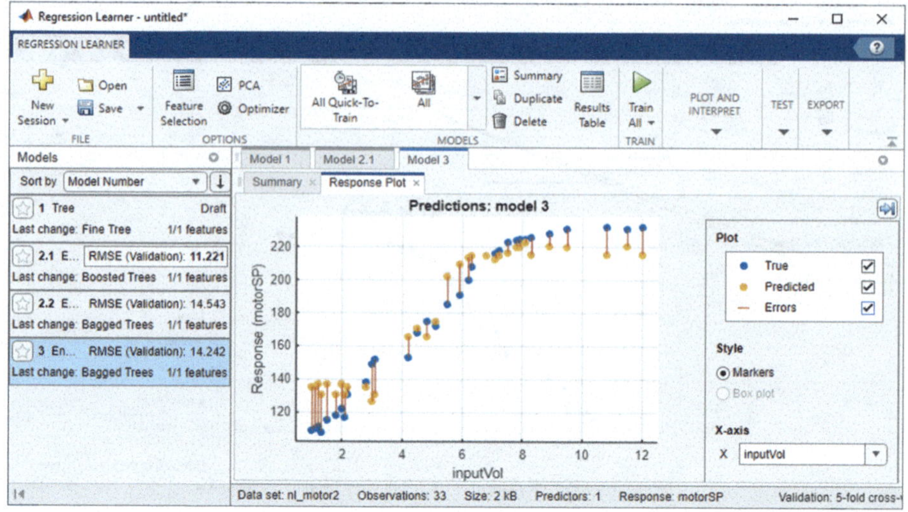

Fig. 5.82 The trained result with Bagged Trees model

```
% Evaluate ensemble and bagged models for motor system
% Name: eval_ensemble_app.m
% Dataset is nl_motor2.xls
% Output: Fitted model performance
% Prior to run this file, the trained model, begged_motor_model, must have been  exported to Workspace.
1  T = readtable("C:\\Artificial Intelligence Book\\Students\\Datasets\\nl_motor2.xls")
2  cVol = T.inputVol;
   cMSP = T.motorSP;
3  inputVol = [1; 3; 5; 7; 9];
4  X = table(inputVol)
5  B = begged_motor_model;
   %B = ensemble_motor_model
6  Y = B.predictFcn(X)
7  plot(cVol, cMSP)
8  hold on
9  plot(inputVol, Y,"k+")
10 hold off
11 grid;
12 legend(["data",""], ...
   "Location","southeast")
   xlabel('Input Voltage (V)');
   ylabel('Motor Speed (RPM)');
```

Fig. 5.83 The codes used to evaluate the begged tree model with new data

The purpose of exporting this trained model tree to the Workspace is because we may need this model to perform an evaluation for some new input data later.

Now let us build a piece of codes to evaluate our trained model tree with some new input data values. Create a new Script file and enter the codes shown in Fig. 5.83 into that file. Save that file as **eval_ensemble_app.m**.

Let us have a closer look at this piece of codes to see how it works.

1. First get the dataset **nl_motor2.xls** by using the **readtable()** function. A point to be noted is that you may need to use your actual path to replace our path if you saved this dataset in your special local folder.

2. Two variables, the input voltage **inputVol** and the output motor speed **motorSP**, are extracted from the dataset, and assigned to two local variables, **cVol** and **cMSP**.

3. A set of new testing data, exactly a new input voltage vector, **inputVol**, is generated and initialized with a set of new input voltage values. A key point to be noted is that the name of this new input vector **MUST** be identical to the name used in the dataset. Otherwise, you may encounter some errors as you run this piece of codes. The MATLAB only knows the name you used in the dataset; thus, the name of the new input vector must be the same.

4. Also this new input vector **inputVol MUST** be converted to a **Table** format even if it is a vector; otherwise, an error would be encountered as the project runs. The converted table is assigned to a local variable **X**.

5. The exported model, **begged_motor_model**, we did in step 10 above is assigned to a local variable **B**, which is exactly a multiple-tree object and it has been trained.

6. The **predictFcn()** function that belongs to the trained model is executed with the new input voltage table **X** as argument. A point to be noted is that when using the exported or trained model to predict the desired output based on new inputs, you cannot use the **predict()** function directly; instead, you must use the function **predictFcn()** that is attached to the trained model. However, you can use **predict()** function directly if the model is trained by using functions, not Apps, and we will show this difference later.

7. First we can plot the original input-output data with desired fitting curve, and the purpose of that plotting is to compare between the original data points and predicted data points.

8. To make that comparison possible, we need to use the **hold on** command to keep the original plotting on the screen.

9. Then we can plot the predicted data based on new input data, and highlight them with + marks.

10. After two plotting has been done in a same graph, the **hold off** command is used to release any holding action.

11. The **grid** command is used to display the grids on two plots.

12. Some **legend()** and **label()** functions are called to make those plots more meaningful.

Now run the project and the running result is shown in Fig. 5.84. It can be found from this figure that the evaluation result is good enough.

You can try to use the trained ensemble tree model to do this evaluation again. Perform the following operations to do this evaluation:

1. Redo the training process for our motor system by using the **Regression Learner** App, and select the **All Ensembles** model with **Boosted Trees** from the **MODELS** pane on the top.

2. Export that trained model to the Workspace with the name of **ensemble_motor_model**.

3. Replace the variable **B** at coding line 5 in Fig. 5.83 with **B = ensemble_motor_model**.

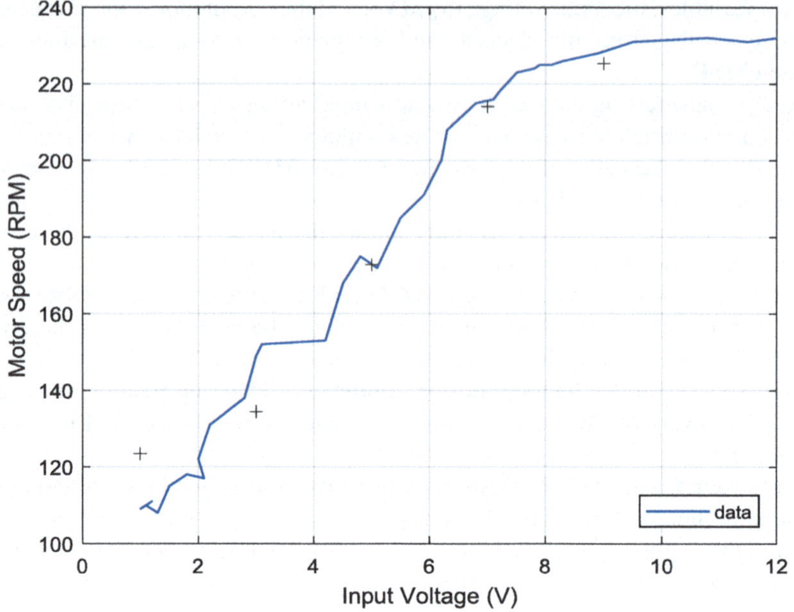

Fig. 5.84 The evaluation running result of using begged model

4. Run the project again, and the running result is shown in Fig. 5.85.

Next let us concentrate our discussions on RegressionBaggedEnsemble related functions, and we like to use those functions to train our motor system with random forest tree model.

5.8.5 The RegressionBaggedEnsemble Related Functions

As we discussed in Sect. 5.8.3, two functions are popular and widely utilized in random forest algorithm, **fitrensemble()** and **TreeBagger()**. The former can only be used for regressions, but the latter can be used for both regressions and classifications. In order to enable us to use the function **TreeBagger()** for the regressions, a **Name-Value** pair for the argument **Method** must be set as **'regression'**.

The function **fitrensemble()** belongs to the Regression Ensemble tree and it can be used to create two kinds of objects for two classes, **RegressionEnsemble** and **RegressionBaggedEnsemble** classes. The function **TreeBagger()** belongs to the Bagged Regression trees. Both functions can be used to perform random forest training processes and predictions.

Now let us have a closer look at both functions to see how they work.

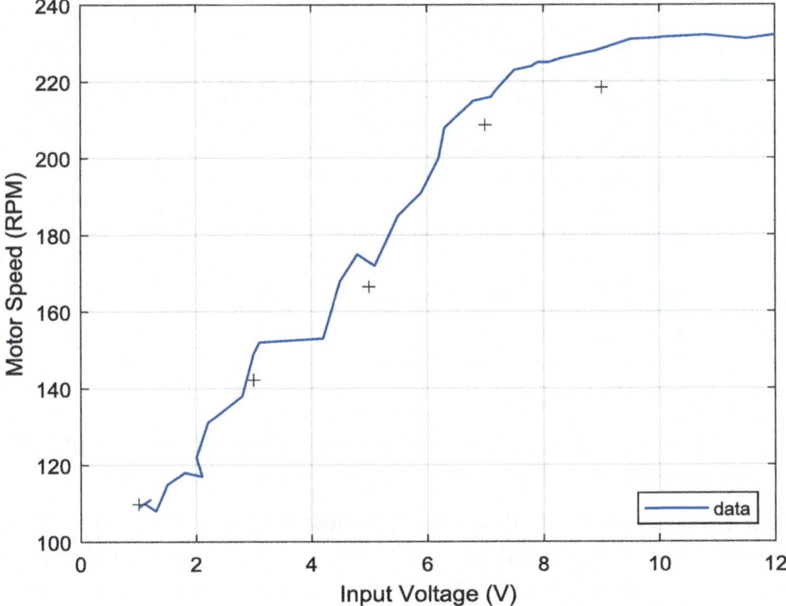

Fig. 5.85 The evaluation running result of using ensemble with boosted tree

5.8.5.1   Introduction to Function Fitresemble()

The **fitrensemble()** function uses bagging **'bag'** with random predictor selections at each split (random forest) by default. You can create an ensemble for regression by using **fitrensemble()** function. To train an ensemble for regression by using that function, use this syntax.

```
ens = fitrensemble(X, Y, Name, Value);
```

1. **X** is the matrix of input data. Each row contains one observation, and each column contains one predictor or input variable.
2. **Y** is the vector of responses or outputs, with the same number of observations as the rows of inputs in **X**.
3. **Name** and **Value** is a Name-Value pair used to specify additional options. For example, you can specify the ensemble aggregation method with the **'Method'** argument, the number of ensemble learning cycles with the **'NumLearningCycles'** argument, and the type of weak learners with the **'Learners'** argument.
4. The **ens** is created ensemble object.

For all nonlinear regression problems, follow these steps to create an ensemble object:

1. Prepare the Predictor Data and the Response Data.

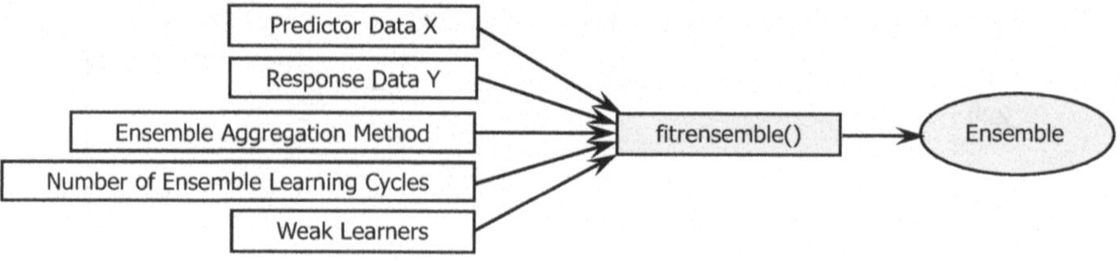

Fig. 5.86 The operational steps to create an ensemble object

2. Choose an Applicable Ensemble Aggregation Method.
3. Set the Number of Ensemble Members.
4. Prepare the Weak Learners.
5. Call the function **fitrensemble()**.

Figure 5.86 shows the steps you need to create a regression ensemble object.In fact, the function **fitrensemble()** has another five constructors, as shown below,

1. **mdl = fitrensemble(Tbl, ResponseVarName):** Return the trained regression ensemble model object (**mdl**) that contains the results of boosting 100 regression trees using **LSBoost** and the predictor and response data in the table **Tbl**. **ResponseVarName** is the name of the response variable in **Tbl**.
2. **mdl = fitrensemble(Tbl, formula):** Apply formula to fit the model to the predictor and response data in the table **Tbl**. The **formula** is an explanatory model of the response and a subset of predictor variables in **Tbl** used to fit mdl. For example, '**Y ~ X1 + X2 + X3**' fits the response variable **Tbl.Y** as a function of the predictor variables **Tbl.X1, Tbl.X2**, and **Tbl.X3**.
3. **mdl = fitrensemble(Tbl,Y):** Treat all variables in the table **Tbl** as predictor or input variables. **Y** is the vector of responses that is not in **Tbl**.
4. **mdl = fitrensemble(X,Y):** Use the predictor data in the matrix **X** and response data in the vector **Y**.
5. **mdl = fitrensemble(___, Name,Value):** Use additional options specified by one or more **Name-Value** pair arguments and any of the input arguments in the previous syntaxes. For example, you can specify the number of learning cycles, the ensemble aggregation method, or implement ten-fold cross-validation.

Table 5.13 shows a list of some popular name-value pair arguments.

Now let us build some real example projects to illustrate how to use this function to perform random forest algorithms to build our desired tree model for our forest fire dataset.

Create a new Script file, name it as **ensemble_fire_func.m**, and enter the codes shown in Fig. 5.87 into that file. Let us have a closer look at this piece of codes to see how it works.

1. Our dataset file, **MFire_Database.xls**, is read out and it is assigned to a local variable **FT**.
2. Seven variables, including the inputs and output, are extracted to a new table **T** as the training data.

Table 5.13 Some popular Name-Value pair arguments

Name-Value Pair	Description
`'Method'`	**LSBoost** – Least squares boosting, **Bag** - Bootstrap aggregation.
`'NumLearningCycles'`	Number of ensemble learning cycles, default = 100.
`'Learners'`	Weak learners to be used in ensemble, 'tree' (default) | tree template object | cell vector of tree template objects.
`'NumBins'`	Number of bins for numeric predictors, Default = [] or Integer number.
`'CategoricalPredictors'`	Categorical predictors list, vector of positive integers
`'PredictorNames'`	Predictor variable names
`'ResponseName'`	Response variable name, default = "Y".
`'CrossVal'`	Cross-validation flag, "on" or "off", default = "off".
`'KFold'`	Number of folds, positive integer > 1, default = 10.
`'Weights'`	Observation weights, numeric vector of positive values.
`'FResample'`	Fraction of training set to resample, positive scalar in (0, 1), default = 1.
`'Replace'`	Flag indicating to sample with replacement, "on" or "off", default = "on".

```
   % Use ensemble functions to set nonlinear model for forest fire dataset
   % Name: ensemble_motor_func.m
   % Dataset is MFire_Database.xls
   % Output: Ensembled model

   % Use fitrensemble() function...
1  FT = readtable("C:\\Artificial Intelligence Book\\Students\\Datasets\\MFire_Database.xls")
2  T = table(FT.FFMC, FT.DMC, FT.DC, FT.temp, FT.wind, FT.BUI, FT.FWI);

   % Use simple syntax to define the model
3  mdl_1 = fitrensemble(T, FT.FWI)

   % Use formula to define the model as FWI = FFMC + DMC + temp + wind + BUI
4  formula = 'FWI ~ FFMC + DMC + temp + wind + BUI';
5  mdl_2 = fitrensemble(FT, formula)

   % Use Name-Value pair to specify additional options
6  X = [FT.FFMC FT.DMC FT.DC FT.temp FT.wind FT.BUI];
7  mdl_3 = fitrensemble(X,FT.FWI,"Method","Bag","NumLearningCycles",120,"NumBins",50,"CrossVal","on")

   % Plot the cumulative 10-fold cross-validated MSE
8  kfloss = kfoldLoss(mdl_3, 'Mode','cumulative');
9  figure;
10 plot(kfloss);
11 grid;
12 ylabel('10-fold cross-validated MSE');
   xlabel('Learning cycle');
```

Fig. 5.87 The detailed codes used to build ensemble tree models

3. The function **fitrensemble()** is executed to estimate the random forest tree model, **mdl_1**. A point to be noted is that the response variable, **FWI**, must be indicated with the original table name **FT**; otherwise, it cannot be recognized by the compiler.

4. Secondly, we want to try to use a formula as an argument to perform this training process. Thus, the formula is defined as **FWI ~ FFMC + DMC + temp + wind + BUI**, which means that we only need five input variables to do this modeling.

5. The second modeling process starts by using that formula with the original table.

6. Thirdly, we can use some **Name-Value** pairs to improve the modeling process to get better modeling results. Still, we use five input variables as predictors to get a partial matrix **X**.

7. The third model is estimated based on some useful **Name-Value** pairs, such as **Method** being **Bag**, **Learning Cycles** being 120, **Number of Bins** being 50, and **CrossVal** being **on**. The output or response variable should be indicated with the original table **FT**.

8. Additionally, we can use the function **kfoldLoss()** to estimate the loss measured by mean squared error, which is obtained by the cross-validated regression model **mdl_3**. For every fold, **kfoldLoss()** function computes the loss for validation-fold observations using a model trained on training-fold observations.

9. To plot these losses, a command **figure** is used to create a blank figure.

10. The loss is plotted with the **plot()** function.

11. The command **grid** is used to set up grids on the plotted graph.

12. Some label functions are used to make the graph more meaningful.

Now run this piece of codes and the running results are shown on the Command window as the project is done. Also, the plotted loss results are shown in Fig. 5.88.

Next let us have a closer look at the function **TreeBagger()**.

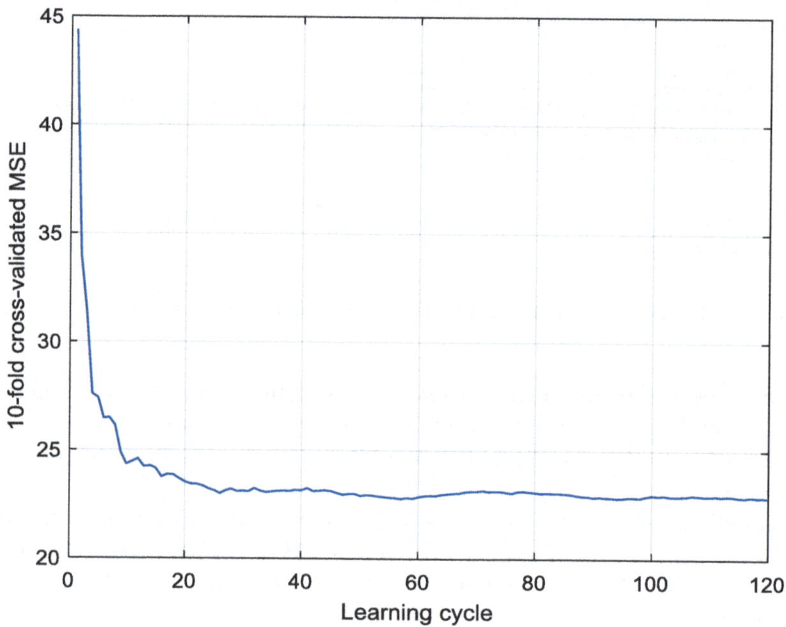

Fig. 5.88 The plotted result for the loss

5.8.5.2 Introduction to Function TreeBagger()

The first algorithm for random decision forests was created in 1995 by Tin Kam Ho using the random subspace method [20, 21], which, in Ho's formulation, is a way to implement the "stochastic discrimination" approach to classification proposed by Eugene Kleinberg [22–24].

An extension of the algorithm was developed by Leo Breiman [25] and Adele Cutler, [26] who registered [27] "Random Forests" as a trademark in 2006 (as of 2019, owned by Minitab, Inc.) [28]. The extension combines Breiman's "bagging" idea and random selection of features, introduced first by Ho [20] and later independently by Amit and Geman [29] in order to construct a collection of decision trees with controlled variance [17].

In fact, the TreeBagger class implements a bagged decision tree algorithm, rather than Random Forests specifically. However, you can get TreeBagger to behave basically the same as Random Forests as long as the **NumPredictorsToSample** parameter in the **Name-Value** pair argument is set appropriately. If you set **NumPredictorsToSample** to any value except **'all'**, the software uses Breiman's random forest algorithm [16]. Fortunately, this parameter is set to an integer for regression trees as the function runs, which is equivalent to random forest algorithm.

The fundamental difference between the Random Forests and Bagging is that in Random forests, only a subset of variables or features are selected at random out of the total and the best split feature from the subset is used to split each node in a tree, unlike in Bagging where all variables or features are considered for splitting a node [30].

In summary, Bagging is an ensemble algorithm that fits multiple models on different subsets of a training dataset, and then it combines the predictions from all models. Random Forest is an extension of bagging that also randomly selects subsets of features used in each data sample.

The **TreeBagger()** function grows every tree in the TreeBagger ensemble model using bootstrap samples of the input data. Observations not included in a sample are considered "out-of-bag" for that tree. The function selects a random subset of predictors for each decision split by using the random forest algorithm [16] since the **NumPredictorsToSample** parameter in the **Name-Value** pair argument will be set to an integer when regression is selected as the function runs.

There are five different constructors for the function **TreeBagger()**, which are listed below:

1. **mdl = TreeBagger(NumTrees,Tbl,ResponseVarName):** Return an ensemble object (**mdl**) of **NumTrees** bagged classification trees, trained by the predictors in the table **Tbl** and the class labels in the variable **Tbl.ResponseVarName**.
2. **mdl = TreeBagger(NumTrees,Tbl,formula):** Return **mdl** trained by the predictors in the table **Tbl**. The input formula is an explanatory model of the response and a subset of predictor variables in **Tbl** used to fit **mdl**. Specify formula using Wilkinson Notation.

3. **mdl = TreeBagger(NumTrees,Tbl,Y):** Return **mdl** trained by the predictor data in the table **Tbl** and the class labels in the array **Y**.
4. **mdl = TreeBagger(NumTrees,X,Y):** Return **mdl** trained by the predictor data in the matrix **X** and the class labels in the array **Y**.
5. **mdl = TreeBagger(___,Name = Value):** Return **mdl** with additional options specified by one or more Name-Value arguments, using any of the previous input argument combinations. For example, you can specify the algorithm used to find the best split on a categorical predictor by using the Name-Value argument **PredictorSelection**.

All of the above five constructors can be used to create related objects used for regression by setting an additional Name-Value pair as **Method = "regression"**, as we mentioned before.

All possible input arguments used for the function **TreeBagger()** are shown in Table 5.14. Some popular Name-Vale pairs used for that function are shown in Table 5.15.

Now let us build some real example projects to illustrate how to use this function to perform random forest process to estimate desired multiple tree models for our forest fire system.

Create a new Script file, name it as **treebagger_fire_func.m**, and enter the codes shown in Fig. 5.89 into the file. Let us have a closer look at this piece of codes to see how it works.

Table 5.14 Some popular input arguments used for TreeBagger() function

Input Argument	Description
NumTrees	Number of decision trees in the bagged ensemble, a positive integer.
Tbl	Sample data used to train the model, specified as a table. • If Tbl contains the response variable, and you want to use all remaining variables in Tbl as predictors, then specify the response variable by using ResponseVarName. • If Tbl contains the response variable, and you want to use only a subset of the remaining variables in Tbl as predictors, then specify a formula by using formula. • If Tbl does not contain the response variable, then specify a response variable by using Y. The length of the response variable and the number of rows in Tbl must be equal.
ResponseVarName	Response variable name, the ResponseVarName must be specified as a character vector or string scalar. E.g., if the response variable Y is stored as Tbl.Y, then specify it as "Y". Otherwise, the software treats all columns of Tbl, including Y, as predictors when training the model.
formula	Explanatory model of the response variable and a subset of the predictor variables, specified as a character vector or string scalar in the form "Y~x1+x2+x3".
Y	**Class labels or response variable,** If you specify Method as "regression", the response variable Y is an n-by-1 numeric vector, where n is the number of observations. Each entry in Y is the response for the corresponding row of X.
X	Predictor data, specified as a numeric matrix.

Table 5.15 Some popular Name-Value pairs used for TreeBagger() function

Name-Value Pair	Description
Method	Type of decision tree, "classification" (default) | "regression"
MinLeafSize	Minimum number of leaf node observations, specified as a positive integer. By default, MinLeafSize is 1 for classification trees and 5 for regression trees.
NumPredictorsToSample	Number of predictor variables (randomly selected) for each decision split, specified as a positive integer or "all". If you set NumPredictorsToSample to any value except "all", the software uses Breiman's random forest algorithm.
NumPrint	Number of grown trees (training cycles) after which the software displays a message about the training progress in the command window.
PredictorNames	Predictor variable names, specified as a string array of unique names or cell array of unique character vectors.
SampleWithReplacement	Indicator for sampling with replacement, specified as "on" or "off". Specify it as "on" to sample with replacement, or as "off" to sample without replacement. If you set it to "off", you must set the name-value argument InBagFraction to a value less than 1.
InBagFraction	Fraction of input data to sample with replacement from the input data for growing each new tree, specified as a positive scalar in the range (0,1].
OOBPrediction	Indicator to store out-of-bag information in the ensemble, specified as "on" or "off". Specify it as "on" to store information on which observations are out-of-bag for each tree. TreeBagger can use this information to compute the predicted class probabilities for each tree in the ensemble.
OOBPredictorImportance	Indicator to store out-of-bag estimates of feature importance in the ensemble, specified as "on" or "off". If you specify it as "on", the TreeBagger function sets OOBPrediction to "on". If you want to analyze predictor importance, specify PredictorSelection as "curvature" or "interaction-curvature".

1. Our dataset file, **MFire_Database.xls**, is read out and it is assigned to a local variable **FT**.
2. Seven variables, including the inputs and output, are extracted to a new table **T** as the training data.
3. The **TreeBagger**() function is executed with the first constructor format, where four arguments are involved and they are, 100 decision trees, the training table **T** contained six inputs and one output variables, response variable and the regression method involved in a **Name-Value** pair. The trained model is returned and assigned to the **mdl_1** with detailed model parameters listed on the Command window as the project runs.
4. A **formula** format used to define the input-output relation is utilized for the second model training process. The formula is represented by using Wilkinson notation with five selected predictors, **FFMC, DMC, temp, wind**, and **BUI**.
5. The second model training process is started by calling the **TreeBagger**() function with the **formula** to indicate the training parameters. One more parameter, **NumPredictorsToSample**, is initialized to 50 to make this function to perform a random forest algorithm.

```
% Use TreeBagger() functions to set random forest model for forest fire dataset
% Name: treebagger_fire_func.m
% Dataset is MFire_Database.xls
% Output: Random forest model

% Use TreeBagger() function...
1   FT = readtable("C:\\Artificial Intelligence Book\\Students\\Datasets\\MFire_Database.xls");
2   T = table(FT.FFMC, FT.DMC, FT.DC, FT.temp, FT.wind, FT.BUI, FT.FWI);

% Use simple syntax to define the model for random forest
3   mdl_1 = TreeBagger(100, T, FT.FWI, Method="regression")

% Use formula to define the model as FWI = FFMC + DMC + temp + wind + BUI
4   formula = "FWI ~ FFMC + DMC + temp + wind + BUI";
5   mdl_2 = TreeBagger(100, FT, formula, Method="regression", NumPredictorsToSample=50)

% Use constrcutor 3 to set the model
6   mdl_3 = TreeBagger(50, FT, FT.FWI, Method="regression")

% Use Name-Value pair to specify additional options
7   X = FT.BUI;
    Y = FT.FWI;
8   mdl_4 = TreeBagger(200, X, Y, Method="regression", OOBPredictorImportance="on", MinLeafSize=4)

9   predX = linspace(min(X), max(X), 10)';
10  YMean = predict(mdl_4, predX);
    YQuartiles = quantilePredict(mdl_4, predX, Quantile=[0.25, 0.5, 0.75]);
    hold on
11  plot(X,Y,"o");
12  plot(predX, YMean);
13  plot(predX, YQuartiles);
    hold off
14  ylabel("Fire Weather Index Values");
    xlabel("Buildup Index Values");
    legend("Data","Mean Response","First Quartile","Median", "Third Quartile");
```

Fig. 5.89 The codes for creating random forest models with TreeBagger() function

6. The third training method is generated by using the third constructor format with number of decision trees, training table **FT**, response **FT.FWI**, and **regression** method. The training results are returned and assigned to the local variable **mdl_3** with the detailed modeling parameters listed on the Command window as this project runs.

7. Finally, we can build this model by using the fourth constructor format with some additional **Name-Value** pair parameters. Here we are using the **OOBPredictorImportance** and **MinLeafSize** pairs to clearly indicate those parameters with some desired values.

8. The **TreeBagger**() function is called to perform that training with number of decision trees as 200.

9. We want to plot the observations, estimated mean responses, and estimated quartiles to show the relations between the input **BUI** and the output **FWI**.

10. We divided 10 equally spaced **BUI** values between the minimum and maximum in-sample **BUI** and predicted conditional mean responses (**YMean**) and conditional quartiles (**YQuartiles**).

11. For the comparison reason, first we plot the original input and output variables.

12. Then plot the mean of the output via equalized-spaced input **X**.

13. Finally plot the **YQuartiles** via equal-spaced input **X**.

14. Some label and legend functions are used to make these plots more meaningful.

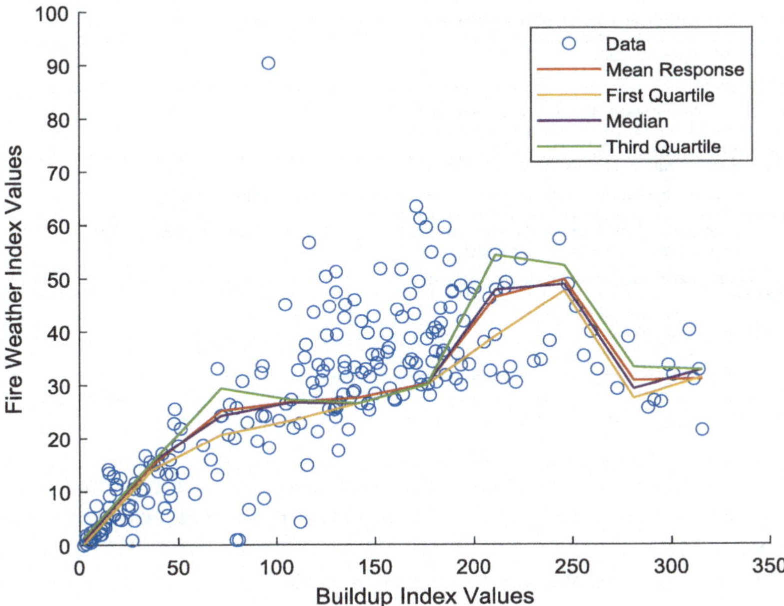

Fig. 5.90 The plotting result

Now run this project and the running results are shown in the Command window with all detailed parameters. The plotting result is shown in Fig. 5.90.

Next let us take care of the evaluations of trained models with the given new testing or input data points.

5.8.5.3   Evaluate the Trained Random Forest Models

To evaluate both **fitrensemble()** and **TreeBagger()** functions with new input data, one can use the function **predict()**. This function provides the ability to predict responses using ensemble of regression models with new predictor or input data. Two constructors can be used to call this function to perform a prediction function:

1. **Yfit = predict(mdl, X):** Return predicted responses to the predictor data in the table or matrix **X**, based on the regression ensemble model **mdl**.
2. **Yfit = predict(mdl, X, Name = Value):** Specify additional options using one or more Name-Value arguments. For example, you can specify the indices of weak learners used for making predictions, and whether to perform computations in parallel.

```
% Evaluate ensemble and treebagger models with predict() function
% Nam: eval_ensemble_treebagger_func.m
% Input: MFire_Databse.xls
1  FT = readtable("C:\\Artificial Intelligence Book\\Students\\Datasets\\MFire_Database.xls");

   % Use ensemble tree algorithm to train the model
2  X = [FT.FFMC FT.DMC FT.DC FT.temp FT.wind FT.BUI];
3  mdl_1 = fitrensemble(X, FT.FWI)
4  FWI_1 = predict(mdl_1, [70 60 715 25 2.9 125])

   % Use formula to define the model as FWI = FFMC + DMC + temp + wind + BUI
5  formula = "FWI ~ FFMC + DMC + temp + wind + BUI";
6  mdl_2 = TreeBagger(100, FT, formula, Method="regression", NumPredictorsToSample=50)
7  FWI_2 = predict(mdl_2, [70 60 25 2.9 125])

8  rsLoss = resubLoss(mdl_1, 'Mode', 'Cumulative');
9  plot(rsLoss);
10 xlabel('Number of Learning Cycles');
   ylabel('Resubstitution Loss');
   grid;
```

Fig. 5.91 The codes used to evaluate some trained random forest models

Now let us build a project to evaluate the performances of some estimated random forest models. Create a new Script file and name it as, **eval_ensemble_treebagger_func.m**, and enter the codes shown in Fig. 5.91 into that file. Let us have a closer look at this piece of codes to see how it works:

1. Our dataset file, **MFire_Database.xls**, is read out and it is assigned to a local variable **FT**.
2. Six predictor or input variables are extracted and assigned to a local variable **X**.
3. The function **fitrensemble()** is called with those selected six predictor variables to train the random forest model, and the trained model is returned and assigned to **mdl_1**.
4. Now let us start the evaluation for that trained model, **mdl_1**, by calling the **predict()** function with a piece of new predictor values. The evaluated result is an estimated FWI value and assigned to the variable **FWI_1**.
5. Next we like to use another way, **formula**, to train a similar model via the **TreeBagger()** function. First, we need to generate a formula with the Wilkinson notation. Five predictors or input variables are involved in this formula.
6. The function **TreeBagger()** is executed with some **Name-Value** pairs, including the **regression** Method and the **NumPredictorsToSample**. The trained model is assigned to a local variable **mdl_2**.
7. Another piece of new predictors or input variables is used with the **predict()** function to estimate an estimated output value **FWI**.
8. Now let us do one more thing, to determine the cumulative resubstitution losses (i.e., the cumulative misclassification error of the labels in the training data). The **rsLoss** is a 100-by-1 vector, where element k contains the resubstitution loss after the first k learning cycles.
9. Plot the cumulative resubstitution loss over the number of learning cycles.
10. Some **label()** functions are used to make the plotting result more meaningful.

Now run this project and the running results are shown on the Command window. Also the plotting result is shown in Fig. 5.92.

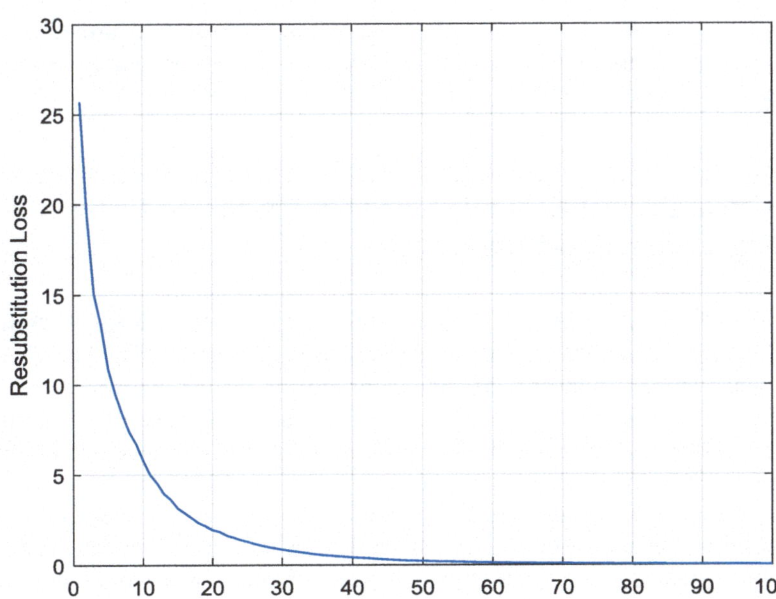

Fig. 5.92   The plotting result

In general, as the number of decision trees in the trained regression ensemble increases, the resubstitution loss decreases.

A decrease in resubstitution loss might indicate that the software trained the ensemble sensibly. However, you cannot infer the predictive power of the ensemble by this decrease. To measure the predictive power of an ensemble, estimate the generalization error by:

1. Randomly partitioning the data into training and cross-validation sets. Do this by specifying **'holdout'**, holdoutProportion when you train the ensemble using **fitensemble()**.
2. Passing the trained ensemble to **kfoldLoss**, which is used to estimate the generalization error.

Regularly, we may need additional testing data to evaluate a model to be trained by using training data, which is a general way to do this kind of evaluation job in machine learning. We will use that strategy to do our evaluations for the following sections in the book.

At this point, we have completed our introductions and discussions about Regression Algorithms under the Supervised Learning category. As for the rest algorithms, such as **Naïve Bayes** and **Similarity Learning**, they are both more popular in Classification implementations. Thus, we leave them to be discussed in the Classifications part later.

Before we can finish this part, we like to build another case study project to illustrate how to use Regression algorithm to predict the current or future stock prices based on the previous stock values.

5.9   A Real Project Example for Stock Prediction Using Regression Algorithm

This case study is based on a stock price dataset, named **Google Stock Dataset**, which provided the stock price data for 5 years from 2012 to 2017 by Google [31]. We like to use some nonlinear regression algorithms to predict the current or future stock prices based on this dataset since most time the stock market has significant variations on the stock prices and those variations are both dynamic and nonlinear functions based on the time.

One of the most important factors or conditions to successfully predict the stock prices based on the history data is that the variations of the stock prices must be or can be approximated as a periodic function based on time. Otherwise, it absolutely does not make any sense to make this kind of prediction.

First let us have a clear picture about this Google Stock Dataset.

5.9.1   *Introduction to Google Stock Dataset*

This dataset is composed of two subdatasets, one of them contained the stock price data for 5 years from 2012 to 2016, which is a 1258 × 6 matrix with stock opening prices, closing prices, high and low prices, and the transaction volumes, and named **Google_Stock_Price_Train.csv**. Another dataset, named **Google_Stock_Price_Test.csv**, contained only 1-month stock price data for 2017, exactly for January 2017, and it is a matrix with a size of 20 × 6.

We can use the first dataset to train our nonlinear regression model and test that model by using the second dataset to check and confirm our estimated model.

To successfully estimate and predict the current or future stock prices based on the stock history data, another import and key issue is that all stock training data or history data must be normalized appropriately since the stock prices are varied and changed at each year, each month, each day, and even each moment. Therefore, we can only predict the relative stock prices, not absolute prices, based on the stock data scale at that time. In other words, we can only predict the trend or the tendency of stock prices, and that is good enough for us to correctly make our investments in a right way.

5.9.2   Build a Sample Project to Train and Test Regression Nonlinear Models

In this project, we like to use three different regression nonlinear algorithms, including the functions **fitrensemble()**, **fitnlm()**, and **TreeBagger()**, to train and build three models to estimate and predict the stock prices based on the history stock data.

Now create a new Script file and name it as **stock_ensemble_treebagger_func.m**, and enter the codes shown in Fig. 5.93 into that file.

```
   % Train and evaluate stock price model with nonlinear regression method
   % Name: stock_ensemble_treebagger_func.m
   % Input: Google_Stock_Price_Train.csv & Google_Stock_Price_Test.csv
1  M = 20;          % testing data with 20 rows
2  T1 = readtable("C:\\Artificial Intelligence Book\\Students\\Datasets\\Google Stock Dataset\\" + ...
                  "Google_Stock_Price_Train.csv");
3  T2 = readtable("C:\\Artificial Intelligence Book\\Students\\Datasets\\Google Stock Dataset\\" + ...
                  "Google_Stock_Price_Test.csv");
   % Use ensemble tree algorithm to train the model
4  X = table(T1.Open/max(T1.Open), T1.High/max(T1.High), T1.Low/max(T1.Low),...
              str2double(T1.Volume)/max(str2double(T1.Volume)));
5  Y = T1.Close/max(T1.Close);
6  mdl_1 = fitrensemble(X, Y);

7  Xnew = table(T2.Open/max(T2.Open), T2.High/max(T2.High), T2.Low/max(T2.Low),...
               str2double(T2.Volume)/max(str2double(T2.Volume)));
8  Y1 = predict(mdl_1, Xnew);
9  Yact = T2.Close/max(T2.Close);

   % Use fitnlm() function to fit the model
10 tbl = table(T1.Open/max(T1.Open), T1.High/max(T1.High), T1.Low/max(T1.Low),...
              str2double(T1.Volume)/max(str2double(T1.Volume)), T1.Close/max(T1.Close));
11 modelfun = @(b,x)b(1) + b(2)*x(:,1).^b(3) + b(4)*x(:,2).^b(5) + b(6)*x(:,3).^b(7) + b(8)*x(:, 4).^b(9);
12 beta0 = [50 500 -1 500 -1 500 -1 500 -1];
13 mdl_2 = fitnlm(tbl, modelfun, beta0);
14 Y2 = feval(mdl_2, Xnew);

15 % Use TreeBagger() and formula to define the model as Close = Open + High + Low + Volume
16 mdl_3 = TreeBagger(100, X, Y, Method="regression", NumPredictorsToSample=50);
17 Y3 = predict(mdl_3, Xnew);

18 XX = 1:M;
   figure(1)
19 plot(XX, Yact, "b-", XX, Y1, "r-", XX, Y2, "g-", XX, Y3, "m-", 'LineWidth', 1);
20 axis([0 20 0.8 1.05]);
   grid;
21 legend(["Actual Stock Price","Ensemble Prediction","fitnlm() Prediction", "TreeBagger() Prediction"], ...
          "Location","northwest");
22 xlabel('Stock Price Per Day - 20 Days');
   ylabel('Stock Prices Predictions');

23 Yact = detrend(T2.Close/max(T2.Close));
   Y1 = detrend(Y1);
   Y2 = detrend(Y2);
   Y3 = detrend(Y3);

24 err_1 = Yact - Y1
   err_2 = Yact - Y2
   err_3 = Yact - Y3

   figure(2)
25 plot(XX, err_1, "b-", XX, err_2, "r-", XX, err_3, "m-");
   grid;
   legend(["Ensemble Prediction Error","fitnlm() Prediction Error", "TreeBagger() Prediction Error"], ...
          "Location","northwest");
   xlabel('Stock Price Per Day - 20 Days');
   ylabel('Stock Prices Prediction Errors');

26 err_sum = abs([max(abs(err_1)) min(abs(err_1)) rmse(Y1, Yact);
                  max(abs(err_2)) min(abs(err_2)) rmse(Y2, Yact);
                  max(abs(err_3)) min(abs(err_3)) rmse(Y3, Yact)])
```

Fig. 5.93 The codes for the project stock_ensemble_treebagger_func.m

Let us have a closer look at this piece of codes to see how it works.

1. Since the testing stock dataset contained only 1 month data with 20 rows, thus the number of rows is first declared.
2. The stock training dataset is loaded into a local table **T1**, which is used to train three different models later. One point to be noted is that path for that training dataset. You need to use a full path that includes both the whole directory and the dataset's name to locate that dataset. You may need to use your real path to replace it if you stored this dataset at different location on your machine.
3. Similarly, the stock testing dataset is also loaded and assigned to another local table **T2**.
4. Organize the input or predictor training data into a table and assigned them to a local variable **X**. One important point when doing this assignment is that all input variables must be normalized by dividing the maximum values for each of them. Since the last column, **Volume**, is a character vector variable, thus the function **str2double()** must be used to convert it to the double type variable. Otherwise, you may encounter some compiling error without using that conversion.
5. Perform the same normalization for the output or response variable **Close** and assign the normalized result to a local variable **Y**.
6. Call the **fitrensemble()** function to train the first model, **mdl_1**.
7. Build a testing stock data matrix **Xnew** by normalizing testing data obtained from the Stock Test dataset, **Google_Stock_Price_Test.csv**.
8. Evaluate the trained model, **mdl_1**, by executing the function **predict()** with the generated testing data matrix, **Xnew**. The predicted response is assigned to a local variable **Y1**.
9. Get the actual stock price values obtained from the stock testing dataset, normalize them, and assign them to a local variable **Yact**.
10. Next using the function **fitnlm()** with model function to perform this model training job, first generate a table **tbl** with the normalized input and the response variables.
11. Build a model function with nine (9) coefficients and four (4) input vectors, where each input vector is indicated by using the column number.
12. Define an initial input vector for all nine (9) coefficients on model function. A trick of setting these initial values is that all of those values should be assumed based on some real system coefficients that could be obtained by building a similar project with MATLAB Curve Fitting App.
13. The function **fitnlm()** is executed to perform the training job for the second model, **mdl_2**.
14. Then the function **feval()** is called to evaluate or predict a set of stock testing data obtained from the stock testing dataset with the trained model, **mdl_2**, to get the evaluated response. The predicted result is assigned to a local variable **Y2**.
15. Now using the function **TreeBagger()** to train our stock price model. First set up a formula as **Close ~ Open + High + Low + Volume** for this model training.

16. The function **TreeBagger()** is executed with some **Name-Value** pairs, including 100 decision trees, the **regression** Method, and the **NumPredictors ToSample**. The trained model is assigned to a local variable **mdl_3**.

17. The trained model, **mdl_3**, is evaluated with a set of testing stock data stored in the variable **Xnew** with the model. The evaluated result is assigned to the variable **Y3**.

18. Now let us use some plots to show and compare those three modeling performances by generating and comparing among three predicted responses with the actual response values. First let us generate a colon variable **XX** to work as the horizontal dataset. Due to the number of stock testing data is 20 (rows), the colon vector is set as 1:20.

19. To plot and display two or more graphs in MATLAB environment, using the MATLAB function **figure(n)** to create multiple (**n**) figure windows. The argument **n** indicates the number of figures to be generated and displayed. Then the **plot()** function is called to plot three predicted responses with the actual response (stock prices) with different colors and styles.

20. The function **axis()** is used to set up all boundary limits on the displayed area.

21. The **legend()** function is used to illustrate all different responses, or who is who, to effectively distinguish among them.

22. Some **label()** functions are used to make plots more meaningful.

23. To remove some DC offsets from three response trajectories, the MATLAB function **detrend()** is used to make comparisons among three predicted responses and the actual response easier and clearer.

24. Also the errors between three predicted responses and the actual response are calculated and displayed in the Command window as the project runs.

25. These errors are also plotted on the second figure, **fig. (2)**, with **legend()** and some **label()** functions.

26. Finally, a summary of errors, including the maximum errors, the minimum errors, and RMSE errors, are displayed on the Command window as the project runs. These errors can be collected and presented by a histogram in Excel.

Now run the project and the running result is shown on the Command window as below,

```
err_sum =

    0.0116    0.0023    0.0067
    0.0214    0.0016    0.0112
    0.0116    0.0005    0.0053
```

The plotting results are shown in Figs. 5.94a, b. A histogram for all errors is shown in Fig. 5.95.

In Fig. 5.94a, a comparison among predictions of three modeling algorithms and the actual response is displayed, where the solid blue color trajectory is the actual response or the actual stock prices variation in January 2017. The green color

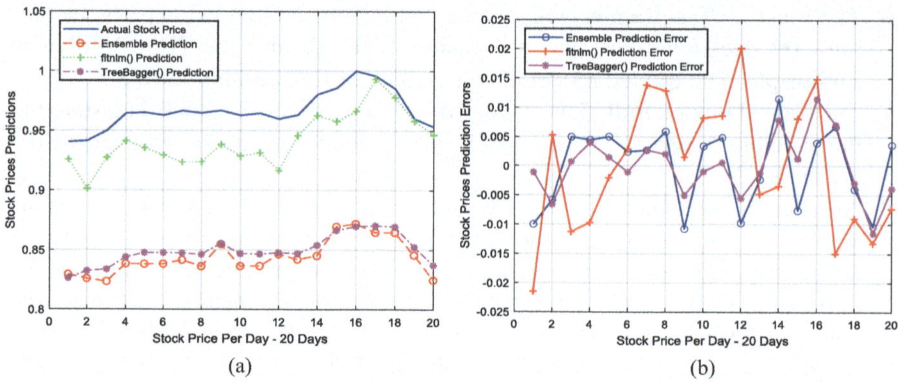

Fig. 5.94 The running results of the project stock_ensemble_treebagger_func.m

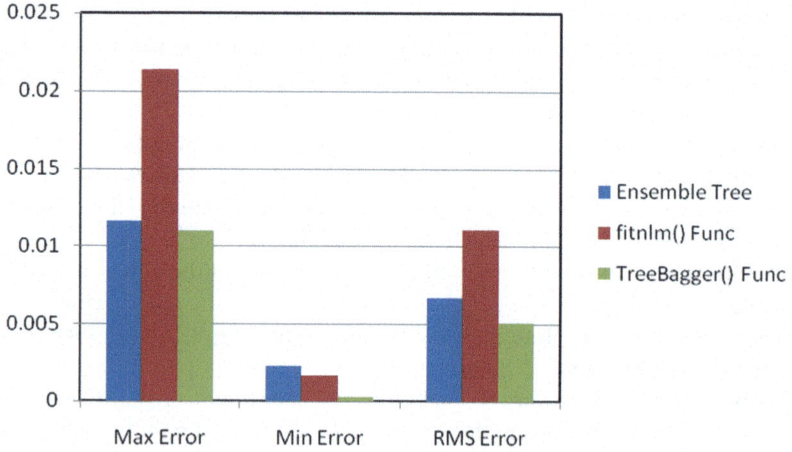

Fig. 5.95 The error comparison among three regression nonlinear algorithms

trajectory with '+' marks is the prediction of stock prices by using the **fitnlm**() function. The red color trajectory with 'o' marks is the estimation of stock prices by using the **fitrensemble**() function, and the pink color trajectory with '*' marks is the prediction of stock prices by using the **TreeBagger**() function.

There are also some DC offsets existed among these responses or predictions, which are not important for our prediction function since we only take care of the trend or the tendency of the stock prices, not the absolute stock price values.

It can be found from Figs. 5.94a, b that comparatively speaking, the prediction of using the **TreeBagger**() function with pink color is the best match to the actual stock prices variation with an RMSE value of 0.005. The prediction of using the function **fitrensemble**() is the second best match to the actual stock prices variation with an RMSE value of 0.0067. The prediction of using the **fitnlm**() function is the worst one with an RMSE value of 0.0112. Therefore, based on those predictions, the best time to do our investment is around the date 15–18 since the closing prices on those days are closer to the peak values.

The same conclusion can also be obtained from the histogram plot shown in Fig. 5.95.

A completed project Script file, **stock_ensemble_treebagger_func.m**, can be found from the Springer ftp site at the folder: **Students\Class Projects\Chapter 5\ Nonlinear Regression Functions**.

Now we have completed our introductions and discussions about regression algorithms, next let us begin our discussions on another supervised learning algorithm, classifications.

5.10   Chapter Summary

The main topic of this chapter is about introductions and discussions for one of the major branches in the machine learning, the regression algorithms.

In fact, MATLAB provided two major Toolboxes, Statistics and Machine Learning Toolbox and Curve Fitting Toolbox, to help users to design and build their machine learning-related projects, especially for regression-related projects, easily and quickly.

Most important and popular regression-related algorithms are discussed in detail in this chapter, which include:

1. Linear regression
2. Multiple linear regression
3. Decision tree-related regression
4. Nonlinear regression
5. Support vector machine (SVM) algorithm
6. K-Nearest Neighbor (KNN) algorithm
7. Random forest algorithm

All of these algorithms are introduced and discussed with some hands-on and real projects in two modes, the MATLAB App mode and MATLAB Function mode. All those Apps and Functions are involved in two Toolboxes, Statistics and Machine Learning and Curve Fitting Toolboxes.

Relatively speaking, the App mode provides an efficient and convenient way to enable students or users, especially the beginners, to build and develop regression-related projects easily and quickly, and it greatly improves the students' study interests and reduces the learning curves. The shortcoming of using those Apps is that some coding and building strategies and techniques are hidden from students, which is not a professional way to study machine learning-related technologies.

To improve that shortcoming, the MATLAB Function mode can be adopted to enable students or users to touch deeper in building more professional AI projects by developing more detailed control functions with more coding processes.

Finally, a hands-on and practical AI project is developed by using some different nonlinear regression models to predict and estimate the stock prices in the stock market. Three classes related functions, **fitrensemble()**, **fitnlm()**, and **TreeBagger()**, are used to build different models to estimate the stock prices. A validation test is performed to confirm the effectiveness of the predictions based on three different algorithms, and a comparison is also performed to evaluate them and to get the optimal algorithm.

Home Works

I. True/False Selections

_______1. MATLAB provides a powerful App, Regression Learner. The function of this App is to interactively train, validate, and tune regression models with different algorithms.

_______2. A regression is a statistical algorithm that relates a dependent variable, which is an input, to one or more independent or explanatory variables, such as outputs.

_______3. The linear regression establishes a nonlinear relationship between two variables, input and output, based on a curve of best fitting.

_______4. Residuals are the difference between the *observed* values of the response (dependent) variable and the values that a model *predicts*.

_______5. The goal of multiple linear regressions is to model a nonlinear relationship between the explanatory (independent) variables and response (dependent) variables.

_______6. A decision tree is a decision support hierarchical model that uses a tree-like model of decisions and their possible consequences, including chance event outcomes, resource costs, and utility.

_______7. A decision tree can be divided into two major types, linear and nonlinear decision tree.

_______8. Nonlinear regression is a statistical technique that helps describe nonlinear relationships in experimental data. Nonlinear regression models are assumed to be parametric.

_______9. Three popular nonlinear fitting functions are, **fit()**, **fitnlim()**, and **nlinfit()**.

_____10. Random forests or random decision forests is an **ensemble learning** method for classification, regression, and other tasks that operates by constructing a multitude of decision trees at training time.

II. Multiple Choices

1. Validation is a process used to examine the predictive accuracy of the fitted models, and three kinds of validation methods are ___________________________.

   (a) Cross-Validation
   (b) Holdout Validation
   (c) Resubstitution Validation
   (d) All of them

2. Regression Learner provides two ways for regression process, they are
   _______________.

   (a) Automated Regression Model Training
   (b) Manual Regression Model Training
   (c) Both a and b
   (d) Either a or b

3. The regression algorithms can be divided into the following categories
   _______________.

   (a) Linear and nonlinear regressions
   (b) Linear, multiple linear, and nonlinear regressions
   (c) Multiple linear and nonlinear regressions
   (d) None of them

4. Regularly, decision trees can be divided into two major types,
   _______________.

   (a) Regression trees and linear trees
   (b) Regression trees and nonlinear trees
   (c) Classification trees and multiple linear trees
   (d) Regression trees and classification trees

5. Some popular algorithms used for fitting a nonlinear regression include
   _______________.

   (a) Decision trees and gradient descent algorithm
   (b) Polynomial regression and Levenberg-Marquardt Algorithm
   (c) Gauss-Newton algorithm and SVM algorithm
   (d) All of them

6. In addition to those nonlinear regression algorithms, there are some popular regression models used for nonlinear regression processes; they are
   _______________.

   (a) Transformable Nonlinear Models and Gaussian Models
   (b) Polynomial Models and Exponential Models
   (c) Logarithmic Models and Fourier Models
   (d) All of them

7. The random forest algorithm can be considered as _______________.

   (a) An extension of bagging that also randomly selects subsets of features
       used in each data sample
   (b) A subset of variables or features selected at random out of the total
       input data
   (c) Both a and b
   (d) None of them

8. Regularly the SVM algorithm can be used for ___________________.

   (a) Only regression issues
   (b) Both classification and regression issues
   (c) Only classification issues
   (d) None of them

9. MATLAB provided three major nonlinear regression fitting functions ______________.

   (a) fit(), fitcsvm(), and fitrsvm()
   (b) fitlm(), fitnlim(), and fitglm()
   (c) fit(), fitrnet(), and fitnlim()
   (d) fit(), fitnlim(), and nlinfit()

10. One can create a new TreeBagger object by calling the function ____________________.

   (a) CreatBagger()
   (b) fitrensemble()
   (c) TreeBagger()
   (d) Bag()

III. Exercises

   1. What is linear regression? And what is multiple linear regression? What is the difference between them?
   2. Provide an explanation for the validation process.
   3. Provide a description about the decision tree and illustrate its function.
   4. Provide a simple description about the nonlinear regression.
   5. List seven popular nonlinear regression models involved in the Curve Fitter App.
   6. Provide a definition about the KNN algorithm.
   7. Provide a definition about the Random Forest algorithm.
   8. Explain the difference between the Random Forests and Bagging.

IV. Lab Projects

   1. Using MATLAB Regression Learner App to build a linear regression project **Car_Regression** to predict the car sale price based on a given dataset **car_prices.csv**. The original dataset can be downloaded at the site: https://www.kaggle.com/datasets/syedanwarafridi/vehicle-sales-data?resource=download. The license for using this dataset can be found in the **Trade Marks and Copyrights** page on the book. We modified this dataset to reduce the data size to 10 K. The modified dataset contains 10 K car sale records with 10 columns, **year, make, model, body, transmission, state, condition, odometer, mmr,** and **sellingprice**. The dataset can be found from the Springer ftp site in a folder: **Students\Datasets\Car Prices Dataset**.

Confirm your trained model with **Predicted vs. Actual Plot** and **Validation Residuals Plot**. Save the trained session as **Car_Price.mat** to a folder in your machine and export the model to the Workspace as **Car_Price_Model**.

Hint1: Refer to Sect. 5.2.4 to build this project with Regression Learner. Select the **sellingprice** as the response column.

Hint2: Use **All Quick-To-Train** to train the model with all possible algorithms.

2. Using MATLAB Regression- related function **fitlm()** to build a car prices model via a Script named **Car_Price_Func.m** with the car prices dataset **car_prices.csv**.

Hint1: Refer to Sect. 5.2.3 to build this project with the function **fitlm()**.

Hint2: Plot the trained model **car_price_mdl** to check and confirm the fitting result.

3. Using MATLAB Regression-related function **fitrsvm()** to build a car prices model via a Script named **Car_Price_SVM.m** with the car prices dataset **car_prices.csv**. Also validate and confirm the trained model **car_price_mdl** with a car prices testing dataset **car_prices_test.csv**, which contained 5 car records in lines 10830–10834 in the original dataset. Compare the predicted and actual selling prices and calculate the RMSE values between them.

Hint1: Refer to Sect. 5.2.3 to build this project with the function **fitrsvm()**.

Hint2: You need to set up both training and testing dataset paths, and use **readtable()** function to read both training and testing data. Setting up training and testing data in 2 Tables and the responses as **Y** and **predY**.

Hint3: Using the **predict()** function and trained model to predict the predicted selling prices based on the testing data.

Hint4: Using **rmse()** function to get RMSE value between the predicted and the actual selling prices.

4. Using MATLAB **fitnlm()** function with **a*log(x) +b** and **a* exp(-((x − b)/c)^2)** functions to train models, **mdl_1** and **mdl_2** shown in Figure 5.72 by using **car_prices.csv** dataset. Then evaluate both models with dataset **car_prices_test.csv**. All data are located under the folder **Car Prices Dataset** and it can be found from the Springer ftp site at the folder **Students\ Datasets\Car Prices Dataset**. The finished Script file is named **car_price_nlfit_func.m**.

Hint1: Refer to Fig. 5.72 on Sect. 5.6.2.3 for the coding developments. You only need two columns, **odometer** and **sellingprice**, on both datasets as inputs and outputs.

Hin2: Using **rmse()** function to get RMSE values between the predicted and the actual selling prices.

5. Build a MATLAB project **car_price_knn.m** with the **kNNeighborsRegressor** method. The trained model should be evaluated with data column **odometer** in the **car_prices_test.csv** dataset. Both training and testing data are stored and located under the folder **Car Prices Dataset** and it can be

found from the Springer ftp site at the folder **Students\Datasets\Car Prices Dataset.**

Hint1: A point to be noticed is that you may need to save the customer-built function file, **kNNeighborsRegressor.m**, at the same folder as this Script file is located to enable the Script file to be run smoothly. This function can be found from the Springer ftp site under the folder, **Students\Class Projects\Chap. 5\KNN Project**.

6. Using **TreeBagger()** function to build four random forest models with **car_prices.csv** dataset. The Script file is named **treebagger_car_price_func.m**. Plot the training and testing results for the last model

   Hint1: Refer to codes developed in Fig. 5.89 at Sect. 5.8.5.2 to build this project.

References

1. https://www.investopedia.com/terms/r/regression.asp#:~:text=A%20regression%20is%20a%20statistical,more%20of%20the%20explanatory%20variables.
2. https://www.investopedia.com/terms/m/mlr.asp#:~:text=Key%20Takeaways-,Multiple%20linear%20regression%20(MLR)%2C%20also%20known%20simply%20as%20multiple,uses%20just%20one%20explanatory%20variable.
3. https://www.investopedia.com/ask/answers/060315/what-difference-between-linear-regression-and-multiple-regression.asp#:~:text=only%20two%20variables.-,Multiple%20regression%20is%20a%20broader%20class%20of%20regressions%20that%20encompasses,regression%20incorporates%20multiple%20independent%20variables.
4. https://archive.ics.uci.edu/dataset/162/forest+fires.
5. Breiman, Leo; Friedman, J. H.; Olshen, R. A.; Stone, C. J. (1984). *Classification and regression trees*. Monterey, CA: Wadsworth & Brooks/Cole Advanced Books & Software. ISBN 978-0-412-04841-8.
6. https://en.wikipedia.org/wiki/Decision_tree_learning#:~:text=Decision%20trees%20where%20the%20target,dissimilarities%20such%20as%20categorical%20sequences.
7. https://www.mathworks.com/discovery/nonlinear-regression.html.
8. https://medium.com/@nathan.myron/gauss-newton-method-7dbb3d4ac32c.
9. https://towardsdatascience.com/gradient-descent-algorithm-a-deep-dive-cf04e8115f21#:~:text=Gradient%20descent%20(GD)%20is%20an,e.g.%20in%20a%20linear%20regression).
10. https://mathworld.wolfram.com/Levenberg-MarquardtMethod.html.
11. https://www.mathworks.com/discovery/support-vector-machine.html.
12. https://en.wikipedia.org/wiki/Hyperplane#:~:text=In%20geometry%2C%20a%20hyperplane%20is,that%20of%20the%20ambient%20space.
13. https://www.geeksforgeeks.org/k-nearest-neighbours/.
14. https://medium.com/@nandiniverma78988/understanding-k-nearest-neighbors-knn-regression-in-machine-learning-c751a7cf516c.
15. David Ferreira (2024). k-Nearest Neighbors (kNN) Regressor (https://github.com/ferreirad08/kNNeighborsRegressor/releases/tag/1.0.1), GitHub. Retrieved January 1, 2024.
16. https://www.mathworks.com/help/stats/treebagger.html.
17. https://en.wikipedia.org/wiki/Random_forest.
18. Gareth James; Daniela Witten; Trevor Hastie; Robert Tibshirani (2013). *An Introduction to Statistical Learning*. Springer. pp. 316–321.

19. https://www.ibm.com/topics/random-forest#:~:text=Random%20forest%20is%20a%20
commonly,both%20classification%20and%20regression%20problems.
20. Ho, Tin Kam (1995). *Random Decision Forests*. Proceedings of the 3rd Intl. Conf. on Document Analysis and Recognition, Montreal, QC, 14–16 August 1995. pp. 278–282.
21. Ho TK (1998). "The Random Subspace Method for Constructing Decision Forests" (PDF). *IEEE Transactions on Pattern Analysis and Machine Intelligence*. **20** (8): 832–844.
22. Kleinberg E (1990). *"Stochastic Discrimination"*. *Annals of Mathematics and Artificial Intelligence*. *1 (1–4): 207–23, 2018-01-18.*
23. Kleinberg E (1996). *"An Overtraining-Resistant Stochastic Modeling Method for Pattern Recognition"*. *Annals of Statistics*. *24 (6): 2319–2349.*
24. Kleinberg E (2000). *"On the Algorithmic Implementation of Stochastic Discrimination"*, *IEEE Transactions on PAMI*. *22 (5): 473–490.*
25. Breiman L *(2001)*. *"Random Forests"*. *Machine Learning*. *45 (1): 5–32.*
26. Liaw A (16 October 2012). "Documentation for R package random Forest". Retrieved 15 March 2013.
27. U.S. trademark registration number 3185828, registered 2006/12/19.
28. *"RANDOM FORESTS Trademark of Health Care Productivity, Inc. - Registration Number 3185828 - Serial Number 78642027 :: Justia Trademarks"*.
29. Amit Y, Geman D (1997). "Shape quantization and recognition with randomized trees" (PDF). *Neural Computation*. **9** (7): 1545–1588.
30. https://stats.stackexchange.com/questions/264129/what-is-the-difference-between-bagging-and-random-forest-if-only-one-explanatory.
31. https://www.kaggle.com/datasets/vaibhavsxn/google-stock-prices-training-and-test-data.

Chapter 6
Introduction to Classification Algorithms

Classifications provided another important branch for supervised learning techniques and generally they are used as classifiers to classify binary or multiclass objects based on predictors. The working principle of classification algorithms is very similar to those regression algorithms, but a key difference between them is that the regression is used to predict or estimate the desired responses represented as a continuous numeric values, and the classification algorithms are used to predict the desired response in terms of classes, either binary or multiclass objects.

6.1 Introduction to Classifications

As shown in Fig. 4.1 in Chap. 4, the classification algorithm is another important and key branch under the Supervised Learning. Essentially, there is no significant difference between the regression and classification algorithm, and the only difference between them, as we mentioned in the previous parts, is that the outputs of regression algorithms are numeric or continuous values, but the outputs of classification algorithms are classes or Boolean values, either true or false.

In fact, classification is a process related to categorization, and the process in which ideas and objects are recognized, differentiated, and understood. Classification is the grouping of related facts into classes. It may also refer to a process which brings together like things and separates unlike things.

As we discussed in Sect. 4.1 in Chap. 4, the AI classification is composed of different algorithms, and each algorithm can be used to perform different functions for the real requirements of various applications. Basically, three algorithms are widely implemented in AI classification applications; they are:

Supplementary Information The online version contains supplementary material available at https://doi.org/10.1007/978-3-031-84423-2_6.

Y. Bai, *AI Foundations and Applications with MATLAB*,
https://doi.org/10.1007/978-3-031-84423-2_6

1. Binary Classifications
2. Multiclass Classifications
3. Multilabel Classifications

Binary Classification

The function of this kind of classification is to group objects or targets into two possible classes, either true or false. One of the most popular applications of using this kind of algorithm is on medical inspection system, to detect some possible diseases, such as diabetes, heart disease, and some cancers.

Multiclass Classification

The so-called multiclass classification is to group objects into more than two classes or two categories. Some popular examples of using this kind of classification are to detect and identify animals, including the animals' types based on sounds and images. It can also be used to distinguish and identify different objects, such as vehicles, plants, merchandise, and products.

Multilabel Classification

Unlike above two classification algorithms, in which each object or target can only be classified or categorized into one class, the multilabel classification algorithm allows each data point or object to belong to several classes. The algorithm assigns multiple descriptive labels to each data point in this scenario.

It can be found that the multilabel classification sounds are more complex compared with the other two algorithms; thus, due to the complication of the multilabel classification, we only concentrate our study on the first two algorithms, binary and multiclass classifications, in this book. Based on Fig. 4.1 in Chap. 4, we limit our study and discussion to the following algorithms:

1. Binary Classification (BC)
2. Multiclass Classification (MC)
3. KNN Algorithm (KNN)
4. Decision Tree (DT)
5. Naive Bayes (NB)
6. Support Vector Machine (SVM)
7. Random Forest (RF)

In fact, another important and popular algorithm used for classification is the neural network. Due to its important property and wide spectrum of applications in AI, we will discuss it in a separate chapter later.

From another point of view, classification is a type of supervised machine learning in which an algorithm *learns* to classify new observations from examples of labeled data. In MATLAB, it also provided two groups of tools to support users to build professional classification-related projects.

- Statistics and Machine Learning Toolbox.
- Deep Learning Toolbox

With the help of those two Toolboxes, all tools can be further divided into two major types for machine learning developers, and they are:

1. User-friendly Apps used to build and develop machine learning and classification-related projects with multiple GUIs in easy and fast ways.
2. Function library used to design, build, and develop machine learning and classification-related projects in more professional and flexible ways.

As we mentioned in Sect. 4.3.1 in Chap. 4, MATLAB Statistics and Machine Learning Toolbox combined nine (9) Apps together, and they contain machine learning and deep learning Apps in one package. One of the Apps, **Classification Learner**, can be used to help users build classification-related projects.

In addition to that App, MATLAB also provided some useful functions with two Toolboxes, Statistic and Machine Learning Toolbox and Deep Learning Toolbox. By using those functions, users can build more professional classification-related applications with more flexibility and functionality.

In this section, we will concentrate on the Classification Leaner App and discuss how to use that App to build classification-related applications.

6.2   Classification-Related Apps in MATLAB

The Classification Learner App is included in the Statistics and Machine Learning Toolbox, and it allows users to choose among various algorithms to train and validate classification models for binary or multiclass problems. After training multiple models, compare their validation errors side-by-side, and then choose the best model.

Users can use Classification Learner App to train models of the following classifiers:

1. Decision Trees
2. Discriminant Analysis
3. Support Vector Machines
4. Logistic Regression
5. K-Nearest Neighbors
6. Naive Bayes
7. Kernel approximation
8. Ensembles
9. Neural Networks

In addition to training models, you can explore your data, select features, specify validation schemes, and evaluate results. You can export a model to the workspace to use the model with new data or generate MATLAB® code to learn about programmatic classification.

A point to be noted is that when training any model, the App does not separate the input data into some sections, such as training data or testing data. Instead, it treats all input data as training data or called full data.

Training a model in Classification Learner App consists of two parts:

1. **Validated Model**: Train a model with a validation scheme. By default, the App protects against over-fitting by applying cross-validation. Alternatively, you can choose holdout validation. The validated model is visible in the App.
2. **Full Model**: Train a model on full data without validation. The App trains this model simultaneously with the validated model. However, the model trained on full data is not visible in the App. When you choose a classifier to export to the workspace, Classification Learner exports the full model.

The App displays the results of the validated model and diagnostic measures, such as model accuracy, and plots, such as a scatter plot or the confusion matrix chart, reflect the validated model results.

When using this App to train different models, you have two options:

1. Automated Classification Training
2. Manual Classification Training

By selecting automatic training, you can train one or more classifiers, compare validation results, and choose the best model that works for your classification problem. However, if you choose for the manual classification training, you need to select a model yourself. When you choose a model to export to the workspace, Classification Learner App exports the full model.

Because Classification Learner App creates a model object of the full model during training, you experience no lag time when you export the model. You can use the exported model to make predictions on new data later.

Figure 6.1 shows a common workflow for training classification models, or classifiers, in the Classification Learner App.

Starting MATLAB R2023a, another powerful App named **Experiment Manager** is provided to enable users to train models under multiple initial conditions and compare the results with multiple different datasets. Due to the space limitation, we will not discuss that App in this book.

Now let's start our discussions with a binary classification project by using the Classification Learner App. We like to use a diabetes dataset in this project to perform a binary classification to train a model that can be used to identify and detect the possibility of being a diabetes patient. The original owner of this dataset is the National Institute of Diabetes and Digestive and Kidney Diseases, and the donor of that dataset is Vincent Sigillito [1].

This diabetes dataset contained 768 records of adult female patients with nine (9) checked or measured medical results, including the number of pregnancies, glucose, blood pressure, skin thickness, insulin, BMI, diabetes pedigree function, age, and

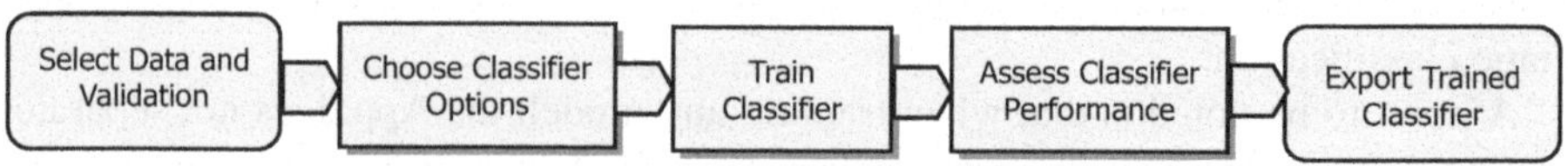

Fig. 6.1 Working flow for training classification models

diabetes result (0, no diabetes, 1 with diabetes). In summary, this dataset is a 768×9 matrix stored in a file, **Diabetes.csv**.

You can find this dataset from the Springer ftp site at a folder **Students\Datasets\ Diabetes Dataset**. You can copy and paste it to one of your local folders to use it later.

6.3   Binary Classification App Example

First perform the following operations to open the Classification Learner App,

1. Open MATLAB and click on the **App** tab on the top of the window.
2. Click on the drop-down arrow on the right to show all models, browse to the **MACHINE LEARNING AND DEEP LEARNING** group, and click on the **Classification Learner** icon to open it.
3. On the **Learn** tab, in the **File** section, click on the **New Session** item and then you can select data from the workspace or from a file. In this study, we prefer to load our dataset file, **Diabetes Dataset.csv**, from our local folder.
4. Click on the **From File** selection to open the **Select File** wizard. Browse to your local folder where this dataset is stored, select it, and click on the **Open** button. The selected dataset with all data is shown in Fig. 6.2.
5. Click on the green color button, **Import Select**, on the upper-right corner to load the data into our App, as shown in Fig. 6.2.
6. The import data wizard is shown in Fig. 6.3. There are two issues to be paid attention to: one issue is that the output (**Outcome**) column in our dataset should be a class or Boolean value, 0 or 1, which means no diabetes (0) and

	A	B	C	D	E	F	G	H	I
	Pregnancies	Glucose	BloodPres...	SkinThick...	Insulin	BMI	DiabetesP...	Age	Outcome
	Number	Number	Number	Number	Number	Number	Number	Number	Number
1	Pregnancies	Glucose	BloodPress...	SkinThickn...	Insulin	BMI	DiabetesPe...	Age	Outcome
2	6	148	72	35	0	33.6	0.627	50	1
3	1	85	66	29	0	26.6	0.351	31	0
4	8	183	64	0	0	23.3	0.672	32	1
5	1	89	66	23	94	28.1	0.167	21	0
6	0	137	40	35	168	43.1	2.288	33	1
7	5	116	74	0	0	25.6	0.201	30	0
8	3	78	50	32	88	31	0.248	26	1
9	10	115	0	0	0	35.3	0.134	29	0
10	2	197	70	45	543	30.5	0.158	53	1
11	8	125	96	0	0	0	0.232	54	1
12	4	110	92	0	0	37.6	0.191	30	0
13	10	168	74	0	0	38	0.537	34	1

Fig. 6.2 The selected dataset with all data in the Diabetes Dataset file

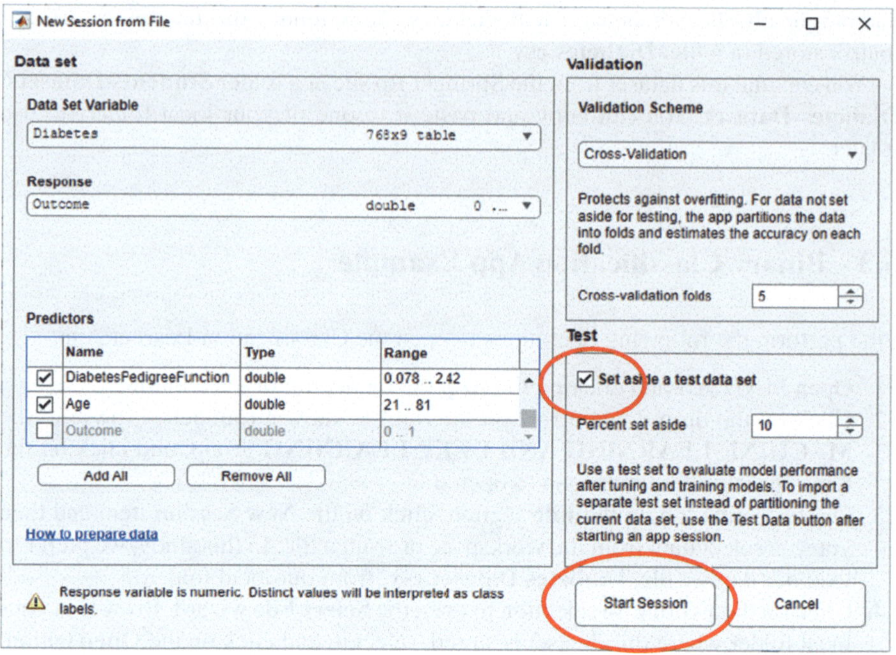

Fig. 6.3 The opened New Session wizard

diabetes (1). But it was a double value in our dataset. Therefore, a warning message is displayed at the bottom of this wizard to indicate that the response variable will be converted to a class label and it is fine for us. The second issue is that we want to use a partial data in this dataset, say 10% of 768 records, as the testing data for this model, so just check the **Set aside a test data set** checkbox to make this clear, as shown in Fig. 6.3.

7. Now click on the **Start Session** button for the next step.
8. To make it easy, first let's perform the automated classification training. In the **Models** section, select the **All Quick-To-Train** item. This option trains all the model presets available for our data set that are fast to fit.
9. Click on the drop-down arrow on the **Train** section, and select the **Train All** item to perform this automated classification process.
10. Totally 12 possible models are trained and the training results are shown in the left pane, as shown in Fig. 6.4.

Among all 12 models, the model 2.10, Efficient Logistic Regression tree, is the best one with the maximum validation accuracy of 76.7% shown in the left pane. To check the training results and model performances, MATLAB provided different ways to enable us to evaluate them. Basically, the following four (4) ways are more popular:

1. Check performance in the Models pane.

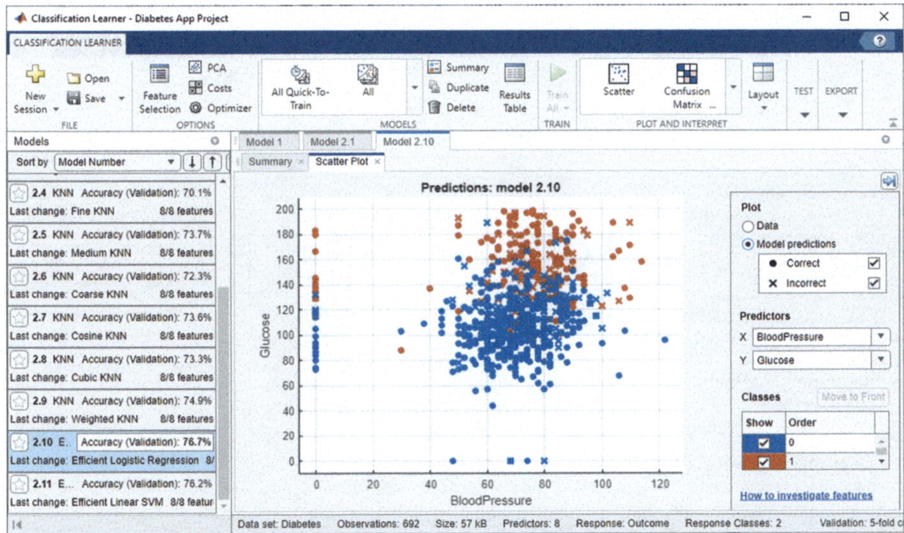

Fig. 6.4 The trained models with results

2. Compare performance by plotting classifier results.
3. Check performance per class in the Confusion Matrix.
4. Check Receiver Operating Characteristic (**ROC**) curve.

Let's do those evaluations and comparisons one by one.

First it can be found from Fig. 6.4 that the key performance parameters of all models, such as Accuracy and Features, are shown in the left pane. The best trained model is 2.10 that had the highest accuracy compared with other models.

Secondly by clicking on the drop-down arrow on the combo box **X** and **Y** under the **Predictors** on the right pane, we can check and compare the predictor results by plotting the prediction results, as shown in Fig. 6.4, where we selected **Blood Pressure** as **X** and **Glucose** as **Y** to get the mapping relation between them. An **x** means incorrect prediction and an **o** means the correct prediction. The blue-color **x** and **o** means 0, and brown-color **x** and **o** means 1.

Next, we can check the performances by opening the confusion matrix, as shown in Fig. 6.5, in which the best model, model 2.10, is displayed. This matrix displays both predicted true class and predicted false class.

To be able to better understand this confusion matrix, we need to familiarize us with some new terminologies listed below:

1. **True Positive Rates** (**TPR**)—The proportion of correctly classified observations per true class.
2. **False Negative Rates** (**FNR**)—The proportion of incorrectly classified observations per true class.
3. **Positive Predictive Values** (**PPV**)—The proportion of correctly classified observations per predicted class.

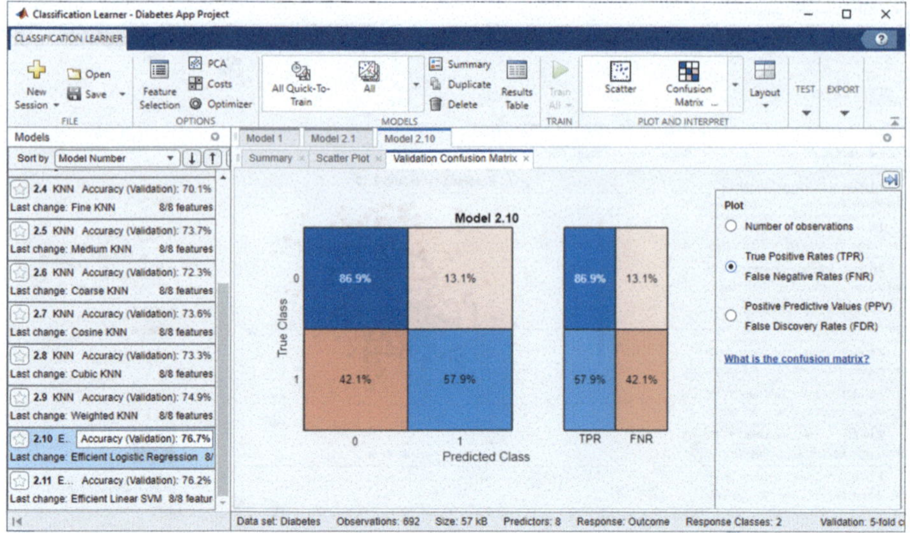

Fig. 6.5 The validation confusion matrix for model 2.10

4. **False Discovery Rates (FDR)**—The proportion of incorrectly classified observations per predicted class.

Based on these terminologies, it can be found from Fig. 6.5 that the correctly predicted or classified observations or outputs are 86.9% for class 0 (no diabetes) in blue color (**TPR**), but the incorrectly predicted observations are 13.1% for class 0 with orange color (**FNR**). Similarly, the correctly predicted observations are 57.9% for class 1 (with diabetes) in blue color (**TPR**), but the incorrectly predicted observations are 42.1% for class 1 with orange color (**FNR**).

On the right of the Fig. 6.5, two columns with blue color and orange color are displayed again based on the **TPR** and **FNR** proportion values. Also it can be found from Fig. 6.5 that the darker the color on a cell, the higher the proportion value on that cell, and vice versa.

Finally you can check the training and estimation results by using the **ROC** curve.

The ROC curve shows the true positive rate (**TPR**) versus the false positive rate (**FPR**) for different thresholds of classification scores, computed by the currently selected classifier. The **Model Operating Point** shows the false positive rate and true positive rate corresponding to the threshold used by the classifier to classify an observation. For example, a false positive rate of 0.4 indicates that the classifier incorrectly assigns 40% of the negative class observations to the positive class. A true positive rate of 0.9 indicates that the classifier correctly assigns 90% of the positive class observations to the positive class.

The Area Under the Curve (**AUC**) value corresponds to the integral of an ROC curve (**TPR** values) with respect to **FPR** from FPR = 0 to FPR = 1. The AUC value is a measure of the overall quality of the classifier. The AUC values are in the range

0 to 1, and larger AUC values indicate better classifier performance. Compare classes and trained models to see if they perform differently in the ROC curve.

To create an ROC curve for our trained model 2.10, click on the drop-down arrow at the **PLOT AND INTERPRET** pane on the top and select the **ROC Curve (Validation)** icon under the **VALIDATION RESULTS** group. Also you can check two classification results, classes 0 and 1, for Positive class, as shown in Fig. 6.6. However, you do not need to examine ROC curves for both classes in a binary classification problem. The two ROC curves are symmetric, and the AUC values are identical. A TPR of one class is a true negative rate (TNR) of the other class, and TNR is 1–FPR. Therefore, a plot of TPR versus FPR for one class is the same as a plot of 1–FPR versus 1–TPR for the other class.

For a multiclass classifier, the App formulates a set of one-versus-all binary classification problems to have one binary problem for each class, and finds an ROC curve for each class using the corresponding binary problem. Each binary problem assumes that one class is positive and the rest are negative. The model operating point on the plot shows the performance of the classifier for each class in its one-versus-all binary problem.

Alternatively, you can choose to train a selected model type if you like to. To do that, perform the following operations:

1. Click the drop-down arrow on the **MODELS** pane on the top and select a desired model type or algorithm, such as **Bagged Trees** under the **ENSEMBLE CLASSIFICATIONS** group in our case, to open the **Model Hyperparameters** wizard.

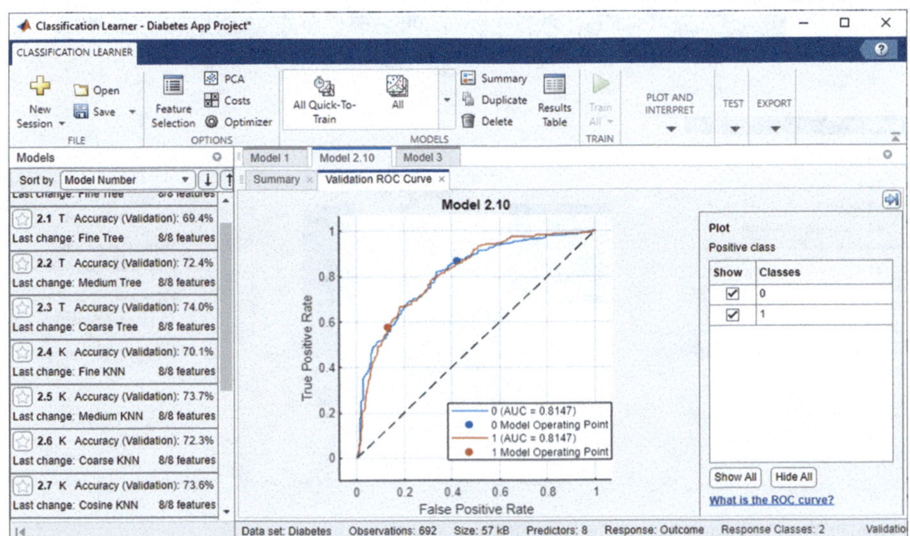

Fig. 6.6 The ROC curve of model 2.10

2. Keep all default parameters with no change and click on the drop-down arrow at the **Train All** item on the **TRAIN** icon on the top, and choose the **Train Selected** item to start the training process.

3. As the training is completed, click on the **Scatter** or **Confusion Matrix** item in the **PLOT AND INTERPRET** pane to check the training results. The Confusion Matrix for this model type is shown in Fig. 6.7.

4. On the same box, select the **ROC Curve (Validation)** icon under the **VALIDATION RESULTS** group to open its ROC curve, which is shown in Fig. 6.8.

Next let's discuss how to use functions to replace that App to perform a similar binary classification model training task for a given or selected model type.

6.4 Classification Classes and Related Functions in MATLAB

First let's have a clear and a complete picture about most popular MATLAB classifiers and related functions used for classifications, which are shown in Table 6.1. Table 6.2 shows typical characteristics of the various supervised learning algorithms in classifications.

In fact, all of these fitting functions belong to related class objects and you can create a desired classification object based on that related function. Table 6.3 lists some popular classification classes and related functions.

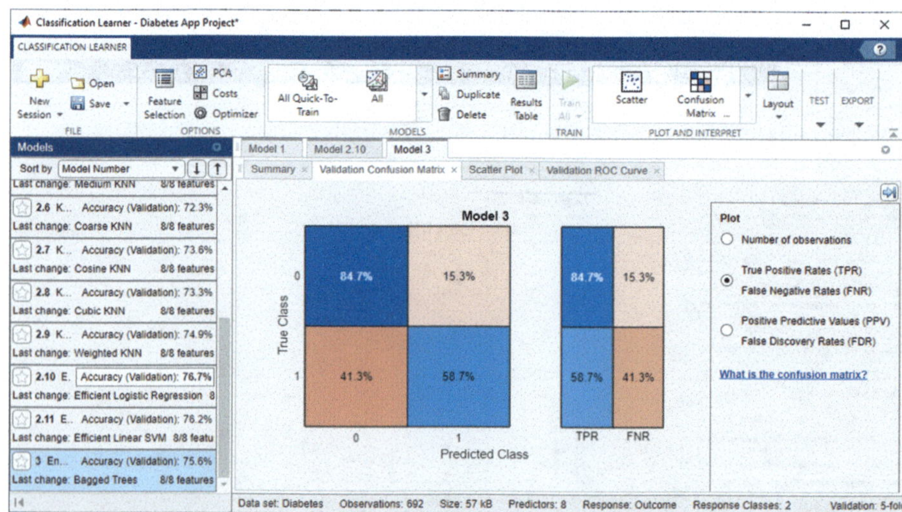

Fig. 6.7 The Confusion Matrix for Bagged Trees algorithm

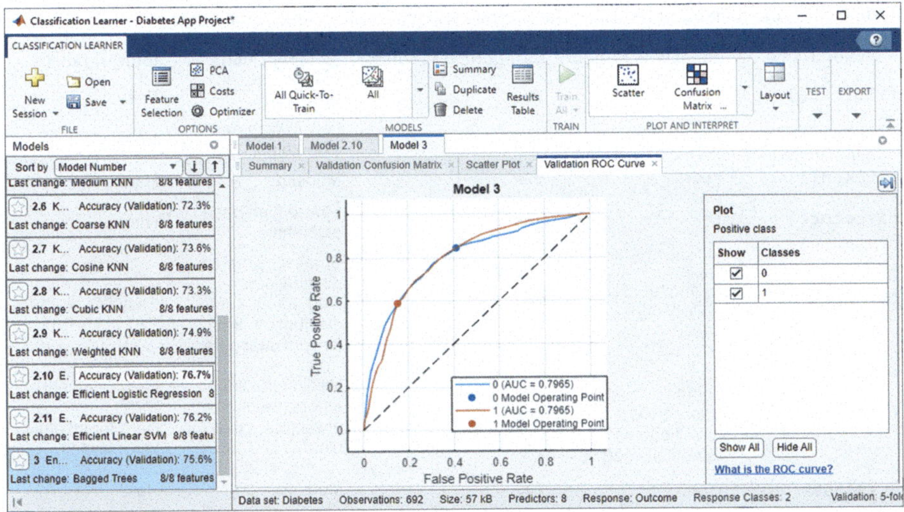

Fig. 6.8 The ROC curve for Bagged Trees algorithm

Table 6.1 Some popular MATLAB fitting functions used for classifications

Classifier	Function Name	Description
Decision Trees	`fitctree()`	Fit binary decision tree for multiclass classification.
Ensemble Classifiers	`fitcensemble()` `TreeBagger()`	Fit ensemble model for classification.
Nearest Neighbor Classifiers	`fitcknn()`	Fit *k*-nearest neighbor classifier.
Efficiently Trained Linear Classifiers	`fitclinear()` `fitcecoc()*`	Fit binary linear classifier to high-dimensional data. * For multiclass.
Kernel Approximation Classifiers	`fitckernel()` `fitcecoc()*`	Fit kernel approximation classifiers to perform nonlinear classification of data with many observations, error-correcting output codes (ECOC) model for SVM or other classifiers. *For multiclass.
Naïve Bayes Classifiers	`fitcnb()`	Train multiclass naive Bayes mode.
Neural Network Classifiers	`fitcnet()`	Train neural network classification model.
Support Vector Machine	`fitcsvm()` `fitcecoc()*`	Train support vector machine (SVM) classifier for one-class and binary classification. * For multiclass.
Logistic Regression Classifiers	`fitglm()` `fitclinear()` `fitcecoc()*`	Fit logistic regression classifier to perform generalized linear or multiclass classifications. *For multiclass.

It can be found from Table 6.3 that the top four classification classes with related functions, **fitclinear()**, **fitckernel()**, **fitcsvm()**, and **fitglm()**, can be used to generate related object to perform one class or binary classifications. Similarly, the lower three classes and related functions, **fitcnb()**, **fitcnet()**, and **fitcecoc()**, can be utilized to create object to perform multiclass classifications.

Table 6.2 Typical characteristics of the various supervised learning algorithms

Function Name	Multiclass Support	Categorical Predictor	Memory Usage	Interpretability
`fitctree()`	Yes.	Yes.	Small.	Easy.
`fitcensemble()`	Yes.	Yes.	Low to high based on algorithms.	Hard.
`fitcknn()`	Yes.	Yes.	Medium.	Hard.
`fitcecoc()`	Yes.	Yes.	Low to high based on algorithms.	Hard.
`fitcnb()`	Yes.	Yes.	Small for simple distributions. Medium for kernel distribution or high-dimensional data.	Easy.
`fitcsvm()`	No. Combine multiple binary SVM classifiers using fitcecoc().	Yes.	Medium for linear. All others: medium for multiclass, large for binary.	Easy for linear SVM. Hard for all other kernel types.
`TreeBagger()`	Yes.	Yes.	Medium.	Hard

Table 6.3 Popular MATLAB classification class/object and related function

Classification Class	Function	Description
ClassificationLinear	`fitclinear()`	Create a *ClassificationLinear* object by using **fitclinear()** for binary classification. The linear model is a support vector machine (SVM) or logistic regression model.
ClassificationKernel	`fitckernel()`	Create a *ClassificationKernel* object using **fitckernel()** for a binary Gaussian kernel classification model using random feature expansion.
ClassificationSVM	`fitcsvm()`	Create a *ClassificationSVM* object by using **fitcsvm()** for one class or binary classification.
GeneralizedLinearModel	`fitglm()`	Create a *GeneralizedLinearModel* object by using **fitglm()** for binary GLM logistic regression in Classification.
ClassificationKNN	`fitcknn()`	Create a KNN model *ClassificationKNN* object using **fitcknn()** for binary and multiclass classification.
ClassificationNaiveBayes	`fitcnb()`	Create a *ClassificationNaiveBayes* object by using **fitcnb()** for multiclass classifications.
ClassificationNeuralNetwork	`fitcnet()`	Create a *ClassificationNeuralNetwork* object by using **fitcnet()**.
ClassificationECOC	`fitcecoc()`	Create a *ClassificationECOC* object by using **fitcecoc()** for multiclass classification model with SVM.

We need to have some basic understanding about all of those classification classes and working principles. Let's start from linear classifications.

6.5   Linear Classifications

A Linear Classifier achieves this by making a classification decision based on the value of a linear combination of the characteristics. An object's characteristics are also known as *feature values* and are typically presented to the machine in a *vector* called a *feature vector* [2].

If the input feature vector to the classifier is a real vector $\vec{x}$, then the output score is

$$S = f\left(\vec{\omega} \cdot \vec{x}\right) = \sum_j \omega_j x_j + b \qquad (6.1)$$

where $\vec{\omega}$ is a real vector of weights or called a weight matrix that is composed of a set of weight values derived from the feature vectors, and b is an initial bias vector.

The weight matrix is composed of m rows and n columns, where m is the total number of input classes that need to be classified and n is the number of features. In other words, each row is for every class that needs to be classified, and one column for every element (feature) of input x.

Let's use some real examples to make these variables clearer. In fact, the linear classification algorithm is a good tool to perform classifications for linear separable observations, and nonlinear classifications are better to handle nonlinear separable observations. Figure 6.9 shows an illustration of linear separable and nonlinear separable observations. Based on that figure, the so-called linear separable elements indicate that all observations can be classified into two separate classes by a straight line (dashed line), but the nonlinear separable observations cannot be divided or classified by a straight line.

Another example of using linear classification is shown in Fig. 6.10. Three different class objects are shown in that figure: cat, dog, and car. Each classifier can be described by a separated straight line, called **Cat Classifier**, **Dog Classifier**, and **Car Classifier**, respectively.

The effect of changing the weight values in the weight matrix will change the angle or slope of each related line, while changing the bias will move the line either left or right based on the polarity of the bias values in the bias vector.

The linear classification is to set up a mapping between the input vectors (feature vector) and class score for a given model with parameters. That parametric model has two important components, they are [3]:

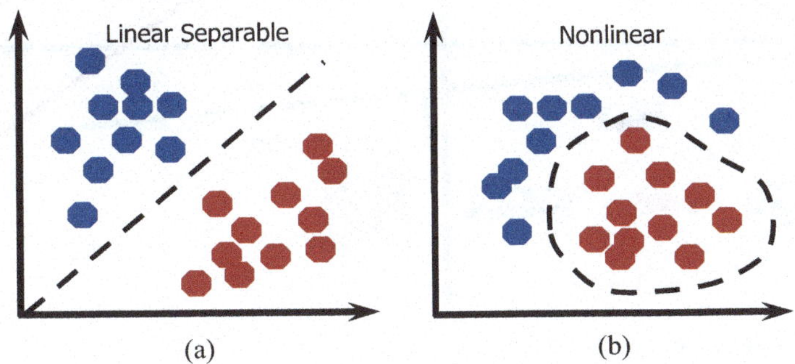

Fig. 6.9 Illustration for linear separable and nonlinear separable classification

1. **Score Function**: Is a function that will map our raw input vector to a score vector, as shown in Eq. (6.1).
2. **Loss Function**: Quantifies how well our current set of weights maps some input x to an expected output y; the loss function is used during training time.

Prior to the classification process, the parametric model must be trained to get the ideal model. The training purpose is to find the optimal weight matrix and initial bias vector, or to find a set of parameters that will change the hypothesis/model to map some input to some of the output class. To make this training process clear, we can use the example shown in Fig. 6.10 to illustrate it. The classification purpose is to classify the **Cat** class from all three input classes, including the **Cat**, **Dog**, and **Car** classes as input vectors.

The training procedure can be divided into the following steps:

1. Prepare the training data—First each input image x in Fig. 6.10 is stretched to a one-dimensional vector, and this operation may lose some spatial information.
2. Construct weight matrix—Set up the weight value for each element on each row starting from the first row **Cat**, exactly for each column on that row, based on the feature of that element in the original image, such as darkness degree, color code, and so on, then the second row **Dog** and continue until the third row **Car**.

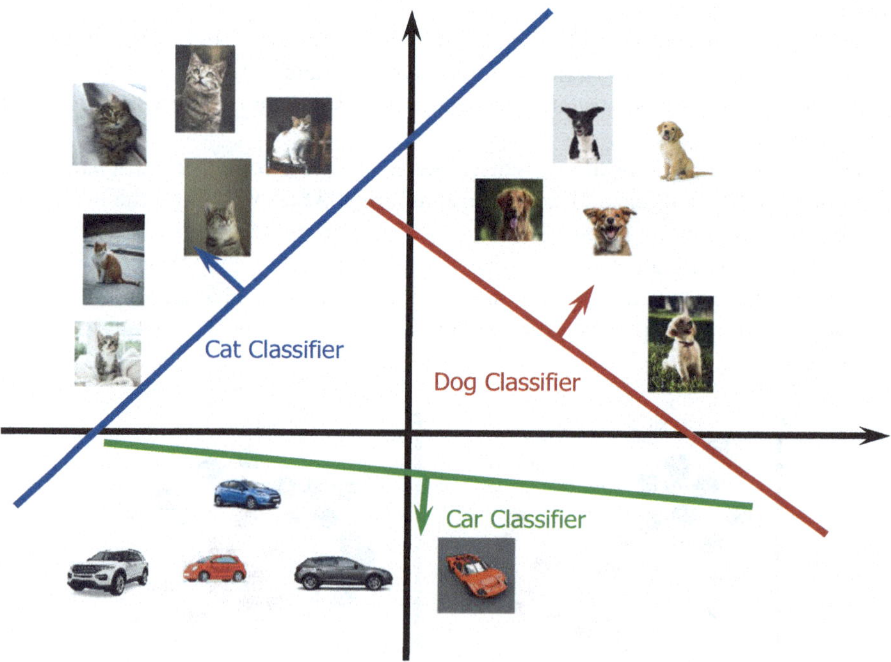

Fig. 6.10 Illustration for linear classification for multiclass

3. Now the resulted weight matrix has three rows, which are equivalent to three output classes or labels. Each row has five elements that are equal to the column number of the input vector.
4. Setup the initial values for the bias vector, which is shown in Eq. (6.1). This vector contains three values for three classes or three labels in one column.
5. During the training process, different input images with three different classes will be fed into the model, and the output score will be compared with the ideal output score represented as a vector. In our case, the ideal output vector score should be $[1\ 0\ 0]^T$ since we just want to classify the cat class with an order [cat dog car] in the weight matrix. The comparison between the actual output score and the ideal output vector score can be considered as a loss function. The training target is to try to make this loss value as small as possible until we get the acceptable error.

Now let's use a set of virtual values to simulate some images to make this training process clearer. As shown in Fig. 6.11, three-class images, Cat, Dog, and Car, are stretched into three vectors as inputs **X**. Each input vector (row) contained 5 elements and here it is transposed to a column vector (5 × 1). The weight matrix is constructed as a 3 × 5 matrix that contained features for each row or input vector.

During the training process, each actual score is calculated based on Eq. (6.1) with the values in weight matrix and input vector. Then the calculated score is compared with the desired score vector **Y (Cat)** whose value is $[1\ 0\ 0]^T$. The Error or the Loss works as an indicator and it is feedback to the weight matrix and bias. Based on that indicator's value, parameters in the weight matrix and bias vector will be adjusted to reduce that Error or Loss until it is smaller enough and can be acceptable.

The MATLAB codes used to simulate these calculations are shown in Fig. 6.12. Let's have a closer look at this piece of codes to see how it works.

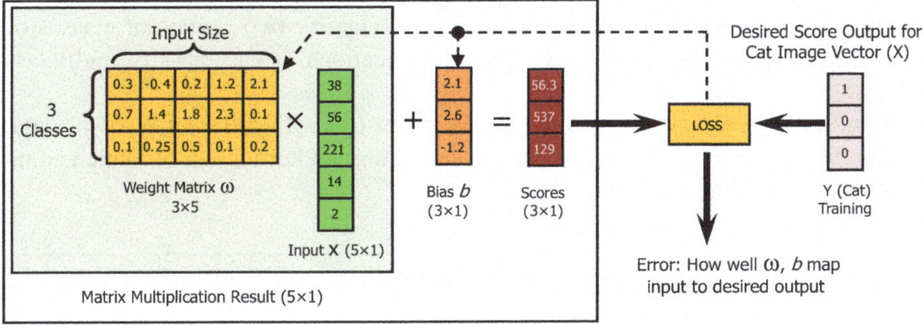

Fig. 6.11 An example of training three-class images to one class image

1. The weight matrix **W** is first generated and it is a 3×5 matrix.
2. The initial bias vector **B** is created and I should be a column vector, thus a transposing operator is attached.
3. One input feature vector **X**, which is a 5×1 vector, is declared and it is a column vector.
4. Finally the actual score **S** is calculated by using Eq. (6.1) and displayed in the Command window as the project runs.

6.5.1 Binary Linear Classifications and Related Functions

After discussion about the linear classifications, let's have a clear idea for linear and binary linear classification algorithm.

A linear classifier makes a classification decision for a given observation based on the value of a linear combination of the observation's features. In a binary linear classifier, the observation is classified into one of two possible classes using a linear boundary in the input feature space [4]. As shown in Table 6.3, the most common binary linear classifiers include:

1. Binary GLM **Logistic Regression** with functions **fitclinear()** and **fitglm()**.
2. Binary Gaussian Kernel Classifier with function **fitckernel()**.
3. **Binary Linear Support Vector Classifier (SVC) with function fitcsvm().**
4. **KNN Binary or Multiclass Classification with function fitcknn().**

The Gaussian Kernel Classification
The kernel function $k(x_n, x_m)$ used in a Gaussian process model tells the model how similar two data points (x_n, x_m) are, an thus the kernel function can be considered as a function used to detect the similarity among a group of data or observations. For binary kernel function model, it can be used to classify two groups or classes of data, either true or false. For multiclass classifications, it can classify multiclass objects.

Several different kernel types are available for use with different types of data, such as linear kernel, polynomial kernel, and Gaussian kernel. An N-Dimensional Gaussian kernel function is defined as:

```
% Simulated calculations for linear classification training process.
% Name: cal_linear_classification.m
% Output: Calculated score for Cat
%
1  W = [0.3 -0.4 0.2 1.2 2.1; 0.7 1.4 1.8 2.3 0.1; 0.1 0.25 0.5 0.1 0.2]
2  B = [2.1 2.6 -1.2]'
3  X = [38 56 221 14 2]'
4  S = W*X + B
```

Fig. 6.12 The sample codes used to calculate the actual scores

$$G_{\mathrm{ND}}\left(\vec{x},\sigma\right) = \frac{1}{\left(\sqrt{2\pi}\,\sigma\right)^{N}} e^{-\frac{|\vec{x}|^{2}}{2\sigma^{2}}} \tag{6.2}$$

The σ determines the width of the Gaussian kernel. In statistics, when we consider the Gaussian probability density function, it is called the standard deviation, and the square of it, σ^{2}, the variance [5].

The idea of using this function is to use a higher-dimension feature space to make the data points almost linearly separable as shown in Fig. 6.9a.

The Support Vector Classifier
We have provided detailed discussions about Support Vector Machine (SVM) algorithm in Sect. 5.5 in Chap. 5. Refer to that section to get more details about that algorithm.

The K-Nearest Neighbor Classifier
We have provided detailed discussions about K-Nearest Neighbor (KNN) algorithm in Sect. 5.7 in Chap. 5. Refer to that section to get more details about that algorithm.

In this section, we like to use some real examples to illustrate how to use binary linear classification functions to train and build models to perform desired classification processes. First let's go through the following binary classification classes and functions:

1. **CalssificationLinear—fitclinear()**
2. **ClassificationKernel—fitckernel()**
3. **ClassificationSVM—fitcsvm()**
4. **ClassificationKNN—fitcknn()**
5. **GeneralizedLinearModel—fitglm()**
6. **Classification with Random Forest—TreeBagger()**

All above top four classes related functions have the same constructors as shown below with only two exceptions, the functions **fitglm()** and **TreeBagger()**.

1. **mdl = fitcxxx(X,Y)** returns a trained linear classification model object **mdl** that contains the results of fitting a binary support vector machine to the predictors **X** and class labels **Y**.
2. **mdl = fitcxxx(Tbl, ResponseVarName)** returns a linear classification model using the predictor variables in the table **Tbl** and the class labels in **Tbl. ResponseVarName**.
3. **mdl = fitcxxx(Tbl, formula)** returns a linear classification model using the sample data in the table **Tbl**. The input argument `formula` is an explanatory model of the response and a subset of predictor variables in **Tbl** used to fit **mdl**.
4. **mdl = fitcxxx(Tbl, Y)** returns a linear classification model using the predictor variables in the table **Tbl** and the class labels in vector **Y**.

5. **mdl = fitcxxx(X, Y, Name, Value)** specifies options using one or more **Name-Value** pair arguments in addition to any of the input argument combinations in previous syntaxes. For example, you can specify that the columns of the predictor matrix correspond to observations, implement logistic regression, or specify to cross-validate. A good practice is to cross-validate using the '**Kfold**' **Name-Value** pair argument. The cross-validation results determine how well the model generalizes.

The function **fitglm()** has the following four different constructors:

1. **mdl = fitglm(tbl)** returns a generalized linear model fit to variables in the table or dataset array **tbl**. By default, fitglm takes the last variable as the response variable.
2. **mdl = fitglm(X,y)** returns a generalized linear model of the responses y, fit to the data matrix **X**.
3. **mdl = fitglm(___,modelspec)** returns a generalized linear model of the type you specify in modelspec.
4. **mdl = fitglm(___,Name,Value)** returns a generalized linear model with additional options specified by one or more **Name, Value** pair arguments.

Next let's build an example project to illustrate how to use these functions to perform binary linear classifications for some real data selected from some datasets. To make thing simple, we still prefer to use the diabetes dataset as the data source.

Create a new Script file and name it as **Diabetes_Func_Project.m** and enter the codes shown in Fig. 6.13 into that file.

Totally, four different binary linear classification algorithms and related functions are discussed, including **fitclinear()**, **fitckernel()**, **fitcsvm()**, and **fitcknn()**. The trained models are also tested with testing data to compare the true label values with the predicted label values to confirm effectiveness of all classification algorithms.

Let's have a closer look at this piece of codes to see how it works.

1. First we need to get the Diabetes dataset and set it up as a table **T**.
2. Then we need to separate all data in the dataset as two parts, the first 668 rows as the training data and the second 100 rows as the testing data, represented by **M** and **N**.
3. Now retrieve the first eight (8) columns as the predictor data with 668 rows as the training data, and assign them to a local variable **X**.
4. Get the last column, **Outcome**, as the response or output variable, and assign it to a local variable **Y**, which is a 688 × 1 vector.
5. Set up the testing output or response data that is a 100 × 1 vector and obtained from the last 100 rows of the Diabetes dataset, and assign it to a local variable **TLabel**. This vector will be used later for testing purpose.

```matlab
% Test to train binary classification model for diabetes
% Name: Diabetes_Func_Project.m
% Input: Diabetes.csv

1  T = readtable("C:\\Artificial Intelligence Book\\Students\\Datasets\\Diabetes Dataset\\" + ...
                "Diabetes.csv");
2  M = 1:668;              % training data - 668 training data
   N = 669:768;            % testing data - 100 testing data
3  X = [T.Pregnancies(M, :), T.Glucose(M, :), T.BloodPressure(M,:), T.SkinThickness(M,:), +...
           T.Insulin(M, :), T.BMI(M, :), T.DiabetesPedigreeFunction(M,:), T.Age(M,:)];
4  Y = T.Outcome(M, :);
5  TLabel = T.Outcome(N, :);

   % Use fitclinear() function to train the model
6  mdl_1 = fitclinear(X, Y);

   % Get testing data XX to perform prediction or classification
7  XX = [T.Pregnancies(N, :), T.Glucose(N, :), T.BloodPressure(N,:), T.SkinThickness(N,:), +...
           T.Insulin(N, :), T.BMI(N, :), T.DiabetesPedigreeFunction(N,:), T.Age(N,:)];
8  label1 = predict(mdl_1, XX);
9  Title = 'The Confusion Chart for fitclinear() function';
10 figure(1)
   plotConfChart(Title, TLabel, label1);      % call confusionchart() to plot confusion chart

   % Use fitckernel() function to fit the model
11 tbl = table(T.Pregnancies(M, :), T.Glucose(M, :), T.BloodPressure(M,:), T.SkinThickness(M,:), +...
              T.Insulin(M, :), T.BMI(M, :), T.DiabetesPedigreeFunction(M,:), T.Age(M,:));
12 [mdl_2, FitInfo] = fitckernel(tbl, Y);
13 label2 = predict(mdl_2, XX);
14 Title = 'The Confusion Chart for fitckernel() function';
15 figure(2)
   plotConfChart(Title, TLabel, label2);      % call confusionchart() to plot confusion chart

   % Use fitcsvm() function to fit the model
16 mdl_3 = fitcsvm(tbl, Y);
17 label3 = predict(mdl_3, XX);
18 Title = 'The Confusion Chart for fitcsvm() function';
19 figure(3)
   plotConfChart(Title, TLabel, label3);      % call confusionchart() to plot confusion chart

   % Use fitcknn() function to fit the model
20 mdl_4 = fitcknn(tbl, Y);
21 label4 = predict(mdl_4, XX);
22 Title = 'The Confusion Chart for fitcknn() function';
23 figure(4)
   plotConfChart(Title, TLabel, label4);      % call confusionchart() to plot confusion chart

24 function[] = plotConfChart(title, tLabel, pLabel)
25 cm = confusionchart(tLabel, pLabel);
26 cm.Title = title;
27 cm.RowSummary = 'row-normalized';
28 cm.ColumnSummary = 'column-normalized';
   end
```

Fig. 6.13 The detailed codes for the binary linear classification algorithms

6. Now call the function **fitcliner**() to train the binary linear model **mdl_1** with the training and response data, **X** and **Y**.

7. To check the trained model, **mdl_1**, we need to get the input testing data that is a 100×8 matrix. Exactly it is the last 100 rows on the original Diabetes dataset without the output column.

8. Now we can do the prediction by using the trained model with the input testing data **XX**. The prediction result **label1** is a binary vector with two possible values, 0 and 1.

9. Next we like to plot a confusion chart to display and check the classification result. Thus we can build a function **plotConfChart()** with three arguments to do that charting. The first argument is a title string that is used to identify different classification methods used. The detailed codes for the function **plotConfChart()** will be discussed later.

10. To display multiple figures, the **figure (1)** command is used to indicate the figure number. Then the user-defined function **plotConfChart()** is called with three arguments, the **title**, the true label values **TLabel**, and the predicted label values **label1**.

11. In order to test the Kernel Classifier with its function **fitckernel()** in a different way, we set up the training data in a table format, **tbl**.

12. Then the function **fitckernel()** is executed to train the model, **mdl_2**, with the input table **tbl** and response **Y**.

13. The **predict()** function is called with the trained model, **mdl_2**, and input testing data **XX** as arguments to get the predicted output label values, **label2**.

14. Similarly, the title string is used to indicate the type of classifier for the Confusion chart.

15. The **figure (2)** command is used to indicate that is the second figure to be displayed. Then the user-defined function **plotConfChart()** is called with three arguments, the **title**, the true label values **TLabel,** and the predicted label values **label2**, to display and confirm the classification result.

16. The SVM Classifier and its related function **fitcsvm()** is executed to train the model, **mdl_3**, with the SVM algorithm.

17. The trained model, **mdl_3**, is also used to derive the predicted label values with the testing label values **XX**.

18. Similarly, the title string is used to indicate the type of classifier for the Confusion chart.

19. The **figure (3)** command is used to indicate that is the third figure to be displayed. Then the user-defined function **plotConfChart()** is called with three arguments, the **title**, the true label values **TLabel**, and the predicted label values **label3**, to display and confirm the classification result.

20. The codes in steps 20–23 are used to perform the training and predicting of the KNN Classifier or model, **mdl_4**, with the function **fitcknn()**.

21. Starting step 24, the user-defined function **plotConfChart()** is built with four (4) coding lines. This function needs to return nothing, so a blank bracket [] following the function name is used. Three (3) arguments are involved with this function: the first is the title used to distinguish different classifiers, the second one is the true or testing label values, and the third one is the predicted label values.

22. The function **confusionchart()** is called with true and predicted label values. The returned function handle is assigned to a local variable **cm**.

23. The passed **title** is assigned to the **title** member data in the **cm** object.

24. The **RowSummary** and **ColumnSummary** are initialized with two **Name-Value** pairs. The first one is used to display the percentages of correctly and incorrectly classified observations for each true class, and the second one is

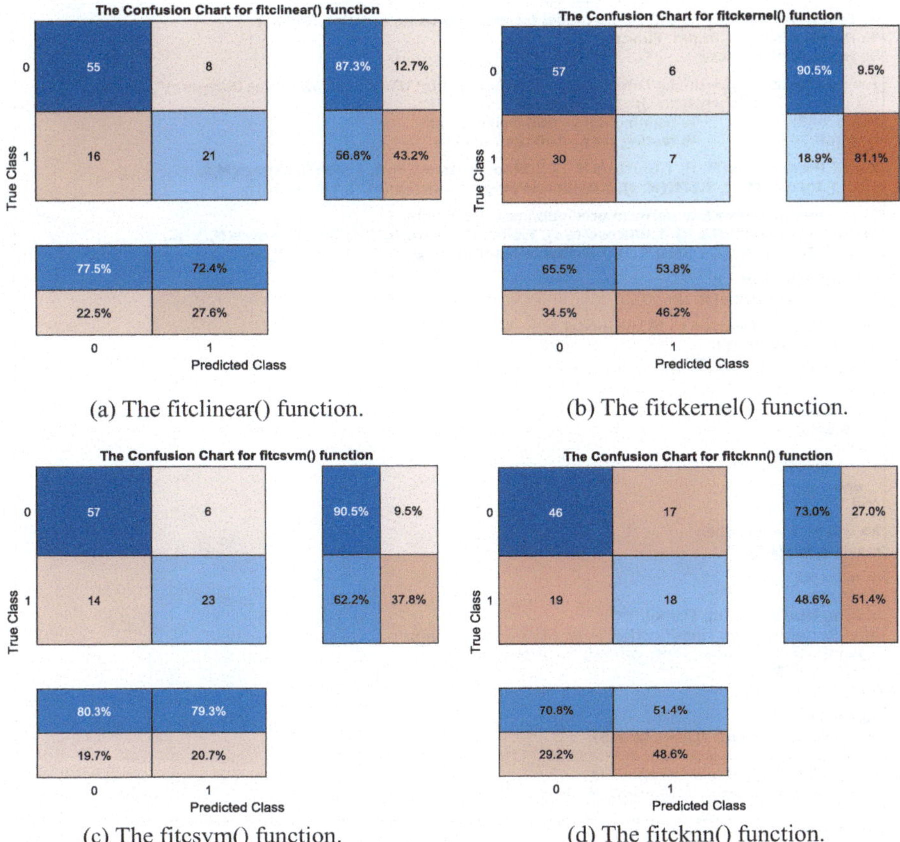

(a) The fitclinear() function.

(b) The fitckernel() function.

(c) The fitcsvm() function.

(d) The fitcknn() function.

Fig. 6.14 The running results of four classification algorithms

used to display the percentages of correctly and incorrectly classified observations for each predicted class.

Now run this project, and four different classification results are displayed as shown in Figs. 6.14a–d. It can be found that the SVM classification algorithm provided the best classification performance, and the KNN algorithm has the worst classification results. The classification results for linear and the kernel classifications are between them, but the former provided better performance.

Next let's build two additional projects to illustrate how to use functions, **fitglm()** and **TreeBagger()**, to perform classifications with generalized linear and random forest algorithms. First let's start with the function **fitglm()**.

Create a new Script file and enter the codes shown in Fig. 6.15 into that file. Name that file as **Diabetes_fitglm_Func.m**. Let's have a closer look at this piece of codes to see how it works.

```matlab
% Test to train binary classification model for diabetes using function fitglm()
% Name: Diabetes_fitglm_Func.m
% Input: Diabetes.csv

T = readtable("C:\\Artificial Intelligence Book\\Students\\Datasets\\Diabetes Dataset\\" + ...
              "Diabetes.csv");
M = 1:668;              % training data - 668 training data
N = 669:768;           % testing data - 100 testing data

X = [T.Pregnancies(M, :), T.Glucose(M, :), T.BloodPressure(M,:), T.SkinThickness(M,:), +...
        T.Insulin(M, :), T.BMI(M, :), T.DiabetesPedigreeFunction(M,:), T.Age(M,:)];

% Get testing data XX to perform prediction or classification
XX = [T.Pregnancies(N, :), T.Glucose(N, :), T.BloodPressure(N,:), T.SkinThickness(N,:), +...
        T.Insulin(N, :), T.BMI(N, :), T.DiabetesPedigreeFunction(N,:), T.Age(N,:)];

Y = T.Outcome(M, :);
TLabel = T.Outcome(N, :);

% Use fitglm() function to fit the model
mdl_5 = fitglm(X, Y);
label5 = predict(mdl_5, XX);

for n = 1:100
   if label5(n) >= 0.5
     label5(n) = 1;
   else
     label5(n) = 0;
   end
end

AA = label5 == TLabel;
result = nnz(AA)

m = 1:100;
figure(1)
plot(m, label5, 'bo', m, TLabel, 'r*');
title('Perform binary classification using fitglm() function');
legend('fitglm() values', 'Test values', 'location', 'west');
axis([0 100 0 1.2]);
grid;

figure(2)
cm = confusionchart(TLabel, label5);
cm.Title = 'Binary Classification With fitglm() Function';
cm.RowSummary = 'row-normalized';
cm.ColumnSummary = 'column-normalized';
```

Fig. 6.15 The codes for the binary classification using the fitglm() function

1. Call **fitglm**() function to train the model, **mdl_5**, using generalized linear algorithm with predictors **X** and response variable **Y**.
2. Then use **predict**() function to predict the desired responses with testing input predictors **XX**, and the predicted results are assigned to **label5**.
3. Convert or normalize the response column **label5** to binary results since the output of the function **fitglm**() is not Boolean values.
4. Compare the binary predicted results with the testing output values with the == function and assign the comparison results to the vector **AA**. In MATLAB, == means an equal operator and **A == B** returns a logical array or a table of logical values with elements set to logical 1 (`true`) where inputs **A** and **B** are equal; otherwise, the element is logical 0 (`false`) bit by bit.
5. The **nnz**() function is utilized to retrieve back all logical 1 (true) in the returned array and assign it to the local vector **result**.
6. Two kinds of results will be displayed in figures. First we need to plot a comparison result between the predicted and the testing values for outputs. The function **axis**() is used to set up both axes in a predefined range.

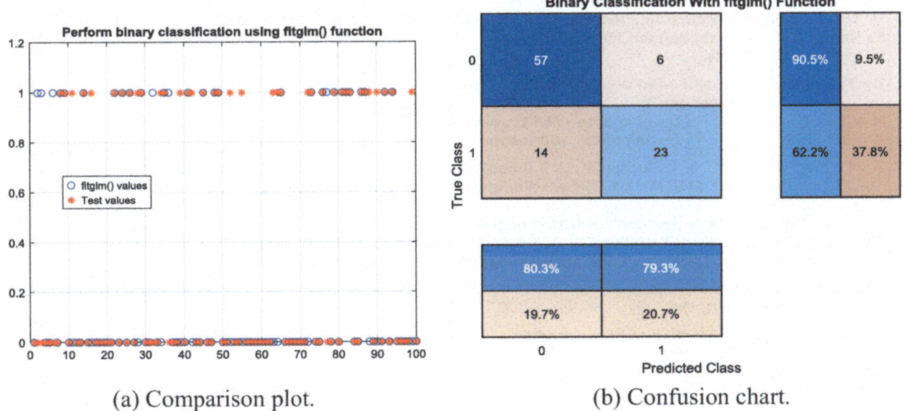

(a) Comparison plot.   (b) Confusion chart.

Fig. 6.16 The running outputs or results of using fitglm() function

7. The second figure is used to plot a confusion chart to show the classification results for both 1 (true) and 0 (false) labels.

Now run this project and the running results are shown in Figs. 6.16a, b.

It can be found from Fig. 6.16 that the classification result of using the function **fitglm()** is similar to that of using the SVM algorithm as shown in Fig. 6.14c, which is the best one.

Now let's take care of the classification with random forest algorithm by using the function **TreeBagger()**. This function has five (5) constructors, as shown below:

1. **mdl = TreeBagger(NumTrees,Tbl,ResponseVarName)** returns an ensemble object (**mdl**) of **NumTrees** bagged classification trees, trained by the predictors in the table **Tbl** and the class labels in the variable **Tbl.ResponseVarName**.
2. **mdl = TreeBagger(NumTrees,Tbl,formula)** returns **mdl** trained by the predictors in the table **Tbl**. The input **formula** is an explanatory model of the response and a subset of predictor variables in **Tbl** used to fit **mdl**. Specify `formula` using Wilkinson Notation.
3. **mdl = TreeBagger(NumTrees,Tbl,Y)** returns **mdl** trained by the predictor data in the table **Tbl** and the class labels in the array **Y**.
4. **mdl = TreeBagger(NumTrees,X,Y)** returns **mdl** trained by the predictor data in the matrix **X** and the class labels in the array **Y**.
5. **mdl = TreeBagger(___, Name = Value)** returns **mdl** with additional options specified by one or more **Name-Value** arguments, using any of the previous input argument combinations. For example, you can specify the algorithm used to find the best split on a categorical predictor by using the **Name-Value** argument **PredictorSelection**. Refer to Table 5.15 in Chap. 5 to get more details about **Name-Value** pairs used by this function.

```
% Test to train binary classification model for diabetes using function TreeBagger()
% Name: Diabetes_TreeBagger_Func.m
% Input: Diabetes.csv

T = readtable("C:\\Artificial Intelligence Book\\Students\\Datasets\\Diabetes Dataset\\" + ...
            "Diabetes.csv");
M = 1:668;              % training data - 668 training data
N = 669:768;           % testing data - 100 testing data

X = [T.Pregnancies(M, :), T.Glucose(M, :), T.BloodPressure(M,:), T.SkinThickness(M,:), +...
        T.Insulin(M, :), T.BMI(M, :), T.DiabetesPedigreeFunction(M,:), T.Age(M,:)];

% Get testing data XX to perform prediction or classification
XX = [T.Pregnancies(N, :), T.Glucose(N, :), T.BloodPressure(N,:), T.SkinThickness(N,:), +...
        T.Insulin(N, :), T.BMI(N, :), T.DiabetesPedigreeFunction(N,:), T.Age(N,:)];

Y = T.Outcome(M, :);
TLabel = T.Outcome(N, :);

% Use TreeBagger() function to fit the model
1   mdl = TreeBagger(100, X, Y);
2   label6 = predict(mdl, XX);
3   label6 =str2double(label6);

AA = label == TLabel;
result = nnz(AA)

m = 1:100;
4   figure(1)
plot(m, label6, 'bo', m, TLabel, 'r*');
title('Perform binary classification using TreeBagger() function');
legend('Prediction values', 'Test values', 'location', 'west');
axis([0 100 0 1.2]);
grid;

5   figure(2)
cm = confusionchart(TLabel, label6);
cm.Title = 'Binary Classification With TreeBagger() Function';
cm.RowSummary = 'row-normalized';
cm.ColumnSummary = 'column-normalized';
```

Fig. 6.17   The codes for the binary classification using function TreeBagger()

Now create a new Script file and name it as **Diabetes_TreeBagger_Func.m**, and enter the codes shown in Fig. 6.17 into that file. Let's have a closer look at this piece of codes to see how it works.

1. The function **TreeBagger**() is executed to train a model, **mdl**, with 100 trees and predictor as well as response variables. By default, this function grows classification decision trees.
2. The trained model is then used to derive the predicted response values based on the testing input data values in matrix **XX**.
3. A point to be noted is that the function **predict**() returns a cell or a string array, not a numeric array. Thus we need to use a function, **str2double**(), to convert the resulted string array to a numeric array.
4. Two kinds of results will be displayed in figures. First we need to plot a comparison result between the predicted and the testing values for outputs. The function **axis**() is used to set up both axes in a predefined range.
5. The second figure is used to plot a confusion chart to show the classification results for both 1 (true) and 0 (false) labels.

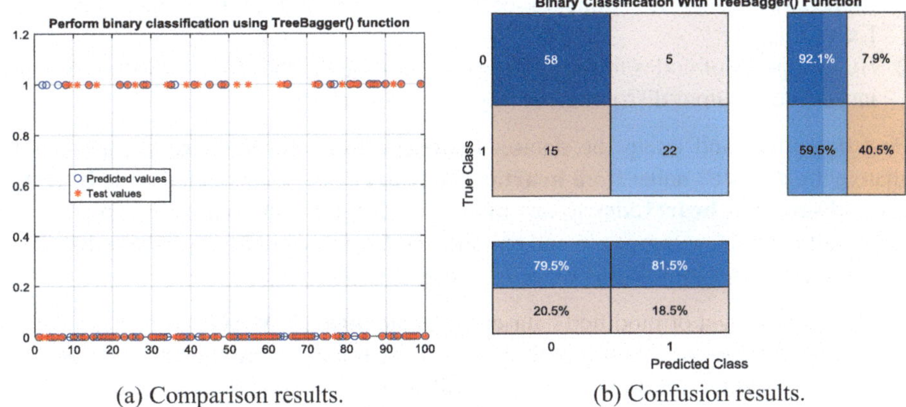

(a) Comparison results.   (b) Confusion results.

Fig. 6.18 The comparison between the predicted and testing classification results

Now run this project and the running results are shown in Figs. 6.18a, b. It can be found from Fig. 6.18 that the classification function **TreeBagger()** also provided a good classification result which is similar to that of SVM classification algorithm.

At this point, we should have a clear picture about binary classification classes and related functions. Next let's use another example project to illustrate how to utilize some functions to perform classification to predict the possibility of suffering the heart disease.

6.5.2   *A Case Study for Binary Classification with Heart Disease Prediction*

The dataset we like to use is a heart dataset named heart.csv, which is located at the **figshare** site [6]. The dataset contained over 280 records of patients who may have heart disease or not. The dataset has 14 columns with all possible parameters, and the last column **output** provided the Boolean value, 1 means heart disease and 0 means no heart disease.

An issue related to this dataset is that the parameters' distribution, especially on the last column **output**, the Boolean values are divided into two parts. The first 165 values are 1 and the rest values are all 0. Thus, we need to make some modifications to make those values distributed evenly in a random order to train our classification models. The modifications include:

1. Convert the format of the dataset from the comma-separated values (**.csv**) to Excel (**.xlsx**) since in the CSV format, the data is in the text format, separated by commas, while in Excel format, information is in the tabular form with rows and columns. The main purpose of this conversion is to enable us to use some

MATLAB-related functions to perform reading and writing operations for Excel file.

2. Build a piece of codes to modify the dataset to make the Boolean values on the last column output distributed evenly in random order.

First download and unzip the dataset **heart.csv** from the **figshare** site [54] and change the dataset's name from **heart.csv** to **heart1.csv**. Then open the file **heart1. csv** and save it as **heart1.xlsx** to one of your local folders in your computer.

Next create a new Script file and enter the codes shown in Fig. 6.19 into that file. Let's have a closer look at this piece of codes to see how it works.

1. Set up our target or modified dataset file name, **heart1_M.xlsx**, with a full path. You need to adjust this full path based on the folder where the target dataset is located in your machine. Then this full name is assigned to a local variable **tDataSet**.

2. Similarly you also need to set up the full name for the source dataset, **heart1. xlsx**, and we need to read that dataset later for the modification purpose. That full name is assigned to another local variable **sDataSet**.

3. By using the **xlsread()** function with the full source name **sDataSet**, we can read all data from the original dataset **heart1.xlsx**, and assign all numeric values to the array **num()**.

4. Next we can define the maximum row number as 289 and desired row number as 280 since we only need to use 280 data as our training, checking and testing data.

5. Create a character array **Tab** that contained the titles for all 14 columns in the target dataset **heart1_M.xlsx**.

6. A **for()** loop is used to repeatedly rewrite each row with a random order into the target dataset **heart1_M.xlsx**.

7. Inside the **for()** loop, first a random integer number is generated by using the **randi()** function that generates a uniformly distributed random number in a range of 1~289, and 1 by 1 matrix with 1D array. The generated random number is assigned to the local variable j. The j row on the original dataset is displayed as the tracking purpose.

8. The title for each column is written into the target dataset in the top row.

```
% Modify heart.xlsx DataSet to make it random distributions

1  tDataSet = 'C:\\Artificial Intelligence Book\\Students\\Datasets\\Heart Dataset\\heart1_M.xlsx'; %target
2  sDataSet = 'C:\\Artificial Intelligence Book\\Students\\Datasets\\Heart Dataset\\heart1.xlsx'; %source

3  [num,txt,raw] = xlsread(sDataSet);
4  M = 289;              % max row number on dataset
   N = 280;              % max row number on (train + check + test) data
5  Tab={'age', 'sex', 'cp', 'trtbps', 'chol', 'fbs', 'restecg', 'thalachh', 'exng', 'oldpeak', 'slp', 'caa', 'thall', 'output'};

6  for i = 1:N
7     j = randi(M,1,1);
      num(j,:)
8     writecell(Tab, tDataSet);
9     writematrix(num(j,:), tDataSet,'WriteMode','append');
   end
```

Fig. 6.19 The codes used to modify the dataset heart1.xlsx

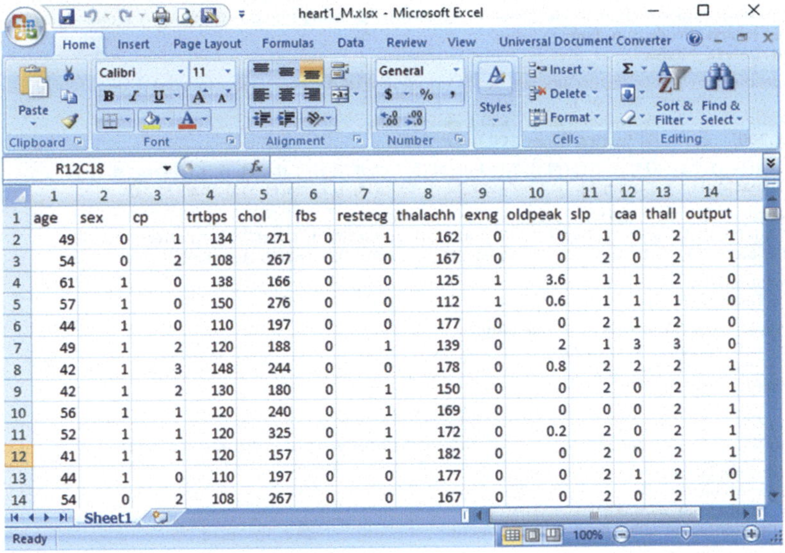

	age	sex	cp	trtbps	chol	fbs	restecg	thalachh	exng	oldpeak	slp	caa	thall	output
1	age	sex	cp	trtbps	chol	fbs	restecg	thalachh	exng	oldpeak	slp	caa	thall	output
2	49	0	1	134	271	0	1	162	0	0	1	0	2	1
3	54	0	2	108	267	0	0	167	0	0	2	0	2	1
4	61	1	0	138	166	0	0	125	1	3.6	1	1	2	0
5	57	1	0	150	276	0	0	112	1	0.6	1	1	1	0
6	44	1	0	110	197	0	0	177	0	0	2	1	2	0
7	49	1	2	120	188	0	1	139	0	2	1	3	3	0
8	42	1	3	148	244	0	0	178	0	0.8	2	2	2	1
9	42	1	2	130	180	0	1	150	0	0	2	0	2	1
10	56	1	1	120	240	0	1	169	0	0	0	0	2	1
11	52	1	1	120	325	0	1	172	0	0.2	2	0	2	1
12	41	1	1	120	157	0	1	182	0	0	2	0	2	1
13	44	1	0	110	197	0	0	177	0	0	2	1	2	0
14	54	0	2	108	267	0	0	167	0	0	2	0	2	1

Fig. 6.20   The modified dataset heart1_M.xlsx

9. Each random row is also written into the target dataset one by one with append format.

Now run this project and you can check the running result by opening the target dataset at the target folder in your machine, which is shown in Fig. 6.20.

Compare the last column **output** for both source and the target dataset, you can find that the Boolean values on the output column in the target dataset are randomly distributed now. We will use this modified dataset as our dataset to train, check, and test our classification model later.

Fourteen columns are involved in this dataset with the first 13 columns as predictors and the last column as the target or response variable. The 13 predictors are exactly input parameters that are composed of all descriptions including the **age** and **sex** as well as other health-measured results. A 0 in the **sex** column represents a male and 1 means a female.

Now create a new Script file and name it as **Hearts_Prediction_Func.m** and enter the codes shown in Fig. 6.21. Due to the similarity between this piece of codes and the previous codes, no explanations are given and they are straightforward and easy to be understood.

Now run this project and the running results are shown in Figs. 6.22a, b.

Two classification algorithms, SVM and TreeBagger, are utilized in this project to compare the classification results of both algorithms. It can be found from Fig. 6.22 that the **TreeBagger()** algorithm provided a better classification result with a 94.6% correct classification rate for the true (1) class, and a 75% correct classification rate for the false (0) class. The SVM classification algorithm provided a

```matlab
% Test to train binary classification model for heart disease using function fitcsvm() & TreeBagger()
% Name: Hearts_Prediction_Func.m
% Input: heart1_M.xlsx

T = readtable("C:\\Artificial Intelligence Book\\Students\\Datasets\\Heart Dataset\\" + ...
            "heart1_M.xlsx");
M = 1:180;              % training data - 180 training data
N = 181:280;           % testing data - 100 testing data

X = [T.age(M, :), T.sex(M, :), T.cp(M,:), T.trtbps(M,:), T.chol(M, :), T.fbs(M, :), T.restecg(M, :) +...
    T.thalachh(M, :), T.exng(M, :), T.oldpeak(M,:), T.slp(M,:), T.caa(M, :), T.thall(M, :)];

% Get testing data XX to perform prediction or classification
XX = [T.age(N, :), T.sex(N, :), T.cp(N,:), T.trtbps(N,:), T.chol(N, :), T.fbs(N, :), T.restecg(N, :) +...
    T.thalachh(N, :), T.exng(N, :), T.oldpeak(N,:), T.slp(N,:), T.caa(N, :), T.thall(N, :)];

Y = T.output(M, :);
TLabel = T.output(N, :);

% Use TreeBagger() function to fit the model
mdl_tree = TreeBagger(100, X, Y);
labelt = predict(mdl_tree, XX);
labelt =str2double(labelt);
Title = 'The Confusion Chart for TreeBagger() function';
figure(1)
plotConfChart(Title, TLabel, labelt);         % call confusionchart() to plot confusion chart

% Use fitcsvm() function to fit the model
mdl_svm = fitcsvm(X, Y);
labels = predict(mdl_svm, XX);
Title = 'The Confusion Chart for fitcsvm() function';
figure(2)
plotConfChart(Title, TLabel, labels);         % call confusionchart() to plot confusion chart

function[] = plotConfChart(title, tLabel, pLabel)
cm = confusionchart(tLabel, pLabel);
cm.Title = title;
cm.RowSummary = 'row-normalized';
cm.ColumnSummary = 'column-normalized';
end
```

Fig. 6.21 The codes used to train two models with heart1_M.xlsx dataset

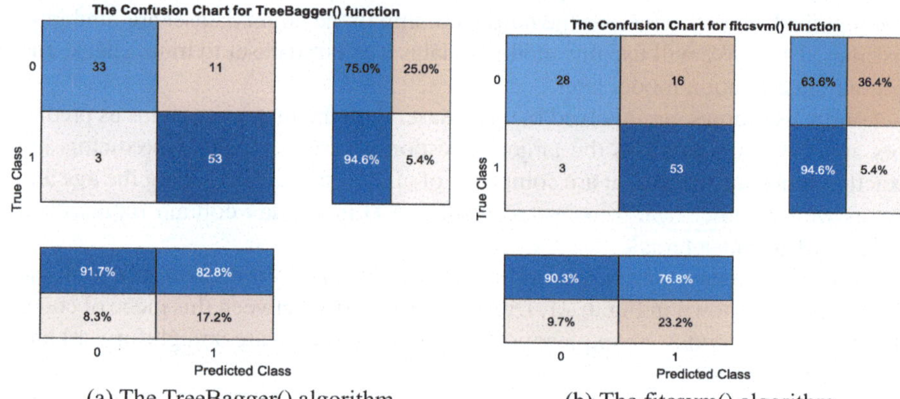

(a) The TreeBagger() algorithm.   (b) The fitcsvm() algorithm.

Fig. 6.22 The classification results with two algorithms

similar correct classification rate for the true (1), 96.4%, but a worse correct classification rate for the false (0) class, which is 63.6%.

Next let's discuss the multiclass classification and its applications. First let's take care of the multiclass classification App built in MATLAB.

6.6 Multiclass Classifications App Example

The so-called multiclass classification means that the classified output classes are more than two levels. For example, multiple different cars can be classified based on some cars' images. Multiple different animals or diseases can be classified based on a collection of different animals' or diagnostics' images. Furthermore, different animals can also be classified based on a collection of recording results of animals' voices or audio files.

As we discussed in Sect. 6.2, similar classification algorithms can be used for both binary and multiclass classifications. In MATLAB, all classification algorithms are involved in an App that is called Classification Learner App, which is included in the Statistics and Machine Learning Toolbox, and it allows users to choose among various algorithms to train and validate classification models for binary and multiclass problems. After training multiple models, compare their validation errors side-by-side, and then choose the best model.

Users can use Classification Learner App to train models of the following classifiers:

1. Decision Trees
2. Support Vector Machines
3. Logistic Regression
4. K-Nearest Neighbors
5. Naive Bayes
6. Kernel approximation
7. Ensembles
8. Neural Networks

Due to the special property and function of the neural networks, we will discuss that algorithm in a later chapter.

Now let's use the MATLAB Classification Learner App and another diabetes dataset to illustrate how to use this tool to train, select, and evaluate some desired models with various algorithms listed above. First let's open that Classification Learner App.

Open MATLAB and click on the **APPS** tab on the top, then click on the drop-down arrow, and click on the **Classification Learner** icon under the **MACHINE LEARNING ANS DEEP LEARNING** group.

Before we can continue, we need to set up our dataset that is a multiclass dataset named **Dataset of Diabetes.csv** and it is located at the site [7] and can be downloaded [8]. Save that dataset to one of your local folders in your machine to make it ready to be used later.

The data in this dataset were collected from the Iraqi society, and were acquired from the laboratory of Medical City Hospital and the Specializes Center for Endocrinology and Diabetes-Al-Kindy Teaching Hospital. The data consist of medical information, laboratory analysis, etc. The data that have been entered initially into the system are: **No. of Patient**, **Sugar Level Blood**, **Age**, **Gender**, Creatinine ratio(**Cr**), Body Mass Index (**BMI**), Urea, Cholesterol (**Chol**), Fasting lipid profile, including total, **LDL**, **VLDL**, Triglycerides(**TG**) and **HDL** Cholesterol, **HbA1c**. The output is the last column, **Class**, or the patient's diabetes disease classes, **Y** = Diabetic, **N** = Non-Diabetic, or **P** = Predict-Diabetic.

On the opened Classification Learner App, perform the following operational steps to perform auto-training process:

1. Click on the **New Session** icon on the upper-left corner and select the **From File** item.
2. Browse to the folder where the diabetes dataset **Dataset of Diabetes.csv** is located, select it, and click on the **Open** button to open this dataset.
3. Click on the **Import Selection** (green color) button on the top to import the dataset. The imported dataset is shown in Fig. 6.23. Check the **Set aside a test**

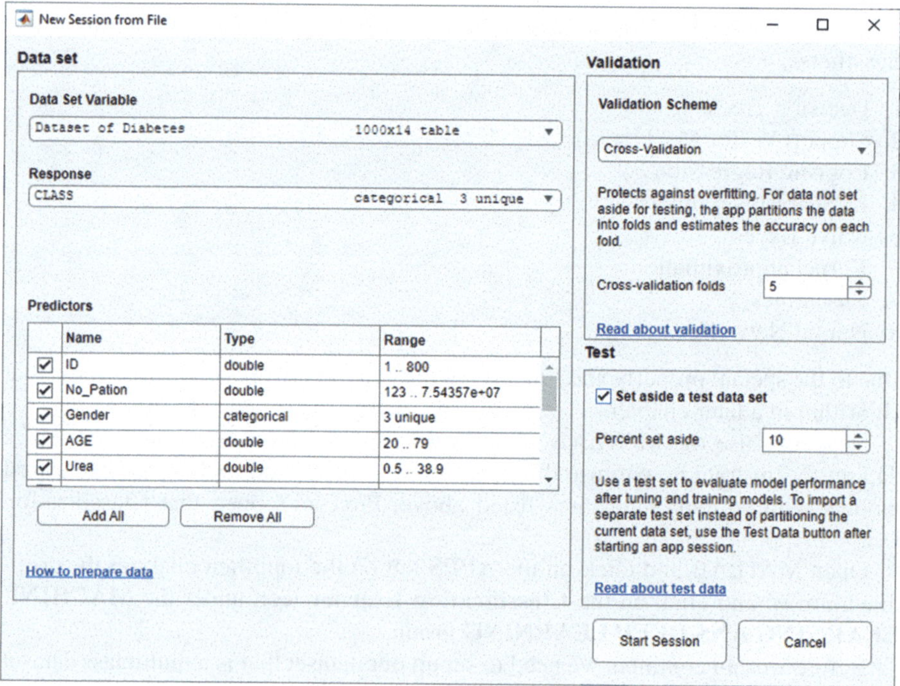

Fig. 6.23 The finished New Session from File wizard

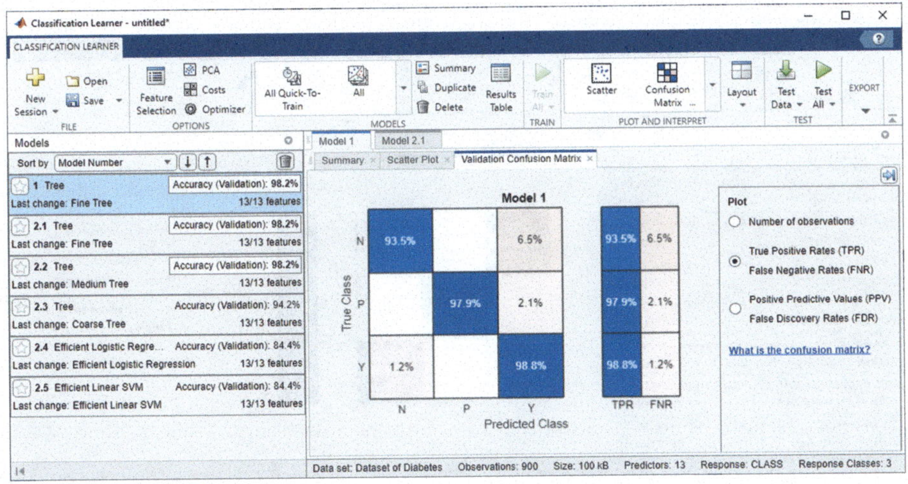

Fig. 6.24 The auto-training results

data set checkbox and keep 10 on the **Percent set aside** box since we want to use 10% data in the dataset as the testing data as the model is trained and checked.

4. Then click on the **Start Session** button to collect and organize the related data from the imported dataset.
5. On the next wizard, click on the **All Quick-To-Train** icon from the **MODELS** pane on the top since we want to perform an auto-training process by using all possible models to select the best model. You could select different classifiers to perform the training process if you like by selecting a desired one.
6. Click on the drop-down arrow on the **Train All** icon on the top and select the **Train All** item to begin the training process.
7. Check the **True Positive Rate** checkbox to see the training results in percent format, and the training results are shown in Fig. 6.24.

It can be found from Fig. 6.24 that the Fine Tree model (**Model 1**) under the **DECISION TREES** algorithm provided the optimal classification result with the validation accuracy as 98.2%, which is much better compared with the Logistic Regression and Linear SVM algorithms (Models 2.4 and 2.5 with accuracy 84.4%).

The correct classification rates for three classes, **N**, **P**, and **Y**, are 95.5, 97.9, and 98.8%, which are pretty good and acceptable for actual applications.

Now let's select a special algorithm to manually train this model with **Bagged Trees** under the **ENSEMBLE CLASSIFIERS** group to see the classification result. Click on the drop-down arrow on the **MODELS** pane on the top and scroll to the **Bagged Trees** icon, select it and keep all default settings for that algorithm, and click on the drop-down arrow on the **Train All** icon, and select **Train Selected** item to begin the training process. The training result is shown in Fig. 6.25.

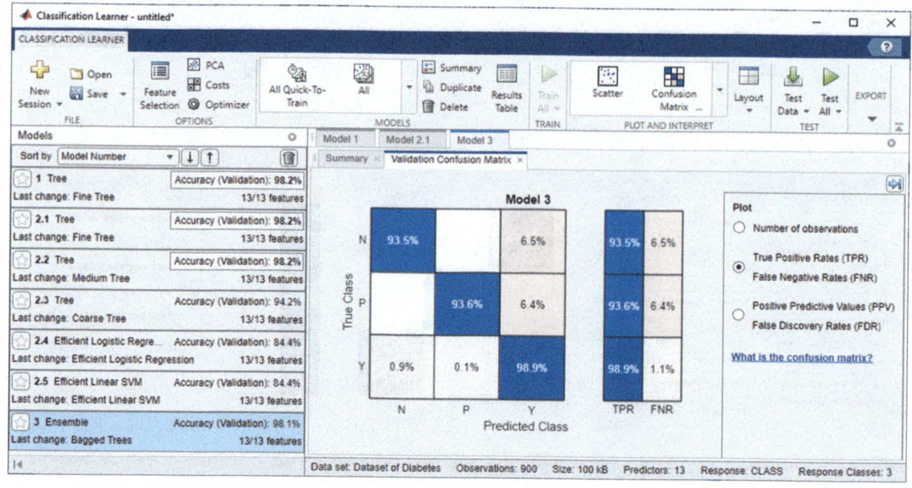

Fig. 6.25 The training result with Bagged Tree algorithm

Table 6.4 Popular MATLAB classification class and function for multiclass classifier

Classification Class	Function	Description
ClassificationKNN	`fitcknn()`	Create a KNN model *ClassificationKNN* object using **fitcknn()** for binary and multiclass classification.
ClassificationNaiveBayes	`fitcnb()`	Create a *ClassificationNaiveBayes* object by using **fitcnb()** for multiclass classifications.
ClassificationNeuralNetwork	`fitcnet()`	Create a *ClassificationNeuralNetwork* object by using **fitcnet()**.
ClassificationECOC	`fitcecoc()`	Create a *ClassificationECOC* object by using **fitcecoc()** for multiclass classification model with SVM.

It can be found from Fig. 6.25 that the training result is similar to that of using the **Decision Tree** or **Model 1** algorithm with three classes with a little lower accuracy of 98.1%.

Now click on the drop-down arrow in the **Save** icon on the top and select the **Save Session** item to save our training result with trained models. Named it as **Diabetes_mclassify_App.mat** and save it to your desired folder and it can be used for evaluation purpose later.

You can also export this model to the Workspace to perform some other prediction or classification functions for some new data in the future. To do that, click on the drop-down arrow from the **EXPORT** icon on the top and **Export Model** item, and select the **Export Model** item. Enter a desired model name, such as **diabeteAppModel**, into the **Variable name** box, and click on the **OK** button to export it. Next let's take care of the multiclass classification process with functions.

6.7   Multiclass Classification Function Example

As we discussed in Sect. 6.5, there are about eight (8) classification classes and objects used for classifiers, as shown in Table 6.3. Among them, four classification algorithms can be used for multiclass classifiers, which are shown in Table 6.4. Due to the specialty of neural networks classifier, we will delay our discussions about it in a later chapter. We only limit our discussions on the rest of the other three classifiers, **ClassificationKNN**, **ClassificationNaiveBayes**, and **ClassificationECOC**, with their related functions in this section.

Prior to starting our discussions, we need to make some adjustments or modifications to make our diabetes dataset suitable for multiclass classification algorithms. The reason for those rectifications is due to requirements of input dataset for some classification algorithm, such as **fitcknn**(), which is: the predictors can be either all numeric values or all text characters or text string, and they cannot be a combination of both numeric values and text string. This requirement may be untrue for some other algorithms. We will discuss more for this issue in the next section. In our original diabetes dataset, **Dataset of Diabetes.csv**, it contained both text strings, such as column 3—**Gender**, and numeric values, such as all other columns except the column **CLASS**. Another reason for this rectification is that the class values on the last column **CLASS** are not randomly distributed, which means that the class values for the first 107 row are all **N**, but the value **Y** occupied more than 800 rows. This distribution is not appropriate for our training process and the training and testing results may be incorrect. Therefore, we need to do some necessary rectifications to make it acceptable to the training process for classifiers.

Perform the following operational steps to make modifications to our dataset:

1. Open our original dataset, **Dataset of Diabetes.csv**, and save it as an Excel file named **Dataset of Diabetes_M.xlsx** since an Excel file is much easier to be modified.
2. Build a small Script file named **MDiabeteDataSet_Excel.m** and enter the codes shown in Fig. 6.26 into that file.

```matlab
% Modify dataset 'Dataset of Diabetes.xlsx' to make it random distributions
% Feb 5, 2024
1  dtPath = 'C:\\Artificial Intelligence Book\\Students\\Datasets\\Diabetes Dataset2\\';
2  tData = 'Dataset of Diabetes_M.xlsx';
3  sData = 'Dataset of Diabetes.xlsx';

4  sDataSet = strcat(dtPath, sData);      % source dataset
   tDataSet = strcat(dtPath, tData);      % target dataset

5  T = readtable(sDataSet);
6  M = 1000;            % max row number on dataset
   N = 1000;            % max row number on (train + check + test) data
7  Tab = {'ID', 'No_Pation', 'Gender', 'AGE', 'Urea', 'Cr', 'HbA1c', 'Chol', 'TG', 'HDL', 'LDL', 'VLDL', 'BMI', 'CLASS'};

8  g_M = strcmp(T.Gender,'M');
9   T.Gender(g_M) = {1};
10  g_F = strcmp(T.Gender,'F');
    T.Gender(g_F) = {0};

11 for i = 1:N
     j = randi(M, 1, 1);
12    writecell(Tab, tDataSet);
      writetable(T(j, :), tDataSet,'WriteMode','append');
   end
```

Fig. 6.26 The codes used to rectify the original dataset

The purpose of this piece of codes is to modify the original dataset to make it suitable for us to use for our model training process. The rectifications include the following components:

- The data in column 3, **Gender**, have been converted to number, **1** for '**M**' and **0** for '**F**'. This conversion would enable MATLAB to use all-numeric dataset to train our classification models.
- All data rows now are randomly distributed to meet the needs of our training process.

Let's have a closer look at this piece of codes to see how it works.

1. First a data path, **dtPath**, is generated and it is a text string used to indicate the location where our original and target dataset are located. The reason we separate the path with our dataset file is to reduce the length of this full address and to make it shorter.
2. The name of the target dataset is assigned to a local string variable **tData**.
3. Similarly, the name of the original or source dataset is assigned to another variable **sData**.
4. The MATLAB function, **strcat**(), is used to concatenate the path with the name of the source dataset together to get a complete or full address for the original dataset, **sDataSet**. In a similar way, we can get the full address for our target dataset, **tDataSet**.
5. Now read all data from the original dataset and assign them to a local variable **T**.
6. The maximized row number for both the original and the target dataset is defined, which are 1000 rows.
7. The title columns for the target dataset are defined since we need to write them into the target or modified dataset later as titles for all columns, which are identical to those in the original or the source dataset.
8. Steps 8~10 are very critical to our modified dataset since we need to use a MATLAB function **strcmp**() to compare all text strings in column 3, **Gender**, to find either '**M**' or '**F**', and assign a 1 for '**M**' and a 0 for '**F**' if a matched letter (string) is found. Two input arguments are column 3, **T.Gender**, and string '**M**'. A true (1) is returned if both are equal and a false (0) is returned if both are not.
9. A trick here is that this string comparison is performed within the entire column 3. In fact, a loop is used to compare each row on that column. When an element in that row is equal to '**M**'. Then a 1 is assigned to that row with {1} format since it is a cell in the table.
10. Similar operations are performed for the comparison result to the string '**F**'. A 0 will be assigned to that row if a match is found.
11. To make all data in the dataset randomly distributed, a **for**() loop and a uniformly distributed random number function, **randi**(), are used to generate a pseudorandom integer j with the maximum row number as the upper bond.
12. Then the titles for all columns are written into the target dataset, and the selected random row from the original dataset is written into the target dataset, too.

```
% Test to train multi class classification model for diabetes
% Name: Diabetes_MC_Func_Project.m
% Input: Dataset of Diabetes_M.xlsx

1  dsPath = 'C:\\Artificial Intelligence Book\\Students\\Datasets\\Diabetes Dataset2\\';
2  sData = 'Dataset of Diabetes_M.xlsx';
3  sDataSet = strcat(dsPath, sData);

4  T = readtable(sDataSet);
5  M = 1:900;                  % training data - 900 training data
   N = 901:1000;               % testing data - 100 testing data

6  X = [T.ID(M, :), T.No_Pation(M, :), T.Gender(M, :), T.AGE(M,:), T.Urea(M,:), T.Cr(M, :), T.HbA1c(M, :) + ...
       T.Chol(M, :), T.TG(M, :), T.HDL(M,:), T.LDL(M,:), T.VLDL(M, :), T.BMI(M, :)];
7  Y = T.CLASS(M, :);
8  TLabels = T.CLASS(N, :);

   % Use fitcknn() function to train the model
9  mdl_1 = fitcknn(X, Y, "ClassNames", ["N", "P", "Y"])

   % Get testing data XX to perform prediction or classification
10 XX = [T.ID(N, :), T.No_Pation(N, :), T.Gender(N, :), T.AGE(N,:), T.Urea(N,:), T.Cr(N, :), T.HbA1c(N, :) + ...
       T.Chol(N, :), T.TG(N, :), T.HDL(N,:), T.LDL(N,:), T.VLDL(N, :), T.BMI(N, :)];
11 [label1, score, cost] = predict(mdl_1, XX);
12 Title = 'The Confusion Chart for fitcknn() function';
   figure(1)
13 plotConfChart(Title, TLabels, label1);        % call confusionchart() to plot confusion chart

   % Use fitcnb() function to fit the model
14 mdl_2 = fitcnb(X, Y);
15 label2 = predict(mdl_2, XX);
16 Title = 'The Confusion Chart for fitcnb() function';
   figure(2)
17 plotConfChart(Title, TLabels, label2);        % call confusionchart() to plot confusion chart

   % Use fitcecoc() function to fit the model
18 mdl_3 = fitcecoc(X, Y, "ClassNames", ["N", "P", "Y"]);
19 label3 = predict(mdl_3, XX);
20 Title = 'The Confusion Chart for fitcecoc() function';
   figure(3)
21 plotConfChart(Title, TLabels, label3);        % call confusionchart() to plot confusion chart

22 function[] = plotConfChart(title, tLabel, pLabel)
   cm = confusionchart(tLabel, pLabel);
   cm.Title = title;
   cm.RowSummary = 'row-normalized';
   cm.ColumnSummary = 'column-normalized';
   end
```

Fig. 6.27 The codes used to perform multiclass classifications

Now run this project and you can find that the original dataset has been modified and all data are now distributed in a random order. With this rectified dataset, **Dataset of Diabetes_M.xlsx**, we can continue to build our multiclass classification projects with different algorithms.

Let's start our discussions based on three classification algorithms one by one, such as **ClassificationKNN**, **ClassificationNaiveBayes**, and **ClassificationECOC** with three functions, **fitcknn()**, **fitcnb()**, and **fitcecoc()**.

Create a new Script file and name it as **Diabetes_MC_Func_Project.m**, and enter the codes shown in Fig. 6.27 into that file. Let's have a closer look at this piece of codes to see how it works.

1. First the data path for the source dataset is established and assigned to a local variable **dsPath**.
2. Then the name of the source dataset is assigned to another local variable **sData**.
3. Both data path and the name of the dataset are combined together by using a concatenating function **strcat()** to get a complete or a full dataset name, and it

is assigned to a local variable **sDataSet**. The reason we used this concatenating is to reduce the length of this command line with no other purpose.

4. The entire source dataset is read out and assigned to a local variable **T**.
5. We selected the first 900 data as the training data and the following 100 data as the testing data.
6. The predictors or the training data are built by extracting the first 13 columns with 900 rows from the source dataset and assigned to a local variable **X**.
7. The response or the output variable, which is the last column in the source dataset **CLASS**, is extracted and assigned to another local variable **Y**.
8. The testing output variables that are the last column with the last 100 rows are extracted and assigned to a local variable **TLabels**.
9. The object of the classification class **ClassificationKNN** is generated by calling the function **fitcknn()** and the training process is performed with the input and the output variables **X**, **Y** and related **Name-Values** pair, **ClassNames**, to indicate three output classes. The trained model is generated and assigned to a local variable, **mdl_1**.
10. The testing data are built by extracting the first 13 columns with the last 100 rows from the source dataset and assigned to a local variable **XX**.
11. To confirm and validate the trained model, **mdl_1**, the function **predict()** with the testing data is executed, and the validation result is assigned to a local variable **label1**.
12. To plot a confusion chart to check the testing result, a **Title** variable is generated to indicate the classification algorithm used.
13. A user-defined function **plotConfChart()** is called with three arguments, **Title**, **TLabels**, and **label1**, which are equivalent to the name of the classification algorithm, testing output, and the predicted output, to plot the confusion chart.
14. Now let's try to create a **ClassificationNaiveBayes** object by using the function **fitcnb()** to train our classification model with predictor **X** and response **Y**. The trained model is returned and assigned to **mdl_2**.
15. The testing data are applied to the trained model to get predicted classes by calling the function **predict()**, and the predicted classes are assigned to the **label2** variable.
16. The **Title** is used to distinguish the classification method, which is passed as an argument to the user-defined function in the next step to plot the confusion chart.
17. The user-defined function is called to plot the classification result.
18. The third classification object, **ClassificationECOC**, is generated by executing its function **fitcecoc()**, and the model is trained by the predictors and response variable.
19. The testing data are applied to the trained model, **mdl_3**, to obtain the predicted classes by calling the function **predict()**, and the predicted results are assigned to the variable **label3**.
20. The **Title** is used to distinguish the classification method, which is passed as an argument to the user-defined function in next step to plot the confusion chart.
21. The user-defined function is called to plot the classification result.

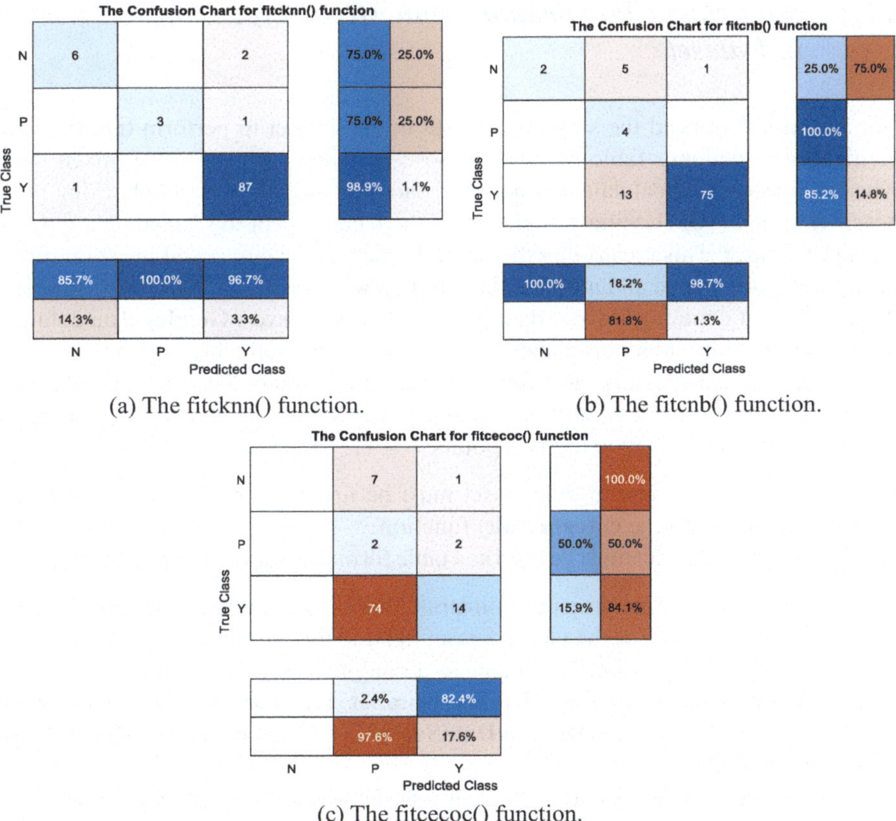

(a) The fitcknn() function.   (b) The fitcnb() function.

(c) The fitcecoc() function.

Fig. 6.28 The running results of three multiclass classification algorithms

22. The detailed codes for the user-defined function, **plotConfChart()**, are shown here.

Now run this project and the running results are shown in Fig. 6.28.

It can be found from Fig. 6.28 that the **fitcknn()** method provided the best classification results, and the **fitcnb()** method had the second result, but the **fitcecoc()** method provided the worst classification results.

As we discussed at the beginning of this section, there is some limitation of using the input or predictor for the class **ClassificationKNN** with its function **fitcknn()**. This means that all data in the training dataset must have the same data type, either all numbers or all test strings or characters. It is prohibited to have mixed or combined data type for all data, such as number combined with text or string together. However, this requirement may not be true for some other classifiers, such as **ClassificationNaiveBayes** and **ClassificationECOC**, with their functions, **fitcnb()** and **fitcecoc()**. Let's have a closer look at this topic in the next section.

6.7.1 Multiclass Classifications with Mixed Type Data in Dataset

Not all classifiers need the single-type data in the dataset to perform training and verification functions. Table 6.5 shows those classifiers who can use mixed data type, such as categorical data and numeric data, for the training and checking purpose. Of all ten (10) classifiers, eight (8) of them can accept the mixed data type in the input dataset. This means that the input dataset can be composed by a combination of categorical data and numeric data. In this way, we do not need to preprocess the dataset that contained mixed data type, such as to convert **Gender** data column to the number column in our Diabetes dataset as we did in the last section.

To use a mixed-typed dataset for the selected classifiers listed in Table 6.5, in addition to select the classifier who can accept the mixed-typed data, we also need to pay attention to the following two points:

1. The string or text data in the dataset must be first converted to the categorical data type by using the **categorical**() function.
2. The converted dataset must be used as a table format, not a vector or matrix format.

Now let's use an example project to illustrate this kind of implementation. We still use our Diabetes dataset as the target dataset to build this example project. First let's revise our target dataset to make all data as a randomly distributed dataset.

Open our existing Script file **MDiabeteDataSet_Excel.m**, modify it, and save it as a new Script file named **MDiabeteDataSet_NCC_Excel.m**. The modified file is shown in Fig. 6.29.

Let's have a closer look at this piece of modified codes to see how it works.

1. The new target or converted dataset is named **Dataset of Diabetes_MM.xlsx**.
2. The codes used to convert text to number for the **Gender** column have been removed.

Table 6.5 The optional dataset selections for different classifiers

Classifier	All predictors Numeric	All predictors Categorical	Categorical & Numeric
Decision Trees	Yes	Yes	Yes
Discriminant Analysis	Yes	No	No
Logistic Regression	Yes	Yes	Yes
Naive Bayes	Yes	Yes	Yes
SVM	Yes	Yes	Yes
Efficiently Trained Linear	Yes	Yes	Yes
K-Nearest Neighbor (KNN)	Euclidean distance only	Hamming distance only	No
Kernel Approximation	Yes	Yes	Yes
Ensembles	Yes	Yes, except Subspace Discriminant	Yes, except Subspace
Neural Networks	Yes	Yes	Yes

```
% Modify dataset 'Dataset of Diabetes.xlsx' to make it random one without convert Gender to number
% Name: MDiabeteDataSet_NCC_Excel.m
% Feb 5, 2024

dtPath = 'C:\\Artificial Intelligence Book\\Students\\Datasets\\Diabetes Dataset2\\';
tData = 'Dataset of Diabetes_MM.xlsx';
sData = 'Dataset of Diabetes.xlsx';

sDataSet = strcat(dtPath, sData);       % source dataset
tDataSet = strcat(dtPath, tData);       % target dataset

T = readtable(sDataSet);
M = 1000;            % max row number on dataset
N = 1000;              % max row number on (train + check + test) data
Tab = {'ID', 'No_Pation', 'Gender', 'AGE', 'Urea', 'Cr', 'HbA1c', 'Chol', 'TG', 'HDL', 'LDL', 'VLDL', 'BMI', 'CLASS'};

for i = 1:N
   j = randi(M, 1, 1);
   writecell(Tab, tDataSet);
   writetable(T(j, :), tDataSet,'WriteMode','append');
end
```

Fig. 6.29 The modified codes for file MDiabeteDataSet_NCC_Excel.m

Now let's build another project to illustrate how to use classifiers, **ClassificationNaiveBayes** and **ClassificationECOC**, with their function, **fitcnb**() and **fitcecoc**(), to perform classification process with the modified dataset **Dataset of Diabetes_MM.xlsx** with mixed-typed data. We can modify an existing project **Diabetes_MC_Func_Project.m**, which can be found from the Springer ftp site at the folder, **Students\Class Projects\Chapter 6\Multiclass Classification Project**, to make it as our new project, **Diabetes_MC_NC_Func_Project.m**.

Open that project and perform the related modifications, and your finished project file should match one that is shown in Fig. 6.30. Let's have a closer look at this piece of modified codes to see how it works.

1. The dataset used for this project has been modified and it is now a mixed-typed dataset.
2. To use this mixed-typed dataset, the text/string column in the dataset (except the last column) must be converted to a categorical type; thus, the **Gender** column is processed by using the **categorical**() function.
3. A set of training data is created, and this set must be presented as a table format, not a vector or a matrix format since the latter cannot be treated as a mixed-typed data set.
4. Similarly, the testing data set must also be in a table format.

Now run this project and the running result is shown in Fig. 6.31.

It can be found from Fig. 6.31 that the classification algorithm **fitcnb**() has better classification performance compared with that of the classification algorithm **fitcecoc**(). Compared with the classification results shown in Fig. 6.28, it can be found that both algorithms provide the similar classification results. Therefore, there is no difference between using a converted dataset in which the string/text is converted to the number and the dataset with mixed data type for these selected classifiers. When using a mix-typed dataset, the only key points to be noted are:

1. All string/text columns must be converted to the categorical data type.

```matlab
% Train multi class classification model for diabetes with mixed dataset - number and categorical data
% Name: Diabetes_MC_NC_Func_Project.m
% Input: Dataset of Diabetes_M.xlsx

dsPath = 'C:\\Artificial Intelligence Book\\Students\\Datasets\\Diabetes Dataset2\\';
sData = 'Dataset of Diabetes_MM.xlsx';
sDataSet = strcat(dsPath, sData);

T = readtable(sDataSet);
M = 1:900;                          % training data - 900 training data
N = 901:1000;                       % testing data - 100 testing data

T.Gender = categorical(T.Gender);   % convert the Gender column to category column

X = table(T.ID(M, :), T.No_Pation(M, :), T.Gender(M, :), T.AGE(M,:), T.Urea(M,:), T.Cr(M, :), T.HbA1c(M, :) +...
        T.Chol(M, :), T.TG(M, :), T.HDL(M,:), T.LDL(M,:), T.VLDL(M, :), T.BMI(M, :));

Y = T.CLASS(M, :);                  % get response variable
TLabels = T.CLASS(N, :);            % get testing data

% Get testing data XX to perform prediction or classification
XX = table(T.ID(N, :), T.No_Pation(N, :), T.Gender(N, :), T.AGE(N,:), T.Urea(N,:), T.Cr(N, :), T.HbA1c(N, :) +...
        T.Chol(N, :), T.TG(N, :), T.HDL(N,:), T.LDL(N,:), T.VLDL(N, :), T.BMI(N, :));

% Use fitcnb() function to fit the model
mdl_1 = fitcnb(X, Y);
label1 = predict(mdl_1, XX);
Title = 'The Confusion Chart for fitcnb() function';
figure(1)
plotConfChart(Title, TLabels, label1);          % call confusionchart() to plot confusion chart

% Use fitcecoc() function to fit the model
mdl_2 = fitcecoc(X, Y, "ClassNames", ["N", "P", "Y"]);
label2 = predict(mdl_2, XX);
Title = 'The Confusion Chart for fitcecoc() function';
figure(2)
plotConfChart(Title, TLabels, label2);          % call confusionchart() to plot confusion chart

function[] = plotConfChart(title, tLabel, pLabel)
cm = confusionchart(tLabel, pLabel);
cm.Title = title;
cm.RowSummary = 'row-normalized';
cm.ColumnSummary = 'column-normalized';
end
```

Fig. 6.30 The modified codes for mixed-typed data in a dataset

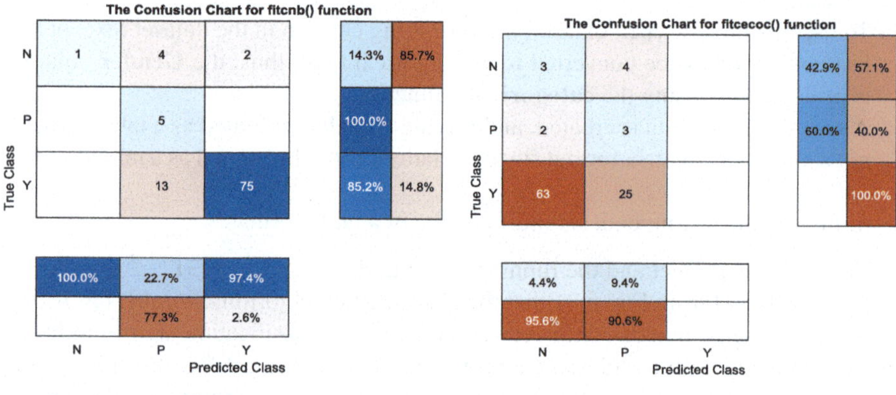

Fig. 6.31 The running results of two classifications

2. All training and validation data in that converted dataset must be treated as a table.

We have illustrated how to use two classifiers, **ClassificationNaiveBayes** and **ClassificationECOC**, with their functions. Now it is time for us to have a clear picture about how those classifiers perform their works to derive the desired classification results.

6.7.2   *Working Principles of Naive Bayes Classifier*

Mathematically, naive Bayes is a conditional probability model and it assigns probabilities $p(Y_k|x_1, x_2, \dots x_n)$ for each of the K possible outcomes or *classes* Y_k for a given problem instance to be classified, represented by a vector $x = (x_1, x_2, \dots x_n)$ encoding some n features (independent variables) [9].

The problem with the above formulation is that if the number of features n is large or if a feature can take on a large number of values, then basing such a model on probability tables is infeasible. The model must therefore be reformulated to make it more tractable. Using Bayes' theorem, the conditional probability can be decomposed as [10]:

$$p(Y_k|x) = \frac{p(Y_k)p(x|Y_k)}{p(x)} \tag{6.3}$$

In practice, we only pay attention to the numerator of that fraction, because the denominator does not depend on Y and the values of the features x_i are given, so that the denominator is effectively constant. The numerator is equivalent to the *joint probability* model with an equation $p(Y_k, x_1, x_2, \dots, x_n)$, which can be rewritten as follows, using the chain rule for repeated applications of the definition of conditional probability:

$$p(Y_k, x_1, x_2, \dots, x_n) = p(x_1, x_2, \dots, x_n, Y_k) == p(x_1|x_2, \dots, x_n, Y_k)$$
$$p(x_2|, x_3|, \dots|, x_n|, Y_k) \cdots p(x_{n-1}|x_n, Y_k) p(x_n|Y_k) p(Y_k) \tag{6.4}$$

Now the so-called *naive* conditional independence assumptions come into play, let's assume that all features in x are mutually independent conditional on the category Y_k. Under this assumption, we have

$$p(x_i|x_{i+1}, \dots, x_n, Y_k) = p(x_i|Y_k) \tag{6.5}$$

This naive independent condition assumption means that each feature in this set or group is independent mutually, and they do not have any dependent relation between and among them. The terminal category Y_k can be determined by each feature separately. Thus, the joint model can be expressed as

$$p(Y_k|, x_1|, x_2|, \dots|, x_n) \propto p(Y_k, x_1, x_2, \dots, x_n) = p(Y_k) p(x_1|Y_k)$$
$$p(x_2|Y_k) p(x_3|Y_k) = p(Y_k) \prod_{i=1}^{n} p(x_i|Y_k) \tag{6.6}$$

where $\propto$ denotes proportionality since the denominator $p(x)$ is omitted.

This means that under the above independence assumptions, the conditional distribution over the class variable Y is:

$$p\left(Y_k|,x_1|,x_2|,\ldots|,x_n\right) = \frac{1}{Z} p\left(Y_k\right) \prod_{i=1}^{n} p\left(x_i|Y_k\right) \tag{6.7}$$

where the evidence $Z = p(x) = \sum p\left(Y_k\right) p\left(x|Y_k\right)$ is a scaling factor dependent only on $x_1, x_2, \ldots, x_n$, that is a constant if the values of the feature variables are known.

The above discussion has derived the independent feature model, that is, the naive Bayes probability model. The naive Bayes classifier combines this model with a *decision rule*. One common rule is to pick the hypothesis that is most probable so as to minimize the probability of misclassification; this is known as the *maximum a posteriori* or *MAP* decision rule. The corresponding classifier, a Bayes classifier, is the function that assigns a class label $\hat{y} = Y_k$ for some k as follows:

$$\hat{y} = \underset{k \in \{1,\ldots,K\}}{\arg\max} \; p\left(Y_k\right) \prod_{i=1}^{n} p\left(x_i|Y_k\right) \tag{6.8}$$

To estimate the parameters for a feature's distribution, one must assume a distribution or generate nonparametric models for the features from the training set [11].

In summary, three popular models, Gaussian naive Bayes, Multinomial naive Bayes, and Bernoulli naive Bayes, are widely implemented in AI studies.

Gaussian Naive Bayes Model

The assumptions on distributions of features are called the *event model* of the naive Bayes classifier. When dealing with continuous data, a typical assumption is that the continuous values associated with each class are distributed according to a Normal or Gaussian distribution.

For example, suppose the training data contains a continuous attribute x. The data is first segmented by the class, and then the mean and variance of x is computed in each class. Let μ_k be the mean of the values in x associated with class Y_k, and let σ_k^2 be the Bessel corrected variance of the values in x associated with class Y_k. Suppose one has collected some observation value v. Then, the probability *density* of v given a class Y_k, i.e., $p(x = v| Y_k)$, can be computed by plugging v into the equation for a normal distribution parameterized by μ_k and σ_k^2. Formally,

$$p\left(x = v|Y_k\right) = \frac{1}{\sqrt{2\pi\sigma_k^2}} e^{-\frac{(v-\mu_k)^2}{2\sigma_k^2}} \tag{6.9}$$

For discrete features like the ones encountered in document classification (including spam filtering), multinomial and Bernoulli distributions are popular. These assumptions lead to two distinct models.

Multinomial Naive Bayes Model

For a multinomial event model, samples or feature vectors represent the frequencies with which certain events have been generated by a multinomial $(p_1, p_2, \ldots, p_n)$ where p_i is the probability that event i occurs or K such multinomials in the multiclass case. A feature vector $x = (x_1, x_2, \ldots, x_n)$ is then a histogram, with x_i counting the number of times event i was observed in a particular instance. This is the event model typically used for document classification, with events representing the occurrence of a word in a single document. The likelihood of observing a histogram x is given by:

$$p\left(x|Y_k\right) = \frac{\left(\sum_{i=1}^{n} x_i\right)!}{\prod_{i=1}^{n} x_i!} \prod_{i=1}^{n} p\left(x_i|Y_k\right) x_i$$

(6.10)

The multinomial naive Bayes classifier becomes a linear classifier when expressed in log-space [12]:

$$\log\left(Y_k|x\right) \propto \log\left(p\left(Y_k\right) \prod_{i=1}^{n} p_{ki}^{x_i}\right) = \log p\left(Y_k\right) + \sum_{i=1}^{n} x_i \cdot \log p_{ki} = b + w_k^T x \quad (6.11)$$

where $b = \log p(Y_k)$ and $w_{ki} = \log p_{ki}$. Estimating the parameters in log space is advantageous since multiplying a large number of small values can lead to significant rounding error. Applying a log transform reduces the effect of this rounding error.

Bernoulli Naive Bayes Model

In the multivariate Bernoulli event model, features are independent Booleans or binary variables describing inputs. Like the multinomial model, this model is popular for document classification tasks [13], where binary term occurrence features are used rather than term frequencies. If x_i is a Boolean expressing the occurrence or absence of the ith term from the vocabulary, then the likelihood of a document given a class Y_k is given by [13]:

$$p\left(x|Y_k\right) = \prod_{i=1}^{n} p_{ki}^{x_i} \left(1 - p_{ki}\right)^{(1-x_i)}$$

(6.12)

where p_{ki} is the probability of class Y_k generating the term x_i. This event model is especially popular for classifying short texts. It has the benefit of explicitly modeling the absence of terms. Note that a naive Bayes classifier with a Bernoulli event model is not the same as a multinomial NB classifier with frequency counts truncated to one.

Next let's use a real example to illustrate how to use this Gaussian naive Bayes model to predict the possibility of having diabetes disease for a given set of predictors.

6.7.3 A Real Example Project for Naive Bayes Classification Model

In this section, we like to use a real example project to illustrate how to use the Gaussian naive Bayes model to predict the possibility of a person having diabetes disease.

The sample dataset we used is the dataset **Diabetes.csv**. However, we did not use the entire dataset, but only selected some data as the training predictors and only one piece of data as the testing data. The selected dataset is shown in Table 6.6.

$$a = \begin{pmatrix} 59 & 51 & 31 & 31 & 26 \\ 189 & 166 & 118 & 107 & 78 \\ 60 & 72 & 84 & 74 & 50 \\ 30.1 & 25.8 & 45.8 & 29.6 & 31.0 \end{pmatrix}$$

$$b = \begin{pmatrix} 26 & 37 & 48 & 42 & 21 \\ 180 & 133 & 106 & 150 & 73 \\ 64 & 84 & 92 & 66 & 50 \\ 34 & 40.2 & 22.7 & 34.7 & 23 \end{pmatrix}$$

Four input predictors are utilized, **Age**, **Glucose**, Blood Pressure (**BP**), and Body Mass Index (**BMI**). The response or the output is **Diabetes**: 1 means that the person has diabetes disease, and 0 means that the person has no diabetes. The top five elements for all features are for Diabetes = 1, and the bottom five elements are for Diabetes = 0 (no diabetes).

To use the Gaussian naive Bayes model to predict the possibility of having Diabetes, first we need to train that model. We can use Eq. (6.9) to calculate all mean and variance values for each given set of predictors, and put the results into related category, as shown in Table 6.7. After this, we can use Eq. (6.7) to get each conditional distribution, and multiply them together to get the numerator. In fact, we

Table 6.6 The sample diabetes dataset

Age	Glucose	BP	BMI	Diabetes
59	189	60	30.1	1
51	166	72	25.8	1
31	118	84	45.8	1
31	107	74	29.6	1
26	78	50	31	1
26	180	64	34	0
37	133	84	40.2	0
48	106	92	22.7	0
42	150	66	34.7	0
21	73	50	23	0

$$a = \begin{pmatrix} 59 & 51 & 31 & 31 & 26 \\ 189 & 166 & 118 & 107 & 78 \\ 60 & 72 & 84 & 74 & 50 \\ 30.1 & 25.8 & 45.8 & 29.6 & 31.0 \end{pmatrix}$$

$$b = \begin{pmatrix} 26 & 37 & 48 & 42 & 21 \\ 180 & 133 & 106 & 150 & 73 \\ 64 & 84 & 92 & 66 & 50 \\ 34 & 40.2 & 22.7 & 34.7 & 23 \end{pmatrix}$$

Table 6.7 The calculated mean and variance values for each predictor

Diabetes	Age (Mean)	Age (Variance)	Glucose (Mean)	Glucose (Variance)	BP (Mean)	BP (Variance)	BMI (Mean)	BMI (Variance)
1	39.6	167.8	131.6	1628.2	68	139.2	32.46	47.6
0	34.8	99.8	128.4	1344.2	71.2	225	30.92	48.0

do not need to take care of the denominator since it is a constant if the values of the feature variables are known. Finally we can get our prediction based on the *maximum a posteriori* or *MAP* decision rule. Let's start our calculations as below:

$$\text{posterior}\left(\text{Diab}\right) = \frac{P\left(\text{Diab}\right)p\left(\text{Age}|\text{Diab}\right)p\left(\text{Glucose}|\text{Diab}\right)p\left(\text{BP}|\text{Diab}\right)p\left(\text{BMI}|\text{Diab}\right)}{\text{Evidence}}$$

$$\text{posterior}\left(\text{NDiab}\right) = \frac{P\left(\text{NDiab}\right)p\left(\text{Age}|\text{NDiab}\right)p\left(\text{Glucose}|\text{NDiab}\right)p\left(\text{BP}|\text{NDiab}\right)p\left(\text{BMI}|\text{NDiab}\right)}{\text{Evidence}}$$

where *Diab* = Diabetes is 1 and *NDiab* = Diabetes is 0. The *Evidence* is the full probability and it equals to P(Diab)p(Age|Diab)p(Glucose|Diab)p(BP|Diab)p(BMI|Diab) + P(NDiab)p(Age|NDiab) p(Glucose|NDiab)p(BP|NDiab)p(BMI|NDiab).

For our case, the P(Diab) and P(NDiat) are equal and both of them equal to 0.5.

To simplify these calculations, two pieces of MATLAB codes can be used to speed up these calculations. Figure 6.32 shows a piece of codes to perform the calculations for means and variances; all of those calculated means and variances are written into the related columns shown in Table 6.7 based on Diabetes = 1 and 0.

These codes have been assembled into a Script file named **Set_Naive_Bayes_Diabetes.m** and it can be found from the Springer ftp site at a folder **Students\Class Projects\Chapter 6\Multiclass Classification Project**. Let's have a closer look at this piece of codes to see how it works.

1. First we need to re-organize data in Table 6.6 and make them into two 4 × 5 matrices, as shown on the right of Table 6.6. For matrix *a*, each row contains five (5) elements for each category (**Age, Glucose, BP, BMI**) when Diabetes = 1. Similarly, for matrix *b*, each row presents five (5) elements for each category when Diabetes = 0. The variable **n** is defined as **5** for 5 columns in matrices *a* and *b*.
2. Vectors **vm**() and **vw**() are defined and they are used to hold the variance values to be calculated later. The former is for the Diabetes = 1 and the latter is for Diabetes = 0.
3. Matrices *a* and *b* are defined as we did in the first step above.
4. A **for**() loop is used to calculate the mean values for 4 rows, where **mm**() is for Diabetes = 1 and **mw**() is for Diabetes = 0.
5. Another **for**() loop is utilized to calculate the variances for Diabetes = 1. This is a nested loop and the outer loop is used to access each row and the inner loop is for the five columns. The running result for the inner loop is to get the numerator

```matlab
% Calculate mean, variance for Diabetes test table
% Name: Set_Naive_Bayes_Diabetes.m
n = 5;                        % number of columns for diabetes
vm = zeros(4);                % variances for 4 rows - diabetes
vw = zeros(4);                % variances for 4 rows - nondiabetes

a = [59 51 31 31 26;          % 4 x 5 matrix for Diabetes = 1, each row = 5 inputs, such as Age, Glucose, ...
     189 166 118 107 78;      % totally 4 rows = 4 inputs, Age, Glucose, BP (Blood Pressure), BMI
     60 72 84 74 50;
     30.1 25.8 45.8 29.6 31];
b = [26 37 48 42 21;          % 4 x 5 matrix for Diabetes = 0, each row = 5 inputs, such as Age, Glucose, ...
     180 133 106 150 73;      % totally 4 rows = 4 inputs, Age, Glucose, BP (Blood Pressure), BMI
     64 84 92 66 50;
     34 40.2 22.7 34.7 23];

for j=1:n-1                   % get mean for 4 rows
    mm(j) = mean(a(j, :))     % mean for Diabetes = 1 (Diabetes)
    mw(j) =mean(b(j, :))      % mean for Diabetes = 0 (Non-Diabetes)
end

% get variances for Diabetes = 1 (Diabetes)
for i = 1:4                                % outer-loop for 4 rows
    for j =1:n                             % inner-loop for 5 columns
        vm(i) = vm(i) + (a(i, j)-mm(i))^2; % calculate variance for each column - numerator only
    end
    vm(i) = vm(i)/n                        % divided by n to get variances
end

% get variances for Diabetes = 0 (Non-Diabetes)
for i = 1:4                                % outer-loop for 4 rows
    for j =1:n                             % inner-loop for 5 columns
        vw(i) = vw(i) + (b(i, j)-mw(i))^2; % calculate variance for each column - numerator only
    end
    vw(i) = vw(i)/n                        % divided by n to get variances
end
```

Fig. 6.32 The MATLAB codes used to calculate means and variances

of the variance for each row, and the outer loop is to divide the numerator of variances by total number of elements to get the final variance values.

6. Similarly, the third **for**() loop is used to calculate the variances for Diabetes = 0.

Now run this project file and the running results are shown in the Command window as:

```
mm = 39.6000   131.6000   68.0000   32.4600
mw = 34.8000   128.4000   71.2000   30.9200


vm = 1.0e+03 *
      0.1678        0        0        0
      1.6282        0        0        0
      0.1392        0        0        0
      0.0476        0        0        0


vw = 1.0e+03 *
      0.0998        0        0        0
      1.3442        0        0        0
      0.2250        0        0        0
      0.0480        0        0        0
```

Now write these means and variance values to the associated column under the related category (**Age mean, Age variance**...) as shown in Table 6.7.

Table 6.8 A piece of testing data (diabetes = 1)

Age	Glucose	BP	BMI	Diabetes
41	187	68	37.7	

```
% Calculate p(Age|Diabetes). p(Glucose|Diabetes), p(BP|Diabetes) and p(BMI|Diabetes)
% Name: CalP_Diabetes.m

1  pAge = (1/sqrt(2*pi*167.8))*exp(-(41-39.6)^2/(2*167.8))
   pGlucose = (1/sqrt(2*pi*1628.2))*exp(-(187-131.6)^2/(2*1628.2))
   pBP = (1/sqrt(2*pi*139.2))*exp(-(68-68)^2/(2*139.2))
   pBMI = (1/sqrt(2*pi*47.6))*exp(-(37.7-32.46)^2/(2*47.6))

   %Classification for Diabetes
2  pDiabetes = 0.5*pAge*pGlucose*pBP*pBMI     % = 8.6421e-08

   %Classification for Non Diabetes
3  pAge = (1/sqrt(2*pi*99.8))*exp(-(41-34.8)^2/(2*99.8))
   pGlucose = (1/sqrt(2*pi*1344.2))*exp(-(187-128.4)^2/(2*1344.2))
   pBP = (1/sqrt(2*pi*225))*exp(-(68-71.2)^2/(2*225))
   pBMI = (1/sqrt(2*pi*48))*exp(-(37.7-30.92)^2/(2*48))

4  pNDiabetes = 0.5*pAge*pGlucose*pBP*pBMI     % = 4.6333e-08

   % Conclusion:
   % Since the posterior numerator on Diabetes is 8.6421e-08, which is greater than the posterior on Non
   Diabetes 4.6333e-08,  Therefore prediction is that the sample is a Diabetes patient.
```

Fig. 6.33 The codes used to test the trained or calculated model

After this model has been set, next we need to test this model with a piece of new or testing data, which is a patient record selected from the Diabetes dataset, as shown in Table 6.8. This testing or sample data is for a person who is diabetes patient (Diabetes = 1).

Let's put testing data shown in Table 6.8 and trained parameters shown in Table 6.7 into the above equations to get all related values as below:

$$p(\text{Age}|\text{Diab}) = \frac{1}{\sqrt{2\pi\sigma^2}}\exp\left(\frac{-(41-\mu)^2}{2\sigma^2}\right) = \frac{1}{\sqrt{2\pi(167.8)^2}}\exp\left(\frac{-(41-39.6)^2}{2(167.8)^2}\right) = 0.0306$$

$$p(\text{Glucose}|\text{Diab}) = \frac{1}{\sqrt{2\pi\sigma^2}}\exp\left(\frac{-(187-\mu)^2}{2\sigma^2}\right) = \frac{1}{\sqrt{2\pi(1628.2)^2}}\exp\left(\frac{-(187-131.6)^2}{2(1628.2)^2}\right) = 0.0039$$

It is too time-consuming to complete all of these calculations. To speed up these calculations, we can build another piece of MATLAB codes to a Script file and run it to do this job. Create a new Script file named **CalP_Diabetes.m** and enter the codes shown in Fig. 6.33 into that file. This piece of codes is used to derive the *maximum a posteriori* or *MAP* value.

Let's have a closer look at this piece of codes to see how it works.

1. As we did in the above two equations, we put each categorized testing data (**Age, Glucose, BP**, and **BMI**) into the related calculated (trained) model's category (mean and variance in Gaussian equation) for Diabetes = 1 assumption to calculate the related value, such as pAge = p(Age|Diabetes), pGlucose = p(Glucose|Diabetes), … The results will be shown in the Command window as no semi-colon is attached at the end of each coding line.

2. Then we calculate the posterior for Diabetes = 1 assumption.
3. Similarly, we calculate prior for Non-Diabetes assumption with the Non-Diabetes model.
4. Calculate the posterior for Diabetes = 0 or Non-Diabetes assumption.

Now run this piece of codes and the running results are displayed in the Command window.

Based on the running results, it can be seen that the calculated posterior value for Diabetes = 1 assumption is **8.6421e-08**, but that value is **4.6333e-08** for Diabetes = 0 assumption. Thus, the posterior value for Diabetes = 1 is greater than that posterior value for Diabetes = 0. Therefore, we can conclude that the testing data is for a Diabetes patient (Diabetes = 1).

The Script files, **CalP_Diabetes.m**, can be found from the Springer ftp site under the folder, **Students\Class Projects\Chapter 6\Multiclass Classification Project**.

Next let's take a look at the working principle of Error-Correcting Output Code (ECOC).

6.7.4   *Working Principles of the Error-Correcting Output Code (ECOC) Algorithm*

The so-called Error-Correcting Output Code (ECOC) method is to use binary classification model to solve or to classify multiclass prediction jobs. In other words, it allows a multiclass classification problem to be reframed as multiple binary classification problems, allowing the use of naive binary classification models directly [14].

Unlike some other classification algorithms that break a multiclass classification job into a piece of binary classification tasks, the error-correcting output codes technique allows each class to be encoded as an arbitrary number of binary classification problems.

In order to convert or reframe a multiclass classification process to multi-binary classification process, two popular methods can be utilized, which are called One-vs-Rest (OvR) and One-vs-One (OvO) techniques.

1. **OvR**: Splits a multiclass problem into one binary problem per class.
2. **OvO**: Splits a multiclass problem into one binary problem per each pair of classes.

A typical ECOC algorithm performs the following operational steps:

1. Encoding and training process.
2. Decoding process.

6.7.4.1   The ECOC Encoding and Training Process

The encoding process is to map a multiclass classification job to multi-binary classifications. In fact, it is to perform a coding design to set up a binary matrix based on the input classes and input elements (classifiers).

For example, still use our Diabetes classification project **MDiabeteDataSet_ NCC_Excel.m** discussed in Sect. 6.7.1, in which the dataset we used is **Dataset of Diabetes_M.xlsx**. To make things simple, we selected a partial of that dataset as shown in Table 6.9. There are ten columns or ten input features (classifiers) with three-class outputs, **Y** (Diabetes), **N** (No Diabetes), and **P** (Predicted Diabetes). We can reframe these three class levels into three binary classification datasets with **OvR** method:

1. Binary classification 1: **Y** vs [**N, P**] → 1 vs [0, 0].
2. Binary classification 2: **N** vs [**Y, P**] → 1 vs [0, 0].
3. Binary classification 3: **P** vs [**Y, N**] → 1 vs [0, 0].

With the **OvO** method, we can reframe these into another three binary classification datasets:

1. Binary classification 1: Y vs N.
2. Binary classification 2: Y vs P.
3. Binary classification 3: N vs P.

Since the **OvR** method is simpler, thus we use this technique as example to discuss how to use it to reframe our dataset to get error-correcting output code table.

As the training process starts, each row of the coding design or the binary matrix corresponds to a distinct class, and each column corresponds to a binary classifier or learner. In a ternary coding design, for a particular column or binary classifier:

- Each row in the corresponding column containing 1 directs the binary classifier to group all observations in the corresponding class into a positive class.
- Each row in the corresponding column containing 0 directs the binary classifier to group all observations in the corresponding class into a negative class.

Table 6.9 A selected partial dataset from dataset of Diabetes_M.xlsx

AGE	Urea	Cr	HbA1c	Chol	TG	HDL	LDL	VLDL	BMI	CLASS
54	3.8	32	10.8	3.4	0.8	1.4	1.6	1.5	27	Y
53	3	48	9.5	7.2	2	1.9	2	0.7	32	Y
30	7.1	81	6.7	4.1	1.1	1.2	2.4	8.1	27.4	Y
50	2.6	106	4	6.3	4.4	1	3.6	1.7	24	Y
60	6	72	10.7	4.4	2.1	1.1	2.5	0.9	26	Y
60	4	63	12	3.6	5.1	0.9	2.5	0.9	30	Y
57	5.3	66	7.9	3.7	2.9	1.1	5.5	1.3	36	Y
50	3.5	59	6	4	2.1	1.4	1.9	0.9	25	P
30	3	45	4.1	4.9	1.3	1.2	3.2	0.5	22	N
61	5.9	56	8.2	4.9	2.1	1.1	2.5	0.9	28	Y
C1	C2	C3	C4	C5	C6	C7	C8	C9	C10	

For our case, each binary classifier (C1~C10) is trained with a group of input or predictor data in the related column to obtain the desired binary encoding string or bit-string for three output classes, **Y[1 0 0 … 1]**, **N[0 0 …1]**, and **P[0 1 … 0]**. Based on Table 6.9, we can get:

- Age = {30~61}, N: 30~40, P: 41~50, Y: 51~61.
- Urea = {3~7.1}, N: 2~3.7, P: 3.8~5.5, Y: 5.6~7.1.
- Cr = {32~81}, N: 28~46, P: 47~64, Y: 65~81.
- HbA1c = {6.7~10.8}, N: 4~6.3, P: 6.4~8.7, Y: 8.8~10.8.
- Chol = {3.4~7.2}, N: 3.4~4.5, P: 4.6~5.9, Y: 6.0~7.2.
- TG = {0.8~2.1}, N: 0.8~1.23, P: 1.24~1.67, Y: 1.68~2.1.
- HDL = {1.1~1.9}, N: 0.8~1.33, P: 1.34~1.87, Y: 1.88~2.4.
- LDL = {1.6~2.5}, N: 1.6~2.31, P: 2.32~3.03, Y: 3.04~3.80.
- VLDL = {0.7~8.1}, N: 0.5~3.1, P: 3.2~5.8, Y: 5.9~8.1.
- BMI = {27~32}, N: 19~23.3, P: 23.4~27.7, Y: 27.8~32.

Here we used a simplified method to avoid the complicated Hamming matrix process. Based on these mappings, Table 6.9 can be reframed to the so-called Error-Correcting Output Code table, as shown in Tables 6.10, 6.11, and 6.12.

A possible issue with using this technique is the large number of classifiers. In our case, we have 10 features, **Age**, **Urea**, **Cr**, …, **BMI**, and each column needs one binary classifier, such as **C1**, **C2**, **C3**, …, **C10**, which are shown at the bottom of Tables 6.9, 6.10, 6.11, and 6.12. These processes are called encoding process.

The encoding results for three classes are: **Y[1 0 0 1 0 0 0 0 0 1]**, **N[1 1 1 1 1 0 0 0 1 0]**, and **P[0 0 1 0 0 1 0 1 0 0]**. All of them come from the three highlighted rows in Tables 6.10, 6.11, and 6.12 shown above.

6.7.4.2   The ECOC Decoding and Testing Process

The classification process is to perform a decoding process, exactly to compare the testing string with our three encoding strings. Table 6.13 shows three testing sample records selected from the Diabetes dataset.

Three samples can be converted to three binary arrays based on the above mapping map:

Table 6.10 Error-correcting output code for binary classification 1

AGE	Urea	Cr	HbA1c	Chol	TG	HDL	LDL	VLDL	BMI	CLASS
1	0	0	1	0	0	0	0	0	1	Y
0	0	1	0	0	1	0	1	0	0	P
1	1	1	1	1	0	0	0	1	0	N
C1	C2	C3	C4	C5	C6	C7	C8	C9	C10	

Table 6.11 Error-correcting output code for binary classification 2

AGE	Urea	Cr	HbA1c	Chol	TG	HDL	LDL	VLDL	BMI	CLASS
0	0	1	0	1	1	0	1	1	0	Y
0	1	0	1	1	0	1	0	1	1	P
1	1	1	1	1	0	0	0	1	0	N
C1	C2	C3	C4	C5	C6	C7	C8	C9	C10	

Table 6.12 Error-correcting output code for binary classification 3

AGE	Urea	Cr	HbA1c	Chol	TG	HDL	LDL	VLDL	BMI	CLASS
0	1	0	0	0	0	1	0	0	1	Y
0	0	1	0	0	1	0	1	0	0	P
0	0	0	0	0	0	0	0	0	1	N
C1	C2	C3	C4	C5	C6	C7	C8	C9	C10	

- SY[1 0 0 1 1 1 1 0 0 1]
- SN[1 1 0 1 0 0 1 0 1 1]
- SP[0 0 1 0 1 1 0 0 0 1]

After the training process is done, a prediction can be made for new examples by using each classifier to make a prediction for the input to create the binary string. Then the created binary string is compared with each class's encoding string (bitstring) obtained from the trained results. The class string that has the smallest distance to the encoding string is then chosen as the output.

Figure 6.34 shows a floating chart of using ECOC to perform multiclass classification process. Next let's build a real example project to use ECOC model to perform multiclass classification task.

In our case, for the comparison between the **SY** sample and the **Y** encoding shown in Table 6.10, **N** encoding shown in Table 6.11, and **P** encoding shown in Table 6.12, we have:

SY[1 0 0 1 1 1 1 0 0 1] to Y[1 0 0 1 0 0 0 0 0 1], the Hamming distance is 3
SY[1 0 0 1 1 1 1 0 0 1] to N[1 1 1 1 1 0 0 0 1 0], the Hamming distance is 6
SY[1 0 0 1 1 1 1 0 0 1] to P[0 0 1 0 0 1 0 1 0 0], the Hamming distance is 7

The smallest Hamming distance is 3, which means that the predicted class is **Y**.

For the comparison between the **SN** sample and the **Y** encoding shown in Table 6.10, **N** encoding shown in Table 6.11, and **P** encoding shown in Table 6.12, we have:

SN[1 1 0 1 0 0 1 0 1 1] to Y[1 0 0 1 0 0 0 0 0 1], the Hamming distance is 4
SN[1 1 0 1 0 0 1 0 1 1] to N[1 1 1 1 1 0 0 0 1 0], the Hamming distance is 3
SN[1 1 0 1 0 0 1 0 1 1] to P[0 0 1 0 0 1 0 1 0 0], the Hamming distance is 8

The smallest Hamming distance is 3, which means that the predicted class is **N**.

For the comparison between the **SP** sample and the **Y** encoding shown in Table 6.10, **N** encoding shown in Table 6.11, and **P** encoding shown in Table 6.12, we have:

SP[0 0 1 0 1 1 0 0 0 1] to Y[1 0 0 1 0 0 0 0 0 1], the Hamming distance is 5

Table 6.13 Three testing sample records

AGE	Urea	Cr	HbA1c	Chol	TG	HDL	LDL	VLDL	BMI	CLASS	Sample
53	3	48	9.5	7.2	2	1.9	2	0.7	32	Y	SY
30	3	45	4.1	4.9	1.3	1.2	3.2	0.5	22	N	SN
31	3.4	55	5.7	4.9	1.6	1	3.2	0.7	24	P	SP

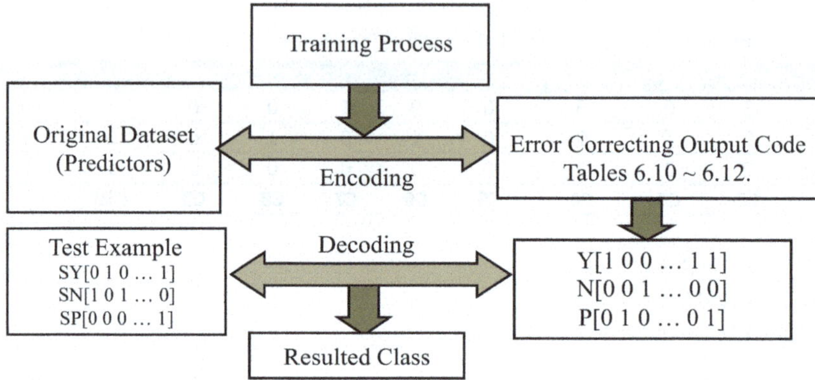

Fig. 6.34 The flowchart of using ECOC to classify multiclass

SP[0 0 1 0 1 1 0 0 0 1] to N[1 1 1 1 1 0 0 0 1 0], the Hamming distance is 6
SP[0 0 1 0 1 1 0 0 0 1] to P[0 0 1 0 0 1 0 1 0 0], the Hamming distance is 3
The smallest Hamming distance is 3, which means that the predicted class is **P**.

It can be found from this example that the ECOC algorithm is a convenient method to simplify multiclass classifications to multiply binary classification processes. More importantly, all binary classification algorithms shown in Tables 6.1 and 6.2 now can be used for multiclass classification process, which provide a broad way and more selections for classification process and enable us to use different algorithms to make multiclass classifications more effectively and easily.

Next let's have a closer look at those algorithms and how to use them in the ECOC method.

6.7.4.3   The ECOC Class ClassificationECOC and Multi-binary Classification Algorithms

As we discussed in Sect. 6.7 with Table 6.4, to perform a multiclass classification by using the class **ClassificationECOC**, a **ClassificationECOC** object should be created by using the function **fitcecoc**() and the default algorithm for that function is SVM. In fact, all of binary classification algorithms shown in Table 6.3 can be selected and used for that function.

The function **fitcecoc**() has the following five different constructors,

1. **mdl = fitcecoc(Tbl, ResponseVarName)** returns a trained multiclass ECOC model using the predictors in table **Tbl** and the class labels in **Tbl. ResponseVarName**. The **fitcecoc()** function uses $K(K − 1)/2$ binary SVM models with the one-versus-one coding design, where K is the number of unique class labels. The **mdl** is a `ClassificationECOC` model.
2. **mdl = fitcecoc(Tbl, formular)** returns an ECOC model using the predictors in table **Tbl** and the class labels. The **formula** is an explanatory model of the response and a subset of predictor variables in **Tbl** used for training.
3. **mdl = fitcecoc(Tbl, Y)** returns an ECOC model using the predictors in table **Tbl** and the class labels in vector **Y**.
4. **mdl = fitcecoc(X, Y)** returns a trained ECOC model using the predictors **X** and class labels **Y**.
5. **mdl = fitcecoc(___, Name, Value)** returns an ECOC model with additional options specified by one or more **Name,Value** pair arguments, using any of the previous syntaxes.

For example, specify different binary learners or classifiers, a different coding design, or to cross-validate. It is good practice to cross-validate using the **Kfold Name,Value** pair argument. The cross-validation results determine how well the model is generalized. Some popular Name-Value pairs are shown in Table 6.14.

Among these Name-Value pairs, the **Learners** is used to indicate which binary classification algorithm should be used and it is presented as a template object. One can create a template object based on each related binary algorithm as below:

1. **templateDiscriminant**, for discriminant analysis.
2. **templateEnsemble**, for ensemble learning. You must at least specify the learning method (**Method**), the number of learners (**NLearn**), and the type of learner (**Learners**). You cannot use the AdaBoostM2 ensemble method for binary learning.
3. **templateKernel**, for kernel classification.
4. **templateKNN**, for k-nearest neighbors.
5. **templateLinear**, for linear classification.
6. **templateNaiveBayes**, for naive Bayes.
7. **templateSVM**, for SVM (default).
8. **templateTree**, for classification trees.

For example, to set up a binary linear algorithm as a **Learner**, it looks like:

```
t = templateLinear;
Mdl = fitcecoc(X,Y,'Learners',t,'ObservationsIn','columns');
```

To set up a SVM binary algorithm as a **Linear** with **onevsall** coding design, it looks like:

Table 6.14 Some popular Name-Value pairs used in fitcecoc() method

Name-Value Pair	Description
Coding **Coding design**	'onevsone' (default)|'allpairs'|'binarycomplete'|'denserandom'|'onevsall'|'ordinal' |'sparserandom'|'ternarycomplete'|
Learners **Binary learner** **templates**	'svm' (default) |'discriminant'|'kernel'|'knn'|'linear'|'naivebayes'|'tree'|'ensemble'|
CrossVal **Flag to train cross-** **validated classifier**	Flag to train a cross-validated classifier, specified as the comma-separated pair consisting of 'Crossval' and 'on' or 'off'. If you specify 'on', then the software trains a cross-validated classifier with 10 folds.
CVPartition **Cross-validation** **partition**	Cross-validation partition, specified as a cvpartition object that specifies the type of cross-validation and the indexing for the training and validation sets. To create a cross-validated model, you can specify only one of these four name-value arguments: CVPartition, Holdout, KFold, or Leaveout.
Holdout **Fraction of data for** **holdout validation**	Fraction of the data used for holdout validation, specified as a scalar value in the range [0,1]. If you specify Holdout = p, then the software completes these steps: 1. Randomly select and reserve p*100% of the data as validation data, and train the model using the rest of the data. 2. Store the compact trained model in the **Trained** property of the cross-validated model. To create a cross-validated model, you can specify only one of these four name-value arguments: CVPartition, Holdout, KFold, or Leaveout.
KFold **Number of folds**	Number of folds to use in the cross-validated model, specified as a positive integer value greater than 1. If you specify KFold = k, then the software completes these steps: 1. Randomly partition the data into k sets. 2. For each set, reserve the set as validation data, and train the model using other k − 1 sets. 3. Store the k compact trained models in a k-by-1 cell vector in the Trained property of the cross-validated model. To create a cross-validated model, you can specify only one of these four name-value arguments: CVPartition, Holdout, KFold, or Leaveout.

```
t = templateSVM('Standardize',true);
Mdl = fitcecoc(X,Y,'Learners', t, 'Coding', 'onevsall');
```

In above sections, we provided detailed introductions and discussions about using multiclass classifiers to classify numeric observations. In the following sections, we like to use some multiclass classifiers to classify different image objects. Due to the different features between the numeric data and image objects, first let's have some basic discussions about some special properties for image classification process.

6.7.5   Special Features and Preprocessing for Image Classifications

Unlike classification process for numeric data or predictors, in which the input predictors and output classes provide one-to-one quantitative mapping relation between them, the image classifications need some special processing prior to performing the classification jobs since there is no quantitative relation existed between each image and the corresponding output label. Therefore, we need to extract or collect some features from input image objects and set up a mapping relation between those features of each input image with its related label to enable the classifier to correctly distinguish or classify each of them easily and effectively.

Feature extraction refers to the process of transforming raw data into numerical features that can be processed while preserving the information in the original data set. It yields better results than applying machine learning directly to the raw data. Feature extraction can be accomplished manually or automatically [15]:

1. Manual feature extraction requires identifying and describing the features that are relevant to a given problem and implementing a way to extract those features. In many situations, having a good understanding of the background or domain can help make informed decisions as to which features could be useful. Over decades of research, engineers and scientists have developed feature extraction methods for images, signals, and text. An example of a simple feature is the mean of a window in a signal.
2. Automated feature extraction uses specialized algorithms or deep networks to extract features automatically from signals or images without the need for human intervention. This technique can be very useful when you want to move quickly from raw data to developing machine learning algorithms. Wavelet scattering is an example of automated feature extraction.

In this section, we are more likely to use ECOC model to build some image classifiers to classify different images; thus, we now pay more attention to image feature extraction. For feature extraction from other predictors, such as time domain signals, frequency domain signal data, and audio signals, we will introduce them in the following related sections later.

Feature extraction for image data represents the interesting parts of an image as a compact feature vector. In the past, this was accomplished with specialized feature detection, feature extraction, and feature matching algorithms. Today, deep learning is prevalent in image and video analysis, and has become known for its ability to take raw image data as input, skipping the feature extraction step. Regardless of which approach you take, computer vision applications such as image registration, object detection and classification, and content-based image retrieval all require effective representation of image features—either implicitly by the first layers of a deep network or explicitly applying some of the longstanding image feature extraction techniques.

Feature extraction techniques provided by Computer Vision Toolbox™ and Image Processing Toolbox™ include:

1. Histogram of oriented gradients (HOG)
2. Speeded-up robust features (SURF)
3. Local binary pattern (LBP) features
4. Bag of Features (BOF)

6.7.5.1 The Histogram of Oriented Gradients (HOG) Method

The histogram of oriented gradients (HOG) is a feature descriptor used in computer vision and image processing for the purpose of object detection. The technique counts occurrences of gradient orientation in localized portions of an image [16].

The essential thought behind the histogram of oriented gradients descriptor is that local object appearance and shape within an image can be described by the distribution of intensity gradients or edge directions. The image is divided into small connected regions called cells, and for the pixels within each cell, a histogram of gradient directions is compiled. The descriptor is the concatenation of these histograms.

To get all gradient values, the most common method is to apply the 1-D centered, point discrete derivative mask in one or both of the horizontal and vertical directions. Specifically, this method requires filtering the color or intensity data of the image with the following filter:

$$[-1,0,1]\,\text{and}\,[-1,0,1]^{T} \tag{6.13}$$

The outputs of using this kind of filter are gradient values for selected pixels in each cell, and the size of selected pixels can be 2×2, 4×4, or 8×8, depending on the actual cases. To account for gradients in illumination and contrast, the gradient strengths must be locally normalized, which requires grouping the cells together into larger, spatially connected blocks. The HOG descriptor is then the concatenated vector of the components of the normalized cell histograms from all of the block regions.

Two main block geometries exist: rectangular R-HOG blocks and circular C-HOG blocks. R-HOG blocks are generally square grids, represented by three parameters: the number of cells per block, the number of pixels per cell, and the number of channels per cell histogram.

6.7.5.2 The Speeded-Up Robust Features (SURF) Method

SURF uses square-shaped filters as an approximation of Gaussian smoothing. Filtering the image with a square is much faster if the integral image is used [17]:

$$S(x,y) = \sum_{i=0}^{x}\sum_{j=0}^{y} I(i,j) \tag{6.14}$$

The sum of the original image within a rectangle can be evaluated quickly using the integral image, requiring evaluations at the rectangle's four corners.

SURF uses a blob detector based on the Hessian matrix to find points of interest. The determinant of the Hessian matrix is used as a measure of local change around the point and points are chosen where this determinant is maximal. SURF also uses the determinant of the Hessian for selecting the scale. Given a point $p = (x, y)$ in an image I, the Hessian matrix $H(p, \sigma)$ at point p and scale σ, is:

$$H(p,\sigma) = \begin{pmatrix} L_{xx}(p,\sigma) L_{xy}(p,\sigma) \\ L_{xy}(p,\sigma) L_{yy}(p,\sigma) \end{pmatrix} \tag{6.15}$$

where $L_{xx}(p, \sigma)$ etc. is the convolution of the second-order derivative of Gaussian with the image $I(x, y)$ at the point p.

The box filter of size 9×9 is an approximation of a Gaussian with $\sigma = 1.2$ and represents the lowest level (highest spatial resolution) for blob-response maps.

Interest points can be found at different scales, partly because the search for correspondences often requires comparison images where they are seen at different scales. In other feature detection algorithms, the scale space is usually realized as an image pyramid. Images are repeatedly smoothed with a Gaussian filter, and then they are subsampled to get the next higher level of the pyramid. Therefore, several floors or stairs with various measures of the masks are calculated:

$$\sigma_{approx} = \text{current filter size} \times \left(\frac{\text{base filter scale}}{\text{base filter size}} \right) \tag{6.16}$$

The scale space is divided into a number of octaves, where an octave refers to a series of response maps covering a doubling of scale. In SURF, the lowest level of the scale space is obtained from the output of the 9×9 filters.

The scale spaces in SURF are implemented by applying box filters of different sizes. Accordingly, the scale space is analyzed by up-scaling the filter size rather than iteratively reducing the image size. The output of the above 9×9 filter is considered as the initial scale layer at scale $s = 1.2$ (corresponding to Gaussian derivatives with $\sigma = 1.2$). The following layers are obtained by filtering the image with gradually bigger masks, taking into account the discrete nature of integral images and the specific filter structure. This results in filters of size 9×9, 15×15, 21×21, 27×27,…. Nonmaximum suppression in a $3 \times 3 \times 3$ neighborhood is applied to localize interest points in the image and over scales. The maxima of the determinant of the Hessian matrix are then interpolated in scale and image space with the method proposed by Brown, et al. Scale space interpolation is especially important in this case, as the difference in scale between the first layers of every octave is relatively large.

6.7.5.3   The Local Binary Pattern (LBP) Features Method

The Local Binary Patterns (LBP) is a type of visual descriptor used for objects classification process in computer vision. An LBP feature vector can be created in the following manner [18]:

1. Divide the examined window into cells (e.g., 16x16 pixels for each cell).
2. For each pixel in a cell, compare the pixel to each of its 8 neighbors on its left-top, left-middle, left-bottom, right-top, etc. Follow the pixels along a circle, in either clockwise or counter-clockwise direction.
3. Where the center pixel's value is greater than the neighbor's value, write "0." Otherwise, write "1." This gives an 8-digit binary number that is usually converted to decimal for convenience.
4. Compute the histogram, over the cell, of the frequency of each "number" occurring (i.e., each combination of which pixels are smaller and which are greater than the center). This histogram can be seen as a 256-dimensional feature vector.
5. Optionally normalize the histogram.
6. Concatenate (normalized) histograms of all cells. This gives a feature vector for the entire window.

The feature vector can now be processed using the SVM, extreme learning machines, or some other machine learning algorithm to classify images. Such classifiers can be used for face recognition or texture analysis.

6.7.5.4   The Bag of Features (BOF) Method

In BOF or Bag of Virtual World (BoVW), the image is broken into a set of independent features. The so-called features consist of related key points and descriptors. Key points can be considered as some interest points, such as color, intensity, brighter or darker, and their spatial locations distributed in an image.

Key points are specific points in an image. Descriptors are the values or description of those key points [19]. Then a dictionary or a codebook could be generated based on these descriptors using clustering algorithms. During the training process, all input images will be inspected and checked to see whether our image has words present in the dictionary. If yes, we increase the count of that particular word. Finally, we create the histogram for this image. From this histogram, we can find similar images or predict the category of this image.

Next let's use a real example project to illustrate how to use the ECOC algorithm to classify some image objects with related feature extraction method.

6.8   A Real Project of Classifying Images by Using the ECOC Algorithm

In MATLAB Statistics and Machine Learning Toolbox, it provides a very useful class, **imageCategoryClassifier**, which can be used to perform image classification function. This class contains a linear support vector machine (SVM) classifier trained to recognize an image category. One can use the system function, **trainImageCategoryClassifier()**, to create that **imageCategoryClassifier** object to perform image classification jobs. The function trains a support vector machine (SVM) multiclass classifier using the error-correcting output codes (ECOC) framework.

In fact, that object runs some related function, such as **fitcecoc()** and **fitcsvm()**, to perform a multiclass classification via multiple binary classifications. By running those functions, either regression or classification algorithm can be executed, but the default type is classification.

6.8.1   Build an Image Classification Project by Using the ECOC Algorithm

In this section, we like to use this object to perform classification job to classify some fruit images, which include **Apple**, **Banana**, **Cherry**, **Chestnut**, **Fig**, and **Walnut**. All of those fruit images are provided by [20], and can be downloaded with no charge. The dataset contains more than 2,350 fruit images with six categories, which are included under two folders, **train_data** and **test_data**. You can find this dataset from the Springer ftp site at a folder **Students\Datasets\Fruit Dataset**, and you can also copy this dataset and save it to your machine to use it later.

To create this new project, generate a new Script file and name it as **Fruits_Classifier.m**, and enter the codes shown in Fig. 6.35 into that file. Let's have a closer look at this piece of codes to see how it works.

1. This coding line is used to list all files and folders under the current full path. In our case, it shows all six folders or classes, **Apple**, **Banana**, **Cherry**, **Chestnut**, **Fig**, and **Walnut**, which contained our training data.
2. This line is used to check and list all subfolders, including all subsymbols, such as (.) and (..), in a logic array format. For each actual subfolder, symbol (.) and (..), it returns a logic 1. In our case, it returns a logic vector [1 1 1 1 1 1 1 1], which includes a symbol (.) and (..) with six subfolders, **Apple**, **Banana**, **Cherry**, **Chestnut**, **Fig**, and **Walnut**.
3. The coding line {**imd(isub).name**} is to display the names for all subfolders with the cell format, but all folders' names are transposed to become column cells.
4. This coding line is used to find all symbols, including (.) and (..), and assign them with blank array []. In other words, the function of this line is to list the names for all six subfolders without any subsymbols (.) and (..).
5. The function **fullfile()** is used to set up a directory by concatenating all paths together, and all paths are separated by comma (,). After this function is done, all six subfolders or six image classes are set up with a full path.
6. The function **imageDatastore()** is a powerful function and it can be used to set up storage to save all images for all six image classes based on the source, **setDir**. Some **Name-Value** pairs can be involved to indicate what kinds of additional options can be adopted with this function. Here the **LabelSource** is to list all **Labels** for all images under six classes, **foldernames** is to show up all

```
% File Name: Fruits_Classifier.m
% Purpose: The complete training codes to train a SVM model with fruit images for 6 classes.
% Inputs: All train data at C:\Artificial Intelligence Book\Students\Datasets\Fruit Dataset\train_data folder.
% Input images classes: 6
% Outputs: Trained SVM model named categoryClassfier.
% Mar 1, 2024
 1  imd = dir('C:\Artificial Intelligence Book\Students\Datasets\Fruit Dataset\train_data\');   % setup the
    source image folder
 2  isub = [imd(:).isdir];                                  % returns logical vector
 3  imgFolds = {imd(isub).name}';
 4  imgFolds(ismember(imgFolds,{'.','..'})) = [];           % get 6 sub folders excluding the . and .. operators.
 5  setDir = fullfile('C:', 'Artificial Intelligence Book', 'Students', 'Datasets', 'Fruit Dataset', 'train_data', imgFolds);
 6  imds = imageDatastore(setDir, 'LabelSource', 'foldernames', 'FileExtensions', {'.jpg'});

    % show all images and labels
 7  imds.Labels
    imds.Files

    % setup training parameters: 70% for training & 30% for validations
 8  numTrain = 0.7;         % 70% car images are used for training, and 30% images are for validations
 9  [trainSet, testSet] = splitEachLabel(imds, numTrain);

    % Create bag of visual words.
10  bag = bagOfFeatures(trainSet);

    % Train a classifier with the training sets.
11  categoryClassifier = trainImageCategoryClassifier(trainSet, bag);

    % Evaluate the classifier using test images. Display the confusion matrix
12  confMatrix = evaluate(categoryClassifier, testSet)

    % Get the average accuracy of the classification.
13  mean(diag(confMatrix))

    % save trained model to Workspace and folder (not necessary here)
14  cd 'C:\Artificial Intelligence Book\Students\Class Projects\Chapter 7\Multiclass Classification Project\'
    save categoryClassifier;
```

Fig. 6.35 The codes used to classify images with ECOC algorithm

classes' names, and **FileExtensions**, **{'.jpg'}** is to select all image files with **.jpg** extension. After this function is executed, all our images are organized under different folders. The folder **Files** contained all images (**.jpg**) categorized under each class's name. The folder **Labels** contained all labels for each image, one image with one related label.

7. To get a direct and clear picture about these folders, we display these two folders with their contents.

8. We try to use 70% training data as our actual training image data and 30% of them as the validation data.

9. The function **splitEachLabel**() is to split training images into two parts, **trainSet** and **testSet**.

10. To set up some features for all training image data, we select the Bag of Feature method to extract useful features from input images.

11. Now we can call the function **trainImageCatagoryClassifier**() with **trainSet** and **bag** feature as arguments to perform classification process for training images. The trained SVM model is returned and assigned to the local variable **categoryClassifier**.

12. To check and evaluate our trained model, the function **evaluate**() is called with our model and testing image data, **testSet**, as arguments. The evaluation results are returned and assigned to another local variable **confMatrix** to be displayed in the Command window.

```
PREDICTED
KNOWN      | Apple  Banana  Cherry  Chestnut  Fig   Walnut
------------------------------------------------------------
Apple      | 0.97    0.00    0.03    0.00     0.00   0.00
Banana     | 0.00    1.00    0.00    0.00     0.00   0.00
Cherry     | 0.01    0.00    0.79    0.20     0.00   0.00
Chestnut   | 0.00    0.00    0.00    1.00     0.00   0.00
Fig        | 0.00    0.00    0.00    0.02     0.97   0.01
Walnut     | 0.00    0.00    0.00    0.00     0.00   1.00

* Average Accuracy is 0.96.

confMatrix =

   0.9730        0   0.0270        0        0        0
        0   1.0000        0        0        0        0
   0.0068        0   0.7905   0.2027        0        0
        0        0        0   1.0000        0        0
        0        0        0   0.0190   0.9668   0.0142
        0        0        0        0        0   1.0000

ans =

   0.9551
```

Fig. 6.36 The running results of ECOC classification algorithm

13. The mean value of the classification accuracy is obtained and displayed in the Command window, too.
14. Finally, the trained model, **categoryClassifier**, can be saved to any folder and it can be used later. In fact, this step is unnecessary if you want to use that model immediately since it has been reserved into the Workspace automatically as it is generated. However, if you want to use this model later without running this project first, this step is necessary and you can directly load this model later as you like to use it at any time.

Now let's run this project and the running results are shown in Fig. 6.36.

It can be found from the running results that the classification accuracy is relatively high (0.96), and most classifiers, except the **Cherry**, have about 100% correct classification results shown in the confusion chart.

Next let's build another project to evaluate the trained model **categoryClassifier** from the above project to check its classification performances.

6.8.2   Build an Evaluation Project to Validate the Trained Model categoryClassifier

Create a new Script file and name it as **Fruits_Eval.m** and enter the codes shown in Fig. 6.37 into that file. Let's have a closer look at that piece of codes to see how it works.

1. To accumulate and track the total number of correct and incorrect image evaluation results, two local variables, **numTrue** and **numFalse**, are generated and initialized first.

```matlab
% File Name: Fruits_Eval.m
% Purpose: Evaluate trained model - categoryClassifier, with test data.
% All test data is in C:\Artificial Intelligence Book\Students\Datasets\Fruit Dataset\test_data folder.
% Input: The trained model - categoryClassifier - is in Workspace. To do it, run the Fruits_Classifier.m first.
% Outputs: Evaluation results.
% Note: Prior to running this project, run the project Fruits_Classifier.m first to get the trained model.
% Mar 1, 2024
1   numTrue = 0;                % number of testing result who is true
    numFalse = 0;               % number of testing result who is false

    % Apply the newly trained classifier to categorize new images.
2   M = 6;         % Number of classes - 6
    N = 80;        % Number of test data, 80 for each class, with 6 classes, totally 480 test data.
3   for i = 1:M   % source folders to get test images
4     path = "C:\Artificial Intelligence Book\Students\Datasets\Fruit Dataset\test_data\" + imgFolds(i);

5     fp = fullfile(path, '*.jpg');
6     f = dir(fp);
7     imgName = {f.name};

8     for k = 1:N
9       fullFileName = fullfile(path, imgName{k});
10      cImage{k} = imread(fullFileName);
11      [labelIdx, score] = predict(categoryClassifier, cImage{k});
12      YPred = categoryClassifier.Labels(labelIdx);
13      fruit_class = imgFolds(i);
14      fruit_result = YPred;
15      result = ['Fruit Class =: ', fruit_class, 'Fruit Result =: ', fruit_result];
16      %disp(result);
17      if cellfun(@isequal, fruit_class, fruit_result)
          numTrue = numTrue + 1;
18      else
          numFalse = numFalse + 1;
        end
      end
19    final = [newline, 'Matched Fruit =: ', num2str(numTrue), ' Unmatched Fruit =: ', num2str(numFalse)];
20    disp(final);
    end
```

Fig. 6.37 The codes used to evaluate the trained model

2. Some local variables, such as the number of classes and number of test data, are declared here.
3. A **for**() loop is used to repeatedly classify each test image and compare it with the class label to check the correctness of the trained model for all six classes.
4. A full path is created first to point to all images based on each class folder. One point to be noted is that each class folder, **imgFolds(i)**, is the name of each class folder and you must run the project **Fruits_Classifier.m** first to get this folder in the Workspace; otherwise, you may get a running error if you run this project without running that one first.
5. Now we need to get each image file by concatenating the full path and **.jpg** extension together and assign it to a *pointer* variable **fp**.
6. All files under the current path are listed by using the **dir**() function.
7. This coding line is used to get all images under each class folder based on the folder name, such as **Apple**, **Banana**, **Cherry**, **Chestnut**, **Fig**, and **Walnut**. Each **imgName** contained 80 test images for each class as the cell format.
8. The second **for**() loop is used to pick up each image under the related class folder, and totally we need to collect 80 images at each class to predict its label by using the trained model.

9. The function **fullfile()** is used to concatenate the path and the selected image file together to provide a complete image file.
10. The selected testing image is read out and assigned to a cell array **cImage{k}**.
11. The **predict()** function is executed with the trained model and the testing image as arguments to classify the label for the testing image. The classified label index is returned and assigned to a local variable **labelIdx** that will be used later.
12. The classified label is derived based on the **Labels()** property of the trained model **categoryClassifier** with the label index **labelIdx** obtained from the last step, and assigned to a local variable **YPred**.
13. To compare between the predicted label and the testing label, the name of the current class folder **imgFolds(i)** is assigned to another local variable **fruit_class**.
14. The classified label **YPred** is assigned to a local variable **fruit_result**.
15. To display the comparison result, a string array **result[]** is generated to display the current class name and the predicted label.
16. The function **disp()** is used to display that comparison result to indicate whether the classification result is correct or not. But we commented it out since we do not need to see the detailed classification result unless if we like to track and check which classifying result is wrong exactly.
17. To count how many correct and incorrect classification results were made by the model, we need to compare the predicted label and the name of the current class folder. However, you cannot use the built-in equal operator (=) to do that comparison, instead you must use a function, **cellfun()** with the **@isequal** argument, to perform this comparison. That function returns a true (1) if both labels are equal, otherwise a false (0) is returned. If a true is returned, the **numTrue** is increased by 1 to indicate that a correct classification is performed.
18. Otherwise, a mismatched label is found and the **numFalse** is increased by 1.
19. Finally another string array **final[]** is generated to be used to display the total number of correct and incorrect classification results. The argument newline is used to get a new line to display this result.
20. The **disp()** function is called to do that displaying.

Prior to running this project, make sure that our previous project, **Fruits_Classifier.m**, has been run at least one time to enable some components, such as the trained model **categoryClassifier** and the cell variable **imgFolds**, to be exported to the Workspace already since we need them to run this project.

Now run this evaluation project and the running result is shown in Command window. It can be found that our classification performance is perfect with a 100% correct classifying rate.

Next let's go a little deeper to build a case study project to illustrate how to get and use HOG features to perform some classification jobs for this Fruit dataset.

6.8.3 A Case Study of Using the HOG Features to Classify Fruits Images

In the last project, we used a bag of features method to extract all features from all images under six classes, and perform both image classifying and evaluation processes. In the following project, we like to use another feature extraction method, histogram of oriented gradients (HOG), to get useful features and use those features to perform image classification and validation process.

Create a new Script file and name it as **HOG_Fruits_Classifier.m** and enter the codes shown in Fig. 6.38 into that file. Let's have a closer look at this piece of codes to see how it works.

1. This piece of codes on the top is identical to that we show in Fig. 6.35, and they are used to collect all image files and related labels.
2. To get a clear picture about all collected image files and labels, we display them here. The **imds.Files** contained all images' names and **imds.Labels** included all related labels.
3. In order to use a part of training image data as the testing image data, the function **splitEachLabel()** is utilized to set 70% of training images as training data and 30% of training images as the testing data.
4. To extract some useful features from training image data, we need first to read a training image. This image selection process could be randomly made and we select any training image. In our case, we just select the 300th image from the training images. The actual number of the training images is 2352 after splitting 3360 training images with a 70% rate. The reason we do this step is to get the length of the returned feature vector.
5. The function **extractHOGFeatures()** is used to extract all useful features from the selected image. The cell size of a featured image should be defined as 2D and measured by pixels. Some popular cell sizes could be [2 × 2], [4 × 4], or [8 × 8]. To make a trade-off between the resolution and memory space, we prefer to use the [4 × 4] as our cell size. This function returns two feature results, the feature vector that is composed of all feature values and feature visualization that is an object. In our case, the **hog_4 × 4** is a 1 × 20,736 vector.
6. This coding line is used to retrieve the length of returned feature vector, which is 20,736.
7. This line is used to get the total number of the actual training images, which is 2352 after the splitting process.
8. Next we need to declare a training feature array, **trainingFeatures[]**, with the total number of actual training images, the length of returned feature vector, and the data type of the returned feature values. This vector is initialized to 0 with **zeros()** function.
9. With steps 9 and 10, a **for()** loop is utilized to read all images one by one, extract features for each of them, and assign them to the training feature array defined in the last step.

```matlab
% File Name: HOG_Fruits_Classifier.m
% Purpose: Training & evaluating a SVM model with fitcecoc() function for fruit images in 6 classes.
% Inputs: All train data at C:\Artificial Intelligence Book\Students\Datasets\Fruit Dataset\train_data folder.
%         All test data is in C:\Artificial Intelligence Book\Students\Datasets\Fruit Dataset\test_data folder.
% Input images classes: 6
% Outputs: Tested results.
% Mar 1, 2024

1  imd = dir('C:\Artificial Intelligence Book\Students\Datasets\Fruit Dataset\train_data\');   % set source.
   isub = [imd(:).isdir];                                      % returns logical vector
   imgFolds = {imd(isub).name}';
   imgFolds(ismember(imgFolds,{'.','..'})) = [];               % get 6 sub folders excluding the . and .. operators.

   setDir = fullfile('C:', 'Artificial Intelligence Book', 'Students', 'Datasets', 'Fruit Dataset', 'train_data', imgFolds);
   imds = imageDatastore(setDir, 'LabelSource', 'foldernames', 'FileExtensions', {'.jpg'});

   % show all images and labels
2  imds.Labels
   imds.Files

   % setup training parameters and CNN model: 70% for training & 30% for validations
3  numTrain = 0.7;          % 70% car images are used for training, and 30% images are for validations
   [trainSet, testSet] = splitEachLabel(imds, numTrain);

   % Extract HOG features & HOG visualization
4  img = readimage(trainSet, 300);
5  [hog_4x4, vis4x4] = extractHOGFeatures(img, 'CellSize', [4 4]);
6  hogFeatureSize = length(hog_4x4);
7  numImages = numel(trainSet.Files);
8  trainFeatures = zeros(numImages, hogFeatureSize, 'single');

9  for m = 1:numImages
      img = readimage(trainSet, m);
10    trainFeatures(m, :) = extractHOGFeatures(img, 'CellSize', [4 4]);
   end

   % Get labels for each image.
11 trainLabels = trainSet.Labels;

   % fitcecoc() uses SVM learners and a 'One-vs-One' encoding scheme.
12 classifier = fitcecoc(trainFeatures, trainLabels);

   % Extract HOG features from the test set. The procedure is similar to get HOG from trainSet above
13 numImages = numel(testSet.Files);
   testFeatures = zeros(numImages, hogFeatureSize, 'single');

14 for n = 1:numImages
      img = readimage(testSet, n);
      testFeatures(n, :) = extractHOGFeatures(img, 'CellSize', [4 4]);
   end
   % Make class predictions using the test features.
15 predictedLabels = predict(classifier, testFeatures);

   % Get labels for each image.
16 testLabels = testSet.Labels;

   % Tabulate the results using a confusion matrix.
17 confMat = confusionmat(testLabels, predictedLabels);
18 helperDisplayConfusionMatrix(confMat);            % call function to format and display confusion matrix
```

Fig. 6.38 The codes used to classify images with the HOG method

11. In order to perform the training process, we get all training labels for all training images.

12. Then we call the function **fitcecoc**() with the **trainFeatures** and **trainLabels** as arguments to perform the training process. The trained model is returned and assigned to a local variable **classifier**.

13. The function of the coding lines 13 and 14 is to extract features for all testing image data, and the returned feature array or vector is assigned to a variable **testFeatures[]**.

15. The function **predict()** is executed to predict the labels for the testing images based on the trained model **classifier** obtained in step 12. The predicted labels are assigned to a local variable **predictedLabels**.
16. This coding line is used to get all actual labels for all testing images, and assign them to a local variable **testLabels**.
17. The function **confusionmat()** is called to tabulate the classifying results in a confusion matrix format.
18. A user-defined function, **helperDisplayConfusionMatrix()**, whose detailed codes will be discussed later, is called to display the classifying results in a professional way.

Now let's generate the codes for the user-defined function **helperDisplayConfusionMatrix()**. Just attach the codes shown in Fig. 6.39 to the end of the codes shown in Fig. 6.38. Let's have a closer look at this piece of codes to see how it works.

1. The prototype of the user-defined function **helperDisplayConfusionMatrix()** is declared first. There is no data type defined in front of the function name since no variable would be returned from this function. The only argument is a confusion matrix.
2. The function **bsxfun()** is a special function and its prototype is: **C = bsxfun(@ fun, A, B)**, and it can be used to apply the element-wise binary operation specified by the function handle **fun** to arrays **A** and **B**. Here the function handle @ **rdivide** indicates to divide the input **confMat** that is a 6×6 confusion matrix by the sum of each row in that matrix to get normalized classification results.
3. Since we have six (6) classes, thus the class number is declared here.
4. The function **arrayfun()** is used to apply a function indicated by @ to each element in that array. Here we use this function to set up titles or header for six confusion columns.
5. The function **repmat()** can be used to define a format to repeat 1 and 11 times with 9 letters (**%-9 s**) aligned to left. The purpose of this function is to define the printing format for all confusion values.
6. The header or the title for the confusion matrix table is displayed first.

```matlab
1  function helperDisplayConfusionMatrix(confMat)     % Display the confusion matrix in a formatted table.

   % Convert confusion matrix into percentage form
2  confMat = bsxfun(@rdivide, confMat, sum(confMat, 2));

3  digits = '1':'6';
4  colHeadings = arrayfun(@(x)sprintf('%d',x),1:6,'UniformOutput',false);
5  format = repmat('%-9s', 1, 11);
6  header = sprintf(format,'digit  |',colHeadings{:});
   fprintf('\n%s\n%s\n',header,repmat('-',size(header)));
7  for idx = 1:numel(digits)
       fprintf('%-9s', [digits(idx) '   |']);
       fprintf('%-9.2f', confMat(idx,:));
       fprintf('\n')
   end
   end
```

Fig. 6.39 The codes for the user-defined function helperDisplayConfusionMatrix()

7. A **for**() loop is used to repeatedly print out all values on the confusion matrix one
   by one in a width of 9 letters aligned to the left.

Now let's run this project and the running result is shown below.

```
digit  | 1          2          3          4          5          6

1      | 0.61      0.00       0.00       0.00       0.39       0.00
2      | 0.00      1.00       0.00       0.00       0.00       0.00
3      | 0.00      0.00       0.99       0.01       0.00       0.00
4      | 0.00      0.00       0.00       1.00       0.00       0.00
5      | 0.00      0.00       0.00       0.46       0.54       0.00
6      | 0.00      0.00       0.00       0.00       0.00       1.00
```

It can be found from this result that the classification results are pretty good.

A completed project file **HOG_Fruits_Classifier.m** can be found from the
Springer ftp site in a folder **Students\Class Projects\Chapter 6\Multiclass
Classification Project**.

6.9   Multiclass Classification Function Example for Audio Sounds Classifications

In this section, we like to discuss how to use multiclass classification algorithms to
classify some audio signals, including digit sounds from 0 to 9 and some popular
animal sounds, including chicken, dog, lion, and sheep.

Similar to image classification processing, to classify any audio signal, some
interesting features related to those audio signals must be extracted first as the classification criteria. To extract features for a piece of audio signal, we must break
down the audio file into a sequence of windows, often between 20 and 100 millisec-
onds, with some intervals or overlaps between those windows. The overlap duration
must be less than the window duration. We then extract these features per window
and can run a classification algorithm for example on each window [21].

In general, all audio signals can broadly be categorized as stationary or nonstationary. Stationary signals have spectrums that do not change over time, like pure
tones. Nonstationary signals have spectrums that change over time, like speech signals. To make machine learning-based tasks easy and tractable, nonstationary signals can be approximated as stationary when analyzed at appropriately small time
scales. Generally, speech signals are considered stationary when viewed at time
scales around 30 ms. Therefore, speech can be characterized by extracting features
from 30 ms analysis windows over time. Figure 6.40 shows an example of converting the audio signals to window-based files.

All our routine audio signals belong to analog signals, which means that the
amplitudes of audio signals are continuous function over time. However, sometimes

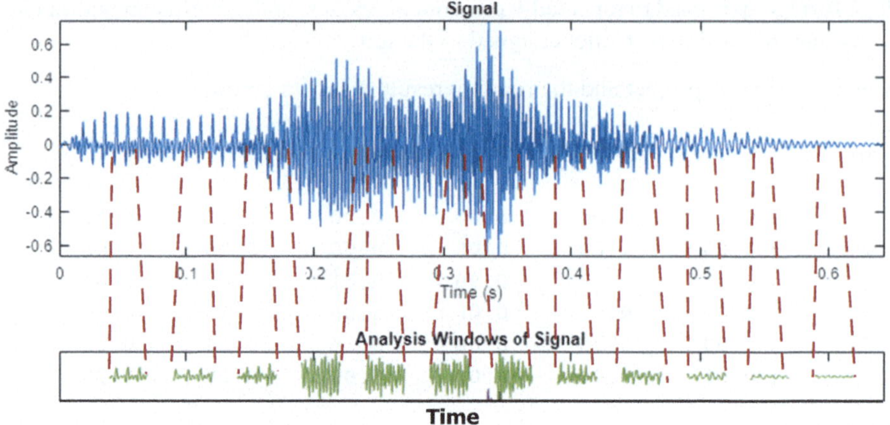

Fig. 6.40 An illustration of converting audio files to window-based files

it is not easy to analyze audio signals in the time domain directly, and it needs to be converted to the frequency domain to form a collection of audio signals over frequency, which is called a spectrum. Some popular algorithms used to perform that conversion include the Fourier Transformation (FT), Fast Fourier Transformation (FFT), Short-Time Fourier Transformation (STFT), and Discrete Fourier Transformation (DFT).

In addition to those conversion algorithms, some other popular algorithms are more popularly used in audio signal processing, including the Mel-Frequency Cepstral Coefficients (MFCCs) and Band Energy Ratio (BER).

Let's have a closer look at these two algorithms and see how they are implemented in audio signal processing.

6.9.1   What Is Mel-Frequency Cepstral Coefficients (MFCCs)

In sound processing, the **Mel-Frequency Cepstrum** (MFC) is a representation of the short-term power spectrum of a sound, based on a linear cosine transform of a log power spectrum on a nonlinear mel-scale of frequency [22].

Regularly, we humans perceive sound logarithmically, which means that we are better at detecting differences in lower frequencies than higher frequencies. For example, we can easily tell the difference between 500 and 1000 Hz, but we will hardly be able to tell the difference between 10,000 and 10,500 Hz, even though the distance between the two pairs is the same. Hence, the **mel scale** was introduced. It is a logarithmic scale based on the principle that equal distances on the scale have the same *perceptual* distance [23].

Conversion from frequency (f) to mel scale (m) is given by

$$m = 2595 \times \log\left(\frac{f}{500}\right)$$

$$(6.17)$$

A melspectrogram is therefore a spectrogram where the frequencies are converted to the mel scale. The information of the rate of change in spectral bands of a signal is given by its cepstrum. A cepstrum is basically a spectrum of the log of the spectrum of the time signal. The resulting spectrum is neither in the frequency domain nor in the time domain, and hence, it was named the **quefrency** (an anagram of the word *frequency*) domain. The Mel-Frequency Cepstral Coefficients (MFCCs) are nothing but the coefficients that make up the mel-frequency cepstrum.

6.9.2 *What Is Band Energy Ratio (BER)*

The Band Energy Ratio is *a ratio of power between defined frequency bands, such as* between the lower and higher frequency bands. Band energy ratio is a relatively common measure, proposed to measure oscillatory or periodic, activity. They are calculated as [24]:

$$BR = \frac{average\left(lowerbandpower\right)}{average\left(higherbandpower\right)}$$

$$(6.18)$$

It can be thought of as the measure of how dominant lower frequencies are. This feature has been extensively used in music-speech discrimination and music classifications.

6.9.3 *MATLAB Classes and Functions Used for Audio Signal Classifications*

MATLAB provides various classes and functions to support audio signal classifications, and these classes and functions are involved in Deep Learning Toolbox™ and Audio Toolbox™, respectively. Among them, the most popular App, classes, and functions are listed in Table 6.15.

The **classifySound()** function is used to classify some unknown audio signals based on a trained audio model **YAMNet**, and the types of those unknown audio signals should be included in the training audio signals for the trained model. Otherwise, you cannot use this function to perform any classification for other types of audio signals.

The **audioFeatureExtractor** is a class and you need to create an object based on this class to collect and encapsulate all input audio features, and then reorganize them into a sequence of streamlined cells to make them easily be picked and selected by users to perform feature selection functions.

Table 6.15 Some popular classes and functions used for audio signal classifications

App/Class/Function	Description
Signal Labeler App	Used to label signal attributes, regions, and points of interest.
audioFeatureExtractor Class	Encapsulate multiple audio feature extractors into a streamlined and modular implementation.
classifySound() Function	Classify audio signals with a pre-trained model **YAMNet**.
read() Function	Read data from a dataset. [data, info] = read(ds);
audioDatastore Class	Use an **audioDatastore** object to manage a collection of audio files, where each individual audio file fits in memory, but the entire collection of audio files does not necessarily fit.
audioread() Function	Read an audio file from an audio signal. [x, info] = audioread(file_name); The function returns an audio file in values with related information.
extract() Function	Extract all selected audio features based on the features selected by the object audioFeatureExtractor.
predict() Function	Predict the class of the tested audio signal based on the trained audio model.
splitEachLabel() Function	Splits datastore according to specified label proportions. [adsTrain, adsTest] = splitEachLabel(ads, 0.85); Split whole audio signal ads into 2 parts in 85%.

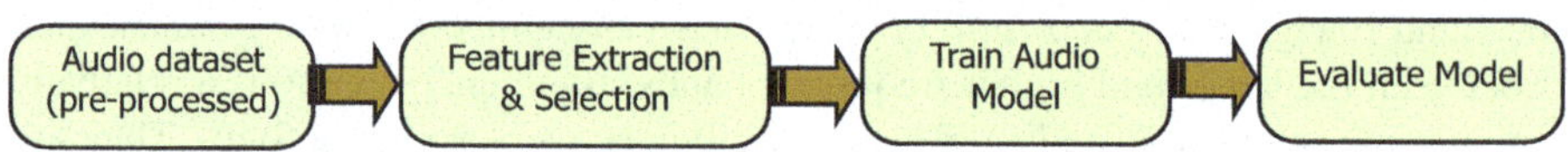

Fig. 6.41 A block diagram of building an audio signal classification project

The **Signal Labeler** is an App, and it is used to assist users to set up and initially configure the input audio dataset, such as selecting interest points for features, setting up the regions for detected audio signals, and mapping related labels to the input audio signals. All of those operations can be performed via some GUIs without any coding process.

The **audioDatastore** is a useful class, especially for multiple audio signal processing. It works like an organized container to set up all input audio signals and audio files to a sequence of list-like module with mapped labels. By using this kind of module or store, you do not need to separate input audio signals into the training data and testing data, and the related function, such as **SplitEachLabel**(), can do that job for you by calling it. Another important advantage of using this module is that some functions can be used to access the module to separate each group or class audio signals based on the folders where each group of audio signals is located, and the related folders can be mapped to the labels for those related groups.

The **extract**() function is used to extract all selected features defined by the class **audioFeatureExtractor**. To do that, an **audioFeatureExtractor** object must be generated first with the selected features. Totally there are 25 features that can be selected for a single audio signal. When an **audioFeatureExtractor** object is first generated, all features are disabled and you need to set up each feature to true to enable it in your program.

A functional block diagram used to perform audio signal classification with MATLAB is shown in Fig. 6.41. In some projects, the feature extraction and feature selection are divided into two steps, but here we combined them into one block.

Next let's build some real projects to use multiclass classification algorithms to classify some audio signals. First let's develop a simple audio classification project to classify digit sounds.

6.9.4   Build Audio Classification Project to Classify Digit Sounds

In this project, the data source we need to use is a Free Spoken Digit Dataset (FSDD) [25]. This dataset can be downloaded from the site in [25] and is free to use with no charge. The dataset contained 10 digit sounds from 0 to 9, and each digit sound includes 200 audio files. We have installed this dataset in our default folder, **C:\ Artificial Intelligence Book\Students\Datasets\FSDD**. A subfolder **recording** under the folder **FSDD** folder controlled all those audio digit files. You can copy and paste it under a folder in your machine if you like.

Now create a new Script named it as **Digit_Audio_Classification.m** and enter the codes shown in Fig. 6.42 into that file. Let's have a closer look at this piece of codes to see how it works.

1. First the path where our audio dataset is located is assigned to a local variable **dataset**. You may need to use your actual path to replace this if you stored this dataset at different folder on your machine.
2. A new object **ads** is generated based on the **audioDatastore** class to collect and manage the **dataset** with all subfolders under that path.
3. A MATLAB function **fileparts**() is used to get all audio signal files' names, such as **'0_jackson_0.wav'**, **'0_jackson_2.wav'**, … and so on. In fact, this function can be used to get the file path, file name, and file extension. Here we only need to get the file names.
4. Each digit audio file contained one digit sound with an extension of **'wav'**, such as **'0_jackson_1.wav'** for digit 1 tone, **'2_jackson_10.wav'** for digit 2 tone. The first digit before the underscore "_" can be considered as a *label* used to indicate its digit. To get all labels for all digit sounds, a function **extractBefore**() is used to get all of those numbers as labels before the first underscore "_", and assign them to the **Labels** property of the object **ads**, or **ads.Labels**.
5. The function **splitEachLabel**() is used to split the datastore where all audio files are stored into two parts in 85% rate, which means that the first 85% audio files are assigned as the training data to the **adsTrain** variable, and the second 15% audio data are assigned as the testing data to the **adsTest** matrix.
6. Now we want to get a sample of audio file to sound and display it to give us a piece of feeling about what sounds like and what looks like for an audio signal. To do that, a **read**() function is used to get the first training audio file, **'0_jackson_0.wav'**. This function returns two results: **x** contained the actual values for that audio signal, which is a 2860 × 1 double array, and **xinfo** is a struct contain-

```matlab
% Name: Digit_Audio_Classification.m
% Input: Digit 0 - 9 audio signals located at folder - C:\Artificial Intelligence Book\Students\Datasets\FSDD\
% Output: Classified digit sounds result - confusion table
% March 14, 2024
1   dataset = 'C:\Artificial Intelligence Book\Students\Datasets\FSDD\';
2   ads = audioDatastore(dataset, IncludeSubfolders=true);
3   [~, filenames] = fileparts(ads.Files);
4   ads.Labels = extractBefore(filenames, "_");

5   [adsTrain, adsTest] = splitEachLabel(ads, 0.85);
6   [x, xinfo] = read(adsTrain);
7   sound(x, xinfo.SampleRate)

8   t = (0:numel(x)-1)/xinfo.SampleRate;
    figure
    plot(t, x)
    title("Label: " + xinfo.Label)
    grid on
    axis tight
    ylabel("Amplitude")
    xlabel("Time (s)")

9   afe = audioFeatureExtractor(SampleRate=xinfo.SampleRate, ...
        Window=hann(round(0.03*xinfo.SampleRate), "periodic"), ...
        OverlapLength=round(0.02*xinfo.SampleRate));

    % Set all feature extractors to true to enable all of them
10  [idx, params] = info(afe, "all");
11  cfeatures = fieldnames(idx);
12  for i = 1:numel(cfeatures)
        afe.(cfeatures{i}) = true;
    end

13  features = extract(afe, adsTrain);
14  N = cellfun(@(x)size(x, 1), features);
15  T = repelem((adsTrain.Labels), N);
16  X = cat(1, features{:});
17  [featureSelectionIdx, featureSelectionScores] = fscmrmr(X, T);

    % Training the model with fitcknn() function
18  numFeatures = 30;
    selectedFeatureIndex = featureSelectionIdx(1:numFeatures);
19  Mdl = fitcknn(X(:, selectedFeatureIndex), T, Standardize=true);

    % Evaluate trained model. Read a sample from the test set.
20  [x, xInfo] = read(adsTest);
    sound(x, xInfo.SampleRate)

    % Extract features from analysis windows.
21  yPerWindow = extract(afe, x);

    % Predict the correct label per window.
22  t = predict(Mdl, yPerWindow(:, selectedFeatureIndex));
23  trueLabel = categorical(xInfo.Label)
24  predictionsPerWindow = categorical(t')

    % Create a file-level prediction by taking the mode of window-level predictions.
25  prediction = mode(predictionsPerWindow)

    % Analyze the whole-word performance over the entire test set.
26  Tfile = categorical(adsTest.Labels);
27  featuresTest = extract(afe, adsTest, UseParallel=canUseParallelPool);
28  Y = cellfun(@(x)mode(categorical(predict(Mdl, x(:,selectedFeatureIndex)))), featuresTest, ...
        UniformOutput=false);
29  Y = cat(1, Y{:});

    figure
30  confusionchart(Tfile,Y,Title="Accuracy = " + 100*mean(Tfile==Y) + " (%)")
```

Fig. 6.42   The codes used to classify digit sounds (0~9)

ing all related properties for that audio file, such as the sampling rate, audio file name, and label.

7. The function **sound**() is executed with the actual values of selected audio file and the sampling frequency as arguments to play the sound or tone for digit 0.

8. The following codes are used to display this audio signal in the time domain with the specified time axis, title, and displaying labels by calling the **figure** command.

9. Then a new audioFeatureExtractor object **afe** is generated to prepare to extract all useful or interesting features for all audio files. Two typical parameters, the window width or duration and the overlaps duration between windows. The former is defined as 30 ms and the latter is set to 20 ms. Both of them are multiplied by sampling rate. The *periodic* property indicates that these windows and overlaps are periodically adopted for each audio file.

10. As mentioned before, as a new **audioFeatureExtractor** object is created, all feature extractors are disabled by default. Thus, to enable all feature extractors, we need to get all feature extractors' names and their active status. For that purpose, the function **info()** is used with the object and the "**all**" property as arguments. The returned feature extractors' names are involved in the returned **idx** struct, and the **params** contains all feature extractors with settable parameters. The latter should contain all blank values by default.

11. Then the **fieldnames()** function is utilized to extract names for all extractors that are formatted as structure array, and are assigned to the **cfeatures** that is a cell array.

12. Then a **for()** loop is used to sequentially set all feature extractors to **true** to enable all of them. A { } symbol is used to select each extractor's name since it is a cell array.

13. Next the function **extract()** is executed to get features from all files in the **audioDatastore**. The output is a (*Number of files*)-by-1 cell array. Each element on the cell array is a (*Number of hops*)-by-(*Number of features*) matrix, where the number of hops depends on the length of the audio file. In this case, there are 306 features and the hops numbers are varied between 13 and 216.

14. To train a machine learning model on window-level features, replicate the file-level labels so that they are in one-to-one correspondence with the features. To do that, we need first to get the total number of labels or total number of audio files. The **features** is a 1700 × 1 cell array, which means that the total files are 1700. Therefore, we need to generate 1700 labels. The function **cellfun()** is used to apply the **size()** to all features.

15. Then the function **repelem()** is used to duplicate 1700 labels and assign them to **T**.

16. The function **cat()** is used to concatenate the features into a single matrix for consumption by machine-learning tools later.

17. Now we need to select the best feature selector. This best feature selector will depend on the intended model. Use **fscmrmr()** function to rank features for classification by using the minimum-redundancy/maximum-relevance (MRMR) algorithm. The MRMR is a sequential algorithm that finds an optimal set of features that is mutually and maximally dissimilar and can represent the response variable effectively. In our case, we pay more attention to the **feature-SelectionIdx**, which determines the elements we need to use from the features.

18. The number of features is defined based on estimations, and the elements in the features are indicated by the **featureSelectionIdx** array with number of features.
19. The function **fitcknn()** is used to train the audio model with the KNN algorithm in a format of **fitcknn(X, Y, Name-Value)**. The Name-Value pair **Standardize = true** is to ask software to center and scale each column of the predictor data **X** by the column mean and standard deviation.
20. After the model is trained, we need to evaluate that model. To do that, first we need to call the **read()** function to get the testing data, exactly to get the testing data stored in a dataset, **adsTest**. The returned audio sound data values are involved in the **x**, and related information, such as the sampling rate, audio file names, and labels, is included in the **xinfo** structure. The sound of the selected testing audio file, it should be a digit 0, is played to confirm its correctness (by default, the audio sound is the first digit sound).
21. To predict the testing audio sound, the **extract()** function is used to get all features for the testing audio sound.
22. Then the function **predict()** is executed based on the trained model **Mdl**, the extracted features from the testing audio sound indicated by the **selectedFeatureIndex**.
23. The actual label is defined based on the real labels of the testing audio sound.
24. The predicted label **t** is a cell column array, and it needs to be converted to a row category array by using the function **categorical()** with the transposed **t**.
25. The function **mode()** is used to return the most frequent value from the predicted labels. In other words, it counts the number of predicted labels and selects the max number of predicted labels.
26. The coding lines in 26 through to 28 are used to analyze and display the entire performance based on all testing audio data.
29. The function **cat()** is used to arrange the predicted labels in a diagonal format.
30. Finally, the entire testing result is displayed in a confusion chart.

Now run the project, and the running result is shown below on the Command window.

```
trueLabel = categorical 0
predictionsPerWindow = 1×33 categorical array
  Columns 1 through 11
   8   3   0   0   0   0   0   3   0   0   0
  Columns 12 through 22
   0   0   0   0   0   0   0   0   0   0   0
  Columns 23 through 33
   0   0   0   0   0   0   0   0   0   0   0
prediction = categorical 0
```

The displayed digit_0 audio sound and the confusion chart are shown in Figs. 6.43a, b.

It can be found from the running result that the true label is digit_0, and the predicted result **predictionPerWindow** shows that among 33 predicted labels, 30 of

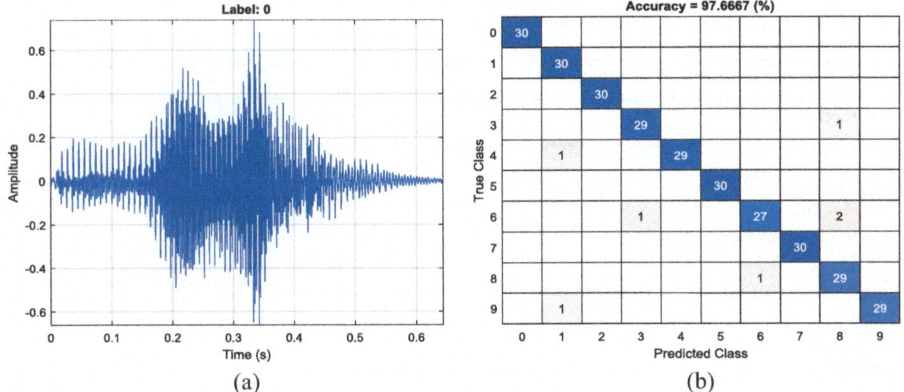

Fig. 6.43 The resulted digit_0 audio sound and the confusion chart

them are label 0 and only 3 of them are other labels (two label 3 and 1 label 8). The correct predicted rate is 91%.

From Fig. 6.43b, which is the resulting confusion chart, it can also be concluded that the overall correct predicted rate is about 98% based on all 10-digit audio sounds. Among them, digits 0–2, 5, and 7 audio sounds are 100% correct rate, and digits 3–4, 8–9 audio sounds are 97% correct rate. The worst case is the predicted rate for digit 6 audio sound, which is 90%.

Next let's perform some evaluations for selected individual digit sounds to see the prediction rate separately.

6.9.5 Evaluate the Classification Result for Individual Digit Audio Sound

Unlike the evaluation for trained digit audio model we did in the last project, in this part we like to evaluate a single digit audio sound with that trained model.

Create a new Script file and name it as **Digit_Audio_Evaluation.m**, and then enter the codes shown in Fig. 6.44 into that file. Let's have a closer look at this piece of codes to see how it works.

1. First we set up our full dataset folder as the current folder to make it ready to get data.
2. The **audioread**() function is used to get a single-digit audio sound file; in this case, we used a digit_6 audio sound file. You may select any other data sound file to do this test. The returned digit sound values and sampling rate are assigned to the local variable **x** and **fs**, respectively. Then the **sound**() function is called to make a sound testing for this file.
3. The label related to that sound is assigned to a local cell variable **Label**.
4. The **plot**() function is executed to plot this sound signal in the time domain.

```matlab
% Digit audio evaluation (digit audio word - 0 ~ 9)
% Name: Digit_Audio_Evaluation.m
% Input: Trained model Mdl obtained from the running result of project Digit_Audio_Classification.m
% Output: Evaluated digit sounds result
% Note: Prior to running this project, run the project Digit_Audio_Classification.m first to get Mdl
% March 15, 2024
1  cd 'C:\Artificial Intelligence Book\Students\Datasets\FSDD\recordings\';

2  [x, fs] = audioread('6_theo_30.wav');
   sound(x, fs)
3  Label = {6};

4  t = (0:numel(x)-1)/fs;
   figure
   plot(t, x)
   title("Label: " + Label)
   grid on
   axis tight
   ylabel("Amplitude")
   xlabel("Time (s)")

5  afe = audioFeatureExtractor(SampleRate=fs, Window=hann(round(0.03*fs), "periodic"), ...
         OverlapLength=round(0.02*fs));

   % Set all feature extractors to true to enable all of them
6  [idx, params] = info(afe, "all");
   cfeatures = fieldnames(idx);
   for i = 1:numel(cfeatures)
      afe.(cfeatures{i}) = true;
   end

   % Extract features from analysis windows.
7  yPerWindow = extract(afe, x);
   numFeatures = 30;
   selectedFeatureIndex = featureSelectionIdx(1:numFeatures);

   % Predict the correct label per window.
   t = predict(Mdl, yPerWindow(:, selectedFeatureIndex));
   trueLabel = Label
   predictionsPerWindow = categorical(t')

   % Create a file-level prediction by taking the mode of window-level predictions.
   prediction = mode(predictionsPerWindow)

   % end
```

Fig. 6.44 The detailed codes for the evaluation of single audio sound

5. A new **audioFeatureExtractor** object **afe** is generated with window and overlaps durations.
6. All feature extractor is enabled by assigning all of them to true.
7. The features of digit_6 audio signal on each window are extracted by calling the function **extract()** with established object **afe** and the collected input digit_6 audio values.

The following codes are similar to those shown in last project in Fig. 6.43, which are used to predict and display the label for the selected input digit_6 audio sound.

One point to be noted is that prior to running this evaluation project, the last project **Digit_Audio_Classifiation.m** must be run first to generate the trained model **Mdl**, and this model would be stored in the Workspace as that project is done. We can directly use that trained model from the Workspace as long as it is created without any problem.

Now run this project, and the running result is shown below:

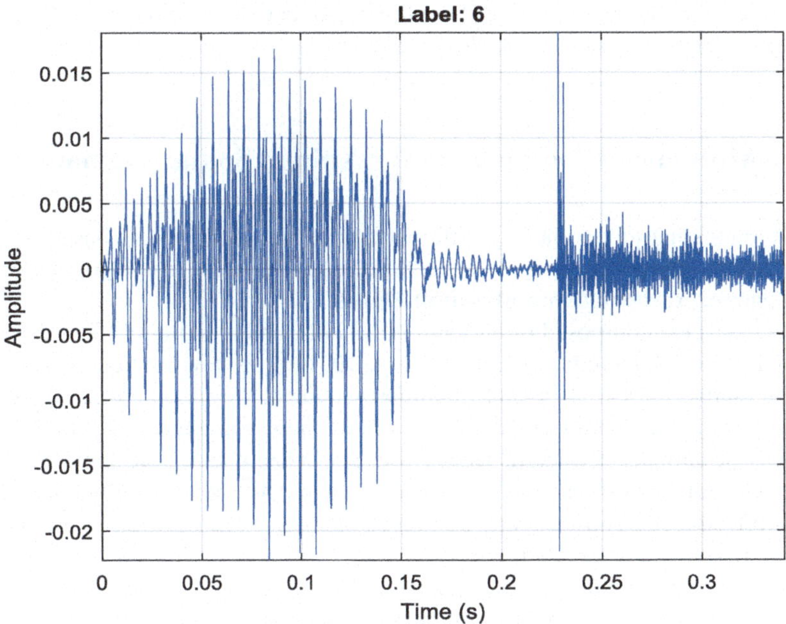

Fig. 6.45 The displayed waveform of digit_6 sound

```
trueLabel = 1×1 cell array {[6]}
predictionsPerWindow = 1×32 categorical array
  Columns 1 through 11
     6    6    6    6    6    6    6    6    6    6    6
  Columns 12 through 22
     6    6    6    6    6    6    6    6    6    6    6
  Columns 23 through 32
     6    6    6    6    6    6    6    6    6    6
prediction = categorical   6
```

The displayed waveform for digit_6 sound is shown in Fig. 6.45.

It can be found from the running result that the rate of predicted label is 100%, and the predicted label (**categorical 6**) is exactly equal to the true label {6}. You can also select other digit sound file to do this testing or evaluation to confirm this classification. If you want to do that testing, what you need to do is just to change the digit number before the first underscore "_" in the audio file name in coding line 2, and the label number in coding line 3 in Fig. 6.44.

The completed digit audio classification project **Digit_Audio_Classification.m** and the related evaluation project **Digit_Audio_Evaluation.m** can be found from the Springer ftp site under the folder **Students\Class Projects\Chapter 6\ Multiclass Classification Project**.

Next let's build one more project to classify some typical animal sounds to make our project more practical and more illustrative.

6.10   Multiclass Classification for Animal Sounds Signals

It is an interesting job to classify different animal sounds or voices based on some real or actual animal sound datasets. In this part, we'd like to use a popular animal sound dataset, YashNita/**Animal-Sound-Dataset** [26].

This dataset is composed of 875 animal sounds, and it contains 10 types of animal sounds, which include 200 cat, 200 dog, 200 bird, 75 cow, 45 lion, 40 sheep, 35 frog, 30 chicken, 25 donkey, and 25 monkey sounds. We modified that dataset by changing the folder names to the actual animals' names and reduced it to hold four (4) kinds of animals, **Chicken**, **Dog**, **Lion**, and **Sheep**, and store them under the folder **Animal Sounds** at our project location **C:\Artificial Intelligence Book\ Students\Datasets**. You can copy and paste them into one of your folders in your machine if you like to use them.

Before we can continue to proceed to build this project, we need to emphasize one important and critical point for features extraction from general audio signals.

6.10.1   *Preprocess the Selected Audio Sounds to Adjust the Multichannel Numbers*

As we mentioned, when running the **extract**() function to try to get all selected features from some input audio sound files, the extracted output is a multiple dimension cell array, exactly it is a *numFiles*-by-1 cell array, where *numFiles* is the number of files in the **datastore**. Each element of the cell array is a *numHops*-by-*numFeatures*-by-*numChannels* array, where the number of hops and number of channels depends on the length and number of channels of the audio file, and the number of features is the requested number of features from the audio data.

In order to use the **extract**() function to get all selected features, it is required that all selected audio files should have the same channel numbers. Regularly most channel numbers should be 2, but occasionally some may be other numbers, such as 3 or higher. The limitation of using the **extract**() function is that all channel numbers must be 2, and an error may be encountered if other numbers are used.

To meet this requirement, we need to preprocess all selected audio files to make sure that all of them have the same channel number. In fact, the channel number for most selected audio files is 2, and only 2% to 3% of them have other numbers. Thus, we can select those small percentage audio files whose channel numbers are higher than 2 and remove them from the selected audio files. We will show those detailed codes later when we perform the preprocessing for our selected audio files in the coding process.

Next let's build our multiclass classification project used to classify some animal sounds.

6.10.2   *Build Audio Classification Project to Classify Animal Sounds*

Create a new Script file and name it as **Animal_Sounds_Classification.m**. Due to large size of this project, we divided this project into two coding parts. The first part is shown in Fig. 6.46 and the second part will be shown in Fig. 6.47. Let's have a closer look at the first part codes to see how they work.

1. First we set up the full path for our modified animal sounds dataset. You may need to replace this path with your actual path if you stored your dataset under a different folder in your machine.
2. The **isdir** field is used to check all subfolders under the current path, and it returns a logical vector to indicate all subfolders: 1 means a subfolder and 0 means not. In this case, it returns a vector as [1 1 1 0 1 0 0 1 0 1], which means that it has 6 subfolders.
3. The **name** property is used to get all subfolders' names in a transposed cell format.
4. The purpose of using the **ismember()** function is to find those subfolders whose names are either "." or "..", and assign them as empty name. If fact, the function of this assignment is to remove those subfolders from the given subfolders. After this operation, we have only four (4) animal sounds subfolders.
5. Set 4 full paths to point to all four subfolders, including the **Chicken, Dog, Lion,** and **Sheep**. This step is to allow all animal sounds files to be stored into the **audioDatastore** that will be executed in the next step.
6. Now we can store all animal sound files into a new **audioDatastore** object **ads**.
7. All stored animal sound labels and sound file names are displayed to show a global picture about the files in the **audioDatastore** object.

```matlab
% Animal sounds feature extract and classifications for selected sounds - 4 animal sounds.
% Name: Animal_Sounds_Classification.m
% Input: Animal sounds data in folder - C:\Artificial Intelligence Book\Students\Datasets\Animal Sounds\
% Output: Classified noise sounds result - confusion table
% March 15, 2024
1  ads = dir('C:\Artificial Intelligence Book\Students\Datasets\Animal Sounds\');   % setup source  folder
2  isub = [ads(:).isdir];                                    % returns logical vector
3  adsFolds = {ads(isub).name}';
4  adsFolds(ismember(adsFolds, {'.','..'})) = [];            % get 4 sub folders excluding the . and .. operators.

5  dataset = fullfile('C:', 'Artificial Intelligence Book', 'Students', 'Datasets', 'Animal Sounds', adsFolds);
6  ads = audioDatastore(dataset, 'LabelSource', 'foldernames', 'FileExtensions', {'.wav'});

   % show all sounds and labels
7  ads.Labels
   ads.Files

8  [adsTrain, adsTest] = splitEachLabel(ads, 0.85);  % Split all input sounds to two parts: training and testing
9  [x, xinfo] = read(adsTrain);
   sound(x, xinfo.SampleRate);

   % Display a sample sound - Chicken sound
10 t = (0:numel(x)-1)/xinfo.SampleRate;
   figure
   plot(t, x)
   title("Label: ", xinfo.Label)
   grid on
   axis tight
   ylabel("Amplitude")
   xlabel("Time (s)")

11 afe = audioFeatureExtractor(SampleRate=xinfo.SampleRate, ...
        Window=hann(round(0.03*xinfo.SampleRate), "periodic"), ...
        OverlapLength=round(0.02*xinfo.SampleRate));

   % Set all feature extractors to true to enable all of them
12 [idx, params] = info(afe, "all");
   cfeatures = fieldnames(idx);
   for i = 1:numel(cfeatures)
      afe.(cfeatures{i}) = true;
   end

   % Extract feature matrix structure
13 featureMatrix = extract(afe, x);
   [numWindows, numFeatures] = size(featureMatrix);
   outputMap = info(afe)

   % Extract features and remove 3D feature cells
14 features = extract(afe, adsTrain, SampleRateMismatchRule="resample");
15 for k = 1: numel(features)
16    [row, col] = size(features{k, :});
17    if col >922   % Normal cell size is 2D in A x 922 columns, but 3D cells show 2D in bigger column number.
18       features{k, :} = [];
      end
   end
```

Fig. 6.46 The first part codes for project Animal_Sounds_Classification

8. To perform both training and testing purpose, we divided the total animal sound files into two parts by using the function **splitEachLabel**(). The training data take 85% and the testing data take the rest of 15% of the whole data.

9. To listen and show a simple of an animal sound, the **read**() function is called to get the first kind of animal sound, which is the chicken's sound. Exactly an entire training dataset that contained 85% of the original sound files, which include the Chicken, Dog, Lion, and Sheep, works as the data source for this function. But the returned data values **x** is just for the first Chicken's sound file. The **sound**() function is executed to play this chicken's sound.

```
1  N = cellfun(@(x)size(x, 1), features);
2  T = repelem((adsTrain.Labels), N);
3  X = cat(1, features{:});
4  [featureSelectionIdx, featureSelectionScores] = fscmrmr(X, T);

   % Training the model with fitcknn() function
5  numFeatures = 30;
   selectedFeatureIndex = featureSelectionIdx(1:numFeatures);
6  Mdl_Animal = fitcknn(X(:, selectedFeatureIndex),T, Standardize=true);

   % Evaluate trained model. Read a sample from the test set. Listen to the sample and plot its waveform.
7  [x, xInfo] = read(adsTest);
   sound(x, xInfo.SampleRate);

   % Extract features from analysis windows.
8  yPerWindow = extract(afe, x);

   % Predict the correct label per window.
9  t = predict(Mdl_Animal, yPerWindow(:, selectedFeatureIndex));
10 trueLabel = categorical(xInfo.Label);
11 predictionsPerWindow = categorical(t');

   % Create a file-level prediction by taking the mode of window-level predictions.
12 prediction = mode(predictionsPerWindow)

   % Analyze the whole-word performance over the entire test set.
13 Tfile = categorical(adsTest.Labels);
   featuresTest = extract(afe, adsTest, SampleRateMismatchRule="resample");
   Y = cellfun(@(x)mode(categorical(predict(Mdl_Animal, x(:,selectedFeatureIndex)))), featuresTest,
   UniformOutput=false);
14 Y = cat(1, Y{:});

15 figure
   confusionchart(Tfile,Y,Title="Accuracy = " + 100*mean(Tfile==Y) + " (%)")
```

Fig. 6.47 The second part codes for project Animal_Sounds_Classification

10. The waveform of this sound is also plotted in the time domain.

11. Now we create a new **audioFeatureExtractor** object **afe** with 30 ms as the window duration and 20 ms as the overlaps between windows.

12. All feature extractors in the new created **audioFeatureExtractor** object are disabled by default. Thus, we need to enable all of those extractors by coding them to true. As we did for the last project, a **for()** loop is used to select and set each of extractors to true to enable all of them.

13. The coding line 13 with the following two lines are used to show up all features of enabled extractors with their values.

14. Now we need to extract features for all training sounds. One issue is that the sampling rate used for the input animal sounds is 48,000 Hz and it is stored in the variable **xinfo.SampleRate**. However, this sampling rate may be different with the sampling rate used in the **audioFeatureExtractor** object. Thus, a **SampleRateMismatchRule = "resample"** Name-Value pair is used to disable any possible mismatching error to be occurred.

15. As we mentioned, the **extrac()** function needs all input sound files to be 2 channels, or at last the channel numbers must be identical for all input sound files. But in this Animal Sound dataset, some sound files contained 3 channels. In order to meet the requirements of using **extract()** function, we need to adjust the channel numbers to make them identical. Since only 2% or 3% of input sound files are 3 channels, thus we can do that adjustment easily by removing those sound files with 3 channels. A **for()** loop is used to check the channel

number for each sound file. The function **numel()** is used to get the length of all features.

16. The function **size()** is used to check the number of rows and columns for each feature cell array, thus a cell array symbol { } is used and we only need to check each row of features.

    The returned number of rows and columns are assigned to variable **row** and **col**.

17. A trick of this detecting is that each 2-channel sound file is a 2D cell, but each 3-channel sound file is a 3D cell. However, that 3D cell is not presented as a 3D cell; instead, it is still presented as a 2D cell. The trick is that it combined the second and the third element as one element by multiplying both together to make them as one element. For example, a 3D cell can be written as row × column × channel, and a 2D cell should be written as row × column without channel number 2 involved by default. But the track is that the 3D cell is still written as a 2D cell and it just multiplies the column and channel together to make them as one body. Thus, you cannot detect any 3D cell in a normal way. To solve this issue, we need to check the columns for all 2D cells; in our case, they are 922. However, for all 3D cells, their column numbers are greater than 922. Hence, we can easily remove those 3D cells when their column numbers are greater than 922.

18. For those cells whose column numbers are greater than 922, just assign them to a blank cell to remove them.

Next let's take look at the second part of this coding process shown in Fig. 6.47.

1. To train a machine learning model on window-level features, replicate the file-level labels so that they are in one-to-one correspondence with the features. To do that, we need first to get the total number of labels or total number of sound files. The **features** is a 183 × 1 cell array, which means that the total features are 183. Therefore, we need to generate related labels. The function **cellfun()** is used to apply the **size()** to all features.

2. Then the function **repelem()** is used to duplicate related labels and assign them to **T**.

3. The function **cat()** is used to concatenate the features into a single matrix for consumption by machine learning tools later.

4. Now we need to select the best feature selector. This best feature selector will depend on the intended model. Use **fscmrmr()** function to rank features for classification by using the minimum-redundancy/maximum-relevance (MRMR) algorithm. The MRMR is a sequential algorithm that finds an optimal set of features that is mutually and maximally dissimilar and can represent the response variable effectively. In our case, we pay more attention to the **feature-SelectionIdx**, which determines the elements we need to use from the features.

5. The number of features is defined based on estimations, and the elements in the features are indicated by the **featureSelectionIdx** array with number of features.
6. The function **fitcknn()** is used to train the sound model **Mdl_Animal** with the KNN algorithm in a format of **fitcknn(X, Y, Name-Value)**. The Name-Value pair **Standardize = true** is to ask software to center and scale each column of the predictor data **X** by the column mean and standard deviation.
7. After the model is trained, we need to evaluate that model. To do that, first we need to call the **read()** function to get the testing data, exactly to get the testing data stored in a dataset, **adsTest**. The returned audio sound data values are involved in the **x**, and related information, such as the sampling rate, audio file names, and labels, are included in the **xinfo** structure. The sound of the selected testing audio file, which should be a chicken sound, is played to confirm its correctness (by default, the audio sound is the first animal sound).
8. To predict the testing animal sound, the **extract()** function is used to get all features for the testing animal sound.
9. Then the function **predict()** is executed based on the trained model **Mdl_Animal**, the extracted features from the testing animal sound indicated by the **selectedFeatureIndex**.
10. The actual label is defined based on the real labels of the testing animal sounds.
11. The predicted label **t** is a cell column array, and it needs to be converted to a row category array by using the function **categorical()** with the transposed **t**.
12. The function **mode()** is used to return the most frequent value from the predicted labels. In other words, it counts the number of predicted labels and selects the max number of predicted labels.
13. These coding lines are used to analyze and display the entire performance based on all testing animal sound data.
14. The function **cat()** is used to arrange the predicted labels in a diagonal format.
15. Finally the entire testing result is displayed in a confusion chart.

Now run the project, and some running result is shown below on the Command window.

The plotting of the sampled chicken sound and the final output of the confusion chart are shown in Figs. 6.48a, b, respectively. The corrected prediction rate is about 84%.

A completed animal sound classification project **Animal_Sound_Classification.m** can be found from the Springer ftp site under the folder **Students\ Class Projects\Chapter 6\Multiclass Classification Project**.

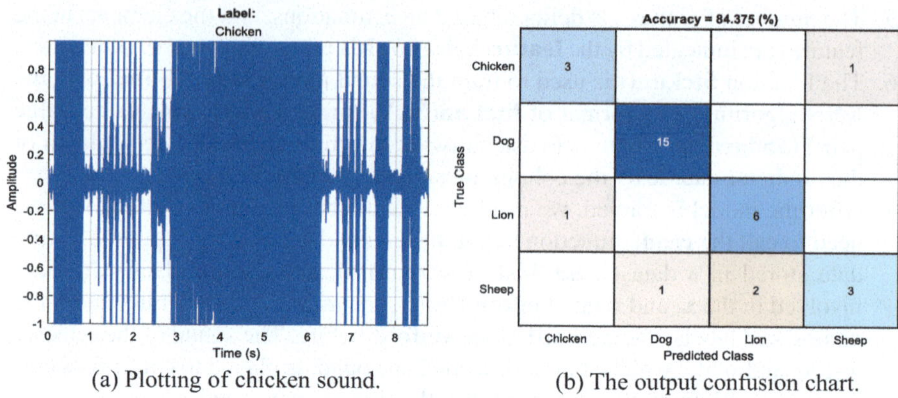

(a) Plotting of chicken sound.   (b) The output confusion chart.

Fig. 6.48 The running result of the project Animal_Sound_Classification

6.11   Chapter Summary

The main topic on this chapter is about introductions and discussions for one of the other major branches in the machine learning, the classification algorithms.

In fact, MATLAB provided two major Toolboxes, Statistics and Machine Learning Toolbox and Curve Fitting Toolbox, to help users design and build their machine learning-related projects, especially for classification-related projects, easily and quickly.

Most important and popular classification-related algorithms are discussed in detail in this chapter, which include:

1. Binary Classifications
2. Multiclass Classifications
3. Multilabel Classifications

Due to the complication of the multilabel classification, we only concentrate our study on the first two algorithms, binary and multiclass classifications, in this book. Based on Fig. 4.1 in Chap. 4, we limit our study and discussion to the following algorithms:

1. Binary Classification (BC)
2. Multiclass Classification (MC)
3. KNN Algorithm (KNN)
4. Decision Tree (DT)
5. Naive Bayes (NB)
6. Support Vector Machine (SVM)
7. Random Forest (RF)

All of these algorithms are introduced and discussed with some hands-on and real projects in two modes, the MATLAB App mode and MATLAB Function mode. All those Apps and Functions are involved in two Toolboxes, Statistics and Machine Learning and Curve Fitting Toolboxes.

Relatively speaking, the App mode provides an efficient and convenient way to enable students or users, especially the beginners, to build and develop classification-related projects easily and quickly, and it greatly improves the students' study interests and reduces the learning curves. The shortcoming of using those Apps is that some coding and building strategies and techniques are hidden from students, which is not a professional way to study machine learning-related technologies.

To improve that shortcoming, the MATLAB Function mode can be adopted to enable students or users to touch deeper in building more professional AI projects by developing more detailed control functions with more coding processes.

Finally, some hands-on and practical AI project is developed by using different classification models to classify actual applications. These projects include:

1. Binary classification project used to classify diabetes patients.
2. Multiclass classification project used to classify mixed data type classes.
3. Using Naive Bayes algorithm to perform multiclass classification for diabetes.
4. Using ECOC binary classification algorithm to perform multiclass image classification for some fruits classes.
5. Multiclass classification project used to classify digit audio signals.

Home Works

I. True/False Selections

______1. Classification is a process related to categorization, and it is the grouping of related facts into classes.

______2. Generally, a classification algorithm can be divided into two categories, binary classification and multiclass classification.

______3. In MATLAB, all classification algorithms and functions are included in two Toolboxes, Statistics and Machine Learning and Deep Learning.

______4. MATLAB divided all classification tools into two components, Machine Learning App and Function library.

______5. When using MATLAB App to train classification models, three possible options can be used: automated training, manual training, and semi-automated training.

______6. When using Confusion Chart to evaluate the classification results, only the correctly classified observations are counted and displayed.

______7. By using the KNN algorithm and its function **fitcknn()**, both binary and multiclass classification process can be performed.

______8. The ECOC algorithm and its function **fitcecoc()** can be used to perform either binary or multiclass classification.

______9. MATLAB provided four class objects to perform multiclass classification job:; **ClassificationKNN**, **ClassificationNaiveBayes**, **Classification ECOC**, and **ClassificaionNeuralNetwork**.

_____10. When using Decision Trees as classifiers, the type of the predictors can be numeric, categorical or categorical and numeric.

II. Multiple Choices

1. Three groups of popular classification algorithms are _____________.

   (a) Nonlinear, linear, and multilinear classifications
   (b) Dynamic, kinematic, and static algorithms
   (c) Binary, multiclass, and multilinear classifications
   (d) Binary, multiclass, and multilabel classifications

2. In MATLAB, all classification tools are included in two Toolboxes, _________.

   (a) Machine Learning and Curve Fitting Toolbox
   (b) Deep Learning and Neural Network Toolbox
   (c) Statistics and Machine Learning, and Deep Learning Toolbox
   (d) Statistics and Machine Learning Toolbox

3. The Classification Learner App can be used to train the following classifiers: ___________.

   (a) Decision Trees, Logistic Regression
   (b) Support Vector Machines, K-Nearest Neighbors
   (c) Naive Bayes, Kernel Approximation
   (d) All of them

4. When using Classification Learner App to train models, you have two options _________

   (a) Linear and nonlinear classification training
   (b) Automated and manual classification training
   (c) Inputs and output classification training
   (d) Binary and multiclass classification training

5. Regularly, a classification process includes the following steps _________________.

   (a) Select data source, choose classifier, train classifier, and evaluate classification
   (b) Select dataset, choose classifier, train, and test classified result
   (c) Select data source, choose classifier, train, and evaluate classification model
   (d) All of them

6. A confusion chart contained the following outputs _______________.

   (a) True positive rates, False negative rates
   (b) Positive predictive values, False discovery rates
   (c) ROC curve and AUC value
   (d) None of them

7. Some popular binary classification algorithms and functions include
   ______________.

   (a) fitctree(), fitcnb()
   (b) fitcensemble(), fitcknn()
   (c) fitclinear(), fitckernel()
   (d) fitcsvm(), fitcecoc()

8. The linear classification is to setup a mapping between _________ and
   __________.

   (a) Inputs and outputs
   (b) Predictors and responses
   (c) Input vectors and class score
   (d) Input vectors and response values

9. The following classifiers can be used for multiclass classifications
   ______________.

   (a) fitclinear()
   (b) fitckernel()
   (c) fitcsvm()
   (d) fitcecoc()

10. To perform image classifications, one needs to use some feature extraction
    methods to extract interesting features from the input predictors or images to
    train selected models to perform image classification. Some popular feature
    extraction methods include ________.

    (a) Histogram of oriented gradients (HOG)
    (b) Speeded-up robust features (SURF)
    (c) Local binary pattern (LBP) features and Bag of features (BOF)
    (d) All of them

III. Exercises

   1. Explain the similarity and dissimilarity between the regression and the clas-
      sification algorithms.
   2. Provide an explanation for four testing parameters, TPR, FNR, PPV,
      and FDR.
   3. Describe how to use Automated Classification Training mode in the
      Classification Learner App to train and check a classification model.

IV. Lab Projects

   1. Refer to Sect. 6.7.3, and use the Gaussian naive Bayes model to predict the
      possibility of a boy or a girl. The inputs or predictors for boys and girls are
      shown in the three tables below (Table 6L.1).
      Hint1: Use Eq. (6.9) to calculate all mean and variance values for each given
      set of predictors with a MATLAB Script file, and put the results into related

Table 6L.1 Inputs or predictors

Person	Height (ft)	Weight (lbs)	Foot size (in.)
Boy	3	90	6.0
Boy	2.96	95	5.5
Boy	2.79	85	6.0
Boy	2.96	83	5.0
Girl	2.50	50	3.0
Girl	2.75	75	4.0
Girl	2.71	65	3.5
Girl	2.88	75	4.5

Table 6L.2 Calculated prior probabilities

Person	Mean (height)	Variance (height)	Mean (weight)	Variance (weight)	Mean (foot size)	Variance (foot size)
Boy	2.93	0.0066	88.25	21.6875	5.63	0.1719
Girl	2.71	0.0186	66.25	104.69	3.75	0.3125

category shown in Table 6L.2 (for your convenience, the calculated results have been there).

Hint2: After this model has been set, next you need to test this model with a piece of new or testing data, which is a person record selected, as shown in Table 6L.3. This testing or sample data is for a Girl (Boy = 0). To speed up these calculations, you can build another piece of MATLAB codes to a Script file and run it to do this job. Create a new Script file named **CalP_ BoyGirl.m** to do it. This piece of codes is used to derive the *maximum a posteriori* or *MAP* value.

2. Using MATLAB Functions to build a speech command classification project to classify three speech commands, '**yes**', '**no**', and '**up**'. The name is **Speech_Command_Classification.m**. The dataset is called kingabzpro/ Speech_Commands_Dataset: "Warden P. Speech Commands: A public dataset for single-word speech recognition, 2017, which is located at the site: http://download.tensorflow.org/data/speech_commands_v0.01.tar.gz".

Hint1: Due to the large size of that dataset, we modified it and used only three speech commands with reduced size of each speech file to 240. The modified dataset is named **Speech Commands**, and it is located at the Springer site: **Students\Datasets\Speech Commands**. Three subfolders or labels, **yes**, **no**, and **up**, are under that folder with all speech files.

Hint2: All speech files were recorded with two channels; thus, no additional channel adjustment is needed.

3. Using MATLAB Functions to build a speech command evaluation project to evaluate each single speech command, such as '**yes**', '**no**', and '**up**'. The project's name should be **Speech_Command_Evaluation.m**. The testing or evaluation speech files are located at the Springer ftp site at the folder:

Table 6L.3 Testing sample person

Person	Height (ft)	Weight (lbs)	Foot size (in.)
Sample	6	130	8

Students\Datasets\Speech Commands Test, with three subfolders or labels, '**yes**', '**no**', and '**up**', with two speech files under each folder.

Hint1: Refer to Fig. 6.44 on Sect. 6.9.5 for the coding developments.

Hint 2: Unlike the project **Speech_Command_Classification.m**, you can directly load each single testing speech file via **audioread**() function with the speech file name. However, prior to doing that, you may need to use **cd** command to set up the full path for that speech file.

Hint3: You may also directly assign the label name (subfolder's name) to the label cell, as **Label = {'yes'};** for the selected speech file.

Hint4: Prior to running this evaluation project, run project **Speech_Command_Classification.m** first to get the trained model, **Mdl_Speech**, to evaluate that model with testing speech data.

4. Using MATLAB **trainImageCategoryClassifier** class to build an image classifier to classify different types of cars, including the **pickup**, **sedan**, and **SUV** three types of cars. The project is named **Cars_Image_Classifier.m**. The dataset used is Boonsirisumpun, Narong; surinta, olarik (2021), "Vehicle Type Image Dataset (Version 2): VTID2", Mendeley Data, V2, https://doi.org/10.17632/htsngg9tpc.2. We modified the dataset and it contained both training data and testing data, and each includes three types of cars with three folders or labels. Each type contained about 100 car images. All data are located under the folder **Car Type Dataset** and it can be found from the Springer ftp site at the folder **Students\Datasets\Car Type Dataset**.

Hint1: Refer to Fig. 6.35 in Sect. 6.8.1 for the coding developments. You can use the coding line **imshow(imds.Files{1});** to display the first car image as a sample image.

Hint 2: You may use the Bag of Features method and split the all training data into 2 parts, 80% for training images and 20% for the validation purpose.

Hin3: The testing data is used later to evaluate single car image in the next project.

Hint4: You may save the trained model to a folder and use it later by loading it.

5. Build a MATLAB project **Cars_Image_Evaluation.m** to evaluate the trained model classified from the last project. The testing data are stored and located under the folder **Car Type Dataset** and it can be found from the Springer ftp site at the folder **Students\Datasets\Car Type Dataset\car_test_data**. Under that folder, three subfolders or labels, **pickup**, **sedan**, and **suv**, are installed and each folder contained 20 testing car images.

Hint1: Refer to codes developed in Fig. 6.37 at Sect. 6.8.2 to build this project.

Hint2: You may add one coding line, **imshow(fullFileName);** to display each tested car image.

Hint3: The evaluation accuracy should be around 75%, and a trade-off should be taken between the evaluation time and the amount of the training data.

6. Using HOG feature extraction method and **fitcecoc()** algorithm to perform classification and testing for the Car Type dataset used in the Project 5. The name of the project should be **HOG_Cars_Classifier.m**.

Hint1: Refer to codes developed in Fig. 6.38 at Sect. 6.8.3 to build this project.

Hint2: You may add one coding line, **imshow(imds.Files{1});** to display each tested car image. Try to split training images and testing images in a ratio of 80% for the former and 20% for the latter.

Hint3: The name of the trained model should be **Car_HOG_Classifier**. The order number of selected image for **readimage()** function should be less than 100, maybe 80, since the total number of training images is about 100.

References

1. (vgs@aplcen.apl.jhu.edu), Research Center, RMI Group Leader, Applied Physics Laboratory, The Johns Hopkins University, Laurel, MD.
2. https://en.wikipedia.org/wiki/Linear_classifier.
3. https://leonardoaraujosantos.gitbook.io/artificial-inteligence/machine_learning/supervised_learning/linear_classification.
4. https://metacademy.org/graphs/concepts/binary_linear_classifiers#:~:text=A%20linear%20classifier%20makes%20a,in%20the%20input%20feature%20space.
5. https://pages.stat.wisc.edu/~mchung/teaching/MIA/reading/diffusion.gaussian.kernel.pdf.pdf.
6. https://figshare.com/articles/dataset/heart_csv/20236848.
7. https://data.mendeley.com/datasets/wj9rwkp9c2/1.
8. Rashid, Ahlam (2020), "Diabetes Dataset", Mendeley Data, V1. https://doi.org/10.17632/wj9rwkp9c2.1.
9. Narasimha Murty, M.; Susheela Devi, V. (2011). Pattern Recognition: An Algorithmic Approach. ISBN 978-0857294944.
10. https://en.wikipedia.org/wiki/Naive_Bayes_classifier.
11. John, George H.; Langley, Pat (1995). Estimating Continuous Distributions in Bayesian Classifiers. Proc. Eleventh Conf. on Uncertainty in Artificial Intelligence. Morgan Kaufmann. pp. 338–345.
12. Rennie, J.; Shih, L.; Teevan, J.; Karger, D. (2003). Tackling the poor assumptions of naive Bayes classifiers (PDF). ICML. Archived (PDF) from the original on 2022-10-09.
13. McCallum, Andrew; Nigam, Kamal (1998). A comparison of event models for Naive Bayes text classification (PDF). AAAI-98 workshop on learning for text categorization, Vol. 752. Archived (PDF) from the original on 2022-10-09.
14. https://machinelearningmastery.com/error-correcting-output-codes-ecoc-for-machine-learning/.
15. https://www.mathworks.com/discovery/feature-extraction.html.
16. https://en.wikipedia.org/wiki/Histogram_of_oriented_gradients#:~:text=The%20histogram%20of%20oriented%20gradients,localized%20portions%20of%20an%20image.
17. https://en.wikipedia.org/wiki/Speeded_up_robust_features#:~:text=In%20computer%20vision%2C%20speeded%20up,feature%20transform%20(SIFT)%20descriptor.
18. https://en.wikipedia.org/wiki/Local_binary_patterns.
19. https://medium.com/analytics-vidhya/bag-of-visual-words-bag-of-features-9a2f7aec7866.
20. https://public.roboflow.ai/classification/fruits-dataset.

21. https://maelfabien.github.io/machinelearning/Speech9/#.
22. https://en.wikipedia.org/wiki/Mel-frequency_cepstrum#:~:text=In%20sound%20
    processing%2C%20the%20mel,collectively%20make%20up%20an%20MFC.
23. https://devopedia.org/audio-feature-extraction.
24. https://fooof-tools.github.io/fooof/auto_motivations/measurements/plot_BandRatios.html.
25. Jakobovski. "Jakobovski/Free-Spoken-Digit-Dataset." GitHub, May 30, 2019. https://github.
    com/Jakobovski/free-spoken-digit-dataset.
26. https://github.com/YashNita/Animal-Sound-Dataset.

Chapter 7
Neural Networks and Deep Learning

As we mentioned in Chap. 4, there is a close relationship between neural networks and deep learning. In fact, a neural network can be considered as a special machine learning model, but it is different from all other machine learning models since it uses a neural network as a body to mimic the human being's brain to improve its analyzing and derivation abilities to provide better performance compared with other models.

The learning algorithm applied to a neural network can be supervised, unsupervised, or reinforced. Thus the neural network block shown in Fig. 4.1 is a floating body.

Deep learning is a subset of machine learning, which is essentially a neural network with three or more layers. These neural networks attempt to simulate the behavior of the human brain, albeit far from matching its ability, allowing it to *learn* from large amounts of data. While a neural network with a single layer, which is called shallow learning, can still make approximate predictions, additional hidden layers can help to optimize and refine for accuracy.

As we mentioned, neural networks are the backbone of deep learning algorithms, which means that if machine learning uses neural networks as models, a deep learning algorithm is coming.

In fact, deep learning involves the use of complex models, neural networks, which exceed the capabilities of all other machine tools or algorithms such as logistic regression and support vector machines, to provide much better performances or some function approximations with smaller errors. First let us concentrate our discussions on neural networks.

Supplementary Information The online version contains supplementary material available at https://doi.org/10.1007/978-3-031-84423-2_7.

Y. Bai, *AI Foundations and Applications with MATLAB*,
https://doi.org/10.1007/978-3-031-84423-2_7

7.1   Introduction to Neural Networks

Neural networks, also called artificial neural networks (ANNs) or simulated neural networks (SNNs), are a subset of machine learning and are the backbone of deep learning algorithms. They are called **neural** because they mimic how neurons in the brain signal one another [1].

Neural networks are composed of nodes and layers—an input layer, one or more hidden layers, and an output layer. Each layer contains multiple nodes and each node is an artificial neuron that connects to the next, and each has a weight and threshold value. When one node's output is above the threshold value, that node is activated and sends its data to the network's next layer. If it is below the threshold, no data passes along.

The basic unit or building block of an artificial neural network (ANN) is an artificial neuron, which can be considered as a node in a network [2]. The general neural neuron is composed of a set of inputs x_j, $(j = 1, 2, \ldots n)$ where the subscript j represents the jth input. Each input x_j has a definite weight factor w_j that is associated to the input x_j, exactly x_j is multiplied by the factor w_j to form a complete input prior to being input to the network. For an ANN with i $(i = 1, 2, \ldots m)$ nodes, an additional subscript i is needed to represent the ith node. Also it has a bias term w_0, a threshold value Θ, a nonlinear function f that acts on the produced signal R, and an output O. A basic model of a neuron i or node i is shown in Fig. 7.1.

The input-output relationship of this neuron can be described by a transfer function as

$$O_i = F_i\left(\sum_{j=1}^{n} w_{ij} x_{ij} - \Theta_i \right) \tag{7.1}$$

and the neuron's firing condition is

$$\sum w_{ij} x_{ij} - \Theta_i \tag{7.2}$$

Regularly a complete ANN contains multiple nodes with multiple layers, including input layers, output layers, and hidden layers. Overall, a feedforward ANN can be considered as a complex brain/machine system that is composed of multilayer with multi-neuron operating in parallel, as shown in Fig. 7.2. In fact, the connections

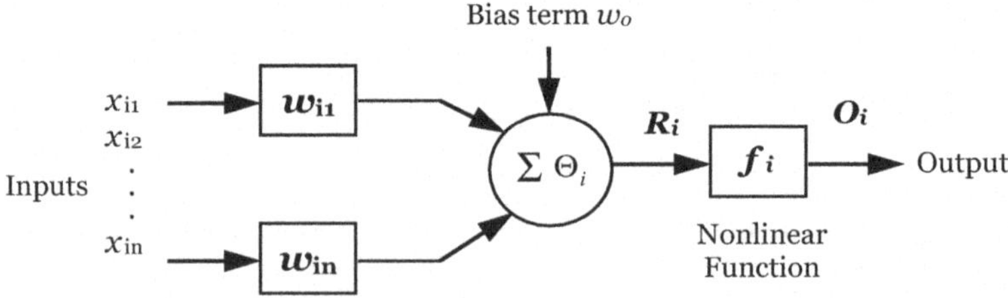

Fig. 7.1 A basic model of a neuron i

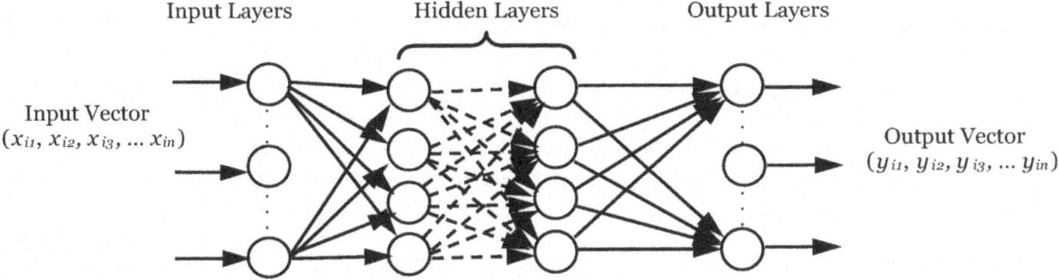

Fig. 7.2 A complete neural network structure

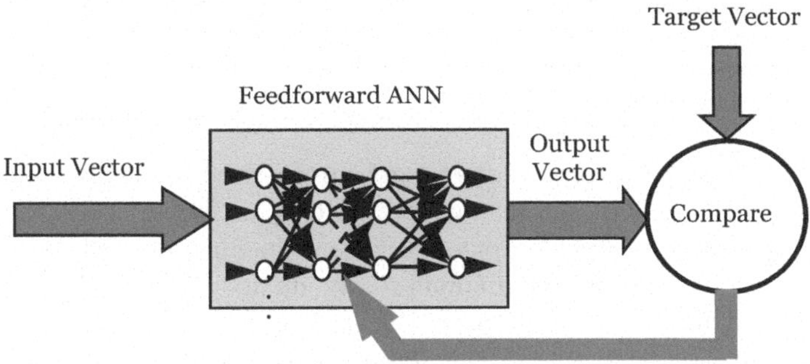

Fig. 7.3 Training process of a feedforward ANN

between nodes largely determine the network function. One can train an ANN to perform a specified function by adjusting the values of the connections (weight factors) between nodes via inputs and desired or target outputs. Figure 7.3 shows an illustration of this training process.

Generally, a neural network can be adjusted or trained, so that a particular input leads to a corresponding target output. In Fig. 7.3, the network is adjusted, based on a comparison between the output and the target, until the network output matches the target. Typically, many such input-target pairs are needed to train a network.

7.2   Structure and Components of a Typical Neural Network System

As shown in Figs. 7.1 and 7.2, a typical neural network is composed of a sequence of layers, such as input layers, hidden layers, and output layers. Each layer is made by a group of neurons or nodes. The connections between each layer are called weight factors with some threshold values. Each input x_j has a definite weight factor w_j that is associated to the input x_j, exactly x_j is multiplied by the factor w_j to form a complete input prior to being input to the network. For an ANN with i ($i = 1, 2, \ldots m$) nodes, an additional subscript i is needed to represent the ith node.

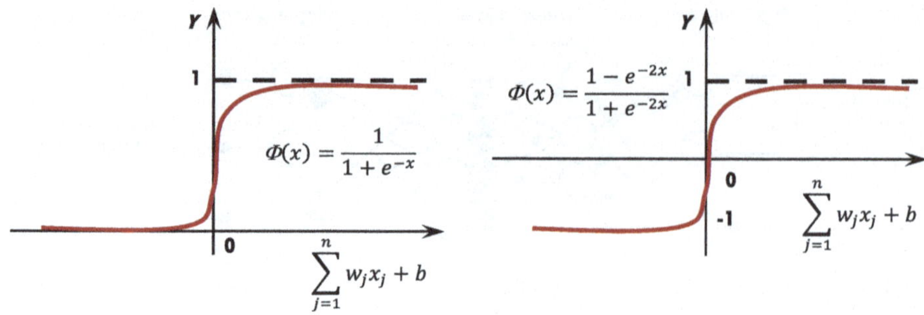

(a) Sigmoid function.   (b) Hyperbolic Tangent Function.

Fig. 7.4 Two popular nonlinear functions used for predicting outputs. (**a**) Sigmoid function. (**b**) Hyperbolic tangent function

The output of each previous node O_i can be transferred to the next node as long as the product of $x_{ij}w_{ij}$ is greater than its threshold Θ_i, as shown in Eq. (7.1). Where the subscript j represents the jth input and i means the ith node, and the same thing to w_{ij}, it is a weight factor for the jth input for the ith node. The final output of each node can be calculated based on a nonlinear function applied to the difference between the sum of $x_{ij}w_{ij}$ and each related threshold value Θ_i.

Generally the nonlinear function used is a sigmoid or a hyperbolic tangent function shown in Fig. 7.4. The advantage of using this kind of function is that it provides a good convergence ability to allow the training result to be converged quickly to get the final solution.

This information transferring process, in which the outputs of previous nodes are transferred to next layers, exactly the related nodes on the next layers, can be mapped to a neural derivation process used by human beings. However, unlike the human being's brain, the neural works cannot perform its normal logic derivation and analysis process until they are trained by supervised learning algorithms with known input and output data pairs.

Generally if neural networks contain two or three layers, it could be called a shallow learning network, but it may be called a deep learning neural network if the number of layers is greater than 3.

7.3   The Types of Neural Networks

The types of neural networks depend on how the inputs can be transferred to outputs and how to map input data to the output data. Regularly, three major types are popularly and widely implemented in neural networks. As we mentioned, a closer relationship exists between neural networks and deep learning. Thus we cannot separate the types of neural works from deep learning. Most time, we call a neural network as a deep learning neural network (DLNN) or deep neural network (DNN). Therefore, the types of neural networks are also categorized based on the deep learning algorithms. Generally, four types of deep neural networks are popularly used today: feedforward neural networks (FNN), backpropagation neural networks (BNN), convolutional neural networks (CNN), and recurrent neural networks (RNN).

7.3.1 Feedforward Neural Networks (FNN)

Feedforward neural networks process data in one direction, from the input layers to the output layers, as shown in Fig. 7.2. Each node in one layer is connected to every node in the next layer. A feedforward network uses a feedback process to improve predictions during the training process. A set of known input and output data pairs is needed to perform the training for the neural networks.

7.3.2 Backpropagation Neural Networks (BNN)

As we mentioned, in a simple term, you can consider that the data are flowing from the input layers to the output layers by passing many different paths in the neural network. However, only one path is the correct one that maps the input node to the correct output node. To find this path, the neural network must be trained by using a feedback loop, which works as follows [3]:

1. Each node makes a guess about the next node in the path with a guessed weight factor.
2. It checks if the guess is correct. Nodes assign higher weight values to paths that lead to more correct guesses and lower weight values to node paths that lead to incorrect guesses.
3. For the next data point, the nodes make a new prediction using the higher weight paths and then repeat Step 1.

This working algorithm means that first a set of input data is applied to the input layers, then goes through the hidden layers, and finally guesses or predicts the output data. After comparing the correct known output data with those predicted data, some errors could be derived. To derive all correct weight factors on the correct path, those errors are back propagated through the networks and all weight factors are adjusted to minimize those errors. The criteria to evaluate these so-called minimized errors are obtained via a cost function. A functional block diagram of the working principle of using backpropagation algorithm is shown in Fig. 7.3.

7.3.3   *Convolutional Neural Networks (CNN)*

The hidden layers in convolutional neural networks perform specific mathematical functions, called convolutions. Based on its definition, the input data are processed by each hidden layer with a convolution algorithm in terms of input data points. They are very useful for image classification because they can extract relevant features from images that are useful for image recognition and classification. For example, an input image that is composed of 256×256 pixels can be divided into a sequence of 8×8 or 16×16 matrices, and a digital convolution algorithm can be applied to those matrices to extract some useful features from that image. The new form is easier to process without losing features that are critical for making a good prediction. Each hidden layer extracts and processes different image features, like edges, color, and depth.

A typical CNN architecture includes the following layers [4]:

1. Input layer
2. Convolutional layer
3. Rectified linear unit (ReLU) layer
4. Pooling layer
5. Fully connected layer (classification layer)
6. Output layer

7.3.3.1   The Input Layer

For the image processing operations, an image input layer inputs images to a CNN and applies data normalization.

7.3.3.2   The Convolutional Layer

Create a feature map to predict the class probabilities for each feature by applying a filter that scans the whole image, few pixels at a time. For image processing, a 2D convolutional layer applies sliding convolutional filters to the input, as normal convolution operations did.

A convolutional layer consists of neurons that connect to sub-regions of the input images or the outputs of the previous layer. The layer learns the features localized by these sub-regions while scanning through an image. When creating a layer, you can specify the size of these regions based on your actual input images.

For each region, the convolution algorithm computes a dot product of the weights and the input and then adds a bias term to it. A set of weights that is applied to a region in the image is called a filter. The filter moves along the input image vertically and horizontally, repeating the same computation for each region. In other words, the filter convolves the input.

When using MATLAB Deep Learning Toolbox, one can adjust the learning rates and regularization options for the layer using name-value pair arguments while defining the convolutional layer.

Regularly, the size of output on the convolutional layer is larger than that of the input image; thus, a feature extraction with reduced size of output is necessary. Then a pooling layer is used to handle this task.

7.3.3.3   The Rectified Linear Unit (ReLU) Layer

Convolutional layer is usually followed by a nonlinear activation function such as a rectified linear unit (ReLU), specified by a ReLU layer. A ReLU layer performs a threshold operation to each element, where any input value less than zero is set to zero, that is,

$$f(x) = \begin{cases} x(x \geq 0) \\ 0(x < 0) \end{cases} \tag{7.3}$$

The ReLU layer does not change the size of its input.

7.3.3.4   The Pooling Layer

It scales down the amount of information from the convolutional layer generated for each feature and maintains the most essential information. In most cases, the process of the convolutional and pooling layers is combined together in a sequence and usually repeats several times to get outputs.

Pooling process can be divided into max polling and average polling. The max polling is to get the max value for a sub-region and the average-polling is to obtain the average value for the selected sub-region. Since pooling layers follow the convolutional layers for down-sampling, it reduces the number of connections to the following layers. They do not perform any learning themselves but reduce the number of parameters to be learned in the following layers. They also help reduce overfitting.

7.3.3.5   The Fully Connected Layer

A fully connected layer multiplies the input or the output of the previous layer by a weight matrix and then adds a bias vector. The convolutional layers are generally followed by one or more fully connected layers.

The so-called fully connected layer means that all neurons in that layer connect to all neurons in the previous layer. This layer combines all of the features learned by the previous layers across the image to identify the larger patterns. When this layer works for the classification problems, the last fully connected layer combines the features to classify the images. This is the reason why the output size of the last fully connected layer of the network is always equal to the number of classes of the dataset.

In fact, the major function of this layer is to apply weights over the input generated by the feature analysis to predict an accurate class label.

7.3.3.6 The Output Layer

The most popular output layers contain softmax and classification layers.

A softmax layer applies a softmax function to the input. A classification layer computes the cross entropy loss for multiclass classification problems with mutually exclusive classes. For classification problems, a softmax layer and then a classification layer must follow the final fully connected layer. The output unit activation function is the softmax function represented as:

$$y_r(x) = \frac{P(x,\theta|c_r)P(c_r)}{\sum_{j=1}^{k} P(x,|\theta,|c_j)P(c_j)} = \frac{e^{a_r(x,\theta)}}{\sum_{j=1}^{k} e^{a_j(x,\theta)}} \tag{7.4}$$

where $0 \leq P(c_r|x, \theta) \leq 1$ and $\sum_{j=1}^{k} P(x,|\theta,|c_j) = 1$. Also $a_r = \ln(P(x, \theta|c_r)P(c_r))$, and $P(x, \theta|c_r)$ is the convolutional probability of the sample given class r, and $P(c_r)$ is the class prior probability.

The softmax function is also known as the normalized exponential and can be considered as the multiclass generalization of the logistic sigmoid function [4]. For typical classification networks, the classification layer must follow the softmax layer. In MATLAB Deep Learning Toolbox, in the classification layer, the variable *trainNetwork* takes the values from the softmax function and assigns each input to one of the K mutually exclusive classes using the cross entropy function for a 1-of-K coding scheme [4]:

$$loss = -\sum_{i=1}^{N} \sum_{j=1}^{k} t_{ij} (\ln y_{ij}) \tag{7.5}$$

where N is the number of samples, K is the number of classes, t_{ij} is the indicator that the ith sample belongs to the jth class, and y_{ij} is the output for sample i for class j, which in this case, is the value from the softmax function. That is, it is the probability that the network associates the ith input with class j.

The functional block diagram of using a CNN to predict and classify some features (true or false) for input images is shown in Fig. 7.5.

Relatively speaking, the convolution algorithm has some powerful functions, such as low-pass filtering, weighted running average, and noise removing.

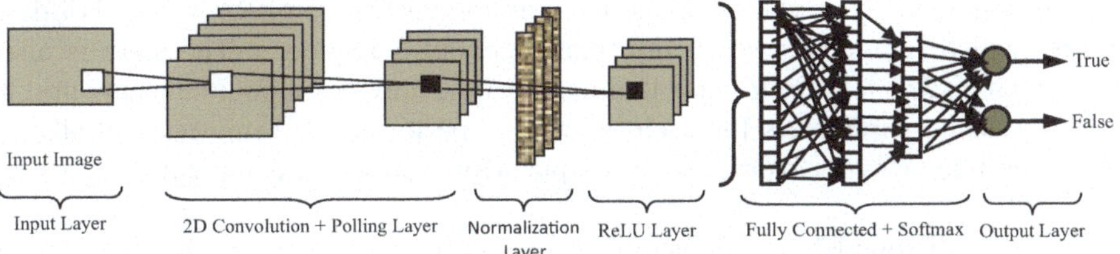

Fig. 7.5 The architecture of using a CNN to classify features for input images

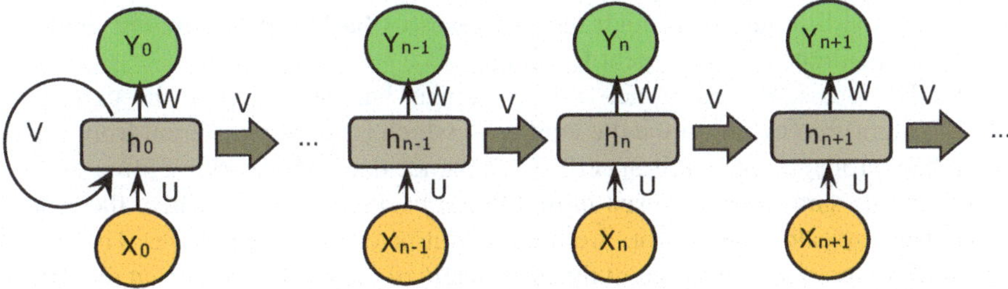

Fig. 7.6 A functional block diagram of recurrent neural networks

7.3.4   Recurrent Neural Networks (RNN)

A recurrent neural network (RNN) is a type of artificial neural network that uses sequential data or time series data [5]. This deep learning algorithm is commonly used for ordinal or temporal problems, such as language translation, natural language processing (NLP), speech recognition, and image captioning; they are incorporated into popular applications such as Siri, voice search, and Google Translate. Similar to feedforward neural networks (FNNs) and convolutional neural networks (CNNs), recurrent neural networks utilize training data to learn. They are distinguished by their *memory* as they take information from prior inputs to influence the current input and output. While traditional deep neural networks assume that inputs and outputs are independent of each other, the output of recurrent neural networks depends on the prior elements within the sequence.

As shown in Fig. 7.6, the inputs X_i and outputs Y_i in RNN are sequences of transferred data via the hidden layers h_i. The output of the current node is not only determined by the current inputs X_i but also determined by the outputs (states) **V** from the previous nodes. Also some outputs are feedback to the input, like the first node shown in Fig. 7.6, to get a so-called recurrent state.

In summary, recurrent neural network (RNN) is a kind of neural network where the output from the previous node is fed as input to the current node. However, the difference is, in traditional neural networks, all inputs and outputs are independent of each other. But in RNN, in cases when it is required to predict the next output of a node, the previous nodes are required, and hence there is a need to remember the previous nodes. Thus RNN came into existence, which solved this issue with the help of a hidden layer. The main and most important feature of RNN is its **Hidden State**, which remembers some information about a sequence. The state is also referred to as *Memory State* since it remembers the previous input to the network. It uses the same parameters for each input as it performs the same task on all the inputs or hidden layers to produce the output. This reduces the complexity of parameters [6].

Another distinguishing characteristic of recurrent networks is that they share parameters across each layer of the network. While feedforward networks have different weights across each node, recurrent neural networks share the same weight parameter **W**, as shown in Fig. 7.6, within each layer of the network. That is, these weights are still adjusted through the processes of backpropagation and gradient descent to facilitate reinforcement learning.

Finally, recurrent neural networks leverage backpropagation through time (BPTT) algorithm to determine the gradients, which is slightly different from traditional backpropagation as it is specific to sequence data. The working principles of BPTT are the same as those used in traditional backpropagation, where the model trains itself by calculating errors from its output layer to its input layer. These calculations allow us to adjust and fit the parameters of the model appropriately. BPTT differs from the traditional approach in that BPTT sums errors at each time step whereas feedforward networks do not need to sum errors since they do not share parameters across each layer.

Based on the number of inputs and the corresponding outputs, the types of RNN can be categorized as:

1. One to one—One input to one output
2. One to many—One input to multiple outputs
3. Many to one—Multiple inputs to one output
4. Many to many—Multiple inputs to multiple outputs

The functional block diagram shown in Fig. 7.6 is for one-to-one type.

Next we will discuss how do neural networks work. In order to provide a clear and complete answer to that question, we need to have more knowledge about deep learning since neural network is only a structure or a backbone for deep learning. In fact, neural networks must be trained by deep learning algorithms to make it a powerful AI tool to solve real problems.

7.4   How Do Neural Networks Work?

To make neural networks work properly, different algorithms are needed to be used to train the neural networks to get different models. The most popular training algorithms are supervised learning algorithms, which role is to feed a set of inputs to a neural network via input layers and go through all hidden layers with various guessed weight factors and threshold values to derive or predict the outputs. The predicted outputs would be then compared with the known actual outputs, and the

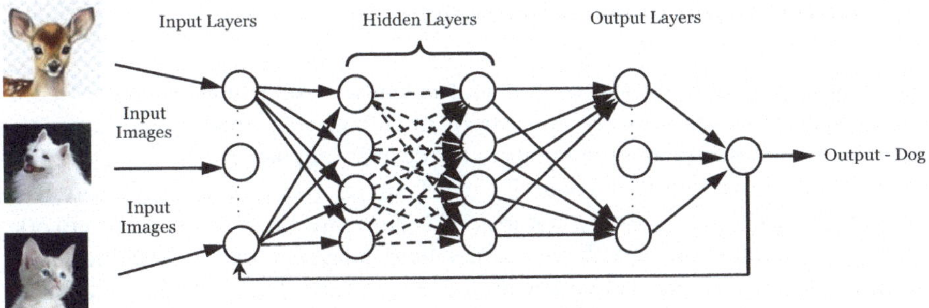

Fig. 7.7 A deep learning training process for a neural network

errors between them are used to feed back to the inputs via neural networks to adjust all weight factors and threshold values to minimize the target errors. To do those, multiple iterations are needed to repeat this kind of backpropagation for maximum accuracy. For each iteration process, the weight factor at every interconnection is adjusted based on the error.

The trained model or neural networks have the optimal weight factors on each node and can be used to automatically estimate or predict the outputs based on new input data.

Let us use a real image processing process to illustrate how a neural network works with a special deep learning algorithm.

As shown in Fig. 7.7, to start the training process for a neural network, some animal images, such as deer, dog, and cat, are fed into a neural network via input layers as the input or prediction data sources. Each kind of image can be considered as a 2D digital matrix composed of 256×256 pixels (for grayscale images). For the color images, they can be thought of as 3D matrix by adding the color codes as the third dimension. Also for each kind of image, a related label is attached to that input as the output of the network. During the training process, all weight factors and threshold values are adjusted based on each input image-output label pair. The output labels will then be fed back to the input layers to compare with the input images. This process will be repeated until the errors between the output label and the input image's label are close to or below an acceptable level.

After the training process is completed, a testing or validation process may be needed to confirm the correctness and accuracy of the trained model by using another group of testing or validation data. Finally, the trained model can be used as an AI system to perform automatic intelligence functions to identify or classify different animal images with new input animal images as inputs.

For example, we can use three labels, **dog**, **deer**, and **cat** to match each related animal image. If the input image is a **dog**, the output label for **dog** is 1, but 0 for both **cat** and **deer** images. Similarly, if input image is a **cat**, the output label for **cat** is 1, but 0 for both other images. This process is very similar to an Error Correcting Output Code (ECOC) algorithm to use multi-binary class classifications for multi-class classifications.

7.5   Introduction to Deep Learning

Deep learning models are based on neural network architectures. Inspired by the human brain, a neural network consists of interconnected nodes or neurons in a layered structure that relates the inputs to the desired outputs. The neurons between the input and output layers of a neural network are referred to as hidden layers. The term *deep* usually refers to the number of hidden layers in the neural network. Generally a neural network with one to three hidden layers is called a shallow learning network, but a neural network with more than three hidden layers is called a deep learning network. Deep learning models can have hundreds or even thousands of hidden layers.

Various deep learning algorithms can be used to train neural networks with large sets of labeled input and output data, and all different features can be learned directly from the data without the need for manual feature extraction. While the first artificial neural network was theorized in 1958, deep learning requires substantial computing power that was not available until the 2000s. Now, researchers have access to computing resources that make it possible to build and train networks with hundreds of connections and neurons.

From another angle to view deep learning, deep learning refers to **computer-simulate** or **automate** human learning processes from a source, such as a collection of images of animals to a learned object (a kind of animal). Therefore, a notion coined as **deeper learning** or **deepest learning** [7] makes sense. The deepest learning refers to the fully automatic learning from a source to a final learned object. Deeper learning thus refers to a mixed learning process: a human learning process from a source to a learned semi-object, followed by a computer learning process from the human learned semi-object to a final learned object [8].

By strict definition, a deep neural network, or DNN, is a neural network with three or more layers. In practice, most DNNs have many more layers. DNNs are trained on large amounts of input-output data pairs to identify and classify phenomena, recognize patterns and relationships, evaluate possibilities, and make predictions and decisions. While a single-layer neural network can make useful, approximate predictions and decisions, the additional layers in a deep neural network help refine and optimize those outcomes for greater accuracy [9].

High-performance GPUs have a parallel architecture that is efficient for deep learning. When combined with clusters or cloud computing, this enables development teams to reduce training time for a deep learning network from weeks to hours or less.

As we discussed in Sect. 7.3, four popular deep learning algorithms are implemented in most applications, feedforward neural network algorithm, backpropagation neural network algorithm, convolutional neural network algorithm, and recurrent neural network algorithm.

To use deep learning algorithms to train neural networks to obtain ideal models, two possible ways can be utilized:

1. **Training from Scratch**: To train a deep learning model from scratch, you select a deep learning algorithm, collect a larger labeled dataset including input and output pairs, and design a network architecture that will learn the features and model. This is a good approach for new or specific applications, or more generally, applications for which preexisting models do not exist. The main disadvantage of this approach is that it requires a large dataset and the training time can take from hours to weeks, depending on your task and computing resources.

2. **Transfer Learning**: In some deep learning applications, such as image identifications or classifications, computer visions, audio processing, and natural language processing, the transfer learning approach is commonly used. It involves fine-tuning a pretrained deep learning model. You start with an existing model, such as **SqueezeNet** or **GoogLeNet** for image classification, and feed in new data containing previously unseen classes. After making some tweaks to the network, you can now perform a new task, such as categorizing only dogs or cats instead of 1000 different objects. This also has the advantage of needing much less data, so the training time drops significantly.

A pretrained deep learning model can also be used as a feature extractor. You can use the layer activations as features to train another machine learning model, such as a support vector machine (SVM) or an error correcting output code (ECOC) algorithm. Or you can use the pretrained model as a building block for another deep learning model. For example, you can use an image classification CNN as the feature extractor for an object detector.

Due to its powerful learning-decision abilities and high accuracy, deep learning algorithms have been widely implemented in most high-tech fields, including

1. Computer visions and image processing
2. Biomedical image identifications and classifications
3. Audio signal identifications and classifications
4. Automatic speech recognitions
5. Neural language processing
6. Virtual art processing
7. Drug discovery and toxicology
8. Natural disasters predictions
9. Medical image analysis
10. Financial fraud detections
11. Materials science
12. Military and battle field applications

As we mentioned in previous sections, deep learning is a subset of machine learning, which means that deep learning is derived from machine learning, but the former provides more powerful functions compared with the latter. This situation is very similar to two classes, a base class and a derived class in C++ programming. The derived class from the base class can access all member data and member functions in base class, but it can also build some additional member data and member functions that cannot be used by base class. This means that the derived class provides more member data and functions and therefore has more functionalities with more powerful ability over the base class.

Next let us concentrate our discussions on these two AI technologies.

7.5.1 *Deep Learning Versus Machine Learning*

Basically, machine learning algorithms leverage structured and labeled data to make predictions, which means that specific features are defined from the input data for the model and organized into tables. This does not necessarily mean that it does not use unstructured data; it just means that if it does, it generally goes through some data preprocessing to organize them into a structured format [9]. In other words, the input and output data pairs used for machine learning algorithms must be managed and preprocessed by human beings by adding appropriate labels to enable the machine learning models to recognize and use them correctly to train the desired models.

Deep learning eliminates some of the data preprocessing that is typically involved with machine learning. These algorithms can ingest and process unstructured data, like text and images, and it automates feature extraction, removing some of the dependency on human experts. As shown in Fig. 7.8, in machine learning, you need to extract all features for the input objects yourself, but for deep learning, the deep learning algorithms can do that for you. Another example is, let us say that we had a set of photos of different pets, and we wanted to categorize them by **cat**, **dog**, **hamster**, and so on. Deep learning algorithms can determine which features, such as ears, are most important to distinguish each animal from another. In machine learning, this hierarchy of features is established manually by a human expert.

Then, through the processes of gradient descent and backpropagation, the deep learning algorithm adjusts and fits itself for accuracy, allowing it to make predictions about a new photo of an animal with increased precision.

Machine learning and deep learning models are capable of different types of learning as well, which are usually categorized as supervised learning, unsupervised learning, and reinforcement learning. Supervised learning utilizes labeled datasets to categorize or make predictions; this requires some kind of human intervention to label input data correctly. In contrast, unsupervised learning does not require labeled datasets, instead, it detects patterns in the data, clustering them by

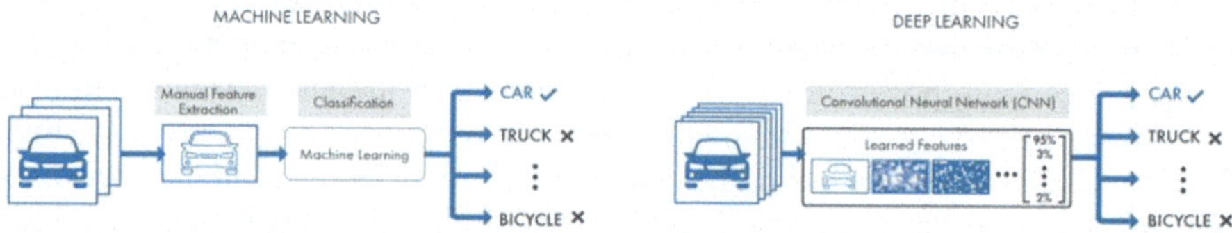

Fig. 7.8 A comparison between machine learning and deep learning [9]

any distinguishing characteristics. Reinforcement learning is a process in which a model learns to become more accurate in performing an action in an environment based on feedback in order to maximize the reward [10].

Now that we have a clear picture about neural networks and deep learning algorithms, next let us have a deep discussion about how neural networks work, or exactly how to use deep learning algorithms to train neural networks to build desired deep learning models to perform actual artificial intelligence predictions or classifications.

7.6 Deep Learning in MATLAB

To handle deep learning-related implementations and applications, MATLAB provided a powerful Toolbox, called Deep Learning Toolbox, to help users design, develop, and build professional and practical implementations by using all kinds of deep learning techniques.

In fact, Deep Learning Toolbox provides Apps, functions, and Simulink® blocks for designing, implementing, and simulating deep neural networks. The toolbox provides a framework to create and use many types of networks, such as convolutional neural networks (CNNs), feedforward and backpropagation networks and transformers. You can visualize and interpret network predictions, verify network properties, and compress networks with quantization, projection, or pruning.

Basically, MATLAB provided two major methods to help users to build their deep learning-related applications:

1. User-friendly Apps that provided some easy-to-use GUIs and tools
2. Detailed coding process by calling various deep learning functions

All Apps provided by MATLAB provided some user-friendly GUIs, combined with various tools and related functions, to set up a good developing and convenient environment for users to seep up and facilitate their designing and developing process for all kinds of deep learning-related implementations in simple and quick ways.

In addition to those Apps, MATLAB also provided various deep learning functions residing in deep learning function library.

Relatively speaking, the Apps method provided an easy and convenient way to enable users to quickly and conveniently design, develop, and build their deep learning-related applications in a more efficient way, which is a good starting point for beginners. But some detailed coding processes using related deep learning functions and methods are hidden for those users, which is a shortcoming for them. Therefore, a more powerful and useful method with a detailed coding process is necessary for them to understand and master real deep learning techniques.

Let us have a closer look at these two methods.

7.6.1   Deep Learning Related Apps

Exactly, MATLAB provided two major types of Apps:

1. Artificial neural networks (ANN)—used for shallow neural networks
2. Deep network designer (DND)—used for deep neural networks

Both ANN and DND Apps can be triggered from the **APPS** icon on MATLAB window and by typing commands from the MATLAB Command window.

With the artificial neural networks (ANN) App and deep network designer (DND) App, users can design, edit, and analyze networks interactively, import pre-trained models, and export networks to Simulink. The toolbox lets you interoperate with other deep learning frameworks. You can import PyTorch®, TensorFlow™, and ONNX™ models for inference, transfer learning, simulation, and deployment. You can also export models to TensorFlow and ONNX.

To open the ANN App, two ways can be adopted:

1. Use **APPS** icon on the MATLAB window
2. Type **nnstart** command in the MATLAB Command window

Let us illustrate this opening with the first way as an example.

Open MATLAB and click on the **APPS** icon, click on the drop-down arrow and browse to **MACHINE LEARNING AND DEEP LEARNING** category. Four tools under that category are ANN-related App, as shown in Fig. 7.9.

1. Neural Net Fitting—**nftool**
2. Neural Net Pattern Recognition—**nprtool**
3. Neural Net Time Series—**ntstool**
4. Neural Net Clustering—**nctool**

Click one of them to open the related App, and most of them belong to two layers of shallow neural networks. One can also open each tool by typing the related command that is attached at the end of each icon shown above. For example, to open the **Neural Net Fitting** tool, just type **nftool** on the MATLAB Command window.

The second way to open these tools is to type **nnstart** command and press the **Enter** key on the keyboard in the MATLAB Command window, and an opened sample is shown in Fig. 7.10.

When clicking one of the tool's buttons, such as **Fitting** shown in Fig. 7.10, the **Neural Net Fitting** tool is opened, as shown in Fig. 7.11. It can be found that the opened tool is a shallow neural network with two layers, one hidden and one output

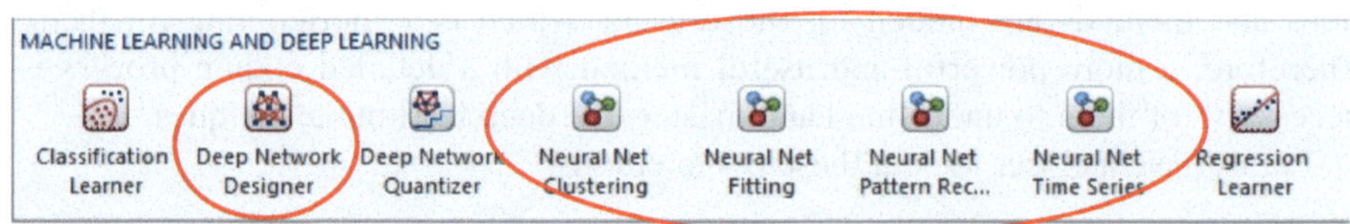

Fig. 7.9 The opened APPS icon for machine learning and deep learning tools

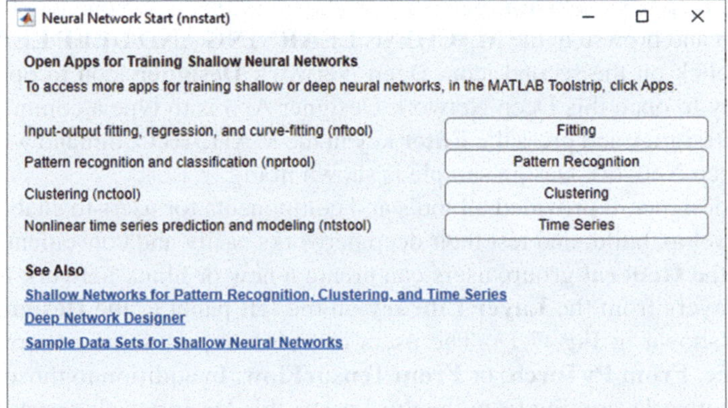

Fig. 7.10 The opened App tools by using the command nnstart

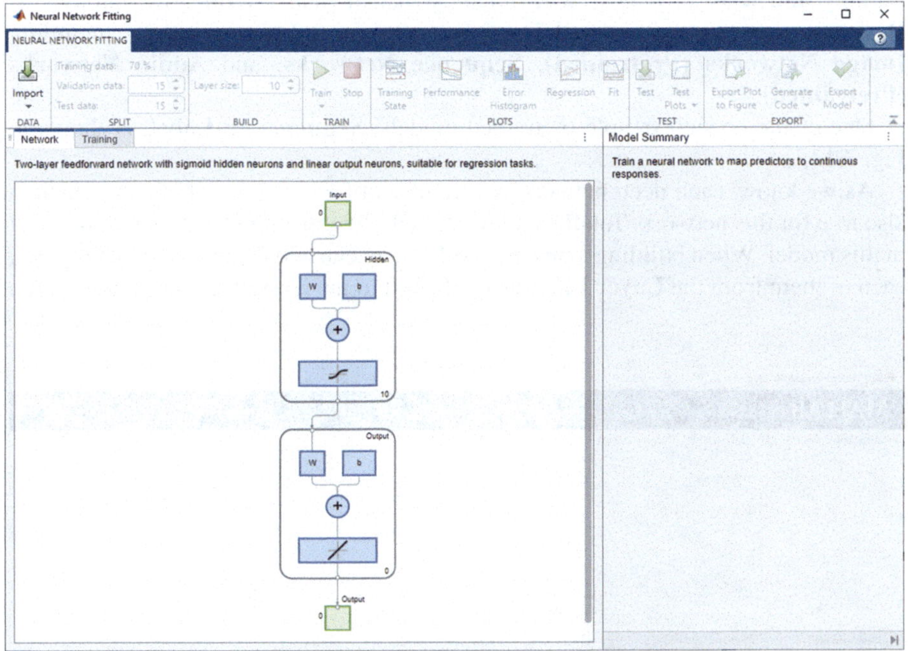

Fig. 7.11 The opened Neural Network Fitting tool

layer. This App allows users to design and build a shallow neural network with the imported or users' data.

Another App, **Deep Network Designer** (DND), provided more powerful App and tools for deep learning algorithms and applications. Similar to the first ANN App, two ways can be used to open that App,

1. Use **APPS** icon on the MATLAB window
2. Type **deepNetworkDesigner** command in the MATLAB Command window

To open a **deepNetworkDesigner** App, as shown in Fig. 7.9, you can click on the **APPS** icon and browse to the **MACHINE LEARNING AND DEEP LEARNING** category, click on the second icon, **Deep Network Designer** icon to open it. The second way to open this Deep Network Designer App is to type a command **deepNetworkDesigner** and press the **Enter** key in the MATLAB Command window. An opened Deep Network Design sample is shown in Fig. 7.12.

In this designer, it provided all tools and components for users to enable them to design, develop, build, and test their deep networks easily and conveniently.

Under the **General** group, users can create a new or blank network by adding different layers from the **Layer Library** on the left panel to the **Designer** on the center, as shown in Fig. 7.13. The users can also import some networks **From Workspace**, **From PyTorch**, or **From TensorFlow**. In addition to those new networks or networks coming from the third party, this **Designer** also provided quite a few pretrained networks to enable users to do some small modifications for those models to speed up their developing process to avoid long time-consuming training and testing process with huge blocks of data. Scroll down along this **Designer**, more pretrained models can be found from different groups, including **Image Networks (Pretrained)**, **Sequence Networks**, and **Audio Networks (Pretrained)**.

One of the opened sample pretrained models, **Sequence-to-Label**, is shown in Fig. 7.13.

As we know, each deep network is composed of a sequence of layers, which is also true for this network. Totally six layers, including the input and the output, exist in this model. When building a new network, users can add these layers by dragging each of them from the **Layer Library** on the left to the **Designer** pane in the center.

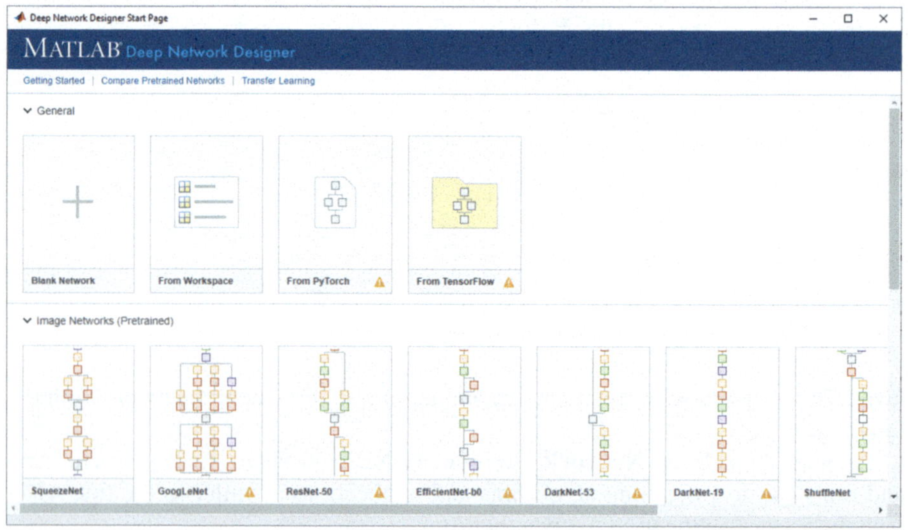

Fig. 7.12 The opened Deep Network Designer

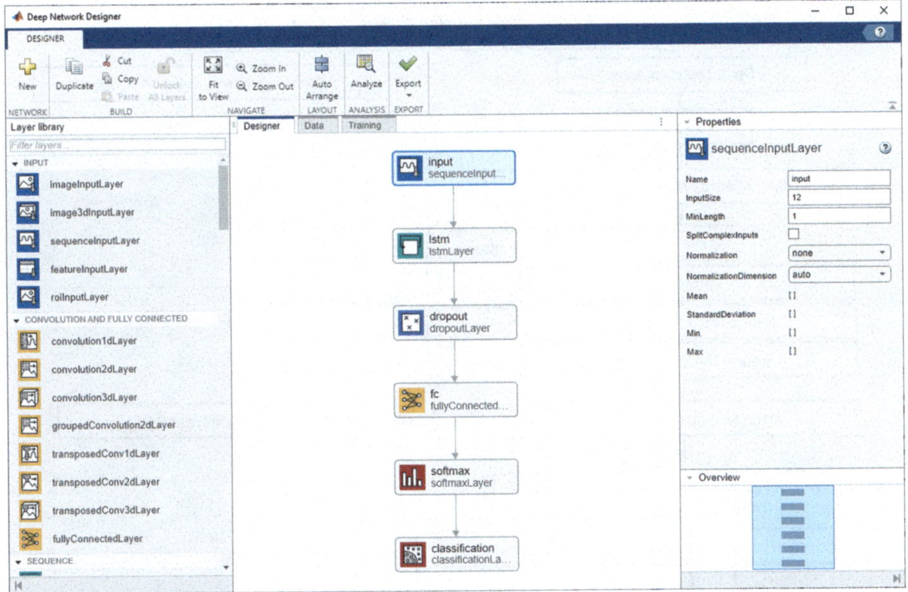

Fig. 7.13 The opened pretrained neural network sequence for classification

The advantages of using this **Designer** to design, build, and develop a user-defined network include but not limited to the following:

1. Some GUIs are provided to facilitate and speed up the designing and building process.
2. A general or predefined network structure is provided to save a lot of designing time.
3. Each layer can be changed or modified based on the users' special requirements.
4. No huge blocks of codes are involved and that is a headache to beginners.

An illustration of the relations between Apps and command functions is shown in Fig. 7.14.

Any good thing must contain some side effects, which is also true for this Designer. One of the most important shortcomings of using this Designer is that the users cannot see what is exactly happening under those layers. This may introduce some inconvenient factors with more details about the networks. To improve this side effect, MATLAB also provided some ways to compensate for this issue:

1. In the related App, users can use **Generate Code** option to convert the users' design and trained networks into MATLAB codes to enable users to have a clear picture and better understanding about what is happening at the coding level with different functions.
2. MATLAB also provided some functional libraries to enable users to design, build, and develop their networks with codes by calling different functions directly. All of those functions are involved in related Toolboxes, such as

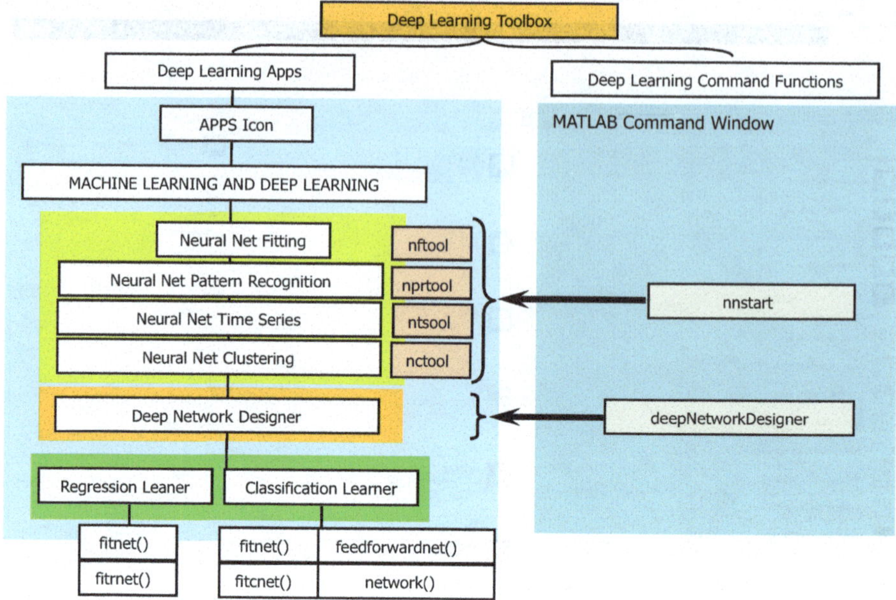

Fig. 7.14 A global description about relations between Apps and functions

Machine Learning and Deep Learning Toolbox and Statistics and Machine Learning Toolbox.

Now let us have a closer look at those functions.

7.6.2　Deep Learning-Related Functions

The functions are included in the MATLAB ANN Function Library. These functions provide a base or foundation for ANN App tool. By using these functions, one can build any ANN model directly with more flexibility but needs more knowledge and better understanding of these functions.

As we discussed in Sect. 5.2.3 in Chap. 5, exactly in Table 5.1, three neural networks-related fitting functions, **fitnet()**, **fitrnet()** and **fitcnet()**, were introduced. The first is used to fit or train a general neural network, and the second and the third are used to fit or train a neural network with regression or classification mode. Regularly, the first one provided a more general function to fit or train most neural networks with better performances, and the others were specially used for networks with regression or classification functions.

Table 7.1 Some popular and important network fitting and training functions

Function Name	Descriptions
fitnet()	Fit and train a general neural network.
fitrnet()	Fit and train a neural network regression model.
fitcnet()	Fit and train neural network classification model.
feedforwardnet()	Generate feedforward neural network.
network()	Create custom shallow neural network.
trainNetwork()	Train shallow or deep neural networks.
trainnet()	Train a deep learning neural network.
trainingOptions()	Create options for training deep learning neural network. To train a neural network, use the training options as an input argument to the **trainnet()** function.
sequenceInputLayer()	Create a sequence input layer to input sequence data to a neural network and apply data normalization.
addLayers()	Add the network layers in `layers` to the `network object net`.
view()	Display the structure of a shallow neural network
perform()	Evaluate the performance of a trained neural network

Generally, function fitting is the process of training a neural network on a set of inputs in order to produce an associated set of target outputs. After you construct the network with the desired hidden layers and the training algorithm, you must train it using a set of training data. Once the neural network has fit the data, it forms a generalization of the input-output relationship. You can then use the trained network to generate outputs for inputs it was not trained on.

Table 7.1 shows some most popular functions used to generate, train, view, and evaluate the trained neural networks. Let us have a closer look at these functions to see how they can be used to perform desired functions.

7.6.2.1   The fitnet() Function

Generally, most neural networks are good at fitting functions. In fact, there is proof that a fairly simple or shallow neural network can fit any practical function. In fact, the **fitnet()** function is good and suitable for shallow network designing and implementation.

Two popular constructors of the function **fitnet()** are:

1. **net = fitnet(hiddenSizes)**
2. **net = fitnet(hiddenSizes, trainFcn)**

The argument **hiddenSizes** is used to define how many nodes or neurons are on each layer, and it is an integer value (default is 10). The argument **trainFcn** is used to define which training method or function to be used (default is **trainlm()**). The generated shallow neural network is assigned to the network object **net**. Table 7.2 shows the most popular training functions.

Table 7.2 Some popular and important training functions

Function Name	Descriptions
`'trainlm'`	Levenberg-Marquardt
`'trainbr'`	Bayesian Regularization
`'trainbfg'`	BFGS Quasi-Newton
`'trainrp'`	Resilient Backpropagation
`'trainscg'`	Scaled Conjugate Gradient
`'traincgb'`	Conjugate Gradient with Powell/Beale Restarts
`'traincgf'`	Fletcher-Powell Conjugate Gradient
`'traincgp'`	Polak-Ribiére Conjugate Gradient
`'trainoss'`	One Step Secant
`'traingdx'`	Variable Learning Rate Gradient Descent
`'traingdm'`	Gradient Descent with Momentum
`'traingd'`	Gradient Descent

7.6.2.2 The fitnet() Function

This function is used to train a feedforward, fully connected neural network for regression.

The first fully connected layer of the neural network has a connection from the network input (predictor data), and each subsequent layer has a connection from the previous layer. Each fully connected layer multiplies the input by a weight matrix and then adds a bias vector. An activation function follows each fully connected layer, excluding the last. The final fully connected layer produces the network's output, namely predicted response values.

Four popular constructors of the function **fitnet()** are:

1. **Mdl = fitnet(Tbl, ResponseVarName)**
2. **Mdl = fitnet(Tbl, Y)**
3. **Mdl = fitnet(X, Y)**
4. **Mdl = fitnet(___, Name, Value)**

The argument **Tbl** includes only the predictors or input data arranged in the table format and the response values in the **ResponseVarName** table variable. The argument **Y** is a vector, and it only contains the response values. The third constructor returns a neural network regression model trained using the predictors or input data in the matrix **X** and the response values in vector **Y**. The **Name, Value** pairs provide additional options for this fitting function.

Some popular and important Name-Value pairs are shown in Table 7.3.

In fact, when calling the **fitnet()** function, a **RegressionNeuralNetwork** object that is a trained, feedforward, and fully connected neural network for regression is generated. The first fully connected layer of the neural network has a connection from the network input or predictor data **X**, and each subsequent layer has a

Table 7.3 Some popular and important Name-Value pairs

Function Name	Descriptions
`'LayerSizes'`	Sizes of the fully connected layers in the neural network model, specified as a positive integer vector. **LayerSizes** does not include the size of the final fully connected layer.
`'LayerWeightsInitializer'`	Function to initialize fully connected layer weights. **'glorot'** (default) | **'he'**.
`'LayerBiasesInitializer'`	Type of initial fully connected layer biases. **'zeros'** (default) | **'ones'**.
`'Activations'`	Activation functions for the fully connected layers of the neural network model, specified as a character vector, string scalar, string array, or cell array of character vectors with values from this table. **'relu'** (default)|**'tanh'**|**'sigmoid'**| **'none'**| string array | cell array of character vectors.
`'ObservationsIn'`	Predictor data observation dimension. **'rows'** (default) | **'columns'**.
`'IterationLimit'`	Maximum number of training iterations. **1e3** (default) | positive integer scalar.
`'LossTolerance'`	Loss tolerance. **1e-6** (default) | nonnegative scalar.
`'StepTolerance'`	Step size tolerance. **1e-6** (default) | nonnegative scalar
`'ValidationData'`	Validation data for training convergence detection. cell array | table.
`'CategoricalPredictors'`	Categorical predictors list. Vector of positive integers | logical vector | character matrix | string array | cell array of character vectors | 'all'.
`'CrossVal'`	Flag to train cross-validated model. **'off'** (default) | 'on'.
`'KFold'`	Number of folds. **10** (default) | positive integer value greater than 1.
`'Lambda'`	Regularization term strength. 0 (default) | nonnegative scalar. Regularization term strength, specified as a nonnegative scalar. The software composes the objective function for minimization from the mean squared error (MSE) loss function and the ridge (L2) penalty term. **Example:** 'Lambda',1e-4.

connection from the previous layer. Each fully connected layer multiplies the input by a weight matrix **LayerWeights** and then adds a bias vector **LayerBiases**. An activation function follows each fully connected layer, excluding the last **Activations** and **OutputLayerActivation**. The final fully connected layer produces the network's output, namely predicted response values.

A typical structure of neural network regression model is shown in Fig. 7.15a.

7.6.2.3 The fitcnet() Function

Use **fitcnet**() function to train a feedforward, fully connected neural network for classification purpose. The first fully connected layer of the neural network has a connection from the network input or predictor data, and each subsequent layer has a connection from the previous layer. Each fully connected layer multiplies the input by a weight matrix and then adds a bias vector. An activation function follows each fully connected layer. The final fully connected layer and the subsequent **softmax** activation function produce the network's output, namely classification scores or posterior probabilities and predicted labels.

The **fitcnet**() function has not only similar four constructors as the **fitrnet**() function did but also has all similar Name-Value pairs as shown in Table 7.3.

The **fitcnet**() function is very similar to the function **fitrnet**(). The only difference between them is due to the last layer prior to the output layer in the former

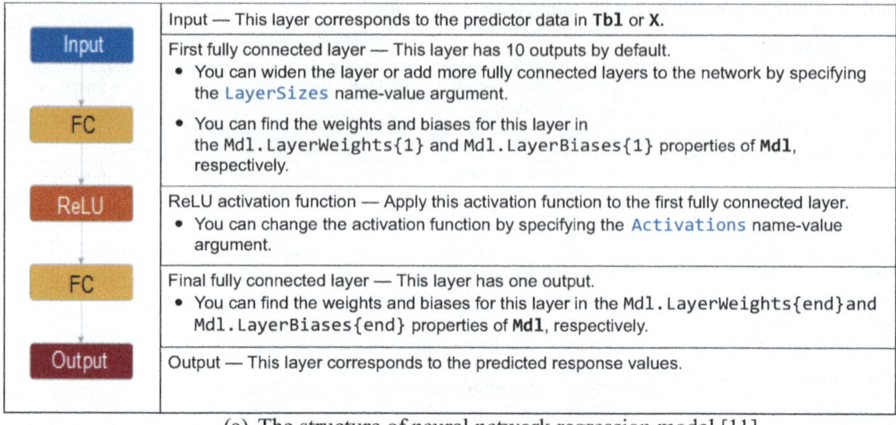

(a) The structure of neural network regression model [11].

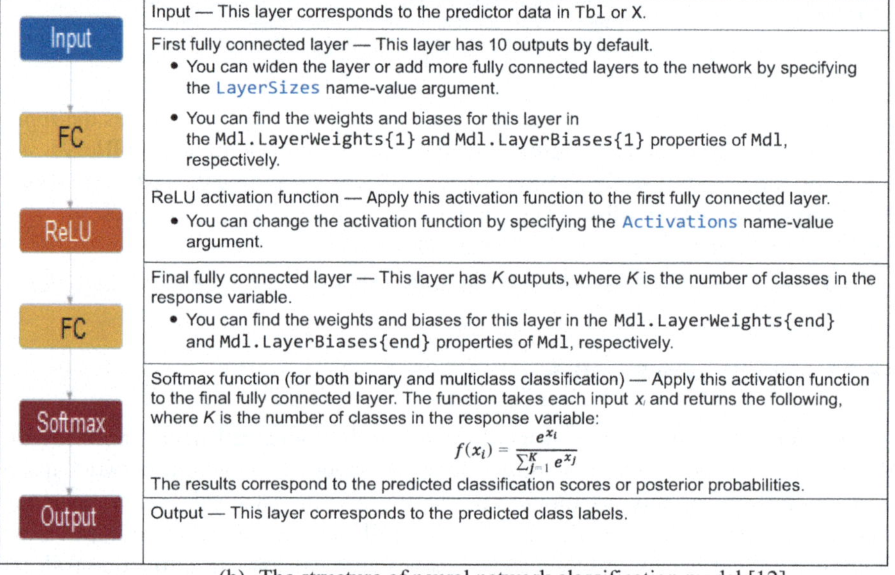

(b) The structure of neural network classification model [12].

Fig. 7.15 The structures of neural network for regression and classification model. (**a**) The structure of neural network regression model [11]. (**b**) The structure of neural network classification model [12]

since it will output a category or a class value, not a continuous value as the **fitrnet()** function did. This layer is called **Softmax**, and it is used to convert the classification results to some category or class level values.

A typical structure of neural network classification model is shown in Fig. 7.15b.

In fact, when calling the **fitcnet()** function, a **ClassificationNeuralNetwork** object that is a trained, feedforward, and fully connected neural network for classification is created. The first fully connected layer of the neural network has a connection from the network input or predictor data **X**, and each subsequent layer has a connection from the previous layer. Each fully connected layer multiplies the input by a weight matrix **LayerWeights** and then adds a bias vector **LayerBiases**. An activation function follows each fully connected layer **Activations** and **OutputLayerActivation**. The final fully connected layer and the subsequent **Softmax** activation function produce the network's output, namely classification scores or posterior probabilities and predicted labels.

Before we can continue to introduce more other functions shown in Table 7.1, we had better have a clear picture about two popular structures used for two major neural network application models, the neural network regression model and the classification model.

7.6.2.4 Two Popular Neural Network Structures

Basically a neural network can be designed and built in either a regression model or a classification model based on its application requirements. Therefore two popular neural network structures, one for regression function and the other for the classification function, are developed. Figures 7.15a, b show these two kinds of structures.

It can be found from Fig. 7.15 that the default neural network regression model and the default neural network classification model have a similar structure except for one more layer of **Softmax** that is located at the bottom and prior to the output layer in the classification model. Besides that layer, two models have the same or identical structure, which includes the following:

1. The input layer—Get input or predictor data from a dataset in **Tbl** or **X** format. The fully connected (**FC**) layer—Set up the **layerSizes**, **layerWeights**, and **layerBiases**.
2. The **ReLU** activation layer—Set and select different nonlinear functions, as shown in Table 7.3, to derive the input and output relations between the neighbored hidden layers.
3. Another fully connected (**FC**) layer that belongs to multiple hidden layers.

4. The **Softmax** layer—Used for the final fully connected layer to convert the last
   layer's outputs to the predicted class scores or possible classification values.
5. Finally the output layer—Create the predicted class labels.

Next let us continue our discussions about those functions used for neural networks.

7.6.2.5   The feedforwardnet() Function

This function is very similar to function **fitnet()**, and both functions have the same
two constructors:

1. **net = feedforwardnet(hiddenSizes)**
2. **net = feedforwardnet(hiddenSizes, trainFcn)**

Exactly, this function returns a feedforward neural network with a hidden layer size
of **hiddenSizes** and training function specified by **trainFcn**. The **haddenSizes** is
used to indicate how many nodes of neurons can be included in layers, and the
trainFcn can be any training function shown in Table 7.2.

7.6.2.6   The network() Function

This function is used to create a customer shallow network with or without arguments, and the related constructors are:

1. **net = network**
2. **net = network(numInputs,numLayers,bias,inputConnect,layerConnect,out
   putConnect)**

The first constructor returns a new shallow neural network with no inputs, layers, or
outputs. The second constructor returns a new neural network with all predefined
arguments. Table 7.4 shows all architecture properties.

   A piece of sample codes used to create a customer neural network by using this
function is:

```
net = network
net.numInputs = 1
net.numLayers = 2
```

   Also you can use the following coding line to replace the above three lines to do
the same thing.

```
net = network(1,2)
```

Table 7.4 Architecture properties

net.numInputs	0 or a positive integer	Number of inputs.
net.numLayers	0 or a positive integer	Number of layers.
net.biasConnect	numLayer-by-1 Boolean vector	If net.biasConnect(i) is 1, then layer i has a bias, and net.biases{i} is a structure describing that bias.
net.inputConnect	numLayer-by-numInputs Boolean vector	If net.inputConnect(i,j) is 1, then layer i has a weight coming from input j, and net.inputWeights{i,j} is a structure describing that weight.
net.layerConnect	numLayer-by-numLayers Boolean vector	If net.layerConnect(i,j) is 1, then layer i has a weight coming from layer j, and net.layerWeights{i,j} is a structure describing that weight.
net.outputConnect	1-by-numLayers Boolean vector	If net.outputConnect(i) is 1, then the network has an output from layer i, and net.outputs{i} is a structure describing that output.
net.numOutputs	0 or a positive integer (read only)	Number of network outputs according to net.outputConnect.
net.numInputDelays	0 or a positive integer (read only)	Maximum input delay according to all net.inputWeights{i,j}.delays.
net.numLayerDelays	0 or a positive number (read only)	Maximum layer delay according to all net.layerWeights{i,j}.delays.

7.6.2.7   The trainNetwork() Function

This function is used to train a neural network with the following constructors:

1. **trainedNet = trainNetwork(images, layers, options)**
2. **trainedNet = trainNetwork(images, responses, layers, options)**
3. **trainedNet = trainNetwork(sequences, layers, options)**
4. **trainedNet = trainNetwork(sequences, responses, layers, options)**
5. **trainedNet = trainNetwork(features, layers, options)**
6. **trainedNet = trainNetwork(features, responses, layers, options)**

Three different input objects can be used to train a neural network, which are images, sequences, and features. This function can be used as either a classification or a regression purpose in training a neural network. The argument **responses** can be either a class or a numeric value for classification or a regression operation. An example coding line is:

```
net = trainNetwork(imdsTrain, layers, options);
```

where the first argument **imdsTrain** is the training input that can be images, sequences, or features.

7.6.2.8 The trainnet() Function

This function is used to train a deep learning neural network with the following constructors:

1. **netTrained = trainnet(images, net, lossFcn, options)**
2. **netTrained = trainnet(images, targets, net, lossFcn, options)**
3. **netTrained = trainnet(sequences, net, lossFcn, options)**
4. **netTrained = trainnet(sequences, targets, net, lossFcn, options)**
5. **netTrained = trainnet(features, net, lossFcn, options)**
6. **netTrained = trainnet(features, targets, net, lossFcn, options)**
7. **netTrained = trainnet(data, net, lossFcn, options)**
8. **[netTrained,info] = trainnet(___)**

The first two constructors are used to train deep learning neural networks for images. The first one is to use a generated net and training images as input data with loss function and options to train a deep learning network. The second constructor is used to classify input images based on the target labels.

The third and fourth constructors are used for sequence or time-series task classifications, and the fifth and sixth constructors are used for feature classifications. The seventh constructor is used to train a neural network with other data layouts or combinations of different types of data. The last constructor is to retrieve information on the training with any of the previous syntaxes.

7.6.2.9 The trainingOptions() Function

This function is used to create options for training deep learning neural network. To train a neural network, use the training options as an input argument to the **trainnet()** function. This function provided the following two constructors:

1. **options = trainingOptions(optionName)**
2. **options = trainingOptions(optionName, Name=Value)**

The argument **optionName** is a collection or a group of options used to specify the training process. An example coding process for this option selection is:

```
options = trainingOptions("sgdm", ...
MaxEpochs=8, ...
Metrics = ["accuracy","fscore"], ...
ValidationData={XTest, labelsTest}, ...
ValidationFrequency=30, ...
Verbose=false, ...
MiniBatchSize=64, ...
Plots="training-progress");
```

The training solver, **sgdm**, is used for this training process with maximum epochs as 8. The **validation data** is used and indicated with testing data and target labels. The **validation frequency** is 30 without verbose. The minimum batch size is 64 and the training process is plotted.

7.6.2.10   The sequenceInputLayer() Function

Neural networks are composed of multiple layers, including the input layers, hidden layers, and output layers. In fact, much more layers exist under those three main layer groups. The **sequenceInputLayer()** function is only one of those input layers under the **Input Layers** group. In fact, under that group, there are some other input layers, which include:

1. **imageInputLayer()**
2. **image3dInputLayer()**
3. **sequenceInputLayer()**
4. **featureInputLayer()**
5. **roiInputLayer()**
6. **inputLayer()**

Different layers shown above are used for different purposes with different functions. The **imageInputLayer()** and **image3dInputLayer()** functions are used to generate image classification–related input layers. The difference between them is that the former is used for grayscale images, but the latter is used for color images as inputs. The **sequenceInputLayer()** function is specially used to build input layers for sequence and time series inputs. The **featureInputLayer()** function is used for tabular and feature data inputs, such as extracted features from audio inputs. The **roiInputLayer()** function is used to generate input images to a **Fast R-CNN** object detection network. The **inputLayer()** function is to input data into a neural network with a custom format.

We will discuss them one by one later when we use it. All of those layers can be used individually or combined with some other layer properties. Now let us only concentrate our discussions on the **sequenceInputLayer()** function first. When it is used individually, the **sequenceInputLayer()** has two constructors, which are:

```
1. layer = sequenceInputLayer(inputSize)
2. layer = sequenceInputLayer(inputSize,Name=Value)
```

Two constructors perform a similar job, which is to input sequence data to a neural network and apply data normalization. The first argument **inputSize** is used to define the number of nodes on the input layer, and the second argument, **Name=Value** pairs, is used to provide additional functions, such as **MinLength, Normalization, Mean,** and **StandardDeviation**.

Some coding examples are:

```
1. layer = sequenceInputLayer(12)
2. layers = [ ...
       sequenceInputLayer(12)
       lstmLayer(100,OutputMode="last")
       fullyConnectedLayer(9)
       softmaxLayer]
```

The first **sequenceInputLayer()** function is used individually to generate a sequence input layer with 12 nodes. The second function is to generate a combined input layer with additional properties, including 100 LSTM hidden layers and 9 output classes.

7.6.2.11   The addLayers() Function

The purpose of this function is to add layers to neural networks. One constructor can be used to create this function:

```
net = addLayers(net,layers)
```

The updated network **net** contains the layers and connections of **net** together with the layers in **layers**, connected sequentially. The second argument, layer names in **layers**, must be unique, nonempty, and different from the names of the layers in **net**. An example is:

```
layers = [ imageInputLayer([32 32 3])
           convolution2dLayer(3,16,Padding="same")
            eluLayer];
net = addLayers(net,layers);
```

7.6.2.12   The view() Function

The **view()** function in MATLAB has multiple purposes. Generally we just use it to display a structure of a trained network or model, such as **view(net)**, where **net** is a trained neural network model.

7.6.2.13   The perform() Function

This function is used to calculate the network performance. The syntax of this function is:

```
perf = perform(net,T,Y,EW)
```

where the **net** is a trained neural network, **T** is the target, **Y** is the output, and **EW** is the error weight factor. This function returns **perf** that is exactly an error between the predicted output and the target output.

7.6.3   The Workflow and Steps of Deep Learning Algorithms on Neural Networks

Regularly, the following workflow for developing and building any deep neural networks should be adopted and used:

1. Collect and preprocess data
2. Create the network and split dataset
3. Train the network
4. Validate the network
5. Deploy and use the network

As we discussed, in MATLAB two ways can be used to perform this kind of deep neural network training and testing; using Apps provided by Machine Learning and Deep Learning Toolbox, or using the library functions. However, no matter which method you like to use, you must first provide a data source or dataset to set up the input-output pairs to train and test the neural networks to get desired models.

7.6.3.1   Collect and Preprocess Data

The first and most important step to successfully build either shallow or deep neural network models is to get good data or dataset as input sources. The so-called good data or dataset means that they have enough available data as source to train and evaluate our network models. These data or dataset include good quality data, and they can be either easily organized or rebuilt into input-output pairs for regressions, or they can be preprocessed into a set of good labeled data for classifications. Fortunately, there are so many potential public datasets available in various sites, such as **Kaggle**, which hosts thousands of large, labeled data sources. Working with these available datasets helps reduce the overhead of starting a deep learning project.

Not all of data or dataset could be used directly to train any of neural networks; instead, some of them need to be preprocessed to meet the needs or requirements of training and testing our neural network models. Regularly, these preprocessing procedures include the following operations:

1. Filtering and cleaning data
2. Normalizing or scaling data
3. Handling categorical and text data

Generally, our original or raw data contain some noises and disturbances, some undesired features, or missing data. Therefore, those data need to be filtered to remove any undesired noises to make them clear.

Also some data items in dataset may have significant larger or smaller values, which are not appropriate to be used to train or test our models. Thus some normalization processes are needed to make them good and appropriate for our usage. Another way to preprocess these data is to standardize them so they have a mean of zero and a variance of one.

In some situations, the data in a dataset may be a combination of different types of data, such as numeric values, categorical data, or text strings. It is prohibited for some neural network models to directly use categorical or text data, instead those data need to be converted to numeric values before they can be used to train and test our models. Another way to handle this issue is that the users can convert categorical or text data to the related data type, such as categorical, by using the **categorical**() function, and that kind of data type can be accepted by most models in MATLAB Deep Learning Toolboxes.

Another issue is that not all data may have the same length, thus we need to add or do some padding jobs to make them the same in length. All of the operations belong to data preprocessing.

7.6.3.2   Create Network Model and Split Dataset

After the input data or dataset have been preprocessed, next we need to create our neural network model based on the deep learning algorithms. For regressions, we can select and create regression-related networks via Apps or functions. For classifications, we would select and use classification-related models.

For a selected model, we need to determine the number of layers, including the input layers, the hidden layers, and the output layers, the number of nodes or neurons in each layer, the initial weight factors and the threshold values, the activation function, and **Softmax** function if it is a classification model. Both Apps and functions can be used to create our desired models.

Prior to training, checking, and evaluating our selected models, we need to take care of another important issue, to reorganize or split the entire data in our dataset into two or three parts: the training data, checking data, and validation data. Regularly, we select 70% of the entire data as training data and 15% as checking and validation data. Some other selections, such as 80% as training data and 20% as checking data, could be used if the validation data could be selected from other data sources.

Some issues may be involved when performing the data splitting, which include the imbalanced data distributed in our dataset. For example, more data of our minority numeric values are concentrated or distributed at either end of our dataset, at starting or ending parts, for regression models, or more instances of our minority classes end up in either the training or the validation dataset. Any of these issues may make our training or checking results not accurate or incorrect.

To solve these possible issues, a stratified split technique should be utilized. The so-called stratified split is to rearrange data in both datasets to make sure that both training and validation datasets have the same proportion of instances from each class.

There is another definition about the imbalanced dataset, where some classes appear much more than others and pose a challenge for deep learning models. If we train neural networks on those imbalanced data, our resulting model will be heavily biased towards predicting those majority classes. This is especially problematic because usually, we care much more about identifying instances of the minority classes [13].

There are two possible ways we can use to solve this issue; under sampling and over sampling, which is:

1. By removing some data or instances from our majority class to make it balance
2. By duplicating more data or instances of our minority class to get it balance

We used the under sampling technique in Sect. 6.10.2 in Chap. 6 to remove some undesired 3D audio feature cells to enable our animal audio classification to be executed correctly.

A point to be noted is to perform over sampling only after the entire dataset has been split to avoid a so-called information leak.

7.6.3.3   Train the Network Models

Prior to train a deep learning model, we need to consider some key points based on the various applications with the following questions:

1. Is any trained model available? If the answer is yes, we can use those models directly or with some modifications via fine training of those models to meet our requirements.
2. If no trained model is available, we need to build some new models and train them with our selected datasets.

For the first situation, it belongs to the so-called transfer learning, and it will be discussed in more detail in the next part. As for the second case, we need to create some new models based on the discussions in the last section and select our data source to perform the training process to get our desired models.

Before starting our training process, the following factors need to be considered:

1. The training method or algorithm to be used
2. The training criterion or terminal condition
3. The maximum number of epochs (full passes of the data) to use for training
4. The initial learning rate
5. The maximum number of iterations to be used for training
6. The training loss and validation loss functions

The first issue is related to the method to be used to compute the states at successive time steps to train the selected model. In fact, a solver applies a numerical method to solve the set of ordinary differential equations that represent the model. Through this computation, it determines the time of the next simulation step. In the process of solving this initial value problem, the solver also satisfies the accuracy requirements that you specify [14].

The following solvers are popular and widely implemented in deep learning model training:

1. **sgdm**—Stochastic gradient descent with momentum (SGDM). SGDM is a stochastic solver. For additional training options, see Stochastic Solver Options.
2. **adam**—Adaptive moment estimation (Adam). Adam is a stochastic solver.
3. **rmsprop**—Root mean square propagation (RMSProp). RMSProp is a stochastic solver.
4. **lbfgs**—Limited-memory   Broyden–Fletcher–Goldfarb–Shanno   (L-BFGS). L-BFGS is a batch solver. Use the L-BFGS algorithm for small networks and datasets that you can process in a single batch.

The second and the sixth issues are the same thing, and they discuss about the training terminal accuracy and the smallest errors that can be accepted. The training process would be terminated as this accuracy is achieved.

Two functions, training loss and validation loss function, are used to describe the error or the difference between the outputs of your trained model and the actual data inputs. In fact, they are different methods used to evaluate how well your algorithm models your dataset. If your predictions are totally off, your loss function will output a higher number. Otherwise, if they are pretty good, it will output a lower number.

In summary, the training loss is used to guide the optimization process during the training process, ensuring that the model fits the training data well. On the other hand, the validation loss is crucial for assessing the model's ability to generalize to new data and helps in preventing over fitting [15].

The maximum number of epochs is used to determine the maximized epochs step number. Regularly it is defined as 30 by default.

The initial learning rate is used to control the rate or speed at which the model learns. If the learning rate is too low, then training can take a long time. If the learning rate is too high, then training might reach a suboptimal result or diverge very soon. The default value is 0.01.

The iteration is defined as the number of batches or steps through partitioned packages of the training data, needed to complete one epoch. In other words, each epoch is composed of a sequence of iterations. The maximum number of iterations to complete one epoch is the upper bound for the iteration number for each single epoch.

After all of these factors have been considered and defined, the neural network can be trained and checked by using predictors and response data. During the training process, some developments may need one or two steps:

1. Training process: It is used to train the model based on the training data (input-output pair data).
2. Checking process: It is used to check the training results by using another separate set of checking data (different input-output pair data).

In some applications, they may use only the training process without the checking process. But in other implementations, they may need to use the two steps above to get better results.

In MATLAB, one can use APPs or library functions to perform model training and checking process. Most training functions and training-related Name-Value pairs are shown in Tables 7.1, 7.2 and 7.3.

To confirm the correctness and accuracy of the trained model, an additional validation or testing process is needed.

7.6.3.4   Validate or Test the Trained Network

To perform validation or testing process for the trained model, a separate set of predictors–responses dataset is needed. Common techniques to evaluate the effectiveness of a trained artificial neural network model include [16]:

1. Measuring the model's performance on a held-out testing dataset
2. Using K-fold cross-validation to estimate the model's performance on unseen data
3. Comparing the model's performance to a baseline or simple model
4. Visualizing the model's decision boundaries and internal representations
5. Evaluating the model's performance on specific tasks or subsets of the data
6. Using metrics such as accuracy, precision, recall, F1 score, and AUC-ROC

When considering performance, it is important to choose the correct evaluation criteria or metric. If the dataset is heavily imbalanced, accuracy and even AUC (Area under the ROC Curve) will be less meaningful. In this case, we likely want to consider metrics like precision and recall. F1-score is another useful metric that combines both precision and recall. A confusion matrix can help visualize what data points are misclassified and what are not [13].

In MATLAB, some popular functions used to perform evaluation or validation process include the **perform()** function shown in Table 7.1 and discussed in Sect. 7.6.2.13.

7.6.3.5   Deploy and Use the Trained Network

Once our model has been trained, checked, and validated, it is the time for us to deploy it in real applications. In MATLAB, two ways can be used to deploy the trained model and utilized in actual applications.

1. When a model is trained by using any App, one can use the **Export** function to export the trained model to the Workspace or a file to keep it to be used in the future.
2. When models are trained by using library functions, the model can also be saved to Workspace or a file by using some special functions or commands.

During the real application process, the exported or saved model can be loaded into the project or Workspace to be used in users' projects.

Additionally, one can use **Export Code** or **Export to Simulink** function to convert a trained model to a set of real codes or a Simulink block to be used later.

Next let us use some actual example projects to illustrate the entire process of building, training, validating, and deploying a deep neural network model. However, before we can do that, it is important for us to have a clear picture of what kinds of building strategies we should adopt to build and use our models.

7.6.4   A Practical Decision Between the Transfer Learning and New Learning Strategy

In deep learning category, two typical applications exist and can be selected by users depending on their actual situations: transfer learning and new learning. The so-called transfer learning means that some application models have been designed and trained, and they can be available to public. The users can easily and conveniently perform some small additional training on those pre-built models to get their desired models by saving a lot of time and energy in the model's building and training process.

In fact, transfer learning lets you take advantage of the expertise in the deep learning community. Popular pretrained models offer a robust architecture and skip starting from scratch. Transfer learning is a common technique for supervised learning because [17]:

- It enables you to train models with less labeled data by reusing popular models already trained on large datasets.
- It can reduce training time and computing resources. With transfer learning, the neural network weights are not learned from scratch because the pretrained model has already learned the weights based on previous learnings.
- You can use model architectures developed by the deep learning research community, including popular architectures such as GoogLeNet and YOLO.

On the other hand, the new learning means that the users need to build, train, check, and validate their models themselves based on their datasets, either public or personal datasets.

However, even with so many advantages of using transfer learning models, the new learning strategy is used in this book, and the main purpose is to enable readers to have a clear picture of the building, training, checking, and validating process for a new model.

7.7 Build Practical Example Projects with Deep Learning Technology

As everybody knows, earthquake is a kind of serious natural disaster, and the damages introduced by earthquakes are terrible and horrible to our society around the world. Thus, it is very critical and vital to predict and evaluate the possibility of earthquakes as correctly and accurately as possible worldwide to effectively avoid possible damages and disasters due to this kind of disaster. The disasters and damages incurred by earthquakes are tremendous and terrible, and sometimes they are invaluable and cannot be estimated in value, just as those huge earthquakes recently occurred in Asia, such as Taiwan and Japan.

In this section, we will discuss some real deep learning projects with various datasets to illustrate how to use Apps and library functions to design, build, train, and test our desired deep neural network models to perform related regressions and classification jobs. First, let us concentrate our discussions on using MATLAB Apps to build and train our desired models.

Prior to building our model, first, let us have some basic understanding about the earthquake parameters and their physical definitions.

7.7.1 Prediction of Energy Level for Earthquake with Deep Learning

In most general earthquake predictions, we need to predict the location, including the latitude, longitude, depth, and time of a possible earthquake to occur, making this a time-series regression problem for every point in the analyzed area. Therefore the data should be divided into a 3D grid (*x-y-z*), as shown in Fig. 7.16.

In fact, a more important factor or variable to be predicted for a possible earthquake is the energy level released by the earthquake. An equation used to describe the energy transformation for a given magnitude is presented as:

$$\log \text{Energy} = 5.24 + 1.44 \times \text{Magnitude} \tag{7.6}$$

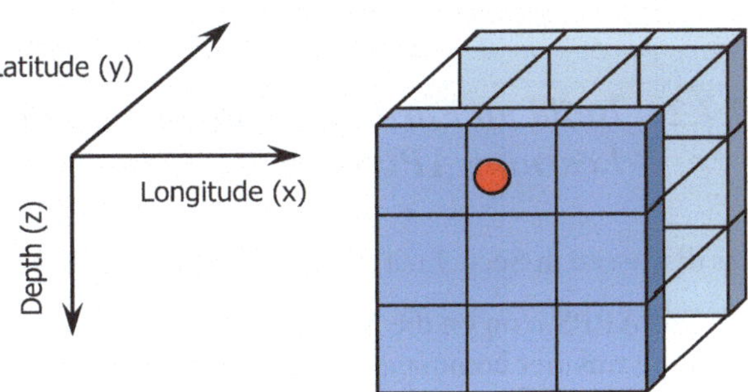

Fig. 7.16 The 3D grid presentation for a location of an earthquake to occur

This equation provides a relation between the energy level and the moment magnitude [18]. Magnitude is the physical size of the earthquake, and the energy released can also be roughly estimated by converting the moment magnitude [19], as shown in Eq. (7.6).

With the equation shown above, the key issue to predict a dangerous degree for an earthquake is the energy level for a given location, which is represented by a log function. Thus, we need to predict the magnitude of an earthquake by using a regression or a deep learning algorithm.

To build our deep neural network model, we need to use a sample dataset as the data source to train and validate our desired model. An existing dataset provided by Kaggle is used.

7.7.1.1   The Sample Earthquake Dataset Provided by Kaggle

In this part, we use a sample earthquake dataset provided by Kaggle, and this dataset contains all global earthquakes that occurred from 1973 until 1997 worldwide [20]. The dataset includes more than 1,048,500 earthquake records that happened during the period between January 1973 and November 1997 with seven columns or categories, **latitude, longitude, depth, mag** (Magnitude), **id, hour**, and **date**. In our applications, we only need to use five of them without using two columns, **id** and **date**.

Go to the site https://www.kaggle.com/datasets/gustavobmgm/earthquakes-for-ml-prediction-new-version to download the dataset, **silver.csv**, and save it to one of your local folders.

Due to the huge size of this dataset, we modified it and use only about 6100 records from that dataset, and the modified dataset is renamed to **silver_m.csv**, which contains the top 6000 records. We used the first 3000 data as the training data, and the second 3000 as the checking data. We also modified that original dataset by collecting 100 data, which are located between 8000 and 8100 rows in the original dataset as our testing or validation dataset, and named it as **silver_test.csv**. Both datasets are available and can be downloaded on the Springer ftp site under the **Students\Dataset\Earthquake Dataset** folder.

With our data source ready, now let us start to build our deep neural network model.

7.7.2   Build an Earthquake Prediction Project with Deep Learning APPS-ANN

As discussed in Sect. 7.6.1, to open the MATLAB ANN App, two ways can be used:

1. Use **APPS** icon on the MATLAB window
2. Type **nnstart** command in the MATLAB Command window

With the artificial neural networks (ANN) App and deep network designer (DND) App, users can design, edit, and analyze networks interactively, import pretrained models, and export networks to Workspace and Simulink. The toolbox lets you interoperate with other deep learning frameworks.

The difference between the **nnstart** and the **deepNetworkDsigner** APPS is that the former only provides a template to build some shallow neural network models, but the latter provides a more complicated or deeper template to enable users to build more powerful deeper models.

Due to the simplicity of our earthquake model, we can use a shallow neural network as a frame to build our model. In this part, we will concentrate on the APPS with ANN method. Based on the workflow discussed in Sect. 7.6.3, let us first collect and preprocess our data in the modified dataset **silver_m.csv**.

7.7.2.1 Collect and Preprocess Data in Our Modified Dataset

In ANN model, the training and checking data are divided into two categories: input and output data, as described below:

1. Both input and output data are M × N matrixes. For example, in our case, if we use four inputs, **LATI, LONG, DEPTH,** and **HOUR**, each of them is a $1 \times N$ array (N = 3000), but the final input matrix is a **4 × N** matrix arranged as four rows: **[LATI; LONG; DEPTH; HOUR]**. The output **MAG** is a **1 × N** array **[MAG]**.
2. Both input and output matrixes are saved into two separate data files.

Create a new Script file, name it **ANN_Data_MAG.m**, and enter the codes shown in Fig. 7.17 into that file. Let us have a closer look at that piece of codes to see how it works.

1. First we need to get our data source by using **readtable()** function and the modified dataset **silver_m.csv** is saved in a folder, in this case, it is **Artificial**

```
% Generate training and target data for earthquake ANN model
% Prior to run this script, make sure the dataset: silver_m.csv, is located at a folder Dataset.
% May 23, 2024

1  T = readtable('C:\\Artificial Intelligence Book\\Students\\Datasets\\Earthquake Dataset\\silver_m.csv');
2  N = 3000;                                    % size of the training data

   % One must use table2array() to convert read values in table to double array to save it as ascii file later.
   % Otherwise one maybe encounter a conversion error for unmatched data type.
3  LATI = table2array(T(1:N, 1))';              % LATI is located at column 1 in database
   LONG = table2array(T(1:N, 2))';              % LONG is located at column 2 in database
   DEPTH = table2array(T(1:N, 3))';             % DEPTH is located at column 3 in database
   HOUR = table2array(T(1:N, 6))';              % HOUR is located at column 6 in dataset
   MAG = table2array(T(1:N, 4))';               % MAG is located at column 4 in database

4  inputData = [LATI; LONG; DEPTH; HOUR];       % the first 3000 data - training data
5  outputData = MAG;                            % the target data

6  save train_data.dat inputData -ascii
   save target_data.dat outputData -ascii
```

Fig. 7.17 The detailed codes for the ANN algorithm

Intelligence Book\Students\Datasets\Earthquake Dataset. You may need to use your folder to replace this if you store this dataset at a different location in your machine.

2. Since we only need to use the first 3000 data records in the dataset as the training data, thus a local variable **N** is defined and initialized to this number.
3. The format of data records read by using the **readtable()** function are table cells, thus the function **table2array()** is needed to convert those cells to double data values since we need to save those data as ASCII files later. A compiling error may be encountered without this conversion since an ASCII file can only contain numeric data values. The **column_number** used in the cell, **T(1:N, column_number)**, is the actual column number of each input column in the modified dataset.
4. An input data array is built by arranging each input variable as a different column and assigned to a local variable array **inputData**.
5. Similarly, a target array, **outputData**, is built by assigning the output **MAG** to it.
6. Finally, we save these two data arrays to the Workspace as two ASCII files, **train_data.dat** and **target_data.dat**.

Run this script file and two data files, **train_data.dat** and **target_data.dat**, are generated.

Next let us select our deep learning algorithm to train our desired model by using the ANN App.

7.7.2.2 Select the Deep Learning Algorithm to Train the Network Model

Perform the following operations to select the deep learning algorithm to train our model:

1. In the MATLAB Command window, load these two data files by typing: **load train_data.dat** and **load target_data.dat**, and press the **Enter** key.
2. Type **nnstart** in the MATLAB Command window and press the **Enter** key to start the ANN App Designer, as shown in Fig. 7.18.
3. Click on the **Fitting** button to begin the fitting process. The Neural Network Fitting wizard is displayed, as shown in Fig. 7.19. The **Network Architecture** is shown up and ten default nodes are selected (you can change it if you like).
4. Click on the drop-down arrow on the **Import DATA** button and select the **Import Data** item to open the **Import Data from Workspace** wizard, as shown in Fig. 7.20.
5. Click on the drop-down arrow from the **Predictors** combo box and select our input data matrix **inputData − [4 × 3000 double]** from the list.
6. Also click on the drop-down arrow from the **Response** combo box and select the item **MAG − [1 × 3000 double]** as the target data. Then click on the **OK** button to continue.
7. The **Neural Network Training wizard** is displayed, as shown in Fig. 7.21. It can be found from the upper-left corner that this training process used 70% of

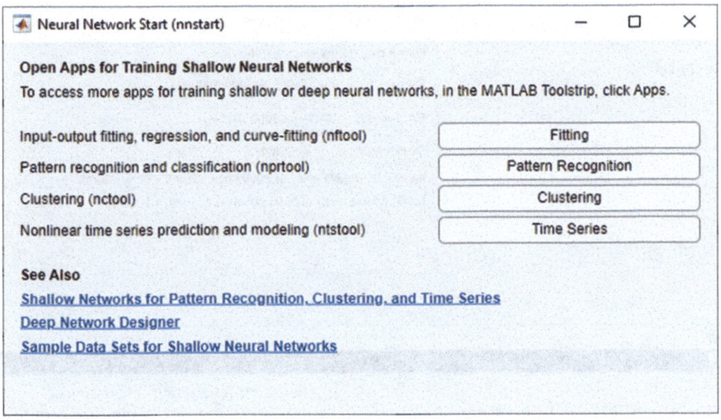

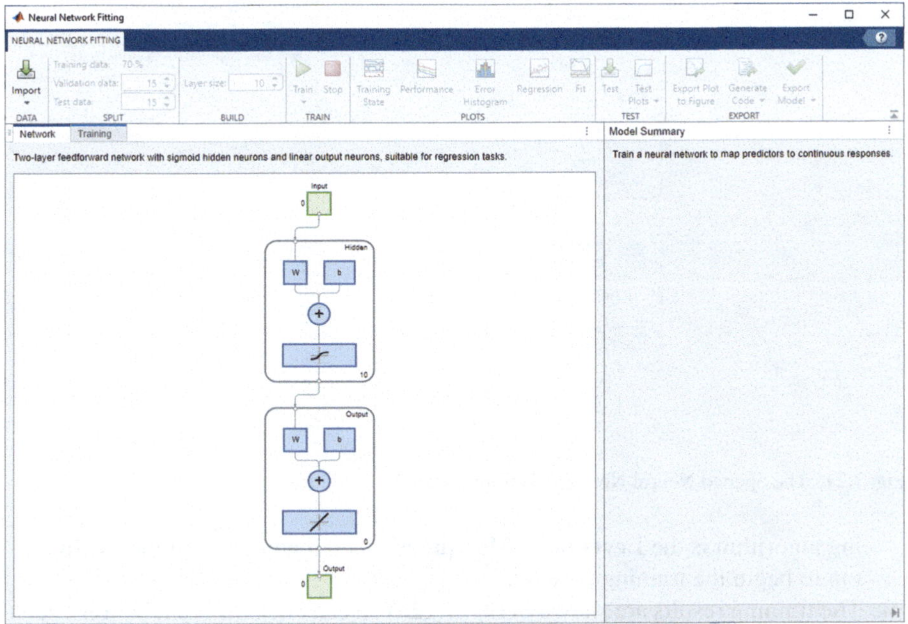

Fig. 7.18 The opened ANN Designer

Fig. 7.19 The opened Neural Network Fitting wizard

our 3000 training data as real training data, and 15% of our 3000 training data as Validation and Testing data. The default layer site is 10, which means that totally ten nodes will be used for this model. You can change it if you like by clicking on the up (increment) or down arrow now.

8. To start the training and testing process, click on the drop-down arrow on the **Train** icon on the top to select one special training algorithm. The default train-

Fig. 7.20 The finished Import Data from Workspace wizard

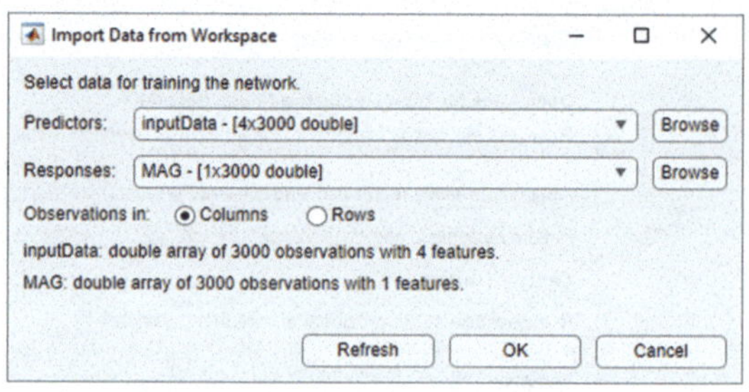

Fig. 7.21 The opened Neural Network Training wizard

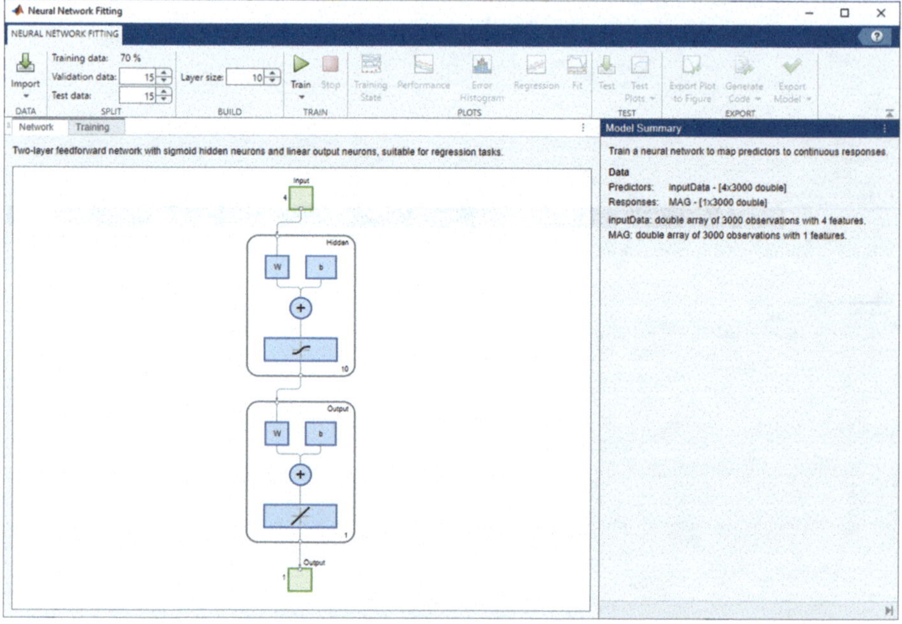

ing algorithm is the **Levenberg-Marquard**, keep it and click on the **Train** button to begin the training process.

9. The training results are shown in Fig. 7.22. You can click different buttons, such as **Training State**, **Performance**, **Error Histogram**, and **Regression**, to plot related result to see each of them. Some of the sample plots are shown in Fig. 7.23.

10. You can also perform additional testing for this trained model by using extra testing data if you like. To do that, you need to prepare the additional testing data and click on the **Test** icon on the top.

11. On this wizard, you can select different ways to save your trained ANN model to MATLAB Workspace or Simulink to be used in the future. To do that, click on the drop-down arrow on the **Export** and **Export Model** icon on the top and select a method. In this case, select the item **Export to Workspace** and enter a

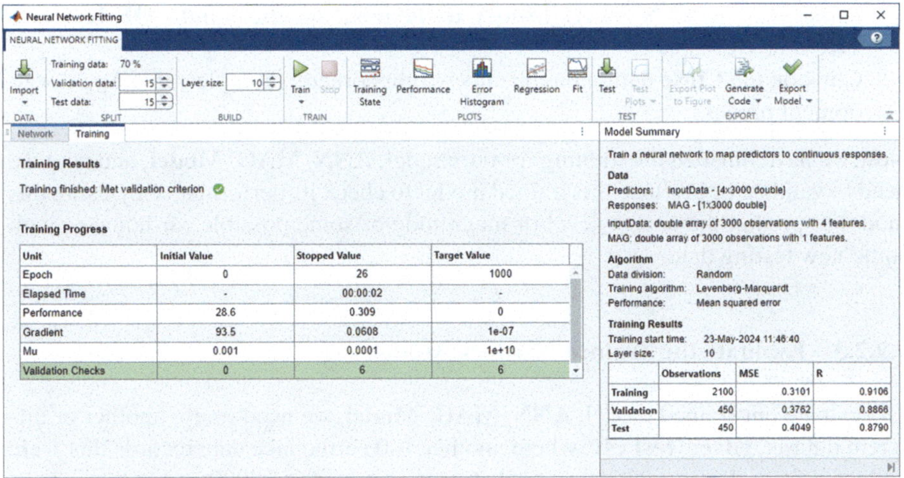

Fig. 7.22 The training results

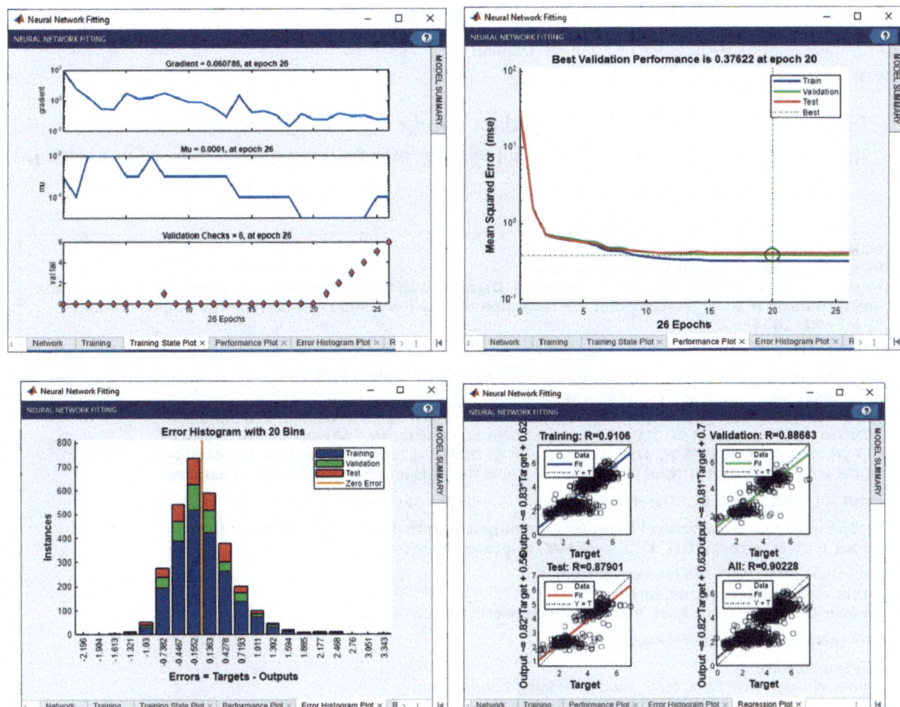

Fig. 7.23 Some plotting results for the model training process

name, such as **ANN_MAG_Model**, to the box, and click on the **OK** button to save it to Workspace.

12. Click on the **Close** button on the upper-right corner to complete this App development process.

Now we have finished the training for our model, **ANN_MAG_Model**, and next we need to validate or evaluate this trained model to check its performance by using this model to predict the energy level or magnitude of some possible earthquakes with some new testing data.

7.7.2.3   Evaluate the Trained Network Model

To evaluate our trained model, **ANN_MAG_Model**, we need to use another or different dataset, **silver_test.csv**, where another 100 earthquake data records that were collected from the original dataset, **silver.csv**, in a range of 8000 and 8100 records, will be used. This dataset can be found and downloaded on the Springer ftp site under the **Students\Dataset\Earthquake Dataset** folder. Save it to one of your folders in your machine.

Create a new Script file, name it **ANN_App_Eval.m,** and enter the codes shown in Fig. 7.24 into that file. Let us have a closer look at that piece of codes to see how it works.

1. First we need to get our testing data source by using **readtable**() function and the testing dataset **silver_test.csv** is saved in a folder, in this case, it is **Artificial**

```
% Name: ANN_App_Eval.m
% Evaluate the trained Deep Learning model.
% Prior to running this script, run Project ANN_Data_MAG.m to get ANN_MAG_Model exported Workspace
% The dataset is silver_test.csv that is a collection of data rows 8000 ~ 8100 from the original dataset
% May 23, 2024
1  T = readtable('C:\\Artificial Intelligence Book\Students\Datasets\Earthquake Dataset\\silver_test.csv');
2  N = 100;                                    % szie of the testing data

   % Assign each column to a related local variable...
3  LATI = table2array(T(1:N, 1))';             % LATI is located at column 1 in database
   LONG = table2array(T(1:N, 2))';             % LONG is located at column 2 in database
   DEPTH = table2array(T(1:N, 3))';            % DEPTH is located at column 3 in database
   HOUR = table2array(T(1:N, 6))';             % HOUR is located at column 6 in dataset

4  input = [LATI; LONG; DEPTH; HOUR];          % The 100 testing data

   % One must use table2array() to convert the output mag to double array to plot it later.
5  target = table2array(T(1:N, 4))';     % MAG is located at column 4 in database

6  outputs = ANN_MAG_Model.Network(input);
7  errors = gsubtract(outputs, target);
8  performance = perform(ANN_MAG_Model.Network, target, outputs)

9  view(ANN_MAG_Model.Network)

   % Plots Results
10 figure, plotperform(ANN_MAG_Model.TrainingResults)
   figure, plottrainstate(ANN_MAG_Model.TrainingResults)
   figure, plotfit(ANN_MAG_Model.Network, input, target)
   figure, plotregression(target, outputs)
   figure, ploterrhist(errors)
```

Fig. 7.24 The codes for the evaluation of the trained model ANN_MAG_Model

Intelligence Book\Students\Datasets\Earthquake Dataset. You may need to use your folder to replace this if you store this dataset at a different location in your machine.

2. Since we only need to use 100 data records in the dataset as the testing data, thus a local variable **N** is defined and initialized to this number.
3. The format of data records read by using the **readtable()** function are table cells, thus the function **table2array()** is needed to convert those cells to double data values since we need to transpose these data records from 100×1 to 1×100 arrays to meet the requirements of testing data used for deep learning method. Do not forget to add each transposing operator, ', on the end of each input variable in step 3 to perform these transposing operations. The **column_number** used in the cell, **T(1:N, column_number)**, is the actual column number of each input column in the testing dataset.
4. An input data array is built by arranging each input variable as a sequence of columns and assigned to a local variable array **input**.
5. Similarly, a target array, **outputs**, is built by assigning the output **MAG** to it. A transposing operation is also needed since we need to plot it as numeric values later.
6. Now it is time for us to calculate the predicted outputs based on the trained model **ANN_MAG_Model** and the testing data stored in the **input** variable. A point to be noted is that the trained model is a network and that object is stored as a property function, **Network()**, under that model. Thus, the completed name for that trained model is: **ANN_MAG_Model.Network**. By calling that function with the **input** as an argument, we can obtain the predicted **outputs**.
7. To evaluate our trained model, a function **gsubtract()** is executed to perform a general subtraction to get the difference between the predicted outputs and the actual outputs stored in the **target** variable. The returned difference is assigned to the variable **errors**.
8. Another evaluation function, **perform()**, which was discussed in Sect. 7.6.2.13, is called to compute the model's performance. The performance is exactly the minimum error among training, checking, and testing data for the trained model.
9. To have a global look at the model structure, the **view()** function that was discussed in Sect. 7.6.2.12 is used.
10. Finally, to compare with those plotting results obtained from the execution of the training and checking by using the App, ANN Designer, in Sect. 7.7.2.2, various plot functions are executed to plot related training and checking results.

Now run this project to get all desired outputs and evaluation results.

A key issue of running this project is that prior to running this project, one must perform all operations shown in Sects. 7.7.2.1 and 7.7.2.2 to generate training data and obtain the trained model **ANN_MAG_Model** and export it to the Workspace.

7.7.3   Build an Earthquake Prediction Project with Deep Learning APPS-DND

As we discussed in Sect. 7.6.2.2 and Fig. 7.15a, a typical structure of a neural network regression model is composed of five layers: input, fully connected (FC), ReLU activation, another FC, and output layer.

In this part, we try to use the deep network designer (DND) App to train our earthquake model. However, prior to doing the training and testing jobs, one key issue was that we had to convert our data source from the regular data array to a kind of data stored in a MATLAB data store, and that is the requirement of using this App.

7.7.3.1   The Functions and Properties of Data Store Used in MATLAB

A datastore is an object for reading a single file or a collection of files or data. The datastore acts as a repository for data that has the same structure and formatting. For example, each file in a datastore must contain data of the same type, such as numeric or text, appearing in the same order, and separated by the same delimiter [21].

Table 7.5 shows all types of popular datastores and related objects used to generate them.

All data items stored in a datastore are formatted to some cell format, and users can create, read out, and modify any data stored in a datastore with related properties and functions. The following properties and functions are the most popular ones and widely used in data operations:

1. **read**() function—Used to read data in a datastore
2. **readall**() function—Used to read out all data items stored in a datastore
3. **reset**() function—Used to reset the datastore to the initial state
4. **preview**() function—Used to preview a subset of data in datastore
5. **combine**() function—Used to combine data from multiple datastores
6. **ReadSize** property—Used to define the size of data to be read from a datastore

Table 7.5 Architecture properties

Type of File or Data	Datastore Object Type
Text files containing column-oriented data, including CSV files.	TabularTextDatastore
Datastore for in-memory data	ArrayDatastore
Image files, including formats that are supported by imread() such as JPEG and PNG.	ImageDatastore
Spreadsheet files with a supported Excel* format such as .xlsx.	SpreadsheetDatastore
Key-value pair data that are inputs to or outputs of mapreduce.	KeyValueDatastore
Parquet files containing column-oriented data.	ParquetDatastore
Custom file formats. Requires a provided function for reading data.	FileDatastore
Datastore for checkpointing tall arrays.	TallDatastore

7. **OutputType** property—Used to define the type of the output data
8. **SelectedVariableNames** property—Used to select desired variables from the datastore

In fact, we have discussed and used quite a few datastores for image and audio data classifications with multiple real projects in Chaps. 6 and 7. However, in this part, we need to discuss and analyze those datastores in different ways or angles.

Now let us use our earthquake dataset as an example to illustrate how to convert our modified earthquake dataset, **silver_m.csv**, to a related datastore and use it for our deep network model training and testing purposes.

7.7.3.2 Convert Our Modified Earthquake Dataset to a Related MATLAB Datastore

As we discussed in the last section, different datastores can be generated and used for the different types of data. For our case, we can use either datastore for in-memory data or Text files containing column-oriented data, including CSV files. In other words, we can use two classes or objects, **ArrayDatastore** or **TabularTextDatastore**, to generate our datastore. In this part, we try to use the first way to generate our datastore due to its simplicity.

Create a new Script file, name it **Generate_Data_Store.m**, and enter the codes shown in Fig. 7.25 into that file. Let us have a closer look at that piece of codes to see how it works.

```
% Name: Generate_Data_Store.m
% Convert training and checking data from silver_m to Data Store used for training DL Model with DND.
% The dataset is silver_m.csv
% May 28, 2024

1  T = readtable('C:\\Artificial Intelligence Book\Students\Datasets\Earthquake Dataset\\silver_m.csv');
2  N = 3000;                        % szie of the training and testing data

   % Assign each column to a related local variable...
3  LATI = T(1:N, 1);                % LATI is located at column 1 in database
   LONG = T(1:N, 2);                % LONG is located at column 2 in database
   DEPTH = T(1:N, 3);               % DEPTH is located at column 3 in database
   HOUR = T(1:N, 6);                % HOUR is located at column 6 in dataset
   MAG = T(1:N, 4);                 % MAG is located at column 4 in database

4  trainDataInput = [LATI LONG DEPTH HOUR];
   trainDataOutput = MAG;

5  trainData = [trainDataInput trainDataOutput];
6  trainStore = arrayDatastore(trainData,"ReadSize", N, "OutputType","same");

7  LATI = T(N+1:2*N, 1);            % LATI is located at column 1 in database
   LONG = T(N+1:2*N, 2);           % LONG is located at column 2 in database
   DEPTH = T(N+1:2*N, 3);          % DEPTH is located at column 3 in database
   HOUR = T(N+1:2*N, 6);           % HOUR is located at column 6 in dataset
   MAG = T(N+1:2*N, 4);            % MAG is located at column 4 in database

8  testDataInput = [LATI LONG DEPTH HOUR];
   testDataOutput = MAG;

9  testData = [testDataInput testDataOutput];
   testStore = arrayDatastore(testData,"ReadSize", N, "OutputType","same");
```

Fig. 7.25 The codes used to convert our dataset to a related datastore

1. First we need to get our data source by using **readtable()** function and the modified dataset **silver_m.csv** is saved in a folder, in this case, it is **Artificial Intelligence Book\Students\Datasets\Earthquake Dataset**. You may need to use your folder to replace this if you store this dataset at a different location in your machine.
2. Since we only need to use the first 3000 data records in the dataset as the training data, and the second 3000 data as the testing or validation data, thus a local variable **N** is defined and initialized to this number.
3. Assign each related column in the reading table T to four input and one output variable, such as **LATI, LONG, DEPTH, HOUR,** and **MAG**, respectively.
4. Now we can define the input data columns for our training data by using the four input columns and make them as a four-column matrix. Similarly, we assign the output or response data to the output data column of the training data.
5. By combining both the input and the output training data columns together, we can get our training data matrix **trainData**.
6. To build our datastore used for the training purpose, **trainStore**, we need to call the function **arrayDatastore()** with our **trainData** matrix to create a new **array-Datastore** object. Two constructors are available for this class, **arrayDatastore(A);** and **arrayDatastore(A, Name-Value);**. Here we used two Name-Value pairs; the size of reading the data is 3000, and the output type is the same as that of the input.
7. In a similar way, we can build our testing datastore. The only issue is that the number of the starting and the ending row should be 3001 and 6000, respectively, to enable us to use the second 3000 data in our modified dataset as the testing data.
8. Arrange our input and output data columns to our **testDataInput** and **test-DataOutput** matrix by assigning related columns to both of them.
9. To build our test datastore, we need to combine the **testDataInput** and the **test-DataOutput** together and call the function **arrayDatastore()** to complete this process.

Now run this project and two datastores, **trainStore** and **testStore**, are generated and saved to the Workspace. Next we can use these datastores to train our deep learning model with the Deep Network Designer (DND) App.

7.7.3.3 Build and Train Our Deep Learning Model with Datastore

Instead of using any existing or pretrained network, in this part we would like to build a new or customer network to enable readers to have a clear and complete picture about the building process of new networks. Open MATLAB and click on the **APPS** icon on the top and select the **Deep Network Designer** item to open the DND App. Perform the following operations to generate and build our model:

1. Click on the + sign in the center of the **Blank Network** to open a new network **Designer** wizard, as shown in Fig. 7.26, since we want to use a brand new network.
2. Refer to Fig. 7.15a in Sect. 7.6.2.2 to get an idea for the structure of a typical regression sequence network. Here the only issue we need to pay attention to is the input and the output layers. Also we may need to add one more layer called Long Short-Term Memory (LSTM). Refer to Sect. 7.3.4 to get more details about the structure of this kind of layer. An LSTM network enables you to input sequence data into a network and make predictions based on the individual time steps of the sequence data.
3. For the input and the output layers, we like to use a **featureInputLayer** as input and the **regressionLayer** as the output layer.
4. Keep those points in mind, now go to the left pane and select related layers, drag each of them one by one, and place them into the Design pane in the center of the wizard. Then connect each of them one by one by dragging the connection line between them. The initial design view of our network is shown in Fig. 7.26.
5. To set related parameters for some layers, we need to change the **OutputSize** for both **Fully Connected** (FC) layers from 10 to 1 since we have only a single output **MAG**. To that, click each FC layer and change the **OutputSize** value to 1 for both of them.
6. Now click on the **Data** tab to open the Import Data wizard to get our datastores to be ready to train our network. Click on the drop-down arrow on the **Import Data** icon on the upper-left corner, and select the **Import Customer Data** item

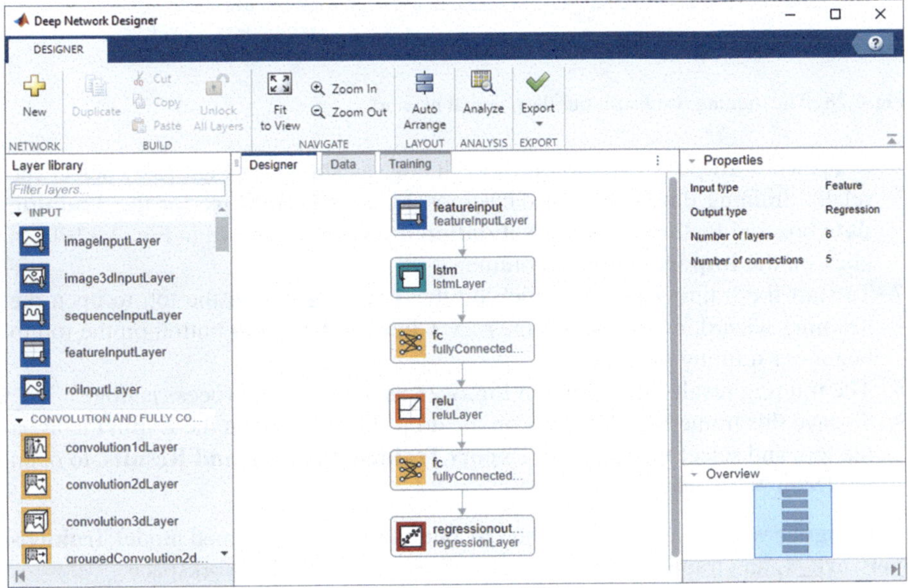

Fig. 7.26 The finished designing view of our network

Fig. 7.27 The completed
data selection process

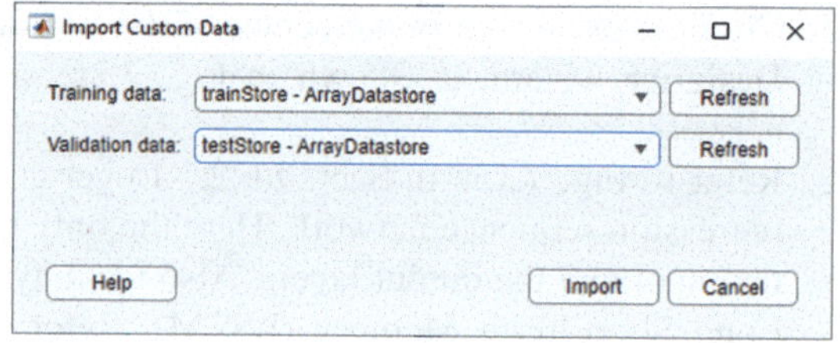

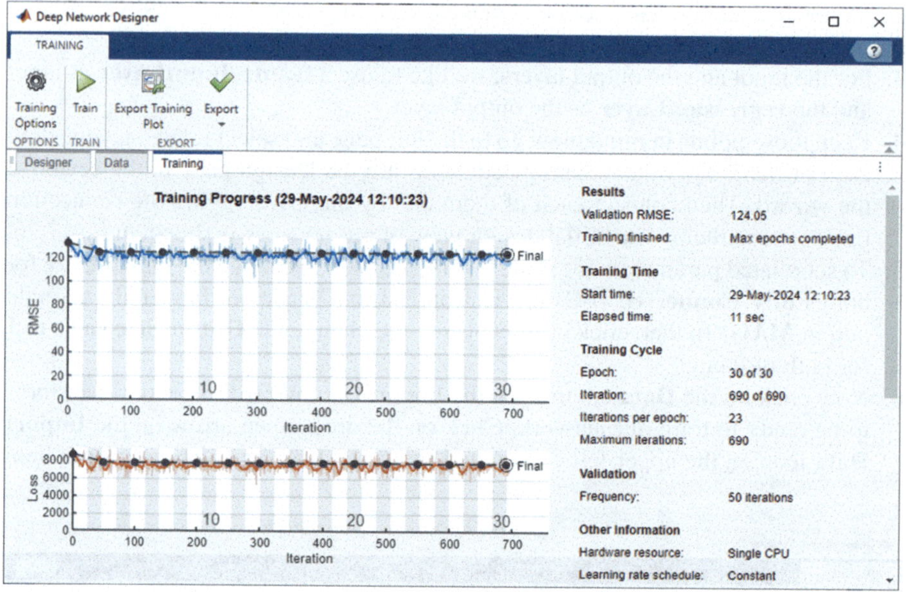

Fig. 7.28 The training results for our deep neural network

to open the **Import Customer Data** dialog, as shown in Fig. 7.27. Select the
related training data from our datastores, such as **trainStore** for the **Training
data** box and **testStore** for the **Validation data** box, as shown in Fig. 7.27. Then
click on the **Import** button to continue.

7. To start the training process, click on the **Training** tab on the top to open the
   Training wizard, as shown in Fig. 7.28. Click on the **Train** button on the top to
   begin our training process.
8. The training results are shown in Fig. 7.28 as the training process is done.
9. To save this trained model, click on the drop-down arrow on the **Export** icon on
   the top, and select the top item, **Export Trained Network and Results**, to open
   the **Deep Network**.

Designer wizard, and click on the **OK** button to save the trained model, **trained-
Network_1**, and trained structure, **trainInfoStruct_1**, to the Workspace.

At this point, we completed our network training and validation process. You can
minimize this App if you like, but do not close it since we need to validate or

evaluate this trained model in the next section, by clicking on **Minimize** button on the upper-right corner on the wizard.

Next let us perform some evaluations to this trained model with our validation data that are retrieved from our testing dataset, **silver_test.csv**.

7.7.3.4   Evaluate and Validate Our Trained Deep Learning Model

To check and confirm our trained model, we need to build another script file.

Open a new Script file, name it **DND_App_Eval.m**, and enter the codes shown in Fig. 7.29 into that file. Let us have a closer look at that piece of codes to see how it works.

1. First we need to get our testing data source by using **readtable**() function and the testing dataset **silver_test.csv** is saved in a folder, in this case, it is **Artificial Intelligence Book\Students\Datasets\Earthquake Dataset**. You may need to use your folder to replace this if you store this dataset at a different location in your machine.
2. Since we only need to use 100 data records in the dataset as the testing data, thus a local variable **N** is defined and initialized to this number.
3. The format of data records read by using the **readtable**() function are table cells, thus the function **table2array**() is needed to convert those cells to double data values. The **column_number** used in the cell, **T(1:N, column_number)**, is the actual column number of each input column in the testing dataset.

```
% Name: DND_App_Eval.m
% Evaluate the trained Deep Learning model with Deep Network Designer.
% Prior to running this script, run Generate_Data_Store.m & generate trained model & export to Workspace
% The dataset is silver_test.csv that is a collection of data rows 8000 ~ 8100 from the original dataset
% May 29, 2024

1  T = readtable('C:\\Artificial Intelligence Book\Students\Datasets\Earthquake Dataset\\silver_test.csv');
2  N = 100;                               % szie of the testing data

   % Assign each column to a related local variable...
3  LATI = table2array(T(1:N, 1));          % LATI is located at column 1 in database
   LONG = table2array(T(1:N, 2));          % LONG is located at column 2 in database
   DEPTH = table2array(T(1:N, 3));         % DEPTH is located at column 3 in database
   HOUR = table2array(T(1:N, 6));          % HOUR is located at column 6 in dataset

4  input = [LATI; LONG; DEPTH; HOUR];      % The 100 testing data

   % One must use table2array() to convert the output mag to double array to plot it later.
5  target = table2array(T(1:N, 4));        % MAG is located at column 4 in database
6  target = repmat(target, 4);
7  outputs = predict(trainedNetwork_1, input);
8  errors = gsubtract(outputs, target);

   % Plots Evaluation Results
9  x = 1:400;
   figure, plot(x, abs(errors'));
   xlabel('The absolute errors between the predicted and the actual outputs');
   grid;
   figure, plot(trainInfoStruct_1.TrainingRMSE);
   xlabel('The training RMSE values');
   grid;
```

Fig. 7.29 The codes used to evaluate the trained model

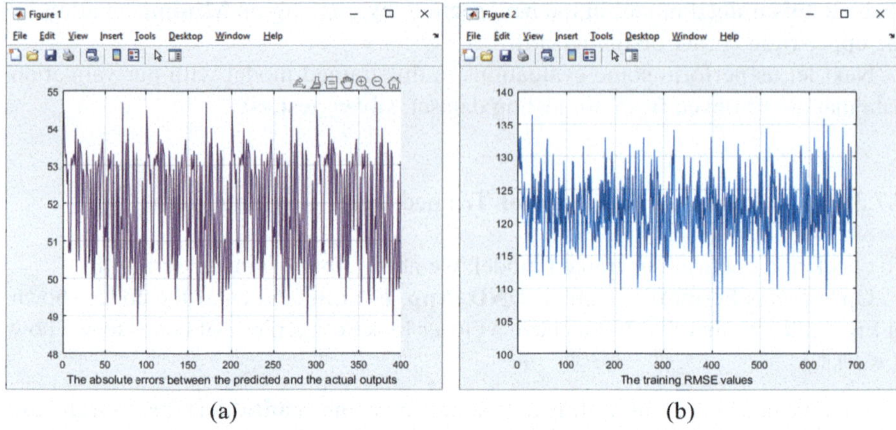

(a)   (b)

Fig. 7.30 The evaluation results for trained model

4. An input data array is built by arranging each input variable as a sequence of columns and assigned to a local variable array **input**.
5. Similarly, we need to get the target data that are actual outputs in our testing dataset. The target data are converted from table cells to double data array to be used later.
6. Since the predicted outputs contained four rows, we also need to make our target data as four rows by using the **repmat()** function.
7. To calculate the predicted outputs, we used function **predict()** with the trained model, **trainedNetwork_1**, which is exported from the Deep Network Designer in the last section, with the input testing data.
8. The errors between the predicted and the actual target data values are compared with the **gsubtract()** function.
9. Some of evaluation results, including the errors and training RMSE, are plotted to display the validation results.

A prerequisite to run this project is that the trained model, **trainedNetwork_1**, must have been generated and exported to the Workspace. Otherwise a running error may be encountered if that model has not been created and exported to the Workspace.

Now run this project and some running results are shown in Figs. 7.30a, b.

Figure 7.30a shows the errors between the predicted and the actual outputs, and Fig. 7.30b shows the training RMSE values.

Next let us discuss how to build our earthquake prediction project by using deep learning-related functions.

7.7.4   Build an Earthquake Prediction Project with Deep Learning Functions

As we discussed in Sect. 7.6.2, various different functions can be used to perform training, testing and validation for a given deep neural network to obtain our desired deep learning model. In this section, we like to use some of them to perform the training and testing functions for our selected deep neural networks with our earthquake dataset.

First let us discuss how to use some functions related to the train and test shallow neural network to get our desired model applied for our earthquake dataset.

7.7.4.1   Build a Shallow Neural Network with Deep Learning Functions

Recall in Sect. 7.6.2, we discussed some popular functions used for deep learning regression and classifications. Especially, most of those functions are shown in Table 7.1. Among them, the top six functions are used for train and test shallow neural networks, and the rest of them are used for deep learning neural networks.

For the regression purpose, two functions, **fitnet()** and **fitrnet()**, are widely used for most regression applications. Comparably speaking, the former provided better performance than that of the latter. Also the function **fitnet()** has more powerful compatibility for both regressions and classifications. In other words, this function can handle both regression and classification operations. But the function **fitrnet()** is mainly used for regressions.

Based on above discussions, in this section we will use the function **fitnet()** to perform our training and testing for a shallow neural network with our earthquake dataset.

Open a new Script file and name it **Shallow_MAG_Func.m** and enter the codes shown in Fig. 7.31 into that file. Let us have a closer look at that piece of codes to see how it works.

1. First we need to get our training and testing data source by using **readtable()** function and the modified dataset **silver_m.csv** that is saved in a folder, in this case, it is **Artificial Intelligence Book\Students\Datasets\Earthquake Dataset**. You may need to use your folder to replace this if you store this dataset at a different location in your machine.
2. Since we only need to use 3000 data records in the dataset as the training and the testing data, thus a local variable **N** is defined and initialized to this number.
3. The format of data read by using the **readtable()** function are table cells; thus the function **table2array()** is needed to convert those cells to double data values. The **column_number** used in the cell, **T(1:N, column_number)**, is the actual column number of each input column in the testing dataset. A transposing operation is needed to convert those records from 3000×1 to 1×3000 arrays to meet the format requirements of deep learning training data, thus a transposing operator, **'**, is attached at the end of each instruction.

```
% Name: Shallow_Train_Func.m
% Using shallow training and testing functions to build a shallow neural network for earthquake dataset
% May 29, 2024
1  T = readtable('C:\\Artificial Intelligence Book\Students\Datasets\Earthquake Dataset\\silver_m.csv');
2  N = 3000;                              % size of the training and testing data

   % Assign each column to a related local variable...
3  LATI = table2array(T(1:N, 1))';        % LATI is located at column 1 in database
   LONG = table2array(T(1:N, 2))';        % LONG is located at column 2 in database
   DEPTH = table2array(T(1:N, 3))';       % DEPTH is located at column 3 in database
   HOUR = table2array(T(1:N, 6))';        % HOUR is located at column 6 in dataset
   MAG = table2array(T(1:N, 4))';         % MAG is located at column 4 in database

4  trainData = [LATI; LONG; DEPTH; HOUR];
5  targetData = MAG;

   % Choose a Training Function, For a list of all training functions type: help nntrain
   % 'trainlm' is usually fastest; 'trainbr' takes longer but may be better for challenging problems.
   % 'trainscg' uses less memory.
6  trainFcn = 'trainlm'; % Levenberg-Marquardt backpropagation.

   % Create a shallow neural network
7  hiddenLayerSize = 10;
8  net = fitnet(hiddenLayerSize, trainFcn);

   % Setup Division of Data for Training, Validation, Testing
9  net.divideParam.trainRatio = 70/100;
   net.divideParam.valRatio = 15/100;
   net.divideParam.testRatio = 15/100;
10 [net, tr] = train(net, trainData, targetData);

   % Test the Network
11 y = net(trainData);
12 e = gsubtract(targetData, y);
13 performance = perform(net, targetData, y)

   % View the Network
14 view(net)

   % Plots training and testing results
15 figure, plotperform(tr);
   figure, plottrainstate(tr);
   figure, ploterrhist(e);
   figure, plotregression(targetData, y);
```

Fig. 7.31 The codes used to train and test a shallow neural network with functions

4. A training data matrix is built by arranging each input variable as a sequence of columns and assigned to a local variable array **trainData**, which is a 4×3000 matrix.

5. Similarly, a target matrix, **targetData**, is built by assigning the output **MAG** to it. The dimension of that matrix is 1×3000.

6. The training function can be selected as any one shown in Table 7.2 based on its algorithm. Here we selected a popular one, **trainlm**(). You can select some other one, such as **trainbr**() or **trainscg**(), based on your preference.

7. We defined the number of nodes or neurons as 10 on each layer.

8. Now we call the function **fitnet**() to create our shallow neural network to be used later.

9. To generate the training, testing, and validation data based on the data records in our earthquake dataset, we can divide those 3000 data into three sets with related ratios.

10. The **train()** function is executed to begin to train our shallow neural work to get our desired model **net**. Three arguments are involved in this calling: our initial model **net**, the training data **trainData**, and the target data **targetData**. This function returned a trained model **net** and related training results stored in the variable **tr**.

11. To test our trained model, we calculate the predicted outputs by using our trained model.

12. To compute the error between the actual output and the predicted outputs, the function **gsubtract()** is used and the results are assigned to a local variable **e**.

13. To evaluate the performance of this trained model, the **perform()** function is executed. In fact, this function returns the RMSE value between predicted and actual outputs.

14. The function **view()** discussed in Sect. 7.6.2.12 is called to display the structure of the trained model.

15. Finally, all training and testing results related to this trained model are displayed.

Now run this project and some running results are shown in Fig. 7.32.

Next let us discuss how to build a deep neural network with deep training functions. Before we can do that, first let us have a more detailed discussion about the structure of a deep neural network or the layers distributions in a network.

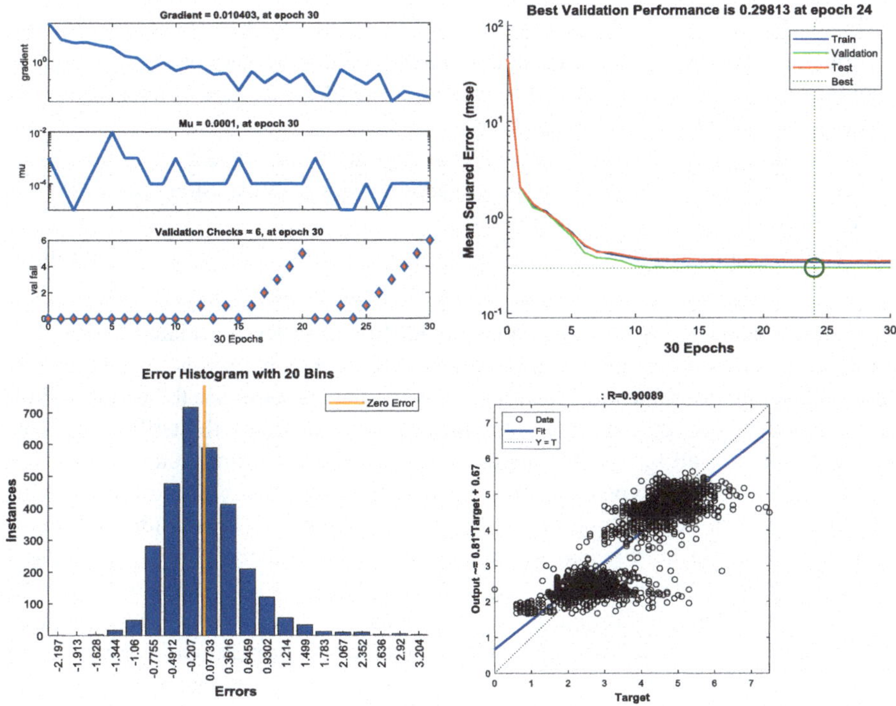

Fig. 7.32 Some running results

7.7.4.2   The List of Deep Learning Layers

In Sects. 7.6.2.2 and 7.6.2.3, we have discussed two typical structures of neural networks, regression and classification structures with different layers. In fact, a popular deep neural network is composed of a sequence of different layers, which include

1. Input layers
2. Convolution and fully connected layers
3. Sequence layers
4. Activation layers
5. Normalization layers
6. Utility layers
7. Resizing layers
8. Pooling and unpooling layers
9. Combination layers
10. Transformer layers
11. Object detection layers
12. Output layers

All of these layers provide different functions, and not all of these layers should be used in most deep neural networks. Regularly, the following layers will be used for most popular neural networks:

1. Input Layers—Used to collect input data and feed to the networks
2. Fully Connected Layers—Used to present all hidden layers to connect from the previous layers to the following layers
3. Sequence Layers—Used to collect and accept input sequence data
4. Activation Layers—Used to perform some nonlinear conversions between outputs of the last layer to the next layer
5. Output Layers—Used to convert the outputs of the last layer to the valid output data values

Some other layers are used to provide special functions to the related layers. For example, to perform convolutional image classifications, one needs to use the convolution layer, to combine multiple related layers, one needs to use the Combination Layer, to apply 2D o 3D cropping to the input, one needs to use the Utility Layer, to normalize some or all inputs, one needs to use the Normalization Later.

If Deep Learning Toolbox does not provide the layer that you require for your task, then you can define your own custom layer using [22] as a guide. After you define the custom layer, you can automatically check whether the layer is valid and GPU compatible, and outputs correctly defined gradients.

In MATLAB Deep Learning Toolbox, you can use the **layers** structure object to define the structure of your selected neural network. Some popular layers used in deep networks are shown in Table 7.6. A point to be noted is that no output layer is needed for any regression network, but it is required by any classification network.

Table 7.6 A popular list of deep learning layers

Type of Network	Layer	Description

LSTM Network	layers = [
	  sequenceInputLayer(numFeatures)
	  lstmLayer(numHidden, OutputMode="last")
	  fullyConnectedLayer(numClass)
	  softmaxLayer];
		This structure is used to build a network by using the sequence layers as input layer, **lstmLayer** and fully connected layers as the RNN and hidden layers, and **softmaxLayer** as the output layer. This kind of neural network is used to classify output based on sequence inputs.

Vector Sequence-to-One Regression Network	layers = [
	  sequenceInputLayer(numFeatures)
	  lstmLayer(numHidden, OutputMode="last")
	  fullyConnectedLayer(numResponse)];
		This structure is to build a sequence regression network with sequence as input layer, one LSTM layer, and one fully connected layer. For a regression network, no output layer should be used.

Vector Sequence-to-Sequence Classification Network	layers = [
	  sequenceInputLayer(numFeatures)
	  lstmLayer(numHidden)
	  fullyConnectedLayer(numClass)
	  softmaxLayer];
		This structure is to build a classification network with the sequence as inputs. One point to b noted is that an output layer, **softmaxLayer**, must be in the last layer to output a class value.

Vector Sequence-to-Sequence Regression Network	layers = [
	  sequenceInputLayer(numFeatures)
	  lstmLayer(numHidden)
	  fullyConnectedLayer(numResponse)];
		This structure is to build a sequence-to-sequence regression network with three layers. The output layer is not necessary for the regression model.

Image Sequence-to-Label Classification Network	layers = [
	  sequenceInputLayer(inputSize)
	  convolution2dLayer(filterSize, numFilters)
	  batchNormalizationLayer
	  reluLayer
	  flattenLayer
	  lstmLayer(numHidden, OutputMode="last")
	  fullyConnectedLayer(numClass)
	  softmaxLayer];
		This structure is used to build an image sequence to label classification network. A convolution layer with 2D and a batchNormalization layer are added to this structure. An output layer is necessary for the classification process.

Image Sequence-to-One Regression Network	layers = [
	  sequenceInputLayer(inputSize)
	  convolution2dLayer(filterSize, numFilters)
	  batchNormalizationLayer
	  reluLayer
	  flattenLayer
	  lstmLayer(numHidden,OutputMode="last")
	  fullyConnectedLayer(numResponse)];
		This structure is to build an image sequence to one regression network. It is very similar to the structure above, image sequence to label. But the difference is that this structure does not need the output layer since it is a regression network.

2-D Image and Feature Classification Network	layers = [
	  imageInputLayer(inputSize)
	  convolution2dLayer(filterSize, numFilters)
	  batchNormalizationLayer
	  reluLayer
	  flattenLayer
	  concatenationLayer(1,2,Name="cat")
	  fullyConnectedLayer(numClass)
	  softmaxLayer];
		This structure is to build a 2D image and feature classification network. More different layers are added into this structure due to its complicity, such as convolution2d layer, batchNormalization layer, flatten layer and concatenation layer. All of those layers are used to help this classification more smooth and effective.

2-D Image Classification Network	layers = [
	  imageInputLayer(inputSize)
	  convolution2dLayer(filterSize, numFilters)
	  batchNormalizationLayer
	  reluLayer
	  fullyConnectedLayer(numClass)
	  softmaxLayer];
		This structure is used to build a 2D image classification network to classify input images to different classes. An output layer, softmaxLayer is needed since this is a classification network.

2-D Image Regression Network	layers = [
	  imageInputLayer(inputSize)
	  convolution2dLayer(filterSize,numFilters)
	  batchNormalizationLayer
	  reluLayer
	  fullyConnectedLayer(numResponse)];
		This structure is used to build a 2D image regression network to map input images to different values. No output layer is needed since this is a regression network.

Keep those layers and structures in mind, and now let us build our deep neural network with deep learning functions.

7.7.4.3   Build a Deep Neural Network with Deep Learning Functions

As we discussed in Sect. 7.6.2, exactly in Table 7.1, some popular functions used for deep learning regression and classifications are displayed. Among them, two of them, **trainnet**() and **trainingOptions**(), are used for deep learning algorithms. In this part, we like to use those functions to build our deep neural network and train it with our earthquake dataset to develop our desired deep learning model and predict the possible earthquake energy level for a given location.

Open a new Script file, name it **Deep_MAG_Func.m**, and enter the codes shown in Fig. 7.33 into that file. Let us have a closer look at that piece of codes to see how it works.

1. As we did before, first we need to build our earthquake training and target data based on our modified earthquake dataset, **silver_m.csv**.
2. Then we need to build our training data and target data matrices.
3. To build our deep neural network, we can use one of the layers discussed in the last section. Since we are using a sequence-to-sequence regression type, we can build our network with five layers: sequence, LSTM, fully connected, RELU, and another fully connected layers. The key points are the arguments' values used for these layers. For the Sequence input layer, the input size is 3000 since we used 3000 data records from our dataset. The number of nodes for the LSTM

```
% Name: Deep_MAG_Func.m
% Using deep training and testing functions to build a deep neural network for earthquake dataset
% May 30, 2024

1   T = readtable('C:\\Artificial Intelligence Book\Students\Datasets\Earthquake Dataset\\silver_m.csv');
    N = 3000;                              % size of the training and testing data

    % Assign each column to a related local variable...
    LATI = table2array(T(1:N, 1))';        % LATI is located at column 1 in database - 1x3000 array
    LONG = table2array(T(1:N, 2))';        % LONG is located at column 2 in database - 1x3000 array
    DEPTH = table2array(T(1:N, 3))';       % DEPTH is located at column 3 in database - 1x3000 array
    HOUR = table2array(T(1:N, 6))';        % HOUR is located at column 6 in dataset - 1x3000 array
    MAG = table2array(T(1:N, 4))';         % MAG is located at column 4 in database - 1x3000 array

2   trainData = [LATI; LONG; DEPTH; HOUR];    % 4x3000 training data matrix
    targetData = MAG;                         % 1x3000 target data matrix

3   layers = [
            sequenceInputLayer(3000, Normalization="zscore")
            lstmLayer(100, OutputMode="last")
            fullyConnectedLayer(3000)
            reluLayer
            fullyConnectedLayer(3000)]

4   options = trainingOptions("sgdm", ...
                        MaxEpochs=150, ...
                        Verbose=false, ...
                        Plots="training-progress", ...
                        Metrics="rmse");

5   net = trainnet(trainData, targetData, layers, "mse", options)
```

Fig. 7.33 The codes used to train and test a deep neural network with functions

layer is 100, and the outputs for two fully connected layers are also 3000 since we need 3000 **MAG** output values.

4. As we discussed in Sect. 7.6.2.9, the **trainingOptions**() function can be used to configure and set up some preferred training methods or options, such as the solver name, the maximum epoch number, the accuracy, the minimum step length for each iteration, and plotting option. Here, we selected the **sgdm** as our solver, 150 as our maximum epoch number, no training process steps displaying with words, displaying training process with plots, and using **RMSE** value as our accuracy criteria.

5. Start to train our neural network with **trainData**, **targetData**, network structure **layers**, accuracy **mse,** and all training **options** defined above. The **mse** means that the target for the loss function is a numeric scalar value.

Now run this project, and the training result is shown in Fig. 7.34. Both training loss and training RMSE values are plotted and displayed for this training process.

7.7.4.4   Evaluate the Trained Model Built with Deep Learning Functions

As we did in Sects. 7.7.2.3 and 7.7.3.4, we like to evaluate the performances for the trained model, **net**, obtained from the last section. In fact, we can do this kind of evaluation by using either project built in those two sections. To make it simple, we like to use the project, **ANN_App_Eval.m**, with a little modification on the coding process. All projects built in this chapter can be found on the Springer ftp site in the

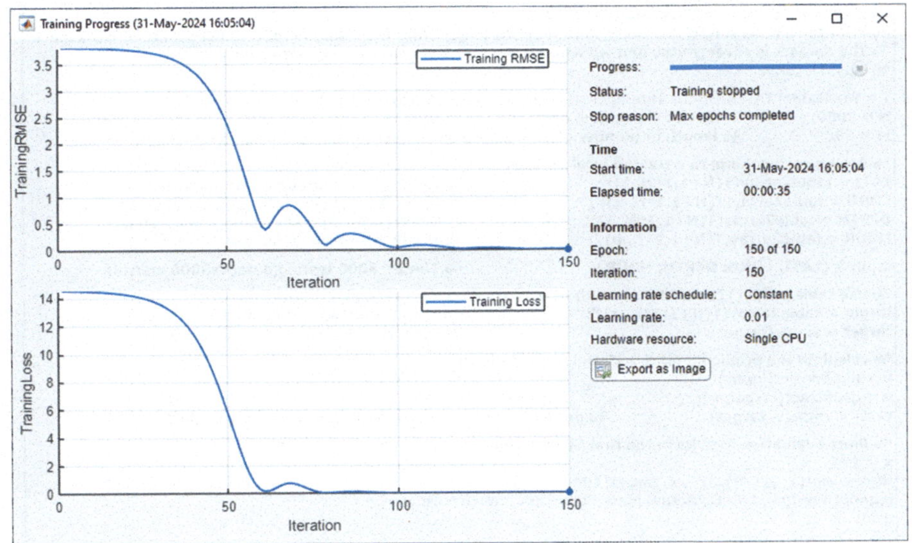

Fig. 7.34 The training results of deep neural network

folder, **Students\Class Projects\Chapter 7**. You can copy any of them and paste it to one of your local folders in your machine.

To build our evaluation project, open the project **ANN_App_Eval.m** and make the following modifications as shown in Fig. 7.35 to make it our new project, **Deep_MAG_Func_Eval.m**.

1. Change the dataset from the testing dataset, **silver_test.csv**, to our modified dataset, **silver_m.csv**, since we like to use the second 3000 data records as our testing data. Also change the variable **N** from 100 to 3000 to match the number of testing data used in this project. Add one more variable **M** that is used to hold the number of data points to be plotted later to check the evaluation results.
2. Change the row numbers for each variable from **1:N** to **N + 1:2*N** or (3001:6000) since we need to use the second 3000 data as our testing data in this project.
3. Perform a similar row-number-changing operation to the target or actual **MAG** vector.
4. This step is not necessary, but it is highly recommended since we like to make sure that both the target or actual **MAG** data and the predicted **MAG** data have the same data type.
5. The property function **predict()** that belongs to the trained model, **net**, is executed to calculate the predicted **MAG** values. Another format, **y = predict(net, input);** is also working.
6. To get the difference between the actual **MAG** values and the predicted **MAG** values, the function **gsubtract()** is used and the result is assigned to a local variable **e**.

```
% Name: Deep_MAG_Func_Eval.m
% Evaluate the trained Deep Learning model.
% Prior to run this script, run Deep_MAG_Func.m to generate the trained model net & export to Workspace
% The dataset is silver_m.csv and we will use the 2nd 3000 data as the testing data
% May 31, 2024

1   T = readtable('C:\\Artificial Intelligence Book\Students\Datasets\Earthquake Dataset\\silver_m.csv');
    N = 3000;              % size of the testing data
    M = 50;               % length or number of data points along the x-axis in plotting later

    % Assign each column to a related local variable...
2   LATI = table2array(T(N+1:2*N, 1))';          % LATI is located at column 1 in database
    LONG = table2array(T(N+1:2*N, 2))';          % LONG is located at column 2 in database
    DEPTH = table2array(T(N+1:2*N, 3))';         % DEPTH is located at column 3 in database
    HOUR = table2array(T(N+1:2*N, 6))';          % HOUR is located at column 6 in dataset

    input = [LATI; LONG; DEPTH; HOUR];           % The 2nd 3000 testing data(4x3000 matrix)

    % Use table2array() to convert the output MAG to double array to plot it later.
3   target = table2array(T(N+1:2*N, 4))';        % MAG is located at column 4 in database
4   target = single(target);                     % convert target to 1x3000 single array

    % calculate the predicted MAG with the trained model
5   y = net.predict(input)
6   e = gsubtract(target, y);
7   error = rmse(y, target)          % get RMSE value between the predicted & actual MAG

    % Plots Evaluation Results in the first 50 data points
8   x = 1:M;
    figure, plot(x, y(1:M),'*r-', x, target(1:M), '.b-');
    legend('Predicted MAG','Actual MAG', 'Location','NorthWest');
    grid;
```

Fig. 7.35 The codes used to evaluate the trained model built with functions

7. The RMSE value between the actual and the predicted **MAG** values is calculated by calling the **rmse()** function.
8. Finally the predicted and the actual **MAG** values are plotted to compare both of them to clarify the effectiveness of our trained model. To make this comparison clear, we only used the first 50 data points. Otherwise the result with all 3000 data points in that plot would be fuzzy.

Now run the project and the running results are shown in Fig. 7.36. The final **RMSE** value between the predicted and the actual **MAG** is **1.8756**, and it is displayed in the Command window.

A point to be noted is that prior to running this project, make sure that you have run the project **Deep_MAG_Func.m** script file to get the trained model **net**.

All projects built in this chapter can be found on the Springer ftp site in the folder, **Students\Class Projects\Chapter 7**. You can copy any of them and paste it to one of your local folders in your machine.

At this point, we have completed our discussions about using deep learning APPS and functions to build and train our deep learning models for earthquake systems. Next we like to provide more real example projects to involve more practical implementations by using deep learning algorithms to illustrate its powerful functions in various AI applications.

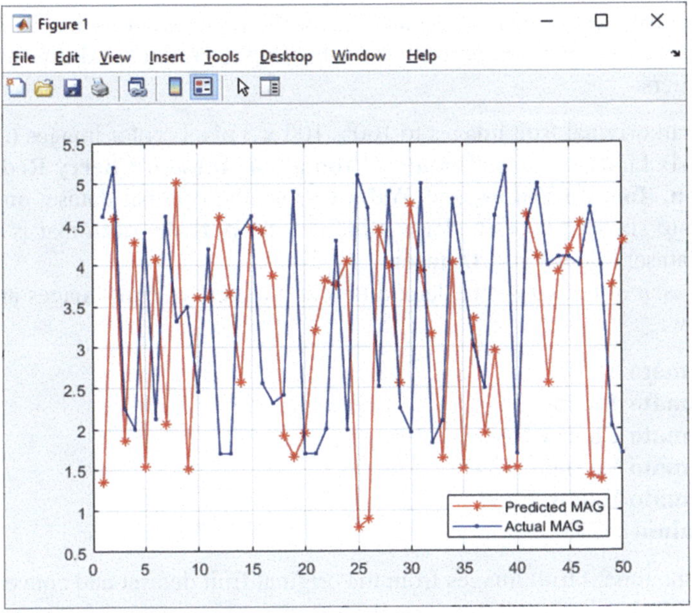

Fig. 7.36 The running result for the evaluation of the trained model

7.8   Using Deep Learning Algorithms to Perform Images Classifications of Fruits

Recall in Sect. 6.8.1, we used a **categoryClassifier** object to classify some fruits by using their images. In this section, we like to use some related deep learning algorithms to perform similar classification jobs. The dataset used in this project is **Fruits Dataset**, which is located at the site: https://public.robo-flow.com/classification/fruits-dataset,and it is a Public Domain license with No Copyright issue.

The dataset contained 2911 different fruit images, including various tomatoes and walnuts. To make this project simple, we only need to select some fruit images as our data source, thus we need to perform a preprocessing for the original fruits dataset to make it our desired one.

First let us have a closer look at this data preprocessing procedure.

7.8.1   Preprocess the Original Fruits Dataset to Get Our Desired Dataset

All original fruit image data are located under the **train_data** folder. To obtain our desired fruit images dataset, perform the following operations and save them to the related folders:

1. Resize all original fruit images to $100 \times 100 \times 3$ pixels color images (**.jpg**) files.
2. Select six kinds of fruits, **Tomato 3**, **Tomato 4**, **Tomato Cherry Red**, **Tomato Maroon**, **Tomato Yellow**, and **Walnut**, from the original dataset and arrange them into six subfolders under a subfolder, **fruit_train_data** that is under the main dataset folder **Fruit Dataset**.
3. Under each of the six subfolders, different numbers of fruit images are defined as below:

    (a) **Tomato 3** = 357
    (b) **Tomato 4** = 458
    (c) **Tomato Cherry Red** = 472
    (d) **Tomato Maroon** = 347
    (e) **Tomato Yellow** = 439
    (f) **Walnut** = 715

4. Select the top 20 fruit images from the original fruit dataset and convert them to $100 \times 100 \times 3$ pixels color images. Save them into another subfolder, **fruit_test_data** that is under the main dataset folder **Fruit Dataset**.

When you complete this preprocessing for desired fruit images dataset, the distributions of related train and test data are shown in Fig. 7.37.

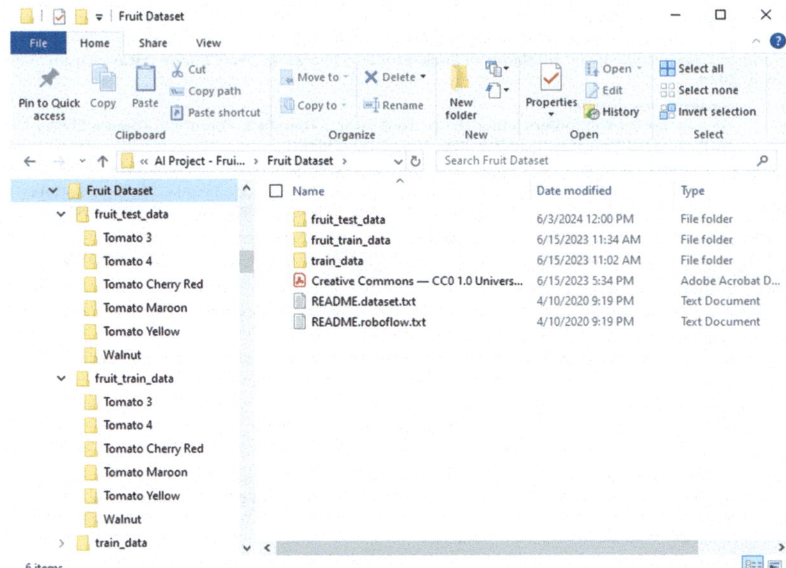

Fig. 7.37 The folder distributions for desired fruits dataset

To build the codes to create our desired fruit images dataset, open a new Script file, name it as **Setup_Images.m**, and enter the codes shown in Fig. 7.38 into that file.

Let us have a closer look at that piece of codes to see how it works:

1. First some local variables are declared and initialized. Where **n** is working as a sequence number to indicate the image sequence to be processed. **M** is the number of subfolders or classes used to store training and testing data. **N** is the number of top 20 data used for testing purposes later.
2. Three directories or folders used to contain training images, testing images, and the original unprocessed images are defined here to make it easy to access them later.
3. Now we need to get all six subfolders that are under the **train_data** folder, where the original or unprocessed images are located, and use them later to access and pick them up, preprocess them, and assign them to the training image subfolders.
4. Next we need to create six subfolders under the **fruit_train_data** folder, and these subfolders will be used to save preprocessed training data later.
5. Starting step 5, we will pick up each original image, preprocess each of them, and save each of them to the related subfolder, such as **Tomato 3**, **Tomato 4**, and so on under the **fruit_train_data** folder.
6. Get the target subfolder, such as **Tomato 3**, **Tomato 4**, … under the **fruit_train_data** folder.

```matlab
% File Name: Setup_Image.m
% Pick up source fruit data & store them to 6 sub folders under the folder, fruit_train_data
% The source folder, C:\Artificial Intelligence Book\Students\Datasets\Fruit Image Dataset\train_data
% June. 3, 2024
1  n = 0;          % sequence number
   M = 6;          % number for sub folders under the current folder: Tomato 3, Tomato 4, Tomato Cherry Red...
   N = 20;         % number of test images in the fruit_test_data folder

2  train_path = "C:\Artificial Intelligence Book\Students\Datasets\Fruit Image Dataset\fruit_train_data\";
   test_path = "C:\Artificial Intelligence Book\Students\Datasets\Fruit Image Dataset\fruit_test_data\";
   orig_path = "C:\Artificial Intelligence Book\Students\Datasets\Fruit Image Dataset\train_data\";

   % Get all 6 sub folders, imgFolds(i), where all original fruit images are stored
3  imd = dir('C:\Artificial Intelligence Book\Students\Datasets\Fruit Image Dataset\train_data\');
   isub = [imd(:).isdir];                          % returns logical vector
   imgFolds = {imd(isub).name}';
   imgFolds(ismember(imgFolds,{'.','..'})) = [];   % get all 6 sub folders excluding the . and .. operators.

   % Create 6 sub folders under the train folder: ..\fruit_train_data
4  cd 'C:\Artificial Intelligence Book\Students\Datasets\Fruit Image Dataset\fruit_train_data\';
   for i = 1:M
      ifold = string(imgFolds(i));
      mkdir(ifold);
   end

5  for i = 1:M
6     impath = train_path + imgFolds(i);            % target folder to save the processed images
7     path = orig_path + imgFolds(i);               % source folders to get original images
8     fp = fullfile(path, '*.jpg');
9     f = dir(fp); imgName = {f.name};

10     for k = 1:numel(imgName)
          n = n + 1;
          fullFileName = fullfile(path, imgName{k});
          sImg = imread(fullFileName);
          imshow(fullFileName);
          imgfullname = fullfile(impath, imgName{k});
          imwrite(sImg, imgfullname);
          n
       end
   end

   % Get top 20 fruit images from the fruit_train_data folder & save them to the fruit_test_data
   % Create 6 sub folders under the target test folder: ..\fruit_test_data
11 cd 'C:\Artificial Intelligence Book\Students\Datasets\Fruit Image Dataset\fruit_test_data\'
   for i = 1:M
      ifold = string(imgFolds(i));
      mkdir(ifold);
   end

12 for j = 1:M
      impath = test_path + imgFolds(j);      % target folder to save the test images
      path = train_path + imgFolds(j);       % source folders to get processed images
      fp = fullfile(path, '*.jpg');
      f = dir(fp); imgName = {f.name};

      for k = 1:N
         fullFileName = fullfile(path, imgName{k});
         cImage{k} = imread(fullFileName);
         delete(fullFileName);
         imgfullname = fullfile(impath, imgName{k});
         imwrite(cImage{k},
         imgfullname);
      end
   end
```

Fig. 7.38 The codes used to preprocess the image data

7. Get the original image subfolder, **Tomato 3**, **Tomato 4**, … under the **train_data** folder.

8. Get each image file handler for the original image files.

9. Obtain the related original image's name.

10. Use a nested **for()** loop to read each original image file, preprocess it, and assign it to the related subfolder under the target training data folder, **fruit_train_data**. An image processing sequence number **n** is also displayed to indicate the progress of this process.
11. Using **cd** command to direct the system to the test data folder, **fruit_test_data**, and then create six subfolders, such as **Tomato3, Tomato 4**, and so on to be ready to transfer the top 20 images on each related subfolder at the subfolder **fruit_train_data** to the test folder.
12. Using another **for()** loop to transfer the top 20 images on each related subfolder under the **fruit_train_data** folder to the related subfolder under the **fruit_test_data** folder as the testing image data.

Now we have finished our preprocessing for both the training and the testing image data, we are ready to perform the classifying process by using those data.

7.8.2   *Perform the Classification Process to Get Our Deep Learning Model*

In Sects. 7.6.2 and 7.7.4.2, we provided very detailed discussions about all deep network-related functions and structures, such as **trainingOptions**, **layers** and **trainNetwork()**. Keep in mind those staff, and we will use some of them to build our image classification project in this part.

Open a new Script file, name it **Fruits_Classify.m**, and enter the codes shown in Fig. 7.39 into that file. Let us have a closer look at that piece of codes to see how it works.

1. First we need to define the location to save our trained model. Due to the over-length of this path, we divided this into two sub strings, **trained_path** and **trained_name**, and use the **append()** function to combine these two strings together to make them shorter in coding.
2. Similarly, we can define another location used to indicate the training image data path, **train_path**.
3. This piece of codes is used to get six subfolders under the image training folder, **fruit_train_data**, where six classes training images are located, such as **Tomato 3, Tomato 4**, and so on without any extension operators.
4. This coding line is used to create an image Datastore used to reorganize and contain all training images in a special structure format. The **fullfile()** function is used to concatenate all strings to get a complete path for the target training images with extension of **.jpg**.
5. These two coding lines are used to display all training images and related labels.
6. To train and validate this model, we need to separate the whole training images into two parts, the training images and the validation images with a ratio of 70% and 30% by using the **splitEachLabel()** function. The training images object is

```matlab
% File Name: Fruits_Classify.m
% Purpose: The complete deep learning training codes to train a DL model with fruit images for 6 classes.
% Inputs: Processed images in C:\Artificial Intelligence Book\...\ Image Dataset\fruit_train_data folders.
% Input images numbers: Tomato 3 = 357, Tomato 4 = 458, ... Walnut = 715
% Outputs: Trained model f_imageNet.mat in C:\Artificial Intelligence Book...\Fruits Image Classifier.
% Notes: The file Setup_Image.m should be run first to prepare all images to be used for this one.
% June 3, 2024

1  trained_path = 'C:\Artificial Intelligence Book\Students\Class Projects\';
   trained_name = 'Chapter 8\Fruits Image Classifier\f_imageNet.mat';
   save_path = append(trained_path, trained_name)
2  train_path = 'C:\Artificial Intelligence Book\Students\Datasets\Fruit Image Dataset\fruit_train_data\';

3  imd = dir(train_path);                          % setup the train source image folder
   isub = [imd(:).isdir];                          % returns logical vector
   imgFolds = {imd(isub).name}';
   imgFolds(ismember(imgFolds,{'.','..'})) = [];   % get 6 sub folders excluding the . and .. operators.

4  imds = imageDatastore(fullfile('C:','Artificial Intelligence Book','Students', 'Datasets', 'Fruit Image Dataset',
                    'fruit_train_data', imgFolds), 'LabelSource', 'foldernames', 'FileExtensions', {'.jpg'});

   % show all images and labels
5  imds.Labels
   imds.Files

   % setup training parameters and CNN model: 70% for training & 30% for validations
6  numTrain = 0.7;        % 70% fruits images are used for training, and 30% images are for validations
   [imdsTrain, imdsValidation] = splitEachLabel(imds,numTrain);
7  inputSize = [100 100 3];
   numClasses = 6;

8  layers = [
       imageInputLayer(inputSize)
       %convolution2dLayer(3,16, 'Padding', 'same')
       batchNormalizationLayer
       reluLayer
       fullyConnectedLayer(numClasses)
       softmaxLayer
       classificationLayer];

9  options = trainingOptions('adam', ...
       'InitialLearnRate',0.001, ...
       'MaxEpochs',10, ...
       'miniBatchSize', 10, ...
       'ValidationData',imdsValidation, ...
       'ValidationFrequency',40, ...
       'Verbose',false, ...
       'Plots','training-progress');

   % training network with input images, layers and options...
10 net = trainNetwork(imdsTrain, layers, options);
11 f_imageNet = net;
   save f_imageNet;
12 save(save_path, 'f_imageNet');
```

Fig. 7.39 The codes used to evaluate the trained model built with functions

stored to a local variable **imdsTrain**, and the validation images body is to **imds-Validation** variable.

7. Some training parameters, such as the input image size, **inputSize**, and the number of classes of images, **numClasses**, are declared first.
8. The training structure, **layers**, which was discussed in Sect. 7.7.4.2, is defined. Since we are using a 2D Image and Feature Classification Network (No. 7 in Table 7.6), six layers are declared here. The **flattenLayer** and **concatenation-Layer** are optional, and we ignore both of them in this project. A point to be noted is that the function **trainNetwork**() is not the updated version of this kind

of training function in MATLAB, but it provided better performances for images training process, thus we keep using it. An updated version is **trainnet()** function. When using that function to replace **trainNetwork()** function, the output layer, **classficationLayer**, which is located at the last layer, is not required. Refer to Table 7.6 to confirm this since that table showed all layers for the updated version.

9. The function **trainingOptions()** is declared with all desired options. In fact, some of the options are not necessary to be selected here, such as **InitialLearnRate**, **miniBatchSize**, and **ValidationFrequency**, since here we are using the default values for all of them. They would be initialized to these default values even they are not selected.

10. Everything is ready, the **trainNetwork()** function is executed to begin the training process for our desired model. The trained model is returned and assigned to a local variable **net**.

11. To save this trained model in a MATLAB format with an extension of **.mat**, a temporary assignment is executed to assign it to a new variable **f_imageNet**. The **save** command is used to save this trained model to the Workspace.

12. Use this format to save this model as a file in one of your folders.

Now we can run this project to start the training and validation process for our input fruits images. The running result is shown in Fig. 7.40. A point to be noted is that prior to running this project, the project **Setup_Image.m** must be run first to finish the image data preprocessing process.

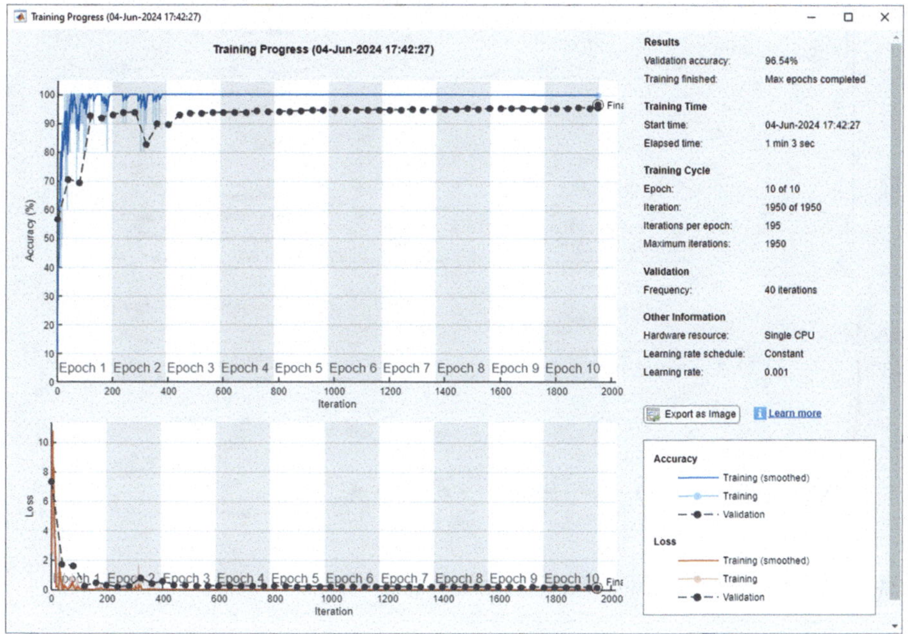

Fig. 7.40 The training result for the fruit images deep learning model without convolution

It can be found that the training and validation results shown in Fig. 7.40 are pretty good, and the validation accuracy is about 97%, which is almost perfect! However, that is not enough. In fact, this training and validation results can be significantly improved by applying a deep learning convolutional layer, as shown in **layers** structure at line 8 in Fig. 7.38, where a convolutional layer has been commented out.

To confirm the effectiveness of using a deep learning convolutional algorithm for this project, recover that coding line by removing the comment out operator, %, and run the project again. The running result is shown in Fig. 7.41. It can be found from Fig. 7.41 that the training and validation results have been significantly improved with the validation accuracy as 100%, which is a perfect result!

Another issue is that the maximum epoch number, **MaxEpochs**, in the **trainingOptions**() function can be reduced to a small one, such as 5 or 6, to make this training and validation process shorter and faster.

After this project is done, a trained model, **f_imageNet.mat**, has been saved to our project folder. Next we can check and confirm this trained model by performing an evaluation process.

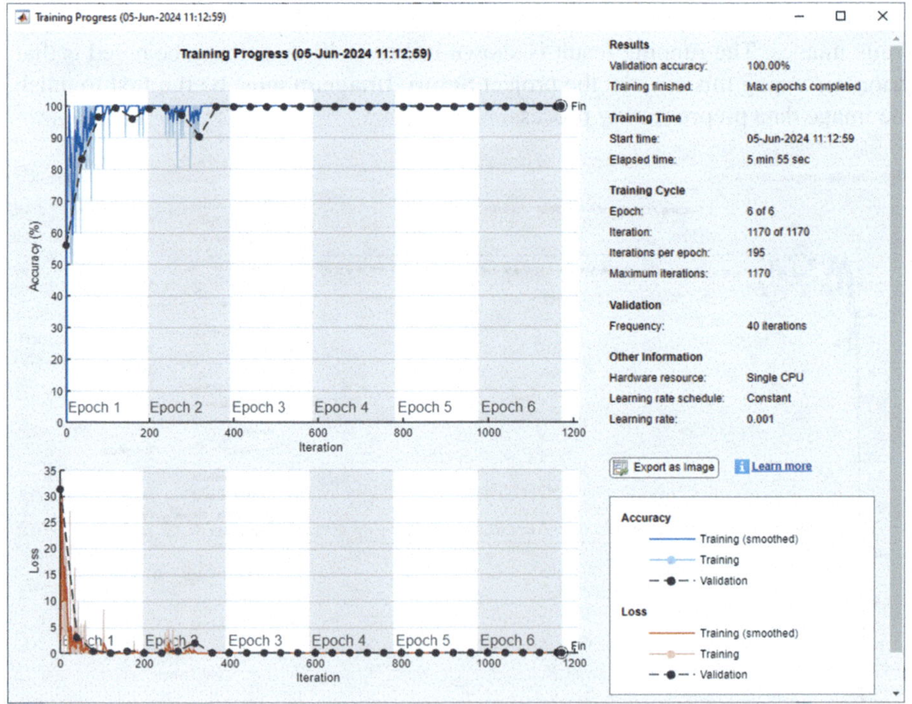

Fig. 7.41 The training result for the fruit images deep learning model with convolution

7.8.3   *Evaluate the Effectiveness of the Trained Fruit Deep Learning Model*

Now we like to evaluate the trained fruit image deep learning model with our testing image data that are saved in the subfolder, **fruit_test_data**, which were saved by us in steps 11 and 12 shown in Fig. 7.38.

Create a new Script file and name it **Fruits_Eval.m** and enter the codes shown in Fig. 7.42 into that file. Save this project file to one of your local folders. You can find

```matlab
% Name: Fruits_Eval.m
% Func: Load the trained net, f_imageNet, and evaluate it with fruits testing data
% June 5, 2024
1  n = 0;                 % loop counter
   M = 6;                 % number of classes is 6: Tomato 3, Tomato 4, ... Walnut
   N = 20;                % number of fruit testing data - images
   T = 120;               % number of total fruit testing data (images)
   numTrue = 0;           % number of result who is true
   numFalse = 0;          % number of result who is false

2  sz = [20 2];
   cTypes = ["string","string"];
   cNames = ["Fruit Class","Fruit Result"];
3  eResult1 = table("Size", sz, "Fruit Type", cTypes, "Fruit Name", cNames);
   eResult2 = table("Size", sz, "Fruit Type", cTypes, "Fruit Name", cNames);
   eResult3 = table("Size", sz, "Fruit Type", cTypes, "Fruit Name", cNames);
   eResult4 = table("Size", sz, "Fruit Type", cTypes, "Fruit Name", cNames);
   csResult = {eResult1, eResult2, eResult3, eResult4};

   % load the trained net model
4  cd 'C:\Artificial Intelligence Book\Students\Class Projects\Chapter 8\Fruits Image Classifier\';
5  load f_imageNet;
6  imgNet = f_imageNet;

   % load all fruits testing data
7  imd = dir('C:\Artificial Intelligence Book\Students\Datasets\Fruit Image Dataset\fruit_test_data\');
   isub = [imd(:).isdir];                      % returns logical vector
   imgFolds = {imd(isub).name}';
   imgFolds(ismember(imgFolds,{'.','..'})) = [];        % get all 6 sub folders excluding the . and .. operators.

   cPath = "C:\Artificial Intelligence Book\Students\Datasets\Fruit Image Dataset\fruit_test_data\" ;
8  for i = 1:M
       path = cPath + imgFolds(i);   % source folders to get test images
       fp = fullfile(path, '*.jpg');
       f = dir(fp);
       imgName = {f.name};

9      for k = 1:N
          n = n + 1;
          fullFileName = fullfile(path, imgName{k});
          %imshow(fullFileName);
          cImage{k} = imread(fullFileName);
10        YPred = classify(imgNet, cImage{k});
11        fruit_class = imgFolds(i);
12        fruit_result = YPred;
13        result(:, 1:4) = ['Fruit Class =: ', fruit_class, 'Fruit Result =: ', fruit_result];
14        csResult(k,:) = cellstr(result(:, 1:4));
15        if k == N
             csResult(k,:) = cellstr(result(:, 1:4))
          end
16        if fruit_class == fruit_result
             numTrue = numTrue + 1;
17        else
             numFalse = numFalse + 1;
          end
       end

18     final = ['Matched Fruit =: ', num2str(numTrue), ' Unmatched Fruit =: ', num2str(numFalse)];
       disp(final);
   end
```

Fig. 7.42 The codes used to evaluate the trained model built with functions

this file in the Springer ftp site under a folder, **Students\Class Projects\Chapter 7\ Fruits Image Classifier**.

Let us have a closer look at that piece of codes to see how it works.

1. Some local variables are declared and initialized first, such as the loop number **n**, the class number **M**, the number of testing images **N**, the total number of testing images **T**, and the number of true or false for the checking results, **numTrue** and **numFalse**.
2. A size structure **sz** is used to indicate the dimension of an element in a table object. A dimension of [20 2] means a 20 × 2 array will be defined. Two arguments, **cType** and **cNames**, will be filled into that size structure in a table later.
3. Four tables generated are used to hold the **Fruit Class** and **Fruit Result** with **sz** size. The former contains the actual fruit images, and the latter includes the predicted fruit images. A string array, **csResult{}**, is declared with four table components.
4. To evaluate our trained model, we need to switch the directory to the folder where our trained model is located by using the **cd** command.
5. Then the **load** command is executed to load our trained model **f_imageNet** into the project.
6. Temporarily save this loaded model to a local variable **imgNet**.
7. Now we need to get our testing image data and change our directory to the folder where the testing images are located by using the **dir** command and retrieve all six subfolders.
8. An outer **for()** is used to get each related subfolder, such as **Tomato 3**, **Tomato 4**, and so on under the **fruit_test_data** folder.
9. Then an inner **for()** loop is used to pick up each fruit image under each related subfolder by using the **imread()** function. The variable **fullFileName** contained the selected fruit image. The retrieved image is assigned to a local variable array **cImage{}** with the related index **k**, which is a loop number for the inner loop.
10. The function **classify()** is used with the trained model and the picked testing image to predict and classify the fruit type, and the classified class label is assigned to a local variable **YPred** that will be used later for comparison purposes.
11. The actual class label of the used testing image is assigned to a local variable **fruit_class**. The variable **imgFolds(i)** is the current selected subfolder's or class's name, such as **Tomato 3**, **Tomato 4**, **Tomato Cherry Red**, and so on.
12. The predicted or classified fruit's label is assigned to another local variable **fruit_result** that will be compared with the actual fruit's label later.
13. A new local cell array, **result[]**, is used to collect the actual class label of the used testing image, **fruit class**, and the predicted or classified fruit's class name, **fruit_result**. Since totally we only need to use four columns for this collection, thus a range on the columns, **1:4**, is used to indicate this range.
14. The local string array, **csResult{}**, defined in step 3 above is used to pick up and collect all elements stored in the **result[]** cell array. Similarly, a range on col-

umns, **1:4**, for the **result[]** cell array is also used to indicate that only four columns are needed for this assignment. A conversion function **cellstr()** must be
used for this assignment since all elements in the **result[]** array are cells, and
they need to be converted to string and assigned to the string array **csResult{}**.
15. If the loop counter **k** is equal to **N**, which is the upper bound of this inner loop,
    it means that the whole inner loop is done, and it can be stopped.
16. Then we need to check the classification result with the actual fruit's label. A
    true result would be obtained if both are equal, and the flag variable **numTrue**
    is increased by one.
17. Otherwise, if both are not equal, a false result is obtained and the flag variable
    numFalse is increased by one.
18. Finally the checking and evaluation results are displayed by using the **disp()**
    function.

Now run this project, and the running result is displayed in the Command window.
It can be found from the last line of those results, it is: **Matched Fruit =: 120
Unmatched Fruit =: 0**, which means that we get a 100% validation and evaluation
result, it is great!

Next let us build another deep learning project to classify some typical animals'
sounds with deep learning functions.

7.8.4 Using Deep Learning Algorithms to Classify Animals' Sounds

In Sect. 6.10, we built a project to classify some typical animals' sounds with
machine learning algorithms. The dataset we used is a typical animal dataset, and it
is located at a site, https://github.com/YashNita/Animal-Sound-Dataset, and this
dataset can be downloaded free of charge. This dataset consisting of 875 animal
sounds contains 10 types of animal sounds, including 200 cat, 200 dog, 200 bird, 75
cow, 45 lion, 40 sheep, 35 frog, 30 chicken, 25 donkey, and 25 monkey sounds. This
dataset can be used without copyright issues with the CC0.10 Universal Deed
license.

7.8.4.1 Modify the Original Animal Sounds Dataset to Get Our Desired Dataset

To make our project simple, we modified the original dataset to cover only five
animals, **Bird**, **Cat**, **Dog**, **Lion**, and **Sheep**, and name the modified dataset as
Animal Sounds Data. Exactly, all five kinds of animals' sounds are stored as five
subfolders or classes named **Bird**, **Cat**, **Dog**, **Lion**, and **Sheep**, and all of these five
classes are under another subfolder, **Train Data**. The structure of this dataset can be
expressed as below:

```
Students
   Datasets
      Animal Sounds Data
         Train Data
            Bird
            Cat
            Dog
            Lion
            Sheep
```

In this project, the modified dataset is located in the folder: **C:\Artificial Intelligence Book\Students\Datasets**. One can find this dataset on the Springer ftp site under the folder: **Students\Datasets**.

In this project, we like to simplify the data processing procedures, convert audio signals to the related spectrogram image files, and process them as 2D images to perform the audio sounds classification process.

7.8.4.2 Build and Evaluate a Deep Learning Model to Classify Animal Sounds

Open a new Script file, name it **Animal_Sounds_Classify.m**, and enter the codes shown in Fig. 7.43 into that file. Save this project file to one of your local folders. You can find this file in the Springer ftp site under a folder, **Students\Class Projects\ Chapter 7\Animal Sounds Classifier**.

Due to the large size of this coding process, we divide the full codes into two parts: The main codes and the function parts. Let us have a closer look at the main codes to see how they work.

1. First we need to set the path for the folder where our animal sounds dataset is located, then connect to our animal sounds dataset folder, **Train Data**, under that folder, all five subfolders or classes marked by five animals are located.
2. The function **audioDatastore()** is executed to create an audioDatastore object **ads**, in which all our training data for five animals' audio signals (**.wav**) are collected and reorganized into a piece of memory. Each animal sound file and label can be retrieved by using the dot operator, for example, the **ads.Files** contained all names of the selected animal sound files and the **ads.Labels** include all related animals' labels or classes.
3. To convert all animals' audio signals to the related spectrogram, or melspectrogram, the sampling frequency for melspectrogram is declared, regularly it is $fs = 44,100\,\text{Hz}$. We have provided a very detailed discussion about the melspectrogram in Sect. 6.9.1 in Chap. 6. Refer to that part to get more details about it. The so-called melspectrogram is to represent an audio signal in the frequency domain with a sequence of windows, but the frequency used is mel-frequency.

```matlab
% Name: Animal_Sounds_Classify.m
% Func: Train the animal-sound net model with deep learning
% Dataset: Animal Sounds Data located at: C:\Artificial Intelligence Book\Students\Datasets
% June 6, 2024
1  mPath = "C:\Artificial Intelligence Book\Students\Datasets\Animal Sounds Data\";
   datafolder = fullfile(mPath, 'Train Data');

2  ads = audioDatastore(datafolder, ...
           'IncludeSubfolders', true, ...
           'FileExtensions', '.wav', ...
           'LabelSource', 'foldernames');

3  fs = 44100;                      % sampling frequency for melspectrogram
4  for i = 1:length(ads.Files)
      [filepath, filename, ext] = fileparts(ads.Files{i});
      audiodata = read(ads);
5     if length(audiodata) < fs
         audiodata = [audiodata; audiodata];
      end
6     spath = fullfile(filepath, filename);
7     spectro(audiodata, fs, spath);   % save spectrogram as image
   end

8  move_file(ads);
9  imds = imageDatastore(fullfile('C:\Artificial Intelligence Book', 'Students', 'Datasets', ...
           'Animal Sounds Data', 'Spectrograms'), 'IncludeSubfolders', true,'LabelSource',...
           'foldernames','ReadFcn',@(f) imresize(imread(f),[100 100]));

10 [trainImgs, valImgs, testImgs] = splitEachLabel(imds, 0.8, 0.1, 0.1,'randomized');
11 numClasses = numel(categories(trainImgs.Labels));
   inputSize = [100 100 3];

12 layers = [
           imageInputLayer(inputSize);
           reluLayer
           fullyConnectedLayer(numClasses);
           softmaxLayer();
           classificationLayer()];

13 options = trainingOptions('adam', ...
           'Plots','training-progress', ...
           'MiniBatchSize', 32, ...
           'ValidationData', valImgs)
14 net = trainNetwork(trainImgs, layers, options);
15 s_trainedNet = net;  save s_trainedNet;

   f_path = 'C:\Artificial Intelligence Book\Students\Class Projects\Chapter 8\';
16 s_path = 'Animal Sounds Classifier\s_trainedNet.mat';
   save_path = append(f_path, s_path);
   save(save_path, 's_trainedNet');

17 % test the classification ressult and accuracy
   predict = classify(net, testImgs)
18 acc = mean(testImgs.Labels == predict)*100;
19 word = ['Classification accuracy = ', num2str(acc), '%'];
20 disp(word);

   % plot confusion matrix
21 confusionchart(testImgs.Labels, predict)
```

Fig. 7.43 The main codes for the animals' sound classification process

4. A **for**() loop is used to get each animal audio signal from our animal dataset with the **read**() function. The **ads** will reorganize all audio signals into a matrix made by M rows and N columns, and each column can be as a frequency channel.

5. The data values on the **audiodata** read from the **ads** are double (64 bits or 8 bytes), and each data item is sampled with the sampling frequency, fs, thus the total length of the **audiodata** in bytes should be less than that of fs. In this case, the **length(audiodata)** is 66,150. The total bytes are: $66,150/8 \approx 8268 < fs = 44,100$. In that case, we need to duplicate the **audiodata**

into two columns to meet the need for the melspectrogram conversions. In other words, if the length of the total samples of **audiodata** is less than 1 s, we need to duplicate them.

6. To be ready to perform melspectrogram processing, set up the target location, including our animal data path and folder.

7. Now the function **spectro()**, which is a customer-built function and its performance will be discussed later, is called to convert all animal sounds files to melspectrogram files.

8. The function **move_file()**, which is another customer-built function, is executed to move all converted melspectrogram files to the same folders as those sound files are located.

9. The **imageDatastore()** function is executed to create an **imageDatastore** object to collect and reorganize all converted melspectrogram image files. Also resize all images to the same size, as 2D matrix with 100×100 pixels.

10. The function **splitEachLabel()** is called to separate our training data into two groups, 80% for the actual training data and 20% for the testing or validation data.

11. To identify how many classes exist in our animal sounds dataset, the **numel()** function is used to get the number of the labels for the training data, **trainImages.Labels**. The input size for the training images is also defined as $100 \times 100 \times 3$.

12. Five layers are defined for this image classification process. Recall that we have provided a detailed discussion about all different layers for deep learning algorithms in Sect. 7.7.4.2. Refer to Table 7.6 in that section, exactly in row 8 on that Table. Six layers are required for a 2D Image Classification Network. However, we do not need the **Convolution2D** and **BatchNormalization** layers in this project, thus only four layers are left for our classification network. One point to be noted is that those six layers are defined for using the training function **trainnet()**, not for the function **trainNetwork()** that we are going to use later in this project. Therefore we need to add one more layer, **classificationLayer()**, at the last line on the **layers** structure.

13. The **trainingOptions()** function is executed to set up all required options for the training process. The **ValidationData** option must be initialized by assigning our validation images, **valImages**, if the validation process is needed.

14. The function **trainNetwork()** is called with training images, layers, and options as arguments to start this training process. The trained model is assigned to a local variable **net**.

15. To save this trained model, it is temporarily assigned to another variable, **s_ traindNet**, and the **save** command is used to save it to the Workspace and to be used later.
16. These three coding lines are used to break a long folder name into two pieces and combine them back as one to make two coding lines shorter.
17. The trained model is also saved in our project folder.
18. Next we need to perform the testing for our trained model by calling the function **classify()** with trained model **net** and the testing image data. The predicted results are assigned to a local variable array **predict**.
19. To get the classification accuracy for this testing, the function **mean()** is used to obtain the final accuracy in percent format.
20. To display that classification accuracy, the **disp()** function is used.
21. A confusion chart is plotted by calling the **confusionchart()** system function to display the classification results for both the true and the false distributions.

The codes for two user-defined functions, **spectro()** and **move_file()**, are shown in Fig. 7.44. No explanations are given due to their simplicity.

Now run the project, and the running results will be displayed after a while due to the larger memory size and long processing time of images. The finished training and validation processes are shown in Fig. 7.45. The final confusion chart is shown in Fig. 7.46.

It can be found that the classification accuracy is 70%, but the validation accuracy is 85%, which is pretty good for our trained model.

```
function spectro(audiodata, fs, path)
   melSpectrogram(audiodata, fs);
   colorbar ('off');
   axis off;
   f = gcf;
   saveas(f,path,'jpg');
   % crop spectrogram data only
   file = [path, '.jpg'];
   img = imread(file);
   crop_im = imcrop(img,[115 50 675 535]);
   imwrite(crop_im, file, "jpg");
end

function move_file(ads)                % move spectrogram image to according folder
   mPath = "C:\Artificial Intelligence Book\Students\Datasets\Animal Sounds Data\";
   num_labels=countEachLabel(ads);
   num_cate = categories(num_labels.Label);
   for i=1:numel(num_cate)
      source_folder = fullfile(mPath, 'Train Data', num_cate{i});
      source_imds = imageDatastore(source_folder, "FileExtensions", '.jpg');
      des_folder = fullfile(mPath,'\Spectrograms' , num_cate{i});
      mkdir (des_folder);
      for j=1: length(source_imds.Files)
         movefile (source_imds.Files{j}, des_folder);
      end
   end
end
```

Fig. 7.44 The codes for two user-defined functions

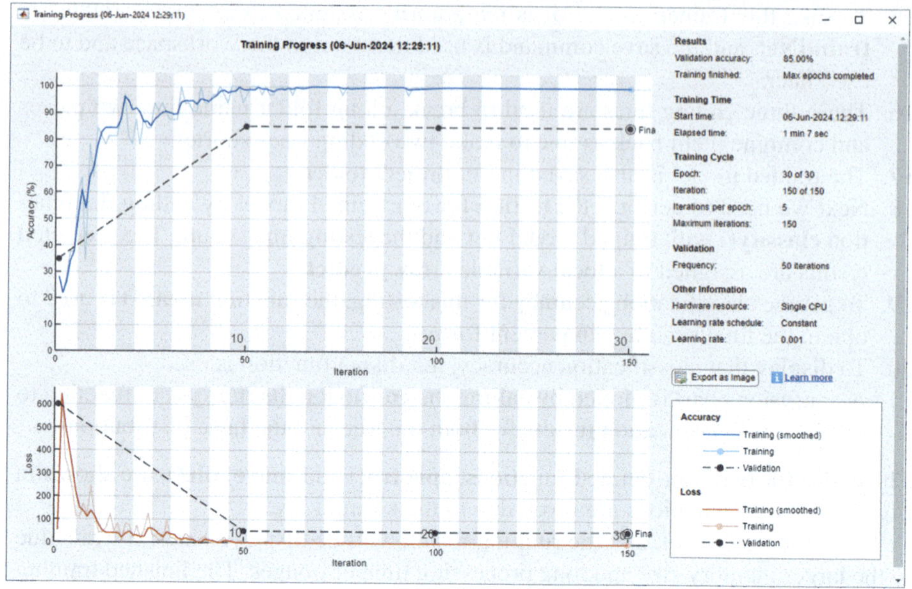

Fig. 7.45　The finished training and the validation processes

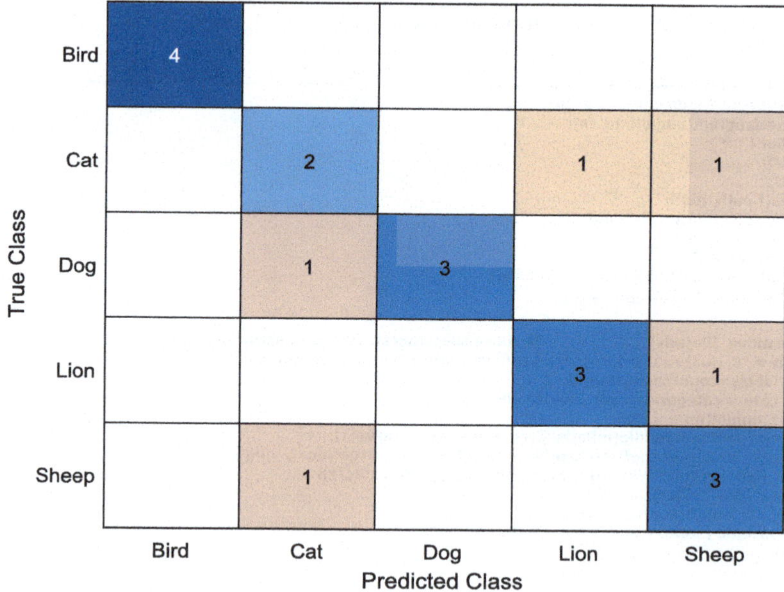

Fig. 7.46　The completed confusion chart for the animal sounds classification

7.9 Chapter Summary

The main topics discussed in this chapter are concentrated on neural networks and deep learning algorithms. The reason we combined these two topics together is due to their closer connections and applications. Applying any machine learning algorithm in a neural network can be considered as a deep learning implementation.

This chapter is divided into two parts, introductions and analyses about neural networks are given in the first part, and the deep learning algorithms are discussed in the second part.

In Sects. 7.1–7.4, detailed introductions and discussions about neural networks are provided, including a basic introduction about neural networks, structures and components of typical neural networks, different types of neural networks, and implementations of neural networks in real applications.

From Sect. 7.5, a fully detailed discussion about deep learning algorithms is given, including the difference between deep learning and machine learning, deep learning in MATLAB, deep learning working flow, and the most popular deep learning-related functions.

Recall that in Sect. 6.8.1 (Chap. 6) where a **bag** of features, BOF, is used to extract images features from fruits images, in Sects. 6.9.4 and 6.10.2 (Chap. 6) where an **audioFeatureExtractor** object is used to extract features of 10 digital sound audio signals and features of some animal sounds. For all above three machine learning projects, the features of predictors, including images and audios, must be extracted by users or human beings before they can input to the machine learning algorithms to enable them to be recognized and utilized to train and validate the desired machine learning models. Those projects provide clear and understandable practical examples to illustrate how to use machine learning algorithms to build intelligent models to meet the needs of AI applications.

Compared with two deep learning projects in Sect. 7.8.2, **Fruits_Classify.m** and in Sect. 7.8.4.2, **Animal_Sounds_Classify.m**, in this chapter, it can be clearly found that no features need to be extracted by users or human beings before they can be fed into the neural networks to be recognized and utilized to train and validate the desired deep learning models. By comparing those projects with real examples, a crisp clear picture could be obtained for the difference between the machine learning and deep learning models.

Sections 7.7 and 7.8 provided three real example projects to use deep learning algorithms to predict the energy levels of possible earthquakes and classify the different fruit images and animals' sounds based on input fruit images and animal sounds datasets. The former project belongs to the deep learning for a regression implementation, and the latter two are classification applications using deep learning to classify the related fruit images and animals' sounds.

After finishing this chapter, students will be able to:

- Understand the basics and fundamentals of neural networks
- Understand foundations and implementations for various deep learning algorithms
- Design and build some real implementations by using deep learning algorithms to train and validate the selected neural network to get the desired model
- Analyze, train, and test some selected deep neural networks with various deep learning algorithms by using some real datasets

Home Works

I. True/False Selections

_______1. Neural networks are the backbone of deep learning algorithms, which means that if machine learning uses neural networks as models, a deep learning algorithm is coming.

_______2. Generally, deep learning is a subset of machine learning, which is essentially a neural network with three or more layers.

_______3. Neural networks are composed of nodes and layers, which include input layers and one or more hidden layers.

_______4. In neural networks, when one node's output is below the threshold value, that node is activated and sends its data to the network's next layer.

_______5. Generally, a neural network can be adjusted or trained, so that a particular input leads to a corresponding target output.

_______6. To make neural networks work properly, different algorithms are needed to be used to train the neural networks to get different models.

_______7. Various deep learning algorithms can be used to train neural networks with large sets of labeled input and output data, but all different features still need to be extracted with manual feature extraction.

_______8. A deeper learning refers to a mixed learning process: a human learning process from a source to a learned semi-object, followed by a computer learning process from the human-learned semi-object to a final learned object.

_______9. Three popular deep learning algorithms are implemented in most applications, feedforward neural network algorithm, backpropagation neural network algorithm, and convolutional neural network algorithm.

_____10. A transfer learning approach is to train a deep learning model from scratch.

II. Multiple Choices

1. Neural networks are composed of ___________________.

   (a) Nodes or neutrons
   (b) Input layers and hidden layers
   (c) Output layers

(d) All of the above

2. Generally, _______ types of deep neural networks are popularly used today, __________.

   (a) 2, Feedforward neural networks, backpropagation neural networks
   (b) 3, Feedforward neural networks, backpropagation neural networks, convolutional neural networks
   (c) 4, Feedforward neural networks, backpropagation neural networks, convolutional neural networks, and recurrent neural networks
   (d) None of the above

3. A typical CNN architecture includes the following layers: __________.

   (a) Input and output layers
   (b) Convolutional layer and ReLU layer
   (c) Fully connected layer
   (d) All of them

4. For classification problems, a ______ layer and then a ______ layer must follow the final fully connected layer.

   (a) Input, output
   (b) ReLU, pooling
   (c) Hidden, fully connected
   (d) Softmax, classification

5. To use deep learning to train neural networks, two ways can be used: __________________.

   (a) Training from scratch and transfer learning
   (b) Training from scratch and training from middle layer
   (c) Transfer learning and training learning
   (d) None of them

6. MATLAB provided two major types of Apps for deep learning ______________.

   (a) Artificial Neural Networks (ANN), machine learning App
   (b) Deep Network Designer (DND), deep learning App
   (c) Artificial Neural Networks (ANN), Deep Network Designer (DND)
   (d) Deep learning App, machine learning App

7. Artificial Neural Networks (ANN) App can be used to design ______________.

   (a) Deeper neural network applications
   (b) Shallow neural network applications
   (c) Deeper neural network and shallow neural network applications

    (d) None of the above

8. Two types of APPS are provided by MATLAB Machine Learning and Deep Learning Toolbox, and they are:

    (a) Regression Learner, Deep Learning Learner
    (b) Classification Learner, Machine Learning Learner
    (c) Tree Learner, Binary Classification Learner
    (d) Regression Learner, Classification Learner

9. Five popular steps to train and validate a deep neural network are ______________.

    (a) Collect data, create network, split input data, train network, validate network
    (b) Preprocess data, create network, train network, validate network, use network
    (c) Collect data, create network, train network, validate network, use network
    (d) Filtering data, create network, train network, validate network, use network

10. Generally three steps are needed to preprocess for input data ________.

    (a) Normalize and scale data
    (b) Filter and clean data
    (c) Handle categorical and text data
    (d) All of them

III. Exercises

1. Explain the similarity and dissimilarity between the machine learning and the deep learning algorithms.
2. Provide a basic description about neural networks.
3. List four types of neural networks.
4. List and illustrate two ways to train neural networks with deep learning algorithms.

IV. Lab Projects

1. Build, train, and validate a shallow neural network model with deep learning algorithm using sample forest fire dataset, **MFire_Database.xls**. Refer to Fig. 7.17 in Sect. 7.7.2.1 to generate both training data and target data. Name the Script as **Generate_FireData_ANN.m**, and use the **nnstart** command to open the ANN APPS. Then use that App to train and validate the forest fire model. Export the trained model to the Workspace with any meaningful name.

    Hint1: Follow Sect. 7.7.2.1 to complete this building and training process.

Hint2: After running this project, need to type two commands on the MATLAB Command window, **load train_data.dat** and **load target_data. dat**, press the **Enter** key for each of them to load these two data items to the Workspace, and they can be used later in the ANN App.

2. Use **Deep Network Designer** to rebuild the project in Lab Project 1 above by using the same dataset **MFire_Database.xls** as data source. To do that, refer to Sect. 7.7.3.2 and Fig. 7.25 in this chapter to create a Script file, **Get_MFire_Datastore.m**, to convert our data source to two datastores, **trainStore** and **testStore**, respectively.

   Hint1: The size for the dataset **MFire_Database.xls** is 500. Thus you can set the number of training and testing data size as **250** to replace the original number **N**.

   Hint2: In the **Designer** panel, modify the **OutputSize** for two **FullyConnectedLayers** from 10 to 1.

   Hint3: In the **Training** panel, change the **MaxEpochs** property in the **Training Options** group from default 30 to 60. Export the trained model and results to the Workspace.

3. Use deep learning functions to design, build, train, and validate a deep learning model **e_imageNet.mat**. Save this model in one of your folders. The model is used to classify three types of cars, such as **pickup**, **sedan**, and **SUV**. Name Script file as **Easy_Car_Classify.m**. A preprocessed dataset, **car_train_data** that contained 102 pickup images, 99 sedans images, and 101 SUVs images, can be found on the Springer ftp site in the folder: **Students\Datasets\Car Type Dataset**.

   Hint1: Refer to Fig. 7.39 in Sect. 7.8.1.2 in this chapter for the coding developments.

   Hint2: You can download the **Car Type Dataset** and store it in one of your folders in your machine. Set up the full path for that dataset in your codes with the **dir()** command.

   Hint3: All preprocessed car images have been configured to the same size, [400 500 3]. Thus set your **inputSize** to that size with **numClasses** as 3 since we have three types of cars.

   Hint4: When setting up the **trainingOptions()** function, the **MaxEpochs** can be set up to 30.

4. Use deep learning functions to evaluate the trained model **e_imageNet.mat** built in the last project with three car testing dataset. The testing dataset **car_test_data** that contained 20 pickup images, 20 sedans images, and 20 SUVs images can be found on the Springer ftp site in the folder: **Students\ Datasets\Car Type Dataset**. Create and name a new Script file as **Easy_ Car_Eval.m** and build your codes based on Fig. 7.42 in Sect. 7.8.1.3 in this chapter.

   Hint1: You can download that testing dataset and save it to one of your local folders. Accordingly, you need to replace the path of that dataset based on your real folder name at line 7 in Fig. 7.42.

Hint2: You also need to load the trained model **e_imageNet.mat** that was built in the last project and saved in one of your local folders.

Hint3: In coding lines 13~15 on Fig. 7.42, use **result** to replace **result(1:4,:)** since you do not need to limit any row on **result** vector in this project.

5. Use deep learning functions to design, train, and validate a deep learning model by using the modified diabetes dataset, **Dataset of Diabetes_M.xlsx**, discussed in Sect. 6.7 in Chap. 6. The reason we selected this kind of dataset is because all input data columns on that modified dataset have been converted to numeric values. The size of that dataset is 1000×14. It is too much trouble to train a deep learning model with predictors that have text or string data involved. You must convert all text, string, or cell values to numeric values before you can use those predictors as inputs to train a deep learning model. Refer to Fig. 6.30 in Sect. 6.7.1 to do your coding development. The Script file is named, **Diabetes_Classifier.m**.

   Hint1: Use the first 900 data as training data and the last 100 data as validation data.

   Hint2: After reading the modified dataset with **T = readtable()** function, set up a new data array **X** as a predictor array by using the **T.column_name(1:900,:)**. The **column_name** is the name of each related column in the dataset. Apply all columns except the second column, **No_Pation**, due to its huge value. The built **X** array is an input array (900×12). The response array **Y**, which is the last column, **CLASS**, in the dataset (900×1) should be converted to a category array by using the **categorical()** function. Perform similar conversion for the 100 validation data predictor array (100×12) and response array (100×1).

   Hint3: When building the **layers** structure, the input layer should be a **featureInputLayer** with an input size of 12 since we have 12 input columns. The number of classes for the **fullyConnectedLayer** should be 3 since we have three output classes, **N**, **P**, and **Y**.

   Hint4: The **MaxEpochs** for the **trainingOptions()** setup could be 500 due to long training time.

6. Use deep learning functions to train and build a deep learning regression model for Google Stock Dataset to predict the future stock's closing prices. Two related Google Stock datasets, **Google_Stock_Price_Train.csv** and **Google Stock_Price_Test.csv**, will be used for this project, one works for training, and the other works for the validation purpose. Both datasets can be found on the Springer ftp site in the folder: **Students\Datasets\Google Stock Dataset**. Create a new Script, name it as **Stock_Price_Predictor.m**, and build your codes to it.

   Hint1: Refer to codes developed in Fig. 5.93 in Sect. 5.9.2 in Chap. 5 to build this project, you only need to refer to the top 22 lines to get the training and validation data.

   Hint2: You need to modify the training dataset, **Google_Stock_Price_Train.csv**, exactly to change the data format of the **Close** column in that dataset. To do that, open that dataset and highlight the **Close** column, then go to the **Format** tab and select the **Format Cells** item. Under the **Number** tab, click on the **Number** item from the **Category** box, and select 2 from the **Decimal places** combo box. Do NOT check **Use 1000 Separator** checkbox. That is all. In this way, we convert all numbers in that **Close** column from the cell format to the number format since MATLAB considers any number in Excel that has a thousand-separator as a cell, not a number. This modification is critical and important; otherwise, you may encounter a **NaN** error when you run this project later.

Hint3: Four layers will be used for this project: The input layer is a **featureInputLayer** with **inputSize** as 1 since we have only one feature for the **Close** price. The second layer is **lstmLayer** with 100 nodes and the **OutputMode** as "last". The **fullyConnectedLayer** is the third layer with four classes since we have four input predictors, **Open**, **High**, **Low**, and **Volume**. The last layer is an output layer, **regressionLayer**, since we are building a deep learning regression model.

Hint4: The **MaxEpochs** for the **trainingOptions**() setup could be 20 due to short training time.

References

1. https://www.ibm.com/blog/ai-vs-machine-learning-vs-deep-learning-vs-neural-networks/.
2. S. V. Kartalopoulos, Understanding Neural Networks and Fuzzy Logic, *IEEE Press*, 1996.
3. https://aws.amazon.com/what-is/neural-network/#:~:text=A%20neural%20network%20is%20a,inspired%20by%20the%20human%20brain.
4. https://missinglink.ai/guides/convolutional-neural-networks/convolutional-neural-network-architecture-forging-pathways-future/#:~:text=CNN%20architecture%20is%20inspired%20by,or%20feature%20of%20the%20image.
5. https://www.ibm.com/topics/recurrent-neural-networks.
6. https://www.geeksforgeeks.org/introduction-to-recurrent-neural-network/.
7. Zhang, W. J.; Yang, G.; Ji, C.; Gupta, M. M. (2018). "On Definition of Deep Learning". *2018 World Automation Congress (WAC)*. pp. 1–5. https://doi.org/10.23919/WAC.2018.8430387. ISBN 978-1-5323-7791-4. S2CID 51971897.
8. https://en.wikipedia.org/wiki/Deep_learning#:~:text=Deep%20learning%20is%20a%20class,digits%2C%20letters%2C%20or%20faces.
9. https://www.mathworks.com/discovery/deep-learning.html.
10. https://www.ibm.com/topics/deep-learning.
11. https://www.mathworks.com/help/stats/fitrnet.html.
12. https://www.mathworks.com/help/stats/fitcnet.html.
13. https://www.codecademy.com/article/deep-learning-workflow.
14. https://www.mathworks.com/help/simulink/ug/choose-a-solver.html.
15. https://www.geeksforgeeks.org/what-is-the-difference-between-val_loss-and-loss-during-training-in-keras/.
16. https://www.quora.com/What-techniques-are-commonly-used-to-evaluate-the-effectiveness-of-a-trained-artificial-neural-network-model.
17. https://www.mathworks.com/discovery/transfer-learning.html#:~:text=Transfer%20learning%20is%20a%20deep,training%20a%20network%20from%20scratch.

18. World Health Organization, Earthquakes, Health topics.
19. United States Geological Survey, Earthquake Magnitude, Energy Release, and Shaking Intensity, Earthquake Hazards.
20. https://www.kaggle.com/datasets/gustavobmgm/earthquakes-for-ml-prediction-new-version.
21. https://www.mathworks.com/help/matlab/import_export/what-is-a-datastore.html.
22. https://www.mathworks.com/help/deeplearning/ug/define-custom-deep-learning-layers.html.

Chapter 8
Introduction to Unsupervised Learning

As we discussed in Sect. 4.2 in Chap. 4, machine learning contains three major components: supervised learning, unsupervised learning, and reinforcement learning. Generally, the first two algorithms, supervised learning and unsupervised learning, play more important roles in most AI applications and implementations in our real world.

Under the unsupervised learning umbrella, three subcomponents, clustering, association rules, and dimensionality reduction, exist and work as three algorithms to perform related unsupervised learning functions. However, regularly speaking, the dimensionality reduction is not necessary until when the number of features or dimensions in a given dataset is too high. It is mainly used to reduce the number of data inputs to a manageable size while also preserving the integrity of the dataset as much as possible [1]. It is commonly used in the preprocessing data stage; thus here we will ignore this algorithm and only consider the first two of them shown in Fig. 8.1 as our major topics in this chapter.

Both neural networks and deep learning can be applied and implemented in either supervised or unsupervised learning models to provide predictions or classifications effectively. In fact, all AI-related models, including supervised, unsupervised, and reinforcement learning, can be trained and built by using deep learning to meet the real needs of practical applications.

We have provided detailed discussions about supervised learning in previous chapters. Starting from this chapter, we will concentrate our discussions on unsupervised learning.

Supplementary Information The online version contains supplementary material available at https://doi.org/10.1007/978-3-031-84423-2_8.

Y. Bai, *AI Foundations and Applications with MATLAB*,
https://doi.org/10.1007/978-3-031-84423-2_8

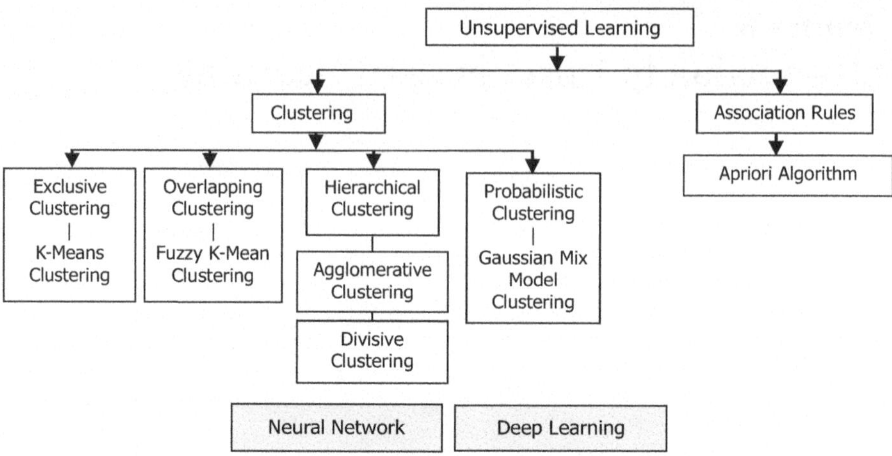

Fig. 8.1 Related algorithms belong to unsupervised learning

8.1   Introduction to Unsupervised Learning Algorithms

Compared with supervised learning, the unsupervised learning is a type of machine learning algorithms that learn from input data without human supervision. Unlike supervised learning, in which a group of input-output data pair (input data and related labels) is needed to train the models, unsupervised machine learning models are given unlabeled data and allowed to discover patterns and insights without any explicit guidance or instruction.

As shown in Fig. 8.1 which is a modified version based on Fig. 4.1 in Chap. 4 two major components or algorithms, clustering and association rules, are included under the unsupervised machine learning category.

According to its definition, clustering is the task of dividing the unlabeled data or data points into different clusters such that similar data points fall in the same cluster than those which differ from the others. In simple words, the aim of the clustering process is to segregate groups with similar traits and assign them into clusters [3]. Each cluster body can be considered as a group body where all objects or data points in a group have similar properties.

A typical example of using clustering algorithm is to classify the customers' purchasing preferences for certain products. For example, a fruit store likes to find the number of customers who like to purchase a special fruit, such as apples, pineapples, bananas, pears, or grapes, and the amount they purchased. Depending on the investigation and survey results, the store can divide customers into different groups or clusters. This is a typical exclusive clustering example, which means that each selected customer can only belong to one group or one cluster without any overlapping or crossing among other groups.

Association rule is a technique used to uncover hidden relationships between variables in large datasets [3].

In fact, association rule is a technique used to identify patterns or relations in large datasets. It involves finding relationships between variables in the data and using those relationships to make predictions or decisions. The goal of association rule is to uncover rules that describe the relationships between different items in the dataset [3].

A typical application of using the association rule is to identify relationships between items that are frequently purchased together. For example, the rule: *If a customer buys a large-screen TV, they are also likely to buy a wall-mount bracket* to install the TV. This is an association rule that could be mined from this dataset. We can use such rules to inform decisions about store layout, product placement, and marketing efforts.

Now let's first concentrate our discussions on the clustering algorithm.

Based on Fig. 8.1, four components or algorithms can be considered as a member of clustering, they are:

(1) Exclusive clustering
(2) Overlapping clustering
(3) Hierarchical clustering
(4) Probabilistic clustering

Let's have a closer look at these clustering algorithms one by one.

8.1.1 *Exclusive Clustering Algorithm*

As we mentioned in the last section, the exclusive clustering is a kind of hard clustering in which data point exclusively belongs to one cluster [4]. The so-called hard clustering is relatively to a soft clustering, in which the data points can belong to either one cluster in some degrees and also belong to another cluster in other degrees. The latter is a typical overlapping clustering. As shown in Fig. 8.1, a typical application of exclusive clustering is the K-means clustering.

This algorithm is to group all the similar data points into K clusters, called or known as K-means clustering. The criterion to group each point into a cluster is based on a centroid-based algorithm in which each group has a centroid. In other words, all data points in a cluster has the desired distance to their center point or centroid. Figure 8.2a shows an illustration of exclusive clustering, where all data points are clustered into $K = 2$ groups, and each data point can only belong to one group A or B without any overlapping.

Here K in K-Means is the number of clusters. To perform a K-Means algorithm, the following procedure should be followed [4]:

(1) First, we randomly assign some centroids to the dataset. And then clusters are formed by assigning data points to the cluster to which the data point is near the corresponding cluster.

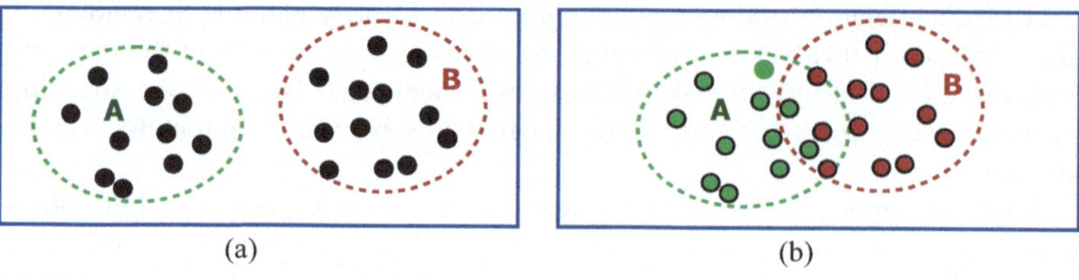

Fig. 8.2 The presentations of exclusive clustering and overlapping clustering

(2) From those clusters, new centroids were formed with the mean data points. This process will continue until the model is optimized which means the final centroids will not change even for the next iteration.

Here a question arises: How can we find the optimal K value during the above process? Two possible methods could be used and they are:

(1) Elbow method
(2) Silhouette method

8.1.1.1   The Elbow Method

The elbow method is one of the most popular methods used to find K for K-Means clustering algorithm. The core of using this method is to find the sum of squared distance (SSD) between each member data point on the related cluster and its centroid by trying to use different K until this SSD keeps a constant or almost no change after some tested K values. The boarding point for that K value is the optimal K value.

To understand this testing process, we need to know a new terminology called Within the Sum of Squares (**WSS**), which is defined as below:

$$\text{WSS} = \sum_{i=1}^{M} d\left(p^{(i)} q^{(i)}\right)^2 = \sum_{i=1}^{M}\sum_{j=1}^{N}\left(p_j^{(i)} - q_j^{(i)}\right)^2 \tag{8.1}$$

where $p^{(i)}$ is the ith data point and $q^{(i)}$ is the ith closest centroid to the data point. The index j indicates the number of cluster K that is initially defined as $N = 10$.

As this testing process starts, the WSS value is generally reduced as the increment of K value and the final K value is chosen after the decrease of **WSS** is almost constant. Figure 8.3 shows an example of this testing process [5].

It can be found from Fig. 8.3 that the K value is 3 and after that point, **WSS** is almost constant. Therefore, 3 is selected as K. In this way, the elbow method is used to find the optimal value of K.

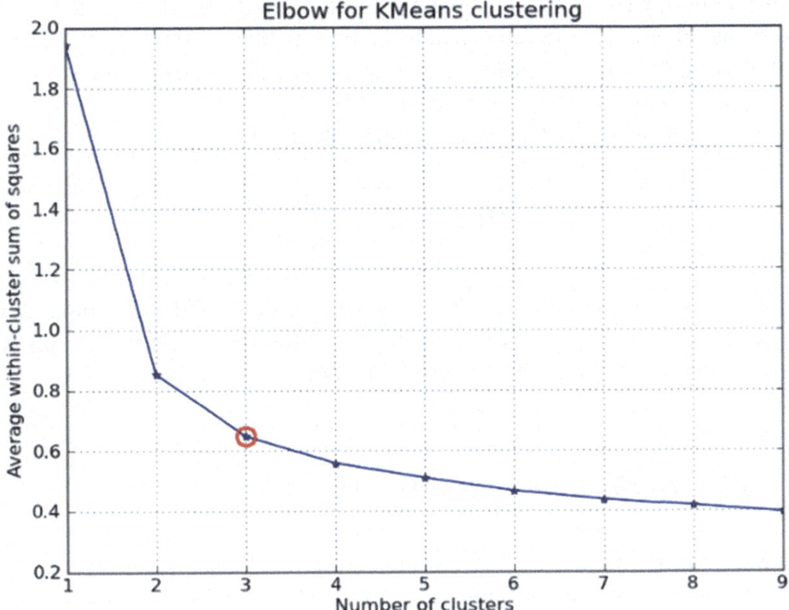

Fig. 8.3 Elbow for K-Means clustering

8.1.1.2   The Silhouette Method

The silhouette method is used to calculate a silhouette value which is a measure of how similar an object is to its own cluster (cohesion) compared to other clusters (separation) [6]. The definition of calculating or measuring the silhouette score is:

(1) For data point $i \in C_I$ (data point i in the cluster C_I), let

$$a(i) = \frac{1}{|C_I| - 1} \sum_{j \in C_I, i \neq j} d(i,j)$$
(8.2)

be the mean distance between i and all other data points in the same cluster, where $|C_I|$ is the number of points belonging to cluster C_I, and $d(i, j)$ is the distance between data points i and j in the cluster C_I (we divide by $|C_I| - 1$ because we do not include the distance $d(i, i)$ in the sum). We can interpret $a(i)$ as a measure of how well i is assigned to its cluster (the smaller the value, the better the assignment).

We then define the mean dissimilarity of point i to some cluster C_J as the mean of the distance from i to all points in C_J (where $C_J \neq C_I$).

For each data point $i \in C_I$, we now define

$$b(i) = \min_{J \neq I} \frac{1}{|C_J|} \sum_{j \in C_J} d(i,j)$$
(8.3)

to be the *smallest* (hence the **min** operator in the formula) mean distance of i to all points in any other cluster (i.e., in any cluster of which i is not a member). The cluster with this smallest mean dissimilarity is said to be the "neighboring cluster" of i because it is the next best-fit cluster for point i.

We now define a *silhouette* (value) of one data point i

$$s(i) = \frac{b(i) - a(i)}{\max\{a(i), b(i)\}} \tag{8.4}$$

Figure 8.4 shows an example of using silhouette score to find the optimal K value [7]. It can be seen from Fig. 8.4 that the highest silhouette coefficient is for $K = 3$. In this way, the silhouette method is used for finding the optimal K value.

8.1.2   The Overlapping Clustering Algorithm

Overlapping clustering is a kind of soft clustering in which a data point can belong to multiple clusters in some different degrees. As shown in Fig. 8.2b, some member data points in the green cluster can also belong to cluster **B**, similarly, some member data points in cluster **B** can belong to cluster **A** in some degrees. In other words, some green points and some red points overlap among two clusters.

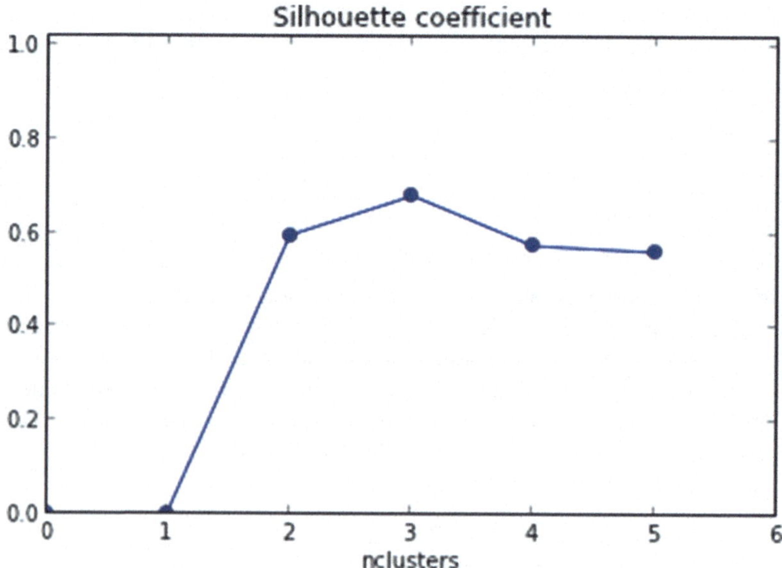

Fig. 8.4 The silhouette coefficient

A typical example of an application using the overlapping clustering algorithm is to collect the number of customers who joined some warehouse clubs. For example, the customers who are Costco club members may also be Sam's club members, and even more, they may be JB club members.

A popular algorithm used for overlapping clustering is called fuzzy clustering, or exactly fuzzy K-means clustering.

In fuzzy clustering, data points can potentially belong to multiple clusters. For example, an apple can be red or green (hard clustering), but an apple can also be red AND green (fuzzy clustering). Here, the apple can be red to a certain degree as well as green to a certain degree. Instead of the apple belonging to green [green = 1] and not red [red = 0], the apple can belong to green [green = 0.5] and red [red = 0.5]. These values are normalized between 0 and 1; however, they do not represent probabilities, so the two values do not need to add up to 1 [8].

The fuzzy K-means (FKM) algorithm is very similar to the K-Means algorithm, and the procedure for using this algorithm includes the following:

(1) Initially choose a number of clusters.
(2) Assign coefficients randomly to each data point for being in the clusters.
(3) Repeat until the algorithm has converged (that is, the coefficients' change between two iterations is no more than ε, the given sensitivity threshold):

    (a) Compute the centroid for each cluster (similar to COG discussed in Sect. 3.6.2 in Chap. 3).
    (b) For each data point, compute its coefficients of being in the clusters.

Any point x has a set of coefficients giving the degree of being in the kth cluster $w_k(x)$, as a membership function (MF $\mu_k(x)$). With fuzzy K-Means, the centroid of a cluster is the mean of all points, weighted by their degree of belonging to the cluster, or, mathematically,

$$K_k = \frac{\sum\limits_x \omega_k\left(x\right)^m x}{\sum\limits_x \omega_k\left(x\right)^m} \tag{8.5}$$

where m is the hyper-parameter that controls how fuzzy the cluster will be. The higher it is, the fuzzier the cluster will be in the end.

In fact, the FKM algorithm attempts to partition a finite collection of n elements $X = \{x_1, x_2, \ldots, x_n\}$ into a collection of K fuzzy clusters with respect to some given criterion [8].

For example, given a finite set of data, the algorithm returns a list of cluster centers $K = \{k_1, k_2, \ldots, k_k\}$ and a partition matrix $W = \omega_{i,j}\epsilon[0, 1]$, $i = 1, 2, \ldots, n$, $j = 1, 2, \ldots, k$, where each element, $\omega_{i,j}$, tells the degree to which element, x_i, belongs to cluster k_j.

The FKM aims to minimize an objective function [8]:

$$J(W,K) = \sum_{i=1}^{n}\sum_{j=1}^{k} \omega_{ij}^{m}\, x_i - k_j^{\,2} \tag{8.6}$$

where

$$\omega_{ij} = \frac{1}{\displaystyle\sum_{k=1}^{K}\left(\frac{x_i - k_j}{x_i - k_k}\right)^{\frac{2}{m-1}}} \tag{8.7}$$

8.1.3   The Hierarchical Clustering

Hierarchical clustering methods, as the name suggests, is an algorithm that builds a hierarchy of clusters. This algorithm starts with all the data points assigned to a cluster of their own. Then two nearest clusters are merged into the same cluster. In the end, this algorithm terminates when there is only a single cluster left [2].

The results of hierarchical clustering can be plotted using a **dendrogram**() function in MATLAB. An example dendrogram for a hierarchical cluster is shown in Fig. 8.5a.

As shown in Fig. 8.5a, at the bottom we start with nine data points, each assigned to separate clusters. The two closest clusters are then merged until we have just one cluster at the top. The height in the dendrogram at which two clusters are merged represents the distance between two clusters in the data space.

The decision of the number of clusters that can best depict different groups can be chosen by observing the dendrogram. The best choice of the number of clusters

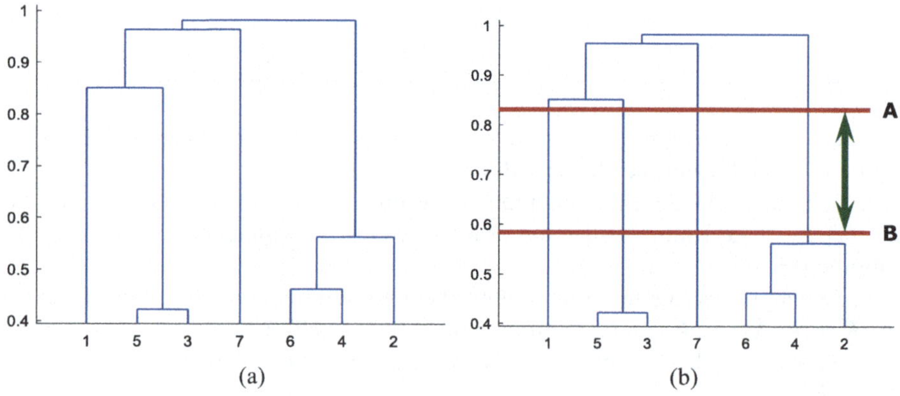

Fig. 8.5 A dendrogram for a hierarchical clustering

is the number of vertical lines in the dendrogram cut by a horizontal line that can transverse the maximum distance vertically without intersecting a cluster (Fig. 8.5b).

In this example, the best choice of number of clusters will be 4 as the red horizontal line in the dendrogram below covers the maximum vertical distance **AB**.

8.1.4   The Probabilistic Clustering

A probabilistic clustering is an unsupervised technique that helps us solve density estimation or *soft* clustering problems. In probabilistic clustering, data points are clustered based on the likelihood that they belong to a particular distribution. The Gaussian mixture model (GMM) is one of the most commonly used probabilistic clustering methods [1].

Similar to fuzzy or fuzzy K-Means clustering, probabilistic clustering allows one data point to belong to more than one cluster, which means a data point is a soft member that can belong to one or another cluster with related degrees simultaneously. In fact, you can use GMM to perform either a *hard* clustering or a *soft* clustering on query data.

To perform *hard* clustering, the GMM assigns query data points to the multivariate normal components that maximize the component posterior probability, given the data. That is, given a fitted GMM, the **cluster**() function can be used to assign query data to the component yielding the highest posterior probability. Hard clustering assigns a data point to exactly one cluster.

Additionally, you can use a GMM to perform a more flexible clustering on data, referred to as *soft* clustering. Soft clustering methods assign a score to a data point for each cluster. The value of the score indicates the association strength of the data point to the cluster. As opposed to hard clustering methods, soft clustering methods are flexible because they can assign a data point to more than one cluster. When you perform GMM clustering, the score is the posterior probability [9].

Figure 8.6 shows an illustration of using probabilistic clustering algorithm to predict the possibility of some data points' distributions for two clusters, A and B, with normal or Gaussian distributions. In Fig. 8.6a, a soft clustering is performed and some data points, such as **a** and **b**, can belong to a shared distribution area. In other words, they can belong to either cluster A or cluster B in some degrees. However, in Fig. 8.6b, a hard clustering is performed and two groups of data points can only belong to each separate cluster with no overlapping.

In summary, GMM clustering can accommodate clusters that have different sizes and correlation structures within them. Therefore, in certain applications, GMM clustering can be more appropriate than other methods such as *k*-means clustering. Like many clustering methods, GMM clustering requires users to specify the number of clusters before fitting the model. The number of clusters specifies the number of components in the GMM [9].

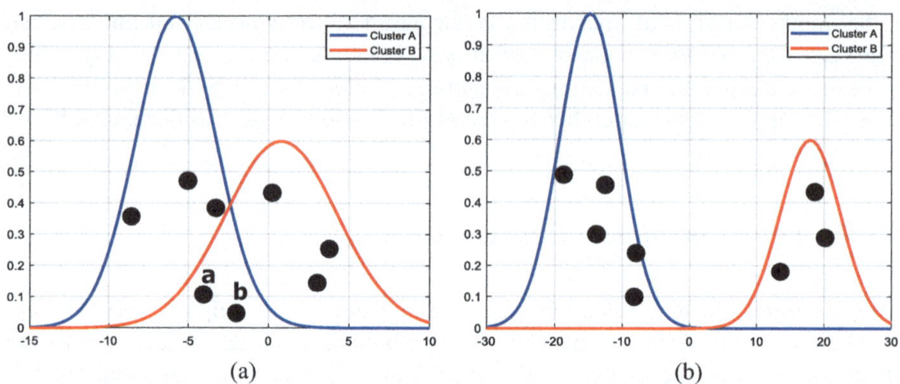

Fig. 8.6 An illustration of hard and soft probabilistic clustering

8.2   The Association Rules

As we mentioned in Sect. 8.1, the association rule is a technique used to uncover hidden relationships between variables in large datasets [3].

The association rules are intended to identify strong rules discovered in databases using some measures of interestingness [10]. In any given transaction with a variety of items, association rules are meant to discover the rules that determine how or why certain items are connected [11].

Association rules are made by searching data for frequent **if-then** patterns and by using a certain criterion under **Support** and **Confidence** to define what the most important relationships are. Support is the evidence of how frequently an item appears in the given data, as Confidence is defined by how many times the **if-then** statements are found to be true. However, there is a third criterion that can be used, it is called **Lift,** and it can be used to compare the expected Confidence and the actual Confidence. Lift will show how many times the **if-then** statement is expected to be found to be true.

For example, if 85% of people who purchased tomato sauces and onions also buy Italian noodles, then there is a discernible pattern in transactional data that customers who buy tomato sauces and onions often buy Italian noodles. An association rule is that there is an association between buying tomato sauces-onions and Italian noodles.

Rules are of the form A $\Rightarrow$ B (e.g., {tomato sauces, onions} - > {Italian noodles}).

The Support and Confidence are used to identify how strong the discovered rules are. To make it simple and straightforward, we can define them as [12]:

(1) Support is the fraction of transactions that contain both A and B: **Support(A,B) = P(A,B)**

(2) Confidence is the fraction of transactions, where items in B appear in transactions that contain A: **Confidence(A,B) = P(B|A).**

We are using Apriori algorithm to identify frequent item sets. It proceeds by identifying the frequent individual items in the database and extending them to larger item sets while the items satisfy the minimum support requirement (frequency of items in the database). The frequent item sets determined by Apriori are then used to determine association rules [12].

According to Apriori probability, the formula for calculating a priori probability is [13]:

$$A\ Priori\ Probability = Desired\ Outcome(s)\ /\ The\ Total\ Number\ of\ Outcomes$$

This is very similar to Confidence and Support defined in the association rules above. In fact, the Confidence is more likely to be a conditional probability as:

$$\mathrm{conf}\left(X \Rightarrow Y\right) = P\{Y|X\} = \frac{\mathrm{support}\left(X \cap Y\right)}{\mathrm{support}\left(X\right)} = \frac{\text{Number of transactions including } X \text{ and } Y}{\text{Number of transactions including } X} \tag{8.8}$$

Now let's use a real example to illustrate how to use Support and Confidence to set up related or association rules among data points or elements.

8.2.1   A Real Example of Using Support and Confidence to Build Association Rules

Assume that we have a small supermarket to provide some life-consuming products. Table 8.1 shows a small supermarket database containing the items. For each entry, the value 1 means the presence of the item in the corresponding transaction, and the value 0 represents the absence of an item in that transaction. The set of items can be presented as:

$$I = \{\textbf{Onions, Tomato Sauce, Italian Noodle, Bread, Milk, Beer}\}$$

An example rule for the supermarket is {**Onions, Tomato Sauce**} ⇒ {**Italian Noodle**} meaning that if onions and tomato sauce are bought, customers also buy Italian noodles.

Table 8.1 Example database with four transactions and six items

Transaction ID	Onions	Tomato Sauce	Italian Noodle	Bread	Milk	Beer
1	1	1	1	0	1	0
2	0	0	1	1	0	0
3	0	1	0	1	1	1
4	1	0	1	0	0	1

In order to select interesting rules from the set of all possible rules, constraints on various measures of significance and interest are used. The best-known constraints are minimum thresholds on support and confidence.

Let X, Y be item sets, $X \Rightarrow Y$ an association rule, and T a set of transactions of a given database.

The **Support** can be described as an indication of how frequently the item set appears in the dataset. In our example, the support can be expressed by writing [11]:

$$\text{Support} = P(A \cap B) = \frac{\text{number of transactions including } A \text{ and } B}{\text{Total number of transactions}} \tag{8.9}$$

where A and B are separate or different item sets that occur at the same time in a transaction.

Let's use Table 8.1 as an example, the item set X = {**Onions, Tomato Sauce, Italian Noodle**} has a support of 1/4 = 0.25, which means that they all occur in 25% of all four transactions. Similarly, the item set Y = {**Bread, Milk**} also has a support of 1/4 = 0.25 since both of them appear in 25% of all four transactions.

Confidence is the percentage of all transactions including X that also satisfy Y [14]. With respect to all transactions T, the confidence value of an association rule can be considered as a ratio of transactions containing both X and Y to the total amount of X values present, where X is the antecedent and Y is the consequent.

As mentioned in Sect. 8.2, the definition of Confidence is very similar to conditional probability, as shown in Eq. (8.8). That equation indicates that Confidence can be calculated by dividing the co-occurrence of transactions X and Y within the dataset by all transactions containing only X. This means that Confidence is a ratio of the number of transactions in both X and Y to those transactions that just happened to X.

For our example, Table 8.1 shows a rule {**Onions, Tomato Sauce**} $\Rightarrow$ {**Italian Noodle**} that has a Confidence of (1/4)/(1/4) = 1.0, which means that all three items, **Onions, Tomato Sauce,** and **Italian Noodle**, can be occurred at the same time that is the nominator is 1 time for 4 transactions. But the **Onions** and **Tomato Source** can occur at the same time that is denominator is also 1 time for 4 transactions. This ratio between the nominator and denominator is defined as a Confidence, and it means that when customers buy Onions and Tomato Sauce, they may also purchase Italian Noodles. This case illustrates the rule being correct 100% of the time for transactions including both Onions and Tomato Sauce.

However, for another rule, {**Bread**} $\Rightarrow$ {**Milk**}, which has a Confidence of (1/4)/(2/4) = 0.25/0.5 = 50%. This means that Milks are bought 50% of the time when Bread is bought. By using this example dataset, it is indicated that the Bread is purchased two times more than Milk.

Finally let's discuss another terminology, Lift. Basically, the Lift of a rule is defined as:

$$\text{Lift}(X \Rightarrow Y) = \frac{\text{Support}(X \cap Y)}{\text{Support}(X) \times \text{Support}(Y)} \tag{8.10}$$

For example, the rule **{Onions, Tomato Sauce}** $\Rightarrow$ **{Italian Noodle}** has a calculated Lift of $(1/4)/(2/4) \times (2/4) = (0.25)/(0.5) \times (0.5) = 1.0$.

If the rule had a lift of 1, it would imply that the probability of occurrence of the antecedent and that of the consequent are independent of each other. When two events are independent of each other, no rule can be drawn involving those two events.

If the lift is >1, that lets us know the degree to which those two occurrences are dependent on one another and makes those rules potentially useful for predicting the consequences in future datasets.

If the lift is <1, which lets us know the items are substitutes for each other. This means that the presence of one item has a negative effect on the presence of another item and vice versa. The value of lift indicates that it considers both the support of the rule and the overall dataset [15].

8.3   Unsupervised Learning in MATLAB

MATLAB provides a set of App and functions to support unsupervised learning, and these App and tools/functions are involved in two major toolboxes:

(1) Statistics and Machine Learning Toolbox
(2) Deep Learning Toolbox

Recall that in Sect. 7.6.1 in Chap. 7, we provided detailed discussions about those Apps under the **MACHINE LEARNING AND DEEP LEARNING** tab under the **APPS** icon in an opened MATLAB window. We redraw Fig. 7.9 here as Fig. 8.7 for your convenience.

The App used to perform related unsupervised learning algorithms is called **Neural Net Clustering,** and it has been highlighted with a red circle in Fig. 8.7. This App is located in the **Deep Learning Toolbox**.

In addition to this App, MATLAB also provides a group of functions used for unsupervised learning algorithms, such as K-Means clustering, hierarchical clustering, fuzzy clustering (works as overlapping clustering), and probabilistic clustering, and these functions are located at the **Statistics and Machine Learning Toolbox.**

There is no special Toolbox available for Association Rules algorithm in MATLAB, but a few supporting tools are provided via some MATLAB File-Exchange Centers.

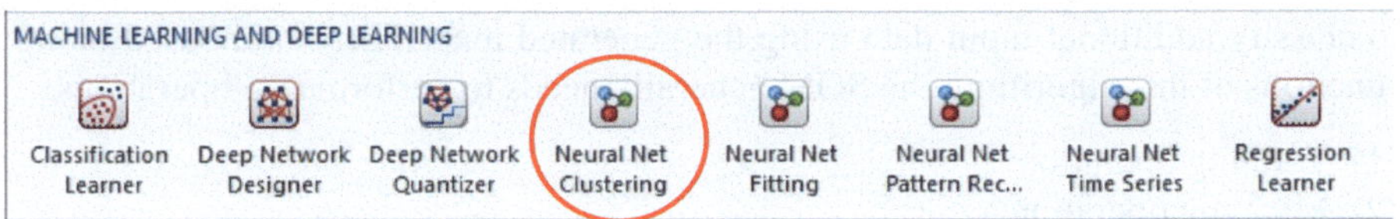

Fig. 8.7 The opened APPS icon for machine learning and deep learning tools

Let's have our discussions for unsupervised learning in the following sequence.

(1) First let's concentrate our discussions on unsupervised learning in the MATLAB App, **Neural Net Clustering**, since it is the only App provided by MATLAB used for unsupervised learning algorithms.
(2) Then we develop our discussions by using four clustering algorithms with real examples.
(3) Finally, we provide our discussions on Association Rules.

8.4 Using Neural Net Clustering App to Build Unsupervised Learning Models

The Neural Net Clustering App provided by MATLAB is located in the Deep Learning Toolbox, and it provides a visual tool to help users perform unsupervised learning tasks via neural networks. This App enables users to create, visualize, and train **Self-Organizing Map** (SOM) networks to solve clustering problems. Using this App, you can:

(1) Import data from file, the MATLAB® workspace, or use one of the example datasets.
(2) Define and train a SOM neural network.
(3) Analyze results using visualization plots, such as neighbor distance, weight planes, sample hits, and weight position.
(4) Generate MATLAB scripts to reproduce results and customize the training process.
(5) Export the trained model and results to the Workspace to make them to be used later.

To better understand this App, first let's concentrate our discussions on a new algorithm, **Self-Organizing Map** (SOM), since it is used for this App.

8.4.1 *Introduction to Self-Organizing Map (SOM) Algorithm*

The SOM is an algorithm used to reduce the dimension of the input data (input space) to a lower dimension for the input data (map space), and then set up a mapping between the input space and the map space. This mapping or map can be used to classify additional input data using the generated map. Therefore, based on the functions of this algorithm, the SOM generally needs to perform two operations:

(1) Training operation
(2) Mapping operation

The training process is to represent an input space with p dimensions as a map space with two dimensions. Specifically, an input space with p variables is said to have p dimensions. A map space consists of components called *nodes* or *neurons*, which are arranged as a hexagonal or rectangular grid with two dimensions [16].

Each node in the map space is associated with a *weight* vector, which is the position of the node in the input space. While nodes in the map space stay fixed, the training process is to move *weight* vectors toward the input data by reducing a distance metric such as Euclidean distance, but without spoiling the topology induced from the map space. After training, the map can be used to classify additional observations for the input space by finding the node with the closest weight vector (smallest distance metric) to the input space vector [17].

In summary, the SOM algorithm follows an unsupervised learning approach and trains its network through a competitive learning algorithm. SOM is used for clustering and mapping (or dimensionality reduction) techniques to map multidimensional data onto lower-dimensional which allows people to reduce complex problems for easy interpretation [18].

For example, a sample clustering dataset named **Earthquake Perception Dataset**, and it is at the site: https://www.kaggle.com/datasets/antoniocola/earthquake-perception-dataset. The dataset is a 472×43 matrix, which contains 43 features and 472 observations. A modified version of this dataset is a 472×10 matrix, and it contains 10 features and 472 observations.

To train a clustering model with the SOM algorithm, the goal of this training is to find the values of all weight vectors, and each of them is related to a node on the grid point in the map space. The node itself cannot be relocated or moved, but the related weight vector can be adjusted or trained to get a desired value that is close or equal to the position (value) of the original data point in the input space.

The values of the weight vector related to each node determined the rate or percentage of contributions of the selected node to the output layer. An architecture illustration of the SOM with two clusters C_0, C_1, and **n** input features is shown in Fig. 8.8.

Now that we have a basic understanding about the SOM algorithm, next let's use this algorithm and **Neural Net Clustering App** to train a clustering model by using the Earthquake Perception Dataset as our data source.

Fig. 8.8 An architecture illustration for SOM with two clusters and n input

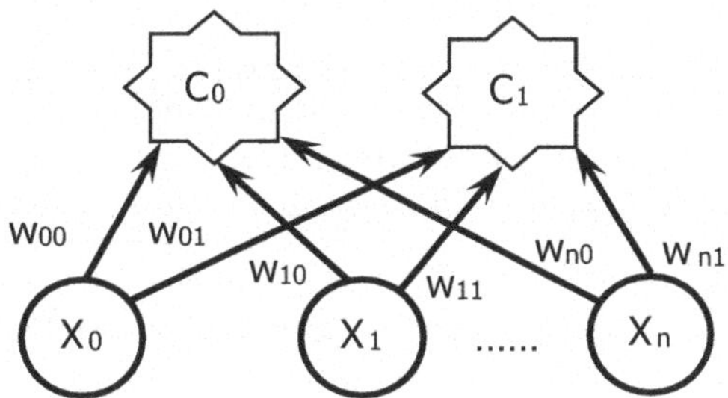

8.4.2   Using *Neural Net Clustering App* to Train a Clustering Model Earthquake_Cluster_App

First let's modify the original earthquake perception dataset to make it our desired data source. This dataset collected a group of physiological data for different groups of persons who have suffered various side effects in their minds and spirits due to serious damage from an earthquake that occurred recently. The modifications include the following parts:

(1) Removing all text columns and some other columns to make the dataset smaller. Only keep the following ten columns:

 (a) age
 (b) family_members
 (c) house_floor
 (d) shocks
 (e) fear
 (f) anxiety
 (g) physiological_symptoms
 (h) decision_timeliness
 (i) insomnia
 (j) seismic_concern

(2) Fill all blank cells under the age column by 40.
(3) Replace all cells with a value of **0 (Ground Floor)** to **0** under the **house_floor** column.
(4) Replace all cells with a value of **Over 5** to **6** under the **house_floor** column.
(5) Replace all cells with a value of **Over 5** to **6** under the **family_members** column.

The modified dataset **Earthquake Emotion.csv** can be found on the Springer ftp site in the folder, **Students\Datasets\Earthquake Emotion**. You need to copy this dataset and save it to one of your local folders in your machine now to do this project.

Now let's use **Neural Net Clustering App** to train a clustering model by using the modified Earthquake Dataset as our data source.

Two ways can be used to open this App; one way is to type **nctool** command in the MATLAB Command window as we discussed in Sect. 7.6.1 in Chap. 7, exactly in Fig. 7.14, and press the **Enter** key. The second way is to click on the **APPS** icon on the opened MATLAB, and click on the **Neural Net Clustering** icon under the **MACHINE LEARNING AND DEEP LEARNING** category.

Perform the following operations to build this project:

(1) Open the **Neural Net Clustering** by using either way shown above. The default clustering neural network is shown in Fig. 8.9. This network has one layer, with neurons organized in a grid. SOM learns to cluster data based on similarity. The map size corresponds to the number of rows and columns in the grid. For this

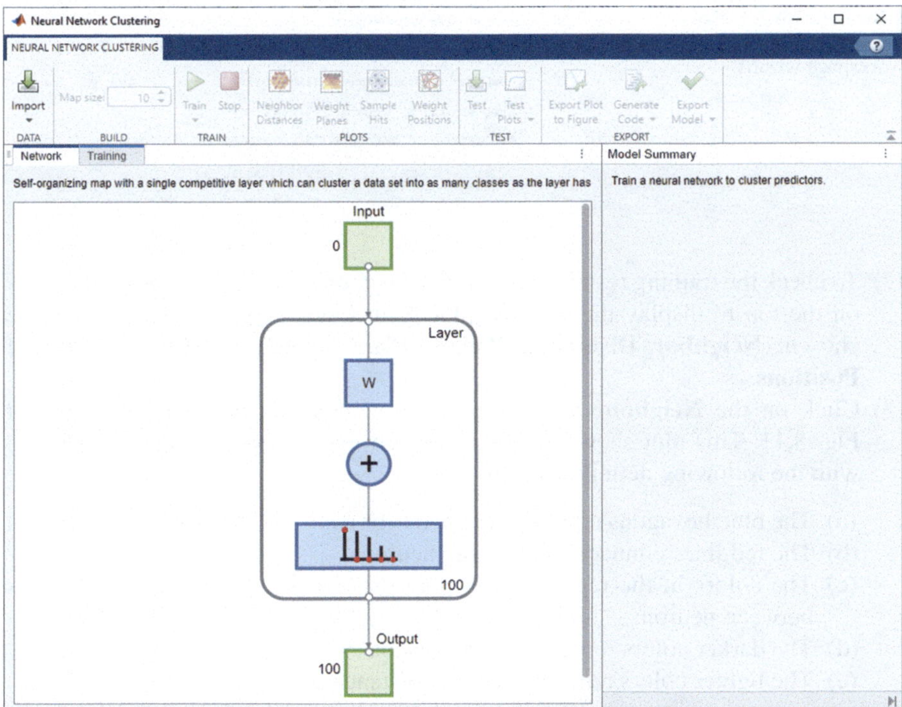

Fig. 8.9 The opened default clustering network

example, the default size value is 10, which corresponds to a grid with 10 rows and 10 columns. The total number of neurons is equal to the number of points in the grid, in this example, the map has 100 neurons. You can see the network architecture in the **Network** pane.

(2) Click on the drop-down arrow on the **Import DATA** icon and select the **Import Data** item.

(3) Click on the **Browse** button to try to find and load our data source. Browse to the folder where you saved our modified dataset, **Earthquake Emotion.csv**, select it and click on the **Open** button to load this dataset into the project.

(4) As the **Import Data from Workspace** wizard appears, as shown in Fig. 8.10, check the **Rows** radio button since we need to transpose the dataset matrix to use them as 10 features with 472 observations. Then click on the **OK** button to continue.

(5) On the next wizard, click on the drop-down arrow on the **Train** icon and select the **Train with Batch SOM** item to start the training process.

(6) As the training process is completed, the training results such as epoch number, stopped value, and target value are shown in the next wizard.

Fig. 8.10 The finished
import data from
workspace wizard

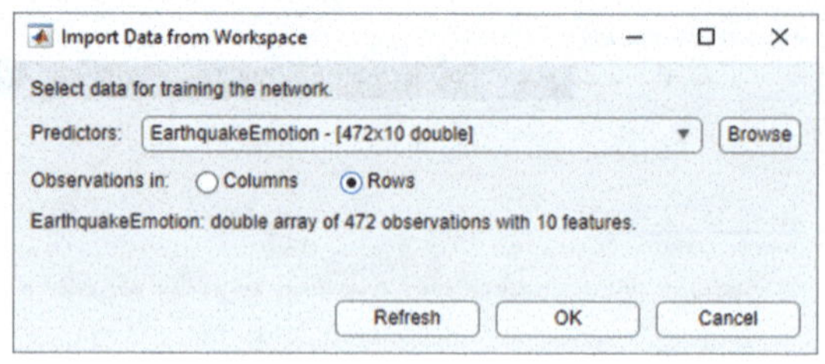

(7) To check the training results in plotting format, one can click on the related icon
on the top to display the selected plot. Four kinds of plotting results can be
chosen: **Neighbor Distances, Weight Planes, Sample Hits,** and **Weight
Positions**.

(8) Click on the **Neighbor Distances** icon to open this plotting, as shown in
Fig. 8.11. This plot shows all distances between or among nodes or neurons
with the following definitions [19]:

(a) The blue hexagons represent the nodes or neurons.
(b) The red lines connect neighboring neurons.
(c) The colors in the regions containing the red lines indicate the distances
between neurons.
(d) The darker colors represent larger distances.
(e) The lighter colors represent smaller distances.

To check the weights of each node, which is related to each input data, click the
Weight Planes icon in the **PLOTS** section. The weight plane for our project is
shown in Fig. 8.12. This figure shows a weight plane for each element of the
input features (ten in this example). The plot shows the weights that connect
each input to each of the neurons, with darker colors representing larger weights.
If the connection patterns of two features are very similar, you can assume that
the features are highly correlated.

It can be seen from Fig. 8.12 that the weights from input 1 (feature 1) are relatively bigger for nodes with small column numbers, but they are gradually reduced
as the column number of nodes increases.

To plot the SOM **Sample Hits**, click **Sample Hits** in the **PLOTS** section. The
resulting plot is shown in Fig. 8.13. This figure shows the neuron locations in the
topology and indicates how many of the observations are associated with each of the
neurons (cluster centers). The topology is a 10-by-10 grid, so there are 100 neurons.
The maximum number of hits associated with any neuron is 13. Thus, there are 13
input vectors (positions) in that cluster.

The **Weight Positions** plot is used to plot the input vectors as green dots and
shows how the SOM classifies the input space by showing blue-gray dots for each
neuron's weight vector
and connecting neighboring neurons with red lines.

To export our trained clustering model to the Workspace, click on the drop-down
arrow on the **Export Model** icon and select the **Export to Workspace** item. Enter

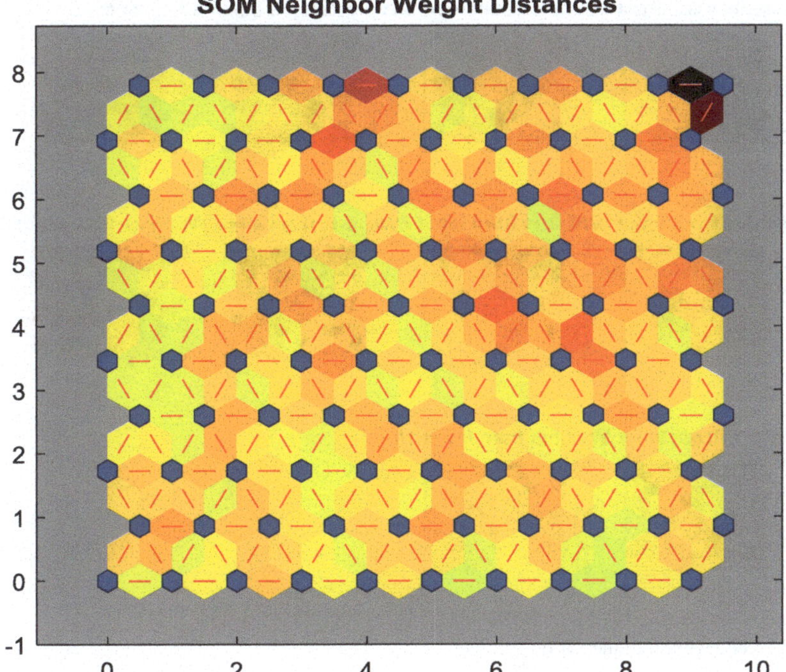

Fig. 8.11 The weight distances plotting

Earthquake_Cluster_App into the box as the name for our model, and click on the
OK button. You can use this trained clustering model to classify new input data later.

   Next let's continue our discussions by following four different clustering algorithms we introduced in Sect. 8.1 with MATLAB functions.

8.5   Using MATLAB Functions to Build Clustering Models

As we discussed in Sect. 8.1, four popular clustering algorithms are widely implemented under the unsupervised learning category, and they are:

(1) Exclusive clustering
(2) Hierarchical clustering
(3) Overlapping clustering
(4) Probabilistic clustering

We will have our discussions based on this sequence and provide details about each
of them with real projects. However, before we can start our discussions, keep in
mind the following key points when studying and developing projects related to
unsupervised learning in MATLAB [20]:

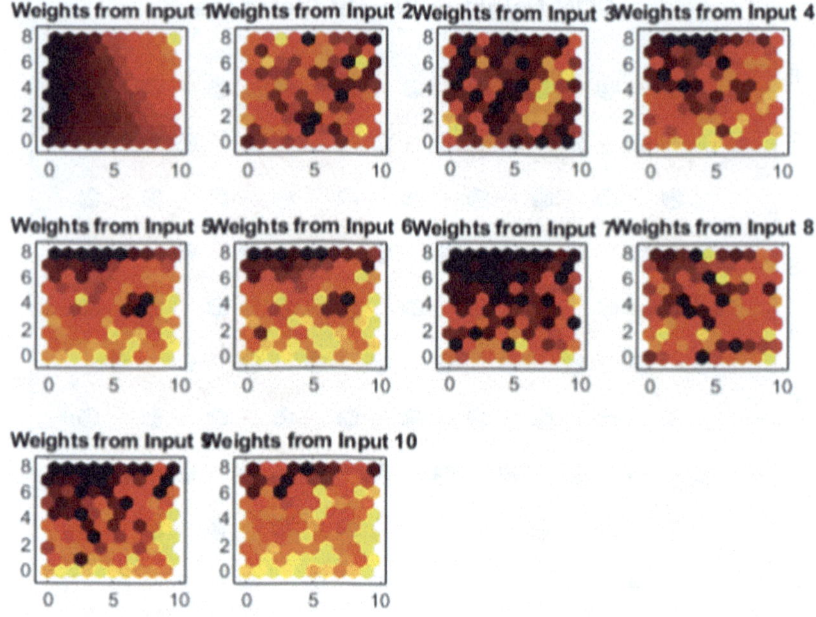

Fig. 8.12 The weight plane for our project

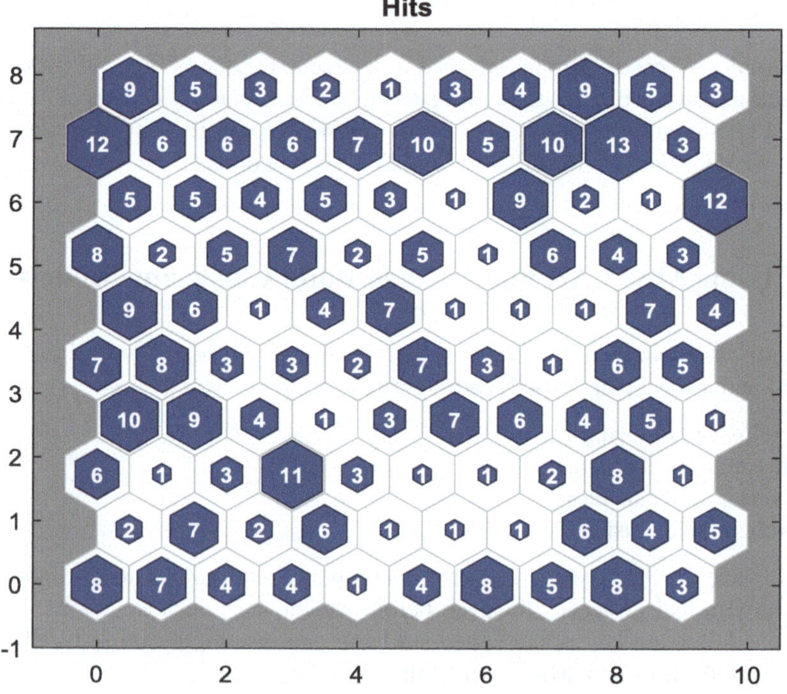

Fig. 8.13 The sample hits plot for our project

- Unsupervised learning is typically applied before supervised learning, to identify features in exploratory data analysis, and establish classes based on groupings.
- K-Means and hierarchical clustering remain popular.
- Unsupervised learning (clustering) can also be used to compress data.
- Unsupervised feature ranking is available to apply distance-based clustering more efficiently to large datasets.

In Sect. 8.1, we have provided some introductions to clustering algorithms and their implementations. In fact, cluster analysis involves applying clustering algorithms with the goal of finding hidden patterns or groupings in a dataset. It is therefore used frequently in exploratory data analysis but is also used for anomaly detection and preprocessing for supervised learning.

Clustering algorithms form groupings in such a way that data within a group or a cluster have a higher measure of similarity than data in any other cluster. Various similarity measures can be used, including Euclidean, probabilistic, cosine distance, and correlation. Most unsupervised learning methods are a form of cluster analysis [21].

Let's start our discussions with the first algorithm, Exclusive Clustering, or a K-Means clustering. We will introduce all related MATLAB functions for each algorithm.

8.5.1   *Exclusive Clustering Algorithm*

As shown in Fig. 8.1, a typical application of exclusive clustering is the K-Means clustering.

This algorithm is to group all the similar data points into K clusters, called or known as K-Means clustering. The criterion to group each point into a cluster is based on a centroid-based algorithm in which each group has a centroid. In fact, this *K-Means* and *K-Medoids clustering* partitions data into k number of mutually exclusive clusters. These techniques assign each observation to a cluster by minimizing the distance from the data point to the mean or median location of its assigned cluster, respectively.

Three functions are built for the K-Means Clustering algorithm in MATLAB, and all of them are shown in Table 8.2.

Both top two functions have five similar constructors, and they are:

(1) **idx = kmeans(X, k)**
(2) **idx = kmeans(X, k, Name, Value)**
(3) **[idx, C] = kmeans(___)**
(4) **[idx, C, sumd] = kmeans(___)**
(5) **[idx, C, sumd, D] = kmeans(___)**

The first constructor performs k-means clustering to cluster the observations of the n-by-p data matrix **X** into **k** clusters and returns an n-by-1 vector (**idx**) containing

Table 8.2 Three popular functions used for exclusive clustering in MATLAB

Function Name	Descriptions
kmeans(X, k)	Perform k-means clustering to partition the observations of the n-by-p data matrix X into k clusters, and returns an n-by-1 vector (idx) containing cluster indices of each observation. Rows of X correspond to points and columns correspond to variables. By default, kmeans() uses the squared Euclidean distance metric and the k-means++ algorithm for cluster center initialization.
kmedoids(X, k)	Perform k-medoids Clustering to partition the observations of the n-by-p matrix X into k clusters, and returns an n-by-1 vector idx containing cluster indices of each observation. Rows of X correspond to points and columns correspond to variables. By default, kmedoids() uses squared Euclidean distance metric and the k-means++ algorithm for choosing initial cluster medoid positions.
mahal(Y, X)	Return the squared Mahalanobis distance of each observation in Y to the reference samples in X

cluster indices of each observation. Rows of **X** correspond to points and columns correspond to variables.

The second constructor is similar to the first one by adding some Name-Value pairs.

The third and the fourth constructors also return the **k** cluster centroid locations in the k-by-p matrix **C** and the within-cluster sums of point-to-centroid distances in the k-by-1 vector **sumd**.

The last constructor also returns distances from each point to every centroid in the n-by-k matrix **D**.

The function **kmedoids**() has similar constructors as those of **kmeans**() function.

The function **mahal**() has only one constructor, which is **d2 = mahal(Y, X)**, and it returns the squared **Mahalanobis distance** of each observation in **Y** to the reference samples in **X**.

The Mahalanobis distance is a measure between a sample point and a distribution. The distance from a vector y to a distribution with mean μ and covariance S is

$$d = \sqrt{\left(y - \mu\right)S^{-1}\left(y - \mu\right)^{T}}$$

(8.11)

This distance represents how far y is from the mean in a number of standard deviations.

The **mahal**() function returns the squared Mahalanobis distance d^2 from an observation in **Y** to the reference samples in **X**, where μ and S are the sample mean and covariance of the reference samples, respectively.

In summary, both k-means clustering and k-medoids clustering partition data into k mutually exclusive clusters. These clustering methods require that you specify the number of clusters k. Both k-means and k-medoids clustering assign every point in your data to a cluster; however, unlike hierarchical clustering, these methods operate on actual observations (rather than dissimilarity measures) and create a single level of clusters. Therefore, k-means or k-medoids clustering is often more suitable than hierarchical clustering for large amounts of data [22].

Let's use some of these functions to build a real project to illustrate how to use the exclusive clustering functions to build an exclusive clustering model with our Earthquake Perception dataset.

8.5.1.1   Build K-Means Clustering Model with Earthquake Perception Dataset

Due to the large size (472 × 10) of the modified Earthquake Perception dataset, to simplify our clustering process, we only use some of the features (columns) on it. Exactly we only use five features, **age, family member, floor, shock,** and **fear**. To make this project complete, we like to use the top 300 observations to do this k-mean clustering but use the rest 172 as testing data.

Now create a new Script file, name it as **Earthquake_KM_Func.m,** and enter the codes shown in Fig. 8.14 into this Script file. Let's have a closer look at this piece of codes to see how it works.

(1) The full path for our modified dataset, **Earthquake Emotion.csv**, is declared in the detailed folder in our machine. You may need to use your actual path to replace this if you saved this dataset in a different folder in your machine.

(2) The modified dataset is read out and saved to a local variable **T** in cells format with the function **readtable()**.

(3) The number of observations or rows in our dataset is defined and assigned to a local variable **N**. The size of our dataset is 472 × 10, and totally there are 472 rows or observations with 10 columns (features).

(4) Due to the large size of our dataset, in this project we only need to use five features, **age, family_member (FAMILY_NO), floor, shock,** and **fear**. Thus we only need to extract those five columns from our dataset and assign them to the related local variables one by one. One point to be noted is that the actual column number in our dataset must be used for this assignment. Also all observations must be converted to numeric values via **table2array()** function since the clustering process needs to use the numeric values as inputs.

(5) An input data array or matrix **X** is established by assigning those variables in an order and making them an input matrix.

(6) To make this clustering more meaningful, we extract the first two features in the first two columns, **age** and **family_member**, and assign them in order to make another matrix **XX**.

(7) To plot some observations, we defined the horizontal variable **M** as a 1 × 472 array.

(8) Now we plot the original observations under two features, **age** and **family_ member** with **M** as horizontal variable. Of course you can plot any other observations for other features if you like. All functions under this **plot()** function are used to support this plotting.

```matlab
% Use K-Means clustering to cluster Earthquake Perception Data
% Name: Earthquake_KM_Func.m
% Input: Earthquake_Emotion.csv dataset
% Output: Clustering results
% Date: June 20, 2024
1   path = 'C:\\Artificial Intelligence Book\Students\Datasets\Earthquake Emotion\\Earthquake Emotion.csv';
2   T = readtable(path);
3   N = 300;                                    % use the first 300 observations as k-means clustering

    % Assign each column to a related local variable...
4   AGE = table2array(T(1:N, 1));               % Age is located at column 1 in database
    FAMILY_NO = table2array(T(1:N, 2));         % Number of persons is located at column 2 in database
    FLOOR = table2array(T(1:N, 3));             % Floor is located at column 3 in database
    SHOCK = table2array(T(1:N, 4));             % Shock is located at column 4 in dataset
    FEAR = table2array(T(1:N, 5));              % Fear is located at column 5 in database

5   X = [AGE FAMILY_NO FLOOR SHOCK FEAR];
6   XX = X(:, 1:2);
7   M = 1:N;

    figure;
8   plot(M, XX(:,1), 'r+', M, XX(:,2),'k*', 'MarkerSize',5);
    title 'Age and Family Member Data';
    xlabel 'Age';
    ylabel 'Family Member';
    legend('Age', 'Family Member');
    grid;

9   [idx, C] = kmeans(X, 3);

    figure;
10  plot(XX(idx==1,1), XX(idx==1,2),'r.','MarkerSize', 12)
11  hold on
12  plot(XX(idx==2,1), XX(idx==2,2),'b.','MarkerSize', 12)
13  plot(XX(idx==3,1), XX(idx==3,2),'g.','MarkerSize', 12)

14  plot(C(:,1), C(:,2), 'kx', 'MarkerSize',15,'LineWidth', 3)
    legend('Cluster 1','Cluster 2', 'Cluster 3', 'Centroids', 'Location','SE')
    title 'Cluster Assignments and Centroids'
    grid;
15  hold off
```

Fig. 8.14 The codes for building a K-Means cluster project

(9) Now we can call the K-Means function **kmeans()** to perform k-Mean cluster-
     ing process. The input is the matrix X with five features, and the value of k is
     3, which means that we like to cluster these input data into three groups or
     clusters. The process returns the indices of the cluster for each observation
     (row) in an array format, and a **k by number of features** matrix **C**, which is a
     (3×5) matrix in this project and contains all clustering centers for each fea-
     ture in a cluster. More discussions about **idx** and **C** will be given later.

(10) Now we can plot three clustering assignments and centroids for all data points
     in the extracted dataset. Where **XX(idx==1, 1)** indicates the observations that
     are clustered to group 1 (cluster 1) for the **age** feature, and **XX(idx==1, 2)**
     includes the observations that are clustered to group 1 (cluster 1) for the **fam-
     ily_member** feature.

(11) A **hold on** command is used to hold the previous plotting unchanged, and then
     continue to plot more other plotting.

(12) Similarly, we can plot observations that are clustered to group 2 for the age
     feature and observations that are clustered to group 2 for the **family member**
     features.

(13) Same plots for observations that are clustered to group 3 for the **age** feature, and observations that are clustered to group 3 for the **family member** feature.
(14) We also like to plot three centers or centroids vectors, each vector containing five feature centers in one of three groups ($k = 3$ clusters). One trick is that you do not need to clearly plot **C(:, 3)** since it has been involved with **C(:, 1)** and **C(:, 2)** together.
(15) Finally, the **hold off** command is executed to turn off the **hold on** command.

Before we can run this project, we like to spend more time to discuss about **idx** and **C** matrix mentioned in step 9 above.

Cluster Indices of Each Observation idx

This **idx** contains the cluster indices or cluster numbers of all observations that are clustered to certain clusters (groups). Exactly this **idx** is an **n×1** vector that contains the cluster numbers for all **n** observations that are grouped into the related cluster. Depending on the total number of clusters, the range of the **idx** value is **1 ~ maximum cluster number**. For our case, we have 472 observations (rows) in our dataset, and each of them should be clustered into one of three groups ($k = 3$ in our case). Thus the range of the **idx** for our case is 1–3, and it is a 472×1 vector composed of a combination of group numbers, 1, 3, 1, 2, … A mapping illustration between the **idx** values and the related observations is shown in Table 8.3.

Cluster Centroid Locations Matrix C

The **C** is a $k \times p$ matrix and it contains all centroids for five features in 3 ($k = 3$) clusters, where p is the number of features. In our case, this **C** should be a 3×5 matrix. Each row represents a cluster and each column represents a feature. In our case, the real values for this matrix **C** are:

```
C =
   41.0357      3.5357      1.6429      2.9464      3.3631
   58.7130      2.9565      2.0609      3.2522      3.3478
   24.9735      3.7460      1.8889      2.4550      2.9153
```

The value in the intersection between each row and each column is the centroid for that feature in that cluster, as shown in Fig. 8.15.

Table 8.3 A mapping illustration between idx value and observations

idx (cluster number)	Observations
3	row 1 (observation 1) with 5 features
1	row 2 (observation 2) with 5 features
1	row 3 (observation 3) with 5 features
2	row 4 (observation 4) with 5 features
⋮	⋮
2	row 472 (observation 472) with 5 features

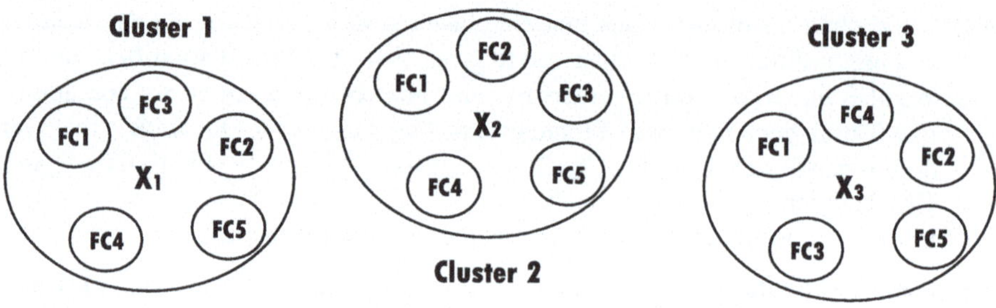

Fig. 8.15 An illustration for cluster centroid location matrix C

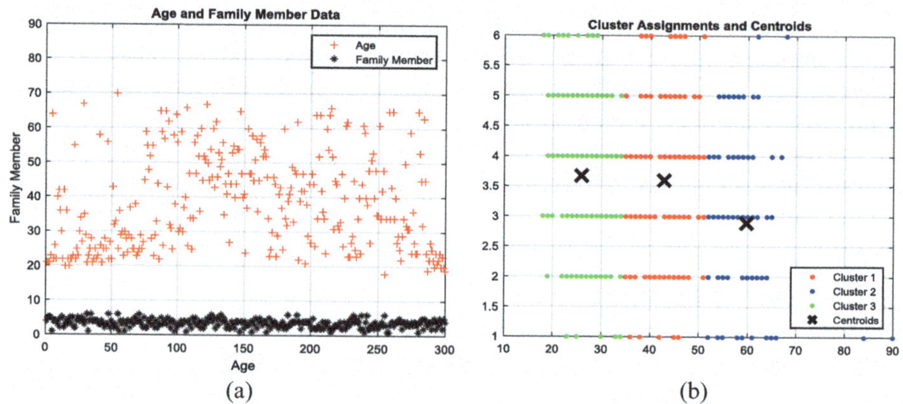

Fig. 8.16 The running result for K-Means clustering algorithm

In Fig. 8.15, each FC_i ($i = 1$–5) represents the related centroid for the ith feature in a cluster, and each X_i represents the centroid for that cluster. For example, the FC_1 in cluster 1 is 41.0375, and FC_2 in cluster 1 is 3.5357, FC_4 in cluster 2 is 3.2522, and FC_5 in cluster 3 is 2.9153. You can set a mapping between FC_i in a selected cluster shown in Fig. 8.15 and the related centroid values shown above easily.

Now let's run the project to see the clustering results. Two figures are plotted based on this running result. The first one, shown in Fig. 8.16a, is the original data point distribution for **age** and **family member** features. Figure 8.16b shows the clustering result that contains three clusters and three centroids for three groups. It can be found from Fig. 8.16b that each cluster contains some observations marked with different colors, red for cluster 1, blue for cluster 2, and green for cluster 3. Three centroids are basically located at the middle or center points of three clusters.

Next let's continue this first project by adding a validation part to test the k-means clustering model to confirm its performance by using the testing observations, which are the rest of 172 observations in our modified dataset.

8.5.1.2   Validate the K-Means Clustered Model with Earthquake Perception Dataset

In this section, we like to use the rest of the 172 observations in our modified dataset, Earthquake Emotion.csv, to test and validate our clustered model to confirm its effectiveness.

Create another new Script file and name it as **Eval_Earthquake_Func.m**, and enter the codes shown in Fig. 8.17 into that Script. The first 10 coding lines are identical to those lines in our last project with no explanations.

Let's have a closer look at this piece of codes to see how it works.

(1) This piece of codes is used to plot the k-means clustering result as we did in coding lines 10–14 in Fig. 8.14 for the last project. The functionality of both pieces of codes are identical even though they look different.
(2) The testing data starts from row 301 and ends at row 171 in our modified dataset, thus these two key numbers are assigned to two new variables, **N** and **P**.
(3) Beginning with this step, a testing observation dataset is built by reading the entire modified dataset starting from row 301 and ending at row 472. The size of that testing dataset is 172×5.

```matlab
% Use testing data to evaluate K-Means clustering for Earthquake Perception Dataset
% Name: Eval_Earthquake_KM_Func.m
% Input: Earthquake_Emotion.csv dataset
% Output: Clustering evaluation results
% Date: June 22,2024

path = 'C:\\Artificial Intelligence Book\Students\Datasets\Earthquake Emotion\\Earthquake Emotion.csv';
T = readtable(path);
N = 300;                              % Size of data for k-means clustering

% Assign each column to a related local variable...
AGE = table2array(T(1:N, 1));             % Age is located at column 1 in database
FAMILY_NO = table2array(T(1:N, 2));       % Number of persons is located at column 2 in database
FLOOR = table2array(T(1:N, 3));           % Floor is located at column 3 in database
SHOCK = table2array(T(1:N, 4));           % Shock is located at column 4 in dataset
FEAR = table2array(T(1:N, 5));            % Fear is located at column 5 in database

X = [AGE FAMILY_NO FLOOR SHOCK FEAR];
[idx, C] = kmeans(X, 3);

figure                                % plot the k-means clustering result
gscatter(X(:,1), X(:,2), idx, 'bgm')
hold on
plot(C(:,1), C(:,2),'kx')
legend('Cluster 1','Cluster 2','Cluster 3','Cluster Centroid', 'Location', 'northeast', 'fontSize', 7.5);
grid;

N = 301;                              % using the rest of 172 observations as testing data
P = 171;

% get the testing data for clustering
AGE = table2array(T(N:N+P, 1));           % Age is located at column 1 in database
FAMILY_NO = table2array(T(N:N+P, 2));     % Number of persons is located at column 2 in database
FLOOR = table2array(T(N:N+P, 3));         % Floor is located at column 3 in database
SHOCK = table2array(T(N:N+P, 4));         % Shock is located at column 4 in dataset
FEAR = table2array(T(N:N+P, 5));          % Fear is located at column 5 in database

Xtest = [AGE FAMILY_NO FLOOR SHOCK FEAR];     % test data matrix

[D, idx_test] = pdist2(C, Xtest, 'euclidean', 'Smallest', 1);

gscatter(Xtest(:,1), Xtest(:,2), idx_test, 'bgm', 'ooo')
legend('Cluster 1', 'Cluster 2', 'Cluster 3', 'Cluster Centroid', 'Test Data-C1','Test Data-C2', 'Test Data-C3');
```

Fig. 8.17   The codes used to evaluate the clustered model

(4) Then a testing data matrix **Xtest** is established by arranging all of those features in an order to make it a testing matrix.

(5) To test and validate the clustered model, the **pdist2()** function is adopted to calculate the distance for a pair of observations in two different datasets. Normally, this function is used to compute the distance between each pair of observations in **X** and **Y** using the metric specified by **D** `(distance)`. Here we use this function to get the cluster numbers **idx_test** (indices) for all observations in the testing dataset Xtest, whose distances to the related clustered centroids are <1 (**Smallest** distance). We will take more time to discuss this function later. The returned indices are assigned to the **idx_test** vector whose size is 1 × 172, and it contains 172 predicted cluster numbers based on those testing data.

(6) Finally, we plot those predicted cluster numbers on the same plotting we did in step 1 above by using the **gscatter()** function (will discuss later) with blue-green-magenta (**bgm**) colors and circle style (**ooo**) to see how many predicted cluster numbers can be located at the related actual clusters or groups to confirm the effectiveness of our k-Means clustering algorithm.

Now run the project and the running result is shown in Fig. 8.18.

It can be found from Fig. 8.18 that those circles with three colors are closely or exactly located at three actual clusters with the same colors. In fact, all blue-color circles belonging to cluster 1 are located in the area (group) where all solid blue-color dots (actual observations) are also located. Same situations are true for green

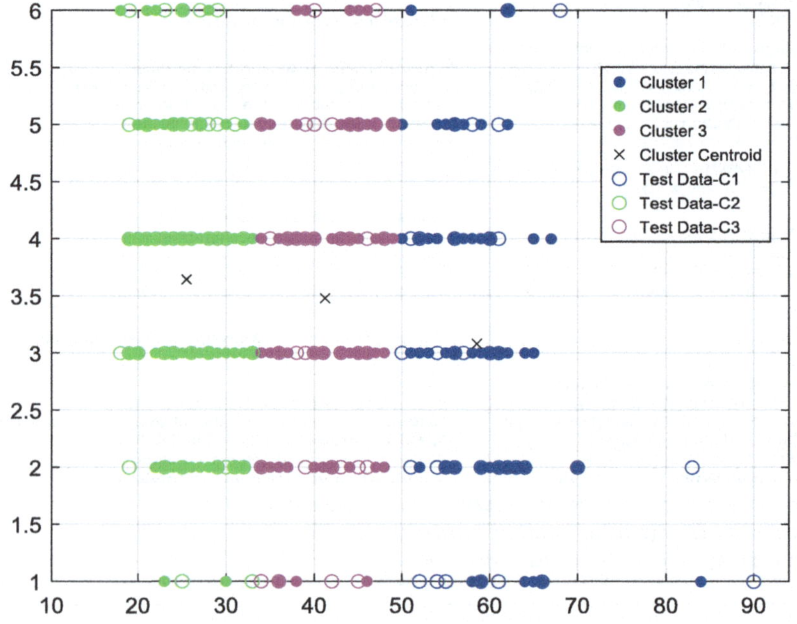

Fig. 8.18 The validation result for our k-means clustering algorithm

color (cluster 2) and magenta color (cluster 3). This means that all predicted observations and all related actual observations are grouped correctly and both of them are located or grouped into the correct clusters. Our k-Means clustering algorithm is very successful.

Next let's have our discussions on functions used for hierarchical clustering algorithms.

8.5.2   Functions Used for Hierarchical Clustering Algorithm

As we introduced in Sect. 8.1.3, hierarchical clustering is to group data over a variety of scales by creating a *cluster tree* or *dendrogram*. The tree is not a single set of clusters, but rather a multilevel hierarchy, where clusters at one level are joined as clusters at the next level. This allows you to decide the level or scale of clustering that is most appropriate for your application. The **clusterdata()** function supports agglomerative clustering and performs all of the necessary steps for you. It incorporates the **pdist()**, **linkage()**, and **cluster()** functions, which you can use separately for more detailed analysis. The **dendrogram()** function plots the cluster tree [23].

A key point of using the above functions to perform any hierarchical clustering is that the function **clusterdata()** combined all necessary functions above to perform all steps for you, and you do not need to execute the **pdist()**, **linkage()**, or **cluster()** functions separately.

In summary, to perform hierarchical cluster analysis on a dataset using Statistics and Machine Learning Toolbox™ functions in MATLAB, follow the procedure shown below:

(1) **Find the similarity or dissimilarity between every pair of objects in the dataset.** In this step, you calculate the *distance* between objects using the **pdist()** function, and this function supports many different ways to compute this measurement.

(2) **Group the objects into a binary, hierarchical cluster tree.** In this step, you link pairs of objects that are in close proximity using the **linkage()** function. The linkage function uses the distance information generated in step 1 to determine the proximity of objects to each other. As objects are paired into binary clusters, the newly formed clusters are grouped into larger clusters until a hierarchical tree is formed.

(3) **Determine where to cut the hierarchical tree into clusters.** In this step, you use the **cluster()** function to prune branches off the bottom of the hierarchical tree and assign all the objects below each cut to a single cluster. This creates a partition of the data. The cluster function can create these clusters by detecting natural groupings in the hierarchical tree or by cutting off the hierarchical tree at an arbitrary point.

To better understand this algorithm, we divide our discussions into three subsections:

(1) Introduction to four functions, **pdist()**, **pdist2()**, **linkage()**, and **cluster()**.
(2) Provide a real example to illustrate how to use these functions to build a hierarchical clustering model.

Due to the similarity between the functions **pdist()** and **pdist2()**, we will discuss them together in this section to make them easy to understand.

8.5.2.1   Introduction to Function pdist()

The function **pdist()** is used to calculate the Euclidean distance between pairs of observations in an input matrix **X**. The **X** must contain a sequence of pairs of observations or objects. Two popular constructors for this function are:

(1) **D = pdist(X)**
(2) **D = pdist(X, Distance)**

The second argument **Distance** in the second constructor indicates what kind of distance model should be used for this action. The default model is Euclidean distance. Some other options include the **squaredeuclidean** distance, **seuclidean** distance, and **minkowski** distance.

For example, we have a matrix X = [2 1; 4.5 2.5; 2 2; 1.5 4; 2.5 4], as shown in Fig. 8.19. The number in the Figure indicates each object's related number.

Now we have five object or data pairs in this X matrix, (2,1), (4.5,2.5), (2,2), (1.5,4), and (2.5,4) represented in (x, y) format. For any two points in a Cartesian plane, the so-called Euclidean distance is calculated as:

$$d = \sqrt{\left(x_i - x_j\right)^2 + \left(y_i - y_j\right)^2} \quad \left(i = 1,2,3,4, j = 2,3,4,5, j \neq i\right)$$

(8.12)

Equation (8.12) means that for an $m \times n$ matrix, we have totally $m \times (m - 1)/2$ Euclidean distances, where m is the total number of observations. In our case, it is

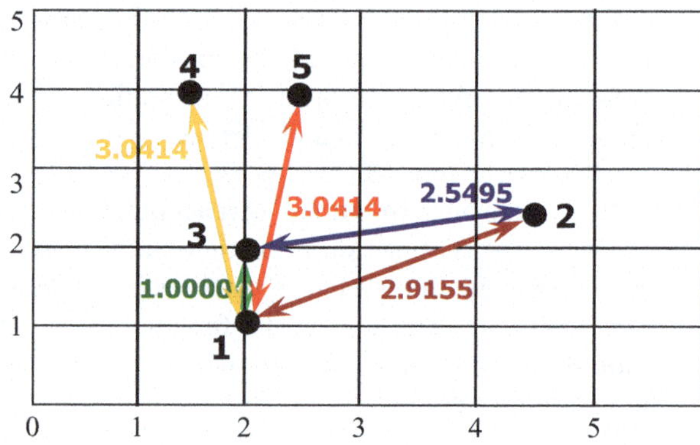

Fig. 8.19 Distribution of all objects in X matrix

5×2 matrix, and it contains five observations, so there are $5 \times 4/2 = 10$ Euclidean distances. In other words, deriving the Euclidean distances is a permutation process, but not a combination process.

If we use the **pdist()** function for this **X** matrix, **D = pdist(X)**, the ten Euclidean distances are:

```
D =     Columns 1 through 10
           2.9155      1.0000      3.0414      3.0414      2.5495
3.3541      2.5000      2.0616    2.0616      1.0000
```

Figure 8.19 shows the first five of these Euclidean distances.

You can reformat the above distance vector into a matrix using the **squareform()** function to make it easier to see the relationship between the distance information generated by **pdist()** function and the objects in the original dataset. In this matrix **X**, element i, j corresponds to the distance between object i and object j in the original dataset. In the following example, element $(1, 1)$ represents the distance between object 1 and itself (which is zero). Element $(1, 2)$ represents the distance between object 1 and object 2, and so on.

```
Y = squareform(D)
Y =
         0          2.9155    1.0000    3.0414    3.0414
         2.9155     0         2.5495    3.3541    2.5000
         1.0000     2.5495    0         2.0616    2.0616
         3.0414     3.3541    2.0616    0         1.0000
         3.0414     2.5000    2.0616    1.0000    0
```

The purpose of using the function **pdist()** is to determine all Euclidean distances between and among all objects or data points in a matrix, and the result is a preparation for the next step in hierarchical clustering, a linkage process that is used to locate and find the smallest or largest distances between or among all of these objects or data points to find the proximity of objects, and furthermore to make grouping of them into different clusters ready.

8.5.2.2   Introduction to Function linkage()

Once the Euclidean distances have been computed, you can determine how objects in the dataset should be grouped into clusters by using the **linkage()** function. This function takes the distance information generated by **pdist()** function and links pairs of objects that are close together into binary clusters which are clusters made up of two objects. The **linkage()** function then links these newly formed clusters to each other and to other objects to create bigger clusters until all the objects in the original dataset are linked together in a hierarchical tree. Two popular constructors for this function are:

(1) **Z = linkage(X)**
(2) **Z = linkage(X, method)**

The second argument **method** in the second constructor indicates how to measure the distance between clusters. The default method is the **single** (shortest distance). Some other optional methods include the **centroid** (centroid distance), **complete** (farthest distance), and **average** (unweighted average distance).

For example, in this case, we have ten Euclidean distances stored in matrix **D**. By using the **linkage()** function, **Z = linkage(D)**, we can get linkage information in a matrix, **Z**, as:

```
Z =
        4.0000      5.0000      1.0000
        1.0000      3.0000      1.0000
        6.0000      7.0000      2.0616
        2.0000      8.0000      2.5000
```

In this result, each row identifies a link between objects or clusters based on their objects' numbers shown in Fig. 8.19. The first two columns identify the objects that have been linked. The third column includes the distance between these objects. For a Cartesian coordinate system, the **linkage()** function begins by grouping objects 4 and 5 since both have the closest proximity (shortest distance = 1). Then the **linkage()** function continues by grouping objects 1 and 3 since both of them also have a shortest distance value of 1.

The third row indicates that the linkage function grouped objects 6 and 7. Here a question arises: If our original matrix **X** contained only five objects, from where objects 6 and 7 are coming? In fact, object 6 is the newly formed binary cluster created by the grouping of objects 4 and 5. When the linkage function groups two objects into a new cluster, it must assign the cluster a unique index value, starting with the value $m + 1$, where m is the number of objects in the original dataset since values 1 through m are already used by the original dataset. Similarly, object 7 is the cluster created by grouping objects 1 and 3.

Linkage uses distances to determine the order in which it clusters objects. The distance vector **D** contains the distance row between the original objects 1 through 5. But linkage must also be able to determine distances involving clusters that it creates, such as objects 6 and 7. By default, linkage uses a method named single linkage (shortest distance). However, there are some other different methods available.

As the final cluster, the function **linkage()** grouped object 8, which is the newly created cluster by grouping objects 6 and 7, with object 2 from the original dataset. Figure 8.20 shows the way linkage groups the objects into a hierarchy of clusters.

To better understand the binary cluster tree created by the **linkage**() function, a graphical view would be a perfect presentation. By using the **dendrogram**() function, this kind of view can be easily obtained. Figure 8.21 shows this view by using function **dendrogram(Z)**.

Now we are ready to complete our hierarchical clustering process by using the function **cluster**() for our example matrix **X**.

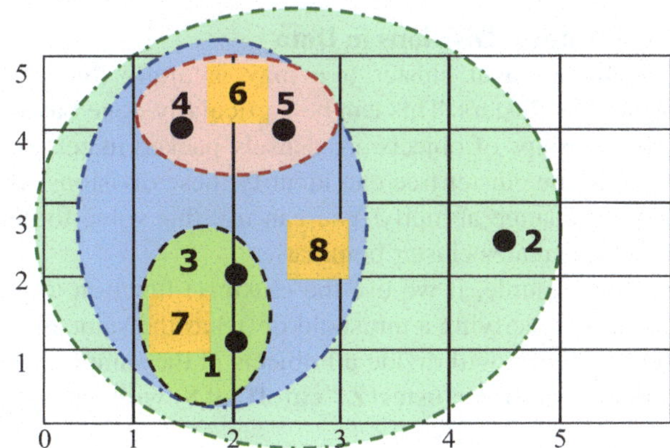

Fig. 8.20 Distribution of all clusters created by linkage() function

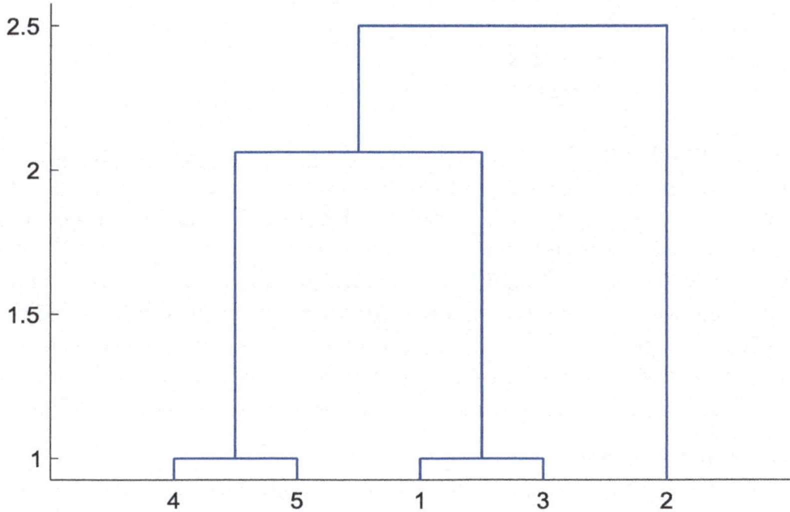

Fig. 8.21 The graphical view of the binary cluster tree

8.5.2.3   Introduction to Function cluster()

After the hierarchical tree of binary clusters is created, you can cut the tree to group your data into clusters using the **cluster()** function. This function allows you to create clusters in two ways to find the division:

(1) Find Natural Divisions in Data
(2) Specify Arbitrary Clusters

Find Natural Divisions in Data
The hierarchical cluster tree may naturally divide the data into distinct, well-separated clusters. This can be particularly done based on the dendrogram diagram where groups of objects are densely packed in certain areas but not in others. The links in the cluster tree can identify these divisions where the similarities between objects change abruptly. You can use this value to determine where the **cluster()** function creates cluster boundaries.

For example, if we use the **cluster()** function to group the sample dataset into clusters, specifying a threshold of 0.8 as the value of the cutoff argument, the **cluster()** function will divide all objects in the sample dataset into three separate clusters, such as **T = cluster(Z,"cutoff",0.8)** with results as:

```
T =

        3           - Object 1
        2           - Object 2
        3           - Object 3
        1           - Object 4
        1           - Object 5
```

This output indicates that objects 1 and 3 are in one cluster, objects 4 and 5 are in another cluster, and object 2 is in its own cluster. There is a mapping between each output or cluster number and its related object number, and the mapped object numbers are listed on the right.

These clusters may, but do not have to, correspond to or meet a horizontal slice across the dendrogram at a certain height. If you want clusters corresponding to a horizontal slice of the dendrogram, you can either use the criterion option to specify that the cutoff should be based on distance rather than inconsistency, or you can specify the number of clusters directly as described in the following section.

Specify Arbitrary Clusters
Instead of allowing the **cluster()** function to create clusters that are determined by the natural divisions in a dataset, we can specify the number of clusters we want to create.

For example, we can specify that we want to use the **cluster()** function to divide the sample dataset into two clusters. In this case, the **cluster()** function will create one cluster containing objects 1, 3, 4, and 5 and another cluster containing object 2. The operational instruction is:

```
T = cluster(Z,"maxclust",2)
```

The running result for this operation is:

```
T =
        2        - Object 1
        1        - Object 2
        2        - Object 3
        2        - Object 4
        2        - Object 5
```

To make this specific arbitrary cluster clearer, Fig. 8.22 shows the hierarchical cluster tree by using the **dendrogram(T)** function. The horizontal dashed line intersects two lines of the dendrogram, corresponding to setting **maxclust** to 2. These two lines divide the objects into two clusters: the objects below the left-hand line, namely 1, 3, 4, and 5, belong to one cluster, while the object below the right-hand line, namely 2, belongs to the other cluster (Fig. 8.22a).

However, if we set **maxclust** to 3, the **cluster()** function groups objects 4 and 5 in one cluster, objects 1 and 3 in a second cluster, and object 2 in a third cluster, as shown in Figure 8.22b.

The instruction to do that operation of setting **maxclust = 3** is:

```
T = cluster(Z,"maxclust",3)
```

The running result for this operation is:

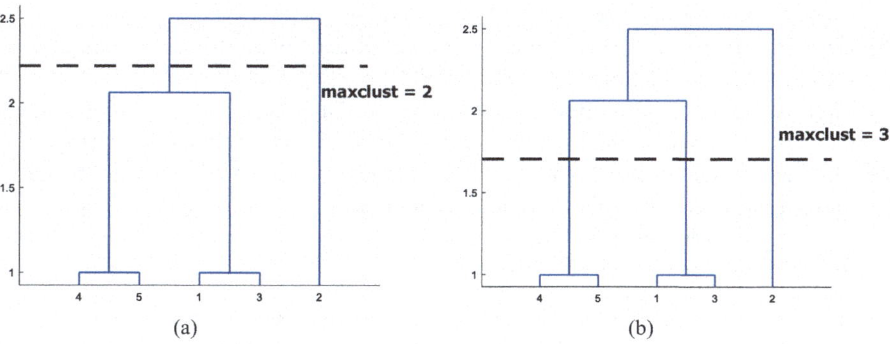

Fig. 8.22 Two specific arbitrary clusters with two selected cluster numbers

```
T =

        2              - Object 1
        3              - Object 2
        2              - Object 3
        1              - Object 4
        1              - Object 5
```

Now we have a basic and fundamental understanding and knowledge about hierarchical clustering algorithms, next let's build a real example project to illustrate how to use it in our actual applications.

8.5.2.4   Build Hierarchical Clustering Model with Diabetes Dataset

In this section, we will build a project to illustrate how to use hierarchical clustering to train a hierarchical clustering model to group persons with or without diabetes disease. The diabetes dataset, **Diabetes.csv**, can be found on the Springer ftp site in the folder **Students\Datasets\Diabetes Dataset**. You can download it and save it to one of your local folders.

Now create a new Script file, name it as **Diabetes_HC_Func.m,** and enter the codes shown in Fig. 8.23 into the file. Let's have a closer look at this piece of codes to see how it works.

(1) First set up our diabetes dataset with a full path. You need to use your actual path to replace this if you saved your dataset in a different folder in your machine.
(2) The dataset is read out and assigned to a local variable **T**. To make this project simple and easy to illustrate the hierarchical clustering function, we reduce the observation numbers to 10, thus setting it to a local variable N = 10.
(3) Retrieve each feature from the dataset variable **T** and convert it to a numeric value. In this project, we selected five features from the original dataset.
(4) Build an input matrix with 10 observations and assign them to a local variable **X**.
(5) Now use the **pdist()** function with default Euclidean distance mode to get all related Euclidean distances for all observations.
(6) Using the function **squareform()** to make all Euclidean distances as a matrix format to make it clearer to be understood.
(7) Then by using the **linkage()** function, the similarity and dissimilarity between those Euclidean distances can be expressed clearly with related distances.
(8) The hierarchical clustering tree can be graphically presented by using the **dendrogram()** function with the figure command.
(9) By using the **cluster()** functions with different cut-off options, we can get different hierarchical clustering results.

Now run the project and the running results are shown in the Command window and the plot is shown in Fig. 8.24. Ten observations mean that we have $10 \times 9 / 2 = 45$ Euclidean distances.

```
   X =
       148.0000      72.0000      33.6000      0.6270      50.0000        -
Object 1
        85.0000      66.0000      26.6000      0.3510      31.0000        -
Object 2
       183.0000      64.0000      23.3000      0.6720      32.0000        -
Object 3
        89.0000      66.0000      28.1000      0.1670      21.0000        -
Object 4
       137.0000      40.0000      43.1000      2.2880      33.0000        -
Object 5
       116.0000      74.0000      25.6000      0.2010      30.0000        -
Object 6
        78.0000      50.0000      31.0000      0.2480      26.0000        -
Object 7
       115.0000            0      35.3000      0.1340
29.0000          - Object 8
       197.0000      70.0000      30.5000      0.1580      53.0000        -
Object 9
       125.0000      96.0000            0      0.2320
54.0000          - Object 10


   Z =
        2.0000       4.0000      10.8758
        7.0000      11.0000      18.6915
        3.0000       9.0000      26.9278
```

```matlab
% Use Hierarchical clustering to cluster Diabetes Dataset
% Name: Diabetes_HC_Func.m
% Input: Diabetes.csv dataset (768 by 9 size), we only need 600 by 5 as clustering data
% Output: Clustering results
% Date: June 25,2024

1  path = 'C:\\Artificial Intelligence Book\Students\Datasets\Diabetes Dataset\\Diabetes.csv';
   T = readtable(path);
2  N = 10;                                  % Size of data for hierarchical clustering

   % Assign each column to a related local variable...
3  GLUCOSE = table2array(T(1:N, 2));        % Glucose is located at column 2 in database
   BLOODP = table2array(T(1:N, 3));         % Blood Pressure is located at column 3 in database
   BMI = table2array(T(1:N, 6));            % BMI is located at column 6 in database
   PEDIGREE = table2array(T(1:N, 7));       % Pedigree is located at column 7 in dataset
   AGE = table2array(T(1:N, 8));            % Age is located at column 8 in database

4  X = [GLUCOSE BLOODP BMI PEDIGREE AGE]    % get input observation matrix: 10 observations (10-row)

5  D = pdist(X)                             % get Euclidean distances
6  Y = squareform(D);
7  Z = linkage(D)                           % get similar clusters
   figure
8  dendrogram(Z)                            % plot hierarchical clustering tree

9  T1 = cluster(Z, "cutoff", 20);
10 T2 = cluster(Z,"maxclust",6);
11 T3 = cluster(Z,"maxclust",2);
```

Fig. 8.23 The codes of the hierarchical clustering algorithm for diabetes

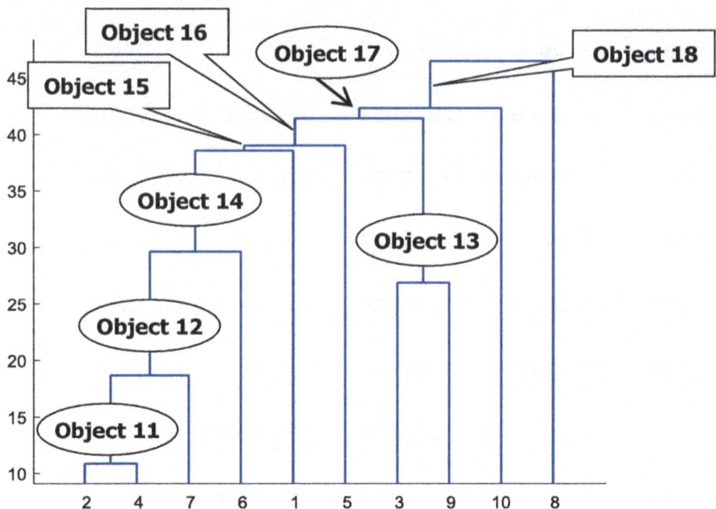

Fig. 8.24 The completed graphical presentation of the hierarchical clustering tree

```
    6.0000      12.0000      29.6690
    1.0000      14.0000      38.6288
    5.0000      15.0000      39.0770
   13.0000      16.0000      41.4619
   10.0000      17.0000      42.3835
    8.0000      18.0000      46.5347
```

$$T1 = [1 \quad 1 \quad 1 \quad 1 \quad 1 \quad 1 \quad 1 \quad 1 \quad 1 \quad 1 \ ]^T$$
$$T2 = [1 \quad 2 \quad 4 \quad 2 \quad 3 \quad 2 \quad 2 \quad 6 \quad 4 \quad 5 \ ]^T$$
$$T3 = [2 \quad 2 \quad 2 \quad 2 \quad 2 \quad 2 \quad 2 \quad 1 \quad 2 \quad 2 \ ]^T$$

First let's discuss the relation or the mapping between the ten observations (objects) $\mathbf{X}$ and the hierarchical clustering tree $\mathbf{Z}$.

It can be found from the tree map $\mathbf{Z}$ that the first two objects, object 2 and object 4 in the $\mathbf{X}$, are grouped together with a distance of 10.8758. Then object 7 is grouped with object 11 which is clustered by combining objects 2 and 4 together to form object 12. This can be found in Fig. 8.24 which has been highlighted. The distance between them is 18.6915.

Next objects 3 and 9 are grouped together to form object 13, and the distance between them is 26.9278. Object 6 is then grouped with object 12 which is clustered by combining objects 11 and 7, as shown in Fig. 8.24. Object 1 is grouped with object 14 which is clustered by combining objects 6 and 12 to form object 15. Then object 5 is grouped with object 15 to form object 16, and object 13 is grouped with object 16 to form object 17. Object 10 is grouped with object 17 to form object 18, and finally object 8 is grouped with object 18 to get one tree.

In operational steps for coding lines 9–11 in Fig. 8.23, we use **cluster()** function to cut off the hierarchical cluster tree at different formats with different values.

First we cut off the tree at 20 and get all objects as one single group **T1**. Then we use **maxclust** argument to cut the tree at the most possible values of 6 and 2, respectively. For the former, the **cluster()** function groups objects 2, 4, 6, and 7 into one cluster, groups objects 3 and 9 into the second cluster, and groups objects 1, 5, 8, and 10 into each individual cluster. Totally they have six clusters. For the latter, the **cluster()** function groups objects 1~7 and 9~10 into one cluster and object 8 as a single cluster.

Compared with the top 10 observations in our dataset, **Diabetes.csv**, it can be found that objects 1, 3, 5, 7, 9, and 10 in the dataset are diabetes patients, but objects 2, 4, and 6 are not diabetes patients. Therefore, the result of our hierarchical clustering algorithm to this diabetes samples' prediction is: the actual number of diabetes patients is 6 over the total 10 samples, and the predicted number of diabetes patients is 9 over the total 10 samples. The error is $0.9 - 0.6 = 0.3$ or 30% in ratio and the prediction accuracy is about 70%. The reason for this low accuracy may be due to the small amount of samples.

A completed project **Diabetes_HC_Func.m** can be found on the Springer ftp sit under the folder: **Students\Class Projects\Chapter 8**.

Next let's concentrate our discussions on overlapping clustering algorithm.

8.5.3   *Functions Used for Overlapping Clustering (Fuzzy-K) Algorithm*

The Fuzzy-K-Mean (FKM) algorithm is also called Fuzzy-C-Mean (FCM) algorithm. As we discussed in Sect. 8.1.2, this algorithm is to cluster a group of data into multiple clusters. It means that some data points can be clustered into more than one cluster, which means that some data points may belong to multiple clusters with different fuzzy degrees represented by some membership functions (MFs).

Anyway, FCM is a data clustering technique where each data point belongs to a cluster to a degree that is specified by a membership grade in a membership function.

The FCM algorithm starts with an initial guess for the cluster centers, which represent the mean location of each cluster. The initial guess for these cluster centers may be likely incorrect. Additionally, FCM assigns every data point with a membership degree for each cluster. By iteratively updating the cluster centers and the membership degrees for each data point, the algorithm iteratively moves the cluster centers to the optimal location within a dataset. This iteration is based on minimizing an objective function that represents the distance from any given data point to a cluster center weighted by the data point membership grade [24].

To cluster a set of data, specify an array of data points, x_i, with N rows. The number of columns for each data point is equal to the data dimensionality.

$$x_j = \begin{bmatrix} x_{j1} \, x_{j2} \, x_{j3}, \ldots, x_{jn} \end{bmatrix}^T \quad (1 \le j \le N)$$

$$(8.13)$$

The FCM algorithm computes the cluster centers, c_i. This array contains one row for each cluster center and the number of columns matches the number of columns in x_i.

$$c_i = \left[c_{i1}\ c_{i2}\ c_{i3}, \ldots, c_{in} \right]^T \quad (1 \le i \le C) \tag{8.14}$$

To specify the number of clusters C, use the **NumClusters** option in MATLAB. The FCM algorithm minimizes the following objective function:

$$J_m = \sum_{i=1}^{C} \sum_{j=1}^{N} \mu_{ij}^m D_{ij}^2 \tag{8.15}$$

where

(1) m is the fuzzy partition matrix exponent for controlling the degree of fuzzy overlap, with $m > 1$. Fuzzy overlap is to indicate how fuzzy the boundaries between clusters are, that is, the number of data points that have significant membership in more than one cluster. To specify fuzzy partition matrix exponent, use the **Exponent** option.
(2) D_{ij} is the distance from the jth data point to the ith cluster.
(3) μ_{ij} is the membership degree of the jth data point in the ith cluster. For a given data point, the sum of the membership values for all clusters is one (1).

In MATLAB, to cluster data using FCM clustering algorithm, use the **fcm()** function. To specify the clustering algorithm options, use an **fcmOptions** object. The **fcm()** function outputs a list of cluster centers and cluster membership degrees for each data point.

You can use cluster information to generate a fuzzy inference system (FIS) that best models the data behavior using a minimum number of fuzzy rules. The rules partition themselves according to the fuzzy qualities associated with each of the data clusters. To automatically generate this type of FIS, use the **genfis()** function. You can generate an initial fuzzy system using clustering results from either FCM or subtractive clustering.

The function **fcm()** supports three types of FCM clustering. These methods differ based on the distance metric used for computing D_{ij}.

Now let's have a clear and basic knowledge and understanding about the **fcm()** function. First let's have a closer look at its constructors.

8.5.3.1   Introduction to Function fcm()

Three popular constructors of this function include:

(1) **[centers, U] = fcm(data)**
(2) **[centers, U] = fcm(data, options)**
(3) **[centers,U,objFcn] = fcm(___)**

The first one is used to compute cluster centers (**centers**) and a fuzzy partition matrix (**U**) using default clustering options. The default options indicate that the **fcm()** function clusters or groups the data ten times, varying the number of clusters from 2 through 11. The reason for starting the cluster number from 2 is that no clustering is necessary for only one cluster. The matrix **U** contained the membership degrees of all data points in two clusters or two rows (μ_{ij} in Eq. (8.15)). Each row is equivalent to a cluster.

The second constructor is similar to the first one, but it specifies clustering options, such as the number of clusters and the distance metric.

The third one returns the objective function values at each optimization iteration for the optimal number of clusters.

One of the most important features or objects related to this **fcm()** function is the **fcmOption**, and it is used to set up and configure some important parameters and properties when using this function. Two constructors are used by this object, which are:

(1) **opt = fcmOptions**
(2) **opt = fcmOptions(Name=Value)**

The first constructor is used to return a default option object for FCM clustering. The second one is to specify options using one or more name-value pair arguments. For example, to set up three clusters, use **NumClusters = 3**.

Table 8.4 lists all popular properties used for this option object.

Regularly, the **NumClusters** property can be set to **auto** to provide more flexibility for the **fcm()** function to enable it to automatically select the optimal number value for the possible clusters. The **Exponent** parameter is used to control how fuzzy or how far away clusters are and how fuzzy between observations or data points. The greater this value, the closer the cluster centers are, and this makes almost all data points to have similar fuzzy degrees in all clusters.

The **ClusterCenters** property is used to indicate all clusters' center positions. Regularly it is presented as a **C × N** matrix, where **C** presents the number of clusters and **N** indicates the number of features on input observations. For example, if we have an input dataset that has 600 × 2 data points, it means that our dataset has 600 observations with 2 features. The **C** will be a 2 × 2 matrix as the number of clusters is 2. The exponent we preferred is 3 and the Euclidean is the distance. Then the following codes and properties can be used with the **fcm()** function:

```
options = fcmOptions(NumClusters=2, Exponent=3.0,
Verbose=true);
  [centers, U] = fcm(X, options);
```

The resulting center matrix is:

$$\mathbf{centers} = \begin{matrix} \mathbf{99.5739} & \mathbf{67.2981} \\ \mathbf{149.0528} & \mathbf{72.8344} \end{matrix} = \begin{pmatrix} C_{11} & C_{12} \\ C_{21} & C_{22} \end{pmatrix}$$

$$(8.16)$$

Table 8.4 All popular properties in fcmOption object

Property Name	Descriptions
NumClusters	Number of clusters C: 1). **auto** – C = 2 ~ 11, 2). **Integer > 1**, C > 1, 3) **Vector of integers > 1**, C > 1. When NumClusters is "auto" or a vector, the fcm function returns cluster centers for the optimal number of clusters, which it determines using a validity index.
Exponent	This is a scale integer > 1. This option controls the amount of fuzzy overlap between clusters, with larger values indicating a greater degree of overlap. Default value is 2. The greater of this value, the closer the cluster centers, and each data point has approximately the same degree of membership in all clusters. Try to reduce this value to avoid this situation.
MaxNumIteration	Maximum number of iterations, specified as a positive integer.
MinImprovement	Minimum improvement in objective function between two consecutive iterations, specified as a positive scalar.
DistanceMetric	Method for computing distance between data points and cluster centers, specified as following values. "euclidean" — Compute distance using a Euclidean distance metric, which corresponds to the classical FCM algorithm. "mahalanobis" — Compute distance using a Mahalanobis distance metric, which corresponds to the Gustafson-Kessel FCM algorithm. "fmle" — Compute distance using fuzzy maximum likelihood estimation (FMLE), which corresponds to the Gath-Geva FCM algorithm.
ClusterCenters	Initial cluster centers, specified as an C-by-N matrix, where C is the number of clusters and N is the number of data features. When ClusterCenters is empty, the FCM algorithm randomly initializes the cluster center values.
Verbose	Flag indicating whether to display the objective value after each iteration. **true** – display, **false**, no display.

Now let's use a simple example project to illustrate how to use this function to perform fuzzy-c-mean clustering process for one of our datasets, **Diabetes.csv**.

Create a new Script file, name it as **Diabetes_FCM_Func.m,** and enter the codes shown in Fig. 8.25 into that file. Let's have a closer look at this piece of codes to see how it works.

(1) The full path for the Diabetes dataset is declared first. You need to use your real path to replace it if you stored your dataset in different folder in your machine.
(2) The Diabetes dataset is read out and assigned to a local variable **T**.
(3) We only need to use the top 600 observations (rows) for this dataset, thus the number of these observations is assigned to a local variable **N**.
(4) In fact, we only need to use two columns or two features from this dataset, **Glucose** and **Blood Pressure**, so we extract and convert them from cells to doubles by using the **table2array**() function since the **fcm**() function needs to use numeric data as inputs.

```matlab
% Using FCM to cluster Diabetes dataset
% Name: Diabetes_FCM_Func.m
% Input: two features in Diabetes.csv dataset
% Output: FCM clustering result
% June 28, 2024
1  path = 'C:\\Artificial Intelligence Book\Students\Datasets\Diabetes Dataset\\Diabetes.csv';
2  T = readtable(path);
3  N = 600;

4  GLUCOSE = table2array(T(1:N, 2));          % Glucose is located at column 2 in database
   BLOODP = table2array(T(1:N, 3));           % Blood Pressure is located at column 3 in database

5  X = [GLUCOSE BLOODP];

6  options = fcmOptions(NumClusters=2, Exponent=3.0, Verbose=true);
7  [centers, U] = fcm(X, options);

   % Classify each data point into the cluster for which it has the highest degree of membership.
8  maxU = max(U);
   index1 = find(U(1, :) == maxU);
   index2 = find(U(2, :) == maxU);

9  plot(X(index1, 1), X(index1, 2),"ob")
   grid;
10 hold on

11 plot(X(index2, 1), X(index2, 2), "or")
12 plot(centers(1, 1), centers(1, 2), "xb", MarkerSize=25, LineWidth=5);
   plot(centers(2, 1), centers(2, 2), "xr", MarkerSize=25, LineWidth=5);
13 xlabel("Glucose")
   ylabel("Blood Pressure")
14 hold off
```

Fig. 8.25 The codes for the FCM clustering for Diabetes dataset

(5) Then compose those two features into an input matrix **X**.

(6) Prior to using the **fcm**() function, we can set the related parameters and properties used by the **fcm**() function. Thus the **fcmOption** object is used for that purpose, which indicates that we want to group our data points into two clusters with a fuzziness of 3 to map those two clusters and data points. The **Verbose** property is set to **true** since we like to monitor the iteration steps during this fuzzy clustering process.

(7) Then the **fcm**() function is executed with the options we selected in step 6. The function returned the cluster centers matrix **centers** and membership degrees of all data points in two clusters, which are stored in the matrix **U**.

(8) To plot all data points with two related clusters to show how fuzzy or how many different degrees of data points belong to related clusters and how closely overlapped between two clusters, we need to do some data preprocessing. These preprocessing steps include:

    a. Find the highest degree of the membership in **U** matrix. Exactly, the **U** matrix is a transpose of input observation matrix **X**, as shown in Eq. (8.13). The input matrix **X** has 600×2 format and the **U** is a 2×600 matrix after that transposing operation. The **maxU** is a 1×600 matrix since it needs to find the maximum membership degree for each column from two rows. The resulting **maxU** is a reduced matrix with 1×600 size.

     b. The **index1** is a one-row array and it contains the positions of the observations located at the first row (cluster) in the **U** matrix and those observations' values are equal to the **maxU** in membership degrees (MD).

     c. Similarly, the **index2** is also a one-row array and it contains the positions of the observations located at the second row in the **U** matrix and whose values are equal to the **maxU** in MD.

     d. The reason to find the observations with the largest membership degree (MD) from the **U** matrix in two rows (clusters) is to collect all of those data points that are closely located in the related cluster to make them ready to be plotted later.

(9) Now we can plot those input observations in **X** by using **index1** and **index2** as the horizontal position indices with the first and the second column (the original **X** is a 600×2 matrix).

(10) A **hold on** command is inserted here, and it is very important since we like to plot multiple graphics in one figure. This command is to hold on all present graphics with no change, otherwise we cannot do this kind of plotting without this command.

(11) As we did in step 9, similarly we can plot the observations with the maximum MD values located at the second cluster.

(12) We also like to plot the centers of two clusters in the same plot. Refer to Eq. (8.16). We can plot each of them with selected **MarkerSize** and **LineWidth**.

(13) Two labels are used to indicate two features in two coordinated systems.

(14) Finally do not forget to use **hold off** command to complete the hold on operation above.

Now run the project, and the running results are shown in the Command window below.

```
Diabetes_FCM_Func
Iteration count = 1, obj. fcn = 353681
Iteration count = 2, obj. fcn = 215281
Iteration count = 3, obj. fcn = 213356
Iteration count = 4, obj. fcn = 207702
Iteration count = 5, obj. fcn = 197281
Iteration count = 6, obj. fcn = 187536
Iteration count = 7, obj. fcn = 182821
Iteration count = 8, obj. fcn = 181334
Iteration count = 9, obj. fcn = 180945
Iteration count = 10, obj. fcn = 180842
Iteration count = 11, obj. fcn = 180813
Iteration count = 12, obj. fcn = 180804
Iteration count = 13, obj. fcn = 180801
Iteration count = 14, obj. fcn = 180800
Iteration count = 15, obj. fcn = 180800
Iteration count = 16, obj. fcn = 180800
```

```
Iteration count = 17, obj. fcn = 180800
Iteration count = 18, obj. fcn = 180800
Iteration count = 19, obj. fcn = 180800
Iteration count = 20, obj. fcn = 180800
Iteration count = 21, obj. fcn = 180800
Iteration count = 22, obj. fcn = 180800
Iteration count = 23, obj. fcn = 180800
Iteration count = 24, obj. fcn = 180800
Iteration count = 25, obj. fcn = 180800
Minimum improvement reached.
```

The plotting for this fuzzy c-mean clustering result is shown in Fig. 8.26.

8.5.3.2   Build a Real Fuzzy C-Mean Clustering Model for Segmentation of Cars

The image segmentation technique involves converting an image into a collection of regions of pixels that are represented by a mask or a labeled image. By dividing an image into segments, we can process only the important segments of the image instead of processing the entire image.

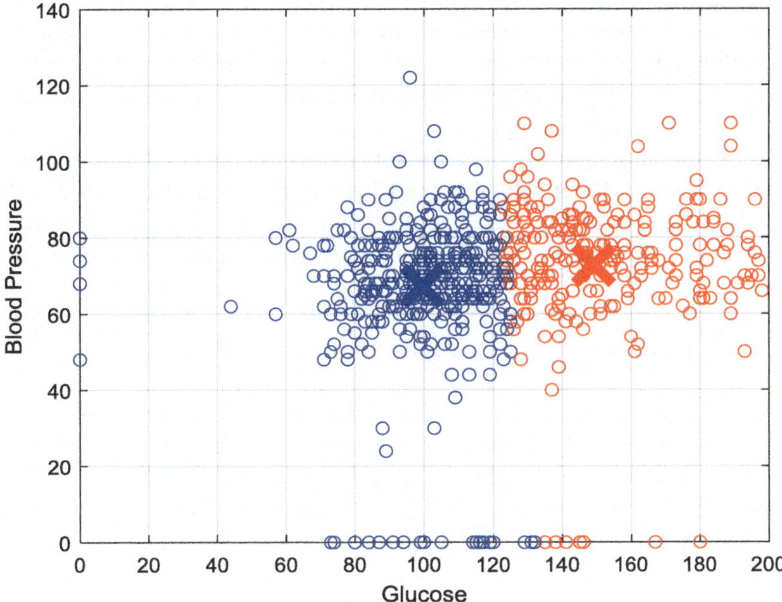

Fig. 8.26   The plotting result of FCM for Diabetes dataset

With functions in MATLAB and Image Processing Toolbox™, you can experiment and build expertise in different image segmentation techniques, including threshold, clustering, graph-based segmentation, and region growth [25].

In this section, we like to use FCM clustering algorithm to build an FCM model to segment some car images. In other words, we try to perform segmentations to some car images to divide an entire car image into different segments to make further study easier.

Before we can start our discussions on this topic, we would like to give a brief introduction about some special input data preprocessing, especially when the inputs are images and a significant difference exists between the numeric data input and image input to an unsupervised or clustering process.

8.5.3.2.1   Preprocessing of Images as Input Data for AI or Clustering Algorithms

Most AI algorithms can handle numeric inputs with no issue, but when the input data are images, either grayscale or color images, we need to perform some preprocessing to convert images to a kind of data format that can be acceptable by the AI algorithms. The main reason for that is due to the different presenting format used between the numeric data and image data. The former is a numeric vector or a matrix, but the latter is a 2D pixel matrix. To enable an AI or a clustering algorithm to process or handle those as input data, we need to do some preprocessing to enable them suitable to be used by the supervised or unsupervised learning methods.

The basic idea for this preprocessing is that each image is presented as a 2D $m \times n$ pixel matrix. To enable the clustering algorithms to recognize that image, each image needs to be converted to one row or one column, which means that one image is equivalent to a single row or a single column as an input in the clustering domain. Figure 8.27 shows this mapping.

In Fig. 8.27a, an image with 4×4 pixels matrix is converted to four rows with each row of 4 pixels. Relatively, this image can also be converted to four columns of 4 pixels, as shown in Fig. 8.27b. In this way, we can finish this preprocessing process to enable a clustering algorithm to recognize each input image.

8.5.3.2.2   Build Clustering Segmentation Model for Car Images

The dataset we like to use is the **Car Type** dataset that contained three major cars, pickup, sedan, and SUV. In this project, we can use two sedans' car images to perform FCM clustering with five clusters to segment them with their similarity.

Create a new Script and name it as **Car_Image_FCM_Func.m**, enter the codes shown in Fig. 8.28 into this script file. Let's have a closer look at this piece of codes to see how it works.

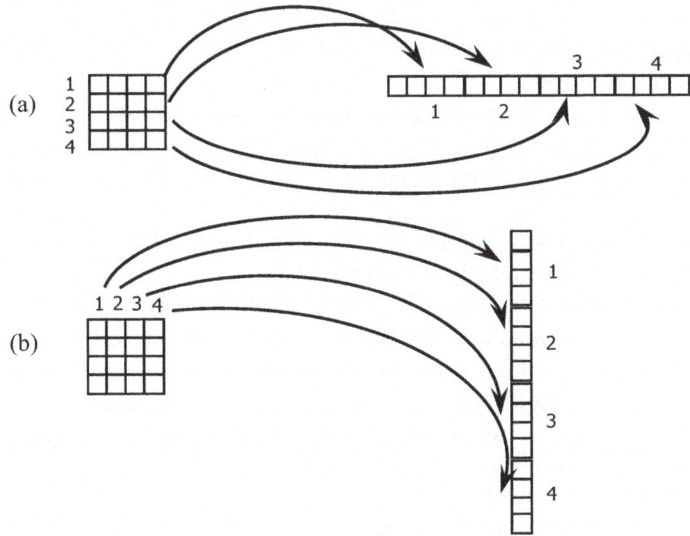

Fig. 8.27 A mapping between an image and its rows or columns

```
% Perform segmentation for two images with FCM algorithm
% Name: Car_Image_FCM_Func.m
% Input: Two car images
% Output: FCM clustering result
1  Img1 = ['C:\Artificial Intelligence Book\Students\Datasets\Car Type Dataset\' ...
            'car_train_data\sedan\PHOTO_7.jpg'];
   Img2 = ['C:\Artificial Intelligence Book\Students\Datasets\Car Type Dataset\' ...
            'car_train_data\sedan\PHOTO_9.jpg'];
2  I1 = im2double(imread(Img1));                   % Load car 1 image
   I1 = rgb2gray(I1);                              % convert to gray scale
   I1 = imadjust(I1);
3  I2 = im2double(imread(Img2));                   % Load car 2 image
   I2 = rgb2gray(I2);                              % convert to gray scale
   I2 = imadjust(I1);
4  subplot(121);
   imshow(I1, []);
5  data = [I1(:) I2(:)];                           % make data array = (400 * 500)) 200000 by 1 array
6  [center, U, obj_fcn, info] = fcm(data, 5);      % Fuzzy C-means classification with 5 classes
7  Noptim = info.OptimalNumClusters;
   % Find the pixels for each class, which have the max MD values
8  maxU = max(U);
9  index = cell(Noptim, 1);
10 for k = 1:Noptim
       index{k} = find(U(k, :) == maxU);
   end

   % Assigning pixel to each class by giving them a specific value
11 N = 1.2;
12 fcmImage(1:length(data))=0;
13 for i = 1:Noptim
       fcmImage(index{i}) = N - 0.2*i;
   end

   % Reshapeing the array to a imag
14 imagNew = reshape(fcmImage, 400, 500);
15 subplot(122);
   imshow(imagNew, []);
```

Fig. 8.28 The codes used to perform FCM clustering for image segmentations

(1) The locations or full paths of two images are declared first. You need to use your real paths to replace these if you saved these two images in different folders in your machine.

(2) Due to the requirements of FCM clustering algorithm, we need to read out, convert, and adjust the first image to double format and appropriate size. After this preprocessing, the first image **I1** is a grayscale image matrix with 400×500 pixels.

(3) Perform similar preprocessing for the second image **I2**, which has the same properties and size as those of the first image.

(4) To take a look at the original first image, a **subplot(121)** function is used, and it means that two plots will be plotted in one-row-two-column format and the first plot will be plotted below first. Then the **imshow()** function is called to plot the first image **I1**. The second argument for this function is a blank [], which means that the actual size of that image can be defined and selected by the system automatically if its size is unknown.

(5) Now to prepare to use FCM clustering algorithm to cluster images, we need first to convert two images from matrix to a two-column vectors, each column representing all pixels for one image. Both images **I1** and **I2** are 400×500 matrices (see step 2 above). After this preprocessing, the **data = [I1($\odot$) I2($\odot$)]** becomes 200000×2 array or matrix. The operator ($\odot$) is to convert each image from an m-row $\times$ n-column matrix to one column with a size of $m \times n \times 1$. For example, in our case, **I1($\odot$)** is converted to $400 \times 500 \times 1 = 200000 \times 1$ vector. Similar operation is performed for I2. The resulted **data** is a 200000×2 matrix with each column mapped to one image.

(6) The **fcm()** function is executed to perform the FCM clustering for these two images with a number of clusters as 5.

(7) To get the optimal cluster number, the **OptimalNummClusters** property is used. In fact, this number should be 5 since we used five clusters for this clustering process.

(8) The matrix **U** contained the fuzzy degrees in distances to the related centers of all five clusters for all observations in five clusters. In fact, the matrix **U** is a transposed version of the preprocessed **data** in step 5 above. The **data** is a 200000×2 matrix, but the **U** is a 5×200000 matrix after the transposing with five clusters. The **maxU** is a 1×200000 matrix since it needs to find the maximum membership degrees for each column from five rows. The resulting **maxU** is a reduced matrix with 1×200000 size.

(9) The **index** is a 5×1 cell vector that contains the number of clusters, which is 5.

(10) A **for()** loop is used to continuously pick up the positions of the observations located at each row (cluster) in the **U** matrix and those observations' values are equal to the **maxU** in membership degrees (MD). Those positions are assigned to the related **index** vector row by row. The reason to find the observations with the largest membership degree (MD) from the **U** matrix in five rows (clusters) is to collect all of those data points that are closely located to the centers in the related clusters to make them ready to be plotted later.

(11) To segment images into different segmentations or parts, we can assign some specific values to observations or pixels in each cluster. These specific values are determined by users to make images with different or desired segmentations or levels in intensities. The smaller the values, the less segmentations or levels can be obtained. In this case, we try to assign an arithmetic sequence {1, 0.8, 0.6,… 0.0} to a new image vector **fcmImage**(). Thus we defined the starting value as **N** = 1.2.

(12) That new image vector is a 1 × 200000 array since the length of the **data** matrix is 200000. First we initialize that new image vector to all 0.

(13) Another **for**() loop is used to assign the new image vector with an arithmetic sequence.

(14) To plot that new image vector, we need to reshape it to a standard size. Recalled in step 2, both images have a size of 400 × 500 pixels, thus we used that size here.

(15) The second graphic is plotted with the **imshow**() function. A blank size, [ ], is used here as the second argument to make this plotting compatible with different sizes of images.

Now run the project and the running sequences are displayed in the command window, and the running result is plotted and shown in Fig. 8.29.

```
Car_Image_FCM_Func
Iteration count = 1, obj. fcn = 10434.6
Iteration count = 2, obj. fcn = 7905.06
Iteration count = 3, obj. fcn = 7902.33
Iteration count = 4, obj. fcn = 7865.93

......
Iteration count = 54, obj. fcn = 766.393
Iteration count = 55, obj. fcn = 766.393
Iteration count = 56, obj. fcn = 766.393
Iteration count = 57, obj. fcn = 766.393
Minimum improvement reached.
```

Next let's discuss how to use functions related to probabilistic clustering algorithm to perform probabilistic clustering for some selected datasets.

8.5.4   *Functions Used for Probabilistic Clustering Algorithm*

As we discussed in Sect. 8.1.4, a probabilistic clustering is an unsupervised technique that helps us solve density estimation or *soft* clustering problems. In probabilistic clustering, data points are clustered based on the likelihood that they belong to a particular distribution. The Gaussian Mixture Model (GMM) is one of the most commonly used probabilistic clustering methods.

By using GMM algorithm, the data points can be either clustered as hard clustering or soft clustering depending on the clustering criterion selected. Similar to other clustering methods, GMM clustering algorithm also requires users to specify the number of clusters before fitting the model. The number of clusters specifies the number of components in the GMM.

8.5.4.1 Introduction to Multivariate Gaussian Distributions

The Gaussian distribution is also called a normal distribution. For a single variable x, the Gaussian distribution or probability density function $f(x)$ can be described as:

$$f(x) = \frac{1}{\sqrt{2\pi\sigma^2}} e^{-\frac{(x-\mu)^2}{2\sigma^2}}$$

$$(8.17)$$

The parameter μ is the mean or expectation of the distribution, while the parameter σ^2 is the variance. The standard deviation of the distribution is σ. A random variable with a Gaussian distribution is said to be normally distributed and is called a normal deviate [26].

However, for multiple variables Gaussian distribution or called multivariate Gaussian distribution, you cannot continue to use a single variance to describe the variance values between or among variables since the variance only works for a single variable. Instead, you need to use covariance matrix to replace the variance variable to describe the covariance values for each pair of input data points.

For a k-dimensional random vector $\mathbf{X} = (\mathbf{X_1}, \mathbf{X_2}, \mathbf{X_3}, \ldots, \mathbf{X_K})$, its probability density function can be written as:

$$f_x\left(X_1, X_2, \ldots, X_k\right) = \frac{1}{\sqrt{(2\pi)^k |\Sigma|}} e^{-\left(\frac{1}{2}(x-\mu)^T \Sigma^{-1}(x-\mu)\right)}$$

$$(8.18)$$

Fig. 8.29 The running result of image segmentations with FCM

where Σ is called a covariance matrix and it contains the variances for each variable if it is a diagonal format or it includes the covariance values between variables if it is off-diagonal format.

The quantity $\sqrt{(x-\mu)^T \Sigma^{-1}(x-\mu)}$ is known as the Mahalanobis distance, which represents the distance of the test point x from the mean μ. Note that in the case when $k = 1$, the distribution reduces to a single normal distribution and the Mahalanobis distance reduces to the absolute value of the standard score [27].

The two major properties of the covariance matrix are [28]:

(1) Covariance matrix is positive semi-definite.
(2) Covariance matrix in multivariate Gaussian distribution is positive definite.

A symmetric matrix M is said to be positive semi-definite if $x^T M x$ is always non-negative for any vector y. Similarly, a symmetric matrix M is said to be positive definite if $x^T M x$ is always positive for any nonzero vector y.

In MATLAB, Statistics and Machine Learning Toolbox™ provides several functions related to the multivariate normal distribution. One of them is used to calculate the probability density function (PDF) at specific values, which is **mvnpdf()**.

Now let's use a simple example to illustrate how to use this function to calculate the PDF values with some data points.

Two variables, x_1 and x_2, are used for this example. The mean values for both of them are 0, and the covariance matrix is a symmetric positive definite one with the values as:

$$\Sigma = \begin{bmatrix} 0.30 & 0.35 \\ 0.35 & 0.98 \end{bmatrix}$$

The related MATLAB codes to perform this calculation and its results are shown in Fig. 8.30. Let's have a closer look at this piece of codes to see how it works.

```
% testing multivariate Gaussian distribution
1  mu = [0 0];
2  Sigma = [0.3 0.35; 0.35 0.98];

3  x1 = -4:0.2:4;
   x2 = -4:0.2:4;
4  [X1, X2] = meshgrid(x1, x2);
   X = [X1(:) X2(:)];

5  y = mvnpdf(X, mu, Sigma);
6  y = reshape(y ,length(x2), length(x1));

7  surf(x1, x2, y)
   axis([-4 4 -4 4 0 0.4])
   xlabel('x1')
   ylabel('x2')
   zlabel('Probability Density')
```

Fig. 8.30 The testing codes for two input variables for GMM algorithm

(1) The mean values for two variables, 0 and 0, are assigned to the local vector **mu**.

(2) The covariance symmetric matrix **Sigma** is declared with initial values.

(3) Two input vectors, x_1 and x_2, are generated with the colon operators (:). Both vectors are 1×31 array starting at -4 and ending at 4 with an interval of 0.2 between each data item.

(4) The function **meshgrid(x_1, x_2)** returns 2D grid coordinates based on the coordinates contained in vectors x_1 and x_2. $\mathbf{X_1}$ is a matrix where each row is a copy of x, and $\mathbf{X_2}$ is a matrix where each column is a copy of x_2. The grid represented by the coordinates $\mathbf{X_1}$ and $\mathbf{X_2}$ has length(x_1) rows and length(x_2) columns. Then both coordinates are assigned to a local variable X.

(5) The function **mvnpdf()** is executed to perform the calculation for PDF values by using GMM algorithm.

(6) To make the plotting easy and symmetry, the **reshape()** function is used to reshape the resulting PDF to be plotted in a symmetry coordinates with **length(x_2)** × **length(x_1)**, which is a 31×31 matrix. However, the total elements in y are not changed after this reshaping.

(7) The function **surf(x_1, x_2, y)** with two vector arguments replacing the first two matrix arguments, must have **length(x_1) = n** and **length(x_2) = m** where **[m, n] = size(y)**. In this case, the vertices of the surface patches are the triples **(x_1(j), x_2(i), y(i,j))**. Note that $\mathbf{x_1}$ corresponds to the columns of **y** and $\mathbf{x_2}$ corresponds to the rows of **y**.

Now run this piece of codes and the running result is plotted in Fig. 8.31.

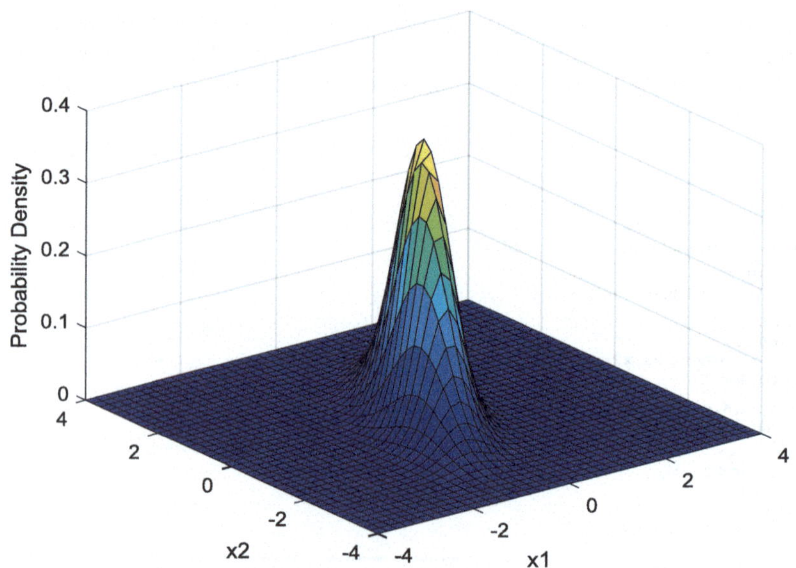

Fig. 8.31 The running result of GMM for two vectors

This piece of codes is stored in a Script file named **testMultivariateGaussian.m,** and it can be found on the Springer site in the folder: **Students\Class Projects\ Chapter 8**.

Now we have gotten some basic idea and understanding about GMM distribution and its implementations, next let's have a deeper discussion about this algorithm and how to use it to perform clustering for real data in datasets.

8.5.4.2   Using GMM Algorithm to Cluster Data in Datasets

In MATLAB, a **gmdistribution** object stores a Gaussian Mixture Distribution (GMD), also called a Gaussian Mixture Model (GMM), which is a multivariate distribution that consists of multivariate Gaussian distribution components. Each component is equivalent to a Gaussian distribution, and it is defined by its mean and covariance. The mixture is defined by a vector of mixing proportions, where each mixing proportion represents the fraction of the population described by a corresponding component [28].

To perform a hard clustering process for real data with GMM object, the following procedures should be adopted:

(1) *Create a GMM object* **gmdistribution** by using two possible ways: (a) using the function **gmdistribution()** to create a **gmdistribution** model object by indicating the distribution parameters and (b) using the function **fitgmdist()** to generate and fit a GMM object with input data.

(2) Then use object functions, such as **cluster()**, **posterior()**, and **mahal()**, to perform related cluster analysis.

(3) As the clustering starts, consider the component covariance structure. You can specify diagonal or full covariance matrices, and whether all components have the same covariance matrix or not.

(4) Specify initial conditions. You can specify your own starting values for the parameters, specify initial cluster assignments for data points, or let them be selected randomly.

(5) Evaluate the model, such as **pdf**, and generate random multivariables or variates.

To do a soft clustering with GMM object, the following procedures should be used [29]:

(1) Assign a cluster membership score or degree to each data point that describes how similar each point is to each cluster's archetype. For a mixture of Gaussian distributions, the cluster archetype is the corresponding component mean, and the component can be the estimated cluster membership posterior probability.

(2) Rank the points by their cluster membership score or degree.

(3) Inspect the scores or degrees and determine cluster memberships.

For algorithms that use posterior probabilities as scores, a data point is a member of the cluster corresponding to the maximum posterior probability. However, if there are other clusters with similar posterior probabilities that are close to the maximum, then the data point can also belong to a member of those clusters. It is good practice to determine the threshold on scores that produce multiple cluster memberships before clustering.

8.5.4.3 Using GMM Algorithm to Cluster Actual Data in Diabetes Datasets

In this section, we like to use a real example project to illustrate how to fit a GMM model to cluster our **Diabetes.csv** dataset with different covariance options and initial conditions.

Prior to starting this project, the following points need to be noted [9]:

(1) The number of components k in a GMM determines the number of clusters. In fact, each cluster is equivalent to a component. In this figure, it is difficult to determine if two, three, or perhaps more Gaussian components are appropriate. Remember, a GMM increases in complexity as k increases.
(2) Each Gaussian component has a covariance matrix. Geometrically, the covariance structure determines the shape of a confidence ellipsoid drawn over a cluster. You can specify whether the covariance matrices for all components are diagonal or full, and whether all components have the same covariance matrix. Each combination of specifications determines the shape and orientation of the ellipsoids.

Keep those points in mind, now let's build our project.

Create a new Script file and name it as **Diabetes_GMM_Func.m** and enter the codes shown in Fig. 8.32 into that file. Let's have a closer look at this piece of codes to see how it works.

(1) First the full path of our **Diabetes.csv** dataset is declared. You need to use your actual path to replace it if you saved this dataset in a different folder in your machine. The whole dataset is read out with **readtable()** function and assigned to a local variable **T**.
(2) Some related local variables are also declared and initialized. The total number of observations on our dataset is 600, and the number of clusters we like to use for this clustering process is 2.
(3) Retrieve two columns from our dataset since we do not want to use all of them and only like to use two of them, **GLUCOSE** and **BMI**. Assign these two features to the input matrix **X**.
(4) Using **statset()** function to create an options structure in which the named fields have the specified values. Here we use this function to set the maximum iteration number as 1000 for the clustering process. Also we like to use **diagonal** covariance matrix in this process. You can change this matrix by selecting the **full** matrix for **Sigma** if you like.
(5) To make the plotting for the clustering result easy, we define the grid length as 500.

```matlab
% Using fitgmdist() function to generate 2 fitted GMM and fit it with 2 clusters
% Name: Diabetes_GMM_Func.m
% Input: Diabetes.csv dataset
% Output: GMM cluster result
% Jul 3, 2024
1  path = 'C:\\Artificial Intelligence Book\Students\Datasets\Diabetes Dataset\\Diabetes.csv';
   T = readtable(path);
2  N = 600;
   k = 2;
3  GLUCOSE = table2array(T(1:N, 2));          % Glucose is located at column 2 in database
   BMI = table2array(T(1:N, 6));              % BMI is located at column 6 in database
   X = [GLUCOSE BMI];
4  options = statset('MaxIter',1000);
   Sigma = 'diagonal';
   %Sigma = 'full';
5  d = 500;                                   % Grid length
6  x1 = linspace(min(X(:,1))-2, max(X(:,1))+2, d);
   x2 = linspace(min(X(:,2))-2, max(X(:,2))+2, d);
7  [x1grid, x2grid] = meshgrid(x1, x2);
8  X0 = [x1grid(:) x2grid(:)];
9  threshold = sqrt(chi2inv(0.99, 2));
10 for j = 1:2
       gmfit = fitgmdist(X, k, 'CovarianceType',Sigma, 'Options', options);
11     clusterX = cluster(gmfit, X);                 % Cluster index
12     mahalDist = mahal(gmfit, X0);                 % Distance from each grid point to each GMM component
       % Draw ellipsoids over each GMM component and show clustering result.
13     figure(j)
14     h1 = gscatter(X(:,1), X(:,2), clusterX);
15     hold on
16     for m = 1:k
           idx = mahalDist(:, m) <= threshold;
17         Color = h1(m).Color*0.75 - 0.5*(h1(m).Color - 1);
18         h2 = plot(X0(idx, 1), X0(idx, 2), '.', 'Color', Color, 'MarkerSize', 1);
19         uistack(h2, 'bottom');
       end
20     plot(gmfit.mu(:,1), gmfit.mu(:,2), 'kx', 'LineWidth', 2, 'MarkerSize', 10)
       title(sprintf('Sigma is %s\n', Sigma), 'FontSize', 8)
21     legend(h1, {'1','2'})
       grid;
22     hold off
end
```

Fig. 8.32 The codes used to perform GMM clustering for Diabetes dataset

(6) The function **linspace(a, b, n)** is used to generate a linearly spaced vector for **n** points. The spacing between the points is **(b-a)/(n-1)**. Here we use it to generate a linearly space starting at the minimum value on the first column on **X**, and ending at the maximum value on the second column on **X** with an offset of ± 2.

(7) For the plotting purpose, we use the function **meshgrid()** to generate 2D grid coordinates based on the coordinates contained in vectors x_1 and x_2. The **x1grid** is a matrix where each row is a copy of x_1, and **x2grid** is a matrix where each column is a copy of x_2. The grid represented by the coordinates **x1grid** and **x2grid** has **length(x_2)** rows and **length(x_1)** columns.

(8) Then both **x1grid** and **x2grid**, having a size of 500×500, are converted to two long columns by using the (:) operator, and both of them now have a size of 250000×1, and are assigned to a new matrix variable **X0**. Refer to Fig. 8.27b to get more details about this kind of conversion. This **X0** will be used later for plotting purpose.

(9) Now we need to set a threshold as our clustering criterion by using the function **chi2inv()**, exactly the square root of this function result. This function returns the inverse cumulative distribution function (**icdf**) of the chi-square distribution with degrees of freedom 2, evaluated at the probability values of 0.99.

(10) Since we used two clusters on this project, thus a **for()** loop is used to generate a GMM and perform clustering process to fit the generated GMM. In this project, we used the second way (refer to Sect. 8.5.4.2 where two ways could be used) to create and fit a new GMM object with input data. Exactly we used the function **fitgmdist()** to do that job. The arguments include the input dataset matrix **X**, the number of clusters k, the **diagonal** as the Sigma type with options. The generated and fitted GMM object is assigned to a local variable **gmfit** to be used later.

(11) Then the **cluster()** function is executed to group the data in **X** into k clusters determined by the k Gaussian mixture components in **gmfit**. The value in **clusterX(i)** is the cluster index of observation i and indicates the component with the largest posterior probability given the observation i. In our case, we have two clusters and it is equivalent to two components. The resulting **clusterX** is a 600×1 vector that contains all cluster indices or the GMM components (1 or 2) with the largest probability values for all 600 observations.

(12) The function **mahal()** is used to calculate the squared Mahalanobis distances, which is discussed in Sect. 8.5.1 with Eq. (8.11), from observations (rows) in **X0** to the class means in **gmfit** object. In our case, since **X0** contained 250000 grid points (see step 8 above), this function returns 250000 squared Mahalanobis distances to the mean values in our **gmfit** object.

(13) Because we need to plot two figures, one for cluster 1 and the other for cluster 2, thus we need to declare two figure objects with **figure(i)** ($i = 1$ or 2).

(14) We like to plot the original input data in two columns with the **gscatter()** function, and this function creates a scatter plot of **X(:, 1)** and **X(:, 2)**, grouped by cluster indices **clusterX**. In our case, we want to plot the **X(:, 1)** values on the x-axis and the **X(:, 2)** values on the y-axis, then group both columns' data points by **clusterX** that contained cluster indices, either 1 or 2, means cluster 1 or cluster 2. Then return the graphic handle corresponding to the groups in **h1**.

(15) Since we need to add some other plots with this plot together, thus a **hold on** command is used to let the system know that keep this graphic and we need to add something soon.

(16) In fact, we need to plot two graphics for two clusters, thus a **for()** loop is used to go through each cluster. First we need to find those grid points at the first column in the **mhhalDist(:, 1)** and whose Mahalanobis distances are less than the **threshold** value. If it is true, a 1 is returned, otherwise a 0 is returned. All of these returned Boolean values are assigned to a local variable vector **idx**, which is a 250000×1 array responding to the first column in **X0**.

(17) This coding line is used to define the different colors for all data points that belong to one cluster since we need to distinguish all data points in two clusters in our plot.

(18) Now we can plot all data points for two columns on **X0** with related indices
      obtained from step 16 in the same color. Then return the graphic handle to **h2**.
(19) Set the above graphic to the bottom. In a similar way, we can plot the same
      graphic for the second cluster.
(20) We also like to plot two mean values for two clusters, indicate the Sigma type,
      and add them to the same plot.
(21) The **legend**() function is used to indicate the number of clusters for both plots.
(22) Finally the hold off command is used to end this hold on operation.

Now run the project and the resulting graphics are shown in Fig. 8.33.

It can be found from Fig. 8.33 that the two-column training data points are basically clustered into two clusters, the **Glucose** with blue color is group 1 and the
BMI with orange color is cluster 2. Two GMM components represented by two
ellipses describe the probability range for all predictor data points. They are basically matched to original data.

Due to space limitations, we can just perform an approximated evaluation for
this clustering result. We consider that the GMM clustered two features into two

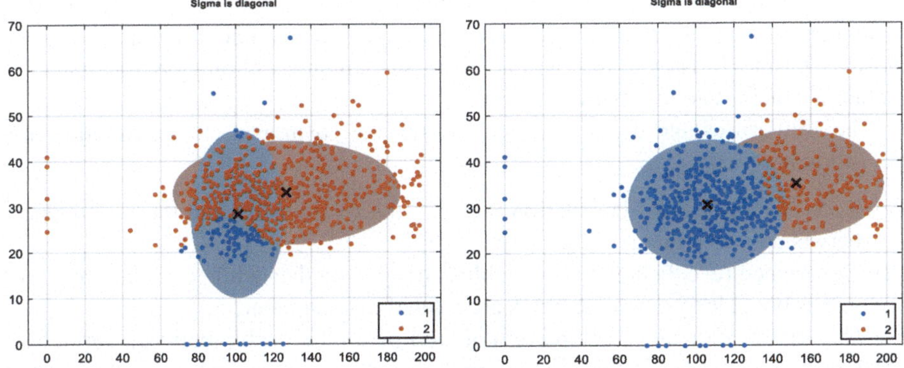

Fig. 8.33   The running results of using GMM to cluster Diabetes dataset

Table 8.5   A comparison for clusterX and output on Diabetes dataset

clusterX Value	The Output Column	Result
1	1	Correct
2	0	Correct
1	1	Correct
2	0	Correct
1	1	Correct
2	0	Correct
2	1	Incorrect
2	0	Correct
1	1	Correct
1	1	Correct

groups based on their properties and covariance, cluster 1 can be mapped to Diabetes and cluster 2 is mapped to non-diabetes, which are equivalent to 1 and 0 in the **Output** column in the Diabetes dataset. The top 10 of these results are listed in Table 8.5.

Based on this comparison, we can get an approximated evaluation result. The correct clustering rate for diabetes patients is about $9/10 \times 100/100 = 90\%$.

8.5.4.4   More About Covariance Matrix in GMM

As we mentioned, the covariance matrix is used for multivariate GMM clustering process. The main function of this matrix is to measure the changing direction of the relationship between two variables. A positive covariance indicates that both variables will change to be high or low in the same direction at the same time. But a negative covariance means that one variable is high but the other tends to be low with opposite direction at the same time. In fact, the covariance matrix

is used to monitor and measure two columns in an input dataset to see their change trends or how much they can change relatively in directions.

A diagonal covariance matrix means that we only measure the change of the symmetry elements on that input variable matrix and keep all nondiagonal elements close to 0. But a full covariance will measure all elements, not only on diagonal elements, for their changing between them.

The similarities and dissimilarities between the covariance and correlation are the following [30]:

(1) Both covariance and correlation measure the relationship and the dependency between two variables.
(2) Covariance reveals how two variables change together while correlation determines how closely two variables are related to each other.
(3) Covariance indicates the direction of the linear relationship between variables.
(4) Correlation measures both the strength and direction of the linear relationship between two variables.
(5) Correlation values are standardized.
(6) Covariance values are not standardized.

Next let's open our discussion to a very interesting topic, how to evaluate the unsupervised learning models.

8.5.5   *Introduction to Evaluation for Unsupervised Learning Models*

As we discussed, all unsupervised learning algorithms do not need to use any label, and exactly they do not use any input-output pair to train and test models. Instead they can use different clustering methods to group or cluster all observations by their similarity or dissimilarity automatically. An interesting question arises here: How to evaluate an unsupervised learning model (clustering) for real applications without output labels? What is the criterion to do this kind of evaluation?

To answer these questions, we need first to have a clear picture about evaluations for unsupervised learning algorithms. This evaluation can be divided into two categories:

(1) Evaluate the clustering number
(2) Evaluate the clustering result

Let's have our discussions about them one by one.

8.5.5.1   Evaluate the Clustering Number

As we discussed in our previous sections with related projects, most times we need to define the clustering number prior to performing any unsupervised learning analysis and building any clustering models to cluster real data in the dataset. However, this kind of definition may introduce some unforeseen or hidden errors, and the predefined clustering number may not be the optimal one. To solve this issue, we can use some MATLAB functions to evaluate an optimal clustering number prior to doing our clustering analysis to make this number more appropriate for our unsupervised learning studies.

In MATLAB, various methods can be used to perform this selection. One of the most popular methods is to use the function **evalclusters()** to do that job. By using this function, different clustering algorithms can be evaluated to obtain the optimal clustering number prior to developing the actual clustering projects.

Two kinds of constructors can be used for this function:

(1) **eva = evalclusters(X, clust, criterion)**
(2) **eva = evalclusters(X, clust, criterion, Name, Value)**

where **X** is the input or predictor matrix, **clust** is the name of the selected clustering algorithm, and the **criterion** is the selected evaluation standard. The **Name-Value** pair on the second constructor is used to provide additional options for this evaluation. Table 8.6 lists some popular input arguments and parameters used by this function.

Now let's use a real example to illustrate how to use this function to find the optimal cluster number for actual dataset.

Still use our Diabetes dataset, **Diabetes.csv**, as input and create a new Script file. Name this file as **Find_Optimal_NCluster.m** and enter the codes shown in Fig. 8.34 into that file. Let's have a closer look at this piece of codes to see how it works.

(1) First set up the full path for our input dataset. You need to use your actual path to replace it if you saved this dataset in a different folder in your machine.
(2) Use **readtable()** function to read the entire dataset and assign them to a local variable **T**.

Table 8.6 Popular input parameters used in evalclusters() function

Parameter	Description	Example
X	Input data matrix	Input data, specified as an *m*-by-*n* matrix, *m* is the number of observations, and *n* is the number of variables or features.
clust	Cluster algorithm name	'kmeans'	'linkage'	'gmdistribution'
criterion	Cluster evaluation criterion	'CalinskiHarabasz'	'DaviesBouldin'	'gap'	'silhouette'
KList	List of number of clusters to be evaluated	Example: 'KList', 1:6. Specifies to test 1, 2, 3, 4, 5 and 6 clusters

```matlab
% Name: Find_Optimal_NCluster.m
% Input: Diabetes.csv dataset
% Output: Optimal cluster number
% Jul 6, 2024
1  path = 'C:\\Artificial Intelligence Book\Students\Datasets\Diabetes Dataset\\Diabetes.csv';
2  T = readtable(path);
3  N = 600;

4  GLUCOSE = table2array(T(1:N, 2));        % Glucose is located at column 2 in database
   BMI = table2array(T(1:N, 6));            % BMI is located at column 6 in database
5  X = [GLUCOSE BMI];

6  eva = evalclusters(X, 'kmeans', 'silhouette', 'KList', 1:6)

7  figure(1)
   plot(eva)
   grid;

8  figure(2)
   clusters = eva.OptimalY;
   gscatter(X(:,1), X(:,2), clusters, [], "xod")
```

Fig. 8.34 The codes used to find the optimal cluster number

(3) Set up the number of observations to 600 since we like to use those numbers of data in our diabetes dataset.

(4) Pick up two columns or two variables, **Glucose** and **BMI**, from that dataset since we only need both of them in this project.

(5) Assign those variables to a local variable **X** to get input matrix.

(6) Call the function **evalclusters()** to try to find the optimal cluster number. In this example, we used **kmeans** as the algorithm, **silhouette** as the cluster criterion, and **KList 1:6** as the range for this cluster searching process.

(7) To take a look at the finding result, we like to plot a graphic to show all possible clusters with the related probability. Regularly, the larger the magnitude of a number, the better the cluster number. Also the **OptimalK** property of the selected cluster number should be displayed on the screen.

(8) Similarly, we also like to plot a graphic to show the distribution of original input data on all clusters to illustrate the power of the selected cluster.

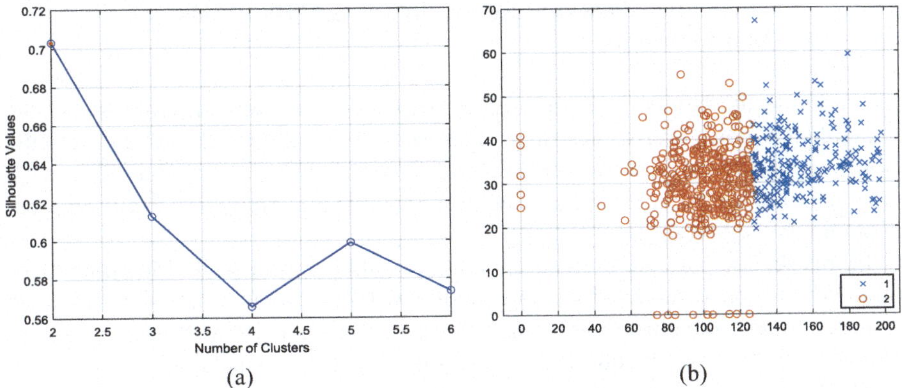

Fig. 8.35   The displayed optimal cluster number and the original data in two clusters

Now run the project and the running results are shown in the Command window. The graphical results are shown in Fig. 8.35.

```
eva =

          SilhouetteEvaluation with properties:


          NumObservations:  600
          InspectedK:  [1 2 3 4 5 6]
          CriterionValues:  [NaN 0.7031 0.6129 0.5659
          0.6039 0.5669]
          OptimalK:  2
```

It can be found from the running results displayed in the Command window that the optimal cluster number is 2, which means that we should use two clusters to perform this diabetes data clustering with GMM object. The type of the clustering object used in the clustering of input data has nothing to do with the algorithm used to find the optimal cluster number. Thus here we used **kmeans** algorithm to search this optimal cluster number and will use it in the GMM distribution object to cluster data in our Diabetes dataset later.

It can also be found from the graphical results that the optimal cluster number is the highest value in magnitude for all possible cluster numbers from 1 to 6, which is 2 with a magnitude of 0.7, as shown in Fig. 8.35a.

Figure 8.35b shows a distribution of all original input data around two clusters. It can be found that the data points with the orange color, which are concentrated around cluster 2, are closer to the center of this plot. Therefore we should select and use two clusters for our GMM clustering project.

Next let's discuss how to evaluate or validate the clustering results for input dataset in which the output labels are available.

8.5.5.2 Evaluate the Clustering Results by Using Output Labels in Input Datasets

As we know, all datasets that are used for supervised learning algorithms are also available for unsupervised learning algorithms. For the former, we can use the input-output pairs or input-output-label pairs to train and test the models. To use the same datasets for the latter, we can only use input data or predictors to cluster them to obtain desired groups.

In this section, we like to illustrate how to evaluate unsupervised learning algorithms or clustering models by using datasets with the output labels since we need them to validate the clustering results. In the next section, we will discuss how to evaluate the clustering results with datasets without output labels.

The basic idea by using the datasets with the output values or labels to evaluate the clustering results is to use the output labels as a reference or an additional criterion to compare the clustering results to confirm the results.

For example, still use our **Diabetes.csv** dataset as the input source, and we can use any unsupervised learning method to cluster desired groups that have the same output label, either cluster 1 (Output Label = 1 – Diabetes) or cluster 2 (Output Label = 0 – No Diabetes). Then by comparing the cluster numbers with the output label number, we can evaluate the clustering result. However, this resulting cluster number may be opposite during the clustering process due to the clustering algorithms and we will discuss this issue later.

Now let's create a new Script file, name it as **Eval_Diabetes_Func.m,** and enter the codes shown in Fig. 8.36 into that file. Let's have a closer look at this piece of codes to see how it works.

(1) Set up the full path to access our **Diabetes.csv** dataset. You may need to replace this path with your actual path if you saved this dataset in a different folder in your machine. The dataset is read out and assigned to a local variable **T**. The total clustering data is 600.

(2) Two columns or features in this dataset are used, **Glucose** and **BMI**, and a local variable **M** is initialized to a 1:600 vector for plotting purpose later.

(3) The original two column data are plotted to provide an image for their distributions.

(4) To find the optimal clustering number, the function **evalclusters()** is executed. This function returns all cluster index **idx** for all observations and the optimal clustering number stored in the property of the object **eva**, exactly as it is in **eva.OptimalK** property. In fact, the **idx** is a vector containing the clustering numbers for all observations. Totally we have 600 input rows (observations), thus the **idx** is a 600×1 array with 600 clustering numbers, either 1 or 2, since the optimal clustering number **eva.OptimalK** is 2 and we need to cluster all observations into two clusters by using **kmeans()** function.

(5) Now we can plot the clustering results with two cluster centers to see the results.

```
% Name: Eval_Diabetes_Func.m
% Input: Diabetes.csv dataset
% Output: Evaluation results
% Date: Jul 8,2024
path = 'C:\\Artificial Intelligence Book\Students\Datasets\Diabetes Dataset\\Diabetes.csv';
T = readtable(path);
N = 600;

GLUCOSE = table2array(T(1:N, 2));          % Glucose is located at column 2 in database
BMI = table2array(T(1:N, 6));              % BMI is located at column 6 in database

X = [GLUCOSE BMI];
M = 1:N;

figure;
plot(M, X(:,1), 'r+', M, X(:,2),'k*', 'MarkerSize',5);
title 'Glucose and BMI Data';
xlabel 'Glucose';
ylabel 'BMI';
legend('Glucose', 'BMI');
grid;

eva = evalclusters(X, 'kmeans', 'silhouette', 'KList', 1:6)
[idx, C] = kmeans(X, eva.OptimalK);

figure;                                           % figure;
plot(X(idx==1,1), X(idx==1,2), 'r.','MarkerSize', 12)
hold on                                           % hold on
plot(X(idx==2,1), X(idx==2,2), 'b.','MarkerSize', 12)
plot(C(:,1), C(:,2), 'kx', 'MarkerSize',15,'LineWidth', 3)
legend('Cluster 1','Cluster 2', 'Centroids', 'Location','SE')
title 'Cluster Assignments and Centroids'
grid;
hold off

MM = 602;
NN = 99;
GLUCOSE = table2array(T(MM:MM+NN, 2));        % Glucose is located at column 2 in database
BMI = table2array(T(MM:MM+NN, 6));            % BMI is located at column 6 in database
XX = [GLUCOSE BMI];
[icdx, C] = kmeans(XX, eva.OptimalK);

OUTPUT = table2array(T(MM:MM+NN, 9));         % Get the Outcome column – Labels
icdx = changem(icdx, 0, 2);

msg =['sum(icdx == OUTPUT) = ', num2str(sum(icdx== OUTPUT))];
disp(msg)
if (sum(icdx == OUTPUT) < 80)
    disp('Enter the if loop...')
    icdx = ~(icdx);
end
CR = ['The correct cluster rate is ', num2str(sum(icdx == OUTPUT)), '%'];
disp(CR)
```

Fig. 8.36   The codes used to evaluate the clustering result

(6) We like to use another 100 data in our **Diabetes.csv** dataset, from rows 601 through 700, to evaluate the clustering result with the help of the output label column, **Outcome**, in the dataset. One point is that the first row in our dataset is used as the title row and the real data starts from the second row. Thus the starting row number is 602 with an offset of 99.

(7) To build a new input testing input matrix, get two columns, **Glucose** and **BMI**, and assign them to another local variable **XX**.

(8) Perform another clustering process for these new testing data by calling the **kmeans**() function. The clustering results, which are a 100 × 1 vector and contain either 1 or 2 cluster numbers, are saved in the variable **icdx**.

(9) Now we need to pick up 100 related output labels for those 100 testing or evaluation rows from the Outcome column and assign them to a local variable **OUTPUT**.

(10) To evaluate the clustering results, we need to compare the clustering results (clustered number) with the real output labels to see how close the clustered number and the output labels. The reason for doing that is, the clustering results should divide all observations into two groups: (1) for Diabetes and (2) for Non-Diabetes. Therefore, if we compare these results with the output labels to see how much they can be matched, we can assess and evaluate the clustering results. However, before we can do that comparison, one issue is that the clustering results contained two clustering numbers, 1 and 2. But the values in the output labels are composed of two Boolean values, 1 or 0. Thus, a conversion is necessary to convert all cluster number 2 in the clustering results to 0. The function **changem()** is used for that purpose. One point to be noted is that the MATLAB Mapping Toolbox needs to be installed if the function **changem()** is used.

(11) These two coding lines are used to display and track the comparison result between the clustering results and the output labels. A decision structure is used for checking the comparison results.

(12) As we mentioned before, the clustering process is a random process. This means that sometimes the clustering uses positive logic, but some other times it uses negative logic. In summary, sometimes the clustering considers the 1 as true and 0 as false. But some other times it considers 1 as false and 0 as true. Thus, sometimes the clustering results would be opposite. To avoid this mix-logic situation to occur, we need to check the comparison results between the clustering results and the output labels. If the matched number is too small, which means that a negative logic has occurred, we use a **disp()** function to indicate that happened. The criterion for this low limit can be obtained from real processing. But generally the clustering result should be better than 80%.

(13) To solve that negative logic issue, we need to inverse the clustering results to make the opposite results back to the correct ones. An inverting operation (~) is performed for that purpose.

(14) Finally, the evaluation result is displayed with percentage number.

Now run this project, and the running results are shown in the Command window. It can be found that the evaluated clustering result is 82%.

```
NumObservations: 600
InspectedK: [1 2 3 4 5 6]
CriterionValues: [NaN 0.7031 0.5995 0.5705 0.6050 0.5624]
OptimalK: 2
sum(icdx == OUTPUT) = 82
The correct cluster rate is 82%
```

For this kind of clustering result evaluation, we need to use the output labels to help us do this job. In other words, we need to use a dataset that is suitable for supervised learning algorithms to meet our needs. In the next section, we discuss how to do this evaluation without using the output labels in a dataset. In other words, we can use a pure dataset that is only suitable for unsupervised learning algorithms.

8.5.5.3   Evaluate the Clustering Results Without Using Output Labels in Input Datasets

In fact, we can still use our **Diabetes.csv** dataset to do this kind of evaluation, and the key point is that we do not need to use the output labels at all. It is equivalent to using a dataset that is only suitable for unsupervised learning algorithms.

The basic idea for doing this kind of evaluation is to compare the mean values of each clustered observation with the mean values of each testing observation. In other words, we need to check the similarity between the clustered observations and the testing observations. If the former is close to the latter, the testing observations belong to the same cluster as those clustered observations. Otherwise they are different clusters.

The operational procedure for this kind of evaluation is:

(1) Evaluate the optimal cluster number by using the **evalclusters()** function.
(2) Then perform clustering process for the training data. In this study, we used the top 600 observations (rows) in our **Diabetes.csv** dataset as the training data. The clustering result contained two clusters, 1 and 2, for all 500 observations.
(3) Now perform the second clustering process for our testing data by using the additional 100 rows in our **Diabetes.csv** dataset, from rows 502 to 601.
(4) Calculate the mean values for all observations in clusters 1 and 2, **m_XC1** and **m_XC2**.
(5) Here a trick is coming, and it is related to the positive logic or negative logic issue for the clustering process. As we mentioned, sometimes the clustering process uses the positive logic (1 = true and 0 = false), but sometimes it uses the negative logic (0 = true and 1 = false). To distinguish this, we need to check the mean values for two clusters. In our case, the mean value for cluster 1 (**m_XC1**) is greater than that of cluster 2 (**m_XC2**). Thus if this is not true, which means that a negative logic is used by the clustering process and we need to exchange the order of both mean values above. We also need to inverse all elements in two clustering results included in the icdx, which contained the clustered numbers for all testing data, due to this negative logic.
(6) Then we can build our testing clustering result **XC** by calculating the mean values for all observations in the testing data. Similarly, we need to take care of the negative logic issue for the clustering results and need to change the cluster number based on the comparison result between the larger mean values (**Glucose**) and small mean values (**BMI**).
(7) Finally we need to evaluate the clustering result by comparing our built XC with the clustering results we did in step 3 above to see how close they are. If the matching result is a small number, it means that a negative logic was used, and we need to inverse the clustering result performed in step 3 above.

(8) The evaluation result is displayed to finish our evaluation process.

Convert the above sequence of pseudo codes to real MATLAB codes (**Eval_Diabetes_NL_Func.m**), as shown in Fig. 8.37. Some explanation steps are provided below:

(1) After obtaining two features from our **Diabetes.csv** dataset, the function **evalclusters**() is used to get the optimal cluster number based on our top 500 rows

```matlab
% Evaluate the clustering result without using the output labels
% Name: Eval_Diabetes_NL_Func.m
% Input: Diabetes.csv dataset
% Output: Evaluated clustering results
% Date: Jul 9,2024

path = 'C:\\Artificial Intelligence Book\Students\Datasets\Diabetes Dataset\\Diabetes.csv';
T = readtable(path);
N = 500;

GLUCOSE = table2array(T(1:N, 2));        % Glucose is located at column 2 in database
BMI = table2array(T(1:N, 6));            % BMI is located at column 6 in database

X = [GLUCOSE BMI];
M = 1:N;

1   eva = evalclusters(X, 'kmeans', 'silhouette', 'KList', 1:6)
    [idx, C] = kmeans(X, eva.OptimalK);

2   figure;                                                          % figure;
    plot(X(idx==1,1), X(idx==1,2), 'r.','MarkerSize', 12)
    hold on                                                          % hold on
    plot(X(idx==2,1), X(idx==2,2), 'b.','MarkerSize', 12)           % plot(C(:,1), C(:,2), 'kx')

    plot(C(:,1), C(:,2), 'kx', 'MarkerSize',15,'LineWidth', 3)
    legend('Cluster 1','Cluster 2', 'Centroids', 'Location','SE')
    title 'Cluster Assignments and Centroids'
    hold off

3   MM = 502;
    NN = 99;
    GLUCOSE = table2array(T(MM:MM+NN, 2));       % Glucose is located at column 2 in database
    BMI = table2array(T(MM:MM+NN, 6));           % BMI is located at column 6 in database

    XX = [GLUCOSE BMI];
4   [icdx, C] = kmeans(XX, eva.OptimalK);

5   m_XC1 = mean(X(idx==1, :));
    m_XC2 = mean(X(idx==2, :));
6   if m_XC1 < m_XC2
        mXC = m_XC1;
        m_XC1 = m_XC2;
        m_XC2 = mXC;
    end

7   XC= zeros(100, 1);
    mup_Limit = 165;

8   for j = 1:NN+1
        m = mean((XX(j, 1) + XX(j, 2)));
9       if m > mup_Limit
            XC(j) = 1;
        else
            XC(j) = 2;
        end
    end

10  if sum(XC == icdx) <50
        disp('Enter the if loop...')
        icdx = reverse(icdx, NN+1);
    end

11  sum(XC == icdx)

12  CR = ['The correct clustering rate is ', num2str(sum(XC==icdx)), '%'];
    disp(CR)
```

Fig. 8.37 The codes used to evaluate the clustering results without using labels

or observations in our dataset. Then the first clustering process is performed by using the **kmeans()** method to group the top 500 observations into two clusters

(2) A figure is used to plot the distributions of all original observations with cluster centers.

(3) A testing dataset is built by taking the **Glucose** and the **BMI** columns with the additional 100 rows from the dataset **Diabetes.csv** and assigned to a local variable **XX**.

(4) The second clustering process is executed by clustering the testing data by using the **kmeans()** method.

(5) Now we need to calculate the mean values for observations clustered in clusters 1 and 2, **m_XC1** and **m_XC2**. Those clustered results are coming from the first clustering process for the top 500 observations.

(6) We need to check the logic used for the first clustering process before we can continue. Based on our case, the mean value for the first cluster (group) should be greater than that of the second cluster. But if it is not, which means that a negative logic was used, then we need to exchange these two mean values.

(7) Now we need to build our testing clustering result **XC** that should be a 100 × 1 array or a vector containing 100 clustered numbers for all 100 observations. Also we can set up the upper limit for the criterion to distinguish the **Glucose** and the **BMI** features. In fact, this limit can be determined by the actual applications with related datasets.

(8) A **for()** loop is used to calculate the sum of mean values for each observation or row by row since a sum of means is more powerful compared with a mean for a single feature.

(9) If the sum of a mean value is greater than the upper limit, it means that this observation should belong to the **Glucose** group, thus a cluster number 1 is assigned to that row on the **XC**. Otherwise a cluster number 2 is assigned to the **XC** since it should belong to the **BMI** feature.

(10) Now we need to perform the evaluation for the clustering result for the testing data by comparing the built clustering result with the actual clustering results in step 4 above to see the similarity or how close they are. The function **sum(XC == icdx)** is equivalent to getting the matched numbers whose values are equal between two vectors. Regularly this number should be >50. But if it is not, it means that the second clustering process used the negative logic. In that case, we need to reverse the clustered numbers between 1 and 2. This can be done by calling a customer-built function **reverse()**, and the detailed codes for that function will be discussed later.

(11) This coding line is used to display the similarity between the built clustering result and the actual clustering result. The larger this number, the better the evaluation result.

(12) Finally this evaluation result is displayed to complete this evaluation process.

The detailed codes for the function **reverse()** are shown in Fig. 8.38.

```matlab
function cidx = reverse(x, n)
  for i = 1:n
   if x(i) == 1
     x(i) = 2;
   else
     x(i) = 1;
   end
   cidx = x;
  end
end
```

Fig. 8.38 The detailed codes for the function reverse()

The function of this piece of codes is simple and straightforward; it just checks all cluster numbers, changes them from 1 to 2 or vice versa, and returns the reversed result.

Now run the project, and the running results are shown in the Command window.

```
NumObservations: 500
InspectedK: [1 2 3 4 5 6]
CriterionValues: [NaN 0.7105 0.6013 0.5753 0.6011 0.5566]
OptimalK: 2
Enter the if loop...
ans = 97
The correct clustering rate is 97%
```

Due to the random property of the clustering process, the evaluation clustering result may be changed in a range, from 77% to 97%.

Three completed MATLAB projects, **Find_Optimal_NCluster.m**, **Eval_Diabetes_Func.m** and **Eval_Diabetes_NL_Func.m** can be found on the Springer ftp site in the folder: **Students\Class Projects\Chapter 8**.

Next let's concentrate our discussions on association rules, which is another unsupervised learning algorithm.

8.6   Using MATLAB Functions to Build Association Rules Models

As we discussed in Sect. 8.2, association rules are made by searching data for frequent **if-then** patterns and by using a certain criterion under **Support** and **Confidence** to define what the most important relationships are. Support is the evidence of how frequently an item appears in the given data, as Confidence is defined by how many times the **if-then** statements are found to be true.

Relatively speaking, using confidence in association rule mining is a great way to bring awareness to data relations. Its greatest benefit is highlighting the relationship between particular items to one another within the set, as it compares co-occurrences of items to the total occurrence of the antecedent in the specific rule. However, confidence is not the optimal method for every concept in association rule mining. The disadvantage of using it is that it does not offer multiple different outlooks on the associations [11].

Unlike support, for instance, confidence does not provide the perspective of relationships between certain items in comparison to the entire dataset, so while bread and milk, for example, may occur 50% of the time for confidence, it only has a support of 0.25 (25%). This is why it is important to look at other viewpoints, such as Support × Confidence, instead of solely relying on one concept incessantly to define the relationships.

In this section, we try to discuss how to use MATLAB functions to perform clustering analysis for some real data by using the association rules. However, unfortunately MATLAB did not provide any function to perform clustering analysis for the association rules algorithm. Therefore we need to build some real example projects by using our codes to illustrate how to use some basic codes to build an association rule model to perform the clustering process.

We still like to use the **Diabetes.csv** as our data source, but some modifications are needed to enable us to use it as our desired dataset to meet the requirement of clustering process with association rules algorithm.

8.6.1   *Modify the Diabetes.csv Dataset to Make It as Our Desired Dataset*

Most datasets used for association rules are binary datasets, thus we need to modify the current **Diabetes.csv** dataset and convert it to an equivalent binary dataset. To perform a reasonable quantitative operation to convert an analog dataset to an equivalent binary dataset, we like to use the median value for each column as the borderline.

In this example project, we like to use three columns, **Glucose, BMI,** and **Outcome**, as a possible association rule, {**Glucose, BMI**} → {**Outcome**}. This rule means that if both **Glucose** and **BMI** occurred, how many percentage of a person can be diagnosed as a Diabetes patient? In this project, we will calculate the **Support, Confidence,** and **Lift** to try to build and check this rule. Relatively speaking, the **Lift** provides a better judgment on the effectiveness of an association rule compared with the Support and Confidence.

Now let's first modify our dataset to make it a binary dataset containing three columns, **Glucose, BMI,** and **Outcome**. Create a new Script file, name it as **Modify_Diabetes.m**, and enter the codes shown in Fig. 8.39. Another two columns, **Blood Pressure** and **Age**, are also involved in this project since we may need them to build and test another association rule later. Let's have a closer look at this piece of codes to see how it works.

```matlab
% Modify the Diabetes.csv to make it as boolean dataset
% Name: Modify_Diabetes.m
% Input: Source dataset - Diabetes.csv
% Output: Target dataset - Diabetes_AR.csv
% Jul 10, 2024
 1 s_path = 'C:\\Artificial Intelligence Book\Students\Datasets\Diabetes Dataset\\Diabetes.csv';
   t_path = 'C:\\Artificial Intelligence Book\Students\Datasets\Diabetes Dataset\\Diabetes_AR.csv';
 2 T = readtable(s_path);
   N = 200;

 3 GLUCOSE = table2array(T(1:N, 2));      % Glucose is located at column 2 in database
   BMI = table2array(T(1:N, 6));          % BMI is located at column 6 in database
   BP =  table2array(T(1:N, 3));          % Blood pressure is located at column 3 in database
   AGE =  table2array(T(1:N, 8));         % Age is located at column 8 in database
   OUTPUT = table2array(T(1:N, 9));       % Outcome is located at column 9 in database

 4 mid = median([GLUCOSE BMI BP AGE])

 5 for i = 1:N
     if GLUCOSE(i, :) >= mid(:, 1)
        GLUCOSE(i, :) = 1;
     else
        GLUCOSE(i, :) = 0;
     end
 6   if BMI(i, :) >= mid(:, 2)
        BMI(i, :) = 1;
     else
        BMI(i, :) = 0;
     end
 7   if BP(i, :) >= mid(:, 3)
        BP(i, :) = 1;
     else
        BP(i, :) = 0;
     end
     if AGE(i, :) >= mid(:, 4)
        AGE(i, :) = 1;
     else
        AGE(i, :) = 0;
     end
   end
 8 Tab = {'Glucose', 'BMI', 'BloodPressure', 'Age', 'Outcome'};
 9 X = [GLUCOSE BMI  BP AGE OUTPUT];
10 writecell(Tab, t_path);
11 writematrix(X, t_path, 'WriteMode', 'append');
```

Fig. 8.39 The codes used to modify the Diabetes.csv dataset

(1) We need to use two full paths, the source and the target path. The former is our original **Diabetes.csv** dataset and the latter is our modified dataset, **Diabetes_ AR.csv**. You may need to use your actual folders to replace them if you saved your dataset in different folders.

(2) Read the original dataset and assign it to a local variable **T**. The number of observations we like to use is 200 (rows) from the original dataset. You can change this number if you like to use a different number.

(3) Pick up five columns (features), **Glucose**, **BMI**, **BP** (Blood Pressure), **Age,** and **Outcome**, from the dataset and assign them to related variables. The function **table2array**() is needed since we need to convert all cells to numeric values.

(4) The function **median**() is executed to find the median values for four input columns since we need them as the criteria to convert four columns to binary values.

(5) A **for**() loop is used to get each feature and convert it to a binary value based on the comparison result between it and the criterion, which is the median

value **mid(:, 1)**. The **GLUCOSE** is a 200 × 1 vector and it has only one column. The result of this first **for()** loop is to convert each row on **GLUCOSE** to either 1 if its value is greater than the median value or 0 if its value is less than the median.

(6) Perform similar operations for **BMI** column to convert it to binary numbers.

(7) For **BP** and **AGE**, do the same process to convert them to Boolean values.

(8) Now we have finished the conversion or modification for our dataset, next we need to write these modified results into our target dataset. First we need to create a title (**Tab**) to include all columns' names.

(9) Also arrange the modified results into a new matrix **X** with five columns by adding the **OUTPUT** column to it.

(10) Use the function **writecell()** to write our title at the top of our target dataset.

(11) Then call the function **writematrix()** to write the modified results into the target dataset.

Now run the project, and the running results are shown on the Command window, which are median values for four columns. The modified dataset is generated, and you can find it in the target folder.

Next let's build our project to cluster some features by using the association rules for the modified Diabetes dataset.

8.6.2 *Build the Diabetes Project to Identify Association Rules for Some Features*

As we mentioned, first we like to use three columns, **Glucose, BMI,** and **Outcome**, as a possible association rule, **{Glucose, BMI}** → **{Outcome}**. This rule means that if both **Glucose** and **BMI** occurred, how many percentage of a person can be diagnosed as a Diabetes patient. In this project, we will calculate the **Support, Confidence,** and **Lift** to try to identify and check this rule. Relatively speaking, the **Lift** provides a better judgment on the effectiveness of an association rule compared with the Support and Confidence.

Create a new Script file, name it as **Diabetes_Association_Rule.m,** and enter the codes shown in Fig. 8.40 into that file. Let's have a closer look at this piece of codes to see how it works.

(1) The target or modified dataset, **Diabetes_AR.csv,** is declared first. The dataset is also read out and assigned to a local variable **T**.

(2) The total number of observations is defined as 200 since we only need to use the top 200 observations from our modified dataset.

(3) Some local variables, such as **A, B,** and **C,** are declared and initialized.

(4) Three columns or features are extracted and converted to numeric values by using the function **table2array()**, which includes the inputs, **GLUCOSE** and **BMI,** as well as the consequent **OUTPUT**.

```matlab
% Find association rules for modified dataset - Diabetes_AR.csv
% Name: Diabetes_Association_Rule.m
% Input: Modified dataset - Diabetes_AR.csv
% Output: Association rules
% Jul 10, 2024
1  path = 'C:\\Artificial Intelligence Book\Students\Datasets\Diabetes Dataset\\Diabetes_AR.csv';
   T = readtable(path);
2  N = 200;
3  A = 0;
   B = 0;
   C = 0;
4  GLUCOSE = table2array(T(1:N, 1));          % Glucose is located at column 1 in database
   BMI = table2array(T(1:N, 2));              % BMI is located at column 2 in database
   OUTPUT = table2array(T(1:N, 5));           % Outcome is located at column 5 in database
5  X = [GLUCOSE BMI OUTPUT];
6  for i = 1:N
     if (X(i, 1) == 1) && (X(i, 2) == 1) && (X(i, 3) == 1)
        A = A + 1;
     end
7    if (X(i, 1) == 1) && (X(i, 2) == 1)
        B = B + 1;
     end
8    if (X(i, 3) == 1)
        C = C + 1;
     end
   end
9  Support = A/N                   % Support calculation
   Confidence = A/B                % Confidence calculation
10 D = B/N;                        % Support X
   E = C/N;                        % Support Y
11 AR = Support * Confidence       % Support * Confidence
12 Lift = Support / (D*E)          % Lift calculation
```

Fig. 8.40 The codes used to identify association rules

(5) The antecedent $\mathbf{X} = \{\mathbf{GLUCOSE, BMI}\}$ and the consequent $\mathbf{Y} = \{\mathbf{OUTPUT}\}$ are arranged to get a matrix X.

(6) A **for**() loop is used to search all three observations (rows) where each feature (column) on that row is 1, or $\{\mathbf{GLUCOSE, BMI, OUTPUT}\} = \{\mathbf{1, 1, 1}\}$, which means that the number of transactions containing $\mathbf{X}$ and $\mathbf{Y}$ and they occurred at the same time. If it is, increase the variable $\mathbf{A}$ by 1. The purpose of this first **if** structure is to detect and record the total number of transactions including $\mathbf{X}$ and $\mathbf{Y}$ that occurred simultaneously.

(7) Similarly, the second **if** structure is to search the first two observations where each column on that row is 1, $\{\mathbf{GLUCOSE, BMI}\} = \{\mathbf{1, 1}\}$, which means that the total number of transactions containing $\mathbf{X}$ only. If it is, increase the variable $\mathbf{B}$ by 1. The purpose of this second **if** structure is to detect and record the total number of transactions including the antecedent or $\mathbf{X}$ only.

(8) The third **if** structure is used to detect and record the third column ($\mathbf{OUTPUT}$) where its value is 1. If it is, increase the local variable $\mathbf{C}$ by 1. The purpose of this third **if** structure is to detect and record the total number of transactions including the consequent or the $\mathbf{OUTPUT}$ only.

(9) When the **for**() loop is done, we can calculate the Support by dividing the $\mathbf{A}$ by the total number of transactions $\mathbf{N}$, and Confidence dividing the variable $\mathbf{A}$

by **B**. Refer to Eqs. (8.9) and (8.8) in Sect. 8.2 to get more details for these calculations.

(10) Similarly, we can calculate the Support **X** and Support **Y** by dividing **B** by **N**, and **C** by **N**, respectively. Refer to Eq. (8.9) in Sect. 8.2 to get details for these calculations.

(11) Now we can get the product of Support and Confidence since it is more meaningful for identifying the association rules compared with by using Support and Confidence individually.

(12) Finally we can get the Lift by dividing the Support by the product of **D** and **E**. Refer to Eq. (8.10) in Sect. 8.2.1 to get more details for this calculation.

Now run the project and the running results are shown in the Command window, as below:

```
Support = 0.1950
Confidence = 0.6290
AR = 0.1227
Lift =1.6774
```

It can be found from these results that it is relatively hard to evaluate a strong relationship between **X** and **Y** only based on Support or Confidence since these two numbers have a significant difference. The product of these two results looks better but still not enough to derive any strong relationship between the antecedent and the consequent. The Lift looks better in detecting a strong relationship since its value is >1, which means that the degree to which those two occurrences are dependent on one another, and makes those rules potentially useful for predicting the consequence in future datasets.

Two completed projects, **Modify_Diabetes.m** and **Diabetes_Association_Rule.m**, can be found on the Springer ftp site in the folder: **Students\Class Projects\Chapter 8**.

We can try to use another set of data, such as Blood Pressure (**BP**) and Age, as antecedent **X = {BP, AGE}** and **Y = {OUTPUT}** as consequent to identify another possible association rule. A completed project named **Diabetes_Association_Rule2.m** can be found on the Springer ftp site in the folder: **Students\Class Projects\Chapter 8**.

8.7   Using MATLAB Functions to Build Apriori Algorithm Models

As shown in Fig. 8.1, under the unsupervised learning category, there are two algorithms, Association Rules and Apriori Algorithm, that are popular and widely implemented in machine learning research.

Similar to association rules, the Apriori algorithm is also used to try to find some strong rules based on the frequency of the item sets to appear in a dataset or a database. If some item sets appear in many times or at high frequency in a dataset, we say that those item sets are highly related together with strong relationships. Otherwise, those item sets are not.

The Apriori algorithm used an iteration sequence to group item sets with different numbers, $k = 1, 2, 3, \ldots$ and group those item sets that have similar properties until no more similarity exists in the dataset. Let's have a closer look at this algorithm.

8.7.1   Introduction to Apriori Algorithm

Apriori algorithm uses a *bottom up* approach, where frequent subsets are extended one item at a time (a step known as *candidate generation*), and groups of candidates are tested against the data (threshold). The algorithm terminates when no further successful extensions are found.

Apriori uses breadth-first search and a Hash tree structure to count candidate item sets efficiently. It generates candidate item sets of length k from item sets of length k - 1. Then it prunes or removes the candidates which have an infrequent subpattern. According to the downward closure lemma, the candidate set contains all frequent k-length item sets. After that, it scans the transaction database to determine frequent item sets among the candidates [31].

Now let's use an example to illustrate how to use the Apriori algorithm to perform its function to group item sets to find the relationships.

Consider a sample dataset shown in Table 8.7 and we will find frequent item sets and generate association rules for them.

We will use Apriori algorithm to determine the frequent item sets of this dataset. To do this, we will say that an item set is frequent if it appears in at least two transactions of the dataset: this means that our selected *support threshold* is 2.

The first step of Apriori is to count up the number of occurrences, called the support, of each member item separately. Let's start our iteration by setting the step as $k = 1$. By scanning the dataset for the first time, we obtain the following result shown in Table 8.8.

Table 8.7 A sample item set with six items

Observation	Item Sets
O1	I1, I2, I3
O2	I2, I3, I6
O3	I2, I4
O4	I1, I3
O5	I1, I2, I4
O6	I1, I3
O7	I1, I2, I3, I5
O8	I1, I2, I5

Table 8.8 The first step on Apriori algorithm

Observation	Support Count
I1	6
I2	6
I3	5
I4	2
I5	2
I6	1

Table 8.9 The item set L1

Observation	Support Count
I1	6
I2	6
I3	5
I4	2
I5	2

Table 8.10 The extended candidate set C2

Observation	Support Count
I1, I2	4
I1, I3	4
I1, I4	1
I1, I5	2
I2, I3	3
I2, I4	2
I2, I5	2
I3, I4	0
I4, I5	0

Table 8.8 contains the number of appearances for each item in the dataset, and this table is called the first extension or the first candidate set C1.

The second step is to compare the item's support count in the first candidate set C1 with minimum support count or support threshold (= 2 in this example). If a support count of candidate set items is less than support threshold, then remove those items since their relations to others can be neglected. It can be found from Table 8.8 that item 6 has a support count 1, which is less than the support threshold 2, thus it can be removed from the dataset and the resulting L1 is shown in Table 8.9.

Now set $k = 2$ to generate candidate set C2 using L1. This step is to generate a list of all pairs of frequent items. The number k indicates how many items should be included in an extended item set. These pairs should be combinations and cover all possible combinations among all items. The sequence should be {I1, I2}, {I1, I3}, {I1, I4}, {I1, I5}, {I2, I3}, {I2, I4}, {I2, I5}, {I3, I4}, {I3, I5}, and {I4, I5}. For this sequence, we get the second candidate set C2, as shown in Table 8.10.

Still comparing each support count of all items in C2 with the support threshold (= 2), we can get the L2, which is shown in Table 8.11.

Now let's continue this process ($k = 3$) by generating the candidate set C3 using L2. This means that we need to combine each three items as an item set and count

Table 8.11 The item set L2

Observation	Support Count
I1, I2	4
I1, I3	4
I1, I5	2
I2, I3	3
I2, I4	2
I2, I5	2

Table 8.12 The extended candidate set C3

Observation	Support Count
I1, I2, I3	2
I1, I2, I5	1
I2, I3, I6	0

its support count. The resulting C3 is shown in Table 8.12. A point to be noted is that no item set on this three-item combination should contain any item that has been removed from C2. For example, the items {I1, I4}, {I3, I4}, and {I4, I5} in C2 have been removed in L2, thus C3 cannot contain any three-item component that includes those combinations {I1, I4}, {I3, I4}, and {I4, I5}.

It can be found in Table 8.7 that five three-item elements exist. However, the item {I1, I2, I4} cannot be counted since it contains {I1, I4} and it has been removed from L2.

Now compare all items in C3 with threshold value 2, two bottom items {I1, I2, I5} and {I2, I3, I6} should be removed since their support count is <2. The final L3 contained only one item {I1, I2, I3}. Thus, we can conclude that the only item {I1, I2, I3} has a value of 2 and it is greater than or equal to the support threshold value 2.

Now let's continue to generate candidate set C4 using L3. Check whether all subsets of these item sets are frequent or not. Here item set {I1, I2, I3, I5} contains {I1, I3, I5}, so it is not frequent. So no item set in C4. We stop here because no frequent itemsets are found further.

The final result is that only item {I1, I2, I3} has a strong support count, and it means that these three items are closely related with strong association rules.

To confirm our result, we can calculate the confidence values for these three items as below:

$$\{I_1 \cap I_2\} \rightarrow \{I_3\} = \frac{\text{Support}\left(I_1 \cap I_2 \cap I_3\right)}{\text{Support}\left(I_1 \cap I_2\right)} = \frac{2}{4} = 50\%$$

$$\{I_1 \cap I_3\} \rightarrow \{I_2\} = \frac{\text{Support}\left(I_1 \cap I_2 \cap I_3\right)}{\text{Support}\left(I_1 \cap I_3\right)} = \frac{2}{4} = 50\%$$

$$\{I_2 \cap I_3\} \rightarrow \{I_1\} = \frac{\text{Support}\left(I_1 \cap I_2 \cap I_3\right)}{\text{Support}\left(I_2 \cap I_3\right)} = \frac{2}{3} = 67\%$$

If 50% is considered as the minimum confidence, then rules 1 and 3 can be considered as strong association rules.

Now we like to use a real project to illustrate how to use Apriori algorithm to identify some rules by using our **Diabetes_AR.csv** dataset.

8.7.2   *Using MATLAB Function to Build Apriori Model for Diabetes Dataset*

Due to relatively complicated Apriori algorithm, MATLAB did not provide any function or App to help users build related application projects. Therefore we like to use a simple example project combined with our modified Diabetes dataset, **Diabetes_AR_csv**, to illustrate how to use this algorithm to develop a real application.

To make it simple and illustrative, we only use three features or three columns, **Glucose, BMI,** and **Outcome**, in the modified dataset **Diabetes_AR.csv** to illustrate how to find hidden relationships or rules among these three features.

Create a new Script file, name it as **Diabetes_Apriori_Rule.m,** and enter the codes shown in Fig. 8.41 into that file. Let's have a closer look at this piece of codes to see how it works.

(1) The full path of our modified dataset, **Diabetes_AR.csv**, is declared and the dataset I read out and assigned to a local variable **T**.

(2) The numbers of the observations (rows) and the features (columns) are defined.

(3) Some local variables, including the subscripts **m12, m13, m23** that means the removed pair of items in dataset, **pairRemove** that is a flag to indicate a pair removing is occurred, **SThreshold** of the Apriori algorithm, and all related Supports, are declared first.

(4) Three columns are extracted from the modified dataset as our item sets in this example, and they are set into a matrix as our dataset **X**. The resulting matrix **X** is a 20 × 3 dimensional matrix.

(5) A **for()** loop is used to check all items for each column whether its value is 1. This operation is to calculate the support for each item.

(6) If it is 1, the candidate set C1, or **SupportC1** is increased by 1.

(7) Then another **for()** loop is utilized to check whether any three items in the C1 is less than the threshold. A negative logic is used for this selection process. If any item in C1 is greater than the threshold, its value is copied and stored into the C2. Otherwise a 0 is stored in the C2.

(8) We only need to keep those items whose values are greater than threshold in C1, which are also stored in C2, thus we need to remove those values of 0 from

```matlab
% Find association rules by using Apriori Algorithm for dataset - Diabetes_AR.csv
% Name: Diabetes_Apriori_Rule.m
% Input: Modified dataset - Diabetes_AR.csv
% Output: Association rules
% Jul 12, 2024
1   path = 'C:\\Artificial Intelligence Book\Students\Datasets\Diabetes Dataset\\Diabetes_AR.csv';
    T = readtable(path);
2   N = 20; M = 3;
3   m12 = 0; m13 = 0; m23 = 0; pairRemove = 0; SThreshold = 5;
    supportC1 = zeros(M, 1); supportC2 = zeros(M, 1);
    supportC12 = 0; supportC13 = 0; supportC23 = 0; supportC123 = 0;

4   GLUCOSE = table2array(T(1:N, 1));          % Glucose is located at column 1 in database
    BMI = table2array(T(1:N, 2));              % BMI is located at column 2 in database
    OUTPUT = table2array(T(1:N, 5));           % Outcome is located at column 5 in database
    X = [GLUCOSE BMI OUTPUT];

5   for i = 1:M
       for j =1:N
6        if (X(j, i) == 1) supportC1(i) = supportC1(i) + 1; end
       end
    end

7   for k = 1:M
        if (supportC1(k, 1) > SThreshold) supportC2(k, 1) = supportC1(k, 1); end
    end
8   supportC2(supportC2 == 0) = []

9   for j =1:N
       if (X(j, 1) == 1) && (X(j, 2) == 1)  supportC12 = supportC12 + 1;  end       % columns 1 & 2
       if (X(j, 1) == 1) && (X(j, 3) == 1)  supportC13 = supportC13 + 1;  end       % columns 1 & 3
       if (X(j, 2) == 1) && (X(j, 3) == 1)  supportC23 = supportC23 + 1;  end       % columns 2 & 3
    end

10  if (supportC12 < SThreshold) m12 = m12 + 1; end
    if (supportC13 < SThreshold) m13 = m13 + 1; end
    if (supportC23 < SThreshold) m23 = m23 + 1; end
11  if ((m12 && m13 && m23) == 0) disp('No double pair removed');
12  else pairRemove = 1; end

13  supportC3 = X;                    % for testing purpose, can set pairRemove = m12=m13=m23 = 1

14  for j =1:N
15     if (m12 ~= 0)
          if (X(j, 1) == 1) && (X(j, 2) == 1) supportC3(j, :, :) = 5
          end
16     else if (m13 ~= 0)
          if (X(j, 1) == 1) && (X(j, 3) == 1) supportC3(j, :, :) = 5
          end
17     else if (m23 ~= 0)
          if (X(j, 2) == 1) && (X(j, 3) == 1) supportC3(j, :, :) = 5
          end
18     else if (X(j, 1) == 1) && (X(j, 2) == 1)  && (X(j, 3) == 1)              % columns 1 & 2 & 3
          supportC123 = supportC123 + 1
       end
     end
    end
    end
    end
19  if (m12 == 1 || m13 == 1 || m23 == 1)
       supportC3(supportC3 == 5) = [];
20     [row, col] = size(supportC3);
21     supportC3 = reshape(supportC3, col/3, 3)
    end
```

Fig. 8.41 The codes to build a simple project with Apriori algorithm

C2 since that means that those values are smaller than the threshold. A blank
symbol [] is assigned to those values that are 0 in C2. In this way, we only keep
those items whose values are greater than threshold in C2.

(9) Next we need to calculate the support for each pair of items in C2, such as
   Glucose & BMI, Glucose & OUTPUT, and **BMI & OUTPUT** (I1 & I2, I1

& I3, and I2 & I3). If one pair or both columns are 1, the related support is increased by 1.

(10) Now we need to check whether any pair of items whose value is less than the threshold value (5 in this project) or not. If it is, the related subscript is increased by 1, and this subscript will work as an indicator to indicate how many pair items whose values are less than the threshold and should be removed on the next candidate set C3.

(11) If all three subscripts are 0, it means that all pair items in C2 are greater than the threshold and no one pair needs to be removed in C3. A message is displayed to indicate this situation.

(12) Otherwise the **pairRemove** flag is set to 1 to indicate that some pairs of items are less than the threshold and have been removed from C3.

(13) Next let's calculate the candidate set C3. To do that, we assign the original matrix **X** to the local variable **supprotC3**.

(14) A **for**() loop is used to check whether any pair of items should be removed from C3.

(15) First check the first pair, **Glucose** and **BMI** or subscript **m12**. If its value is not 0, which means that some pair should be removed. To indicate this, a value 5 is assigned to the first row on **supportC3**.

(16) Similarly, we check the second pair, **Glucose** and **OUTPUT** or subscript **m13**. If its value is not 0, it means that some pair should be removed. To indicate this, a value of 5 is assigned to the first row on **supportC3**.

(17) Perform a similar operation to the third pair, **BMI** and **OUTPUT**.

(18) If all the above situations do not match, it means that no pair is less than the threshold and no pair removal has happened. The final support, **supportC123**, is increased by 1.

(19) If any of the pair of items is removed, we need to remove all values of 5 from the **supportC3** since we put 5 when an item removal occurred in steps 15–17.

(20) After this removing operation, the **supprotC3** changed to a 1 × 42 vector.

(21) But we need it to be an **m × 3** matrix, where **m** is the number of rows after those values of 5 are removed. Thus we need to get the size, or exactly to get the column (col) from the changed **supportC3** and the value of **m** should be equal to **col/3**. Finally we need to reshape the **supportC3** to a **col/3 × 3** matrix.

It is a little weird why the **supportC3** is changed to a 1 × 42 vector after that blank symbol [] is

sent to it to clean up all rows whose values are 5 in **supportC3** matrix. This means that the blank operation may have some issues and need special attention when using it.

Now run the project and the running results are shown in the Command window.

```
supportC2 =
    13
     8
    13
No double pair removed
supportC123 = 1
supportC123 = 2
supportC123 = 3
supportC123 = 4
supportC123 = 5
```

Thus, we can see that all three items occurred at the same time which is 5, and this means that this is the final result and no further candidate set could be involved. Let's calculate the confidences for these final five item sets.

$$\{\text{Glucose} \cap \text{BMI}\} \rightarrow \{\text{OUTPUT}\} = \frac{\text{Support}(\text{Glucose} \cap \text{BMI} \cap \text{OUTPUT})}{\text{Support}(I_1 \cap I_2)} = \frac{5}{6} = 83\%$$

$$\{\text{Glucose} \cap \text{OUTPUT}\} \rightarrow \{\text{BMI}\} = \frac{\text{Support}(\text{Glucose} \cap \text{BMI} \cap \text{OUTPUT})}{Support(\text{Glucose} \cap \text{BMI})} = \frac{5}{10} = 50\%$$

$$\{\text{BMI} \cap \text{OUTPUT}\} \rightarrow \{\text{Glucose}\} = \frac{\text{Support}(\text{Glucose} \cap \text{BMI} \cap \text{OUTPUT})}{\text{Support}(\text{BMI} \cap \text{OUTPUT})} = \frac{2}{5} = 40\%$$

The items relations 1 and 2, and 1 and 3, can be considered two stronger relations if 50% confidence and above can be thought of as the bottom line for good relations.

A completed project, **Diabetes_Apriori_Rule.m**, can be found on the Springer ftp site in the folder: **Students\Class Projects\Chapter 8**.

8.7.3 *Unsupervised Learning and Supervised Learning*

As we discussed at the beginning of this chapter, unsupervised learning is a type of machine learning algorithms that learn from input data without human supervision. Unlike supervised learning, in which a group of input-output data pair (input data and related labels) is needed to train the models, unsupervised machine learning models are given unlabeled data and allowed to discover patterns and insights without any explicit guidance or instruction.

Regularly, unsupervised learning algorithms can solve the following problems [32]:

(1) **Discovering patterns within unlabeled data:** Unlabeled data or data that do not have identifying characteristics or properties can be tricky for humans to understand. However, unsupervised learning algorithms can be used to evaluate

the data and detect patterns that might be too subtle or complex for humans to detect on their own.

(2) **Finding and constructing rules for current or future datasets:** Based on the patterns it releases, unsupervised learning algorithms can be used to build rules between or among the data. As more data are used, it can apply those same rules to the new data or even improve upon its existing rules based on the new data.

(3) **Understanding and categorizing data:** Today with a huge of block data available, it can sometimes be too challenging for humans to extract valuable findings from the data. With the help of unsupervised learning, we can think of unique categories for the data that might help professionals implement the data findings in useful ways.

(4) **Determining new and unique approaches:** When an unsupervised learning algorithm is used, it can think of radically original ideas or solutions. For example, unsupervised learning can play a human game and create strategies for that game unlike anything ever thought of or used by human players.

Based on these properties and specialties, unsupervised learning can be implemented in the following popular applications:

- Customer segmentations
- Anomaly detection and market segmentations
- Disease studies and detections
- Cyber security
- Speech recognition
- Recommender systems
- Dimensionality reduction

As with anything in this world, one good side always brings some bad points, or in other words, nothing is perfect. Unsupervised learning is also true for that word. The following factors need to be considered when using unsupervised learning compared with supervised learning [33]:

(1) The results provided by unsupervised learning models may be less accurate as input data do not contain labels as answer keys.

(2) The method requires output validation by humans, internal or external experts who know the field of research.

(3) The training process is relatively time-consuming because algorithms need to analyze and calculate all existing possibilities.

(4) More often than not unsupervised learning deals with huge datasets, which may increase the computational complexity.

(5) Unsupervised learning may cost organizations more than supervised learning because it can take longer for the former to develop rules or extract insights from the data. It might also become more costly if a company needs to hire experts to review the accuracy of unsupervised learning's algorithms, rules, or findings.

8.8 Chapter Summary

The main topic of this chapter is about introductions and discussions for another major and important branch involved in AI, unsupervised learning algorithms. Two main categories are included under the unsupervised learning umbrella, clustering and association rules.

The clustering algorithm is the task of dividing the unlabeled data or data points into different clusters such that similar data points fall in the same cluster than those which differ from the others. In simple words, the aim of the clustering process is to segregate groups with similar traits and assign them into clusters. Each cluster body can be considered as a group body where all objects or data points in a group have similar properties.

The Association rule is a technique used to uncover hidden relationships between variables in large datasets. In fact, association rule is a technique used to identify patterns or relations in large datasets. It involves finding relationships between variables in the data and using those relationships to make predictions or decisions. The goal of association rule is to uncover rules that describe the relationships between different items in the dataset.

Four components or algorithms can be considered as a member of clustering; they are:

(1) Exclusive clustering
(2) Overlapping clustering
(3) Hierarchical clustering
(4) Probabilistic clustering

MATLAB provides a set of App and functions to support unsupervised learning, and these App and tools/functions are involved in two major toolboxes:

(1) Statistics and Machine Learning Toolbox
(2) Deep Learning Toolbox

In addition to theoretical introductions and discussions, ten real example projects are involved in the first part, clustering algorithm and its implementations.

Besides the clustering, two other algorithms related to association rules and Apriori algorithm are also introduced and discussed with real example projects. Due to some reasons, MATLAB did not provide any App or direct function for the association rules and Apriori algorithms. To make these two algorithms clear and understandable, three real example projects are developed in MATLAB codes and used to illustrate how to use those algorithms to detect and identify possible hidden rules among item sets with a modified Diabetes dataset.

Unlike supervised learning, it is a difficult and even challenging task to evaluate an unsupervised learning algorithm effectively and accurately due to no output label available. A new approach is discussed in this chapter to help users quickly and easily evaluate some unsupervised learning algorithms with MATLAB codes.

Home Works

I. True/False Selections

______1. Machine learning contains two major components: supervised learning and unsupervised learning.

______2. Under the unsupervised learning umbrella, three subcomponents, clustering, association rules, and dimensionality reduction, exist.

______3. Both neural networks and deep learning can be applied and implemented in either supervised or unsupervised learning models to provide predictions or classifications.

______4. Clustering is the task of dividing the labeled data or data points into different clusters.

______5. Association rule is a technique used to uncover hidden relationships between variables in large datasets.

______6. The exclusive clustering is a kind of hard clustering in which data point exclusively belongs to one cluster.

______7. The K-Means clustering is a kind of soft clustering.

______8. Overlapping clustering is a kind of soft clustering in which a data point can belong to multiple clusters in some different degrees.

______9. The Gaussian Mixture Model (GMM) is one of the most commonly used probabilistic clustering methods.

_____10. The frequent item sets determined by Apriori algorithm are then used to determine association rules.

II. Multiple Choices

1. Four algorithms can be considered as a member of clustering, they are: ____________.

   a. Overlapping, Hierarchical, Probabilistic, Random
   b. Overlapping, Hierarchical, Probabilistic, Exclusive
   c. Overlapping, Random, Hierarchical, Exclusive
   d. None of the above

2. The following algorithm belongs to hard clustering, _________.

   a. Overlapping
   b. Probabilistic
   c. Fuzzy C-Mean
   d. Hierarchical

3. The Confidence is more likely to be a _________________ probability.

   a. Theoretical
   b. Classical
   c. Conditional
   d. Subjective

4. The Apriori algorithm can be used to identify _______________.

   a. Input and output relations
   b. Connections between inputs and outputs by labels
   c. Frequent item sets
   d. Random datasets

5. The Gaussian Mixture Model (GMM) is one of the commonly used
   __________ methods:

   a. Hard clustering
   b. Soft clustering
   c. Fuzzy K-Means clustering
   d. Random clustering

6. The results of hierarchical clustering can be plotted using a ______________
   function in MATLAB.

   a. dendrogram()
   b. hierarchical()
   c. plot()

7. A popular algorithm used for overlapping clustering is ______________.

   a. Hierarchical clustering
   b. Probabilistic clustering
   c. Gaussian mixture model
   d. Fuzzy clustering

8. To find the optimal K value on a K-Means clustering, two methods used are
   __________.

   a. Least square method, combination method
   b. Center of gravity method, Gaussian method
   c. Elbow method, least square method
   d. Silhouette method, elbow method

9. Which of the following statement is true? ______________.

   a. Association rule is to uncover hidden relationships between variables in
   large datasets
   b. Association rules intended to identify strong rules discovered in databases using some measures of interestingness
   c. Support is the evidence of how frequent an item appears in the given data
   d. All of them

10. Which of the following equation is correct to calculate the Confidence?
    ________.

   a. $Support(X \cap Y)/Support(Y)$
   b. $Support(X \cap Y)/Support(X)$
   c. $Support(X \cup Y)/Support(Y)$
   d. $Support(X \cup Y)/Support(X)$

III. Exercises

1. Explain the differences between the supervised learning and the unsupervised learning.
2. Provide a basic description about hard clustering and soft clustering.
3. List four types of clustering algorithms.
4. Provide a basic description about the Association rules algorithm.
5. Provide a basic description about the Apriori algorithm.

IV. Lab Projects

1. Using Neural Net Clustering App to build an unsupervised learning model with **Diabetes.csv** dataset. Refer to Sect. 8.4.2 to get more details for this building process. Name the Script as **Diabetes_Cluster_App.m**, and use the **nctool** command to open the ANN APPS. Then use that App to build the model. Export the trained model to the Workspace with any meaningful name.

Hint1: Follow Sect. 8.4.2 to complete this building and training process.

Hint2: As the **Import Data from Workspace** wizard appears, check the **Rows** radio button since we need to transpose the dataset matrix to use them as 9 features with 768 observations. Then click on the **OK** button to continue.

2. Using MATLAB function **kmeans()** to build a K-Means clustering model with **Diabetes.csv** dataset. Refer to Sect. 8.5.1.1 in this chapter to create a Script file, **Diabetes_KM_Func.m**, and build your codes for that file.

Hint1: To make it simple, only use five columns, Glucose, Blood Pressure (BP), BMI, PEDIGREE, and Age, and set cluster number as 3. Plot the results with three clusters and cluster centers.

3. Validate the K-Means clustered model built above with **Diabetes.csv** dataset. Refer to Sect. 8.5.1.2 in this chapter to create a Script file, **Eval_Diabetes_KM_Func.m**, and build your codes for that file.

Hint1: Still use five columns in the Diabetes.csv dataset, but using the first 100 observations (rows) as the training data, and another 100 (rows) data located between rows 602 and 703 as the testing data.

4. Build a hierarchical clustering model with heart dataset. Refer to Sect. 8.5.2.4 in this chapter to create a Script file, **Heart_HC_Func.m**, and build your codes for that file.

Hint 1: Use five features (columns), Age, Chol, Oldpeak, Trtbps, and Fbs, with the top 10 observations (rows) as training data.

5. Use the Fuzzy-C-Mean function **fcm()** to perform overlapping clustering for the modified Heart dataset, **heart1_M.xlsx**. The Script file is named: **Heart_FCM3_Func.m**.

Hint1: Refer to Sect. 8.5.3.1 and the sample project Diabetes_FCM3_Func.m to get details about the coding process for this project.

Hint2: Use three features (columns), chol, oldpeak, and trtbps, with the total 280 observations as input data.

6. Use GMM algorithm to cluster the modified Heart dataset, **heart1_M.xlsx**. Use only two features from that dataset, **chol** and **trtbps**, with 280 observations as input data. The Script file is named: **Heart_GMM_Func.m**.

   Hint1: Refer to codes developed in Fig. 8.32 in Sect. 8.5.4.3 in Chap. 8 to build this project.

7. Build an association rule model to find the strong relations among three item sets, **age, chol,** and **oldpeak** in the modified dataset, **Heart_AR.csv**. The modified dataset can be found on the Springer ftp site in the folder: **Students\ Datasets\Heart Dataset**. The name of the project Script file is: **Heart_Association_Rule.m**.

   Hint1: Refer to Sect. 8.6.2 and the sample project Diabetes_Association_ Rule.m shown in Fig. 8.40 to get details about the coding process for this project.

   Hint2: Use three features (columns), age, chol, and oldpeak, with a total 280 observations as the input data.

References

1. https://www.ibm.com/topics/unsupervised-learning.
2. https://www.analyticsvidhya.com/blog/2016/11/an-introduction-to-clustering-and-different-methods-of-clustering/#:~:text=Clustering%20is%20the%20task%20of,and%20assign%20them% 20into%20clusters.
3. https://www.datacamp.com/tutorial/association-rule-mining-python.
4. https://www.analyticsvidhya.com/blog/2022/02/clustering-machine-learning-algorithm-using-k-means/#:~:text=Exclusive%20Clustering%3A%20Exclusive%20Clustering%20is,all%20similar%20datapoints%20are%20clustered.
5. https://machinelearninginterview.com/wp-content/uploads/2020/12/elbow.png.
6. https://en.wikipedia.org/wiki/Silhouette_(clustering)#:~:text=Silhouette%20refers%20to%20 a%20method,statistician%20Peter%20Rousseeuw%20in%201987.
7. http://i.stack.imgur.com/iAWnF.png.
8. https://en.wikipedia.org/wiki/Fuzzy_clustering.
9. https://www.mathworks.com/help/stats/clustering-using-gaussian-mixture-models.html.
10. Piatetsky-Shapiro, Gregory (1991), Discovery, analysis, and presentation of strong rules. In: Piatetsky-Shapiro Gregory, Frawley, William J. (eds.), Knowledge Discovery in Databases, AAAI/MIT Press, Cambridge, MA.
11. https://en.wikipedia.org/wiki/Association_rule_learning.
12. https://www.mathworks.com/matlabcentral/fileexchange/42541-association-rules/#:~:text=Association%20Analysis%20is%20a%20method,%7D%20%2D%20%3E%20 %7Bburger%7D).
13. https://www.investopedia.com/terms/a/apriori.asp#:~:text=A%20priori%20probability%20 stipulates%20that,cannot%20deduce%20the%20next%20outcome.
14. Wong, Pak (1999). "Visualizing Association Rules for Text Mining" (PDF). *BSTU Laboratory of Artificial Neural Networks*. Archived (PDF) from the original on 2021-11-29.
15. Hahsler, Michael (2005). "Introduction to arules – A computational environment for mining association rules and frequent item sets" (PDF). *Journal of Statistical Software*. doi: https://doi.org/10.18637/jss.v014.i15. Archived from the original (PDF) on 2019-04-30. Retrieved 2016-03-18

16. Jaakko Hollmen (9 March 1996). "Self-Organizing Map (SOM)". *Aalto University*.
17. https://www.geeksforgeeks.org/self-organising-maps-kohonen-maps/.
18. https://www.mathworks.com/matlabcentral/answers/1973769-how-to-interpret-and-determine-clusters-in-som-neighbour-distance-plot.
19. https://www.mathworks.com/discovery/unsupervised-learning.html.
20. https://www.mathworks.com/discovery/cluster-analysis.html.
21. https://www.mathworks.com/help/pdf_doc/stats/stats.pdf.
22. https://www.mathworks.com/help/stats/hierarchical-clustering.html.
23. https://www.mathworks.com/help/fuzzy/fuzzy-clustering.html.
24. https://www.mathworks.com/discovery/image-segmentation.html#:~:text=How%20Image%20Segmentation %20Works,of%20processing%20the%20entire%20image.
25. https://en.wikipedia.org/wiki/Normal_distribution.
26. https://en.wikipedia.org/wiki/Multivariate_normal_distribution.
27. https://www.mathworks.com/help/stats/gmdistribution.html.
28. https://www.mathworks.com/help/stats/cluster-gaussian-mixture-data-using-soft-clustering.html.
29. https://builtin.com/data-science/covariance-vs-correlation#:~:text=Covariance%20 reveals%20how %20two%20variables,the%20linear%20relationship%20between%20 variables.
30. https://en.wikipedia.org/wiki/Apriori_algorithm.
31. https://www.indeed.com/career-advice/career-development/unsupervised-learning.

Chapter 9
Introduction to Reinforcement Learning

As we discussed in Sect. 4.2 of Chap. 4, machine learning contains three major components: supervised learning, unsupervised learning, and reinforcement learning. Unlike the first two algorithms, reinforcement learning is a special algorithm, and it provides some special controllability with more flexibility.

Under the reinforcement learning umbrella, two subcomponents, positive reinforcement and negative reinforcement, exist and work as two algorithms to perform related reinforcement learning functions.

Reinforcement learning differs from supervised learning in not needing labeled input-output pairs to be presented, and in not needing suboptimal actions to be explicitly corrected. Instead the focus is on finding a balance between exploration (of uncharted territory) and exploitation (of current knowledge) with the goal of maximizing the long-term reward, whose feedback might be incomplete or delayed [1].

The environment is typically stated in the form of a Markov Decision Process (MDP), because many reinforcement learning algorithms for this context use dynamic programming techniques [2]. The main difference between the classical dynamic programming methods and reinforcement learning algorithms is that the latter do not assume knowledge of an exact mathematical model of the Markov decision process and they target large Markov decision processes where exact methods become infeasible [3].

In this chapter, we will provide a detailed introduction and discussion about this algorithm with some real example projects. First let's have a clear and better understanding of the reinforcement learning algorithm and related components.

Supplementary Information The online version contains supplementary material available at https://doi.org/10.1007/978-3-031-84423-2_9.

Y. Bai, *AI Foundations and Applications with MATLAB*, https://doi.org/10.1007/978-3-031-84423-2_9

9.1   Introduction to Reinforcement Learning Algorithms

Reinforcement learning is a goal-directed computational approach where an agent learns to perform a task by interacting with an unknown dynamic environment. During training, the learning algorithm updates the agent policy parameters. The goal of the learning algorithm is to find an optimal policy that maximizes the long-term reward received during the task.

Depending on the type of agent, the policy is represented by one or more policy and value function representations. You can implement these representations using deep neural networks. You can then train these networks using Reinforcement Learning Toolbox™ software [4].

In other words, reinforcement learning (RL) is the science of decision-making. It is about learning the optimal behavior in an environment to obtain maximum reward. In RL, the data is accumulated from machine learning systems that use a trial-and-error method. Data is not part of the input that we would find in supervised or unsupervised machine learning.

Reinforcement learning uses algorithms that learn from outcomes and decide which action to take next. After each action, the algorithm receives feedback that helps it determine whether the choice it made was correct, neutral, or incorrect. It is a good technique to use for automated systems that have to make a lot of small decisions without human guidance.

Reinforcement learning is an autonomous, self-teaching system that essentially learns by trial and error. It performs actions with the aim of maximizing rewards, or it is learning by doing in order to achieve the best outcomes [5].

Basically, reinforcement learning is based on the Markov decision process and dynamic programming methods but provides more flexibility on the control target or environment, which means that the control target or environment may not be defined by any mathematical model. To get a better understanding of this algorithm, let's start from a classical or a modern control system.

9.1.1   Components Involved in Reinforcement Learning Control Systems

As we know, a classical control system is a closed-loop control system with the following components, as shown in Fig. 9.1a:

(1) Input state or variable **Set**
(2) Control plant or target **G**
(3) Controller **D**
(4) Control output or command **U**

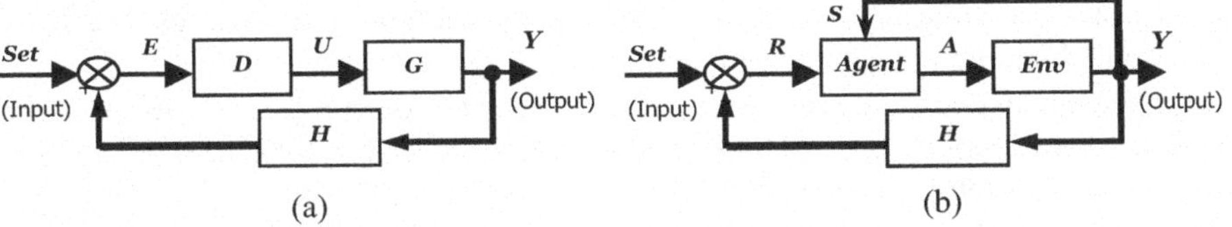

Fig. 9.1 Block diagram for classical and reinforcement learning control system

(5) Feedback system **H**

Compared with a reinforcement learning control system shown in Fig. 9.1b, it can be found that the only difference between a classical and a reinforcement learning control system is, one more variable or variable set called State (**S**) is added and fed back to the controller. Also different terminologies are used for the latter as described below:

- The Error **E** between the input **Set** and the feedback **H** is changed to Reward **R** in the reinforcement learning system.
- The controller **D** is changed to the **Agent**.
- The output of the controller **U** is changed to Action **A**.
- The control target or plant is changed to the Environment **Env**.
- One more feedback, State **S**, is added and feedback to the Agent.

The identical thing is that both control systems are still closed-loop control systems.

Like a modern control system, five key components are involved in the reinforcement learning control system:

(1) Environment system
(2) Agent control system
(3) State system
(4) Action system
(5) Reward system

By comparing a classical and a reinforcement learning system, it looks like they are very similar in components. However, they are significantly different on the working principle and process. A classical controller works as a digital controller for a known dynamic model or plant, but reinforcement learning works as a discrete control agent for an unknown environment based on a Markov decision process.

Now let's have a clear picture about the Markov decision process.

9.1.2   *The Markov Decision Process*

In mathematics, a **Markov Decision Process (MDP)** is a discrete-time stochastic control process. It provides a mathematical framework for modeling decision-making in situations where the outcomes are partly random and partly under the control of a decision-maker. MDPs are useful for studying optimization problems solved via dynamic programming [6].

Unlike a classical control system, some variables or components used in the reinforcement learning control system are a sequence of variables or a set of variables. For example, the state, action, agent, and reward variables are all sets. Therefore, a Markov decision process is composed of the following sets:

(1) S is a set of states called the state space, $S = \{s_1, s_2, \ldots, s_n\}$.
(2) A is a set of actions known as action space, $A = \{a_1, a_2, \ldots, a_n\}$. A_S is a set of actions for the state s, such as $A_{s1}, A_{s2}, \ldots, A_{sn}$.
(3) $P_a(s, s') = P_r(s_{t+1} = s' \mid s_t = s, a_t = a)$ is the probability that action a in state s at time t will lead to the next state s' at time $t + 1$.
(4) $R_a(s, s')$ is the immediate reward received after transitioning from the current state s to the next state s' due to action a.

Reinforcement learning uses this kind of Markov decision process to select the optimal action a based on the current state s. This selection process can be considered as a mapping from the state to the action, and this map is called a *policy* and it can be expressed as:

$$\pi : A \times S \rightarrow [0,1]$$

$$\pi(a,s) = P_r\left(A_t = a \mid S_t = s\right) \tag{9.1}$$

Two kinds of mapping or policies are available and they are:

- **Stochastic (Stationary) Policy:** A policy is called *stochastic* or *stationary* if it returns a probability distribution over actions in a given (last) state. This process is called Partially Observable Markov Decision Process (POMDP). The policy is determined without using a value function.
- **Deterministic Policy:** A *deterministic* policy deterministically selects actions based on the current state without any uncertainty. It happens when we have a deterministic or definite environment like a robot's workspace. Since any of such policies can be identified with a mapping from the set of states to the set of actions, these policies can be identified with such mappings with no loss of generality.

Equation (9.1) describes a stochastic policy π, which maps the last state to the next action with a probability distribution in a range of [0, 1]. The mapping or policy returns a target strategy $\pi(a, s)$ for the agent.

The goal of a Markov decision process is to find a good or an optimal *policy* for the decision-maker: a function π that specifies the action $\pi(s)$ that the decision-maker will choose an action $a = \pi(s)$ for the state s. Once a Markov decision process is combined with a *policy* in this way, this fixes the action for each state and the resulting combination behaves like a Markov chain since the action chosen in the state is completely determined by $\pi(s)$ and $Pr(s_{t+1} = s' \mid s_t = s, a_t = a)$ reduces to $Pr(s_{t+1} = s' \mid s_t = s)$, a Markov transition matrix.

To realize the above MDP to find the optimal policy, two kinds of functions, the state-value function and the action-value function, should be used. To avoid the duplication, we discussed these functions and selection process with details in the next section.

9.1.3 The Reinforcement Learning Process

The purpose of reinforcement learning is for the agent to learn an optimal, or nearly optimal, policy that maximizes the *reward function* or other user-provided reinforcement signal that accumulates from the immediate rewards.

A reinforcement learning agent interacts with its environment in discrete time steps. At each time t, the agent receives the current state S_t and reward R_t. It then chooses an action A_t from the set of available actions, which is subsequently sent to the environment. The environment moves to a new state S_{t+1} and the reward R_{t+1} associated with the *transition* (S_t, A_t, S_{t+1}) is determined. The goal of a reinforcement learning agent is to learn a *policy*; $\pi: S \times A \rightarrow [0, 1]$, $\pi(S, A) = P_r(A_t = a \mid S_t = s)$ that maximizes the expected cumulative reward [8].

The policy map gives the probability of taking action a when in state s [8]. How to get the optimal policy and which kind of criterion should be adopted? To answer these questions, we need to have a clear picture about some evaluation functions. First let's take a closer look at the state-value function.

9.1.3.1 The State Value Function

The state-value function $V_\pi(s)$ is defined as, *expected discounted return* starting with state s, i.e., $S_0 = s$, and successively following policy π. Hence, roughly speaking, the value function estimates *how good* it is to be in a given state [7].

$$V_\pi\left(s\right) = E\left[G \mid S_0 = s\right] = E\left[\sum_{t=0}^{\infty} \gamma^t R_{t+1} \mid S_0 = s\right] \tag{9.2}$$

where the random variable G denotes the **discounted return** and is defined as the sum of future discounted rewards:

$$G = \sum_{t=0}^{\infty} \gamma^t R_{t+1} = R_1 + \gamma R_2 + \gamma^2 R_3 + \ldots, \tag{9.3}$$

where R_{t+1} is the reward for transitioning from state S_t to S_{t+1}, $0 \leq \gamma < 1$ is the discount rate. γ is less than 1 but greater than or equal to 0, so rewards in the distant future are weighted less than rewards in the immediate future.

The so-called discounted return is exactly a sum of the discounted rewards for a given state.

The algorithm must find a *good* policy with maximum expected discounted return. From the theory of Markov decision processes it is known that, without loss of generality, the search can be restricted to the set of so-called *stationary* policies. The policy can be either *stationary* or *deterministic* depending on the state used.

9.1.3.2   The Action Value Function

The action-value function approach attempts to find a policy that maximizes the discounted return by maintaining a set of estimates of expected discounted returns $E[G]$ for some policy, usually either the *current* (on-policy) or the *optimal* (off-policy) one.

These methods rely on the theory of Markov decision processes, where optimality is defined in a sense stronger than the one above: A policy is optimal if it achieves the best-expected discounted return from *any* initial state (i.e., initial distributions play no role in this definition). Again, an optimal policy can always be found among stationary policies.

To define optimality in a formal manner, define the state-value of a policy π by

$$V^{\pi}\left(s\right) = E\left[G\mid s\mid,\pi\right] \tag{9.4}$$

where G stands for the discounted return associated with the following π from the initial state s. Defining $V^{*}(s)$ as the maximum possible state-value of $V^{\pi}(s)$, where π is allowed to change,

$$V^{*}\left(s\right) = \max_{\pi} V^{\pi}\left(s\right) \tag{9.5}$$

A policy that achieves these optimal state-values in each state is called *optimal*. Clearly, a policy that is optimal in this sense is also optimal in the sense that it maximizes the expected discounted return, since $V^{*}\left(s\right) = \max_{\pi} E\left[G\mid s\mid,\pi\right]$, where s is a state randomly sampled from the distribution μ of initial states, so $\mu(s) = P_r(S_0 = s)$.

Although state-values suffice to define optimality, it is useful to define action-values. Given a state s, an action a and a policy π, the action-value of the pair (s, a) under π is defined by

$$Q^{\pi}\left(s,a\right) = E\left[G\mid s\mid,a\mid,\pi\right] \tag{9.6}$$

where G now stands for the random discounted return associated with first taking action a in state s and following π, thereafter.

The theory of Markov decision processes states that if π^{*} is an optimal policy, we act optimally (take the optimal action) by choosing the action from $Q^{\pi^{*}}\left(s,.\right)$ with the highest action-value at each state, s. The *action-value function* of such an optimal policy ($Q^{\pi^{*}}$) is called the *optimal action-value function* and is commonly denoted by Q^{*}. In summary, the knowledge of the optimal action-value function alone suffices to know how to act optimally.

Assuming full knowledge of the Markov decision process, the two basic approaches to compute the optimal action-value function are value iteration and policy iteration. Both algorithms compute a sequence of functions Q_k ($k = 0, 1, 2, \ldots$) that converge to Q^*. Computing these functions involves computing expectations over the whole state-space, which is impractical for all but the smallest (finite) Markov decision processes. In reinforcement learning methods, expectations are approximated by averaging over samples and using function approximation techniques to cope with the need to represent value functions over large state-action spaces [6].

To perform those optimal searching processes, different methods or algorithms are utilized in reinforcement learning. First let's have a quick review of some basic algorithms used in RL.

9.2   Some Basic Algorithms Used in Reinforcement Learning

Now we have a basic understanding about reinforcement learning and its algorithm, as well as its components. Next we need to know how to combine these together to perform actual reinforcement learning process to find the optimal actions for the selected state with the reward. Prior to doing this, we like to emphasize some important terminologies.

(1) **Agent**—The learner and decision-maker for the reinforcement learning.
(2) **Action**—A setoff action issued by the agent based on the state and the reward set.
(3) **State**—A set of states related to the environment at different time intervals.
(4) **Reward**—A set of feedback errors returned by the environment (discounted return).
(5) **Policy**—The decision-making function of the agent, which is a mapping from the states to the actions.
(6) **Environment**—A control target or a plant, where the agent learns and decides what kind of actions to take to control it.
(7) **Value Functions**—Search to find the optimal policy that maximizes the discounted return by maintaining a set of estimates of expected discounted returns $E[G]$.
(8) **Markov Decision Process** (MDP)—A discrete-time stochastic model used to select the optimal action a based on the current state s. This selection process can be considered as a mapping from the state to the action, and it is called a *policy*. The goal of a Markov decision process is to find a good or an optimal *policy* for the decision-maker.

Keeping the above terminologies in mind, let's have a closer look at some popular algorithms used in reinforcement learning. One point to be noted is that a selected algorithm is directly related to a kind of agent to be used; thus, the chosen algorithm sometimes is also called a related agent.

9.2.1 *Monte Carlo Methods*

Monte Carlo algorithm can be used for policy iteration. Policy iteration consists of two steps: *policy evaluation* and *policy improvement*.

Monte Carlo is used in the policy evaluation step. In this step, given a stationary or a deterministic policy π, the goal is to compute the function values $Q^\pi(s, a)$ or a good approximation to them for all state-action pairs (s, a). Assume (for simplicity) that the Markov decision process is finite, that sufficient memory is available to accommodate the action-values, and that the problem is episodic and after each episode a new one starts from some random initial state. Then, the estimate of the value of a given state-action pair (s, a) can be computed by averaging the sampled returns that originated from (s, a) over time. Given sufficient time, this procedure can thus construct a precise estimate Q of the action-value function Q^π. This finishes the description of the policy evaluation step.

In the policy improvement step, the next policy is obtained by computing a *greedy* policy with respect to Q: Given a state s, this new policy returns an action that maximizes $Q(s, .)$. In practice, lazy evaluation can defer the computation of the maximizing actions to when they are needed. Problems with this procedure include the following [7]:

(1) The procedure may spend too much time evaluating a suboptimal policy.
(2) It uses samples inefficiently in that a long trajectory improves the estimate only of the *single* state-action pair that started the trajectory.
(3) When the returns along the trajectories have *high variance*, convergence is slow.
(4) It works in episodic problems only.
(5) It works in small, finite Markov decision processes only.

9.2.2 *The Temporal Difference Method*

Temporal Difference Learning is very commonly used for the purpose of predicting the total reward expected over the future. They can also be used to predict other quantities as well. It is essential to learn how to predict a quantity that is dependent on the future values for a given signal. This method can be used to compute the long-term utility of a pattern of behavior from a series of intermediate rewards.

In fact, Temporal Difference (TD) learning is used to predict a variable's future value for a given sequence of states. TD learning is a major technology in solving the problem of reward prediction. It could be said that TD learning utilizes a mathematical trick that allows it to replace complicated reasoning with a simple learning procedure that can be used to generate the very same results.

The trick is that rather than attempting to calculate the total future reward, TD learning just attempts to predict the combination of immediate reward and its own reward prediction at the next moment in time. Now when the next moment comes and brings fresh information with it, the new prediction is compared with the expected prediction. If these two predictions are different from each other, the TD learning algorithm will calculate how different the predictions are from each other and make use of this temporal difference to adjust the old prediction toward the new prediction [9].

9.2.3   The Function Approximation Method

The function approximation method uses a *linear function approximation* that starts with a mapping ϕ that assigns a finite-dimensional vector to each state-action pair. Then, the action values for a state-action pair (s, a) are obtained by linearly combining the components of $\phi(s, a)$ with some *weights* θ:

$$Q(s,a) = \sum_{i=1}^{d} \theta_i \varnothing_i (s,a) \qquad (9.7)$$

The algorithm then adjusts the weights, not the values associated with the individual state-action pairs. A method based on ideas from nonparametric statistics has been explored.

Value iteration can also be used as a starting point, giving rise to the Q-Learning algorithm and its many variants [10]. The Deep Q-Learning (DQL) algorithm is involved when a neural network is used to represent Q with various applications in stochastic search problems [11].

The problem with using action-values is that they may need highly precise estimates of the competing action-values that can be hard to obtain when the returns are noise involved, though this problem is mitigated to some extent by temporal difference methods. Using the so-called compatible function approximation method compromises generality and efficiency [7].

To perform those optimal searching process, different methods or algorithms are utilized in reinforcement learning. All of these algorithms are divided into two categories, model-based and model-free algorithms.

9.3   Model-Based Reinforcement Learning

Like a classical control system, in which a dynamic model of the plant should be established based on experience or testing, and this model provides a mathematic equation that gives the input and the output relation in either a frequency domain or a discrete time domain. For a reinforcement learning system, a similar model is also needed, and it is used to describe the relations among states, rewards, and actions for a given environment.

On the other hand, similarly to the modern control system, such as fuzzy logic control systems, in which no dynamic model is needed. This is very similar to the model-free reinforcement learning systems, in which it uses experience to learn directly one or both of two simpler quantities (state-action values or policies) which can achieve the same optimal behavior but without estimation or use of a world model. Given a policy, a state has a value, defined in terms of the future utility that is expected to accrue starting from that state.

Based on these two models, quite a few RL algorithms are categorized, developed, and used. Figure 9.2 shows a global picture about these algorithms and their distributions under two model categories.

Basically, the model-based reinforcement learning (RL) algorithms use experience to construct an internal model of the transitions and immediate outcomes in the environment. Appropriate actions are then chosen by searching or planning based on this world model.

In fact, a model-based algorithm needs first to learn a model of the Markov Decision Process.

In other words, a model-based method aims to understand the environment's rules. It builds a policy network that guides their choices. This requires extra steps, like creating a model that needs updates as new information comes in.

Thus, it takes less time for it to learn from experiences since they have a head start with some knowledge of how things work. But making decisions can be slower because they always check against their internal model first—kind of like double-checking your work on a math problem before turning it in [12].

As shown in Fig. 9.2, under the model-based RL, there are two ways to build a model: learn the model and give the model. Let's have a closer look at these two ways to see how they can develop or build a model in different ways.

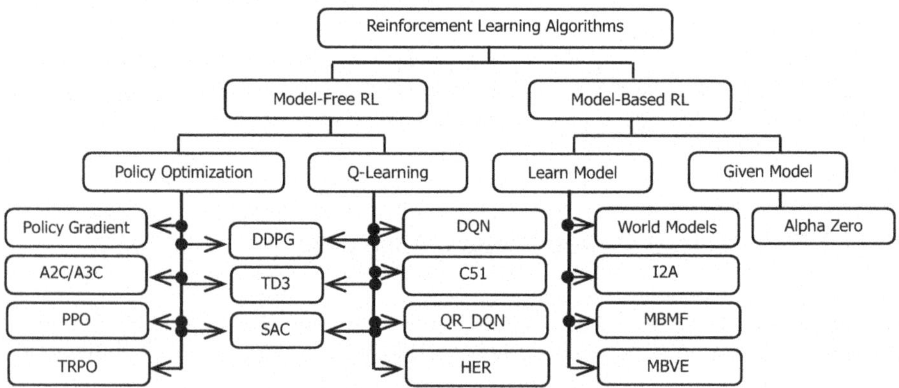

Fig. 9.2 A global picture about all RL algorithms under two models

9.3.1   *Learn the Model*

To learn the model, a base policy is run, like a random or any educated policy, while the trajectory is observed. The model is fitted using the sampled data with the following procedure:

(1) Perform the basic policy based on the current state s_t and action a_t, $\pi_0(a_t \mid s_t)$ to collect and derive the next state $D = \{(s, a, s')_i\}$.
(2) Learn the dynamic model $f(s, a)$ to minimize the cost function $\sum_i f\left(s_i, a_i\right) - s_i'^2$.
(3) Search the plan through $f(s, a)$ to get related actions.

Supervised learning is used to train a model to minimize the least square error from the sampled data for the control function. Optimal trajectory using the model and a cost function is used in step three. The cost function can measure how far we are from the target location and the amount of effort spent [13].

The following models are involved under this category:

(1) **World models**: the agent can learn from its own "dreams" due to the Variable Auto-encoders.
(2) **Imagination-Augmented Agents (I2A)**: learns to interpret predictions from a learned environment model to construct implicit plans in arbitrary ways, by using the predictions as additional context in deep policy networks. Basically it's a hybrid learning method because it combines model-base and model-free methods.
(3) **Model-Based Priors for Model-Free Reinforcement Learning** (MBMF): aims to bridge the gap between model-free and model-based reinforcement learning.
(4) **Model-Based Value Expansion** (MBVE): this method controls for uncertainty in the model by only allowing imagination to fixed depth [14]. By enabling wider use of learned dynamics models within a model-free reinforcement learning algorithm, we improve value estimation, which, in turn, reduces the sample complexity of learning.

9.3.2   *Given the Model*

A typical example of the given model is the AlphaZero algorithm, in which it learns the model without any human input by playing games against itself and discovering new strategies, becoming its own teacher.

Let's use a real example to illustrate how a model-based RL controls a robot in real time. The following operational procedures should be adopted:

(1) **Modeling the Environment**: The agent builds a model for the robot's working environment, which includes other robots, workspace, objects, and so on.

(2) **Planning**: The agent uses the model to simulate various scenarios, like the movements of the current robot, other robots, and objects, and plans its actions accordingly (e.g., when to pick up, when to drop down, when to accelerate, when to rotate, when to move forward and back).

(3) **Sample Efficiency**: The model helps the agent learn to operate safely with fewer actual working trials, as it can learn a lot through simulation.

(4) **Challenges**: The complexity of the real world can make it extremely difficult to create an accurate model. Any discrepancy between the model and the real world (model bias) can lead to poor decision-making.

In summary, model-based learning attempts to model the environment and then chooses the optimal policy based on its learned model and related environment parameters, such as states, rewards, or discounted returns.

9.4 Model-Free Reinforcement Learning

As we mentioned, model-free learning does not need to try to understand the environment dynamics. In other words, it does not need a model at all. It builds a policy for itself that tells what the optimal behavior is for a given state and issues the related optimal action for the state. This is accomplished by using error-and-trial methods by the agent.

In fact, the agent cannot predict or guess the future output of its action, instead it will try to derive it by using the error and trial only in real time. Exactly, in model-free RL, the key is to learn by observing the consequences of actions rather than attempting to understand the dynamics of the environment.

The agent does not estimate the transition probability distribution and the reward function associated with the environment. This approach is particularly useful in situations where the underlying model is either unknown or too complex to be accurately developed.

It looks that the model-free RL can save time to avoid building or developing any model. However, anything cannot be perfect, and one good thing may bring some side effects. This is also true for model-free RL. Model-free algorithms are statistically less efficient than model-based methods because information from the environment is combined with previous, and possibly erroneous, estimates or beliefs about state values, rather than being used directly.

In conclusion, while model-based methods can be more sample-efficient and capable of planning, they suffer from model bias and complexity. On the other hand, model-free methods are simpler and potentially more robust but require more interactions with the environment to learn effectively. The choice between model-based and model-free RL often depends on the specific requirements and constraints of the application at hand.

Let's use a real example to illustrate how a model-free RL controls a robot in real time. The following operational procedures should be adopted [12]:

(1) **Learning from Interaction**: The agent learns to move the robot by interacting with the robot's environment, receiving feedback in the form of rewards (e.g., positive for tracking of positions of robots, negative for losing the robots).
(2) **Direct Policy or Value Function Estimation**: Without modeling the environment, the agent directly learns the policy (what kind of action to take in each state) or value function (how good each state or action is).
(3) **Sample Inefficiency**: The agent might need to experience many tracking hours to learn a good policy, as it learns purely from interaction.
(4) **Robustness**: The approach is potentially more robust to the complexities of the real-world robots' environment since it does not rely on a possibly flawed model.

Next let's have more discussions about popular algorithms used under the model-free RL.

9.4.1   Policy Optimization and Policy Iteration

As shown in Fig. 9.2, there are two components to represent agents under the model-free RL category, Policy Optimization and Q-Learning.

As we discussed in Sect. 9.2.1, the so-called policy iteration consists of two steps: *policy evaluation* and *policy improvement*, and these two steps are repeated iteratively until the policy converges. Let's have a closer look at these terminologies.

- **Policy Evaluation** uses value functions to compute and compare different policies with the function value $Q^\pi(s, a)$ for all state-action pairs (s, a). Then, the estimate of the value of a given state-action pair (s, a) can be computed by averaging the sampled returns that originated from (s, a) over time. Given sufficient time, this procedure can thus construct a precise estimate Q of the action-value function Q^π. This finishes the description of the policy evaluation. The value function of a policy is calculated by solving the Bellman equations. This step involves estimating the value of each state under the current policy.
- Similarly to policy evaluation, **Policy Improvement** is also to estimate the value for each state under the current policy, but a new policy is derived based on the current value function. The new policy is typically a greedy policy with respect to the value function. Policy iteration guarantees convergence to the optimal policy as long as the algorithm is run to convergence.
- **Policy Optimization** is also called **Policy Search**, which is a wide class of RL algorithms where the purpose is to directly search for an optimal policy without explicitly estimating value functions. In policy search methods, parameterized policies are used, and the goal is to find the best set of parameters that define the policy. Policy search methods can use various optimization techniques such as evolutionary algorithms, gradient descent, or other black-box optimization methods to search for the optimal policy parameters [15].

A point to be noted is that there are two types of policies: deterministic and stochastic. Deterministic policy maps state to action without uncertainty. It happens when you have a deterministic environment like a robot's workspace. Stochastic policy outputs a probability distribution over actions in a given state. This process is called Partially Observable Markov Decision Process (POMDP).

Next let's have a brief and quick discussion about some popular algorithms used under the policy optimization category. Four algorithms are involved and they are: Policy Gradient (PG), Asynchronous Advantage Actor-Critic (A3C), Proximal Policy Optimization (PPO), and Trust Region Policy Optimization (TRPO).

9.4.1.1 Policy Gradient (PG) Algorithm

This algorithm belongs to the class of Monte Carlo method. The agent does not learn during an episode but only after an episode is finished. To reduce the variance of the parameter updates, one can use a baseline network that estimates the expected discounted cumulative long-term reward [16].

The policy gradient algorithm can be used to model and optimize the policy directly with any model. The policy is usually defined with a parameterized function with respect to θ, $(a|s)$. The value of the reward or objective function depends on this policy and then various algorithms can be applied to optimize θ for the best reward.

Similar to Monte Carlo methods, the policy gradient algorithm uses a function $J(\theta)$ that is defined as:

$$J(\theta) = E\left\{\sum_{k=0}^{H} c_k r_k\right\} \tag{9.8}$$

to update and optimize this function to get the optimal policy.

Where θ is the policy parameter to be estimated or optimized, c_k is a weight factor and it can be defined as $c_k = \gamma^k$ ($0 \leq \gamma \leq 1$) for discounted return or $c_k = 1/H$ for the average reward as Monte Carlo method did. The return $r_k = r(s_k, a_k)$, s_k is the state at time k and a_k is the action at time k.

This method updates the policy parameterization according to the gradient update rule:

$$\theta_{k+1} = \theta_k + \alpha_k \nabla_\theta J\big|_{\theta=\theta_k} \tag{9.9}$$

where α_k is the learning rate and $k = \{0, 1, 2, \ldots\}$ is the updating number.

The operational procedure of the PG algorithm includes:

(1) Measure and monitor the accuracy of the policy by using the policy score function $J(\theta)$

(2) Using policy gradient ascent to find the optimal parameter θ that improves the policy until the optimal one is obtained

9.4.1.2   Asynchronous Advantage Actor-Critic (A3C) Algorithm

This algorithm is one of the newest algorithms to be developed in the field of Deep Reinforcement Learning Algorithms.

Asynchronous: Unlike Deep Q-Learning (DQL) which uses a single agent and a single environment, A3C algorithm uses multiple agents with each agent having its own network parameters and a copy of the environment. This agent interacts with their respective environments asynchronously, learning with each interaction. Each agent is controlled by a global network. As each agent gains more knowledge, it contributes to the total knowledge of the global network. The presence of a global network allows each agent to have more diversified training data.

Actor-Critic: This item stands for two neural networks—**Actor** and **Critic**. The goal of the **Actor** is to optimize the policy, like ***How to act?*** The **Critic** aims at optimizing the value, such as ***How good action is?*** Thus, it creates a complementary situation for an agent to gain the best experience of fast learning.

Advantage: In the implementation of Policy Gradient algorithm, the value of discounted returns ***G*** tells the agent which of its actions were rewarding and which ones were penalized. By using the value of **Advantage** instead, the agent also learns how much better the rewards were compared with its expectations. This gives a new-found insight to the agent into the environment and thus the learning process becomes better. The advantage metric is given by the following expression:

$$A(s,a) = Q(s,a) - V(s) \tag{9.10}$$

where $Q(s, a)$ stands for the expected future reward of taking action at a particular state, and $V(s)$ stands for the value of being in a specific state or the average value of that state.

9.4.1.3   Proximal Policy Optimization (PPO) Algorithm

Proximal Policy Optimization (PPO) algorithm can be considered as a state-of-the-art algorithm because it seems to strike a balance between performance and comprehension. Compared with other algorithms, the three main advantages of PPO are simplicity, stability, and sample efficiency [17].

PPO is classified as a policy gradient (PG) method for training an agent's policy network. The policy network is the function or a map that the agent uses to make decisions. Essentially, to train the right policy network, PPO takes a small policy update (step size), so the agent can reliably reach the optimal solution.

Essentially, the PPO used the Advantage function as we discussed in Sect. 9.4.1.2, to try to learn how much better the rewards were compared with its expectation and answer the question of whether a specific action of the agent is better than the other possible action in a given state or worse than the other action. The positive output of the advantage function means that the chosen action is better than the average return, so the possibilities of that specific action will increase, and vice versa.

In PPO, the ratio function $rt(\theta)$ calculates the probability of taking action a at state s in the current policy network divided by the previous old version of policy.

In this function, $rt(\theta)$ denotes the probability ratio between the current and old policy:

- If $rt(\theta) > 1$, the action a at state s is more likely based on the current policy than the old policy.
- If $rt(\theta)$ is between 0 and 1, the action a at state s is less likely based on the current policy than the old policy.

This ratio function can easily estimate the divergence between old and current policies [18, 19].

9.4.1.4   Trust Region Policy Optimization (TRPO) Algorithm

Trust Region Policy Optimization (TRPO) algorithm updates its policies by taking the largest step possible to improve performance while satisfying a special constraint on how close the new and old policies are allowed to be. The constraint is expressed in terms of KL-Divergence, a measure of (something like, but not exactly) distance between probability distributions.

This is different from normal policy gradient (PG) algorithm, which keeps new and old policies close in parameter space. But even seemingly small differences in parameter space can have very large differences in performance, so a single bad step can collapse the policy performance. This makes it dangerous to use large step sizes with vanilla policy gradients, thus hurting its sample efficiency. TRPO nicely avoids this kind of collapse and tends to quickly and monotonically improve performance [20].

9.4.2   *Q-Learning or Action Value Iteration Algorithm*

Basically, the Q-Learning (Quality-Learning) algorithms are used to learn the action-value function $Q(s, a)$ to check how good to take an action at a particular state. *Generally a scalar value is assigned* over an action a given the state s.

As we discussed in Sect. 9.1.3.2, given a state s, an action a , and a policy π, the action-value of the pair (s, a) under π is defined by Eq. (9.6):

$$Q^{\pi}\left(s,a\right) = E\left[G|,s|,a|,\pi\right]$$

where G now stands for the random discounted return associated with first taking action a in state s and following π, thereafter.

The theory of Markov decision processes states that if π^* is an optimal policy, we act optimally (take the optimal action) by choosing the action from $Q^{\pi^*}(s,.)$ with the highest action-value at each state, s. The *action-value function* of such an optimal policy (Q^{π^*}) is called the *optimal action-value function* and is commonly denoted by Q^*. In summary, the knowledge of the optimal action-value function alone suffices to know how to act optimally.

Q-learning uses this Q-function, or action-value function, to model optimal behavior. The Q-function measures the expected discounted reward for taking a specific action in a given state. Q-learning iteratively updates Q-values for each state-action pair using the Bellman Equation until the Q-function converges to Q^*.

The Q-function is similar to a value function, which is also a quality measure. However, the value function is tied to a policy and measures the expected return of trajectories.

The key difference between Q-Function and value function is: the Q function takes both the state and the action as input, while the value function only takes the state as input. This means that the Q function can be used to learn an optimal policy, while the value function can only be used to evaluate different policies.

The learning rate or *step size* determines to what extent newly acquired information overrides old information. A factor of 0 makes the agent learn nothing (exclusively exploiting prior knowledge), while a factor of 1 makes the agent consider only the most recent information (ignoring prior knowledge to explore possibilities). In fully deterministic environments, a learning rate of $r_t = 1$ is optimal. When the problem is stochastic, the algorithm converges under some technical conditions on the learning rate that require it to decrease to zero. In practice, often a constant learning rate is used, such as $r_t = 0.1$ for all t [21].

The operational procedure of a typical Q-Learning algorithm is a loop operation:

(1) Set a 2D Q-Learning table of states by actions and initialize all cells to 0. Each cell can be updated during the training process later.
(2) Then, at each time t the agent selects an action A_t, gets a reward R_{t+1}, enters a new state

S_{t+1} that may depend on both the previous state S_t and the selected action and Q is updated.

This procedure will be continued until an optimal Q is obtained.

9.4.2.1   Deep Q Neural Network (DQN) Algorithm

A possible or potential problem when using Q-Learning is that some challenges including the complexity of Q-table and very time-consuming processing may be encountered when the state is a huge set. To solve this issue, the Deep Q Network (DQN) algorithm appeared.

The Deep Q-Network (DQN) algorithm is a model-free, online, off-policy reinforcement learning method. A DQN agent is a value-based reinforcement learning agent that trains a critic to estimate the expected discounted cumulative long-term reward when following the optimal policy. DQN is a variant of Q-Learning that features a target critic and an experience buffer.

DQN is Q-Learning combined with neural networks. Instead of using a Q-table, neural networks replace Q-values with each action based on the state. In fact, the DQN algorithm follows a deep neural network-based approach to learn and optimize action-value functions. The working process can be summarized as follows [22]:

(1) **State Representation**: Convert the current state of the environment into a suitable numerical representation, such as raw pixel values or preprocessed features.
(2) **Neural Network Architecture**: Design a deep neural network, typically a convolutional neural network (CNN), which takes the state as input and output action-values for each possible action.
(3) **Experience Replay**: Store the agent's experiences consisting of state, action, reward, and next state in a replay memory buffer.
(4) **Q-Learning Update**: Sample mini-batches of experiences from the replay memory to update the neural network weights. The update is performed using the loss function derived from the Bellman equation, which minimizes the discrepancy between the predicted and target action-values.
(5) **Exploration and Exploitation**: Balance exploration and exploitation by selecting actions either greedily based on the current policy or stochastically to encourage exploration.
(6) **Target Network**: Use a separate target network with the same architecture as the main network to stabilize the learning process. Periodically update the target network by copying the weights from the main network.
(7) Repeat Steps 1–6: Interact with the environment, gather experiences, update the network, and refine the policy iteratively until convergence.

9.4.2.2   C51 Algorithm

C51 is an RL algorithm very similar to Q-Learning algorithm, and it is based on DQN. Like DQN, it can be used in any environment with a discrete action space. But the returns are not values of state and discounted rewards, instead they are the distributions. The algorithm is able to stay more stable during training, leading to improved final performance. This is particularly true in situations with bimodal or even multimodal value distributions, where a single average does not provide an accurate picture. The number 51 represents the use of 51 discrete values to parameterize the value distribution $Z(s, a)$.

In fact, the main difference between C51 and DQN is that rather than simply predicting the Q-value for each state-action pair in DQN, C51 predicts a histogram model for the probability distribution of the Q-value. In C51 algorithm, it uses a so-called *Distributional Bellman Equation*, which is very similar to the Bellman Equation and expressed as:

$$Z(s,a) = R + \gamma Z(s',a')$$

(9.11)

where Z represents the distribution of future rewards, not like Q which is a scalar. R is the reward distribution and γ is the discounted rate. According to [23], *the distributional Bellman equation states that the distribution of Z is characterized by the interaction of three random variables: the reward R, the next state-action (s',a'), and its random return $Z(s',a')$. By analogy with the well-known case, we call this quantity the value* value distribution.

9.4.2.3   Distributional RL with Quantile Regression (QR-DQN) Algorithm

Like the C51 algorithm, the QR-DQN also estimates a distribution of values during the learning process. The distribution of the values, rather than just the average, can improve the policy. This means that quantiles are learned which threshold values are attached to certain probabilities in the cumulative distribution function.

The similarity and dissimilarity between the C51 and the QR-DQN include the following:

(1) With a nearly identical neural network architecture as DQN, the output layer is changed to be of size $|A| \times N$, where N is a hyper-parameter giving the number of quantile targets.
(2) The Huber loss function used by DQN is replaced with a quantile Huber loss function. The Huber loss is a loss function used in robust regression, which is less sensitive to outliers in data than the squared error loss. See [24] to get more details for the Huber loss function.
(3) The Root Mean Square Propagation (RMSProp) is replaced by Adam which is a method used for stochastic optimization [25].

9.4.2.4   Hindsight Experience Replay (HER) Algorithm

In the HER algorithm, suppose the agent performs an episode of trying to reach goal state G from initial state S, but fails to do so and ends up in some state S' at the end of the episode. We cache the trajectory into our replay buffer [26]:

$$\left\{ \left(S_0, G, a_0, r_0, S_1 \right), \left(S_1, G, a_1, r_1, S_2 \right), \dots, \left(S_n, G, a_n, r_n, S' \right) \right\} \tag{9.12}$$

where r with subscript k is the reward received at step k of the episode, and a with subscript k is the action taken at step k of the episode. The idea in HER is to **imagine that the goal has actually been S' all along**, and that in this alternative reality, the agent has reached the goal successfully and got a positive reward for doing so. Thus, in addition to caching the real trajectory as seen before, we also cache the following trajectory:

$$\left\{ \left(S_0, S', a_0, r_0, S_1 \right), \left(S_1, S', a_1, r_1, S_2 \right), \dots, \left(S_n, S', a_n, r_n, S' \right) \right\} \tag{9.13}$$

This trajectory is the imagined one and is motivated by the human ability to learn useful things from failed attempts. It should also be noted that in the imagined trajectory, the reward received at the final step of the episode is now a positive reward gained from reaching the imagined goal.

By introducing the imagined trajectories to the replay buffer, we ensure that **no matter how bad our policy is, it will always have some positive rewards to learn from**.

The HER algorithm gives us a very similar outcome without requiring us to adapt the problem or design a curriculum. We can think of HER as an implicit curriculum learning process, in which we always supply our agent with problems that it is indeed capable of solving, and gradually increase the spectrum of those problems.

9.4.3 Hybrid RL Algorithms

In addition to the above popular RL algorithms, there are also some mixed or hybrid algorithms that combine some advantages from the above algorithms, and these algorithms can be categorized into a third one called hybrid algorithms.

The following algorithms can be considered as the hybrid algorithm:

(1) **Deep Deterministic Policy Gradients (DDPG)**: The DDPG algorithm is a model-free, online, off-policy reinforcement learning method. A DDPG agent is an actor-critic reinforcement learning agent that searches for an optimal policy that maximizes the expected cumulative long-term reward. DDPG agents can be trained in environments with continuous or discrete states and continuous action spaces. DDPG agents use the Q-value function as critic $Q(S, A)$ with deterministic policy actor $\pi(S)$ [27].

(2) **Soft Actor-Critic (SAC)**: The SAC algorithm is a model-free, online, off-policy, actor-critic reinforcement learning method. The SAC algorithm computes an optimal policy that maximizes both the long-term expected reward and the entropy of the policy. The policy entropy is a measure of policy uncertainty given the state. A higher entropy value promotes more exploration. Maximizing both the expected cumulative long-term reward and the entropy helps to balance between exploitation and exploration of the environment. SAC agents can be trained in environments with discrete or continuous states and continuous action spaces. SAC agents use the Q-value function as critics $Q(S,A)$ with stochastic policy actor $\pi(S)$ [28].

(3) **Twin Delayed Deep Deterministic Policy Gradients (TD3)**: The TD3 algorithm is an extension of the DDPG algorithm. DDPG agents can overestimate value functions, which can produce suboptimal policies. To reduce value function overestimation, the TD3 algorithm includes the following modifications of the DDPG algorithm [29].

(a) A TD3 agent learns two Q-value functions and uses the minimum value function estimate during policy updates.

(b) A TD3 agent updates the policy and targets less frequently than the Q functions.

(c) When updating the policy, a TD3 agent adds noise to the target action, which makes the policy less likely to exploit actions with high Q-value estimates.

You can use a TD3 agent to implement one of the following training algorithms, depending on the number of critics you specify.

(a) TD3—Train the agent with two Q-value functions. This algorithm implements all three of the preceding modifications.

(b) Delayed DDPG—Train the agent with a single Q-value function. This algorithm trains a DDPG agent with target policy smoothing and delayed policy and target updates.

TD3 agents can be trained in environments with continuous or discrete states with continuous action spaces. TD3 agents use one or more Q-value function critics $Q(S,A)$ with deterministic policy actor $\pi(S)$.

Now we have some basic understanding about popular RL algorithms, next let's have a closer look at how to use them to develop real applications with MATLAB.

9.5   Reinforcement Learning in MATLAB

MATLAB provides a Toolbox named **Reinforcement Learning Toolbox** to direct and help users to build all kinds of popular RL-related applications with associated algorithms or agents. In fact, three kinds of tools are involved and included in this toolbox:

(1) APP—Reinforcement Learning Designer
(2) Functions—All RL-related functions
(3) Simulink Block

With the help of the Reinforcement Learning Toolbox, the user is able to

(1) Represent policies and value functions using deep neural networks or look up tables and train them through interactions with environments modeled in MATLAB® or Simulink.

(2) Evaluate the single- or multiagent reinforcement learning algorithms provided in the toolbox or develop your own.

(3) Experiment with hyperparameter settings, monitor training progress, and simulate trained agents either interactively through the App or programmatically.

(4) Improve training performance or simulations by running your implementations in parallel on multiple CPUs, GPUs, computer clusters, and the cloud (with Parallel Computing Toolbox™ and MATLAB Parallel Server™).

(5) Existing policies can be imported from deep learning frameworks such as TensorFlow™ Keras and PyTorch (with Deep Learning Toolbox™) through the ONNX™ model format.
(6) Generate optimized C, C++, and CUDA® code to deploy trained policies on microcontrollers and GPUs. The toolbox includes reference examples to help you get started.

Next let's take a look at the operational sequence and procedure in building a real RL-related application with MATLAB.

9.5.1   Procedure to Build Real RL-Related Applications in MATLAB

The general workflow for training an agent using reinforcement learning toolbox includes the following steps [30].

(1) **Formulate Problem**—Define the task for the agent to learn, including how the agent interacts with the environment and any primary and secondary goals the agent must achieve
(2) **Create Environment**—Define the environment within which the agent operates, including the interface between agent and environment and the environment dynamic model
(3) **Define Reward**—Specify the reward signal that the agent uses to measure its performance against the task goals and how to calculate this signal from the environment
(4) **Create Agent**—Create the agent, which includes defining a policy approximator (**actor**) and value function approximator (**critic**) and configuring the agent learning algorithm
(5) **Train Agent**—Train the agent approximators using the defined environment, reward, and agent learning algorithm
(6) **Simulate Agent**—Evaluate the performance of the trained agent by simulating the agent and environment together
(7) **Deploy Policy**—Deploy the trained policy approximator using, for example, generated GPU code

A point to be noted is that training an agent using reinforcement learning is an iterative process. Decisions and results in later stages can require you to return to an earlier stage in the learning workflow. For example, if the training process does not converge to an optimal policy within a reasonable amount of time, you might have to update some of the following before retraining the agent:

- Training settings
- Learning algorithm configuration
- Policy and value function (actor and critic) approximators

- Reward signal definition
- Action and observation signals
- Environment dynamics

Let's start our project developments by using the Reinforcement Learning APP, Reinforcement Learning Designer.

9.6   Using MATLAB Reinforcement Learning APP to Build Real Applications

By using the Reinforcement Learning Designer App provided by MATLAB, you can design, train, and simulate agents for some existing or predefined environments built by MATLAB. After finishing this section, you will be able to:

(1) Import an existing environment from the MATLAB® workspace or create a predefined environment
(2) Automatically create or import an agent for your environment, including DQN, DDPG, TD3, SAC, and PPO agents
(3) Train and simulate the agent against the environment
(4) Analyze simulation results and refine your agent parameters
(5) Export the final agent to the MATLAB workspace for further use and deployment

A point to be noted is that if you like to use the following features to build your agents, they will not work since these agents are not supported in the Reinforcement Learning Designer App:

(1) Multiagent systems
(2) Q, SARSA, PG, AC, and SAC agents
(3) Custom agents
(4) Agents relying on table or custom basis function representations

In that case, you need to use functions to build your projects.
   Let's start our journey with a quick review of this App.

9.6.1   Introduction to Reinforcement Learning Designer

Two possible ways can be used to open this Reinforcement Learning Designer:

(1) Clicking on the **APPS** icon on the MATLAB window, browse to the **MACHINE LEARNING AND DEEP LEARNING** group, and select the **Reinforcement Learning Designer** icon.
(2) Typing command **reinforcementLearningDesigner** in the MATLAB Command window.

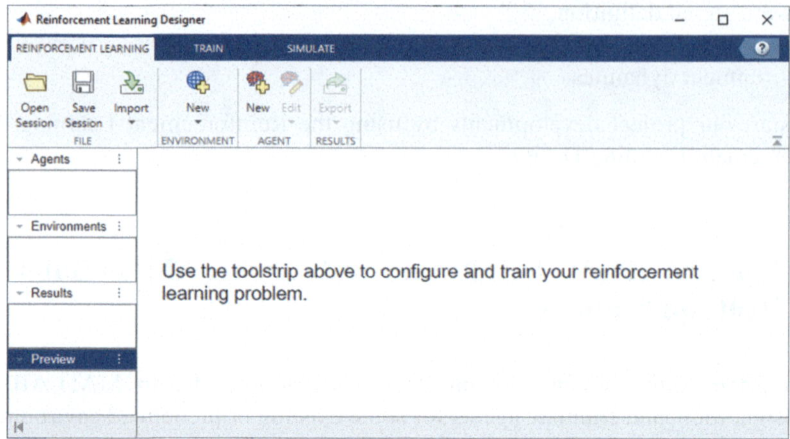

Fig. 9.3 The opened reinforcement learning designer

This App provides a completed GUI with all related function wizards to enable users to design, set up, build, and evaluate users' projects easily and conveniently. When you open this App, a major GUI is displayed as shown in Fig. 9.3.

Let's have a closer look at these icons or wizards to see how they work. First let's concentrate on all icons on the menu bar.

(1) **Open Session**: This icon is to enable users to open an existing session built by the user.
(2) **Save Session**: It is used to save the user's project with environments together with a session format.
(3) **Import**: It is used to allow users to import an agent or environment built by users before. You can import an environment from the MATLAB® workspace or create a predefined environment. You can also create a customer environment and use it. To use a custom environment, you must first create the environment object at the MATLAB command line and then import the object into Reinforcement Learning Designer.
(4) **New Environment**: It allows users to generate a new MATLAB Environment or a Simulink Environment based on some existing MATLAB environments or Simulink environments built by MATLAB.
(5) **New Agent**: It enables users to create a new agent based on the selected environment. The point to be noted is that you must first generate an environment and then generate a predefined agent based on the environment. The types of agents are limited to DQN, DDPG, TD3, SAC, and PPO in this Designer.
(6) **Edit Agent**: It allows users to edit an agent.
(7) **Export Result**: It enables users to export their project to the Workspace.

Initially, no agents or environments are loaded in the App.

On the left-hand side, four real-time monitors, **Agents**, **Environments**, **Results**, and **Preview**, are used to display the current components for each element created or selected by the users.

Next let's use a real example project to illustrate how to use this Designer to design and build a real RL-related project.

9.6.2   Build an RL Project with Reinforcement Learning Designer

In this section, we like to use a Cart Pole system as an example to illustrate how to build an RL project to control its function. Let's have a basic understanding about the Cart Pole system.

9.6.2.1   A Real Cart Pole System

A Cart Pole system is composed of a pole called pendulum attached by an un-actuated joint to a cart moving along a frictionless track. The pendulum starts upright, and the goal is to prevent it from falling over by increasing or reducing the cart's moving speed.

The cart with the pole can be considered as an environment, and the forces can be thought as actions. The running speeds can be considered as states, and the degrees between the pole and the vertical line could be thought of as rewards.

The pole starts upright, and the goal of the agent is to prevent it from falling over by applying a force of -10N or +10N to the cart to set moving speed. A reward of +1 is given for every time step the pole remains upright. An episode ends when: (1) the pole is more than 15 degrees from vertical or (2) the cart moves more than 2.4 units from the center.

The policy is a mapping function set between the current state and the next action, and it is a deterministic policy since all states can be detected without any uncertainty.

9.6.2.2   Build an RL Project to Control a Cart Pole System

Perform the following operations to build this project with Reinforcement Learning Designer and use a predefined environment:

(1) On the opened Reinforcement Learning Designer, click on the **New Environment** icon and select the **Continuous Cart Pole** environment (Fig. 9.4) by clicking on it. The selected environment is loaded and displayed in the **Environment** pane on the left side. Click on it and click on the **Preview** pane, a description about that environment is shown in that panel, as shown in

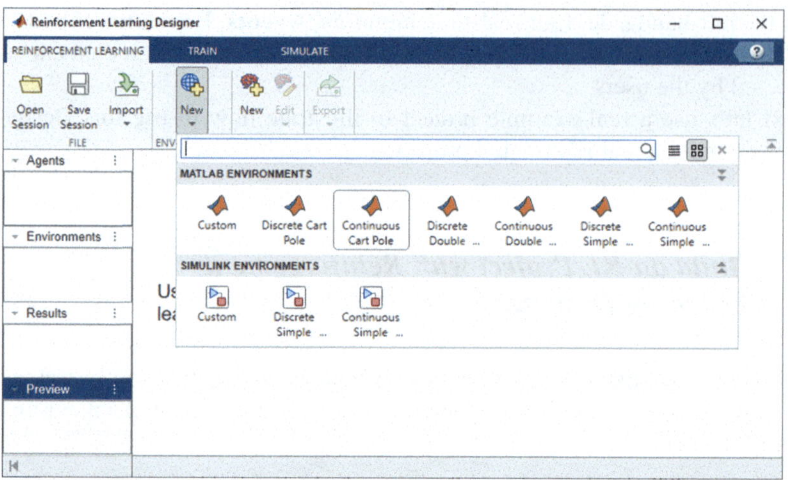

Fig. 9.4 Select the continuous cart pole environment

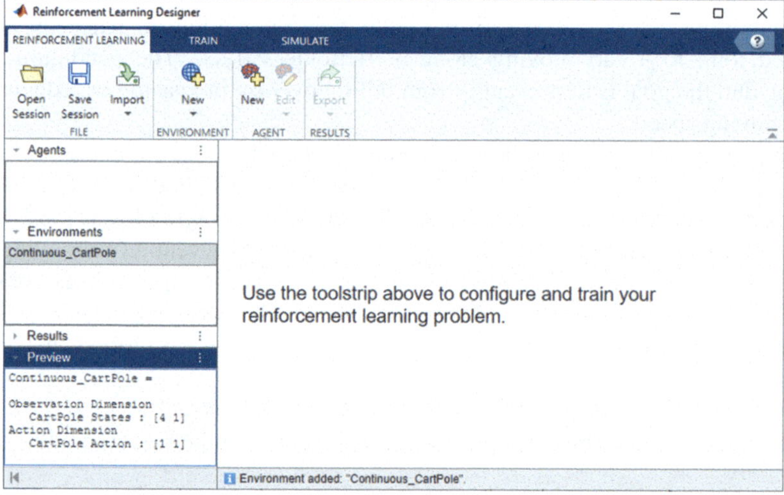

Fig. 9.5 The selected Continuous Cart Pole environment

Fig. 9.5. This environment has a continuous four-dimensional state space (the positions and velocities of both the cart and pole) and a discrete one-dimensional action space consisting of two possible forces, −1N or 1N.

(2) Click on the **New Agent** icon to create a predefined agent PPO. On the opened **Create agent** wizard shown in Fig. 9.6, select the agent or algorithm type, **PPO**, from the **Compatible algorithm** box and keep the default agent name **agent1**. Your finished **Create agent** wizard should match the one that is shown in Fig. 9.6. Click on the **OK** button to complete this step.

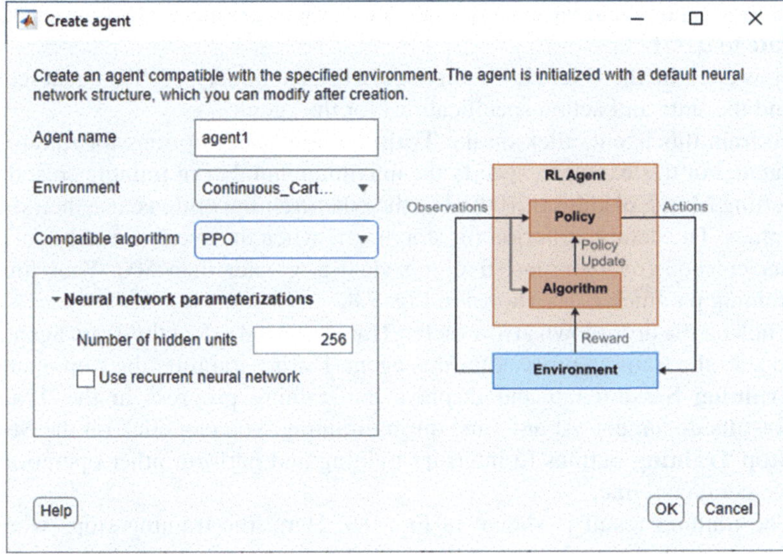

Fig. 9.6   The selected agent—PPO

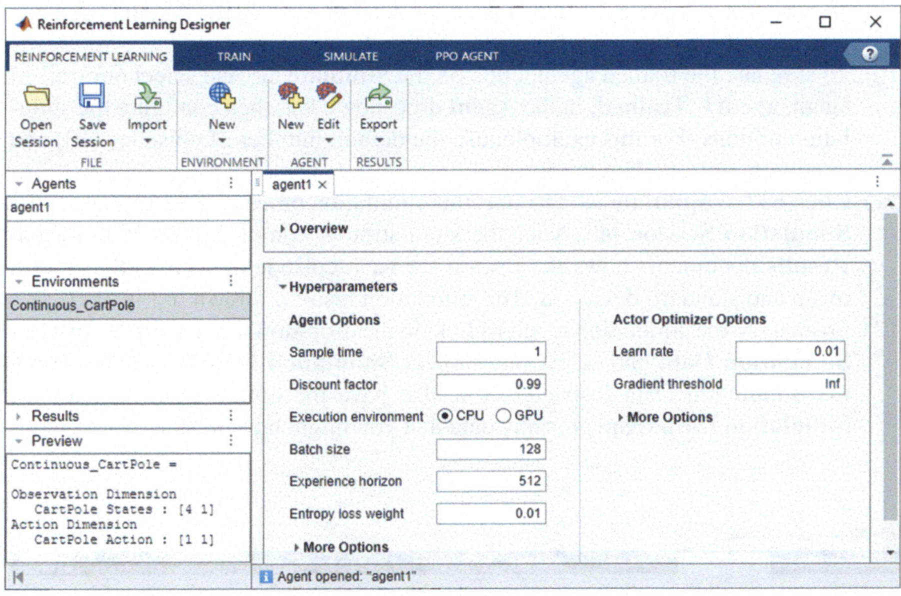

Fig. 9.7   The overview of the created agent PPO

(3) An **Overview** and **Hyperparameter** summary for this created agent is shown
in the next wizard, as shown in Fig. 9.7.

(4) In the **Hyperparameter** section, under **Critic Optimizer Options** set **Learn rate** to 0.001.

(5) Now click on the **Overview** item to get a brief summary of PPO agent features and the state and action specifications for the agent.

(6) To train this agent, click on the **Train** tab to specify options for training the agent. For this example, specify the maximum number of training episodes by setting **Max Episodes** to 1000. For the other training options, use their default values. The default criterion for stopping is when the average number of steps per episode (over the last five episodes) is greater than 500. Your finished training parameters are shown in Fig. 9.8.

(7) Click on the drop-down arrow on the Train icon and select the Train agent item to start the training process for this agent. During training, the App opens the **Training Session** tab and displays the training progress in the **Training Results** document. At any time during training, you can click on the **Stop** or **Stop Training** buttons to interrupt training and perform other operations on the command line.

(8) The training result is shown in Fig. 9.9. Here, the training stops when all agents reach the stop training criteria. The trajectory with light blue color is the episode reward, and the trajectory with dark blue color is the average reward. The trajectory with yellow color is episode Q0.

(9) To accept the training results click on the **Accept** icon. In the **Agents** pane, the App adds the trained agent, **agent1_Trained**.

(10) To simulate the trained agent, click on the **Simulate** tab and select our trained agent, **agent1_Trained**, in the **Agent** drop-down list, then configure the simulation options. For this example, use the default number of episodes (10) and maximum episode length (500).

(11) Click on the **Simulate** icon to start this simulation process. The App opens the **Simulation Session** tab. After the simulation is completed, the **Simulation Results** document shows the reward for each episode as well as the reward mean and standard deviation. The simulation result is shown in Fig. 9.10.

(12) To analyze the simulation results, click on the drop-down arrow on the **Inspect Simulation Data** and select the **Inspect Simulation Data** item. This opens the Simulation Data Inspector. You also have the option to clear from the **Simulation Data Inspector** any data that you might have loaded in a previous

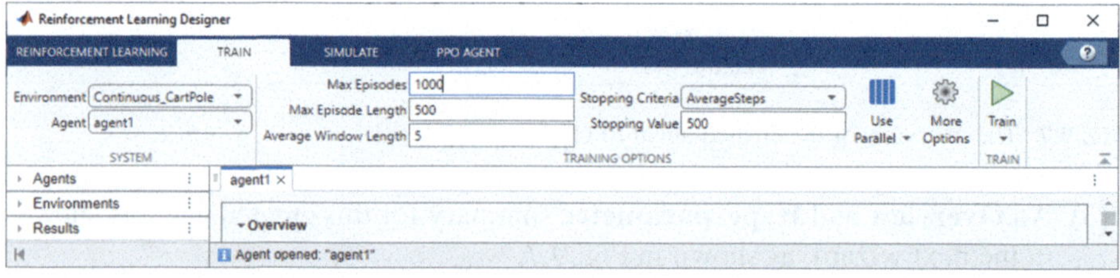

Fig. 9.8 The modified training parameters

Fig. 9.9 The training result for the PPO agent under the Continuous_CartPole environment

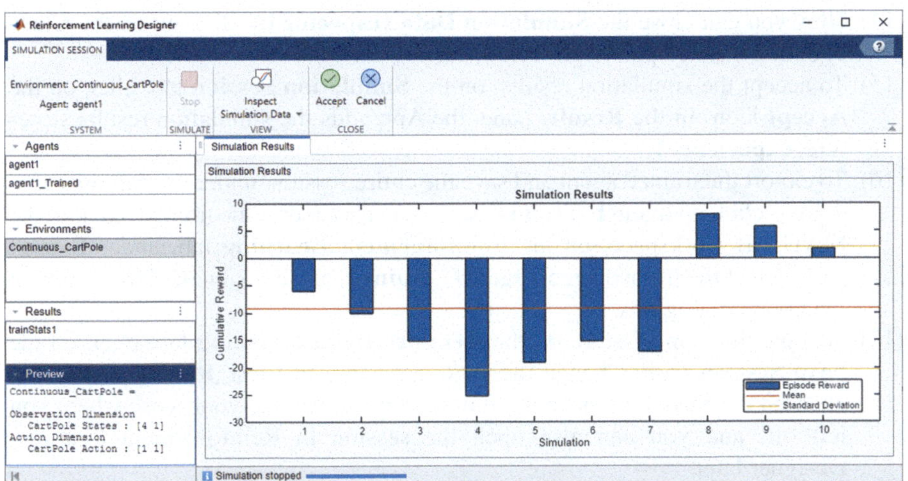

Fig. 9.10 The simulation result

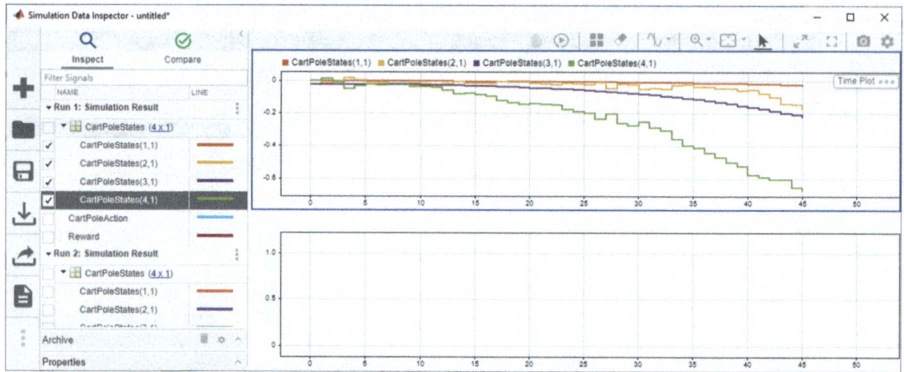

Fig. 9.11 The plotting results for four states

session. To do so, under **Inspect Simulation Data**, select **Clear and Inspect Simulation Data** item.

(13) In the **Simulation Data Inspector** you can view the saved signals for each simulation episode. By default, the upper plot area is selected. To show the first state (the cart position), during the first episode, under **Run 1: Simulation Result**, expand the **CartPoleStates** variable, and select all states, **CartPoleStates(1,1) ~ CartPoleStates(4,1)**. All four states are plotted in the top plot, as shown in Fig. 9.11.

(14) Now you can close the **Simulation Data Inspector** by clicking on the **Close X** button on the upper-right corner on this wizard.

(15) To accept the simulation results, on the **Simulation Session** tab, click on the **Accept** icon. In the **Results** pane, the App adds the simulation results structure, **experience1**.

(16) To export the trained agent and save the entire session, under the **Agents** pane, double-click on **agent1_Trained** item. Then, to export the trained agent to the MATLAB workspace, on the **Reinforcement Learning** tab, under **Export** icon, select the trained agent, **agent1_Trained**, and click on the **OK** button on the popup dialog box.

(17) To save the App session, on the **Reinforcement Learning** tab, click on the **Save Session** icon. Change the session's name to **First_RL_App.mat**, and click on the **Save** button. In the future, you can resume your work where you left off, and you can also open the session in Reinforcement Learning Designer later.

A complete session for this project, **First_RL_App.mat**, can be found on the Springer ftp site in the folder, **Students\Class Projects\Chapter 9**. You can use the **Open Session** icon on the App to load, retrain, and simulate it based on your options.

Next let's discuss some popular methods used to build a customer environment.

9.6.3 Three Popular Methods Used to Build Customer Environments

In the above project, we used a predefined environment **CartPole** built by MATLAB. In this section, we like to show users how to create a customer environment based on a real DC motor control system that we discussed in Chap. 5.

Reinforcement Learning Toolbox defines environments as objects. Such objects interact with agents using object functions or methods such as **step()** or **reset()**. Specifically, at the beginning of each training or simulation episode, the **reset()** function is called by a training or simulation function to set the environment's initial condition. Then at each training or simulation time step, the **step()** function is called to update the state of the environment and return the next state along with a reward.

After you create an environment object in the MATLAB workspace, you can extract observation and action specifications from the variable. Then you can use these specifications to create an *agent* that works within your environment. You can also use both the environment and agent variables as arguments for the built-in functions **train()** and **sim()**, to train or simulate the agent within the environment, respectively.

Three types of customer environments can be generated with MATLAB codes.

9.6.3.1 Custom Function Environments

For the purpose of calculations of the state transition, reward, observation, and initial state, the custom function environments use two functions, **step()** and **reset()**.

For single-agent environments, if you define your action and observation specifications and write them to your custom **step()** and **reset()** functions, you can use the function **rlFunctionEnv()** to return an environment object that can interact with your agent in the same way as any other environment does.

You can also create two different kinds of custom *multiagent* function environments:

(1) Multiagent environments with universal sample time, in which all agents execute in the same step.
(2) Turn-based function environments, in which agents execute in turns. Specifically, the environment assigns execution to only one group of agents at a time, and the group executes when it is its turn to do so.

For both kinds of multiagent environments, the observation and action specifications are cell arrays of specification objects in which each element corresponds to one agent.

For custom multiagent function environments with universal sample time, use the function **rlMultiAgentFunctionEnv()** to return an environment object. For custom turn-based multiagent function environments, use the **rlTurnBasedFunctionEnv()** function.

To specify options for training agents in multiagent environment, create and configure the **rlMultiAgentTrainingOptions** object. With the help of this object, it allows you to specify whether different groups of agents are trained in a decentralized or centralized manner. In a group of agents subject to decentralized training, each agent collects its own set of experiences and learns from its own set of experiences. In a group of agents subject to centralized training, each agent shares its experiences with the other agents in the group and each agent in the group learns from the collective shared experiences.

You can train and simulate your agents within a multiagent environment by using the functions **train()** and **sim()**, respectively. You can visualize the training progress of all the agents using the Reinforcement Learning Training Manager.

9.6.3.2   Custom Template Environments

The custom template environments are based on a modified class template.

To create a custom template environment, you can use the **rlCreateEnvTemplate()** function to open a MATLAB script that contains a template class for an environment, then modify the template by specifying environment properties, required environment functions, and optional environment functions.

Using the **rlCreateEnvTemplate()** function to create custom template environments is more elaborate than just writing custom **step()** and **reset()** functions, it gives you more flexibility in adding properties or methods that might be needed for your applications. For example, you can write a custom plot method to plot a visual representation of the environment at a given time.

9.6.3.3   Custom Simulink Environments

Custom Simulink environments are based on a Simulink model that you designed.

You can also use Simulink to design multiagent environments. In particular, Simulink allows you to model environments with multirate execution, in which each agent may have its own execution rates.

Now let's use a real project to illustrate how to use these methods to build customer environments for a DC motor control system. Let's start from the first method.

9.6.4   Build an RL Project to Control DC Motor with Customer Function Environment

As we discussed above, to create a customer environment, basically we need to configure two built-in functions, **reset()** and **step()**. The first function is used to set the environment's initial conditions and the second one is used to update the state of the environment and return the next state with a reward. Let's concentrate on our first function, **reset()**.

For our DC motor control system, the control target is to set a desired voltage (**Action**) to the DC motor and track and check the motor rotation speed to make sure that it is close to our setup or desired speed (not velocity since it is a vector). The environment related to this motor system includes the rotation angle θ (theta), the rotation angle speed $d\theta$ (dtheta), and the initial state.

The initial conditions for this motor environment contain a random starting rotation theta value, an initial value for the dtheta, and an initial state value.

Create a new Script file, name it as **mResetFunction.m,** and enter the codes shown in Fig. 9.12 into that file. Let's have a closer look at this piece of codes to see how it works.

(1) The customer reset function, **mResetFunction()**, is defined with two returned objects, the **InitialObservation** and **InitialState**.
(2) The initial theta **T0** is set to a random value with ± 0.05 rad.
(3) The dtheta is initialized to 0.
(4) The **InitialState** is defined by a column vector **[T0; Td0]**, which is a 2 × 1 array.
(5) The **InitialObservation** is set to equal to the **InitialState**.

Next let's build our customer function **step()**.

Create a new Script file, name it as **mStepFunction.m,** and enter the codes shown in Fig. 9.13 into that file. Let's have a closer look at this piece of codes to see how it works.

(1) The customer function head is declared first with two input arguments and four returned objects, **NextObs, Reward, IsDone,** and **NextState**. Two input arguments include the current **Action** and **State**.
(2) The maximum rotation speed of the motor is defined as 50 RPM.
(3) The sampling time is defined as 20 ms.
(4) The threshold angle is defined as 2 degree or 0.035 *rad*, and this is the maximum tolerated angle error for the rotating motor.

```matlab
% Customer reset() function definition
% This function is used to help to build a customer function environment
% July 27, 2024

1  function [InitialObservation, InitialState] = mResetFunction()
   % Reset function to place custom motor environment into a random initial state.

   % Theta (randomize)
2  T0 = 2 * 0.05 * rand() - 0.05;
   % Thetadot
3  Td0 = 0;

   % Return initial environment state variables as logged signals.
4  InitialState = [T0; Td0];
5  InitialObservation = InitialState;

   end
```

Fig. 9.12 The detailed codes for the customer reset() function

```matlab
% Customer step function for motor system environment
% This function is used to help to build a customer function environment
% July 27, 2024
1  function [NextObs, Reward, IsDone, NextState] = mStepFunction(Action, State)
   % Custom step function to construct motor environment for the function name case.

   % Max speed the input can apply
2  MaxSpeed = 50;
   % Sample time
3  Ts = 0.02;
   % Motor rotation angle at which to fail the episode - 2 degree
4  AngleThreshold = 2 * pi/180;
   % Reward each time step as the motor is in normal rotating range
5  RewardNormalMotor = 1;
   % Penalty when the motor fails to rotating in normal range
6  PenaltyErrorMotor = -1;

   % Check if the given action is valid.
7  if (Action < -MaxSpeed) || (Action > MaxSpeed)
       error('Action must be %g for going left and %g for going right.', -MaxSpeed, MaxSpeed);
   end

8  U_Voltage = Action;            % Assign the Action to the output command U_Voltage

   % Unpack the state vector from the logged signals.
9  Theta = State(1);
   ThetaDot = State(2);

   % Calculate output based on PD controller.
10 P = 80;
   D = 120;
   ThetaDotDot = U_Voltage + P*Theta + D*ThetaDot;

   % Perform Euler integration to calculate next state.
11 NextState = State + Ts.*[ThetaDot; ThetaDotDot];

   % Copy next state to next observation.
12 NextObs = NextState;

   % Check terminal condition.
13 Theta = NextObs(1);
14 IsDone = abs(Theta) > AngleThreshold;

   % Calculate reward.
15 if ~IsDone
       Reward = RewardNormalMotor;
16 else
       Reward = PenaltyErrorMotor;
   end

   end
```

Fig. 9.13 The detailed codes for the customer step() function

(5) If the motor is rotating under the normal tolerated error (<0.035 *rad*), the Reward is returned as 1.

(6) Otherwise the Reward returns a penalty value of -1.

(7) Here we need to check whether the action is in our required range, between -50 and 50 RPM. A warning message is displayed if it is not.

(8) Assign the **Action** to the control output command **U_Voltage**. Here the agent output or **Action** is equivalent to the control's output to the DC motor.

(9) Configure or assign two components in the state to two variables, **Theta** and **ThetaDot**.

(10) Calculate the next output or **dtheta** based on a PD control equation.

(11) Calculate the next state by adding the current state with the Euler integration.

(12) The next state is assigned to the next observation since they are the same in this project.

(13) Assign the first component in the state object, which is the current theta angle value, to the local variable **Theta**. The purpose of this assignment is for the next instruction.
(14) Check whether the current Theta is greater than the angle threshold value defined in step 4. If it is, set **IsDone** to true or 1 to help define the Reward value in the next step.
(15) If the **IsDone** returns a false, which means that the current rotation error is less than 2 degree, the Reward returns **RewardNormalMotor** (1).
(16) Otherwise it returns **PenaltyErrorMotor** (−1).

Finally let's build the codes to call those two functions to create our customer environment.

Create a new Script file, name it as **motor_function_env.m,** and enter the codes shown in Fig. 9.14 into that file. Let's have a closer look at this piece of codes to see how it works.

(1) Assign all two states from the environment, including the motor rotation angle and motor rotation speed, to the **Observation Information** object.
(2) Define the action space where an agent can apply one of two possible voltages, -50V or 50V, to the DC motor.
(3) Set up the initial action to 0.
(4) Call the customer reset function, **mResetFunction**(), to set the environment's initial conditions.
(5) Execute the customer step function, **mStepFunction**(), to update the state of the environment and return the next state with a reward.
(6) Use the system function **rlFunctionEnv**() to create a custom environment object with the observation and action specification, and the names of your customer step and reset functions. A customer environment object named **motor_func_env** is generated after this function calling.

An interesting point is that the function **rlFunctionEnv**() can automatically call another system function **validateEnvironment**() to verify the operation of the environment it created. The function

```
% Generate a customer environment for a DC motor system
% This piece of codes is to call 2 functions to build a customer environment
% July 27, 2024
1  ObsInfo = rlNumericSpec([2 1]);
   ObsInfo.Name = "Motor States";
   ObsInfo.Description = 'theta, dtheta';
2  ActInfo = rlFiniteSetSpec([-50 50]);
   ActInfo.Name = "Motor Action";
3  Action = 0;
4  [InitialObservation, Info] = mResetFunction()
5  [NextObservation, Reward, IsDone, UpdatedInfo] = mStepFunction(Action, Info)
6  motor_func_env = rlFunctionEnv(ObsInfo, ActInfo, "mStepFunction", "mResetFunction")
```

Fig. 9.14 The codes used to call two functions to create customer environment

validateEnvironment() will reset the environment, generate an initial observation and action, and simulate the environment for one or two steps. If there are no errors during these operations, validation is successful, and the **validateEnvironment**() function returns nothing. If errors occur, these errors will be displayed in the Command window. Based on error information, you can determine how to change or modify your observation specification, action specification, custom functions, or Simulink model in your customer reset and reset functions.

Now we have completed all coding jobs for creating a customer environment. However, before we can run this piece of codes to create our customer environment, a key issue is that you must save or store this Script file with both reset and step functions we built in the last section into the same folder or location at your computer to enable MATLAB compiler to find and locate them, and to call them to do this environment creation.

Run the project file **motor_function_env.m** and the running results are displayed in the Command window as shown below, and an environment object, **motor_func_env**, is created and added into the MATLAB Workspace.

motor_function_env
InitialObservation = -0.0239 0
Info = -0.0239 0
NextObservation =
-0.0239
-0.0382
Reward = 1
IsDone = logical 0
UpdatedInfo =
-0.0239
-0.0382
motor_func_env = rlFunctionEnv with properties:
StepFcn: "mStepFunction"
ResetFcn: "mResetFunction"
Info: [2×1 double]

Later on, we can import this customer environment to the Reinforcement Learning Designer, create a new agent for that imported environment, train the agent by using different algorithms, and check the training results.

9.6.4.1   Create a New Agent for Customer Generated Environment in App

Now let's create a new agent based on the customer-created environment object to perform and check the training results for the new agent. Perform the following operations to complete these operations:

(1) Open the Reinforcement Learning Designer by using either the APP or the command.

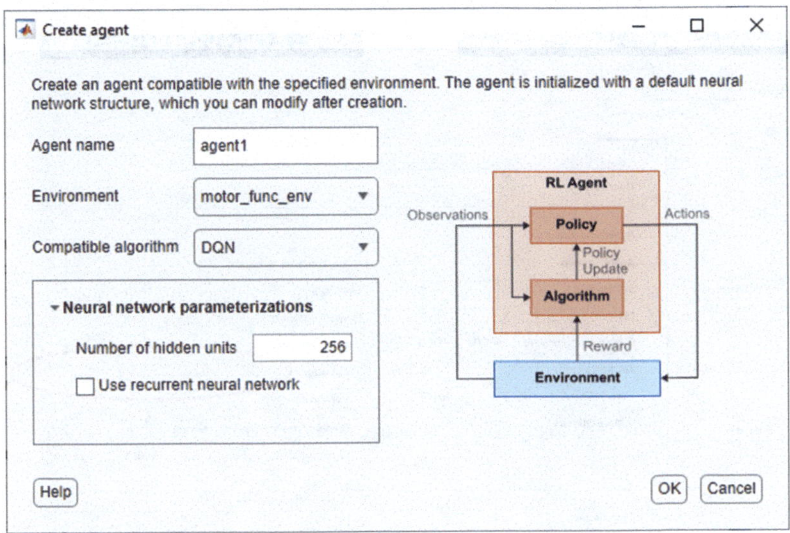

Fig. 9.15 The opened Create agent wizard

(2) On the opened Designer, click on the drop-down arrow on the **Import** icon, you can find that our created environment, **motor_func_env**, is located under the Select Environment category. Click and select it, then you can find that our customer-created environment, **motor_func_env**, is imported and displayed under the **Environments** tab on the left pane.

(3) Now click on the **New** from the **AGENT** icon to create a new Agent.

(4) On the opened Create agent wizard, as shown in Fig. 9.15, keep the Agent name **agent1**, and Environment name, **motor_func_env**, with no change. For the RL algorithm, you can select any one from the default three if you like. For this example, just keep the DQN algorithm and click on the **OK** button to go to the next wizard.

(5) On the next wizard shown in Fig. 9.16, expand the **More Options** under the **Critic Optimizer Options** group, and select the **sgdm** algorithm. Then click on the **TRAIN** item on the top menu bar.

Next let's perform the training operation for this newly created Agent.

9.6.4.2   Train and Export the New Agent for Customer-Generated Environment

Perform the following operations to train our newly created Agent **agent1**:

(1) On the opened TRAIN wizard as shown in Fig. 9.17, change the **Max Episodes** from 500 to 300, and this number is used to indicate how many episodes can be used for this training process.

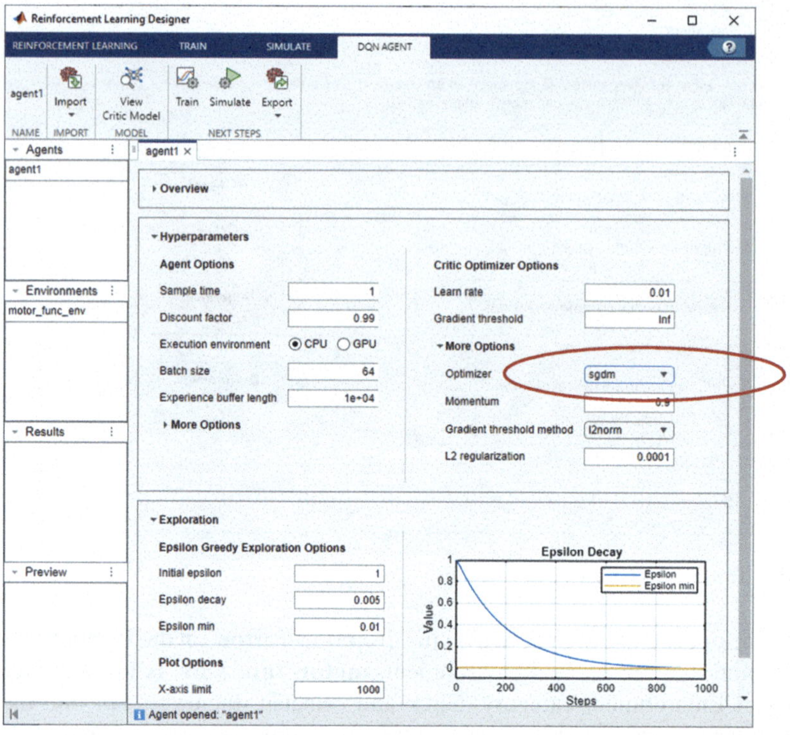

Fig. 9.16 Select the Optimized wizard

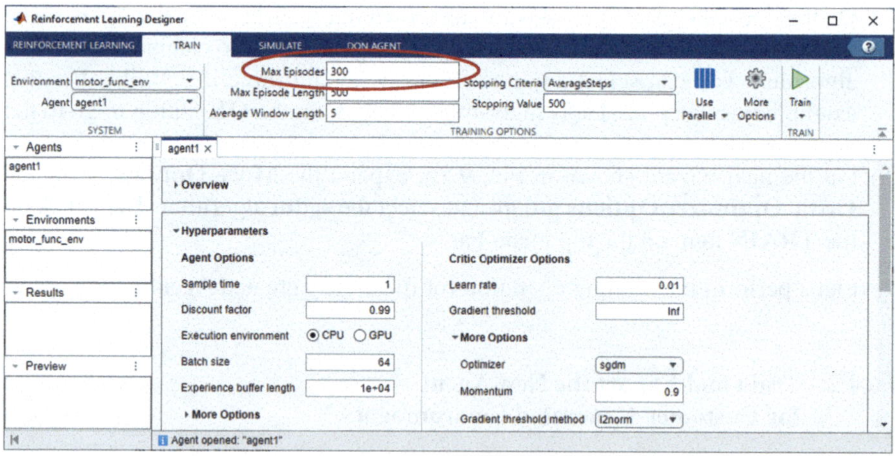

Fig. 9.17 The opened TRAIN wizard

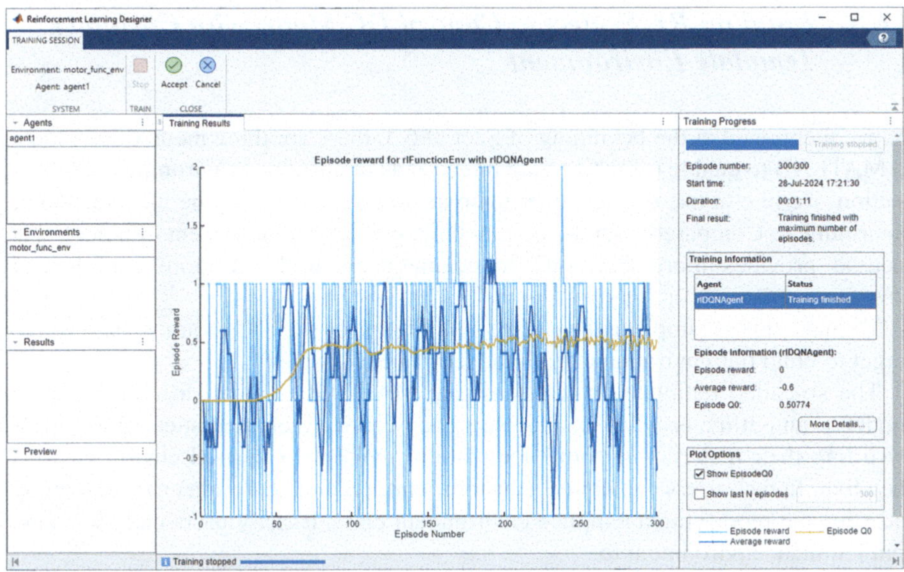

Fig. 9.18 The training result

(2) Click on the drop-down arrow on the Train icon and select the Train agent item
     to start the training process.

The training starts and the training result is shown in Fig. 9.18.

If you are satisfied with this training result, click on the green-color **Accept** circle on the top to accept it. To save and export this training result with the entire
session, perform the following operations:

(1) Click on the **REINFORCEMENT LEARNING** tab on the top to return to the
     home wizard.
(2) Click on the drop-down arrow on the **Export** icon and select the trained agent,
     agent1_Trained.
(3) Click on the **OK** button on the popup dialog to export it to the MATLAB
     Workspace.
(4) To save this entire session, click on the **Save Session** icon on the top, change the
     session name to **trained_func_env.mat**, and click the **Save** button to save it to
     your folder.

Next let's try to create a customer environment by using another method, the
Customer Template Class.

9.6.5   Build an RL Project to Control DC Motor with Customer Template Environment

As we mentioned at the beginning of Sect. 9.6.3, there are three methods available in MATLAB to enable users to create three kinds of customer environments. In this section, we like to use the customer template class method to generate a customer environment. Compared with the first method, customer function environment, this method provides more flexibility to enable users to build more complicated environments.

To make things simple and easy, we still like to use our DC motor system as the target to build our environment with the template class method.

The so-called template class means that MATLAB provides a template class model as an outline or a frame, and users can generate their customer environment by filling their materials into that template to make their desired environment as they like. In fact, users can define a custom reinforcement learning environment by creating and modifying a template environment class. Exactly users can use a custom template environment to:

- Implement more complex environment dynamics
- Add custom visualizations to your environment
- Create an interface to third-party libraries defined in languages such as C++, Java®, or Python®

By default, the MATLAB sample template class contained a simple cart-pole balancing model similar to the cart-pole predefined environments described in Sect. 9.6.2.1.

This template class includes three major parts with predefined codes for car-pole model:

(1) Environment properties
(2) Required environment methods
(3) Optional environment methods

Table 9.1 Some popular required environment methods used in MATLAB

Function Name	Descriptions
getObservationInfo()	Return information about the environment observations.
getActionInfo()	Return information about the environment actions.
sim()	Simulate the environment with an agent
validateEnvironment()	Validate the environment by calling the reset() function and simulating the environment for one time step using step() function.
reset()	Initialize the environment state and clean up any visualization.
step()	Apply an action, simulate the environment for one step, and output the observations and rewards; also, set a flag indicating whether the episode is complete.
Constructor()	A method with the same name as the class that creates an instance of the class.

The environment properties part can be used to declare your physical constants and variables, environment variables and constraints. The required environment methods are the methods defined by users, and they are required for basic and normal operations for building environment. The optional environment methods are not required but highly recommended to assist users to make the building process smooth and easy.

Some popular required environment methods provided by MATLAB are shown in Table 9.1. More optional methods are not listed in this table but we will discuss them later in our real example project as they are needed.

Three methods, **reset()**, **step()** and **constructor**, in Table 9.1 are required methods, all others are not required or optional methods. MATLAB provides a function, **rlCreateEnvTemplate()**, to enable us to create a template environment object based on its class.

First let's create a template environment class file based on the template environment class by calling the function **rlCreateEnvTemplate("Environment_Name")**. Perform the following operation steps to create our customer environment class file:

(1) Open the MATLAB and type **rlCreateEnvTemplate("motor_template_env")** in the Command window to create our customer environment class file. The name of our customer environment is **motor_template_env**.
(2) Press the **Enter** key on the keyboard to create and open this object file. The name of this opened template object file is **untitled.m**.

The opened template file contained all codes for the sample cart-pole model with three parts. We need to fill and replace all of those codes with our codes in three parts to build our customer environment. First let's pay attention to the first part, environment properties part.

```matlab
1  classdef motor_template_env < rl.env.MATLABEnvironment
       %MOTOR_TEMPLATE_ENV: Template for defining custom environment in MATLAB.

       %%% Properties (set properties' attributes accordingly)
       properties
           % Specify and initialize environment's necessary properties
2          MaxSpeed = 50;
           % Sample time
3          Ts = 0.02;
           % Motor rotation angle at which to fail the episode: 2 degree
4          AngleThreshold = 2 * pi/180;
           % Reward each time step as the motor is in normal rotating range
5          RewardNormalMotor = 1;
           % Penalty when the motor fails to rotating in normal range
6          PenaltyErrorMotor = -1;
       end

       properties
           % Initialize system state [theta, dtheta]'
7          State = zeros(2, 1)
       end

       properties(Access = protected)
           % Initialize internal flag to indicate episode termination
8          IsDone = false
       end
```

Fig. 9.19 The codes for the environment properties part

Environment Properties

On the opened object template file, replace all codes under the **properties** part with the codes shown in Fig. 9.19. Let's have a closer look at this piece of codes to see how it works.

(1) This class definition header line is created by MATLAB automatically as a new template environment object is generated. This coding line means that our template object belongs to the **rl.env.MATLABEnvironment** class or namespace. All constants, variables, or environment constraints declared under the properties category belong to this class, and they can be considered as family members for this class. All of these member data (family members) are private, and they can only be accessed by using some related optional methods defined later.

(2) The first member data is the environment constraint, maximum angular speed of the DC motor, **MaxSpeed**, which is defined as 50 RPM.

(3) The sampling time T_S is defined as 20 ms.

(4) The tolerated rotation angular error is 2 degree or 0.035 rad.

(5) The reward for motor rotating in the normal status is defined s 1.

(6) The reward for motor rotating in the abnormal status is defined as -1.

(7) The **State** contains two variables, **theta** and **dtheta**, respectively with a 2 × 1 vector.

(8) Initially we set the **IsDone** flag as false to indicate that the process has not been done.

Next let's continue on **Required Environment Methods** part.

Required Environment Methods

All user-defined methods used to assist in building this customer environment should be covered under this category. On the opened template environment object file, continue to replace the original codes with the codes shown in Fig. 9.20.

Let's have a closer look at this piece of codes to see how it works.

(1) The system keyword, **methods**, is declared here first to indicate that all the following codes are used to define the required methods used for this template environment class.

(2) The constructor is declared here with the same name as that of this object file. Regularly a constructor never returns anything, but here we use this to indicate that this constructor returns to this class itself. In fact, **this** should be a pointer, but here it is used as a reference.

(3) Create specifications object for a numeric action or observation channel. Exactly the function **rlNumericSpec()** is used here to create an object specified for our action with predefined number of observations or states. In this project, we have two states, the motor rotation angle theta and derivative of theta. The name and the description about these states are also defined.

(4) Create specifications object for a finite-set action or observation channel. The function **rlFiniteSetSpec()** is used to create a finite-set object for our action. The difference between this function and the last one is that this function generates a finite-set action object with the upper and lower bounds, from -50

```matlab
%% Necessary Methods
   methods
       % Constructor method creates an instance of the environment
       function this = motor_template_env()
           % Initialize Observation settings
           ObservationInfo = rlNumericSpec([2 1]);
           ObsInfo.Name = "Motor States";
           ObsInfo.Description = 'theta, dtheta';

           % Initialize Action settings
           ActInfo = rlFiniteSetSpec([-50 50]);
           ActInfo.Name = "Motor Action";

           % The following line implements built-in functions of RL env
           this = this@rl.env.MATLABEnvironment(ObservationInfo, ActInfo);

           % Initialize property values and pre-compute necessary values
           updateActionInfo(this);
       end

       % Apply system dynamics and simulates the environment with the given action for one step.
       function [NextObs, Reward, IsDone, NextState] = step(this, Action)

           % Get action – call optional method
           U_Voltage = getVoltage(this, Action);

           % Unpack state vector
           Theta = this.State(1);
           ThetaDot = this.State(2);

           % Calculate output based on PD controller.
           P = 80;
           D = 120;
           ThetaDotDot = U_Voltage + P*Theta + D*ThetaDot;

           % Perform Euler integration to calculate next state.
           NextState = this.State + this.Ts.*[ThetaDot; ThetaDotDot];
           % Copy next state to next observation.
            NextObs = NextState;

           % Check terminal condition.
           Theta = NextObs(1);
           IsDone = abs(Theta) > this.AngleThreshold ;

           % Get reward – call optional method
           Reward = getReward(this);

           % (optional) use notifyEnvUpdated to signal that the environment has been updated
           notifyEnvUpdated(this);

       end
       % Reset environment to initial state and output initial observation
       function InitialObservation = reset(this)
           % Theta (randomize)
           T0 = 2 * 0.05 * rand() - 0.05;
           % Thetadot
           Td0 = 0;

           % Return initial environment state variables as logged signals.
           InitialObservation = [T0; Td0];
           this.State = InitialObservation;

           % (optional) use notifyEnvUpdated to signal that the environment has been updated
           notifyEnvUpdated(this);
       end
   end
```

Fig. 9.20 The codes for the required methods

RPM to 50 RPM, with counter-clockwise as positive and clockwise as negative values in direction.

(5) This coding line is used to create a new RL environment with our customer-defined observations, states, and action by calling and implementing a built-in

MATLAB environment function, **MATLABEnvironment()**. The created environment is returned and assigned to this class.

(6) The function **updateActionInfo()** is used to update the default environment with our customer-defined properties or parameters.

(7) The customer **step()** function starts from here with four returned components and two input arguments. First we need to get the input argument, **Action**, which belongs to a private member data in this class, by using the optional method **getVoltage()** whose body will be discussed later, and assign it to our real Action **U_Voltage**.

(8) Next we need to get two states, theta in **State(1)** and dtheta in **State(2)**, and assign them to two local variables, **Theta** and **ThetaDot**, respectively. Since the **State** is defined in the **properties** part (Fig. 9.19) and they belong to private data, the class name, **this**, must be prefixed before those state variables.

(9) Three coding lines starting from step 9 are used to calculate the next action with a PD controller, and the result is assigned to the local variable **ThetaDotDot**.

(10) The next state is calculated by adding the current state with the Euler integration. Since both **State** and $\mathbf{T_S}$ are defined in the properties part and they belong to private member data, thus the current class, **this**, must be prefixed before them.

(11) The next state is assigned to the next observation to update the observation.

(12) Assign the next observation, **NextObs(1)**, whose value is the updated theta, to the local variable **Theta**, and then compare it with the angular threshold value to see whether it is greater than the threshold. If it is, the flag **IsDone** is set to true to indicate that upper or the lower bound of the rotation angle has been met, and the project should be terminated.

(13) To get the current reward, the optional method, **getReward()** whose detailed codes will be discussed later, is called.

(14) An optional notification for this environment updating can be sent to the system by calling the function **notifyEnvUpdated()**.

(15) The customer function **reset()** starts from here. The current class name, **this**, must be forwarded into this function since this function needs to initialize all variables that belong to private member data in that class. The motor rotation angle theta or **T0** is initialized to a random degree, and the dtheta, **Td0**, is set to 0.

(16) Both initialized theta and dtheta are assigned to a 2 × 1 vector **InitialObservation**, and this vector is then assigned to the current **State** with the current class, **this**, being prefixed.

(17) Another optional notification is sent to the system to indicate that the current environment has been updated.

Next let's continue on **Optional Environment Methods** part.

Optional Environment Methods

All optional methods used in this project are shown in Fig. 9.21. Let's have a closer look at this piece of codes to see how it works.

```matlab
%%% Optional Methods (set methods' attributes accordingly)
1    methods
       % Helper methods to create the environment
2      function U_Voltage = getVoltage(this, action)
         if ~ismember(action, this.ActionInfo.Elements)
            error('Action must be %g for going left and %g for going right.', -this.MaxSpeed, this.MaxSpeed);
         end
3         U_Voltage = action;
       end
       % update the action info based on max force
4      function updateActionInfo(this)
         this.ActionInfo.Elements = this.MaxSpeed*[-1 1];
       end

       % Reward function
5      function Reward = getReward(this)
         if ~this.IsDone
            Reward = this.RewardNormalMotor ;
         else
            Reward = this.PenaltyErrorMotor;
         end
       end

       % (optional) Visualization method
6      function plot(this)
       % Initiate the visualization

       % Update the visualization
7         envUpdatedCallback(this)
       end

       % (optional) Properties validation through set methods
8      function set.State(this, state)
         validateattributes(state,{'numeric'},{'finite', 'real','vector', 'numel', 2},'','State');
         this.State = double(state(:));
         notifyEnvUpdated(this);
       end

9      function set.Ts(this, val)
         validateattributes(val,{'numeric'},{'finite','real','positive','scalar'},'','Ts');
         this.Ts = this.Ts;
       end
10     function set.AngleThreshold(this,val)
         validateattributes(val,{'numeric'},{'finite','real','positive','scalar'},'','AngleThreshold');
         this.AngleThreshold = val;
       end
    end

11   methods (Access = protected)
       % (optional) update visualization everytime the environment is updated
       % (notifyEnvUpdated is called)
12     function envUpdatedCallback(this)
       end
    end
end
```

Fig. 9.21 The codes for the optional methods

(1) All optional methods are defined under the **Optional Methods** category.
(2) The first optional method is **getVoltage(),** and it is used to get the current
    Action. Prior to doing this pickup, we need to check whether the current action
    is in the normal operational range by using the function **~ismember()**. This
    function will check if the current action value is between -50 and 50 RPM
    (**-MaxSpeed** and **MaxSpeed**). If it is not, a true is returned since a not (~)
    operator is applied in front of this function, a warning message is displayed to
    indicate this situation.

(3) Otherwise if it is normal, the current action is assigned to the local variable **U_Voltage**.

(4) The function **updateActionInfo()** is used to update the bound conditions for the action, which are two bound speed values.

(5) The optional function **getReward()** is used to pick up the current reward value. Based on the IsDone flag, the reward returns the related value, either the **RewardNormalMotor** for the normal running of the motor or **PenaltyErrorMotor** for running error of the motor.

(6) The optional function **plot()** can be used to plot the current running status of the environment. Here we leave it blank without any coding for it.

(7) The **evnUpdatedCallback()** function is used to update any visualization if we have one.

(8) The function **set.State()** is used to set a set of definite states for the **State** vector. The only point to be noted for this function is the number of states that is 2 in our current project.

(9) The function **set.Ts()** can be used to set up a specified value for the sampling time.

(10) Similarly, the function **set.AngleThreshold()** can be used to set a new or an updated threshold value for the motor rotating angle.

(11) No method in this part is defined as a protected one, so leave this space as a blank.

(12) No codes are developed for the **evnUpdatedCallback()** function since we do not need it in this project.

Now click on the **Save** icon on the top menu to save this customer template class as a Script file with the same name of this class, **motor_template_env.m**, to your selected folder.

At this point, we have completed the building and development of our customer template environment class. Next we need to create a new object based on this class to enable us to use it in the Reinforcement Learning Designer to train and test our customer environment.

9.6.5.1 Instantiate the Custom Template Environment

To use this customer environment, we need to instantiate this class. To do that, perform the following operations:

(1) Running this project by clicking on the green-color **Run** button on the menu bar. The running result **ans** is displayed in the Command window as below:

```
   motor_template_env
    ans = motor_template_env with properties:
         MaxSpeed: 50    Ts: 0.0200   AngleThreshold: 0.0349
 RewardNormalMotor: 1
          PenaltyErrorMotor: -1     State: [2×1 double]
```

(2) On the Command window, type: **motor_template_env = ans**, and press the
    Enter key from the keyboard. Immediately our customer environment object
    motor_template_env is added and displayed in the Workspace.

Next we can use this customer environment in the Reinforcement Learning Designer
to train and test our customer environment with selected agent or algorithm.

9.6.5.2   Create and Train a New Agent for Customer Template Environment

Now open the Reinforcement Learning Designer, and we like to create and train a
new agent under that App. Perform the following operations for this kind of creation
and training:

(1) Click on the drop-down arrow from the **Import** icon and select our customer
    template environment **motor_template_env** under the **Select Environment**
    item. The selected environment is added to this project and shown under the
    Environments pane.

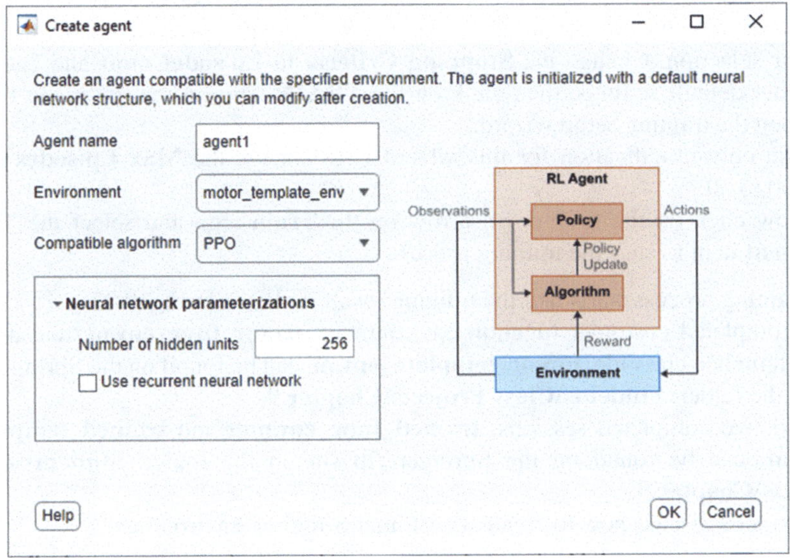

Fig. 9.22 The opened Create agent wizard

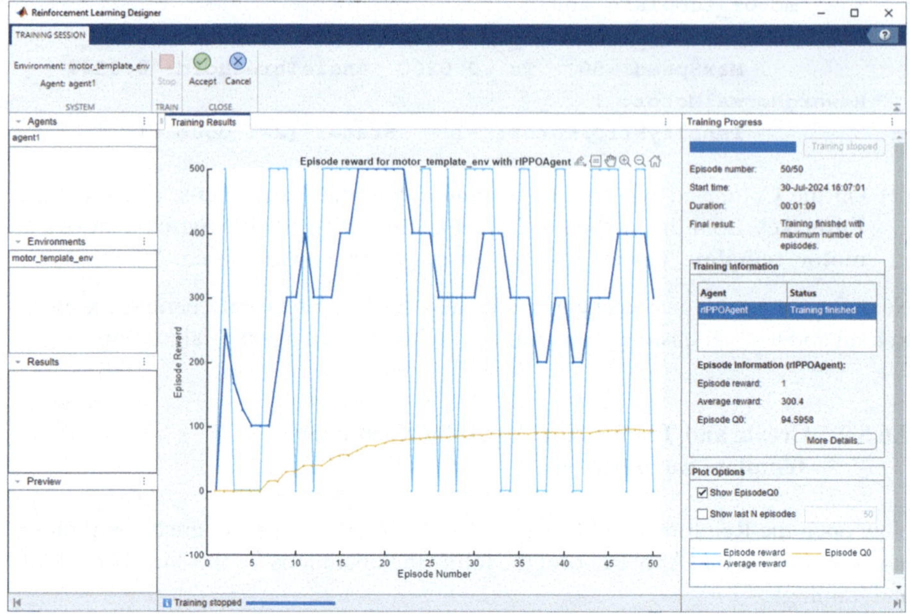

Fig. 9.23 The training result for customer template class environmen

(2) Click on the **New** from the **AGENT** icon to open the **Create agent** wizard, as shown in Fig. 9.22.
(3) Select the **PPO** from the **Compatible algorithm** box as our agent or algorithm for this environment, as shown in Fig. 9.22. Click on the **OK** button to continue.
(4) On the next wizard, all possible Agent options are displayed to enable us to do our selection. Change the **Stopping Criteria** to **EpisodeCount** and keep all other default settings, then click on the **TRAIN** item on the top menu bar to open the training setup wizard.
(5) The only modification for this wizard is to change the **Max Episodes** from 500 to 50.
(6) Now click on the drop-down arrow on the **Train** icon and select the **Train agent** item to start the training process.

The training process starts and the training result is shown in Fig. 9.23.

A completed customer function environment, **motor_func_env.m**, and a customer template class file, **motor_template_env.m**, can be found on the Springer ftp site in the folder: **Students\Class Projects\Chapter 9**.

Also two completed sessions, **trained_func_env.mat** and **trained_template_ env.mat**, can be found on the Springer ftp site in the folder, **Students\Class Projects\Chapter 9**.

Next let's discuss how to create a customer Simulink environment.

9.6.6   Build an RL Project to Control DC Motor with Customer Simulink Environment

To create a custom Simulink environment, first create a traditional Simulink model that represents a closed-loop control system with input, forward controller, the plant, the feedback channel, and error signal derivation parts. Then this traditional closed-control system model can be modified to a Simulink environment model.

Your environment model must have an input signal, the action, which influences (through some discrete, continuous, or mixed dynamics) its next internal state and its outputs, which are the observation, the reward, and the Is-Done signals. The Is-Done signal is a scalar that indicates the termination of an episode, causing the simulation to stop when its value is true.

Regularly the observations include reward and state that are feedback to the agent to enable the latter to make an action decision to the environment. Generally to build a customer Simulink environment, the following operational steps are needed:

(1) Build a traditional closed-loop control model with Simulink blocks
(2) Remove the feedforward controller and replace it with an RL Agent block
(3) Build all other components, including the Generating Observations block, Calculating Reward block, and Stop Simulation block
(4) Connect all of the above blocks with the RL Agent and the original plant (either in the continuous domain or in the discrete domain)
(5) Build the MATLAB codes to initialize and configure most blocks above to enable the Simulink environment model can be used or called by Reinforcement Learning Designer to perform training or simulation job

In this section, we still like to use our DC motor control system as an example to build our customer Simulink environment. Due to its simplicity, we can skip the first step and directly build our closed-loop model. Let's start our building process from step 3. Build all other components, including the Generating Observations block, Calculating Reward block, and Stop Simulation block. Finally we can connect them together with an Agent.

A key point to be noted when building these components is that all components are subsystem blocks in Simulink library.

9.6.6.1   Build All Subsystems for the Simulink Environment Model

Let's build this Simulink environment model starting from each subsystem one by one. The first subsystem is the Generating Observation block.

Build the Generating Observations Subsystem Block

Open the Simulink by typing **simulink** in the MATLAB Command window and press the **Enter** key on the keyboard. Perform the following operations to build this subsystem:

(1) Click on the **Blank Model** to open a blank model frame.
(2) Click on the **Library Browser** icon on the top to open the Simulink Library since we need to use all desired blocks from that Library.
(3) On the opened library, expand the **Simulink** and the **Commonly Used Blocks** item, drag the **Subsystem** block, and add it to the work panel on the right.
(4) Double-click on the added **Subsystem** block to open it, remove the connection line between the **In1** and the **Out1** ports by clicking on that line, and press the **Delete** key on the keyboard. To get the part's name, just click the port and the name is displayed under it.
(5) Click and select the name under the port **In1**, change it to **speed_error**.
(6) Go to the Library again, locate and add another port **In1** by dragging and placing it to the right panel. Locate it just under the **speed_error** port. Then change its name to **speed**.
(7) In the Library, expand the **Math Operations** group, select and add the **Add** to the panel.
(8) Go to the Library again, locate and add a **Bus Creator** to the right panel.
(9) Go to the Library, expand the **Discrete** group and select and add the **Discrete Derivative** block to the right panel. Set the value for the derivative **Gain value** K = 0.05.
(10) Change the name of the port **Out1** to **observation**.
(11) Make connections between these blocks by dragging between the arrows.

Your finished **generate observations** subsystem block is shown in Fig. 9.24a.

Click on the upper arrow ⬆ on the menu bar to get a subsystem block, as shown in Fig. 9.24b. Change this block's name to **generate observations**, as shown in Fig. 9.24b.

Next let's continue for the next subsystem, Calculating Reward block.

Build the Calculating Reward Subsystem Block

Perform the following operations to build this subsystem:

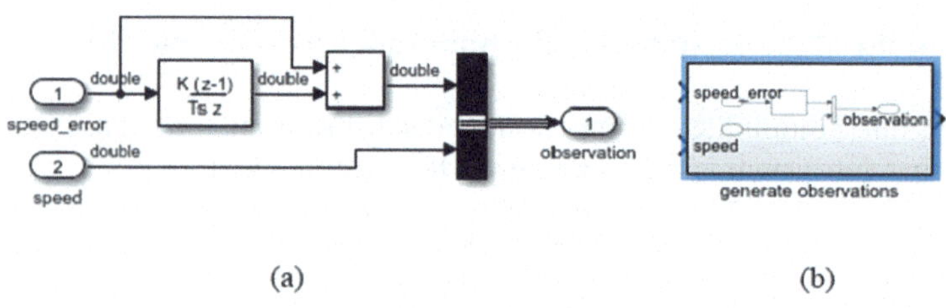

(a)   (b)

Fig. 9.24 The completed subsystem for generate observations block

(1) Go to the Library and select the **Subsystem** block and add it to the right panel. Locate it just under the **generate observation** block.

(2) Double-click on the new added subsystem block to open it. Remove the connection line between the port **In1** and the port **Out1**.

(3) Add another port **In1** into the panel and locate it just under the first port **In1**. Change the name of the first port **In1** to **speed_error**, and the name of the second port **In1** to **exceeds bounds**, respectively. Also change the name of the port **Out1** to **reward**.

(4) Go to the Library and expand the **Logic and Bit Operations** item and locate the **Compare To Constant** block. Drag and add two of them to the right panel.

(5) Double-click on the first compare block to open it, click on the drop-down arrow on the **Operator** combo box, and select the < operator and enter 0.035 into the **Constant value** box. Click on the **OK** button to make this setup effective.

(6) Double-click on the second compare block and change its settings to > 0.035.

(7) Go to the Library, re-expand the **Commonly Used Blocks,** and add three **Gain** blocks. Change three gain values for three added Gain blocks to: 28.57, −28.57, and −100, respectively since we like the **reward** to be 1 when the **speed_error** is less than 0.035, otherwise to be −1 when the **speed_error** is greater than 0.035. A product of 28.57 and 0.035 is equal to 1. The third gain is −100 to indicate that an error has occurred and the project needs to be terminated.

(8) Go to the Library, expand the **Math Operations** item, and locate and add the **Add** block to the right panel. Double-click on that **Add** block to open it, and type one more + after the first two ++ in the **List of signs** box to make it a three-input adder. Click on the **OK** button to make this setup effective.

(9) Make connections between arrows and your finished calculated reward subsystem block should match one that is shown in Fig. 9.25a.

(10) Click on the upper arrow ⬆ on the menu bar to get a subsystem block, as shown in Fig. 9.25b. Change this block's name to **calculate reward**, as shown in Fig. 9.25b.

Build the Stop Simulink Subsystem Block

Perform the following operations to build this subsystem:

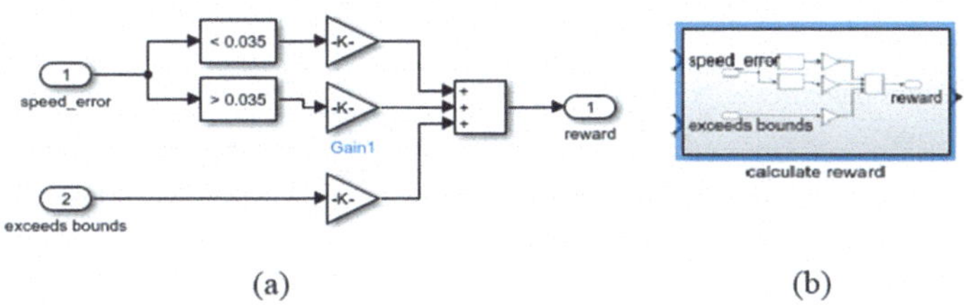

Fig. 9.25 The completed subsystem for calculating reward block

(1) Go to the Library, select the **Subsystem** block, and add it to the right panel. Locate it just under the **calculate reward** block.
(2) Double-click on the new added subsystem block to open it. Remove the connection line between the port **In1** and the port **Out1**.
(3) Change the name of the port **In1** to **speed**, and the name of the port **Out1** to **stop**.
(4) Go to the Library, expand the **Logic and Bit Operations** item, and locate the **Compare To Constant** block. Drag and add two of them to the right panel.
(5) Double-click on the first compare block to open it, click on the drop-down arrow on the **Operator** combo box, select the >= operator, and enter 50 into the **Constant value** box. Click on the **OK** button to make this setup effective.
(6) Double-click on the second compare block and change its settings to <= -50.
(7) Still in the **Logic and Bit Operations** item locate the **Logic Operator** block. Drag and add it to the right panel.
(8) Double-click on that Logic Operator block to open it, select **OR** from the **Operator** box, and keep 2 in the **Number of input ports**. Then click on the **OK** button to make these settings effective.
(9) Make connections between arrows and your finished stop simulation subsystem block should match one that is shown in Fig. 9.26a.
(10) Click on the upper arrow ⬆ on the menu bar to get a subsystem block, as shown in Fig. 9.26b. Change this block's name to **stop simulation**, as shown in Fig. 9.26b.

Next we need to add four more blocks into this Simulink model to complete this model. These blocks include the input constant speed (we assumed the input is a speed to simplify this simulation) block, a signal Sum block, the DC motor dynamic model block that is a discrete transfer function, and the RL Agent block.

Perform the following operation steps to add these blocks:

(1) Go to Library and expand the **Commonly Used Blocks** group, and locate and add the **Constant** block to the right panel. Set its value to 10.
(2) Go to the Library and still in the **Commonly Used Blocks** group, and locate and add the **Sum** block to the right panel.
(3) Double-click on the Sum block to open its property wizard, and replace the second + symbol with the symbol on the **List of signs** box. Click on the **OK** button to make this setup effective.

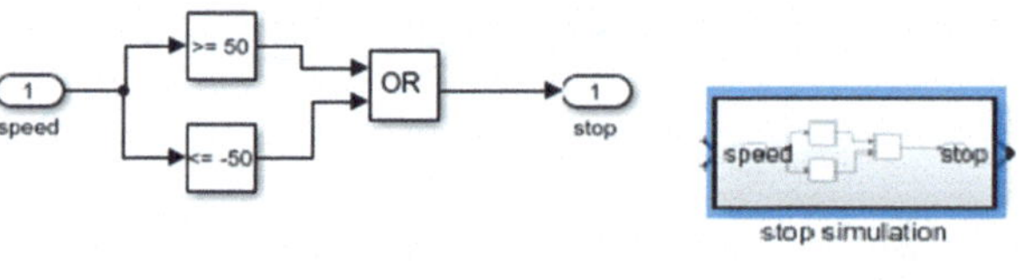

Fig. 9.26 The completed subsystem for stop simulation block

(a)   (b)

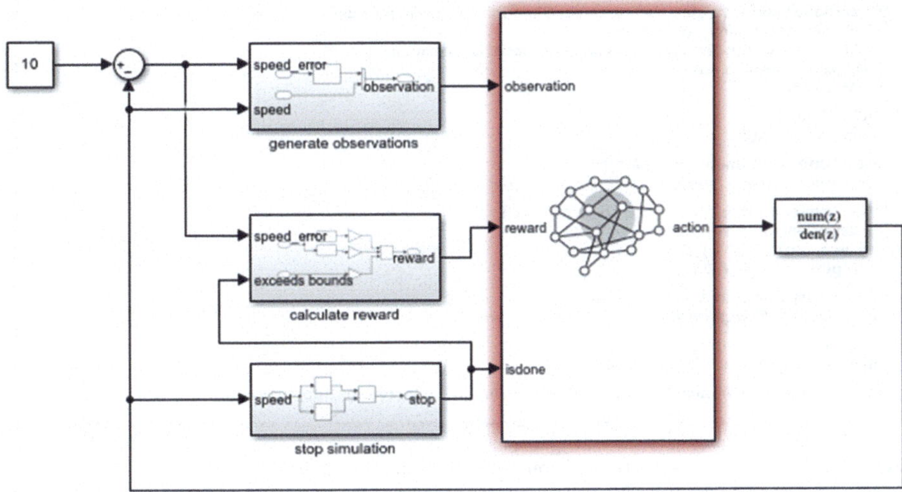

Fig. 9.27 The finished Simulink environment model for DC motor system

(4) Go to Library and expand the **Discrete** group, locate and add the **Discrete Transfer Fcn** block to the right panel.

(5) Double-click on the Transfer block to open its property wizard, and enter 20 into the **Numerator** box and [1 430 0] into the **Denominator** box. These are parameters used to set a dynamic model for our DC motor in the discrete domain. Click on the **OK** button.

(6) Go to Library and expand the **Reinforcement Learning** group, and locate and add the **RL Agent** block to the right panel.

(7) Make connections between arrows to make it a closed-loop control system, as shown in Fig. 9.27.

(8) The **RL Agent** block is surrounded by a red color box, and this indicates that some error has occurred about that block. One of the most popular possible errors is that the block may contain an incorrect Agent object. To fix it, double-click on that block, and enter **rlACAgent** to the **Agent object** box. Then click on the **Apply** and the **OK** button to complete this correction. Now the red color box has disappeared.

Before we can move to the next step, click on the **Save** button to save this Simulink environment model to one of your folders. On the opened **Save** dialog box, enter **motor.slx** as the name for this model and click on the **Save** button.

We complete the building process for our Simulink environment model in the Simulink domain, now let's move to the next step, developing and building the MATLAB codes to initialize and configure this Simulink environment model to make it ready for the training and simulation process for this DC motor system.

```
% Initialize and configure the customer Simulink environment model
% Name: motor_simulink_env.m
% The name of the Simulink environment model is: motor.slx
% Output: motor_s_env
% Aug 2, 2024
1  mdl = "motor";
2  open_system(mdl)

3  actionInfo = rlNumericSpec([1 1]);
   actionInfo.Name = "speed";

4  open_system(mdl + "/generate observations");
   observationInfo = rlNumericSpec([2 1],...
5     LowerLimit=[-50 0]',...
      UpperLimit=[50 0.035]');

6  observationInfo.Name = "observations";
   observationInfo.Description = "speed, speed_error";

7  open_system(mdl + "/calculate reward");
   open_system(mdl + "/stop simulation");

8  motor_s_env = rlSimulinkEnv(mdl, mdl + "/RL Agent", observationInfo, actionInfo)
```

Fig. 9.28 The codes to initialize and configure the Simulink environment model

9.6.6.2   Build MATLAB Codes to Set Up and Configure the Simulink Environment

To initialize and configure the Simulink environment model, we need to develop a piece of MATLAB codes to perform those functions.

To do that, create a new MATLAB Script file, name it as **motor_simulink_env.m**, and enter the codes shown in Fig. 9.28 into that file.

Let's have a closer look at this piece of codes to see how it works.

(1) The name of our Simulink environment model is **motor.slx**, which was built by us in the last section. A key issue is that you cannot use or assign that model with the full name when using it. Instead, you can only use the model name without the extension **.slx**. Otherwise a compiling error may be encountered since this is a requirement in MATLAB. In this way, the model is assigned to another local variable **mdl**. Another issue is that you must store this file with your Simulink environment model in the same folder to enable the compiler to locate and use it.

(2) To configure the Simulink environment model, we need to open it first. A system function **open_system()** is used for that purpose.

(3) We need to generate all configuration structures and parameters with a system function **rlNumericSpec()** for our Action object, which is a single object or motor speed.

(4) Similarly, to configure the **generate observations** block, we need to open it first. The full path for that block is our model name + block name. The same system function **rlNumericSpec()** is used for that purpose. Our observation object contains two variables, **speed** and **speed_error**, which is a 2×1 vector.

(5) The upper and the lower bounds are set for those two variables, the lower bound for the motor rotation speed is -50 RPM, and for the rotation speed error is 0.

The upper bound for the motor rotation speed is 50 RPM, for the rotation speed error is 0.035 rad.

(6) Set up and define the observations' name and descriptions with related properties.

(7) For both **calculate reward** and the **stop simulation** blocks, we do not need to perform any configuration job. Just open them to confirm both of them.

(8) Finally create our completed Simulink environment model by adding our Agent with the system function **rlSimulinkEnv()** and the configuration parameters involved in **observationInfo** and the **actionInfo** structures that work as arguments.

Now run this project and the Simulink environment model will be opened during the running process, and the running result is displayed in the Command window as shown below:

```
motor_simulink_env
motor_s_env = SimulinkEnvWithAgent with properties:


        Model : motor
        AgentBlock : motor/RL Agent
        ResetFcn : []
    UseFastRestart : on
```

Also a Simulink environment model, **motor_s_env**, has been added to the Workspace on the right of the MATLAB Command window. We will use it to perform our training process for this model in the Reinforcement Learning Designed in the next section.

At this point, we have completed building our Simulink environment model. Next let's try to train and simulate it for our DC motor system.

9.6.6.3   Train and Simulate the Simulink Environment Model

To train and simulate our Simulink environment model, open the Reinforcement Learning Designer first. You can do that by opening either from the **APP** tab or from the Command window.

On the opened Reinforcement Learning Designer, perform the following operational steps to perform the training process:

(1) Click on the drop-down arrow from the **Import** icon on the top and select our Simulink environment model **motor_s_env** that is under the **Select Environment** group by clicking on it to add it to our RL trainer.

(2) Then click on the drop-down arrow on the **New** located at the **AGENT** icon to open the Create agent wizard, as shown in Fig. 9.29.

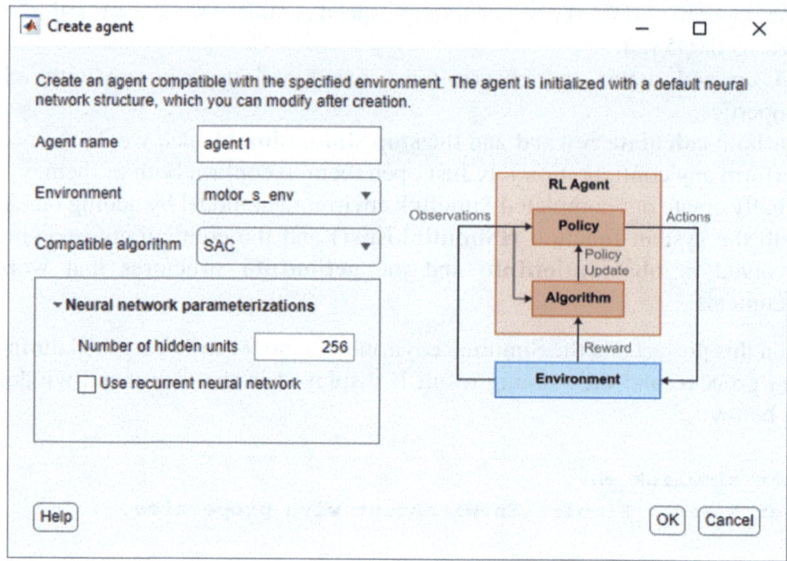

Fig. 9.29 The opened Create agent wizard

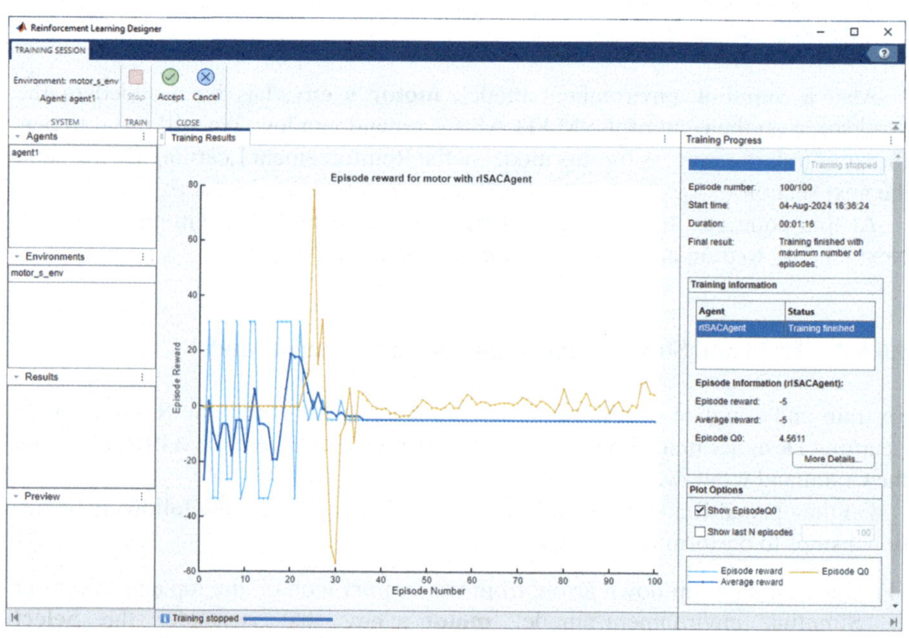

Fig. 9.30 The training result for the Simulink environment model

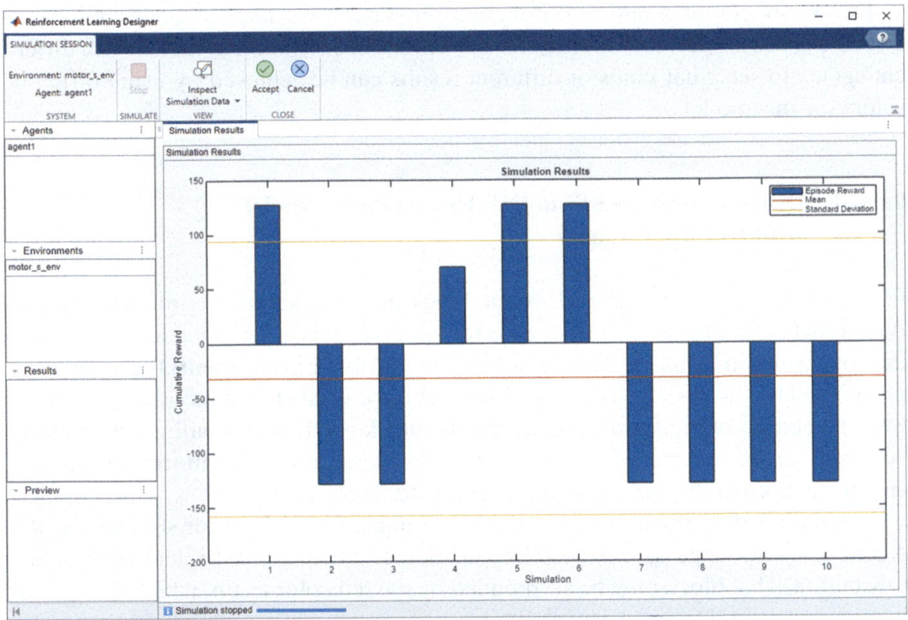

Fig. 9.31 The simulation result for the Simulink environment model

(3) Keep the default agent, **agent1**, with no change for the agent name, and select the **SAC** from the **Compatible algorithm** box as the training algorithm. Click on the **OK** button to continue to the next step.

(4) Keep all default options for this algorithm with no change, and click on the **TRAIN** tab on the top to open the **TRAIN** wizard. The only modification is to change the number of the **Max Episodes** from 500 to 200 since we do not need so many episodes for this training.

(5) Now click on the drop-down arrow on the **Train** icon and select the **Train agent** to start the training process.

The training result is shown in Fig. 9.30.

To perform the simulation study for this model, just click on the **SIMULATE** tab on the top, and click on the **Simulate** icon to start the simulation process. The simulation result is shown in Fig. 9.31.

If this simulation result is acceptable, click on the green-color Accept icon. You can also Export this trained model to the Workspace and save this entire model with the training and simulation results with a single session file. To do that, click on the **Save Session** icon under the **ENVIRONMENT LEARNING** tab to save it in one of your preferred folders. The name of this saved session can be: **trained_sinulink_env.mat**.

A completed Script file **motor_simulink_env.m** and a Simulink environment model **motor.slx** can be found on the Springer ftp site in the folder: **Students\Class Projects\Chapter 9**.

Due to its complex and multifunction properties, we like to spend more time discussing some special properties for the Simulink environment model with different agents to see what kinds of different results can be achieved by using different actors for this model.

9.6.6.4   Discussions About Simulink Environment Model with Different Agents

First let's pay attention to the RL Agent block in our Simulink environment model. As we mentioned in Sect. 9.6.6.1, an error appeared when we first added this block into our Simulink model. The default name for this block, **agentObj**, cannot be recognized by the system. The reason for that error is due to the incorrect initialization and configuration for the block, and the block needs to be configured correctly before it can be recognized. In fact, the codes in our Script file **motor_simulink_ env.m** can correct this error when it is executed later.

To confirm that, open our Simulink environment model **motor.slx** and the **RL Agent** block. Change its name back to the default one, **agentObj**, and click on the **OK** button. That block may be surrounded by the red color again, which means that it has some error. Do not worry about it.

Now run our Script file **motor_simulink_env.m** and you can find that the error is gone as the Simulink environment model is opened. That it is!

The reason why we entered the Soft Actor-Critic (SAC) algorithm as the agent to that block in Sect. 9.6.6.1 is to speed up our building process and leave the solution to this section.

Secondly, you can use any other different algorithms, including the TD3, PPO, DDPG, and TRPO, as an agent to train and simulate our Simulink environment model. To do that, on the opened Reinforcement Learning Designer, after importing our Simulink environment model, **motor_s_env**, select any other algorithm to train and simulate our model. You can even compare the training and simulation results among different agents with different algorithms to see the differences among them.

To keep all training results for comparison purposes, you can add more agents one by one and compare their training results with the same Simulink environment model. In Figs. 9.32, 9.33, 9.34, 9.35 and 9.36, different training results of using different agents, such as the agent TD3, DDPG, PPO, SAC, and TRPO, are displayed. It can be found that the training result of using the SAC agent is better compared with the results of using other agents, and it provides smoother control output or actions to the control environment or DC motor plant.

One issue may have been noticed by some readers, and it is the difference in the training results between one shown in Fig. 9.30 and the other one shown in Fig. 9.35. Both training results are obtained by using the same agent SAC. One possible reason for that difference is due to the actual and nominal agent used for those trainings. The result shown in Fig. 9.30 is trained by an actual SAC agent with a hard copy of the SAC Agent, **rlACAgent**, which was assigned to the **RL Agent** block directly in Simulink. But the result shown in Fig. 9.35 is a nominal agent, which

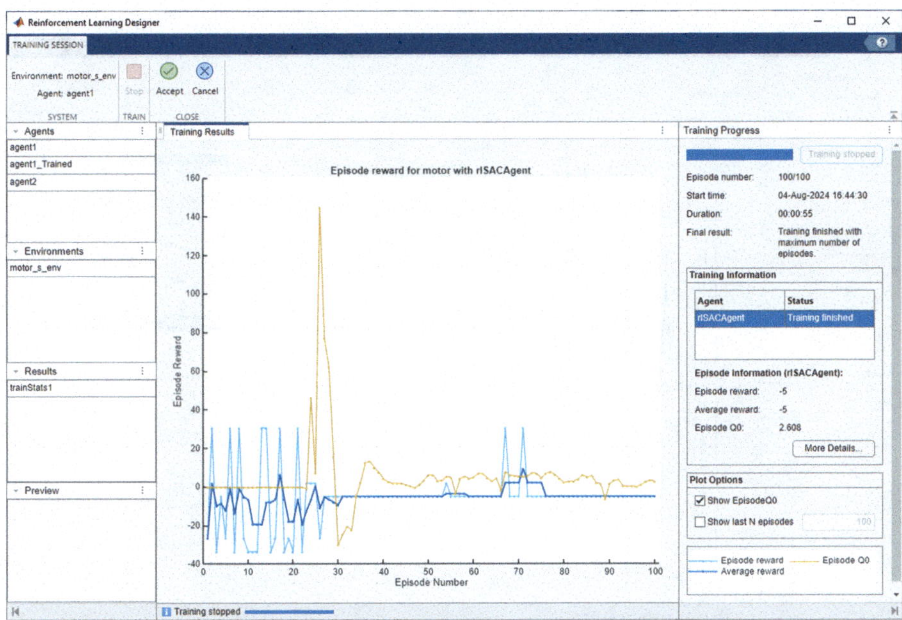

Fig. 9.32 The training result by using the TD3 Agent

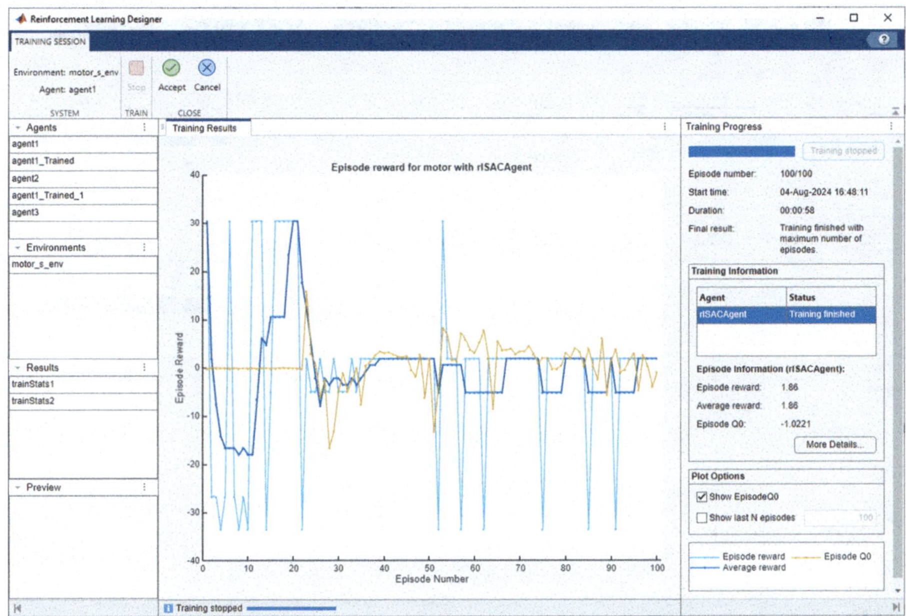

Fig. 9.33 The training result by using the DDPG Agent

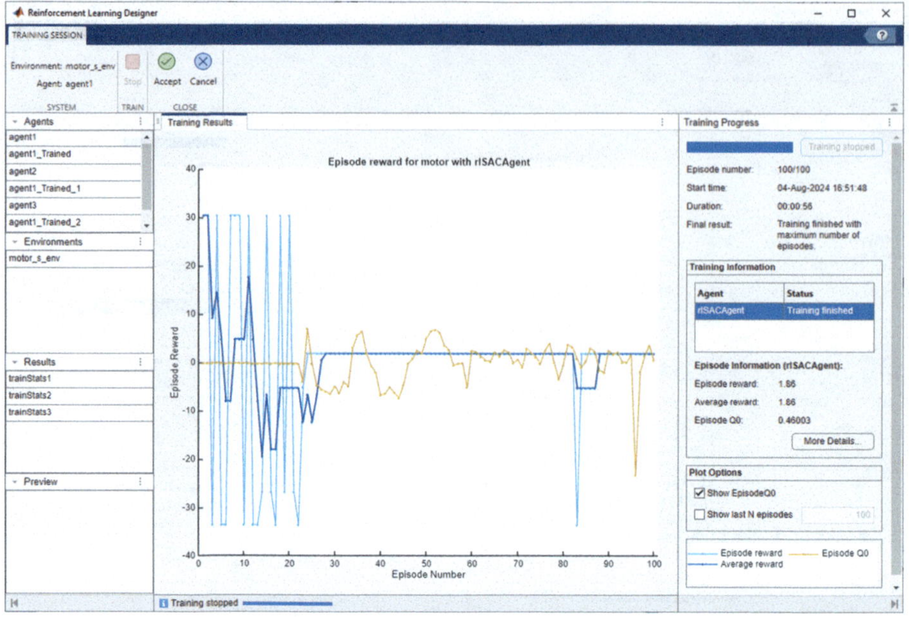

Fig. 9.34 The training result by using the PPO Agent

Fig. 9.35 The training result by using the SAC Agent

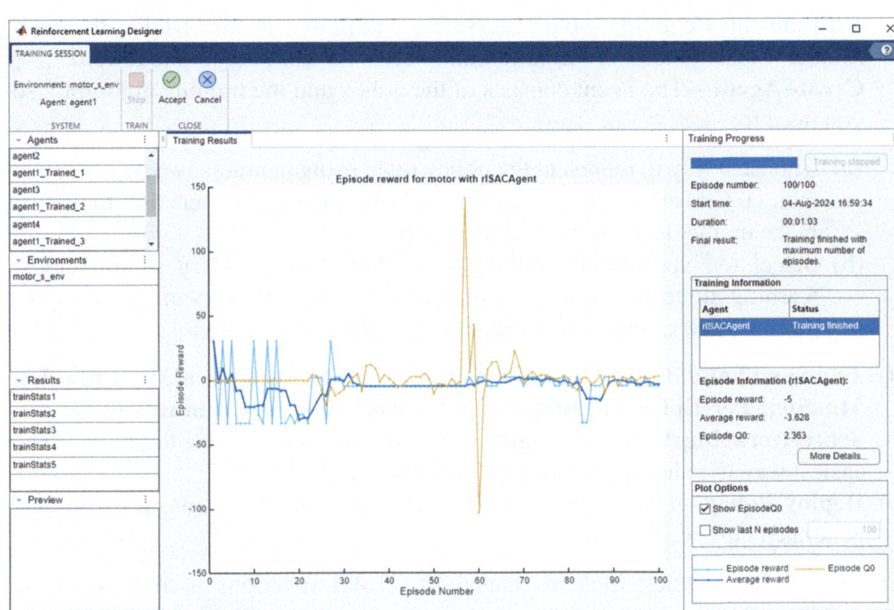

Fig. 9.36 The training result by using the TRPO Agent

means that the default agent in the **RL Agent** block is replaced by a SAC agent dynamically during the Script running process.

By comparing the above training results, it looks like the training results of using the DDPG, PPO, and TD3 are relatively better compared with other agents.

In the following sections, we concentrate our discussions on how to build real applications by using MATLAB reinforcement learning-related functions.

9.7 Using MATLAB Reinforcement Learning Functions to Build Real Applications

As we mentioned in Sect. 9.5.1, the general workflow for building an RL-related project using reinforcement learning toolbox functions can be simplified with the following steps:

(1) **Create Environment**—Define the environment within which the agent operates, including the interface between agent and environment and the environment dynamic model.
(2) **Define Reward**—Specify the reward signal that the agent uses to measure its performance against the task goals and how to calculate this signal from the

environment. Regularly, both observation and policy are involved in the environment. Thus you do not need to build additional codes to define reward.

(3) **Create Agent**—The agent consists of the policy and the training algorithm, so you need to:

(a) Choose a way to represent the policy (e.g., using neural networks or lookup tables). Consider how you want to structure the parameters and logic that make up the decision-making part of the agent.

(b) Select the appropriate training algorithm. Most modern reinforcement learning algorithms rely on neural networks because they are good candidates for large state/action spaces and complex problems.

(4) **Train and Validate the Agent**—Set up training options, such as **MaxEpisodes**, **MaxStepsPer-Episode**, **StopTrainingCriteria**, **StopTrainingValue**, and **ScoreAveraging-WindowLength**, and train the agent to tune the policy. The easiest way to validate a trained policy is via simulation.

(5) **Deploy Policy**—Deploy the trained policy approximator using, for example, generated GPU code.

First let's have a quick look at related MATLAB functions used to play the above roles.

9.7.1 Introduction to MATLAB Reinforcement Learning Functions

Let's start with creating the customer environment with MATLAB functions.

9.7.1.1 MATLAB Functions Used to Create Customer Environments

In Sects. 9.6.4, 9.6.5, and 9.6.6, we have discussed in detail how to use functions, **rlFunctionEnv()**, **rlCreateEnvTemplate()**, and **rlSimulinkEnv()** to create and build three different types of customer environments. Review those sections with detailed codes to understand better the MATLAB functions used to create customer environments. To save time and energy, you can also directly use some predefined environments built by MATLAB.

9.7.1.2 MATLAB Functions Used to Define Rewards

To provide a guide to the learning process, reinforcement learning uses a scalar reward signal generated from the environment. This signal measures the performance of the agent with respect to the task objectives. In fact, for a given observation (state), the reward works as feedback and measures the immediate effectiveness of taking a specified action. During training, an agent updates its policy based on the rewards received for different state-action combinations.

Regularly, you should provide a positive reward to encourage the agent's actions and a negative reward or a penalty to discourage the other related actions. A well-designed reward signal guides the agent to maximize the expectation of the (possibly discounted) cumulative long-term reward.

For example, when an agent performs a task for as long as possible, a common strategy is to provide a small positive reward for each time step when the agent successfully performs the task and a large penalty when the agent fails. This approach encourages longer training episodes while heavily discouraging actions that lead to episodes in which the agent fails.

You can specify either continuous or discrete reward signals. In either case, you must provide a reward signal that provides rich information when the action and observation signals change.

For control system applications in which cost functions and constraints are already available, you can also generate reward functions from such specifications.

Regularly, both observation and policy are involved in the environment at the MATLAB Reinforcement Learning. Thus you do not need to build additional codes to define reward.

9.7.1.3   MATLAB Functions Used to Create Agents

Various and different built-in agents are included and available in the MATLAB RL Toolbox, and you do not need to create a new agent in most applications. Instead, you can directly use those agents to save time and energy. Regularly those agents provide better performances compared with those developed by users or called customer agents.

Due to multiple agents available, different functions should be used to create related built-in agents. Table 9.2 lists some popular functions used to create different agents.

Table 9.2 Some useful methods used to create popular built-in agents

Function Name	Descriptions
rlDQNAgent()	Create a DQN Agent.
rlSARSAAgent()	Create a SARSA agent.
rlPGAgent()	Create a PG agent
rlACAgent()	Create an AC agent.
rlPPOAgent()	Create a PPO agent.
rlTRPOAgent()	Create a TRPO agent.
rlDDPGAgent()	Create a DDPG agent.
rlTD3Agent()	Create a TD3 agent.
rlSACAgent()	Create a SAC agent.
rlMBPOAgent()	Create a MBPO agent (not good for image environment).

In addition to using these built-in agents, you can also build your customer agent based on your algorithm. To define your custom agent, first create a class that is a subclass of the **rl.agent.CustomAgent** class. As a starting point for your own agent, you can open and modify this custom agent class later.

Basically MATLAB provides two possible ways to create a built-in agent, and they are:

(1) Create a default DQN agent based on the observation and action specifications from the environment.
(2) Create an actor and critics and use these objects to create your agent.

For example, to create a DQN agent in the first way, perform the following operations:

(1) Create observation specifications for your environment. If you already have an environment object, you can obtain these specifications by using the function **getObservationInfo**() from your environment.
(2) Create action specifications for your environment. If you already have an environment interface object, you can obtain these specifications using the function **getActionInfo**().
(3) If needed, specify agent options using an **rlDQNAgentOptions** object.
(4) Create the agent using the **rlDQNAgent**() object method.

Alternatively, you can create actors and critics and use these objects to create your agent. In this case, ensure that the input and output dimensions of the actor and critic match the corresponding action and observation specifications of the environment.

To create a DQN agent in the second way, perform the following operations:

(1) Create a critic using an **rlQValueFunction**() or **rlVectorQValueFunction**() object method.
(2) Specify agent options using an **rlDQNAgentOptions** object. Alternatively, you can also create the agent first and use the dot notation to access its option object and modify the options.
(3) Create the agent using the **rlDQNAgent**() object method.

Exactly, the so-called actor is the agent, and the critics are the value functions, either state-value functions or action-value functions.

9.7.1.4   MATLAB Functions Used to Train Agents

After you have created an environment and reinforcement learning agent, it is ready for you to train the agent in the environment using the **train**() function. To configure your training, you can use the **rlTrainingOptions**() function. For example, create a training option set **opt**, and train an agent named **agent** in environment **env**, using the following piece of codes:

```
opt = rlTrainingOptions(MaxEpisodes=200,...
        MaxStepsPerEpisode=500,...
            StopTrainingCriteria="AverageReward",...
            StopTrainingValue=480);
```

If the **env** is a multiagent environment, specify the agent argument as an array. The order of the agents in the array must match the agent order used to create **env**. For multiagent training, use **rlMultiAgentTrainingOptions**() to replace **rlTrainingOptions**(). Using **rlMultiAgentTrainingOptions** gives you access to training options that are specific to multiagent training.

Training can be terminated automatically when the conditions you specify in the **StopTrainingCriteria** and **StopTrainingValue** options of your **rlTrainingOptions**() object are met. You can also terminate training before any termination condition is reached by clicking **Stop Training** button in the Reinforcement Learning Training Monitor.

When training terminates the training results are stored in the **trainResults** object.

Because the training updates the agent at the end of each episode, and the **trainResults** store the last training results along with data to correctly recreate the training scenario and update the Reinforcement Learning Training Monitor, you can later resume the training from the exact point at which it stopped. To begin a training process, at the command line, type:

```
trainResults = train(agent, env, trainResults);
```

This starts the training from the last values of the agent parameters, and the training results object is obtained after the previous train call.

Next let's use a real example to illustrate how to use MATLAB functions to create an agent and train it with our environment.

9.7.2   Build a Real RL Project with Reinforcement Learning Functions

In this section, we like to use a real example to show users how to use MATLAB functions included in the MATLAB Reinforcement Learning Toolbox to create a desired agent and train it with our environment.

To make things simple and easy, we still use our DC motor environment, **motor_func_env**, which was built in Sect. 9.6.4, to create a built-in agent and perform training process for that agent. The training result will be displayed in the Reinforcement Learning Training Monitor.

As we mentioned in Sect. 9.7.1.3, two ways can be used to create a built-in agent. For our application, we prefer to use the first way to do that since we have already had our customer environment available. In this project, we like to create a DQN

```
% Create a built-in DQN Agent with MATLAB Functions
% Name: Create_DQN_Agent.m
% Input: Customer environment motor_func_env, which has been added in Workspace
% Output: Created and trained DQN agent
% Aug 6, 2024
1  obsInfo = getObservationInfo(motor_func_env);
2  actInfo = getActionInfo(motor_func_env);

3  agent = rlDQNAgent(obsInfo, actInfo);

4  getAction(agent, {rand(obsInfo(1).Dimension), rand(obsInfo(1).Dimension)})

5  policy = getExplorationPolicy(agent)

6  trainOpts = rlTrainingOptions;

7  trainOpts.MaxEpisodes = 100;
   trainOpts.MaxStepsPerEpisode = 500;
   trainOpts.StopTrainingCriteria = "AverageReward";
   trainOpts.StopTrainingValue = 495;
   trainOpts.ScoreAveragingWindowLength = 20;
   trainOpts.Verbose = false;
   trainOpts.Plots = "training-progress";

8  trainingInfo = train(agent, motor_func_env, trainOpts);
```

Fig. 9.37 The codes used to create and train a built-in DQN agent

agent. Of course you can select and use any other built-in agent as those shown in Table 9.2 if you like.

To use our created customer environment, **motor_func_env**, first we need to run the project file **motor_function_env.m** and the running results are displayed in the Command window, and an environment object, **motor_func_env**, is created and added into the MATLAB Workspace.

Now create a new Script file, name it as **Create_DQN_Agent.m,** and enter the codes shown in Fig. 9.37 into that file. Let's have a closer look at this piece of codes to see how it works.

(1) Since we have already had our customer environment, **motor_func_env**, available, we can create our agent in the first way as we discussed in Sect. 9.7.1.3. To do that, we can use the function **getObservationInfo**() to pick up the observation information stored in our customer environment and assign it to a local object **obsInfo**.

(2) Similarly, we can get the action-related information by calling the **getActionInfo**() function, and assign it to another local object **actInfo**, respectively.

(3) Then create our built-in DQN agent by calling the function **rlDQNAgent**() with two arguments, **obsInfo** and **actInfo**.

(4) This coding line is just used to confirm the correctness of the execution of the function **rlDQNAgent**() with some collected pieces of observation and action information. The collected information should be displayed in the Command window.

(5) Similarly, we can also get the created policy by calling the function **getExplorationPolicy**(), and the retrieved policy should be displayed in the Command window.

(6) Prior to performing the training process, we can use the **rlTrainingOptions** object to set up and configure some desired options for this training process.

(7) Some popular training options are set and initialized to our desired values. The **Verbose** option is reset to **false** since we do not want to display any tracking

and processing descriptions during the training process. The **Plots** option is set to **training-progress** since we like to display the training process with a plotting format to see or monitor the real training trajectory.

(8) Finally the **train**() function is called with three arguments, the newly created agent, our customer environment, and the training options to start this training process.

Now run this project file, and the agent is created and some collected information for that created agent is displayed in the Command window, as shown below.

Create_DQN_Agent
    ans =
    1×1 cell array {[50]}
    policy =
    rlEpsilonGreedyPolicy with properties:
    QValueFunction: [1×1 rl.function.rlVectorQValueFunction]
    ExplorationOptions: [1×1 rl.option.EpsilonGreedyExploration]
    Normalization: "none"
    UseEpsilonGreedyAction: 0
    EnableEpsilonDecay: 1
    ObservationInfo: [1×1 rl.util.rlNumericSpec]
    ActionInfo: [1×1 rl.util.rlFiniteSetSpec]
    SampleTime: 1

The training result is shown in Fig. 9.38.

Fig. 9.38 The training result for DQN agent with motor_func_env

9.7.3   *Comparison of Different Agents Based on Their Training Results*

You can try to create some other built-in agent to see the training result for our customer environment by doing a little modification for this Script file if you like.

The training results for PG and AC agents with our customer environment are shown in Figs. 9.39 and 9.40, respectively. For these training processes, one of the Options parameters, **MaxEpisodes**, has been changed to 200.

However, you cannot create all types of agents listed in Table 9.2 by modifying this Script file due to some special functional properties for some agents.

The following agents cannot be created by modifying the Scrip file:

(1) SAC and TD3: Since they need to use two Q-value function critics for them
(2) SARSA: Since SARSA agents do not use an actor
(3) DDPG: Since it works only for continuous actions
(4) MBPO: Since it uses a network environment, **rlNeuralNetworkEnvironment** object

Compared with the three sample training results shown in Figs. 9.38, 9.39, and 9.40 with three agents, DQN, PG, and AC, it can be found that the AC agent provides the best training result. The training performance for PG agent is between the training results of DQN and AC, but better than that of the DQN agent. The training result of the DQN agent is the worst one among them.

Three completed MATLAB Script files, **Create_DQN_Agent.m**, **Create_AC_Agent.m,** and **Create_PG_Agent.m**, can be found on the Springer ftp site in the folder: **Students\Class Projects\Chapter 9**.

9.8   Chapter Summary

The main topic discussed in this chapter is about reinforcement learning (RL) algorithm which is one of the three major technologies used in machine learning.

Reinforcement learning differs from supervised learning in not needing labeled input-output pairs to be presented, and in not needing suboptimal actions to be explicitly corrected. Instead the focus is on finding a balance between exploration (of uncharted territory) and exploitation (of current knowledge) with the goal of maximizing the long-term reward, whose feedback might be incomplete or delayed.

The environment is typically stated in the form of a Markov Decision Process (MDP), because many reinforcement learning algorithms for this context use dynamic programming techniques. The main difference between the classical dynamic programming methods and reinforcement learning algorithms is that the latter do not assume knowledge of an exact mathematical model of the Markov decision process and they target large Markov decision processes where exact methods become infeasible.

Fig. 9.39 The training result for PG agent with motor_func_env

Fig. 9.40 The training result for AC agent with motor_func_env

Starting with Sect. 9.1, a basic introduction about RL algorithm is provided with a comparison between a traditional closed-loop control system and an RL control system. Following that introduction, some fundamental and important components involved in the RL algorithm are introduced, including environment, agent, state, action, and reward. Some important processes and functions are also involved, such as

(1) The Markov decision process
(2) The state-value function
(3) The action-value function
(4) The policy
(5) Some basic algorithms used in reinforcement learning

Two typical types of RL algorithms, model-based and model-free reinforcement learning, are discussed in Sects. 9.3 and 9.4.

Reinforcement learning algorithms implemented in MATLAB are discussed in Sect. 9.5. All RL-related algorithms are involved in a toolbox called Reinforcement Learning Toolbox that provides two major tools, Reinforcement Learning Designer that is an APP with a GUI, and an RL function library that provides all required RL-related functions.

Detailed introductions and discussions about the APP with some real projects are given in Sect. 9.6, which includes

(1) Create customer environment with three methods
(2) Create a built-in agent with policy
(3) Train and simulate the selected agent

In Sect. 9.7, a real project used to control a DC motor is developed by using the RL-related functions, including how to create a customer environment, create agent, and train and simulate the agent. Also a comparison of training results by using different agents to that DC motor control system is provided.

Home Works
  I. True/False Selections

      1. A reinforcement learning system contains environment, agent, reward, and policy.
      2. Reinforcement learning can handle known dynamic environments.
      3. Reinforcement learning uses algorithms that learn from outcomes and decide which action to take next.
      4. The only difference between a classical and a reinforcement learning control system is, one more variable or variable set called State (**S**) is added and fed back to the controller.
      5. The reinforcement learning uses the Markov decision process to select the optimal action a based on the current state s.
      6. Two kinds of policies are implemented in RL algorithms: stationary and deterministic.

_____7. The so-called on-policy means the optimal policy, but the off-policy means the current policy.

_____8. A state-value function is used to estimate how good the actor is to provide a given action for a given state. It can be used to evaluate a state or a state-action pair.

_____9. The action-value function is used to find a policy that maximizes the discounted return by maintaining a set of estimates of expected discounted returns E[G] for some policy.

_____10. All RL algorithms can be divided into two categories: model-based and model-free.

II. Multiple Choices

1. To train an RL algorithm, the following data and method will be used: _____________.

   a. Input-predictor and output-label pairs
   b. Policy mapping
   c. Markov decision process
   d. None of the above

2. Which of the following components is not involved in an RL? _________.

   a. Environment
   b. Policy
   c. State
   d. Control output

3. An observation set is equivalent to _________.

   a. Action
   b. State
   c. Environment
   d. Policy

4. Almost all RL algorithms use the _______________ to estimate how good for an agent to create an action based on the states.

   a. Reward
   b. Policy
   c. State
   d. Value function

5. The action value function is generally used to _______________________.

   a. Find the optimal policy
   b. Identify the optimal state
   c. Find the optimal action
   d. Find the optimal reward

6. In the actor-critic framework, an agent (actor) learns a policy to make decisions, and a critic or a _______________ evaluates the actions taken by
   the actor.

   a. Approximator
   b. Decision-maker
   c. Controller
   d. Value function

7. MATLAB provides a Reinforcement Learning Toolbox and it contains
   _______ parts, they are ___________ and _______________.

   a. 2, model-based RL, model-free RL
   b. 2, App, function library
   c. 2, Reinforcement Learning Designer, Markov Decision Process
   d. 2, actor, critic

8. Markov decision process is used to select the optimal _______ based on the
   ___________.

   a. Optimal action, current state
   b. Optimal policy, current reward
   c. Optimal state, current environment
   d. Optimal action, current reward

9. Which of the following statements is true? _______________.

   a. On policy is to estimate and improve the same policy used by the actor
   b. Off policy is to estimate and improve the policy by using different policy
   c. Value function is used by the actor to find the optimal action based
      on state
   d. All of them

10. Which of the following algorithms does not belong to model-free one?
    _______.

    a. SAC, DDPG
    b. MBVE
    c. PPO, TRPO
    d. TD3

III. Exercises

   1. Provide a basic description about the Markov Decision Process.
   2. Provide a basic description about the state value function and the action
      value function.
   3. List some popular components used in RL.
   4. Explain the differences between the on policy and off policy method.
   5. List three methods used to create customer environments.

IV. Lab Projects

1. Modify the Script file **motor_function_env.m**, exactly the file **mStepFunction.m**, to create a customer environment **pid_func_env** with a PID controller to replace the original PD controller to control the DC motor.

   Hint1: Follow Sect. 9.6.4 and Fig. 9.13 to complete this modification process. Just add an integration gain **I = 10;** under the gain **D**.

   Hint2: Add one more item on the right of the coding line, **ThetaDotDot**, as **I*ThetaDot**.

   Hint3: Name the modified Script files as, **pid_function.env.m** and **mpidStepFunction.m**. Also rename the original Script file, **mResetFunction.m** to **mpidResetFunction.m**. You also need to change the function header for two function files.

   Hint4: Rename the returned or created environment to **pid_func_env** in the last line.

2. Use MATLAB App, Reinforcement Learning Designer, to import the above customer environment and create an agent to train and simulate the selected agent. Set the **Max Episodes** as 100 in the **TRAIN** panel.

   Hint1: Follow Sect. 9.6.4.1 to complete these creating and training processes.

3. Create a new Script file and use MATLAB functions to create an agent, AC or PG, based on the above customer environment. Train and simulate those two agents and compare the training results for two agents. Name the Script file as **PID_AC_Agent.m**.

   Hint1: Refer to Sect. 9.7.2 to complete these training and comparison processes.

4. Modify the Simulink model **motor.slx** in the Simulink, exactly modify the subsystem **generate observations** by replacing the PD blocks and Adder with a continuous PID controller block. Set $K_P = 1.0$, $K_I = 0.005$, and $K_D = 0.80$ to that block. Replace the original plant with a continuous transfer function block and set the numerator as [6.74] and denominator as [1 1.79]. Replace the input constant block with a **Step** block, and set its **Finale value** to 20 with **Sample time** as 0.01. Rename this model as **pid_motor.slx**.

   Hint 1: Refer to Sect. 9.6.6.1 to complete these modification processes.

5. Modify the Script file **motor_simulink_env.m** to enable it to use the Simulink model file **pid_motor.slx** built above to generate a customer Simulink environment, **pid_slink_env**. Rename this Script file as **pid_simulink_env.m**.

   Hint1: You need to change the model name for the variable **mdl** to **pid_motor**.

   Hint2: When running this Script file, make sure that the **pid_motor.slx** is in the same folder.

6. Use MATLAB App, Reinforcement Learning Designer, to import the above customer environment, **pid_slink_env**, and create an agent to train and simulate the selected agent.

7. Use MATLAB functions to create an agent, AC or PG, to train the above customer environment, **pid_slink_env**. Compare the training results for two agents. Name the Script file as **pid_slink_func.m**.

   Hint1: Refer to Sect. 9.7.2 to complete these training and comparison processes.

References

1. Kaelbling, Leslie P., Littman, Michael L., Moore, Andrew W. (1996). Reinforcement Learning: A Survey. Journal of Artificial Intelligence Research. 4: 237–285. doi: https://doi.org/10.1613/jair.301. S2CID 1708582. arXiv:*cs/9605103*. Archived from the original on 2001-11-20

2. van Otterlo, M., Wiering, M. (2012). Reinforcement Learning and Markov Decision Process. Reinforcement Learning Adaptation, Learning, and Optimization. 12. 3–42. doi: https://doi.org/10.1007/978-3-642-27645-3_1. ISBN 978-3-642-27644-6.

3. Li, Shengbo (2023). Reinforcement Learning for Sequential Decision and Optimal Control (1st ed.) Springer Verlag, Singapore. pp. 1–460. doi: https://doi.org/10.1007/978-981-19-7784-8. ISBN 978-9-811-97783-1. S2CID 257928563

4. https://www.mathworks.com/help/deeplearning/reinforcement-learning.html.

5. https://www.geeksforgeeks.org/what-is-reinforcement-learning/.

6. https://en.wikipedia.org/wiki/Markov_decision_process.

7. https://en.wikipedia.org/wiki/Reinforcement_learning#:~:text=Reinforcement%20learning%20(RL)%20is%20an,to%20maximize%20the%20cumulative%20reward.

8. "Reinforcement learning: An introduction" (PDF). *Archived from* the original (PDF) on 2017-07-12. Retrieved 2017-07-23.

9. https://www.engati.com/glossary/temporal-difference-learning.

10. Watkins, Christopher J.C.H. (1989). Learning from Delayed Rewards (PDF) (PhD thesis). King's College, Cambridge, UK.

11. Matzliach Barouch, Ben-Gal Irad, Kagan Evgeny. Detection of Static and Mobile Targets by an Autonomous Agent with Deep Q-Learning Abilities. Entropy 2022;24(8): 1168. doi: https://doi.org/10.3390/e24081168. Bibcode:2022Entrp..24.1168M. PMC 9407070. PMID 36010832

12. https://medium.com/@kalra.rakshit/the-difference-between-model-based-and-model-free-reinforcement-learning-9499af3770db.

13. https://jonathan-hui.medium.com/rl-model-based-reinforcement-learning-3c2b6f0aa323.

14. https://doi.org/10.48550/arXiv.1803.00101.

15. https://www.quora.com/Whats-the-difference-between-policy-iteration-and-policy-search.

16. https://www.mathworks.com/help/reinforcement-learning/ug/pg-agents.html.

17. T. Simonini, "Proximal Policy Optimization (PPO)," Hugging Face – The AI community building the future, https://huggingface.co/blog/deep-rl-ppo.

18. W. Heeswijk, "Proximal Policy Optimization (PPO) explained," Medium, https://towardsdatascience.com/proximal-policy-optimization-ppo-explained-abed1952457b.

19. Edan Meyer. "Proximal Policy Optimization Explained," *YouTube*, May 20th, 2021 [Video file]. Available: https://www.youtube.com/watch?v=HrapVFNBN64.

20. https://spinningup.openai.com/en/latest/algorithms/trpo.html.

21. Sutton, Richard; Barto, Andrew (1998). Reinforcement Learning: An Introduction. MIT Press.

22. https://medium.com/@shruti.dhumne/deep-q-network-dqn-90e1a8799871.

23. https://arxiv.org/abs/1707.06887.

24. https://en.wikipedia.org/wiki/Huber_loss.

25. https://arxiv.org/abs/1412.6980.

26. https://towardsdatascience.com/reinforcement-learning-with-hindsight-experience-replay-1fee5704f2f8.

27. https://www.mathworks.com/help/reinforcement-learning/ug/ddpg-agents.html.

28. https://www.mathworks.com/help/reinforcement-learning/ug/sac-agents.html.

29. https://www.mathworks.com/help/reinforcement-learning/ug/td3-agents.html.

30. https://www.mathworks.com/help/pdf_doc/reinforcement-learning/rl_ug.pdf.

Chapter 10
Introduction to Adaptive Neuro Fuzzy Inference System

An Adaptive Neuro-Fuzzy Inference System or adaptive network-based fuzzy inference system (ANFIS) is a kind of artificial neural network that is based on the Takagi–Sugeno fuzzy inference system. The technique was developed in the early 1990s [1, 2]. Since it integrates both neural networks and fuzzy logic principles, it has the potential to capture the benefits of both in a single framework [3].

Compared with pure fuzzy inference systems and neural networks, the ANFIS technology provides some special advantages over either of them. In fact, this technology combines both fuzzy inference principles and the derivation process of neural networks to enable it to have both artificial decision ability and intelligent control ability on one system.

Refer to Sect. 3.8.1.2 in Chap. 3 and Sect. 10.7.1 to get more details about Sugeno FIS and neural networks since we have discussed both of them in those chapters.

In this chapter, we will provide a detailed introduction and discussion about this technique with some real example projects. First, let's have a clear and better understanding about this ANFIS algorithm and related components.

10.1 Introduction to Adaptive Neuro Fuzzy Inference Algorithm

The Adaptive Neuro Fuzzy Inference System (ANFIS), which integrates the merits of fuzzy inference systems (FIS) and neural networks (NN), was proposed by Jang [1]. Using the neural network learning mechanism, it automatically extracts rules

Supplementary Information The online version contains supplementary material available at https://doi.org/10.1007/978-3-031-84423-2_10.

Y. Bai, *AI Foundations and Applications with MATLAB*,
https://doi.org/10.1007/978-3-031-84423-2_10

from input and output sample data, and thus constitutes a self-adaptive neural fuzzy controller [4].

It works in Sugeno fuzzy inference system and its structure is similar to a multilayer feedforward neural network structure, except that the links in ANFIS indicate the signals' flow direction and there are no associated weights with the links [5].

Relatively speaking, the approximate knowledge reasoning and uncertainties could be modeled by a fuzzy logic process but it lacks learning rules whereas neural network has learning capabilities to strengthen the adaptive learning rules. On the other hand, neural network lacks representation of knowledge as compared to fuzzy logic. A neuro fuzzy system (NFS) combines the main features of both neural networks and fuzzy logic [6].

In summary, the FIS provides good decision-making ability for uncertain or vague input data, and the neural networks provide some advantages on learning ability for unknown system. Thus, the ANFIS technique combines the advantages of both the FIS and the NN and provides some special merits used for AI applications.

10.2 The Components and Architecture of an Adaptive Neuro Fuzzy Inference System

As we discussed, an ANFIS is composed of two parts, an FIS and an NN. In more detail, the architecture is composed of five layers.

1. The first layer takes the input values and determines the membership functions belonging to them. It is commonly called the fuzzification layer. The membership degrees of each function are computed by using the premise parameter set, namely $\{a, b, c\}$.
2. The second layer is responsible for generating the firing strengths for the rules. Due to its task, the second layer is denoted as the rule layer.
3. The role of the third layer is to normalize the computed firing strengths, by dividing each value for the total firing strength.
4. The fourth layer takes as input the normalized values and the consequence parameter set. The values returned by this layer are the defuzzificated ones and those values are passed to the last layer.
5. The fifth layer is called the output layer and it is used to organize the received defuzzificated values coming from the fourth layer and send them as the final output [7].

An architecture block diagram of a typical ANFIS is shown in Fig. 10.1.

After the input layer, the first layer, MFs, is used to convert the crisp input data to a set of fuzzy data, or called membership functions (MFs) data, which means that each data point value belongs to some membership function in certain degree with a full scale of degree as 1.0. The purpose of performing this fuzzification process is to enable all input data to meet the requirements of FIS for further processing.

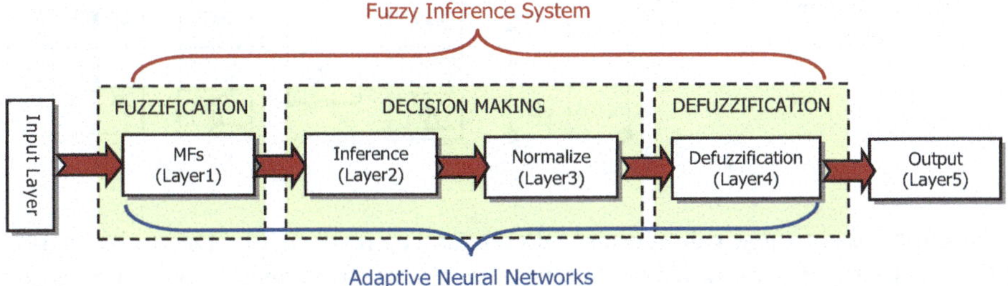

Fig. 10.1 A typical architecture of an ANFIS

The second layer, as we mentioned, is a rule layer, exactly it should be a rule inference layer. The function of this layer is to find the MFs data based on related rules to derive the corresponding fuzzy output.

In the third layer, some normalization processing may be needed to make fuzzy outputs normalized in case the outputs contain some significant differences in magnitudes. In other words, this layer normalizes the firing strengths to ensure that the sum of all rule activations equals one, facilitating weighted summation in the next layer. Generally, this step is to normalize the data to ensure that the outputs have a consistent distribution and reduces the internal covariate shift problem that can occur during training.

The defuzzification processing happens in the fourth layer. The purpose of this step is to convert the fuzzy outputs back to the crisp and normal outputs. This step needs to use some defuzzification methods, such as Center of Gravity (COG), to complete this conversion.

The fifth layer or the output layer is generally used for a neural network to produce the desired final prediction. It has its own set of weights and biases that are applied before the final output is derived.

In Fig. 10.1, the blocks in yellow colors are used to indicate that those components belong to fuzzy inference system (FIS) with three processing stages. The layers in white colors used to indicate the components used in neural networks with the layer structure.

Now we like to use a two-input ANFIS to illustrate the components and structures as well as a detailed calculation process to provide readers with a complete and detailed picture about ANFIS.

10.3   A Real Example of a Two-Input Adaptive Neuro Fuzzy Inference System

In this section, we like to use a 2D robot modeless calibration process as an example to illustrate how to use an ANFIS to improve the robot's calibration accuracy. In a 2D robotic workspace, robots can be moved either in x or y direction with some

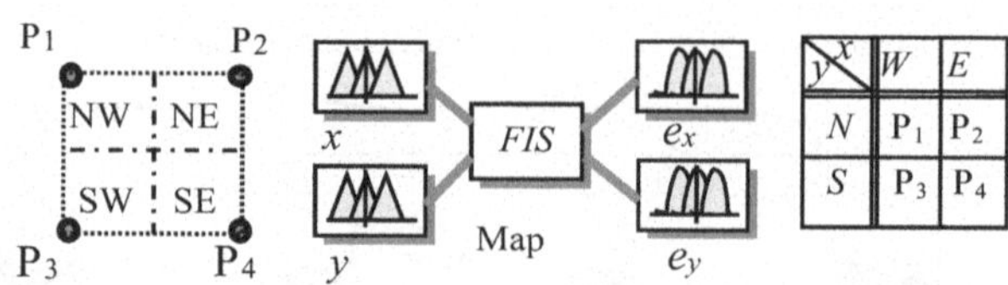

Fig. 10.2 Definition of the fuzzy error interpolation system

position errors in x and y directions, called e_x and e_y, respectively. Due to its uncoupling property, moving in either direction has no effect on another direction at all. Thus we can separate these two-direction moving and treat them individually with no issue.

A mapping between positions and position errors is shown in Fig. 10.2.

All errors are divided into four areas: NW, NE, SW, and SE. If an error is located in the SW area, its actual error value can be interpolated based on the error on point P_1. If an error is located in the SE area, its actual error value can be interpolated based on the error on point P_4 and so on.

The ANFIS architecture for this robotic model is shown in Fig. 10.3 with five layers.

Layer 1: Input membership function and its output membership grade.

$$\text{Input } x: \quad O_{1,1} = \mu_W(x), \quad O_{1,2} = \mu_E(x)$$
$$\text{Input } y: \quad O_{1,1} = \mu_N(y), \quad O_{1,2} = \mu_S(y) \tag{10.1}$$

Layer 2: All nodes in this layer are fixed, and the output of the node is the outcome of multiplying the signals coming into the node and carried out to the next node.

$$\text{Input } x: \quad O_{2,1} = w_{11} = \mu_W(x), O_{2,2} = w_{12} = \mu_E(x)$$
$$\text{Input } y: \quad O_{2,1} = w_{21} = \mu_N(y), O_{2,2} = w_{22} = \mu_S(y) \tag{10.2}$$

Layer 3: All nodes in this layer are fixed. The output of every node is the normalized firing strength, which is the ratio between the ith rules' firing strength and the total sum of all the rules' firing strengths.

$$\text{Input } x: \quad O_{3,1} = \bar{w}_1 = w_{11} / (w_{11} + w_{12})$$
$$O_{3,2} = \bar{w}_2 = w_{12} / (w_{11} + w_{12})$$
$$\text{Input } y: \quad O_{3,1} = \bar{w}_1 = w_{21} / (w_{21} + w_{22})$$
$$O_{3,2} = \bar{w}_2 = w_{22} / (w_{21} + w_{22}) \tag{10.3}$$

Layer 4: Output membership function. The nodes on this layer are adaptive nodes. Its outputs are defined as:

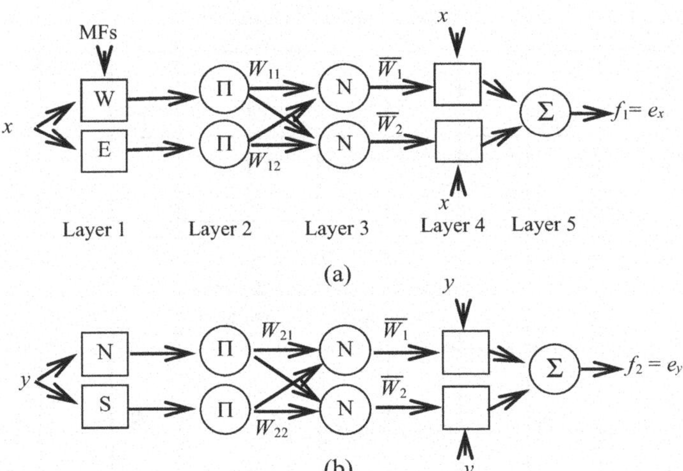

Fig. 10.3 The actual architecture of our ANFIS

$$\text{Input } x: \quad O_{4,1} = \overline{w}_1 f_{11} = \overline{w}_1 \left(p_1 x + r_1 \right)$$
$$O_{4,2} = \overline{w}_2 f_{12} = \overline{w}_2 \left(p_2 x + r_2 \right)$$
$$\text{Input } y: \quad O_{4,1} = \overline{w}_1 f_{21} = \overline{w}_1 \left(p_3 y + r_3 \right)$$
$$O_{4,2} = \overline{w}_2 f_{22} = \overline{w}_2 \left(p_4 y + r_4 \right)$$

$$(10.4)$$

Layer 5: This layer contains only a single node, which is fixed and calculates the total summation of all the arriving signals from the previous node to find the final output.

$$\text{Input } x: \quad O_5 = \left(\overline{w}_1 f_{11} + \overline{w}_2 f_{12} \right) / \left(\overline{w}_1 + \overline{w}_2 \right)$$
$$\text{Input } y: \quad O_5 = \left(\overline{w}_1 f_{21} + \overline{w}_2 f_{22} \right) / \left(\overline{w}_1 + \overline{w}_2 \right)$$

$$(10.5)$$

Four control rules are defined and can be interpreted as follows:

(1) If x is W, $f_{11} = p_1 x + r_1$.
(2) If x is E, $f_{12} = p_2 x + r_2$.
(3) If y is N, $f_{21} = p_3 y + r_3$.
(4) If y is S, $f_{22} = p_4 y + r_4$.

During the training process, 20 sets of position data, including 20 sets of positions and related 20 sets of position errors, are used to estimate optimal internal parameters for this ANFIS. Another 20 sets of position data are used to check and validate this ANFIS for the calibration results.

The membership functions used for the input in the x and y directions are shown in Fig. 10.4.

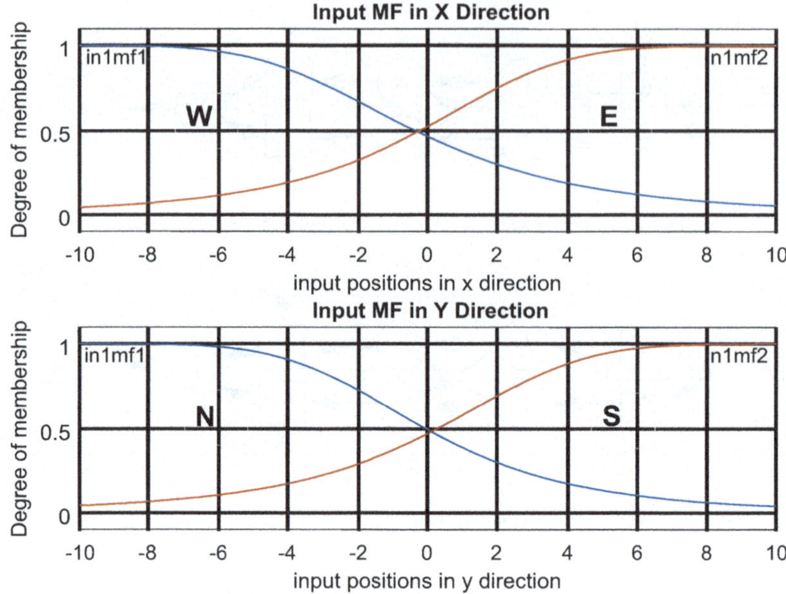

Fig. 10.4 The membership functions used for both x and y inputs

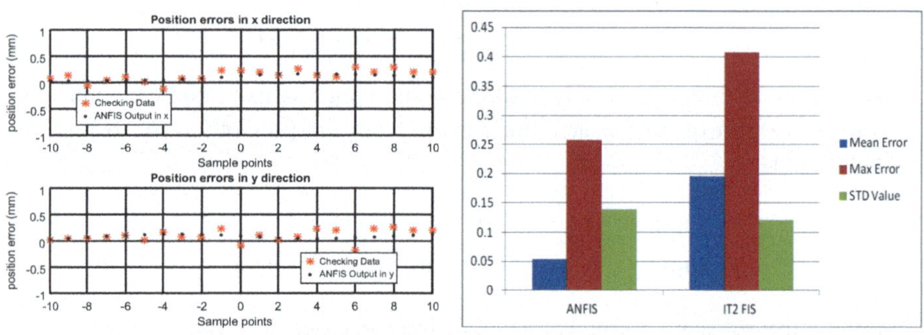

Fig. 10.5 The norm errors comparison

Extensive simulation has been performed in order to illustrate the effectiveness of the proposed ANFIS calibration method in comparison to an interval type-2 FIS calibration method.

Figure 10.5 shows the simulation results of comparing the actual position errors with the outputs of a trained ANFIS calibration method. Table 10.1 shows the simulation results for mean position errors, max position errors, and STD values between the ANFIS calibration method and an IT2 FIS method. The unit is mm.

In these figures, the simulated target (testing) positions on the standard calibration board are spaced from -10 mm to 10 mm within each cell with a step size of 1 mm.

Table 10.1 Comparison between ANFIS and IT2

	Mean Err	Max Err	STD		Mean Err	Max Err	STD
x	0.0471	0.1673	0.0833	*x*	0.1305	0.2015	0.08035
y	0.0271	0.1956	0.1120	*y*	0.1458	0.3548	0.09125
Norm	0.0544	0.2574	0.1396	*Norm*	0.1957	0.4080	0.1216

It can be seen that both the mean errors and maximum errors of the ANFIS technique are smaller than those of IT2 FLS methods. For errors in both directions, the mean errors of the ANFIS calibration method are approximately 30% smaller compared with those of the IT2 FIS method.

The simulated results show the effectiveness of the ANFIS calibration technique in reducing the position errors in the modeless robot compensation process.

Next, let's have a closer look at how the ANFIS is implemented with MATLAB.

10.4   Adaptive Neuro Fuzzy Inference System in MATLAB

MATLAB provides support for ANFS and these supports are involved in the Fuzzy Logic Toolbox in two ways. One is an App and it is a good tool for beginners in ANFIS, called **Neuro-Fuzzy Designer**. This tool provided not only a collection of functions with a friendly GUI but also some interfaces to enable users to quickly and easily build an ANFIS project. The second way is a set of functions used to support and help experienced users to build more professional ANFIS-related projects.

In summary, two kinds of tools can be used by users to build ANFIS-related projects:

(1) An APP named **Neuro-Fuzzy Designer** that provides friendly user GUI and interfaces.
(2) A set of ANFIS-related functions.

Now let's start our discussion from the APP or Neuro-Fuzzy Designer. We prefer to use a real example project to illustrate how to use this APP to build a real application.

However, like any other AI-related project, a collection of user data is needed prior to building any AI-related project. This is also true for ANFIS projects. So we need to first get familiar with some data generation and pre-processing procedures before we can start our journey to our target. However, fortunately, today we have so many datasets available on websites, and we can collect those data and use them for our project.

In most cases, things are not as easy as you expected, and in some situations you need to perform some pre-processing for the raw data to make them suitable for the ANFIS projects.

Now let's use an example to illustrate how to generate a good dataset and make it suitable for our desired ANFIS applications.

10.5 Generate Input Datasets and Make It Suitable for ANFIS Projects

For most algorithms applied in machine learning, especially in supervised learning, a set of input-predictor and output-response data pairs or a dataset is needed for the training purpose. That dataset can also be divided into two or three parts for training, testing, or validating purposes for the desired models. Some datasets may need pre-processing and then they can be applied as input-out pairs to the ANFIS project. This is also true for ANFIS-related project.

In this section, we like to use a real example project to illustrate how to use and pre-process a dataset, **Google Stock Price** dataset, to make it ready to be used for our example ANFIS project to estimate and predict the current stock price on the stock market.

10.5.1 Introduction to Google Stock Price Dataset

This dataset **Google_Stock_Price.csv** contained the stock price records for 5 years, from 2012 to 2016, and it includes those records in six columns:

(1) (transaction) **Date**
(2) (stock opening price) **Open**
(3) (stock high price) **High**
(4) (stock low price) **Low**
(5) (stock closing price) **Close**
(6) (stock total volume) **Volume**

In addition to this dataset, two additional datasets, **Google_Stock_Prive_Train.csv** and **Google_Stock_Price_Test.csv**, are also provided. The first one is identical to the original dataset, **Google_Stock_Price.csv**, but the second one contained only one month stock records for 2017.

An illustration of the first dataset is shown in Fig. 10.6.

This dataset is free of use without any license, and you can download this dataset from the site, https://www.kaggle.com/datasets/vaibhavsxn/google-stock-prices-training-and-test-data. All of the above three datasets can also be found on the Springer ftp site under the folder, **Students\Datasets\Google Stock Dataset**.

In this project, you need to familiarize and use some MATLAB functions to build the data sets, which include the training data and checking data arrays.

In an ANFIS system designing process, the following operation sequences are adopted:

(1) A set of **training data**, which includes a group of inputs and related outputs, is used first to train the ANFIS with some Neuro-Fuzzy and Back Propagation

	Date	Open	High	Low	Close	Volume
Row 1	1/3/2012	67.3	65	5.68	12.785	
Row 2	1/4/2012	35.8	86	7.32	23.674	
	⋮	⋮	⋮	⋮	⋮	⋮
Row 1258	12/30/2012	79.5	58	8.77		......
	Col 1	**Col 2**	**Col 3**	**Col 4**	**Col 5**	**Col 6**

Fig. 10.6 An illustration of the Google stock price dataset

algorithms provided by MATLAB Fuzzy Logic Toolbox to get an optimal system,

(2) Then another set of **checking data**, which contains another group of inputs and
related outputs, is utilized to check the performances of the obtained optimal
system in step 1.

Two sets of data, training data and checking data, must not be identical but must
have logical or functional relationships to correctly reflect the input-output relationship of the designed system or model.

10.5.2   *Generate the Training Dataset and Checking Dataset*

In MATLAB, the format or structure of a set of training or checking data is represented as an **M × N** array or matrix. **M** represents the number of rows and **N** indicates the number of columns that only contained numeric values (**Data** column
contained Text value). Each column represents one piece of input data and the last
column is the output data. In other words, **M** is the number of input or output data
rows (points) and **N** is the sum column number of both inputs and outputs.

For example, in our project, we need to use five sets of data (Date column is not
a number column), Open Price (**OPEN**), High Price (**HIGH**), Low Price (**LOW**),
and Total Volume (**VOLUME**), as inputs, and use the Close Price (**CLOSE**) as the
output. These data can be found in the related columns in our Google Stock Dataset,
Google_Stock_Price.csv. Totally that dataset provided 1258 data points, which
means that each column has 1258 rows and each row in each column represents a
single data point, as shown in Fig. 10.6.

We need to organize this **1258 × N** data matrix into one data matrix, the first **600
× N** data works as training data, and the second **600 × N** data contains checking data.

In our case, we need to use four (4) columns as our input data, **OPEN, HIGH,
LOW,** and **VOLUME**, and one column **CLOSE** as the output. Thus you can determine the **N**.

As we mentioned, in some situations you need to perform some normalization process for the dataset to make it meet the needs of the related applications. This is true for our project, too. The reason for that is that we do not care about the absolute stock prices for a period of time, but instead, we only pay attention to the relative stock prices for a period of time. We can start to sell or buy some stocks based on their peak or valley values for that period.

Keep those points in mind, and let's build our MATLAB codes to generate the desired dataset based on the Google Stock Price dataset for our ANFIS project.

Create a new Script file, name it as **Generate_Stock_Data.m**, and enter the codes shown in Fig. 10.7 into that file. Let's have a closer look at this piece of codes to see how it works.

(1) The full path for our dataset is defined first and a system function **readtable()** is executed to read our dataset as a cell matrix and assign it into a local variable **T**. You may use your actual path to replace this if you stored this dataset at different folder in your machine.

(2) For both training and checking data, we need to make them as 600 rows, thus a local variable **N** is used for that purpose.

```
% Program used to generate the training/checking data for the stock project
% (Normalized dataset)
% Name: Generate_Stock_Data.m
% Aug 10, 2024

1  path = 'C:\Artificial Intelligence Book\Students\Datasets\Google Stock DataSet\Google_Stock_Price.csv';
   T = readtable(path);
2  N = 600;             % max row number in dataset - 600

3  OPEN = table2array(T(1:N, 2));
   HIGH = table2array(T(1:N, 3));
   LOW = table2array(T(1:N, 4));
   VOLUME = str2double(table2array(T(1:N, 6)));
   CLOSE = table2array(T(1:N, 5));

   % Normalize all columns for training data
4  OPEN = OPEN/max(OPEN);
   HIGH = HIGH/max(HIGH);
   LOW = LOW/max(LOW);
   VOLUME = VOLUME/max(VOLUME);
   CLOSE = CLOSE/max(CLOSE);

5  TrainData = [OPEN HIGH LOW VOLUME CLOSE];        % the first 600 data - training data

6  M = 601;      % Start the second or checking data row from 601

7  OPEN = table2array(T(M:2*N, 2));
   HIGH = table2array(T(M:2*N, 3));
   LOW = table2array(T(M:2*N, 4));
   VOLUME = str2double(table2array(T(M:2*N, 6)));
   CLOSE = table2array(T(M:2*N, 5));

   % Normalize all columns for checking data
8  OPEN = OPEN/max(OPEN);
   HIGH = HIGH/max(HIGH);
   LOW = LOW/max(LOW);
   VOLUME = VOLUME/max(VOLUME);
   CLOSE = CLOSE/max(CLOSE);

9  ChkData = [OPEN HIGH LOW VOLUME CLOSE];        % the second 600 data - checking data

10 save train_data.dat TrainData -ascii
   save check_data.dat ChkData -ascii
```

Fig. 10.7 The codes used to generate the desired dataset for training of the ANFIS

(3) Now we need to retrieve each column from readout dataset **T**. Two issues are involved in this retrieving. First, we need to use a system function **table2array**() to convert each element to a double numeric value since the data readout and stored in the variable **T** is a cell, not a number. The second issue is that the data values in the **VOLUME** column are nested cell arrays even after using the function **table2array**(), and you need to convert them to double values by calling another function **str2double**(). Otherwise you may encounter some compiling errors when you run this piece of code later. The colon operator, **1:N**, is used to get the first 600 rows for the training data.

(4) A normalization processing is performed for all five columns by dividing each element in each column by the maximum value in that column to make them as relative values.

(5) Then assign those five normalized columns to the training data matrix **TrainData**.

(6) Similarly, we need to build our checking data and start to get data from row **601**.

(7) Now retrieve each column for our checking data one by one. One trick is that even if we start from row 601 to perform this reading, the **readtable**() function will start from row 602 and end at row 1,202 due to its functionality.

(8) Perform similar normalizations for the checking data to make it as relative values.

(9) Arrange the second 600 rows as our checking data, **ChkData**.

(10) Finally, save both datasets as two ascii data files to make them to be used later.

Now run the project and two datasets are generated. Next, we can use the MATLAB APP, Neuro-Fuzzy Designer, to build our ANFIS project to train, check, and validate this model to confirm the correctness of our ANFIS model to predict the stock prices correctly.

10.6   Use Neuro-Fuzzy Designer to Build Our Stock Price Prediction Project

This Neuro-Fuzzy Designer APP allows users to design, train, test, and validate any ANFIS project by providing some GUI and interfaces. It is so easy to make an ANFIS-related project in just a few minutes without touching any codes. Yes, no matter you believe it or not, it is a true fact. Now let's build our stock price prediction project with this APP to familiarize us with the environments and functions of this APP.

There are two ways to open this APP, by using the APP icon on the MATLAB menu bar or by typing the command in the Command window, as shown below:

(1) On the opened MATLAB, click on the **APP** tab on the top and select the **Neuro-Fuzzy Designer** under the **CONTROL SYSTEM DESIGN AND ANALYSIS** group.
(2) Type the command, **neuroFuzzyDesigner**, in the Command window.

Perform the following operational steps to build this project:

(1) Open the MATLAB Command window and load our two dataset files into the Workspace with two following commands:

      load train_data.dat
      load check_data.dat

(2) In the MATLAB Command window, type command: **neuroFuzzyDesigner** to open the Neuro-Fuzzy Designer, as shown in Fig. 10.8a.
(3) In the opened **Neuro-Fuzzy Designer**, keep the **Training** radio button selected and choose the **worksp** radio button under the **Load data** group box.
(4) Click on the **Load Data** button to load the training data. Enter **train_data** into the Messagebox, as shown in Fig. 10.8b, to give the data file name. Click the **OK** button to load this data.
(5) Now select the **Checking** radio button under the **Load data** group box, and click on the **Load Data** button again to complete loading the **check_data** file.
(6) Both **train_data** and **check_data** are displayed on the Neuro-Fuzzy Designer, as shown in Fig. 10.9a.
(7) Next, we need to create the initial model for our FIS system. Click on the **Generate FIS** button under the **Generate FIS** group box. Keep the default radio button, **Grid partition**, with no change since we want to create an initial model with this mode.

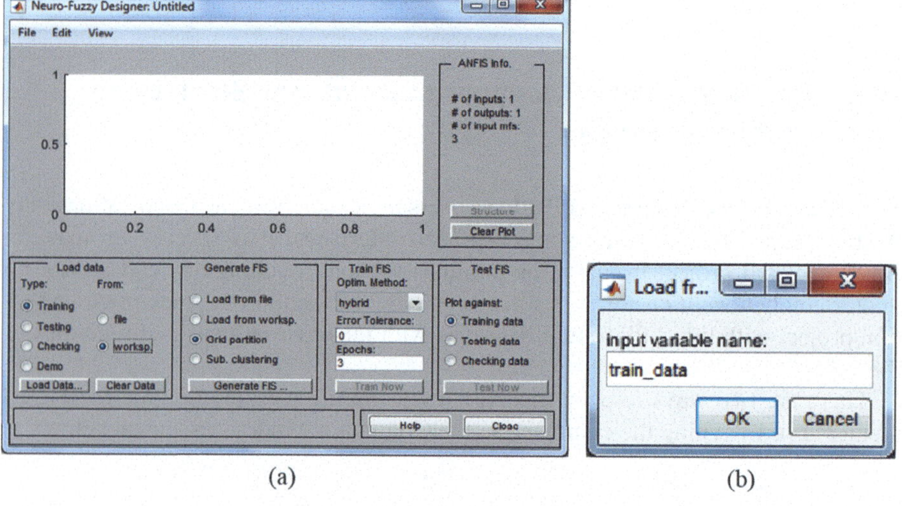

(a)   (b)

Fig. 10.8 The opened neural-fuzzy designer

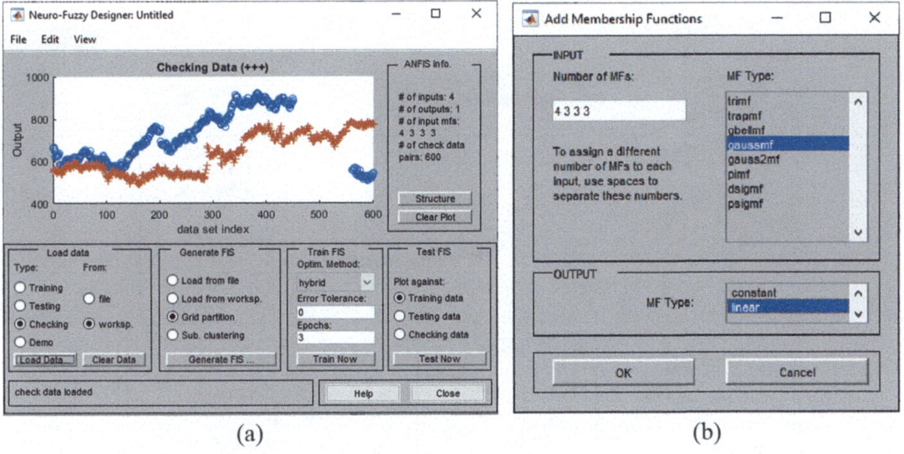

(a)   (b)

Fig. 10.9   The loaded data shown in the Neural-Fuzzy Designer

(8) Select the number of membership functions (MFs) for four inputs, **OPEN, HIGH, LOW,** and **VOLUME**. Here we used four MFs and kept the default **4 3 3 3** in the **Numbers of MFs** box, as shown in Fig. 10.9b. Select the type of MFs as **gaussmf** and the type of the output MF as **linear**, as shown in Fig. 10.9b. Click on the **OK** button to generate this FIS.

(9) Now you may need to save this FIS to the workspace or a file by going to **File|Export|To Workspace** or **File|Export|To File**. The name can be **ANFIS_Stock_App**.

(10) Then we can begin to train this FIS with Adaptive Neural Network. Change the epoch times from 3 to 60 in the **Epochs** box, select **backpropa** from the **Optim Method** box, as shown in Fig. 10.10, and click on the **Train Now** button to start this training process. The training process is shown in Fig. 10.10 (Minimal training RMSE = 0.104949, Minimal checking RMSE = 0.0681429).

(11) To confirm this training result, we can check this FIS model by comparing the original output **CLOSE** and the **CLOSE** that is the output of the trained FIS model. To do that, click the **Checking data** radio button under the **Test FIS** group box, as shown in Fig. 10.10, and click on the **Test Now** button. The testing result is shown in Fig. 10.11.

(12) You can check the structure of this ANFIS by clicking on the **Structure** button located in the **ANFIS Info** group box. The structure of this ANFIS is shown in Fig. 10.12. Four inputs, **OPEN, HIGH, LOW,** and **VOLUME**, each of them has three or four MFs indicated with three lines pointed to three white circles. All internal rules and output rules are generated by this ANFIS automatically.

(13) You can go to **View|Rules** menu item to open the Rules View, as shown in Fig. 10.13, to dynamically test the ANFIS to get real-time output, **CLOSE,**

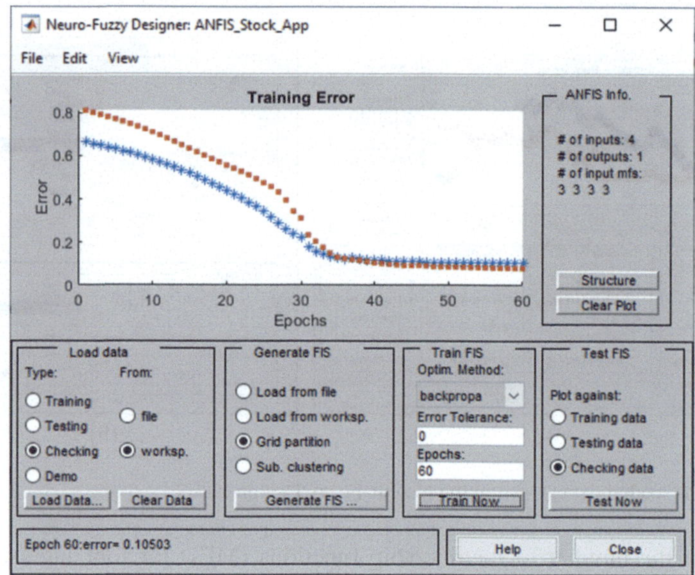

Fig. 10.10 The training process is going on

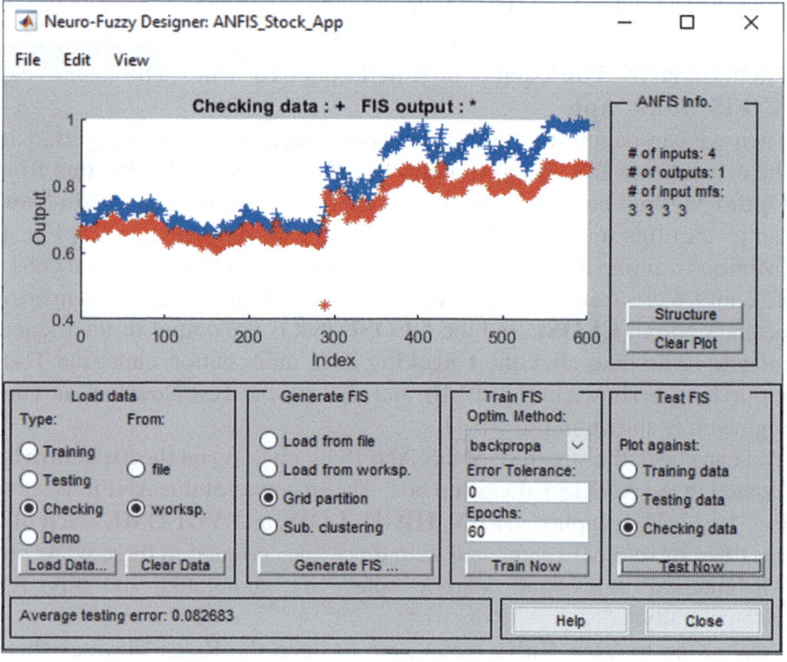

Fig. 10.11 The testing result

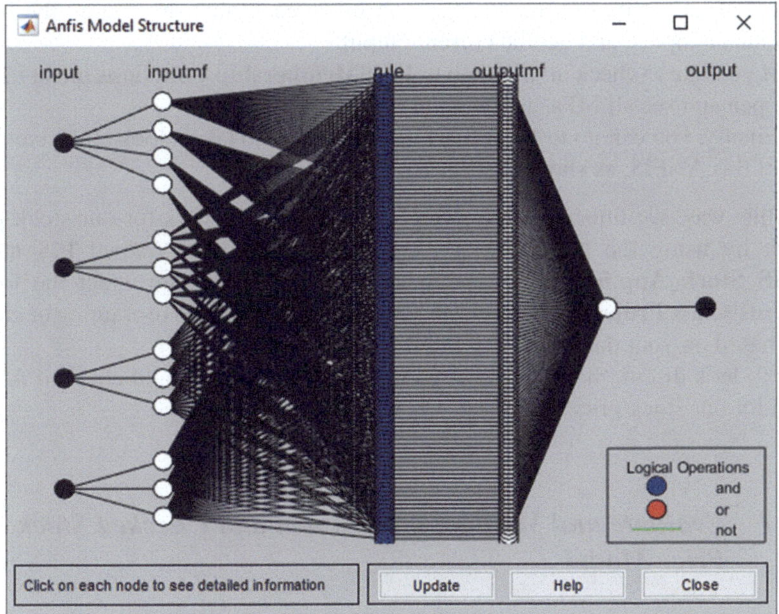

Fig. 10.12 The structure of the ANFIS

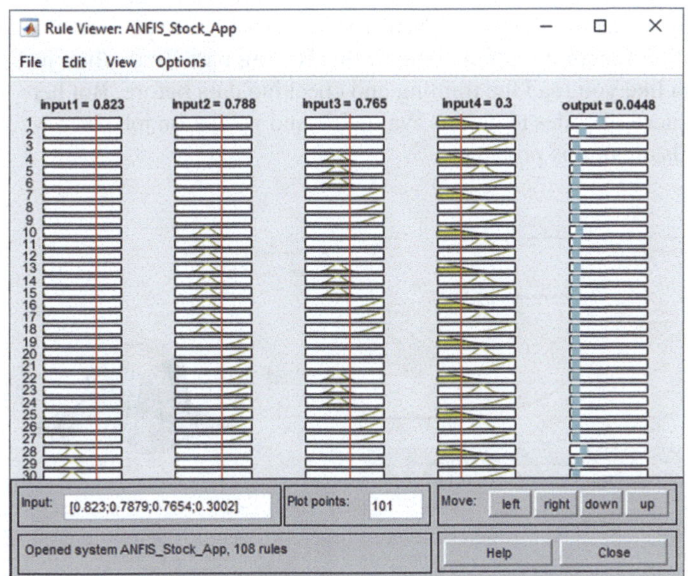

Fig. 10.13 The rule-view of the generated ANFIS

based on the current inputs. You can move each input by sliding the bar to change inputs, and get the current output.

(14) If you like to check all MFs, go to **Edit|Membership Functions** menu item to open and see all MFs, as shown in Fig. 10.14a.

(15) Finally, you can go to **View|Surface** menu item to take a look at the envelope of this ANFIS, as shown in Fig. 10.14b.

In this way, we finished the training and checking process for our stock price project by using the Neuro-Fuzzy Designer. A completed trained FIS named **ANFIS_Stock_App.fis** can be found on the Springer ftp site under the folder, **Students\Class Projects\Chapter 10**. You can load this FIS to re-training or checking it based on your dataset later if you like.

Next, let's discuss how to evaluate or validate our trained and checked ANFIS model for our stock price project.

10.6.1   *Evaluate and Validate the Trained and Checked Stock Price Model*

Now that we finished the training and checking of our stock price model built by ANFIS, how to evaluate and validate that model to confirm its correctness and availability? In this section, we need to build some codes to answer those questions.

In fact, you can use the APP—Neuro-Fuzzy Designer to perform this evaluation and validation function with the help of the **Testing** part. To do that, just load your testing data like you load the training and checking data before. But here we like to develop a piece of codes to do this evaluation and validation job since we like to see more details about this process.

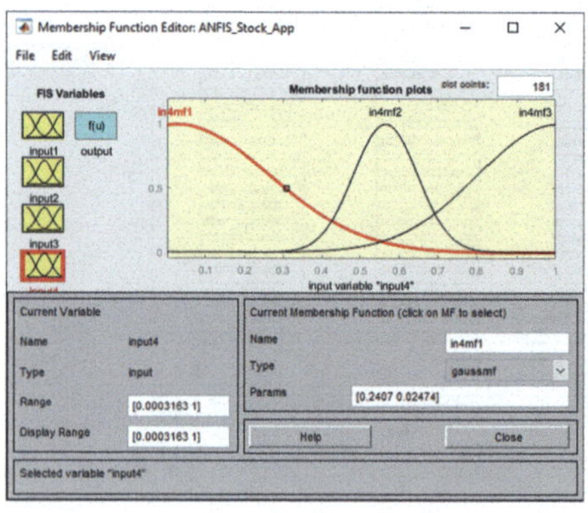

(a) One of Membership functions.

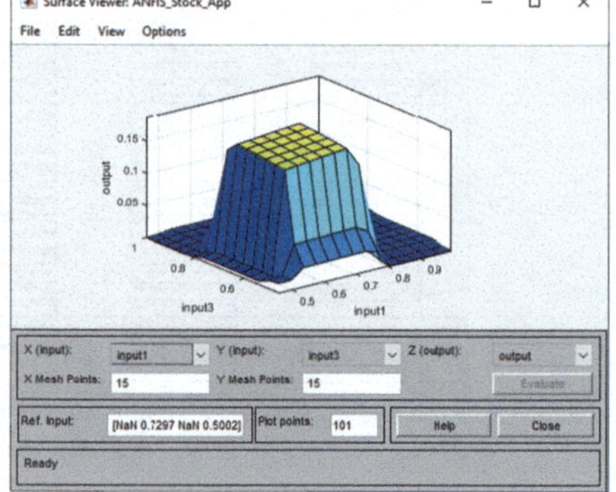

(b) The surface of envelop of ANFIS.

Fig. 10.14 The MFs and surface of envelop of the trained ANFIS. (**a**) One of the Membership functions. (**b**) The surface of the envelop of ANFIS

In the last project, we used the ANFIS App to build our model based on training and checking data. In this project, we need to evaluate the ANFIS stock model with another dataset collected from the Google Stock Price dataset, **Google_Stock_Price_Test.csv**, which provided a collection of stock price records during the period between January 3 and January 31, 2017.

With the help of this dataset, we can use it as our testing data to compare the predicted stock prices, exactly the closing prices, to evaluate our trained ANFIS model we got from the last project. We can use all four columns, **OPEN**, **HIGH**, **LOW**, and **VOLUME**, in that testing dataset as the inputs to our trained model to produce the predicted stock closing prices, and then compare them with the actual closing prices on the **CLOSE** column in our testing dataset.

Create a new Script file and name it as **Stock_Eval_App.m**, and enter the codes shown in Fig. 10.15 into that file. Let's have a closer look at this piece of codes to see how it works.

(1) Due to an over-long path name for our testing dataset, we use two strings and concatenate them with the **strcat()** function.
(2) The function **readtable()** is used to read out our testing dataset, and the reading result is assigned to a local variable **T**.

```
% Evaluation the ANFIS model performance - normalized dataset
% Name: Stock_Eval_App.m
% The stock model, ANFIS_Stock.fis, should have been built prior to running this one.
% Aug 11, 2024

1  path = 'C:\Artificial Intelligence Book\Students\Datasets\Google Stock DataSet\';
   name = 'Google_Stock_Price_Test.csv';
   fname = strcat(path, name);

2  T = readtable(fname);
3  N = 20;            % max row number in testing dataset - 20

4  OPEN = table2array(T(1:N, 2));
   HIGH = table2array(T(1:N, 3));
   LOW = table2array(T(1:N, 4));
   VOLUME = str2double(table2array(T(1:N, 6)));
   CLOSE = table2array(T(1:N, 5));

   % Normalize all columns for training data
5  OPEN = OPEN/max(OPEN);
   HIGH = HIGH/max(HIGH);
   LOW = LOW/max(LOW);
   VOLUME = VOLUME/max(VOLUME);
   CLOSE = CLOSE/max(CLOSE);

6  stock_fis = readfis('ANFIS_Stock_App');

7  finput = [OPEN HIGH LOW VOLUME];
8  foutput = evalfis(stock_fis, finput);
   save foutput;

9  x = 1:N;
   p = plot(x, foutput, 'b-o', x, CLOSE, 'r-o');
   p(1).LineWidth = 2;
   p(2).LineWidth = 2;
   axis([1 20 0.4 1.1]);
   legend('Predicted Stock Price','Actual Stock Price','Location','SouthWest')
   grid;
```

Fig. 10.15   The codes used to evaluate the trained ANFIS model

(3) Unlike the original Google Stock Price dataset, the testing dataset only contained 20 stock price records during the period of January 2017. Thus the number of the total rows is assigned to another local variable **N**.

(4) Each testing data column is retrieved, converted to double values by calling the function **table2array()**, and assigned to the related variable. The colon operator, **1:N**, is used to indicate the range of the rows to be retrieved.

(5) Each read column is normalized by dividing its maximum value on that column. This normalization process is very important and it is a key to make this project successful. The reason for that is due to the significant variations of the stock prices on the markets. We cannot, or we do not need to, estimate or predict the absolute stock prices on a period, instead, we only need to predict a group of relative stock prices for a certain period of time. In other words, we only take care of changing trends or tendency of the stock prices in a period, and that is good enough for us to make our decisions to sell or to buy stocks at some valley or peak points.

(6) To predict the stock prices (closing prices—**CLOSE**) based on the input data in the testing dataset, **OPEN, HIGH, LOW,** and **VOLUME**, we need to load our trained model we built in the last project, **ANFIS_Stock_App.fis**. Here you do not need to use any extension **.fis** for that model's name. A system function **readfis()** is needed to perform this loading action when you load a FIS model.

(7) Then we organize all four input columns to an input matrix **finput**.

(8) A system function **evalfis()** is executed with two arguments, the trained model and our input matrix, to calculate the predicted output, the closing prices. The predicted closing prices are stored in a local variable **foutput**. You can save it to Workspace if you like.

(9) Finally, we can plot both the predicted and the actual stock prices on one plotting to compare them to evaluate our trained model, exactly to evaluate its function. The **foutput** is our predicted prices and the **CLOSE** is the actual prices.

Prior to running this project, make sure that the last project has been run at least one time since we need that trained model **ANFIS_Stock_App.fis** in this project.

Now run the project and the plotting result is shown in Fig. 10.16.

It can be found from Fig. 10.16 that the absolute values of predicted and the actual stock prices are different; exactly there is an offset existed between them. But that is not important at all for our purpose. We do not pay any attention to those absolute stock prices since we do not need those values. What we want is the trend or tendency of these prices. In other words, we only care about the peak or valley points on these trajectories.

The key issue is that these two price trajectories have a very similar changing trend or tendency, and we can clearly obtain the peak and valley points on any trajectory since they are very similar or identical in changing trends. That is all we need!

A completed Script file, **Stock_Eval_App.m**, can be found on the Springer ftp site under the folder, **Students\Class Projects\Chapter 10**.

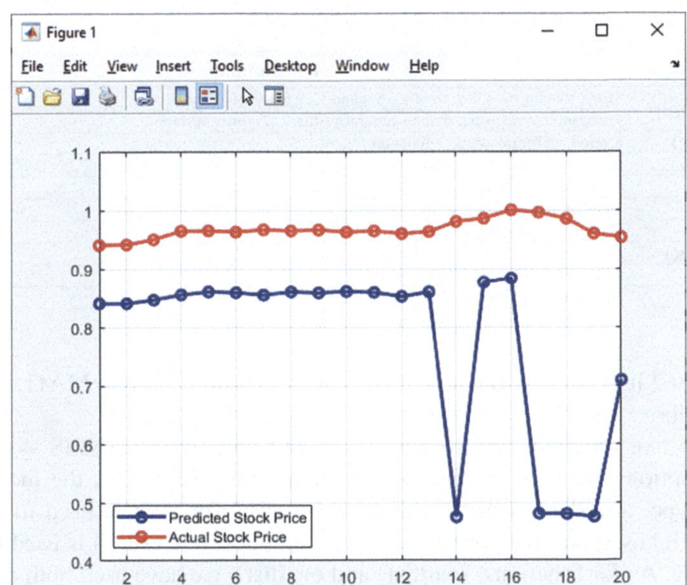

Fig. 10.16 The evaluating result for the trained ANFIS model

Next, let's discuss how to use MATLAB functions to build and evaluate our stock price model with the ANFIS algorithm.

10.7   Use MATLAB ANFIS Functions to Build Our Stock Price Prediction Project

As we mentioned in Sect. 10.4, in addition to APP, MATLAB also provides a set of ANFIS-related functions to help users build their ANFIS-related projects. In this section, we will discuss how to use those functions to build and evaluate our stock price model.

10.7.1   Introduction to ANFIS Functions

In MATLAB, all ANFIS-related functions are involved in the Fuzzy Logic Toolbox. Instead of using the APP or Neuro-Fuzzy Designer, you can use those functions to build, train, and evaluate a customer-built ANFIS model. This method provides more flexibilities and controllability to enable users to develop more professional applications.

Table 10.2 Some popular functions used for ANFIS model

Function Name	Descriptions
anfis()	Using training data to generate a Sugeno-type fuzzy inference system.
genfis()	Using input and output data to generate a fuzzy inference system.
genfisOptions()	Option set for `genfis()` function.
anfisOptions()	Option set for `anfis()` function.
sugfis()	Generating a Sugeno-type fuzzy inference system.
readfis()	Read or load a fuzzy inference system.
tunefis()	Tuning a fuzzy inference system.
tunefisOptions()	Option set for `tunefis()` function.
evalfis()	Evaluating a fuzzy inference system.
writeFIS()	Saving a fuzzy inference system to a file.

Table 10.2 lists some popular ANFIS-related functions used in MATLAB Fuzzy Logic Toolbox.

The top five functions are used to create and initialize an ANFIS system with different options, such as the number of membership functions, the membership function type, and the partition type. The **tunefis()** function is used to tune and adjust a FIS based on given parameters. The function **writeFIS()** is used to save a FIS to a file. As for functions, **readfis()** and **evalfis()**, we have used both of them in our previous projects to perform loading and evaluating a FIS model, respectively. The difference between the function **anfis()** and **genfis()** is that the former can be used to generate and tune a Sugeno-type FIS based on input parameters, but the latter can only generate an FIS.

Next let's use these MATLAB ANFIS-related functions to generate, train, and evaluate our stock price model.

10.7.2   Design and Evaluate Our Stock Price Model with MATLAB Functions

Create a new Script file, name it as **Stock_Train_Eval_Func.m**, and enter the codes shown in Fig. 10.17 into that file. Let's have a closer look at this piece of codes to see how it works.

(1) The epoch number is defined first with 25 for the training process.
(2) Both the training and checking data are loaded into the workspace to enable us to use them. A key issue is that the Script file **Generate_Stock_Data.m** must be run prior to running this project since we need to have those two datasets.
(3) The function **genfisOptions()**sets up all necessary parameters for the structure of this FIS. The Grid Partition, 4 MFs with 3 MFs for each input, Gaussian2 waveforms and Linear, are used for this FIS.

```
% Train and evaluate the ANFIS model performance - normalized dataset via ANFIS Functions
% Name: Stock_Train_Eval_Func.m
% The Script file: Generate_Stock_Data.m should be run prior to running this program
% Aug 11, 2024
1  M = 25;                      % Epoch number
2  load train_data.dat;
   load check_data.dat;

3  genOpt = genfisOptions('GridPartition');
   genOpt.NumMembershipFunctions = [3 3 3 3];
   genOpt.InputMembershipFunctionType = 'gauss2mf';
   genOpt.OutputMembershipFunctionType = 'linear';

4  inFIS = genfis(train_data(:, 1:4), train_data(:, 5), genOpt);

5  opt = anfisOptions('InitialFIS', inFIS, 'EpochNumber', M);
   opt.DisplayANFISInformation = 1;
   opt.DisplayErrorValues = 1;
   opt.DisplayStepSize = 0;
   opt.DisplayFinalResults = 1;
   opt.ValidationData = check_data;

6  [fis, trainError, stepSize, chkFIS, chkError] = anfis(train_data, opt);

7  x = 1:M;
   figure
   plot(x, trainError, '.b', x, chkError, '*r');          % plot RMS value on ANFIS for each epoch
   legend('Training Error', 'Checking Error', 'Location', 'NorthWest');
   grid;

8  path = 'C:\Artificial Intelligence Book\Students\Datasets\Google Stock DataSet\';
   name = 'Google_Stock_Price_Test.csv';
   fname = strcat(path, name);
   T = readtable(fname);
   N = 20;          % max row number in testing dataset - 20

9  OPEN = table2array(T(1:N, 2));
   HIGH = table2array(T(1:N, 3));
   LOW = table2array(T(1:N, 4));
   VOLUME = str2double(table2array(T(1:N, 6)));
   CLOSE = table2array(T(1:N, 5));

   % Normalize all columns for training data
10 OPEN = OPEN/max(OPEN);
   HIGH = HIGH/max(HIGH);
   LOW = LOW/max(LOW);
   VOLUME = VOLUME/max(VOLUME);
   CLOSE = CLOSE/max(CLOSE);

11 finput = [OPEN HIGH LOW VOLUME];
   foutput = evalfis(fis, finput);
   save foutput;

   figure
12 x = 1:N;
   p = plot(x, foutput, 'b-o', x, CLOSE, 'r-o');
   p(1).LineWidth = 2;
   p(2).LineWidth = 2;
   grid;
   legend('Predicted Stock Price', 'Actual Stock Price', 'Location', 'NorthWest')
```

Fig. 10.17 The codes used to build, train, and evaluate the stock price model

(4) The **genfis**() function is executed to create this FIS with the first four columns
      in the **train_data** dataset as inputs, the fifth column as the output and **genOpt**
      as additional structure parameters.

(5) Then the **anfisOptions**() function is called to set up the training structure and
      parameters for this FIS, which includes an initial FIS, epoch time, **check_
      data,** and other related displaying configuration parameters.

(6) The function **anfis**() is executed to generate and train this FIS with the training
      dataset and configured structure parameters. This function returns the training

results, including the trained **fis** model, training error, training step size, validation error (**chkFIS**), and checking error.

(7) To display those training and checking errors, a **plot()** function is executed.

(8) Steps 8 through 10 are used to get and normalize the testing dataset.

(9) Retrieve all columns from the testing dataset.

(10) Normalize all columns.

(11) Set the input matrix with the testing columns and evaluate the trained model to get the predicted output, stock closing prices, and assign them to a local variable **foutput**.

(12) To compare the predicted stock prices with the actual stock prices, another **plot()** function is executed, and the comparison result is displayed as the project is done.

Now run the project and the training process is displayed in the Command window with the minimum training and checking RMSE values, as shown below:

```
ANFIS info:
    Number of nodes: 193
    Number of linear parameters: 405
    Number of nonlinear parameters: 48
    Total number of parameters: 453
    Number of training data pairs: 600
    Number of checking data pairs: 600
    Number of fuzzy rules: 81


Start training ANFIS ...



1        0.0349168        0.118568
2        0.0344528        0.110141
3       0.0340689        0.124388
4       0.0341518        0.125312
5       0.0339088        0.126783
6       0.0340545        0.132331
7       0.0338031        0.127672
8       0.0339655        0.144869
9       0.0336455        0.134538
10      0.0338687        0.168224
11       0.0335817        0.149263
12      0.0337826       0.19761
13      0.0334469       0.174659
14       0.0336539       0.246374
15       0.0333016        0.225196
16       0.0334935        0.29801
17       0.0331985        0.36074
```

```
18      0.0333221      0.262363
19      0.0330952      0.499653
20      0.0332186      0.368111
21      0.0330151      0.511792
22      0.0331392      0.430997
23      0.0329551      0.562023
24      0.0330662      0.486861
25      0.0328741      0.602802
```

```
Designated epoch number reached. ANFIS training completed at
epoch 25.
```

```
Minimal training RMSE = 0.0328741
Minimal checking RMSE = 0.110141
```

Two plots, including the comparison between the training and the checking errors, and the comparison between the predicted stock prices and the actual stock prices, are also displayed, as shown in Figs. 10.18a, b.

It can be found from Fig. 10.18b that the trend or tendency of the predicted and the actual stock prices is basically identical with similar peak and valley values. This trend is good enough for us to make our decisions for the current stock market to obtain the maximum benefits.

A completed Script file, **Stock_Train_Eval_Func.m**, can be found on the Springer ftp site under the folder, **Students\Class Projects\Chapter 10**.

With this example project, we can finish this chapter since all other projects can be developed and built in a similar sequence or steps as this one.

10.8   Chapter Summary

This chapter focuses on another popular AI technology, adaptive neuro fuzzy inference system or called ANFIS.

ANFIS is a powerful and efficient method that enables users to quickly and easily develop and build professional applications to predict and estimate the desired or optimal outputs based on the current inputs. Like other AI-related algorithms, ANFIS also needs the training and testing process to make it to be applied in practical implementations to provide optimal outcomes.

Compared with pure fuzzy logics and traditional neural networks, ANFIS has some advantages over both of them to provide both better decision-making strategy with the vague inputs and learning ability to improve its learning-decision functions.

An introduction to ANFIS is given in Sect. 10.1, and the major components and architecture of a typical ANFIS model are provided in the following part.

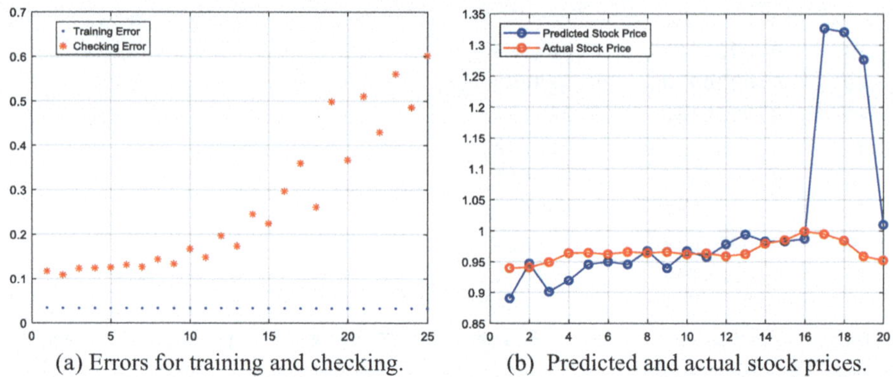

(a) Errors for training and checking.   (b) Predicted and actual stock prices.

Fig. 10.18 Comparison of training-checking error and predicted-actual stock prices. (**a**) Errors for training and checking. (**b**) Predicted and actual stock prices

To make things more sense, a real ANFIS example project with two inputs is discussed in Sect. 10.3 with detailed illustrations. This part provides a clear and detailed picture about ANFIS to readers to enable them to have a better understanding and knowledge of ANFIS. This part also builds a solid foundation for their further study on this topic in the future.

Starting Sect. 10.4, an introduction to ANFIS applied in MATLAB is given with two styles, the Neuro-Fuzzy Designer which is an APP with good GUIs and a set of ANFIS-related functions involved in the Fuzzy Logic Toolbox. The first way enables users to quickly and easily build a professional ANFIS project, and the latter allows users to develop more flexible and advanced projects with more codes.

A real and practical example project is provided in Sects. 10.5 and 10.6 with a dataset, which is a Google Stock Price dataset. A detailed developing process with the APP is given to direct users to build this project with real datasets step by step with detailed illustrations.

Section 10.7 provides another way to build the same project, which is to use a set of ANFIS-related functions. This way provided more flexibility to enable users to use more codes to build their applications. With the help of functions, users can develop more professional applications and implement them in most real situations.

Both above sections also provide introductions and discussions about the evaluations and validations for the trained ANFIS models with real projects.

Home Works

I. True/False Selections

  ______1. An ANFIS can be considered as an Adaptive Neural Fuzzy Inference System.

  ______2. ANFIS combines fuzzy inference system with neural networks to provide a better environment for building AI-related applications.

  ______3. Fuzzy inference system provided a good way to make decision-making based on vague and uncertain input data.

_____4. Neural networks provided a good way for the learning rules and thinking ability.

_____5. A complete ANFIS is composed of five layers, namely, the input layer, fuzzification layer, rule layer, normalization layer, and output layer.

_____6. The function of the fourth layer or the defuzzification layer is to convert the fuzzy outputs to the related crisp outputs.

_____7. The fuzzy inference system used in an ANFIS is a Mamdani fuzzy inference system.

_____8. MATLAB provided two ways for ANFIS development, an APP with GUIs and a set of ANFIS-related functions, both are located in the Fuzzy Logic Toolbox.

_____9. A set of training and checking data is needed to train and test an ANFIS model.

____10. You do not need to provide any training and checking data to the Neuro-Fuzzy Designer to build an ANFIS model and the APP can handle those data itself.

II.   Multiple Choices

1. The ANFIS combined the following two kinds of technologies together: ___________.

   a. Neuro networks, fuzzy logic
   b. Fuzzy inference system, neuro networks
   c. Neural networks, fuzzy inference system
   d. A Sugeno fuzzy inference system, neuro networks

2. Which of the following components is not involved in an ANFIS? _________.

   a. Fuzzy inference system
   b. A Mamdani fuzzy inference system
   c. A Sugeno fuzzy inference system
   d. Neural network

3. An ANFIS is also called a(n) _________.

   a. Artificial neural fuzzy inference system
   b. Adaptive neural fuzzy inference system
   c. Adaptive network fuzzy inference system
   d. Adaptive network-based fuzzy inference system

4. All links used to connect different layers in an ANFIS _____contain some weight factors.

   a. Do not
   b. Do
   c. May
   d. Seemingly

5. In Equation (10.1), which is used for the first layer, the symbol μ is to
   __________________.

   a. Indicate the transfer degree for variable x
   b. Indicate the transfer degree for variable y
   c. Indicate the membership degree for variable x
   d. Indicate the working degree for variable y

6. The links connected between layers in an ANFIS are used to
   __________________.

   a. Indicate the weight for each link
   b. Indicate the relationship for each link
   c. Indicate the function for each link
   d. Indicate the flow direction for each link

7. Which of the following methods is not used for defuzzification process?
   __________________.

   a. Mean of Maximum (MOM) method
   b. Center of Gravity (COG) method
   c. Average Magnitude (AM) method
   d. The Height Method (HM)

8. MATLAB provides _____ tools to support ANFIS developments, they are
   __________________.

   a. 3, Fuzzy logic toolbox, APP, and ANFIS-related functions
   b. 3, Membership functions, APP, and GUIs
   c. 2, Fuzzy logic toolbox and ANFIS-related functions
   d. 2, APP and ANFIS-related functions

9. Which of the following statements is not true? __________________.

   a. A set of input and output data is needed to train, check, and test an
      ANFIS model
   b. MATLAB provided two tools for ANFIS projects, an APP and a set of
      functions
   c. Both tools provided by MATLAB are involved in the Fuzzy Logic Toolbox
   d. The anfis() function is used to create a new FIS model

10. What is the purpose of functions, genfis() and anfis()? _________.

    a. Both functions are used to generate a new FIS model
    b. The first one is used to create a new model, but the second is to train a model
    c. Both functions are used to train a new FIS model
    d. The first one is used to train a model, but the second is to create a model

III.   Exercises

1. Provide a basic description about ANFIS.
2. Provide a basic description about components and architecture of a typical ANFIS.
3. Explain why we need to use links between layers in an ANFIS model.
4. Explain the function for each layer in a five-layer ANFIS model.
5. Explain the advantages and disadvantages of using the APP and functions provided by MATLAB for ANFIS developments.

IV.   Lab Projects

1. Generating the training and checking data for an ANFIS model to be built by using a forest fires dataset, **forestfires.csv**, located at: https://archive.ics.uci.edu/ dataset/162/forest+fires. However, that original dataset cannot be used directly and you need to use a modified version, **MFire_Database.csv**, which can be found on the Springer ftp site under the folder: **Students\Datasets**. Use four columns, **temp**, **RH**, **wind,** and **rain**, as inputs, and **FWI** as the output column. **FWI** is a key index used to indicate how dangerous a forest fire may occur. The MATLAB Script file should be named as **Generate_Fires_Data.m**.

   Hint1: Refer to Sect. 10.5.1 to generate the training and checking data, and save them to two datasets. Then load them into Workspace to be used by APP later.

   Hint2: No normalization processing is needed for all data in this project. Use the first 200 rows as training data, and next 200 as checking data, and the rest 117 rows as testing data.

   Hint3: Set the **Number of MFs** on **INPUT** for the **Generate FIS** wizard to: **3 3 3 3**, and select the **gbellmf** as the **INPUT MF Type** and **constant** as the **OUTPUT MF Type**.

   Hint4: Select the **hybrid** as the **Optim Method** for the **Train FIS** group, and set **Epochs** to 100.
2. Using the MATLAB APP, Neuro-Fuzzy Designer, to design, train, and check an ANFIS model, **ANFIS_Fires_App.fis**, for a modified forest fires dataset, **MFire_Database.csv**, which can be found on the Springer ftp site under the folder: **Students\Datasets**. Use four columns, **temp**, **RH**, **wind,** and **rain**, as inputs, and **FWI** as the output column. **FWI** is a key index used to indicate how dangerous a forest fire may occur.

   Hint1: Load two datasets generated by Project 1 above into Workspace by typing two commands in the Command window to enable them to be used by APP later.

   Hint2: Save or export the trained ANFIS model into a file, **ANFIS_Fires_App.fis**.

3. Develop a MATLAB Script file, **Fires_Eval_App.m**, to evaluate the trained ANFIS model **ANFIS_Fires_App.fis**. Plot the comparison result between the predicted and the actual FWI values. Using the last 117 rows in the dataset **MFire_Database.csv** as the testing data.

   Hint1: Refer to Sect. 10.6.1 to build your codes for this evaluation project.

4. Using MATLAB Functions to build, train, and evaluate the ANFIS model built by using the forest fires dataset above. The Script file is named as **Fires_Train_Eval_Func.m**.

   Hint1: Follow Sect. 10.7.2 to complete these creating, training, and evaluation processes.

5. Generating the training and checking data for an ANFIS model to be built by using a flood dataset, **AEGISDataset.csv**, which is a Metro Manila Flood Landscape Data and located at https://www.kaggle.com/datasets/giologicx/ aegisdataset/data. However, that original dataset cannot be used directly and you need to use a modified version, **AEGISDataset_M.csv**, which can be found on the Springer ftp site under the folder: **Students\Datasets\Flood Dataset**. Use the column **flood_height** as the output and all others as inputs. The MATLAB Script file should be named as **Generate_Flood_Data.m**.

   Hint1: Refer to Sect. 10.5.1 to complete these training and checking data, and save them to two datasets. Then load them into Workspace to be used by APP later.

   Hint2: Using the top 2000 data rows as training + checking data, and the last 1000 as testing data.

   Hint3: Select the **gbellmf** as the **INPUT MF Type** and **linear** as the **OUTPUT MF Type**.

   Hint4: Select the **backpropa** as the **Optim Method** for the **Train FIS** group, and set **Epochs** to 60.

6. Using the MATLAB APP, Neuro-Fuzzy Designer, to design, train and check an ANFIS model, **ANFIS_Flood_App.fis**, for a modified flood dataset, **AEGISDataset_M.csv**, which can be found on the Springer ftp site under the folder: **Students\Datasets\Flood Dataset**. Use the column **flood_height** as the output and all others as inputs.

   Hint1: Load two datasets generated by Project 5 above into Workspace by typing two commands in the Command window to enable them to be used by APP later.

   Hint2: In the opened App, select the **gbellmf** as the **INPUT MF Type** and **constant** as the **OUTPUT MF Type**. Select the **hybrid** as the **Optim Method** for the **Train FIS** group, and set **Epochs** to 60.

   Hint3: Save or export the trained ANFIS model into a file, **ANFIS_Flood_App.fis**.

7. Develop a MATLAB Script file, **Flood_Eval_App.m**, to evaluate the trained ANFIS model **ANFIS_Flood_App.fis**. Plot the comparison result between the predicted and the actual **flood height** values.

   Hint1: Refer to Sect. 10.6.1 to build your codes for this evaluation project.

Hint2: Plot only the first 100 predicted and the actual data points for comparison purpose, otherwise the density on the plot is too high to be viewed.

8. Using MATLAB Functions to build, train, and evaluate the ANFIS model built by using the flood dataset.

    Hint1: Follow Sect. 10.7.2 to complete these creating and training processes.

    Hint2: Select the **gbellmf** as the **InputMembershipFunctionType** and **linear** as the **Output-MembershipFunctionType**.

    Hint3: Plot only the first 50 predicted and the actual data points for comparison purpose, otherwise the density on the plot is too high to be viewed, and set **Epochs** to 30.

References

1. Jang, Jyh-Shing R (1991). Fuzzy modeling using generalized neural networks and Kalman Filter Algorithm (PDF). In: Proceedings of the 9th National Conference on Artificial Intelligence, Anaheim, CA, USA, July 14–19. Vol. 2. pp. 762–767.
2. Jang, J.-S.R. (1993). ANFIS: adaptive-network-based fuzzy inference system. IEEE Transactions on Systems, Man, and Cybernetics, 23(3): 665–685. doi: https://doi.org/10.1109/21.256541.
3. https://en.wikipedia.org/wiki/Adaptive_neuro_fuzzy_inference_system#cite_note-1.
4. Yi Yang, Yanhua Chen, Yachen Wang, Caihong Li, Lian Li, Modelling a combined method based on ANFIS and neural network improved by DE algorithm: A case study for short-term electricity demand forecasting, Applied Soft Computing, 2016;49:663–675. doi: https://doi.org/10.1016/j.asoc.2016.07.053.
5. Mohammed Imran, Sarah A. Alsuhaibani, Chapter 7 - A Neuro-Fuzzy Inference Model for Diabetic Retinopathy Classification, In: D. Jude Hemanth, Deepak Gupta, Valentina Emilia Balas, Intelligent Data-Centric Systems, Intelligent Data Analysis for Biomedical Applications, Academic Press, 2019, Pages 147-172, ISBN 9780128155530, doi: https://doi.org/10.1016/B978-0-12-815553-0.00007-0.
6. Poras Khetarpal, Madan Mohan Tripathi, A critical and comprehensive review on power quality disturbance detection and classification, Sustainable Computing: Informatics and Systems, Volume 28, 2020, 100417, ISSN 2210-5379, doi: https://doi.org/10.1016/j.suscom.2020.100417.
7. Karaboga, Dervis; Kaya, Ebubekir (2018). Adaptive network based fuzzy inference system (ANFIS) training approaches: a comprehensive survey. *Artificial Intelligence Review* 52 (4): 2263–2293. doi:https://doi.org/10.1007/s10462-017-9610-2. ISSN 0269-2821. S2CID 40548050.

Appendix A: Download and Install MATLAB 2023a Software

Try to download and install the updated version of MATLAB software at any time.

Perform the following steps to download and install the MATLAB R2023a 30-day free trial version as well as related Toolboxes:

(1) After submitting your request for free 30-day trial version of MATLAB, go to the site https://www.mathworks.com/downloads/web_downloads/11526200 to open the free downloading wizard, as shown in Fig. A.1.
(2) Click on the **Download for Windows** button as shown in Fig. A.1 to start the download process.
(3) As the download process id is done, open the File Explorer and go to the **Downloads** folder on your machine. Then double click on the downloaded file, **matlab_R2023a_win64.exe**, to begin the installing process.
(4) Complete the login process with your username and password that were used when you sent request to MathWorks, and click on the **Next** button to go to the **PRODUCTS** wizard, as shown in Fig. A.2.
(5) On this wizard, we can select all related Toolboxes we like to use in our projects. Select the following components and Toolboxes by clicking on each of them:

   (a) Simulink
   (b) Audio Toolbox
   (c) Computer Vision Toolbox
   (d) Control System Toolbox
   (e) Data Acquisition Toolbox
   (f) Deep Learning Toolbox
   (g) DSP System Toolbox
   (h) Fuzzy Logic Toolbox
   (i) Image Processing Toolbox
   (j) Signal Processing Toolbox

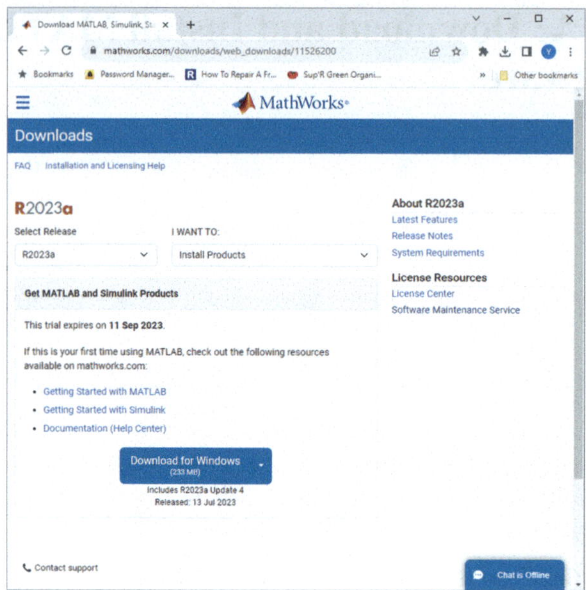

Fig. A.1 The opened downloading wizard

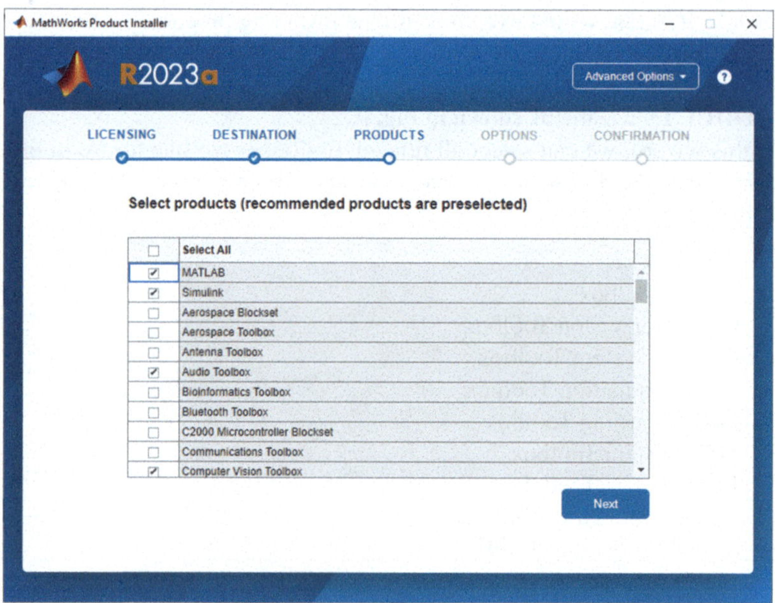

Fig. A.2 The opened PRODUCTS wizard

   (k) Simulink Control Design
    (l) Statistics and Machine Learning Toolbox
 (m) System Identification Toolbox

(6) Then click on the **Next** button to continue.
(7) In the next wizard, the **OPTIONS** wizard, check the **Add shortcut to desktop** checkbox to add a shortcut of this MATLAB to the desktop to make it easy to be used. Then click on the **Next** button to continue.
(8) Click on the **Begin Install** button to start this installation process. The installation process begins, as shown in Fig. A.3.
(9) When the installation process is done, as shown in Fig. A.4, click on the **Close** button to complete this process.

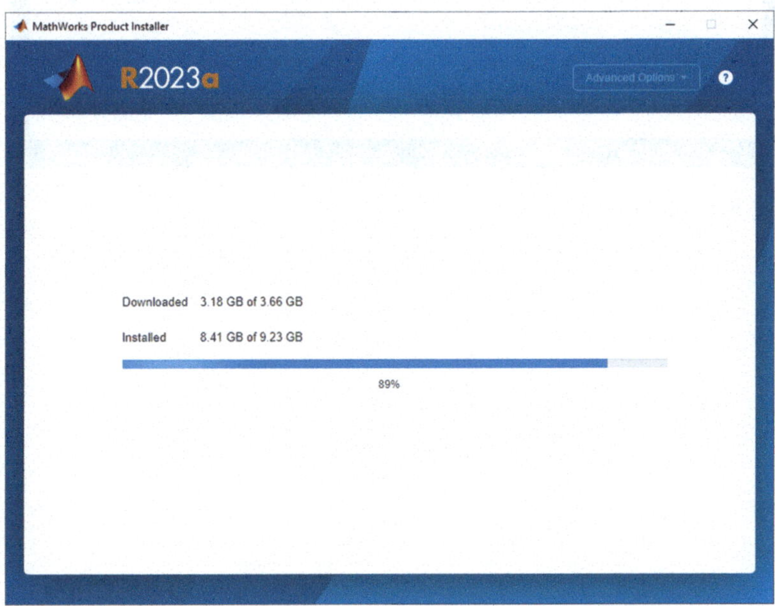

Fig. A.3 The installation starts

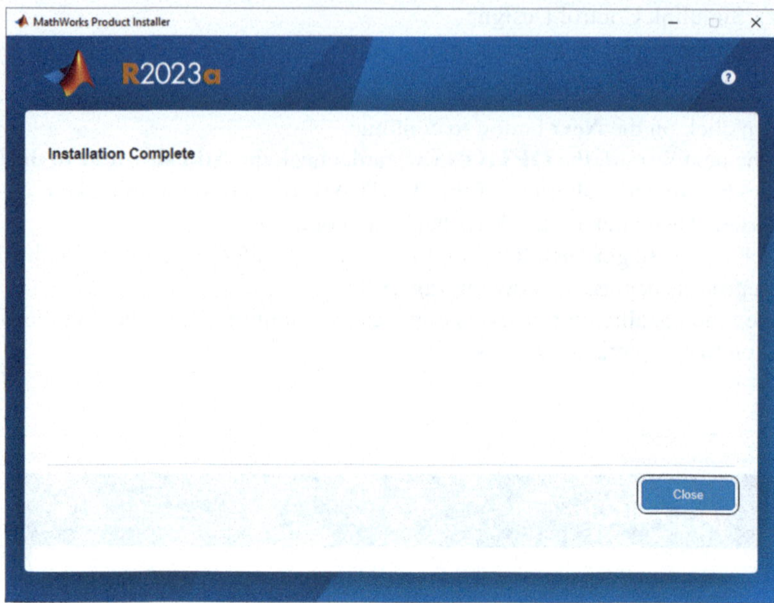

Fig. A.4 The installation ends

Index

A

Abductive inference system, 28

Action, 3, 8–12, 15, 31, 36, 37, 42, 68, 74, 136, 154, 155, 197, 238, 240, 251, 267, 280, 399, 498, 557–577, 579, 581, 582, 584, 587, 589–592, 599–602, 605, 610, 611, 614, 618–620, 622–624, 626, 648

Action system, 559

Action-value function, 15, 561–564, 569, 572–574, 620, 626

Adam, 418, 450, 457, 475

Adaptive neural fuzzy inference system (ANFIS), 2, 16, 631–659

Add All Possible Rules, 93

addInput(), 96, 98

addLayers(), 414

addMF(), 96, 99, 100, 127

addOutput(), 96

addRule(), 96, 105, 129

Agent control system, 559

Agent policy, 558

Aggregate type-2 fuzzy set, 114, 115

AI knowledge cycle, 29–30, 51

All Quick-To-Train, 176, 188, 289, 298, 323

anfis(), 650, 651

ANFIS functions, 649–653

anfisOptions(), 651

Animal image dataset, 395

Apriori algorithm, 470, 479, 541–550

ArrayDatastore, 431, 432

arrayfun(), 358

Artificial neural networks (ANNs), 36, 156, 217–219, 386, 387, 393, 396, 400, 401, 404, 419, 422–429, 443, 444, 631

Association rule, 15, 25, 150, 153, 154, 469–471, 478–482, 536–542, 544–546, 550

Asynchronous advantage actor-critic (A3C), 566, 570, 571

Attended stimulus, 7

audioDatastore, 362–365, 371, 372, 456, 457

audioFeatureExtractor, 361, 362, 364, 365, 368, 372, 373, 461

audioread(), 367, 381

Automated classification training, 296, 298

Automated regression model training, 176, 177, 188, 189

B

Backpropagation neural networks (BNN), 15, 389, 396

Backward chaining, 31–33, 51

Bagged trees, 263–266, 268, 273, 301–303, 315, 323, 324

Bagging, 260–263, 269, 273, 288

Bag of features (BOF), 15, 347, 350, 352, 356, 381, 461

Band energy ratio (BER), 360–361

BatchNormalization, 458

Bernoulli naïve bayes model, 334, 335

beta0, 247–252, 254, 281

Bi-conditional, 26, 28

Binary classification, 15, 25, 150, 152, 294, 296–303, 309, 311, 314, 316–321, 340–346, 351, 376, 377

Binary linear classifications, 308–317

Blank model, 134, 606

© The Editor(s) (if applicable) and The Author(s), under exclusive license to
Springer Nature Switzerland AG 2025
Y. Bai, *AI Foundations and Applications with MATLAB*,
https://doi.org/10.1007/978-3-031-84423-2

Boosting, 260, 270
Bootstrap aggregation, 260–261, 263
bsxfun(), 358
Build Up Index (BUI), 185, 196, 197, 223,
       271, 275, 276, 278

C
C51 algorithm, 566, 574–575
Canadian Forest Fire Danger Rating System
       (CFFDRS), 184
Case-based systems, 30
cat(), 365, 366, 374, 375
categorical(), 330, 331, 366, 375, 416, 466
cellfun(), 214, 355, 365, 374
cellstr(), 455
Center of gravity (COG) method, 14, 29, 32,
       33, 59, 71, 72, 74, 75, 109–111, 140,
       475, 633
Cepstrum, 360, 361
cfit, 242, 243
changem(), 532
chi2inv(), 524
Classical set, 60–66, 68, 140
Classification, 15, 18, 25, 40, 41, 150–153,
       156, 158–161, 167, 199–200, 218–220,
       225, 254–257, 260–263, 268, 273, 278,
       279, 285, 293–382, 390, 392, 395, 397,
       399, 403, 404, 407–413, 415–417, 421,
       431, 437, 440, 442, 446–461, 469, 515
ClassificationECOC, 325, 327–329, 331,
       332, 344–346
ClassificationKNN, 309, 325, 327–329, 377
classificationLayer(), 457
Classification Learner, 160, 295–297, 321,
       322, 404
ClassificationNaiveBayes, 325, 327–329, 331,
       332, 377
Classification trees, 199, 200, 261, 273,
       315, 345
classifySound(), 361
Clear all rules, 93
cluster(), 477, 497, 498, 501–504, 506, 507,
       521, 524
Cluster analysis, 153, 489, 497, 521
ClusterCenters, 509
Cluster centroid locations matrix, 493, 494
clusterdata(), 497
Clustering, 15, 25, 150, 151, 153–154, 156,
       158, 160, 220, 350, 398, 400, 404,
       469–478, 481–537, 550
Coefficient of determination, 169, 172, 173,
       183, 204

Cognitive process, 2
CompactRegressionEnsemble class, 263
Complement, 61, 63–65, 111, 112, 571
Conditional, 26–28, 40, 198, 276, 333, 334,
       336, 479, 480
Confidence, 159, 227, 242, 253, 478–481, 522,
       536–537, 539–541, 544, 545, 548
confusionchart(), 311, 312, 320, 327, 332, 459
Conjunction, 26
Control surface, 93, 94, 125, 126, 144
convertfis(), 96
convertToSugeno(), 96
convertToType1(), 96, 133
convertToType2(), 96
Convolution2D, 458
Convolutional layer, 390–391, 452
Convolutional neural networks (CNN), 15,
       159, 357, 389–393, 396, 397, 399, 413,
       450, 574
Cross-validation, 151, 168, 175, 176, 187,
       188, 200, 210, 213–215, 225, 261, 263,
       270, 279, 296, 310, 345, 419
Curve fitting toolbox, 14, 17, 158–160, 162,
       240–250, 285, 376, 377
Custom FIS, 83, 85
Custom function environments, 587–588
Custom Simulink environments, 588, 605
Custom template environments, 588,
       596, 602–603

D
Dancing robot, 2
Decision making process, 23–54, 154
Decision nodes, 40, 41, 44, 198
Decision/leaf nodes, 40, 41, 43, 44, 49,
       198, 199
Decision trees, 14, 15, 25, 40–51, 150, 151,
       158, 160, 161, 177, 178, 198–215, 217,
       260–263, 273, 275, 276, 279, 283, 285,
       294, 295, 316, 321, 323, 324, 376
Deductive inference system, 28
Deductive reasoning, 3, 4
Deep learning neural network (DLNN), 388,
       389, 412, 437
Deep learning toolbox, 14, 17, 158, 159, 162,
       294, 295, 361, 391, 392, 399, 404, 415,
       416, 440, 481, 482, 550, 578, 661
Deep network designer app, 159, 401,
       402, 404
Deep network designer (DND), 16, 159, 160,
       400–402, 404, 423, 430–436, 465
Deep network quantizer, 160

Deep neural network (DNN), 37, 159, 389, 393, 396, 399, 400, 415, 420–422, 434, 437, 439, 440, 442–443, 462, 558, 574, 577
Deep Q-Learning (DQL) algorithm, 565, 571
Deep Q Network (DQN) algorithm, 566, 573–576, 579, 580, 593, 620–624
defuzz(), 96
Defuzzifications, 14, 58–60, 66, 70–76, 80, 86, 111, 115–116, 140, 633
dendrogram(), 476, 497, 501, 504
Density threshold, 153
DESIGN BROWSER, 86
Deterministic policy, 560, 564, 570, 576, 577, 581
Direct policy/value function estimation, 569
Discounted return, 561–563, 568, 570–572
Discrete Fourier Transformation (DFT), 360
Disjunction, 26, 27
Distance function, 153
Driving-car-study process, 3
Drought code (DC), 185
Duff moisture code (DMC), 185, 192–194, 271, 275, 276, 278

E
ECOC decoding and testing process, 342–344
ECOC encoding and training process, 340–342
Elbow method, 472–474
End nodes, 40, 198
ENSEMBLE CLASSIFIERS, 323
Ensemble learning method, 260
Entropy, 41–49, 51, 392, 576
Environmental stimulus, 7
Environment system, 559
Error-checking function, 3
Error correcting output code (ECOC), 15, 340–347, 350–359, 377, 395, 397
Euro-US currency exchanging rates, 16
evalclusters(), 527, 528, 530, 533, 534
evalfis(), 96, 105, 106, 113, 131, 132, 648, 650
Evaluate the clustering number, 527–529
Evaluate the clustering result, 525, 527, 529–536
Evenly distribute MFs, 87–89, 122
evnUpdatedCallback(), 602
Exclusive clustering, 15, 470–474, 487, 489–497, 550
Experiment manager, 159, 160, 296
Expert systems, 31, 33, 79

Exponential models, 218, 227, 228, 234, 235, 238
Export fuzzy inference system to workspace, 89
extract(), 362, 365, 366, 368, 370, 373, 375
extractHOGFeatures(), 356

F
False positive rate (FPR), 300, 301
Fast Fourier Transformation (FFT), 360
fcm(), 508–511, 516, 553
fcmOptions, 508–511
Feature bagging, 261, 262
featureInputLayer, 413, 433, 466, 467
featureInputLayer(), 413
feedforwardnet(), 404, 410
Feedforward neural networks (FNN), 15, 389, 393, 396, 410, 632
Fine fuel moisture code (FFMC), 185, 192–194, 271, 275, 276, 278
Fire weather index (FWI), 184, 185, 187, 192–194, 196, 197, 271, 276, 278, 657, 658
First-order Sugeno systems, 81
FIS from data, 83
fit(), 219, 243–247, 251, 259
fitcecoc(), 303, 327–329, 331, 332, 344–346, 351, 357, 377, 382
fitcensemble(), 262
fitckernel(), 303, 308–313
fitcknn(), 308–313, 325, 327–329, 364, 366, 373, 375
fitclinear(), 303, 308–311, 313
fitcnb(), 303, 327–329, 331, 332
fitcnet(), 303, 404, 407–409
fitcsvm(), 303, 308–313, 320, 351
fitglm(), 303, 308–310, 313–315
fitgmdist(), 521, 523, 524
fitlm(), 181, 182, 192, 194–196, 289
fitnet(), 404–405, 410, 437, 438
fitnlm(), 219, 246–248, 251, 252, 281, 282, 284, 286, 289
fitoptions(), 234, 246, 251
fitrensemble(), 262, 263, 268–271, 277, 278, 281, 282, 284, 286
fitrlinear(), 192
fitrnet(), 404, 406–407, 409, 437
fitrtree(), 210, 213, 214
fittype, 241–245, 251
Footprint of uncertainty (FOU), 108
Forward chaining, 31–33, 51

Fourier models, 218, 227, 229, 235, 236
Fourier Transformation (FT), 270–272, 275, 276, 278, 360
Fraud bank checks, 16
fscmrmr(), 365, 374
Full model, 168, 296
Fully connected layer (Classification layer), 390–392, 406–410, 440, 442, 443
Function approximation method, 565
Fuzzy control rule, 60, 68–70, 77, 83, 90
Fuzzy idea, 58, 140
Fuzzy implication rule, 69, 70
Fuzzy inference system (FIS) plot, 2, 14, 16, 18, 23, 25, 28, 29, 33, 50, 57–59, 66, 68, 76, 77, 79–82, 85, 89, 91, 97, 98, 102–140, 149, 508
Fuzzy K-Means (FKM), 470, 475, 477, 507
Fuzzy logic designer app (FLDA), 79, 81–95, 99, 107, 113, 120, 123, 125, 132, 136, 140
Fuzzy logic designer app mode, 79, 81, 82, 140
Fuzzy logic functions, 79, 82, 95, 140
Fuzzy logic idea, 58–60
Fuzzy mapping rule, 69–70
Fuzzy-neural-works, 2
Fuzzy set, 2, 60–68, 76, 77, 79, 83, 86, 107, 109–117, 119, 140
Fuzzy variables, 29, 59, 66, 70

G

gauss2mf, 651
Gaussian kernel classification, 308–309
Gaussian mixture model (GMM), 160, 477, 517–526, 529, 551, 552, 554
Gaussian models, 218, 227, 229, 235, 237, 252, 287
Gaussian naive Bayes model, 334–336, 379
gaussmf, 100, 101, 104, 127, 128, 130, 643
Gauss-Newton algorithm, 217, 218, 287
gbellmf, 100, 101, 104, 657–659
Generative adversarial networks (GANs), 159
genfis(), 96, 113, 141, 142, 508, 650, 651, 656
genfisOptions(), 650
gensurf(), 96, 105, 130
getExplorationPolicy(), 622
getObservationInfo(), 620, 622
getReward(), 600, 602
gmdistribution object, 521
Goal-based agents, 10
Goal-directed computational approach, 558

Google Stock datasets, 19, 280–281, 466, 638–640, 647, 651
Gradient descent algorithm, 217, 218, 287
gscatter(), 496, 524
gsubtract(), 429, 436, 439, 444

H

The Height method (HM), 71, 72, 140, 141, 656
Hidden Markov models (HMM), 160
Hierarchical clustering, 15, 25, 150, 160, 470, 471, 476–477, 481, 487, 489, 490, 497–507, 550, 552, 553
Hindsight experience replay (HER) algorithm, 566, 575–576
Histogram of oriented gradients (HOG), 15, 347–348, 355–359, 379, 382
Hold off, 196, 197, 239, 240, 254, 266, 267, 276, 492, 493, 511, 512, 523, 525, 531, 534
Hold on, 170, 196, 197, 239, 240, 254, 266, 267, 276, 492, 493, 495, 511, 512, 522–525, 531, 534
Holdout validation, 168, 175, 176, 187, 286, 296
Hybrid RL algorithms, 576–577
Hyperplane, 219–221
Hypertext manipulation systems, 31

I

im2double(), 515
imadjust(), 515
image3dInputLayer(), 413
imageDatastore(), 351, 458
imageInputLayer(), 413
Image on the retina, 7
Imagination-augmented agents, 567
Import DATA, 297, 424, 426, 433, 482, 485, 486, 553
Import Data from workspace, 424, 426, 485, 486, 553
imread(), 454
imshow(), 516, 517
Inductive inference system, 28
Inductive reasoning, 3–5
Inference engine, 30, 31, 33, 38, 39, 52, 111
Inference system by analogy, 28
Inferential knowledge, 38, 39, 51, 52
Information gain, 44–47, 49, 51, 53, 54
Inheritable knowledge, 38, 39, 51

Initial spread index (ISI), 185
inputLayer(), 413
Input layer, 156, 386, 387, 389, 390, 393–395,
     409, 413, 414, 416, 440, 442, 462, 466,
     467, 632, 633, 655
Inspect simulation data, 584, 586
Intelligent agents, 8–12, 154
Intelligent software, 2
Intelligent tutoring systems, 31
Intersection, 61, 63–65, 70, 74, 108, 111, 112,
     117, 123, 220, 493
Interval type-1 fuzzy set, 115
Interval type-2 FIS (IT2FIS), 108–117, 126,
     131, 139–141, 143, 145, 146, 636
Interval type-II FIS, 16, 25, 50, 57
IS-A relation, 35, 36
ismember(), 371, 601

K
Karnik-Mendel algorithm, 111
Kind-of-relation, 35, 36
kmeans(), 490, 492, 530, 531, 535, 553
K-Means clustering, 25, 150, 156, 163,
     470–473, 475, 477, 481,
     489–497, 551–553
K-Nearest Neighbor (KNN), 14, 15, 25, 150,
     151, 160–163, 254–260, 285, 288–290,
     294, 295, 308–313, 321, 325, 345,
     366, 375–378
K-Nearest Neighbor classifier, 309–313
kNNeighborsRegressor(), 257, 259
Knowledge base, 9, 29–33, 39, 40, 51, 52, 68
Knowledge based systems, 29–33, 51, 52
Knowledge representations, 26, 29, 33–40, 52

L
Language understanding, 3, 8
Laser tracking system (LTS), 16
lbfgs, 418
Learning, 1–3, 6, 9, 11–19, 21, 23–54, 68, 82,
     95, 149–164, 167, 174, 181, 185, 186,
     199, 210, 217–220, 248, 254, 259–262,
     264, 269–272, 278, 279, 285, 286,
     293–295, 297, 302, 304, 321, 345, 347,
     350, 359, 361, 365, 374, 376–378,
     385–467, 469–554, 557–632, 638, 653,
     655, 661, 663
Learning ability, 1, 9, 12, 16, 632, 653
Learning agents, 11, 12, 561, 574, 576, 620
Learning from interaction, 569
The learning element, 11

legend(), 139, 240, 267, 283, 525
Levenberg-Marquardt algorithm, 217,
     218, 287
Library Browser, 134, 606
Linear-Fold AI algorithm, 2
Linear function approximation, 565
Linear regression, 14, 25, 150, 152, 161,
     167–198, 217–219, 285, 286, 288
LineWidth, 106, 131, 138, 281, 492, 511, 512,
     523, 531, 534, 647, 651
Linguistic variables, 59, 66–69, 71, 119
linkage(), 497–501, 504
linspace(), 276, 523
Live script file, 95–97
Local binary pattern (LBP) features, 15, 347,
     349–350, 379
Logarithmic models, 218, 227–229, 235, 236,
     251, 287
Logical intelligence, 24–29, 51, 57
Logical representation, 34–35, 52
Logistic regression, 24, 25, 150–152, 156,
     161, 167, 295, 298, 308, 310, 321,
     323, 378, 385
Long short-term memory (LSTM) networks,
     159, 414, 433, 442
Long-term reward, 557, 558, 570, 574, 576,
     619, 624
Lookup table, 14, 58, 59, 70–77, 140, 141,
     227, 618
Loss function, 151, 218, 272, 306, 307, 412,
     417, 418, 443, 574, 575
Lower membership function, 109, 110, 113,
     116, 140

M
mahal(), 490, 521, 524
Mamdani fuzzy inference systems,
     79–80, 655
Mamdani systems, 79, 80, 96, 97, 112–114
Mamdani type-I FIS, 83, 85
Mamdani type-II FIS, 83
mamfis(), 96, 97, 133, 141
mamfistype2, 112, 116, 125–128, 130
Manual classification training, 296, 378
Manual regression model training, 176,
     188, 287
MarkerSize, 492, 511, 512, 523, 531, 534
Markov chain, 560
Markov decision process (MDP), 15, 154,
     557–564, 566, 570, 573, 624, 626–628
Markov transition matrix, 560
MATLABEnvironment(), 600

MATLAB functions used to create
agents, 619–620
MATLAB functions used to define
rewards, 618–619
maxclust, 503, 505, 507
MaxEpochs, 412, 442, 450, 452, 465–467
Maximum-margin hyperplane, 220, 221
Mean Of Maximum (MOM), 14, 59, 71, 74,
140, 141, 656
Mel-Frequency Cepstral Coefficients
(MFCCs), 360–361
Melspectrogram, 361, 456–459
Membership degrees (MD), 109, 507–509,
511, 512, 516
Membership Function (MF) Editor,
89–90, 121
Membership functions (MFs), 14, 29, 30, 33,
58–60, 64, 66–68, 70–74, 76, 77,
80–93, 96, 99–105, 107–117, 119,
121–123, 127–130, 133, 140, 143–146,
475, 507, 632–636, 643, 646, 650,
656, 657
meshgrid(), 523
minkowski distance, 255, 258, 498
mode(), 366, 375
Model-based priors for model-free
reinforcement learning, 567
Model-based reflex agents, 10
Model-based reinforcement learning,
565–568
Model-based value expansion, 567
Modeless robot calibration, 637
Model-free reinforcement learning,
566–577, 626
modelfun, 247–254, 281
Model hyperparameters, 223, 301
Model operating point, 300, 301
Monte Carlo methods, 15, 564, 570
Multi-class classifications, 15, 25, 150
Multinomial naïve Bayes model, 335
Multiple linear regression function
modeling, 192–198
Multiple linear regression (MLR), 14,
184–198, 285, 286, 288
mvnpdf(), 519, 520

N

Naïve Bayes, 15, 25, 150, 151, 161–163, 279,
294, 295, 321, 333–340, 345, 376–379
NASDAQ dataset, 16, 19
nctool, 400, 404, 484, 553

Negation, 26, 27
Negative reinforcement, 25, 150, 155, 163, 557
Neighbor Distances, 482, 486
network(), 404, 410–411, 429
Neural net clustering, 160, 400, 404,
481–487, 553
Neural net clustering App, 482–487, 553
Neural net fitting, 160, 404
Neural net pattern recognition, 160, 400, 404
Neural net time series, 160, 400, 404
Neural processing, 7
neuroFuzzyDesigner, 642
Neuro-Fuzzy Designer, 637, 641–649, 654,
655, 657, 658
New script, 96, 97, 126, 133, 138, 181, 192,
194, 208, 213, 252, 258, 266, 270, 274,
278, 281, 310, 313, 316, 318, 319, 327,
330, 339, 351, 353, 356, 363, 367, 371,
380, 423, 428, 431, 435, 437, 442, 447,
449, 453, 456, 465, 466, 491, 495, 504,
510, 514, 522, 527, 530, 537, 539, 545,
589, 591, 622, 629, 640, 647, 650
nftool, 400, 404
nlinfit(), 219, 243, 248–252, 286, 288
nnstart, 400, 401, 404, 422–424, 464
Nonlinear regression, 14, 25, 150, 161, 184,
217–219, 226–254, 256, 269, 280,
281, 285–288
notifyEnvUpdated(), 600
nprtool, 400, 404
ntstool, 400
numel(), 374, 458

O

Off-policy, 562, 573, 576, 627
One-vs-One (OvO), 340, 341, 357
One-vs-Rest (OvR), 340, 341
On-policy, 562, 627
Ontologies, 30, 34, 52
OptimalK property, 528, 530
Optimal policy, 558, 560–564, 568–570, 573,
574, 576, 578, 627, 628
Optimization results, 223
Optional environment methods, 596, 597
Ordinary least-squares (OLS) regression, 184
Output layer, 33, 37, 156, 386, 387, 389, 390,
392–396, 407, 409, 410, 413, 416, 430,
433, 440, 451, 462, 463, 467, 483, 575,
632, 633, 655
Overlapping clustering, 15, 470–472,
474–476, 481, 487, 507–517, 550–553

P

Partially observable Markov decision process (POMDP), 560, 570

pdist(), 497–499, 504

pdist2(), 496, 498

Perception, 3, 6–9, 11, 37, 38, 483, 484, 491–497

perform(), 414, 419, 429, 439

The performance element, 11

plotfis(), 96

plotmf(), 96

Policy evaluation, 564, 569

Policy gradient (PG), 566, 570–572, 576–577

Policy optimization, 566, 569–572

Polynomial models, 218, 227–228, 252, 287

Pooling layer, 390, 391, 440

Positive reinforcement, 25, 150, 155, 163, 557

posterior(), 521

predict(), 210, 214, 259, 267, 277, 278, 282, 289, 312, 314, 316, 328, 355, 358, 366, 375, 436, 444

Principal component analysis (PCA), 158, 160, 208

Probabilistic clustering, 15, 470, 471, 477–478, 481, 487, 517–526, 550–552

probability density function (PDF), 309, 518–521

The problem generator, 11

Problem solving, 3, 6, 9, 14, 37, 38

Procedural knowledge, 38, 40, 51, 52

PROPERTY EDITOR: INPUT, 87, 121

PROPERTY EDITOR: OUTPUT, 88

PROPERTY EDITOR: RULE, 91–93, 123

Propositional logic, 25–28, 33, 51–54

Proximal policy optimization (PPO), 566, 570–572, 579, 580, 582–585, 604, 614, 616, 617, 628

Pruning, 41, 399

Pure-sub-split, 43

Q

Q-function, 573

Q-Learning (Quality-Learning) algorithms, 565, 566, 572–576

Quantile regression (QR-DQN) algorithm, 575

R

Random forest (RF), 14, 15, 25, 150, 151, 161, 162, 259–280, 285–288, 290, 294, 309, 313, 315, 376

Random subspace, 262, 273

Rational models, 218, 219, 227, 230, 235, 237, 238

readfis(), 96, 105, 130, 132, 133, 142, 648, 650

readtable(), 181, 240, 266, 289, 423, 424, 428, 429, 432, 435, 437, 466, 491, 522, 527, 640, 641, 647

Reasoning, 3–5, 9, 25, 28–33, 35, 37, 38, 52, 564, 632

Reasoning by analogy, 4, 5

Recognition, 2, 6, 8, 12, 30, 152, 153, 160, 162, 219, 350, 380, 390, 393, 397, 400, 404, 549

Rectified linear unit (ReLU) layer, 390, 391, 393, 408, 409, 430, 442, 463

Recurrent neural networks (RNN), 15, 389, 393, 394, 396, 463

Regression, 14, 18, 24, 25, 41, 150–153, 156, 158–162, 164, 167–290, 293, 295, 298, 308, 310, 321, 323, 351, 378, 379, 385, 404, 406–409, 411, 415, 416, 421, 422, 426, 430, 433, 437, 440, 442, 461, 464, 466, 467, 575

Regression algorithm, 14, 18, 24, 151–152, 167–290, 293

RegressionBaggedEnsemble class, 263, 268

RegressionEnsemble class, 263

regressionLayer, 433, 467

Regression learner, 158, 160, 167, 168, 173, 174, 176, 180–186, 188, 191, 200, 208, 210, 222, 223, 225, 264, 267, 286–289, 464

Regression trees, 167, 199–217, 221, 261, 263, 268, 270, 273, 287, 298

reinforcementLearningDesigner, 579

Reinforcement Learning Designer, 577, 579–586, 592, 602, 603, 605, 611, 614, 626, 628, 629

Reinforcement learning process, 23, 24, 50, 561–563

Reinforcement learning (RL), 14–18, 23–25, 50, 51, 150, 154–155, 157, 161–163, 394, 398, 399, 469, 557–630

removeInput(), 96

removeMF(), 96

removeOutput(), 96

repmat(), 358, 436

reshape(), 520

Residuals, 172, 173, 180, 184, 189, 191, 193, 207, 208, 227, 248, 250, 286, 289

Resubstitution validation, 175, 187, 286

Reward function, 155, 163, 561, 568, 601, 619

Reward system, 559
rgb2gray(), 515
rlDQNAgent(), 620, 622
rlFiniteSetSpec(), 598
rlFunctionEnv(), 587, 591, 618
rlMultiAgentTrainingOptions(), 621
rlNeuralNetworkEnvironment object, 624
rlNumericSpec(), 598, 610
rlQValueFunction(), 620
rlTrainingOptions(), 620, 621
rlVectorQValueFunction(), 620
rmsprop, 418, 575
RNA sequence, 2
Robotic process automation (RPA), 2
Robotic vacuum cleaner, 2
Robustness, 569
roiInputLayer(), 413
Root node, 40, 41, 44–47, 52
Rule-based machine learning method, 154
Rule-based systems, 31, 34
Rule firing, 80, 81, 113, 114
Rule inference, 93, 94, 124, 125, 144, 633
Rule inference viewer, 93, 94
Rule output value, 80
Rules editor, 91–93, 123, 124

S
Sample hits, 482, 486, 488
Sample inefficiency, 569
Samples-trees, 261
Score function, 306, 570
Script file, 95–97, 99, 105, 126, 129, 131, 133,
    138, 139, 181, 182, 192, 194, 197, 208,
    213, 215, 239, 250, 252, 254, 258, 259,
    266, 270, 274, 278, 281, 285, 289, 290,
    310, 313, 316, 318, 319, 325, 327, 330,
    337, 339, 340, 351, 353, 356, 367, 371,
    379, 380, 423, 424, 428, 431, 435, 437,
    442, 445, 447, 449, 453, 456, 465, 466,
    491, 495, 504, 510, 514, 521, 522, 527,
    530, 537, 539, 545, 553, 554, 589, 591,
    592, 602, 610, 613, 614, 622, 624, 629,
    630, 640, 647, 648, 650, 651, 653,
    657, 658
Self-organizing map (SOM) networks, 482
Semantic networks, 34–36, 52
Semi-supervised learning process, . 51, 24
sequenceInputLayer(), 413–414
seuclidean distance, 498
sfit, 237, 238, 242
sgdm, 412, 413, 418, 442, 443, 593

Short-Time Fourier Transformation
    (STFT), 360
showrule(), 96
sigmf, 127, 128, 130
Silhouette method, 472–474, 552
Similarity learning, 25, 150, 151, 279
Simple reflex agents, 10
Simple relational knowledge, 38, 51, 52
Simulated neural networks (SNNs), 156, 386
Simulation Data Inspector, 584, 586
Simulink Library Browser, 134
Smart homes, 2
softmax layer, 392, 410
Speech recognition, 2, 12, 380, 393, 397, 549
Speeded-up robust features (SURF), 15,
    347–349, 379
splitEachLabel(), 352, 356, 362, 363, 372,
    449, 458
squaredeuclidean distance, 498
squareform(), 499, 504
State-action spaces, 563
State system, 559
State value function, 561–562, 620, 626, 627
Stationary policies, 562
Statistics and machine learning toolbox, 14,
    17, 158, 160, 162–164, 181, 199, 248,
    264, 285, 294, 295, 321, 350, 376, 378,
    404, 481, 497, 519, 550, 663
statset(), 249, 552
Stochastic (stationary) policy, 560, 570, 576
Stop Time, 137
Sub-tree, 41
Sugeno fuzzy inference systems, 80–81, 112,
    631, 632, 655
Sugeno systems, 79–81, 96, 112–114, 116
Sugeno type-I FIS, 81
Sugeno type-II FIS, 81, 85
sugfis(), 96, 141
sugfistype2, 112, 116
Sum of squared distance (SSD), 472
Supervised learning, 14, 16, 23–25, 32, 33, 41,
    50, 51, 150, 151, 154, 156, 161–164,
    167, 218, 219, 279, 285, 293, 302, 304,
    388, 394, 398, 420, 469, 470, 489, 530,
    532, 548–551, 553, 557, 567, 624, 638
Support, 14, 20–21, 25, 31, 35, 40, 45, 67, 68,
    71, 79, 89, 116, 140, 150, 151, 156,
    158–160, 163, 167, 198, 217–225, 229,
    230, 285, 286, 294, 295, 308, 309, 321,
    350, 351, 361, 376, 378, 385, 397,
    478–481, 491, 497, 508, 536, 537,
    539–548, 550, 552, 637, 656

Support vector classifier, 308, 309
Support vector machines (SVMs), 14, 25, 150, 151, 156, 158, 160, 167, 217–225, 285, 294, 295, 309, 321, 350, 351, 376, 378, 385, 397
Switch points, 110, 111, 115, 116, 142
SYSTEM BROWSER, 86, 121

T
TabularTextDatastore, 431
Takagi-Sugeno-Kang fuzzy inference, 80, 631
templateDiscriminant, 345
templateEnsemble, 345
Template Fuzzy Inference System (FIS), 85
templateKernel, 345
templateKNN, 345
templateLinear, 345
templateNaiveBayes, 345
templateSVM, 345, 346
templateTree, 345
Temporal difference method, 564–565
Test nodes, 41, 49, 52
Test/chance nodes, 40, 41, 198
To Workspace blocks, 135, 138
trainImageCatagoryClassifier(), 352
Training from scratch, 397, 463
trainingOptions(), 412–413, 442, 443, 451, 452, 458, 465–467
trainnet(), 412, 442, 451, 458
trainNetwork(), 411, 449–451, 458
Transfer Fcn, 134, 137, 609
Transfer learning, 159, 397, 400, 417, 420, 462, 463
Transformable nonlinear models, 218, 219, 227, 287
TreeBagger(), 262, 263, 268, 272–278, 281–284, 286, 288, 290, 309, 313, 315–317, 319, 320
True negative rate (TNR), 301
True positive rate (TPR), 299–301, 323, 378, 379
Trust region policy optimization (TRPO), 566, 570, 572, 614, 617, 628
tunefis(), 113, 650

Type-2 fuzzy set, 107, 109, 111, 112, 114, 115
Type-2 Mamdani systems, 112, 114
Type-2 Sugeno systems, 112, 113
Type-I FIS, 25, 50, 57, 58, 83, 85
TypeReductionMethod parameter, 112
TypeReduction property, 116

U
Union, 61, 63–65, 72, 111
Universe of discourse, 60, 62, 64, 65
Unsupervised learning, 14–16, 18, 23–25, 50, 51, 150, 151, 153–154, 161–164, 220, 398, 469–554, 557
Unsupervised learning process, 24, 150
updateActionInfo(), 600, 602
Upper membership function, 109, 110, 113, 114, 116, 140
User interface, 30, 52, 227
Utility-based agents, 10

V
Validated model, 168, 212, 215, 296
validateEnvironment(), 591, 592
Vehicle Type Image Dataset, 381
view(), 214, 414, 429, 439

W
Weight planes, 482, 486, 488
Weight positions, 482, 486
Within the Sum of Squares (WSS), 472
World models, 566, 567
writecell(), 539
writeFIS(), 96, 105, 141, 650
writematrix(), 539

X
xlsread(), 182, 318

Z
Zero-order Sugeno system, 80

The manufacturer's authorised representative in the EU is Springer
Nature Customer Service Centre GmbH, Europaplatz 3, 69115 Heidelberg,
Germany. If you have any concerns regarding our products, please
contact ProductSafety@springernature.com

Printed and bound by CPI Group (UK) Ltd, Croydon, CR0 4YY

02/01/2026

02028189-0015